CHAOS

中央之帝爲渾沌
（莊子：內篇卷七）

The Emperor of Center was called Hundun (Chaos).

(Zhuangzi, Part I, Chapter VII.)

CHAOS

HAO BAI-LIN

THE INSTITUTE OF THEORETICAL PHYSICS, BEIJING

World Scientific

Published by
World Scientific Publishing Co Pte Ltd.
P O Box 128, Farrer Road, Singapore 9128

The author and publisher are indebted to the original authors and the following publishers of the various journals and books for their assistance and permission to reproduce the selected articles found in this volume:

Academic Press Inc. *(J. Combinatorial Theory);* American Association for the Advancement of Science *(Science);* American Meteorological Society *(J. Atmos. Sci.);* American Physical Society *(Phys. Rev.* and *Phys. Rev. Lett.);* Cambridge University Press *(J. Fluid Mech.);* John Wiley & Sons Inc. *(Commun. Pure and Appl. Math.);* Kyoto University *(Prog. Theor. Phys.);* Macmillian Journals Ltd. *(Nature);* Mathematical Association of America *(Amer. Math. Monthly);* North-Holland Physics Publishing *(Physica* and *Phys. Lett.);* Pergamon Press *(Collected Papers of L. D. Landau);* Plenum Press Publishing Corporation *(J. Stat. Phys.);* Springer-Verlag *(Commun. Math. Phys.* and *Lect. Notes Phys.).*

ISBN: 9971-966-50-6
9971-966-51-4 pbk

Printed in Singapore by Singapore National Printers (Pte) Ltd.

FOREWORD

Chaos is a rapidly expanding field of research to which mathematicians, physicists, hydrodynamists, ecologists and many others have all made important contributions, but in the end it is a newly-recognized and ubiquitous class of natural phenomena and thus belongs to the realm of physics. This volume is designed mainly for physicists with a standard mathematics education, and does not pretend to mathematical rigour in the Introduction but instead tries to rely more or less on physical intuition. However, a demanding reader can easily find precise formulations in the reprinted papers or trace them through the bibliography.

The better understood part of chaos is essentially classical. Quantum mechanics, in spite of its probabilistic interpretation, happens to be more deterministic, i.e. less chaotic, compared to its classical counterpart. In this volume we shall put aside the problem of quantum chaos except for citing a number of references in the bibliography.

Stochastic behaviour in classical Hamiltonian, in particular, conservative systems has become a well-shaped chapter of mathematics and there have been several excellent reviews and books. Chaotic phenomena in dissipative systems are closer to the heart of the physicist, but we need some notions formulated in the study of Hamiltonian systems. That is why a short chapter on the KAM theorem and stochasticity in classical dynamical systems is included. But in the main, this volume deals with dissipative systems.

The idea to compile an Introduction and Reprints Volume on chaos was suggested by Drs. K. K. Phua and K. Young. Dr. K. Young gave valuable advice on the organization of this book. However, the author alone takes responsibility for any possible mistakes in the Introduction and any bias in the choice of reprinted papers. It is absolutely impossible to include all important publications on chaos in a single volume. The author apologizes to all those whose papers did not find a place in this book.

The bibliography in this volume is based on a computerized bibliography list maintained by Ms. Zhang Shu-yu of the Institute of Physics, Academia Sinica, from which preprints and papers published after December 1983 were deleted. If any papers have been overlooked, we hope the authors would send us their papers for inclusion in future editions of the bibliography.

The author expresses his gratitude to all persons mentioned above and to many more colleagues not mentioned who have helped and taught him so much on life in the vast Empire of Chaos.

TO THE READER

This volume consists of three parts. The introduction and the reprinted papers are divided into chapters under the same headings indicating their rough correspondence. The bibliography is subdivided into "Books and Conference Proceedings" and "Papers including Review". At the end of each introductory chapter there are "References and Guide to the Bibliography", which contain not only entries cited in the text, but also a classified index to the bibliography. A reprinted paper is referred to as, e.g. "Hénon (1976), Paper 13 in this volume", while a title in the bibliography is referred to as, e.g. "Feigenbaum (1980a)". Entries in the "Books and Conference Proceedings" section are numbered from B1 to B40.

Unlike a textbook, this volume is not supposed to be read from the beginning. Those coming across chaos for the first time are recommended to skip papers of historical significance and to start with "May (1976), Paper 6 in this volume".

Good luck!

CONTENTS

PART THREE: BIBLIOGRAPHY

Part One

INTRODUCTION

Chapter 1
WHAT IS CHAOS?

There is no generally accepted definition of chaos. We begin with a few historical remarks.

In 1963 E. N. Lorenz published his numerical observations on a simplified model of thermal convection[1], a model which now carries his name. He discovered that in this completely deterministic system of three ordinary differential equations, all nonperiodic solutions were bounded but unstable, i.e. they underwent irregular fluctuations without any element of randomness introduced from the outside. In 1971 D. Ruelle and F. Takens[2] coined the term "strange attractor" for dissipative dynamical systems, though they were unaware of the Lorenz model as the first example having strange attractors. They further suggested a new mechanism for the onset of turbulence. Li and Yorke[3] seemed to be the first to introduce the word "chaos" into the mathematical literature to denote the apparently random output of some mappings, although the use of "chaos" in physics dates back to L. Boltzmann in another context, not unrelated to its present usage. In an excellent review published in 1976[4], R. May called attention to the very complicated dynamics including period-doubling and chaos in some very simple population models. Then came the discovery by Feigenbaum[5] of the scaling properties and universal constants in one-dimensional mappings, as well as the introduction of renormalization group idea into this field by him. The work of Feigenbaum triggered an upsurge of research interest among physicists, as witnessed by the bibliography in this volume.

There has been another course of events leading to the domain of chaos, namely, the study of nonintegrable Hamiltonian systems in classical mechanics. By the end of the last century the development of celestial mechanics and the foundation of statistical mechanics had posed a number of deep problems in classical dynamics. Since the success of relativity and quantum theory as well as the rapid progress of modern technology absorbed the attention of almost all physicists, these difficult problems were left to mathematicians for calm study for more than half a century. Their efforts crystallized in the formulation of the so-called KAM theorem on nearly integrable Hamiltonian systems in the early 60's (see Chapter 2). Numerical studies on what happens when the conditions of the KAM theorem fail have revealed an abundance of random motion in nonintegrable systems[6].

The phenomena related to the occurrence of randomness and unpredictability in completely deterministic systems have been called "dynamical stochasticity", "deterministic chaos", "self-generated noise", "intrinsic stochasticity", "Hamiltonian stochasticity" and so on by various authors. We prefer the word "chaos" as the shortest and use it mainly in the context of dissipative systems.

Actually there are two distinct classes of deterministic equations, and at least two different levels of stochasticity in physics.

On the fundamental "microscopic" level we have the law of dynamics, represented by Newton's equations in classical mechanics. In the absence of external time-dependent forces, they describe reversible movement of conservative systems. Does stochasticity and the necessity of a statistical description occur spontaneously, when the system gets complicated enough? This is the old problem of the relation between mechanics and statistical mechanics. In the hands of mathematicians, as part of ergodic theory, dynamical system theory and the qualitative theory of differential equations, significant progress has been reached in understanding this problem. We shall touch briefly on the essentials in Chapter 2, because in studying dissipative systems many concepts are borrowed from the well-shaped mathematics of stochasticity in Hamiltonian systems.

On the other hand, there are many macroscopic equations in physics, the first example being the Navier-Stokes equations describing the velocity field $\mathbf{V}$ of a fluid:

$$\frac{\partial \mathbf{V}}{\partial t} + (\mathbf{V}\nabla)\mathbf{V} = -\frac{1}{\rho}\nabla p + \nu\nabla^2\mathbf{V} \tag{1-1}$$

where p, ν and ρ denote respectively pressure, kinematic viscosity and density. For another example one may take the reaction-diffusion equations describing chemical reactions in an inhomogeneous medium:

$$\frac{\partial x_i}{\partial t} = f_i(x_1, \ldots, x_n) + D_i\nabla^2 x_i \quad , \qquad i = 1, \ldots, n \tag{1-2}$$

where the reaction kinetics is represented by functions f_i and the diffusion caused by spatial inhomogeneity is described by the second term, D_i being the diffusion constant of the i-th component x_i.

These nonlinear (due to the $(\mathbf{V}\nabla)\mathbf{V}$ term or the functions f_i) and dissipative (due to the ν or D_i terms) evolution equations are irreversible in nature. Although they can be derived, in principle, from microscopic dynamics by making statistical assumptions or coarse-graining at certain steps of the derivation, the equations themselves, supplemented with appropriate boundary and initial conditions, are apparently deterministic. It is a basic experimental fact that under certain conditions a fluid or a reacting system may undergo transitions into states of more and more erratic motion, which in turn necessitates a statistical description. This leads to another long-standing problem in physics: the problem of turbulence. Does the onset of turbulence follow from such macroscopic deterministic equations as (1-1) or (1-2)? And furthermore, can one characterize the state of developed turbulence starting from these same equations? To a certain extent the recent upsurge of interest in chaos was roused by the hope to understand these questions. We shall devote Chapter 3 to the problem of turbulence.

While equations (1-1) and (1-2) are partial differential equations, i.e. systems with infinite degrees of freedom, most models studied so far are low-dimensional, e.g., one- or two-dimensional mappings, ordinary differential equations with three or more variables, etc. A moral drawn from recent studies on chaos in low-dimensional systems consists of the belief that at least the onset of turbulent behaviour in dissipative systems, no matter how large the original phase space, may be described by motions in subspaces of much lower dimension, called attractors. In other words, it is dissipation that realizes the contraction of description in a natural way: a vast number of modes die out due to dissipation, only those spanning the attractors need be taken into account in modelling the system. This kind of simplification cannot take place in Hamiltonian systems owing to the preservation of phase volume (Liouville's theorem). Moreover, dissipation is unavoidable in most experiments. This explains why chaos in dissipative systems has attracted the attention of more and more physicists.

To conclude this introductory chapter we emphasize that chaos is not to be equated simply with disorder. It is more appropriate to consider chaos as a kind of order without periodicity. Within generally chaotic regimes one can discover patterns of ordered motion interspersed with chaos at smaller scales, provided sufficiently high resolving power is reached in numerical or laboratory experiments. Instead of the usual spatial or temporal periodicity, there appears some kind of scale invariance which opens the possibility for renormalization group considerations in studying chaotic transitions.

References and guide to the bibliography

1. Lorenz (1963), Paper 14 in this volume, see also book B32 (Sparrow, 1982).

2. Ruelle and Takens (1971), Paper 5 in this volume.
3. Li and Yorke (1975), Paper 10 in this volume.
4. May (1976), Paper 6 in this volume. This is a good starting point for a first acquaintance with chaos.
5. Feigenbaum (1978, 1979), Papers 7 and 8 in this volume.
6. See, e.g., Walker and Ford (1969), Paper 2 in this volume, and references given at the end of Chapter 2.
7. For semi-popular articles on chaos see, e.g., Ruelle (1980b), Feigenbaum (1980), GBL (1981), Hofstadter (1981), and Kadanoff (1983b).
8. Among recent books on chaos we draw attention to B31 (Lichtenberg and Lieberman, 1982) and B36 (Guckenheimer and Holmes, 1983), both from the Springer Applied Mathematics Series. B31 deals basically with conservative systems and is more readable for physicists, while B36 distinguishes itself in mathematical clarity.
9. To our knowledge no book on the physics of chaotic phenomena has appeared yet, but there are many reviews in physics journals, e.g., Eckmann (1981), Ott (1981), Shaw (1981), Hu (1982a), Tomita (1982a), Hao (1983), Huberman (1983a), and especially, Swinney (1983) for a review on experimental observations of chaos. One may also consult the proceedings of many conferences listed in the first part of the bibliography.
10. Since we shall not touch at all the problem of quantum chaos in this volume, we mention here the book B4 (Casati and Ford, 1977) and two reviews: Chirikov *et al.*, (1981), Zaslavsky (1981), as well as a few recent papers: Berman and Zaslavsky (1977, 1982), Shepelyansky (1981, 1983), Bellissard *et al.*, (1982), Fishman *et al.*, (1982), Hogg and Huberman (1982, 1983), Jaffe and Reinhardt (1982), Shapiro and Child (1982), Weissman and Jortner (1982), Berman and Kolovsky (1983), Casati and Guarneri (1983), and Graham (1983b).

Chapter 2

KAM THEOREM AND STOCHASTICITY IN CLASSICAL HAMILTONIAN SYSTEMS

As mentioned before, stochasticity in Hamiltonian systems has become a well-shaped chapter of mathematics and there are many excellent reviews and books on this subject[1]. This chapter is included since some notions formulated for Hamiltonian systems remain useful for understanding chaotic behaviour in dissipative systems.

To begin with we recall a few concepts from analytic mechanics. The motion of a classical conservative system of N degrees of freedom with Hamiltonian function

$$H = H(p_1, \ldots, p_N; q_1, \ldots, q_N) \tag{2-1}$$

is described by the Hamilton's canonical equations

$$\dot{q}_i = \frac{\partial H}{\partial p_i}, \qquad \dot{p}_i = -\frac{\partial H}{\partial q_i} \qquad i = 1, \ldots, N \tag{2-2}$$

If there exist successive canonical transformations changing $\{p_i, q_i\}$ into a new set of canonical variables $\{J_i, Q_i\}$ such that in terms of these new variables the Hamiltonian function depends only on the J_i's, but not on the Q_i's, i.e., all the Q_i's become cyclic variables:

$$H = H(J_1, \ldots, J_N) \quad , \tag{2-3}$$

then the corresponding canonical equations

$$\dot{Q}_i = \frac{\partial H}{\partial J_i} = \Omega_i(J_1, \ldots, J_N)$$

$$\dot{J}_i = -\frac{\partial H}{\partial Q_i} = 0 \tag{2-4}$$

can readily be integrated to give

$$Q_i(t) = \Omega_i t + Q_i(0) \quad ,$$

$$J_i(t) = J_i(0) \quad . \tag{2-5}$$

Now going back to the old variables, one gets $2N$ combinations of $\{p_i, q_i\}$ and t:

$$Q_i(0) = Q_i(p_1(t), \ldots, p_N(t); q_1(t), \ldots, q_N(t); t) \quad ,$$

$$J_i(0) = J_i(p_1(t), \ldots, p_N(t); q_1(t), \ldots, q_N(t)\ ) \quad , \tag{2-6}$$

which do not change with time. In other words, we have solved the equations of motion completely and obtained $2N$ constants (or integrals, or invariants) of the motion. Such Hamiltonian systems are said to be ***integrable.*** In fact, the existence of N independent integrals of motion suffices to make the system integrable.

A qualitative picture of the motion of an integrable system looks very much like that of a system of coupled oscillators, as can be seen from (2-5). To keep the motion in a finite region of the phase space, the linearly growing $Q_i(t)$'s must appear as arguments of periodic functions. When $N = 1$ the motion can be visualized as rotation around a circle of radius $\sqrt{2J(0)}$ with constant angular velocity Ω; therefore the bounded motion of a system with one degree of freedom is always periodic. When $N = 2$, there are two radii determined by $J_1(0)$ and $J_2(0)$ with angular velocities Ω_1 and Ω_2, so the motion is confined to a two-dimensional toroidal surface or a 2-torus. Besides periodic motion a new possibility appears. When the ratio Ω_1/Ω_2 happens to be an irrational number, the motion can no longer be periodic: the trajectory winds up the 2-torus densely and endlessly. This kind of motion is called ***quasiperiodic*** (or conditionally periodic in the Russian literature). In general, the motion of an integrable Hamiltonian system with N degrees of freedom is quasiperiodic and is confined to an N-torus. Therefore, integrable systems have nothing to do with the requirements of statistical mechanics, since the dimension of the constant energy surface $(2N - 1)$ is larger than that of the torus whenever $N > 1$, and the trajectory can in no way fill up the energy surface, not to mention the equal probability assumption of microcanonical ensemble.

Two questions arise immediately:

1. Are there many integrable systems among all Hamiltonian systems?

2. What happens with the qualitative picture of motion when the system is made slightly nonintegrable, i.e., when the Hamiltonian becomes

$$H = H_0 + V \quad , \tag{2-7}$$

where H_0 is integrable and V contains a small parameter[2]?

The answer to the first question is definitely negative[2]. It was found that integrability is an exceptional property for Hamiltonian systems whenever the number of degrees of freedom gets larger than two. Integrable systems are so rare that in general it is impossible to approximate a nonintegrable Hamiltonian system by a series of integrable ones. This statement is to be compared with irrational numbers which can always be approached from both sides by sequences of rationals, because rational numbers are dense on the number axis, though having zero measures.

The answer to the second question is provided by the KAM theorem[3], first enunciated by Kolmogorov in 1954 and completely proved by Arnold and Moser in the early 60's. The proof required a successful treatment of the small divisor problem in perturbative solutions to the classical many-body problem. The mathematical prerequisite goes beyond that of an average physicist, so we confine ourselves to a loose formulation of the theorem and then turn to its physical implications.

KAM proved that provided the following two conditions hold:

(a) The perturbation V causing nonintegrability in (2-7) is small (we ignore the precise formulation for smallness);

(b) The frequencies Ω_i of the unperturbed integrable Hamiltonian satisfy the noncorrelated or nonresonance condition

$$\frac{\partial(\Omega_1, \ldots, \Omega_N)}{\partial(J_1, \ldots, J_N)} = 0 \quad , \tag{2-8}$$

then the motion is still confined to an N-torus except for a negligible set (of measure zero) of initial conditions which may lead to wandering motion on the energy surface. These N-tori, now called *KAM surfaces*, or *KAM curves* if seen in plane sections, may be slightly distorted compared to that of the $V = 0$ case; nevertheless the qualitative picture of the motion remains much the same as the unperturbed integrable system.

If we follow the exceptional trajectories mentioned in the KAM theorem, then a qualitatively new phenomenon appears for systems with N degrees of freedom, $N > 2$. This follows from the fact that the boundary of the constant energy surface must be of dimension $2N-2$ and when $N < 2N-2$, the N-dimensional KAM torus cannot serve as boundaries dividing the energy surface into regions impenetrable for the wandering trajectories. Therefore, when $N > 2$ these trajectories may wander along the whole energy surface and give rise to a new mechanism of randomness called Arnold diffusion[4]. Physically this new possibility remains unobservable as far as the KAM conditions hold.

It is appropriate to summarize briefly at this point: Hamiltonian systems with $N = 1$ are all integrable; the overwhelming majority of systems with $N \geqslant 2$ becomes nonintegrable; for $N > 2$ Arnold diffusion may show up.

What happens when one violates the condition of KAM theorem? This appears to be a very difficult problem, just to cite such a competent mathematician as Arnold: "Nonintegrable problems of dynamics appeared inaccessible to tools of modern mathematics"[5]. Still mathematicians were able to tell the qualitative picture of how the KAM tori are destroyed and modern computers are of much help to visualize this process[6].

Let us try to grasp the essentials without involving mathematics. We start with the mathematical pendulum described by the simple nonlinear differential equation

$$\ddot{\varphi} + \omega^2 \sin\varphi = 0 \qquad (2\text{-}9)$$

The angle φ and the angular velocity $\dot{\varphi}$ span a two-dimensional phase plane. Due to periodicity of the motion we can consider only an infinite strip $-\pi \leqslant \varphi < \pi$ in the phase plane (Fig. 2-1). Among the equi-energy curves shown in Fig. 2-1, there is one connecting the points $(-\pi, 0)$ and $(\pi, 0)$. This separatrix divides the strip into three regions with different types of motion: oscillation in the central region and rotation in opposite directions in the upper and lower regions.

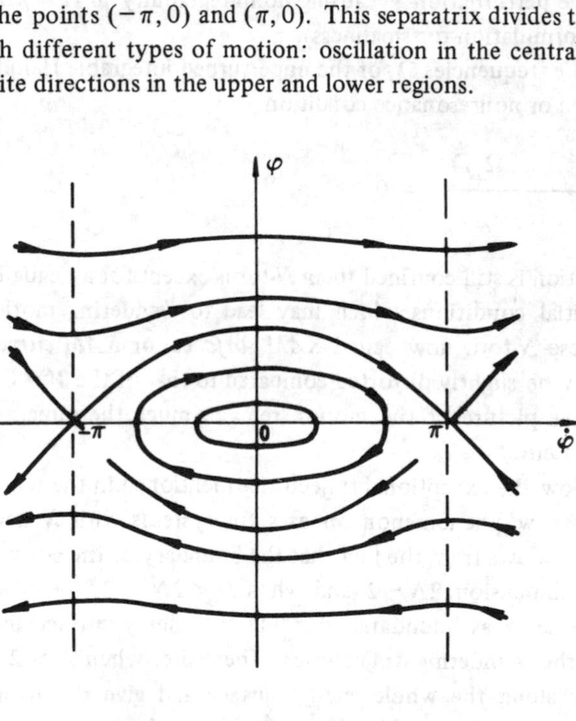

Fig. 2-1 Phase plane of the mathematical pendulum.

In Fig. 2-1 the stable equilibrium point $(0, 0)$, surrounded by ellipses, is an elliptic point or a center, while the unstable equilibrium point $(\pi, 0)$ is a hyperbolic or saddle point. We see that the separatrix leaves one saddle point along the "unstable" direction and enters the other saddle point along the "stable" direction and vice versa. Actually, $(\pi, 0)$ and $(-\pi, 0)$ correspond to the same point and the "stable" and "unstable" directions of the separatrix intersect at the saddle point.

Now take a system of two coupled oscillators in its nearly integrable regime. A section of the phase space in terms of J_i, Q_i would appear as if assembled from Fig. 2-1, i.e., there are regions corresponding to motion of different frequencies called *resonance zones*, see Fig. 2-2. When the coupling gets stronger and the conditions of KAM theorem begin to break down, these resonance zones tend to overlap and the original separatrices become stochastic layers of finite width (see Fig. 2-3). At the same time, some of the originally simple closed curves split into successions of elliptic and hyperbolic points at small scales. The overlap of resonance zones can be cast into a quantitative criterion for the appearance of stochasticity (the *Chirikov criterion*[7]).

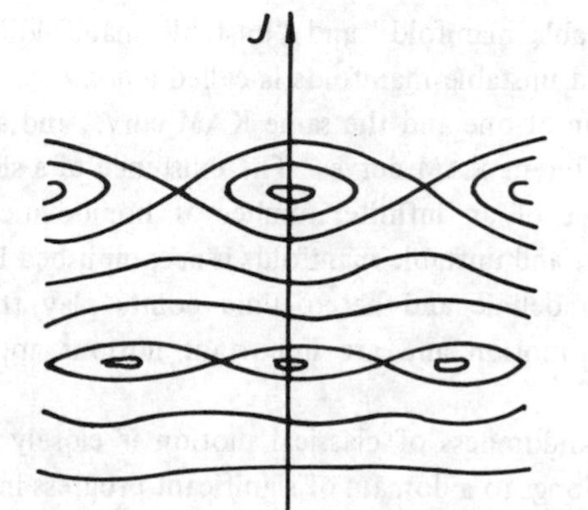

Fig. 2-2 Resonance zones of coupled oscillators.

In general, the KAM torus looks much like these figures if intersected by planes in the phase space. The KAM curves can be classified according to the "distance" of their underlying frequency ratio from rationals. The more irrational a curve is, the longer it persists during the violation of the KAM conditions. The destruction of KAM curves shares the general feature of going from Fig. 2-2 to Fig. 2-3. The intersection of stable and unstable directions seen in Fig. 2-1 now generalizes to

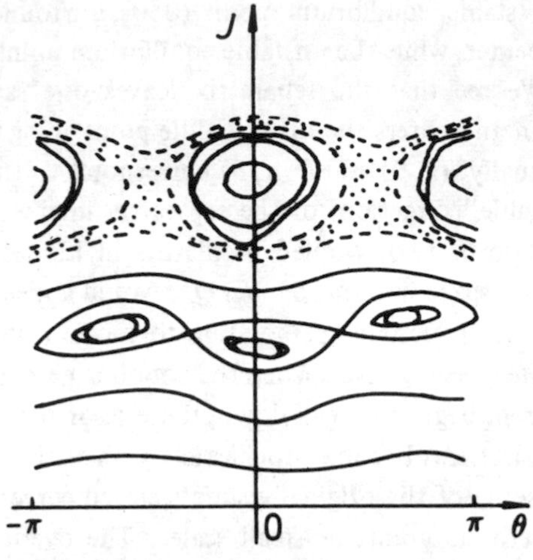

Fig. 2-3 Stochastic layers.

intersections of "stable manifold" and "unstable manifold". A transversal intersection of stable and unstable manifolds is called a *homoclinic point*, if they result from the destruction of one and the same KAM curve, and a *heteroclinic point* if they come from different KAM curves. The existence of a single homoclinic point implies the presence of an infinite number of homoclinic points, because the intersection of stable and unstable manifolds is accomplished in a very intricate way (see Fig. 2-4). Homoclinic and heteroclinic points play the role of organizing centers for chaotic motion and are important notions applicable to dissipative systems as well.

The degree of randomness of classical motion is closely related to its ergodic property, which belongs to a domain of significant progress in last 20 years[8]. Since one encounters the concepts of ergodicity when characterizing the attractors in dissipative systems, a few words may be in order.

A system is said to be ergodic on its energy surface if time averages along a trajectory are equal to ensemble averages over the whole energy surface. Ergodicity alone means very little in randomness: two neighbouring points may remain correlated all the time. The next step on the ladder of randomness is called mixing: any initial region on the energy surface evolves into filaments which cover the whole surface when time goes on. Correlation between initial neighbours must decay with time, but nothing is required as regards the decay rate. If the motion suffers local orbital instability, i.e. any nearby trajectories go apart exponentially, then the correlation also decays exponentially. Such system is said to be a *K-flow*

Fig. 2-4 Intersections of stable and unstable manifolds (schematic).

(after Kolmogorov). The separation rate of neighbouring points averaged along the trajectory determines the *K-entropy* of the system. A system is a *K*-flow if it possesses positive *K*-entropy. This serves as another criterion for stochasticity along with the Chirikov resonance overlap criterion, which often give identical results in simple cases. We shall return to the notion of entropy in Chapter 8. Skipping the highest level of randomness, namely the Bernoulli flow[8], we only recall that every higher step on the ladder of ergodicity implies the lower ones, but not vice versa. K-flow is the normal case one encounters in "Hamiltonian chaos".

To conclude this chapter let us emphasize that instability does not mean collapse of the system, but opens the way to intrinsic stochasticity. A deeper thought on the schematic parallel

stability —— determinism

instability —— randomness

would help the reader to get rid of the traditional prejudice that classical mechanics is fully deterministic and to recognize stochasticity as an ubiquitous and intrinsic property of nonintegrable Hamiltonian systems.

References and guide to the bibliography

1. Books on stochasticity in Hamiltonian systems: B2 (Moser, 1973), B4 (Casati and Ford, 1977), B6 (Arnold, 1978), B31 (Lichtenberg and Lieberman, 1982); Reviews: Chirikov (1969, 1979), Ford (1973, 1974, 1975), Whiteman (1977), Berry (1978), Helleman (1980), and Wightman (1981).
2. Siegel (1941, 1954).
3. Kolmogorov (1954). We include this paper as the first one in this volume for its historical significance; Arnold (1963a) and Moser (1962) were purely mathematical, but Arnold (1963b) is readable for physicists.
4. For Arnold diffusion see Arnold (1964), Chirikov *et al.* (1979), Tennyson *et al.* (1979), Lieberman (1980), Chirikov and Shepelyansky (1982), Holmes and Marsden (1982a, 1983).
5. Arnold (1963a).
6. The first computer study was reported by Hénon and Heiles (1964) and then continued by Ford and coworkers. See Walker and Ford (1969), Paper 2 in this volume, Ford and Lunsford (1970, 1972), Lunsford and Ford (1972). Renormalization group ideas have been applied to the destruction of KAM surfaces recently, see Kadanoff (1981a, b, 1983a), Escande and Doveil (1981a, b), Shenker and Kadanoff (1982), Escande (1982a, b, c).
7. For stochastic layer and the resonance overlap criterion see the reviews by Chirikov (1969, 1979).
8. An elementary introduction to modern ergodic theory can be found in J. L. Lebowitz and O. Penrose, *Physics Today*, 1973, February, p. 23, but nothing was said about *K*-systems. We recommend another *Physics Today* paper by Ford, April 1983, p. 40. The standard reference on ergodic theory is B1 (Arnold and Avez, 1968).

Chapter 3
THE PROBLEM OF ONSET OF TURBULENCE

As we mentioned in Chapter 1, turbulence has been a long-standing problem in physics. The difficulty is rooted in the simultaneous presence of many, many length scales, or in other words, in the lack of a single characteristic length. This can be seen from the intuitive picture of a turbulent fluid: nested and interpenetrated eddies of all scales, from macroscopic down to "molecular". In this respect the problem of turbulence bears similarity to the problem of continuous phase transitions, where length scales ranging from the correlation length, which approaches infinity at the transition temperature, down to the atomic scale all play an essential role. Perhaps this explains why people like L. D. Landau and K. G. Wilson who contributed so much to the understanding of phase transitions also thought about turbulence.

Just as in the case of phase transitions, the key to understanding turbulence may be hidden in the onset mechanism, as pointed out by L. D. Landau forty years ago: "... the problem may be in a new light if the process of initiation of turbulence is examined thoroughly"[1]. We would like to make it clear from the very beginning that chaotic phenomena in dissipative systems, at least for the time being, are relevant only to the onset mechanism of turbulence, i.e. to the stage of weak turbulence. It has nothing to do with the fully developed turbulence which is of primary importance in engineering. In addition, chaos as it is treated in this volume concerns mainly erratic behaviour in time evolution, whereas turbulence necessarily involves stochasticity in the spatial distribution as well.

Real turbulence occurs in three-dimensional space. However, most mathematical models and experimental situations studied so far are confined to finite or low-dimensional geometry, e.g., fluid instability between rotating cylinders (the Taylor instability) or thermoconvective instability in small boxes (the Rayleigh-Bénard instability). Developed turbulence must involve a great number of fluid motion modes, but the onset of turbulence may stem from the loss of stability of only a few modes. Finite geometry together with dissipation just provides the mechanism to suppress many irrelevant modes what makes the experimental results closer to predictions based on simple theoretical models.

It has been realized for a long time that turbulence might be a sophisticated regime of nonlinear oscillation in continuous media. L. D. Landau[1] and E. Hopf[2] attributed the onset of turbulence to the appearance of an increasing number of quasiperiodic motions resulting from successive bifurcations in the system. Being an extension of the Hopf bifurcation idea, the first steps of this process can easily be understood geometrically.

When the Reynolds' number R, which represents the relative importance of nonlinear to dissipative terms in the Navier-Stokes equations (1-1), remains small enough, the fluid motion is laminar and stationary, corresponding to a stable *fixed point* in its phase space. A stable fixed point acts as an *attractor*, i.e., it attracts all nearby initial points towards itself (see Fig. 3-1(a)). Now, let the Reynolds' number be increased infinitesimally larger than the first critical value R_{c1} where the fixed point loses stability and begins to repel all nearby trajectories (Fig. 3-1(b)). Since a small change in R cannot cause such drastic consequence as inverting the direction of all flows on the whole phase space, the neighbourhood of the fixed point may become repelling, but it must remain attracting with respect to regions located far enough. Local repulsion and global attraction of the flow implies the formation of a closed curve around the now unstable fixed point and the curve attracts all nearby

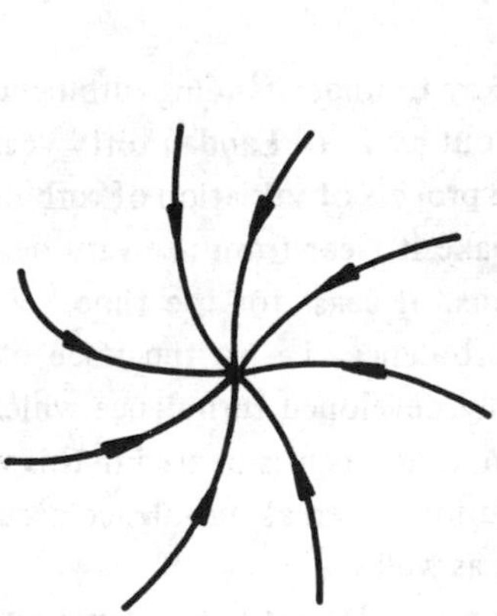

Fig. 3-1(a)
A stable fixed point.

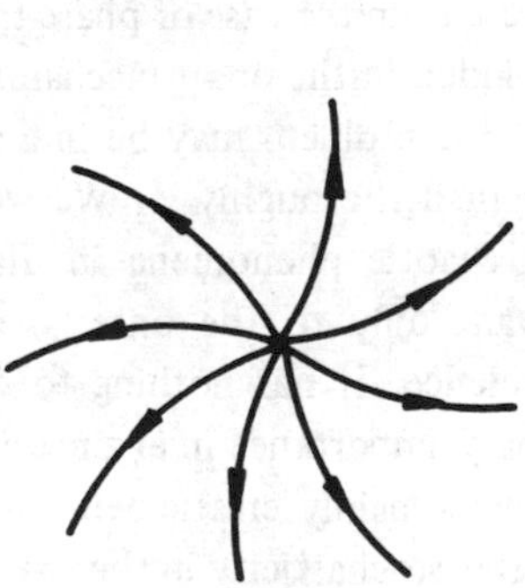

Fig. 3-1(b)
An unstable fixed point.

flows (see Fig. 3-1(c)). This closed curve is called a *limit cycle* and corresponds to a periodic motion of the system.

Fig. 3-1(c)
A compromise appearance of a limit cycle.

The process of generating a limit cycle from a fixed point is called a *Hopf bifurcation.* Repeated use of the above arguments would reveal the nature of the next bifurcation when the limit cycle loses stability and becomes repelling at R_{c2}. There would appear an attracting closed tube, i.e. a 2-torus, around the unstable limit cycle. The motion becomes quasiperiodic if the two frequencies on the torus are incommensurable. Landau and Hopf allowed this process to continue infinitely and identified the final state with an infinite number of incommensurable frequencies as fully turbulent. This was called the *Landau-Hopf route* (or scenario) to turbulence.

At present, we do not know any reasonable mathematical model which follows the Landau-Hopf route to turbulence. The Landau-Hopf route requires successive appearance of new incommensurable frequencies in the power spectrum, which remains discrete all the time as Reynolds' number increases. Turbulent spectra in laboratory experiments do develop a few independent frequencies, but then turn into broad noisy bands. There is no mechanism for sensitive dependence on initial conditions in Landau-Hopf scheme, but the details of turbulent states do depend on initial conditions sensitively. In addition, the Landau-Hopf picture ignores an important physical phenomenon — frequency locking. In fact, in nonlinear systems new incommensurable frequencies cannot appear infinitely without interacting with each other. Nearby frequencies tend to get locked, which will diminish the number of independent frequencies. The above remarks exclude the Landau-Hopf route as an onset mechanism for turbulence from both the theoretical and the experimental points of view.

In 1971 Ruelle and Takens[3] showed that the Landau-Hopf route is unlikely to occur in nature. It is enough to have four consecutive bifurcations to get into a state of erratic motion described by interweaving trajectories attracted to a low-dimensional manifold in the phase space called a ***strange attractor.*** They identified the motion on strange attractors with turbulence. Their scheme may be summarized as: fixed point → limit cycle → 2-torus → 3-torus → strange attractor (turbulence). A few years later, in collaboration with Newhouse, these authors succeeded in reducing the scheme to: fixed point → limit cycle → 2-torus → strange attractor, i.e., quasiperiodic motion on a 2-torus may lose stability and give birth to turbulence directly. This so-called *Ruelle-Takens route* to turbulence seems to be more consistent with recent hydrodynamical experiments, but it is less well understood in theoretical models except for some deep results on circle mappings. We shall devote Chapter 7 to transitions from quasiperiodic motion to chaos.

The work of Ruelle and Takens played an eye-opening role. More and more people now believe that at least the problem of turbulence onset can be settled within the framework of Navier-Stokes equations and many new routes to turbulence have been suggested. Now we are facing a situation of "all routes lead to turbulence", among which the most thoroughly studied are the *period-doubling route* of Feigenbaum[4] and the *intermittent route* of Pomeau-Manneville[5]. These two routes are actually twin phenomena. The next Chapter 4 on one-dimensional mappings is at the same time an introduction to the period-doubling route and Chapter 6 will deal with the intermittent route, so we shall not go into details here.

In conclusion we would like to point out that turbulence has become a general concept, related to many branches of natural sciences and not less important than the concept of order. New terms such as solid state turbulence, chemical turbulence, acoustic turbulence or optical turbulence are emerging into the literature. Subtle measurements at liquid helium temperatures, laser Doppler velocimetry and modern data acquisition technique[6] have brought the study of turbulence back into physics laboratories. The idea of scaling and universality as well as renormalization group arguments which have proved so successful in understanding phase transitions are now being adapted to chaos and turbulence. In one word, turbulence should not be considered as a specific problem in hydrodynamics. It should attract the attention of physicists, because "That is the central problem which we ought to solve some day, and we have not"[7].

References and guide to the bibliography

0. Review articles on the onset mechanism of turbulence in light of chaos and strange attractors: Monin (1978), Rabinovich (1978, 1980), Iooss and Lanford (1980), Lanford (1981, 1982a), Boon (1981), Coullet and Vanneste (1983),

and especially Eckmann (1981).

1. Landau (1944), Paper 3 in this volume.
2. Hopf (1948), Paper 4 in this volume.
3. Ruelle and Takens (1971), Paper 5 in this volume, Newhouse *et al.*, (1980), Ruelle (1978c, 1979b, 1980a, 1981a, 1983).
4. Feigenbaum (1980a, b, c, d), and reviews listed at the end of Chapter 4.
5. Manneville and Pomeau (1979), Pomeau and Manneville (1980), Paper 20 in this volume, and reviews at the end of Chapter 6.
6. Swinney and Gollub (1978), and Swinney *et al.*, (1977a, b).
7. R. P. Feynman, R. B. Leighton and M. Sands (1963). *The Feynman Lectures on Physics*, Vol. I, pp. 3–10, Addison-Wesley.

Chapter 4

UNIVERSALITY AND SCALING PROPERTIES OF ONE-DIMENSIONAL MAPPINGS

It often happens in physics that one-dimensional models are either too trivial or too specific to be extended to higher dimensions, but chaos in dissipative systems offers a lucky and instructive exception. The reason is very simple: dissipation plays a global stabilizing role against local orbital instability and causes the volume representing the initial states in phase space to contract in the process of evolution. This contraction makes the phase volume approach one-dimensional objects in some of its sections and thus enables higher dimensional systems to enjoy the universal properties of one-dimensional mappings.

One-dimensional mappings of an interval into itself are simple enough to be accessible to certain analytical tools and are not very time-consuming in numerical study. At the same time they are rich enough to show many of the scaling and universal properties of chaotic transitions observed in higher dimensional systems. Therefore, we shall treat them in more detail than we did in previous chapters.

Consider a real interval I and a nonlinear function f which transforms any point x of I into some point x' in the same interval I. This is called *a map of the interval*

$$f\colon\ I \rightarrow I \tag{4-1}$$

In general, the function f may depend on a parameter μ. We can choose an arbitrary initial point $x_0 \in I$ and iterate it using f:

$$x_n = f(\mu, x_{n-1}) \quad , \qquad n = 1, 2, 3, \ldots \tag{4-2}$$

Formula (4-2) can be rewritten as

$$x_n = f^{(n)}(\mu, x_0) \quad , \qquad n = 1, 2, 3, \ldots \tag{4-3}$$

where $f^{(n)}$ denotes the n-th iterate (not derivative!) of f and sometimes it is useful to have another specific notation for it:

$$F(n,\mu,x) \equiv f^{(n)}(\mu,x) \equiv \underbrace{f(\mu,f(\mu,\ldots f(\mu,x)}_{n\ \text{times}}\ldots)) \quad . \tag{4-4}$$

In general, the property of the sequence $\{x_i,\ i=0,1,2,\ldots\}$ depends on the function f and on the choice of x_0 and μ. We confine ourselves to those functions f which have only one maximum on the interval I. Without loss of generality one can rescale f and x in such a way that

(1) the maximum is located at $x=0, f'(\mu,0)=0$, and $f(\mu,0)=1$;

(2) $f(\mu,x)$ is monotonically increasing when $x<0$, and monotonically decreasing when $x>0$.

This kind of mappings has been called *unimodal*. We further require that

(3) in the neighbourhood of $x=0$, f can be expanded as

$$f(\mu,x)=1-ax^z+\ldots \tag{4-5}$$

where $z=2,4,6,\ldots$ etc.

For concreteness we shall refer to the *logistic mapping*

$$x_n=1-\mu x_{n-1}^2 \tag{4-6}$$

as the representative of unimodal mappings. In this case $I=[-1,+1]$, $\mu\in(0,2)$, and $z=2$. We see that even when x runs over the whole interval I, $f(\mu,x)$ does not necessarily fill up the interval I, hence the name *endomorphism* (endo = internal) of the interval in some mathematical literature.

Now let us return to the sequence $\{x_i,\ i=0,1,2,\ldots\}$. Usually after a few hundred transient points, it settles into one of two kinds of stationary patterns: periodic or aperiodic. A periodic pattern of period p: $x_{i+p}=x_i, x_{i+k}\neq x_i$ for all $k<p$ and all i larger than certain N, is also called an *orbit* of period p or a *p-cycle* for the mapping f. In what follows we shall assume that the transients have died away and omit the phrase "for all i larger than certain N". The particular case $p=1$ corresponds to a fixed point for f:

$$x^*=f(\mu,x^*) \quad . \tag{4-7}$$

By definition, all points from a p-cycle of f must be fixed points of the p-th iterate of f, i.e.

$$x_i=F(p,\mu,x_i) \qquad i=1,2,\ldots,p \quad . \tag{4-8}$$

A standard question to be asked about a fixed point is its stability, i.e. if x_n

is chosen very close to the fixed point x^*:

$$x_n = x^* + \epsilon_n$$

what happens with the next iterate

$$x_{n+1} = x^* + \epsilon_{n+1} \quad ?$$

If

$$\left|\epsilon_{n+1}/\epsilon_n\right| < 1 \quad , \tag{4-9}$$

we say that the fixed point x^* is stable. The stability condition (4-9) is equivalent to the requirement

$$|f'(\mu, x^*)| < 1 \quad . \tag{4-10}$$

Similarly, a p-cycle is stable if

$$\left|F'(p, \mu, x_i)\right| = \left|\prod_{j=1}^{p} f'(\mu, x_j)\right| < 1 \ , \qquad i = 1, 2, \ldots, p \quad . \tag{4-11}$$

where the chain rule of differentiation has been used.

The most favourable case for stability appears when

$$F'(p, \tilde{\mu}, x_i) = 0 \ , \qquad i = 1, 2, \ldots, p \ , \tag{4-12}$$

then one has quadratic convergence towards the fixed point. The stability condition (4-11) holds for a finite interval called a *periodic window* on the μ-axis, whereas (4-12) takes place only at one particular value $\tilde{\mu}$ somewhere in the middle of the periodic window. This $\tilde{\mu}$ value corresponds to a *superstable period*, which serves as the representative of all p-cycles from the same periodic window. With our conventions on f we can say that any cycle containing the point $x = 0$ must be superstable, since $f'(\mu, 0) = 0$ leads to $F'(p, \mu, 0) = 0$ for all p. This suggests an idea to determine the superstable value $\tilde{\mu}$ by solving the fixed point equation $F'(p, \mu, 0) = 0$, but we shall mention a better method later.

Equipped with a desk calculator one can easily find the first periodic windows for the logistic mapping (4-6):

$p = 1$	$0 < \mu < \mu_1 = 0.75$
$p = 2$	$\mu_1 < \mu < \mu_2 = 1.25$
$p = 4 = 2^2$	$\mu_2 < \mu < \mu_3 = 1.3680989 \ldots$
$p = 8 = 2^3$	$\mu_3 < \mu < \mu_4 = 1.3940461 \ldots$
$\ldots$	$\ldots$

This is a *period-doubling bifurcation cascade* with period $p = 2^n$ which quickly converges to an aperiodic orbit at $n = \infty$, the value $\mu_\infty = 1.401155\ldots$ being approached as a geometric progression, namely

$$\mu_n \approx \mu_\infty - \frac{A}{\delta^n} \quad , \qquad \text{as} \quad n \to \infty \tag{4-13}$$

where $\delta = 4.66920\ldots$ is a universal (for unimodal mappings with $z = 2$) constant, first discovered by M. J. Feigenbaum[1].

What happens when μ gets larger than μ_∞ is more interesting from the viewpoint of chaos. In the parameter range $(\mu_\infty, 2)$ there exists an infinite number of periodic windows immersed in the background of aperiodic regime. If one examines more carefully the distribution of points in the aperiodic regime, then one sees the iterates jumping between 2^n subintervals (or islands) of the interval I with n decreasing from ∞ to 0 when μ goes from μ_∞ to 2. This is the so-called *reversed*, or *period-halving bifurcation sequence* of chaotic bands, which looks much like a washed-out mirror image of the direct period-doubling bifurcation sequence with respect to μ_∞. This overall structure can be seen clearly from the bifurcation diagram, obtained by plotting a few hundred stationary outputs of the iteration (4-6) versus the parameter μ (see[2] for figures).

If what described above is restricted only to a specific mapping, e.g. (4-6), it would be worth no more than a rare bird in the mathematical zoo. However, this kind of bifurcation structure with its universal numerical characteristics (such as the Feigenbaum constant δ) appears more and more in nonlinear mathematical models and real experiments. In this introductory chapter we confine ourselves to listing some useful notions and rigorous results related to the universal and scaling properties of one-dimensional mappings.

1. A unimodal mapping can have at most one stable period for each parameter value μ. It may have no stable period at all for many μ values. The necessary condition for f to have at most one stable period was found in 1978 and consists in the *Schwarzian derivative* Sf of f being negative on the interval I[3]:

$$Sf(x) \equiv \frac{f'''(x)}{f'(x)} - \frac{3}{2}\left(\frac{f''(x)}{f'(x)}\right)^2 < 0 \tag{4-14}$$

(see[4] for more on the Schwarzian derivative). The condition (4-14) is not sufficient for the stable period to exist. It is just this insufficiency that opens the possibility for chaotic orbit to appear: even when $Sf < 0$ one can get different aperiodic sequences starting from different x_0 and never reach a stable period.

2. The set of all parameter values which give birth to chaotic orbits possesses a positive measure on the μ-axis, although no such μ values form an interval. Moreover, to classify an orbit as chaotic one must show that all consecutive points approach a continuous distribution with respect to dx. Computers are of no use in proving this kind of statements, and one must appeal to rigorous mathematics. Proofs for some particular parameter values have been known since long[5], but a general proof for certain classes of mappings appeared only in 1981[6].

3. The classification and ordering of periodic orbits has been a well-studied mathematical problem with an extensive literature. In this and the next subsections we list a few results, frequently encountered in the literature. First of all, there was a theorem proved in 1964[7] which became known to physicists much later:

Sarkovskii theorem: consider the following ordering of integers

$$3 \to 5 \to 7 \to 9 \to \ldots \to 3*2 \to 5*2 \to 7*2 \to 9*2 \to \ldots \to 3*2^2 \to 5*2^2$$

$$\to 7*2^2 \to 9*2^2 \to \ldots \to 3*2^n \to 5*2^n \to 7*2^n \to 9*2^n \to \ldots\ldots 2^m$$

$$\to \ldots \to 32 \to 16 \to 8 \to 4 \to 2 \to 1\,.$$

(the symbol $\to$ means "precede"). If f is an unimodal mapping and has a point x leading to a p-cycle, then it must have a point leading to a q-cycle for every $q \leftarrow p$ in the sense of the above ordering.

It must be emphasized that the Sarkovskii theorem is only a statement concerning different x's at fixed parameter value μ. It says nothing about the stability of the periods, nor about the measure, i.e. the observability of these periods. Unaware of the work of Sarkovskii, many equivalent formulations or corollaries of this theorem were suggested in the intervening years, among which the Li-Yorke theorem "Period 3 implies chaos"[8] and the Oono's statement "Period $\neq 2^n$ implies chaos"[9], etc.

4. The *U-sequence* and *symbolic dynamics.* The order of appearance of various stable periods in an unimodal mapping $f(\mu, x)$ when the parameter μ varies is certainly more important for physics. We introduce here the notion of *U*-sequence, called also *MSS* sequence in the literature according to the authors of[10] or kneading sequence according to the procedure how it is constructed.

With our conventions on the scale of $f(\mu, x)$ and x in mind, let us take $x = 0$ as the "Centre" and initial point of the mapping and then follow the subsequent iterations. We mark a point by the letter R or L according to whether it falls to the Right or Left of the Centre. Therefore, each periodic sequence corresponds to a word of finite length made of R's and L's. Then the important result of *MSS*[10] says that it is possible to introduce an ordering for all admissible words, i.e. to compile a dictionary of these words and there is a correspondence between the position of a word in the dictionary and the place of occurrence of a period on the parameter axis for a wide class of unimodal mappings. In particular, if the word P corresponds to a period, we can construct another word $H(P) = PxP$, where $x = R$ if there is an even number of R's in P and $x = L$ otherwise. H is called the harmonic of P and represents the period-doubled orbit adjacent to P. All admissible words corresponding to periods equal to or less than 11 were given in the Appendix of[10].

Acutally the construction of harmonics happens to be a particular case of a more general composition rule, introduced and denoted by $*$ in the paper of Derrida *et al.*[11]. Using the $*$ composition one can express the harmonic of P as $P{*}R$, and a period-doubling sequence starting with P corresponds to the symbolic sequence $P{*}R^{*n} = P{*}R{*}\ldots{*}R$ (n times of R), etc. Moreover, one can select period-tripling sequence $(RL)^{*n}$, period-quadrupling sequence $(RL^2)^{*n}$, period-quintupling sequences $(RLR^2)^{*n}$, $(RL^2R)^{*n}$ and $(RL^3)^{*n}$, etc. from the infinite number of periods embedded in the chaotic bands. Periods in these sequences are not adjacent on the parameter axis, but they have their own scaling factors and convergence rates.

Exploring the properties of the $*$ composition, Derrida *et al.*[11] succeeded in proving the self-similarity of the U-sequence, i.e. a part of it appears to be similar to the whole sequence. This important result is the mathematical manifestation of the hierarchy structure of chaotic bands, seen in bifurcation diagrams. U-sequence has been observed in real[12] and computer experiments[13].

Classification of periods by using combinations of two letters is an application of *symbolic dynamics*, a notion which originated in topological dynamical systems theory in the 30's and later adapted to differential dynamical systems by Bowen and others. For physicists, symbolic dynamics is nothing but a mathematical realization of coarse-graining. Instead of detailed information on the phase trajectory at all times, one simply labels a reasonable partition of the phase space by a set of letters and then follows the succession of letters. If properly chosen, this symbolic dynamics will reflect the essential feature of the evolution process, e.g. its periodicity. The method of symbolic dynamics is expected to play more role in physics (see the Preface of J. Ford to Ref. 14).

5. How to determine the superstable parameter values? Again with our conventions on f in mind we define a sequence of polynomials $P_n(\mu)$ using recursive relations based on the original mapping $f(\mu, x)$:

$$P_0(\mu) = 0 \quad ,$$

$$P_n(\mu) = f(\mu, P_{n-1}(\mu)) \quad . \tag{4-15}$$

The $P_n(\mu)$'s as functions of μ describe the dark lines seen in the chaotic region of bifurcation diagrams (see figures cited in Ref. 2), because the iterates of $x=0$ always correspond to extremes in islands of the chaotic bands. Actually, $P_n(\mu)$ is nothing but the n-th iterate of f taken at point $x=0$, i.e.

$$P_n(\mu) \equiv F(n, \mu, 0) \quad . \tag{4-16}$$

Therefore, real zeros of $P_n(\mu)$ in the appropriate interval give the parameter values for superstable orbits of period n. For example, $P_5(\mu)$ in the case of the logistic mapping (4-6) has three zeros 1.62541 . . . , 1.86078 . . . and 1.98542 . . . , which correspond to superstable orbits of types RLR^2, RL^2R and RL^3. However, when zeros get very close to each other most numerical algorithms to solve $P_n(\mu)=0$ lose stability, which makes this method impractical. Nevertheless, there exists a nice method to calculate the superstable parameter value separately, given the corresponding word in the U-sequence. Take for example the word RLR^2 which represents the first superstable period 5. Write it as a 5-cycle starting and ending at $x=0$, i.e.

$$f_R(\mu, f_L(\mu, f_R(\mu, f_R(\mu, f(\mu, 0))))) = 0 \tag{4-17}$$

where the subscript R or L indicates which branch, Right or Left, of the map has been used at a given step. (The first f maps into the Centre and thus has no subscript.) Once the branches have been marked, we can invert (4-17) to yield

$$1 = f(\mu, 0) = f_R^{-1}(f_R^{-1}(f_L^{-1}(f_R^{-1}(0)))) \quad . \tag{4-18}$$

Now define

$$R(x) = f_R^{-1}(x) \ , \qquad L(x) = f_L^{-1}(x) \tag{4-19}$$

and rewrite (4-18) using the function composition symbol $\circ$. We see that the word RLR^2 has been "lifted" to an equation for μ:

$$R \circ L \circ R \circ R(0) = 1 \quad , \tag{4-20}$$

which determines the superstable parameter μ for RLR^2. In the case of the logistic map (4-6) taking the square of (4-18) a few times to get rid of the radicals would bring us back to $P_5(\mu)=0$ with all other zeros present. So we retain (4-20) and multiply it by μ to get

$$\sqrt{\mu + \sqrt{\mu - \sqrt{\mu - \sqrt{\mu}}}} = \mu$$

H. Kaplan[15] suggested to solve this equation by iteration, i.e. let

$$\mu_{n+1} = \sqrt{\mu_n + \sqrt{\mu_n - \sqrt{\mu_n - \sqrt{\mu_n}}}}$$

with a suitable μ_0, say, $\mu_0 = 2$. One can easily read out the general rule to write down the iteration equation for any word in the U-sequence.

6. The renormalization group (RG) equation and the scaling factor. Generally speaking, behind any RG arguments there always figures some geometry with infinitely-nested self-similar internal structure (a kind of fractal geometry, to use a term now in fashion). One-dimensional mappings provide us with a simple example.

The very instructive discussion on scaling property of one-dimensional mapping in Feigenbaum's first paper[1] was illustrated by several not well-proportioned figures, so we redraw them in Fig. 4-1. Only the vicinity of certain fixed points is shown in these figures. A part of $f^{(2^n)}$ at the exact superstable parameter μ_n is given in Fig. 4-1(b). If we draw $f^{(2^{n-1})}$ at the same parameter value μ_n, it represents the period-doubled regime shown in Fig. 4-1(a), where locally one has a few superstable 2-cycles as outlined by the square boxes. Now increase μ_n from the situation of Fig. 4-1(b), until a new period-doubled superstable regime appears at the next parameter value μ_{n+1}, as shown by the smaller boxes in Fig. 4-1(c). Comparing the two hatched boxes in Figs. 4-1(c) and 4-1(a) shows what happened was a rescaling and change of signs in both the x and y directions. This process repeats itself with μ increasing.

To put the above geometrical observation into a mathematical frame, let us introduce an operator T to represent this period-doubling, μ-shifting and rescaling procedure starting from some fixed μ_n, i.e.

$$Tf(\mu_n, x) = -\alpha f(\mu_{n+1}, f(\mu_{n+1}, -x/\alpha)) = -\alpha F(2, \mu_{n+1}, -x/\alpha)$$

$$T^2 f(\mu_n, x) = (-\alpha)^2\, F(2^2, \mu_{n+2}, x/(-\alpha)^2)$$

$$\ldots \qquad (4\text{-}21)$$

Feigenbaum gave some plausible arguments to conjecture the existence of an universal, i.e. independent on the starting f, limit

$$\lim_{k\to\infty} (-\alpha)^k\, F(2^k, \mu_{n+k}, x/(-\alpha)^k) = g(x) \qquad (4\text{-}22)$$

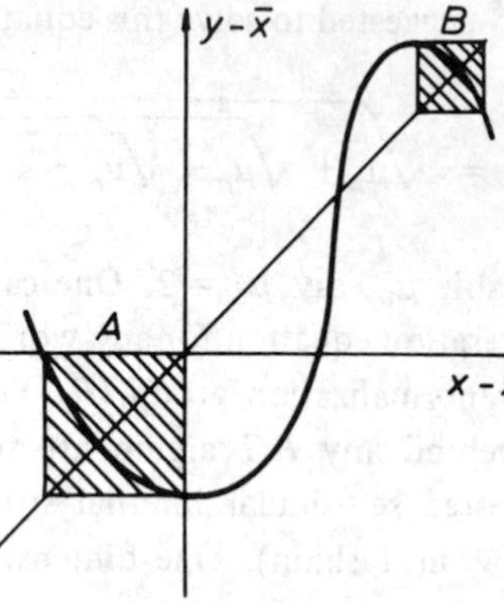

Fig. 4-1(a)
A segment of $f^{(2^{n-1})}(\tilde{\mu}_n, x)$,
$\tilde{\mu}_n$ is the superstable point for $f^{(2^n)}$.

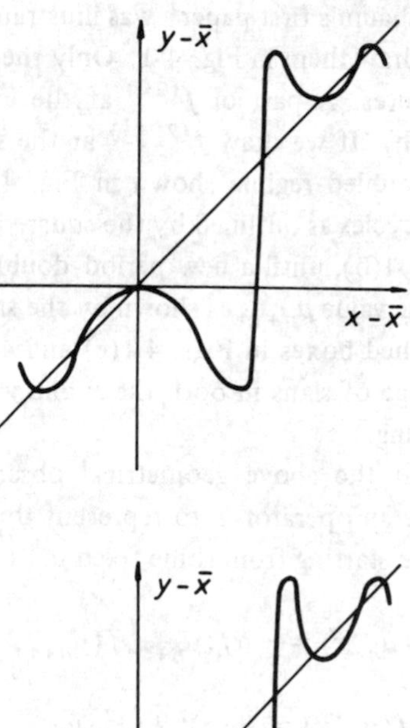

Fig. 4-1(b)
A segment of $f^{(2^n)}(\tilde{\mu}_n, x)$.

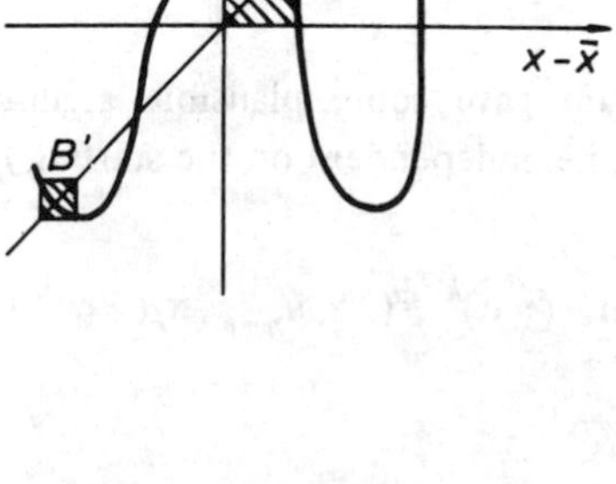

Fig. 4-1(c)
A segment of $f^{(2^n)}(\tilde{\mu}_{n+1}, x)$ to be rescaled and inverted to get (a) again.

This conjecture was proved later[16]. From the definition (4-4) of the composite function F it follows

$$F(2^k, \mu_{n+k}, x/(-\alpha)^k) = F(2^{k-1}, \mu_{n+k}, F(2^{k-1}, \mu_{n+k}, x/(-\alpha)^k)) \quad (4\text{-}23)$$

Taking the $k \to \infty$ limit on both sides of (4-23) results in the *RG* equation of Feigenbaum:

$$g(x) = -\alpha g(g(-x/\alpha)) \quad . \quad (4\text{-}24)$$

The normalization of $f(\mu, 0) = 1$ and the superstable condition (4-12) imply the following boundary conditions for (4-24):

$$g(0) = 1 , \qquad g'(0) = 0 \quad . \quad (4\text{-}25)$$

It is worth mentioning that conditions (4-25) alone do not determine the solution of (4-24) uniquely. At least the behaviour of $g(x)$ near $x = 0$ should be given, e.g. series expansion

$$g(x) = 1 + Ax^z + Bx^{2z} + Cx^{3z} + \ldots \quad (4\text{-}26)$$

will give different solutions to (4-24) for different choice of z.

The *RG* equation (4-24) can be generalized to period-n-tupling sequences:

$$\alpha^{-1} g(\alpha x) = -g^{(n)}(-x) \quad . \quad (4\text{-}27)$$

Equation (4-24) defines a saddle point in the space of all functions leading to $g(x)$[17]. Therefore, one must be careful in devising numerical procedure to calculate $g(x)$, because the approximations may get away from the true solution after converging to it initially. This situation resembles the *RG* theory of continuous phase transitions where the critical point also corresponds to an unstable saddle point.

Actually the simplest way to solve equations (4-24) or (4-27) consists in substituting the expansion (4-26) and then solving the resulted equations for $\alpha, A, B, \ldots$ etc. by combined use of algebraic manipulation and numerical languages[18].

7. Convergence rate δ and other universal constants. The Feigenbaum constant $\delta = 4.66920\ldots$ only determines the convergence rate of period-doubling bifurcation sequences for quadratic mappings, i.e. for $z = 2$ in (4-5). There is a smooth dependence of δ on z and its values for $z = 4, 6, 8$ are 7.248, 9.296 and 10.048 respectively[19]. In general, δ is defined as the eigenvalue in excess of 1 of a linearized functional equation obtained from the *RG* equation. Therefore, different solutions of the *RG* equation lead to different α's and δ's for various period-n-tupling sequences[18].

There are many other universal constants related to one-dimensional unimodal mappings. Not being able to go into details in this Introduction, we simply list some of them: the height ratio of two adjacent period-doubled peaks in the power spectra is 13.2 db[20], the critical slowing down exponent at μ_n is $\Delta = 1$ [21], the fractal dimension of the attractor at μ_∞ equals 0.538 . . .[22] etc. Another exponent related to external noise will be mentioned in Chapter 8.

8. The crisis of chaotic attractors[23]. The "microscopic" structure of a chaotic attractor does not depend continuously on parameter μ, whereas the general shape of the attractor changes smoothly with μ except for certain values where abrupt changes take place. This phenomenon was called crisis of the attractor and explained by Grebogi *et al.*[23] as the result of collisions of the attractor with an unstable orbit. Actually all the band-merging points in the bifurcation diagram occur at intersections with unstable orbits left by the main period-doubling bifurcations from the other side of μ_∞.

The sudden change or disappearance of chaotic attractors at "collisions" with other unstable objects certainly plays a more important role in higher dimensional systems.

References and guide to the bibliography

0. There is a monograph on one-dimensional mappings: B14 (Collet and Eckmann, 1980); For reviews see May (1976), Paper 6 in this volume, Grassmann and Thomae (1977), Feigenbaum (1980a, 1983).
1. Feigenbaum (1978), Paper 7 in this volume.
2. For bifurcation diagrams of the logistic mapping see Fig. I.19 in B14, Figs. 13–15 in Hao (1983), or Fig. 1 in Grebogi *et al.* (1983a).
3. Singer (1978).
4. E. Hille (1976). *Ordinary Differential Equations in the Complex Domain*, Chap. 10, Wiley.
5. Ulam and von Neumann (1947), Ruelle (1977).
6. Jacobson (1981).
7. Sarkovskii (1964), Stefan (1977).
8. Li and Yorke (1975), Paper 10 in this volume; Li *et al.* (1982).
9. Oono (1978b).
10. Metropolis *et al.* (1973), Paper 9 in this volume.
11. Derrida *et al.* (1978), Paper 11 in this volume.
12. Simoyi *et al.* (1982).
13. Hao *et al.* (1983).
14. Alekseev and Yakobson (1981); we list a few more papers on symbolic dynamics: Bowen (1973), Parry (1976), Guckenheimer (1980), Crutchfield

and Packard (1982, 1983), and Aizawa (1983).

15. Kaplan (1983).
16. Collet *et al.* (1980), Lanford (1982b).
17. Feigenbaum (1979), Paper 8 in this volume.
18. W.-Z. Zeng, B.-L. Hao, G.-R. Wang and S.-G. Chen (1984). "Scaling property of period-n-tupling sequences in one-dimensional mappings", *Commun. Theor. Phys.*, to appear.
19. For the dependence of Feigenbaum's δ on z see Mendes (1981), McGuive and Thompson (1981), Hu and Satija (1983).
20. Nauenberg and Rudnick (1981); see also Feigenbaum (1980b, c, d, 1981), Collet *et al.* (1981c).
21. Hao (1981).
22. Grassberger (1981), Hu and Hao (1983); Y.-Q. Wang and S.-G. Chen (1984), "Metric properties of chaotic region in one-dimensional maps", *Acta Physica Sinica*, **33**, 341.
23. For crises of chaotic attractors see Grebogi *et al.* (1982), Paper 12 in this volume, and (1983a), Jeffries and Perez (1983), Yamaguchi and Sakai (1983).

Chapter 5
BIFURCATION AND CHAOS IN HIGHER DIMENSIONAL SYSTEMS

Most physical processes are described by ordinary or partial differential equations, so we are more concerned with chaotic behaviour in such higher dimensional systems. Discrete mappings of a plane or an annulus occupy an intermediate place and often provide the clues for understanding chaotic transitions in higher dimensional systems. In what follows we first list some mathematical models encountered frequently in the current literature. Then we shall say a few words about analytical and numerical methods to study these systems, with emphasis on the latter, since one has to rely heavily on numerical experiments in studying higher dimensional systems. A table summarizing the findings in computer experiments will be given at the end of this chapter.

Higher dimensional systems may be divided into three categories: discrete mappings of two or more dimensions (difference equations of second order or higher); ordinary differential equations (ODE's, autonomous, nonautonomous and time-delayed); and partial differential equations (PDE's). We cite examples from each category.

1. Two- and higher-dimensional mappings

In contradistinction to one-dimensional mappings, higher-dimensional mappings may be conservative (volume-preserving) as well as dissipative (volume-contracting), invertible as well as noninvertible, depending on the parameter range in the model. As the first example one should mention the well-studied *Hénon mapping*[1] :

$$\begin{aligned} x_{n+1} &= 1 - \mu x_n^2 + y_n \\ y_{n+1} &= b\, x_n \end{aligned} \tag{5-1}$$

with the Jacobian

$$J = \frac{\partial(x_{n+1}, y_{n+1})}{\partial(x_n, y_n)} = -b \quad .$$

It is an invertible transformation when $b \neq 0$ and goes back to the logistic mapping (4-6) when $b = 0$. It preserves area for $b = 1$ and corresponds to a dissipative system for $b < 1$. The case $b = 0.3$ was studied in great detail by Hénon and many others. The essential difference from one-dimensional mappings may be summarized as follows. First, even for one and the same parameter value, the character of stationary output of (5-1) depends on initial point taken. In other words, the (x, y) plane divides into basins, and starting from different basins one may be led to different periodic or aperiodic orbits. In contrast, there exists at most only one stable orbit for one-dimensional unimodal mappings. Second, for certain parameters and initial points, the iteractions of (5-1) converge to an attractor with self-similar internal structure (see Figs. 3–6 in Paper 13). This was the first strange attractor known to have fractal dimension. We shall return to this point in Chapter 9. Third, it was proved that there exist intersections of stable and unstable manifolds in Hénon's model, i.e. homoclinic points give rise to chaos.

Another well-known two-dimensional mapping is the so-called *standard mapping*[2]:

$$x_{n+1} = x_n + y_{n+1}$$

$$y_{n+1} = y_n - \frac{\mu}{2\pi} \sin 2\pi x_n \quad . \tag{5-2}$$

Being an example of two-dimensional Hamiltonian system, this conservative mapping occurs in many physical applications such as the motion of charged particle in toroidal magnetic field, or the Frenckel-Kontorova model of modulated structures in solid state physics (see Chapter 10).

Instead of listing other two- and higher-dimensional mappings we confine ourselves to a few remarks. First, some mathematical results on one-dimensional mappings can be "lifted" to mappings on R^n as outlined in Ref. 4. Higher dimensional mappings are richer in their behaviour, e.g., basin dependence and transitions from quasiperiodic motion to chaos may take place. Second, period-doubling bifurcations do occur in area-preserving mappings of the plane, but the universal convergence rate happens to be $\delta = 8.7210\ldots$[5]. Third, complex mappings[6] present a new class of two-dimensional mappings. While no physical relevance is known for the time being, their graphic representations are beautiful creations of computer science.

2. Ordinary differential equations

ODE's displaying chaotic behaviour may be subdivided into three groups: autonomous systems with three or more variables, nonautonomous systems with two or more variables, and time-delayed systems of at least one variable.

A system of ODE's is called *autonomous*, when there is no explicit time dependence on the right hand side of the standard form:

$$\frac{dx_i}{dt} = f_i(x_1, \ldots, x_N) , \qquad i = 1, \ldots, N . \tag{5-3}$$

The classical example of an autonomous system exhibiting chaos is the *Lorenz model*[7]:

$$\begin{aligned} \dot{x} &= -\sigma(x-y) , \\ \dot{y} &= -xz + rx - y , \\ \dot{z} &= xy - bz . \end{aligned} \tag{5-4}$$

It was obtained by truncating the PDE's describing thermal convection between two infinite plates. There are three parameters in this model: r is the ratio of the Rayleigh number to its first critical value, σ is the Prandtl number, and b.

The Lorenz model is one of the simplest truncated hydrodynamical systems. If one considers planar flow of a fluid with periodic boundary conditions imposed in both directions, then the original PDE's (the Navier-Stokes equations) can be transformed into an infinite system of ODE's for the Fourier coefficients of various hydrodynamical quantities. By truncating this latter system one gets various finite systems of ODE's. Truncated equations with four, five, six, seven or more modes retained have been suggested[8], among which the following two-parameter family of seven-mode equations[9]

$$\begin{aligned} \dot{x}_1 &= -2x_1 + 4\sqrt{5}\, x_2 x_3 + 4\sqrt{5}\, x_4 x_5 \\ \dot{x}_2 &= -9x_2 + 3\sqrt{5}\, x_1 x_3 \\ \dot{x}_3 &= -5x_3 - 7\sqrt{5}\, x_1 x_2 + 9\epsilon x_1 x_7 + R \\ \dot{x}_4 &= -5x_4 - \sqrt{5}\, x_1 x_5 \\ \dot{x}_5 &= -x_5 - 3\sqrt{5}\, x_1 x_4 + 5\epsilon x_1 x_6 \\ \dot{x}_6 &= -x_6 - 5\epsilon x_1 x_5 \\ \dot{x}_7 &= -5x_7 - 9\epsilon x_1 x_3 \end{aligned} \tag{5-5}$$

permits a continuous transition from a five-mode ($\epsilon = 0$) to a seven-mode ($\epsilon = 1$) model, both well-studied in the literature[8].

The behaviour of all the above-mentioned models is very complicated, so Rossler tried to construct models as simple as possible, but still exhibiting chaotic behaviour. The simplest of his models contains only one quadratic nonlinear term[10]:

$$\begin{aligned} \dot{x} &= -(y+z) \\ \dot{y} &= x + ay \\ \dot{z} &= b + xz - cz \end{aligned} \tag{5-6}$$

Among other autonomous systems displaying chaotic behaviour we mention the model of coupled unstable and damping waves in plasma[11], the double-diffusive or two-component Lorenz model[12], a 32-mode system supposed to describe the Taylor instability of rotating fluid[13], the 40-mode system for Gunn instability in semiconductors[14], and the coupled Brusselators[15]. (The Brusselator is the name for a hypothetical model of trimolecular reaction with an autocatalytic step, which shows rich temporal and/or spatial structures when a diffusion term is added[16].)

We digress to explain why at least three variables are required to allow chaotic behaviour. This can best be seen from the example of period-doubling as the first step leading to chaos. Since the trajectory has no reason to change drastically at an infinitesimal increase of the parameter value, and yet the period T doubles to $2T$, the only imaginable picture is an almost imperceptible splitting of the original orbit. If this splitting takes place in a plane, then there must be at least one point where the trajectory intersects itself and thus violates the uniqueness of the solution. However, usually the ODE systems under consideration are good enough to ensure the validity of the uniqueness theorem. Therefore, splitting of the orbit without self-intersection can take place only in three- and higher-dimensional space. The subsequent bifurcations and the development of chaotic trajectories follow the same principle.

Now let us return to *nonautonomous* ODE's. It is well-known that a nonautonomous system can be made autonomous by adding one or more variable, so at least two variables are required for a nonautonomous system to show chaotic behaviour. In fact, the only kind of nonautonomous systems studied in detail up to now consists of nonlinear oscillators driven by external periodic force.

The forced van der Pol equation[17]

$$\ddot{x} - k(1-x^2)\dot{x} + x = b\lambda k \cos(\lambda t + \varphi) \tag{5-7}$$

seemed to be the first example exhibiting stochastic behaviour reported long before the modern jargon of chaos and strange attractor has emerged.

The forced anharmonic oscillator

$$\ddot{x} + k\dot{x} - \beta x + \alpha x^3 = b\cos\omega t \quad , \tag{5-8}$$

being a particular case of the Duffing's equation[18]

$$\ddot{x} + k\dot{x} + f(x) = g(t) \tag{5-9}$$

where $g(t)$ is a periodic function of t, and $f(x)$ a nonlinear function of x, appears in many physical applications, e.g., motion of dislocation line in supersonic field[19].

The forced mathematical pendulum

$$\ddot{x} + k\dot{x} + \sin x = \alpha\cos\omega t \tag{5-10}$$

and the parametrically excited pendulum[20]

$$\ddot{x} + k\dot{x} + (A + \alpha\cos\omega t)\sin x = 0 \tag{5-11}$$

represent another class of physical models, encountered, say, in describing Josephson junctions in microwave cavity (see References at the end of Chapter 10).

The forced Brusselator[21]

$$\dot{x} = A - (B+1)x + x^2 y + \alpha\cos\omega t$$

$$\dot{y} = Bx - x^2 y \tag{5-12}$$

provides at the present time the most thoroughly studied model by combined use of various numerical methods. All well-known routes to chaos and the *U*-sequence of Metropolis-Stein-Stein (see Chapter 4) have been shown to exist in this model.

In principle, every autonomous system can be extended to a nonautonomous one by adding a periodic force. From such models we mention only the forced Lorenz model[22].

Most periodically forced systems can be viewed as coupled systems of one nonlinear and one linear oscillators. This point of view allows for a more intuitive interpretation of various regimes of oscillation, including period-doubling and chaos. Having the driving frequency as a control parameter at hand opens the possibility to reach very high frequency resolution by using various stroboscopic sampling techniques (see below). These merits distinguish the periodically driven equations from purely autonomous systems in numerical studies.

From recent studies on forced nonlinear oscillators there is a short paper worthy of note[23]. An exactly solvable nonlinear oscillator was subjected to periodic kicks.

Using the known solution of the free oscillator this model was transformed into a discrete mapping of two variables which enables a detailed study using a reasonable amount of computer time. Further investigation on this model is desirable, because there is a good chance of discovering U-sequences similar to that observed in the forced Brusselator and of exploring how the Faray sequences describing locking frequencies immersed in quasiperiodic regime are replaced by the U-sequences describing periodicities embedded in the chaotic region when nonlinearity gets stronger. We believe these issues are universal for a wide class of nonlinear systems and are relevant to the transition from quasiperiodicity to chaos.

Now comes the last group of ODE's — the *time-delayed equations*. A seemingly simple equation like

$$\tau \dot{x}(t) + x(t) = \mu x(t-1)(1-x(t-1)) \tag{5-13}$$

may possess very complicated bifurcation and chaos structure except for the limiting case $\tau \ll 1$ when it can be approximated by the logistic mapping (4-6). Formally the time delay may be written as an infinite order differential operator acting on the function $x(t)$:

$$x(t-1) = \exp\left(-\frac{d}{dt}\right)x(t) = \sum_{n=0}^{\infty} \frac{(-1)^n}{n!}\frac{d^n}{dt^n}x(t) .$$

Now it becomes clear that a time-delayed equation in one variable may correspond to an infinite system of ODE's. One meets this kind of equations in models describing chaos in optical bistability devices (see references at the end of Chapter 10).

3. Partial differential equations

Many ODE's listed above are truncations of PDE's. However, direct study of chaotic behaviour in PDE's is still in its beginning stage. We mention briefly a few lines of research on this topic.

Numerical study of PDE's is a time-consuming task. A recent paper reported the first example of period-doubling and chaos in a system of PDE's for thermosolutal convection[24]. Actually the equations in (t, x, z) were replaced by a finite difference scheme on a mesh and the essential nature of the solution was shown not to depend upon altering the mesh interval. Oscillations, period-doublings and chaos were observed in the time evolution of certain global characteristics of the solution. Nothing essentially new was reported compared to ODE's, although one should expect some new features in chaotic behaviour in PDE's, e.g., the interplay between spatial structure and temporal chaos.

Another interesting development connects the two extremes of nonlinear equations: those exhibiting chaos and those having solitons. It is well-known that PDE's solvable by the inverse scattering technique are integrable systems. The existence of an infinite number of integrals of motion guarantees the creation of such "stable" objects as solitons and precludes the possibility for chaotic behaviour. However, integrability is a very subtle property which can easily be broken by periodic perturbations. Chaotic behaviour has been observed in the perturbed sine-Gordon equation[25], the perturbed nonlinear Schrodinger equation[26] and the closely related Ginsburg-Landau equation[27]. In particular, the interplay of coherent spatial structure and temporal chaos and the evidence for spatial period halving reported in Ref. 25 may have far-reaching consequence, e.g., in understanding cell division in developmental biology.

Now we turn to the methodological aspects in studying higher dimensional systems. As regards analytical tools, there are very few mathematical results ready for physicists to use: no *a priori* criterion for chaos, no classification and enumeration theorem for attractors, etc. Still we would like to call attention to two methods due to Melnikov and Silnikov respectively, both centered on the existence of homoclinic points or homoclinic orbits.

The *Melnikov's method*[28] applied to near-integrable systems subjected to dissipative time-dependent (usually periodic) perturbations. We mentioned in Chapter 2 that a separatrix becomes a stochastic layer under perturbation which makes the system nonintegrable, but still Hamiltonian. This is not the general rule for dissipative perturbations, as it can be seen from Fig. 5-1. An unperturbed separatrix shown in (a) may be perturbed into one of other three cases, among which only case (d) leads to an infinite number of homoclinic intersections. The criterion for case (d) consists in the alternating sign of the Melnikov function — the distance between stable and unstable branches of the trajectory.

The *Silnikov's method* applies to stationary points of saddle-focus type, when there is a homoclinic trajectory passing through the saddle-focus. The theorem says roughly that if the only real eigenvalue and the real part of a pair of complex eigenvalues of the linearized system are of opposite sign and the absolute value of the former is larger, then there exists a set of chaotic trajectories near the original homoclinic trajectory. This criterion has been used for the Rossler model (5-6) and a few other systems.

Before going to numerical methods we discuss a plausible objection against any numerical study of chaos using digital computers. The point is since one always works with a finite field of rational numbers representable on a digital computer of finite word length, it is impossible to realize even a true quasiperiodic process, not to mention chaotic orbits. Actually, things are not so hopeless.

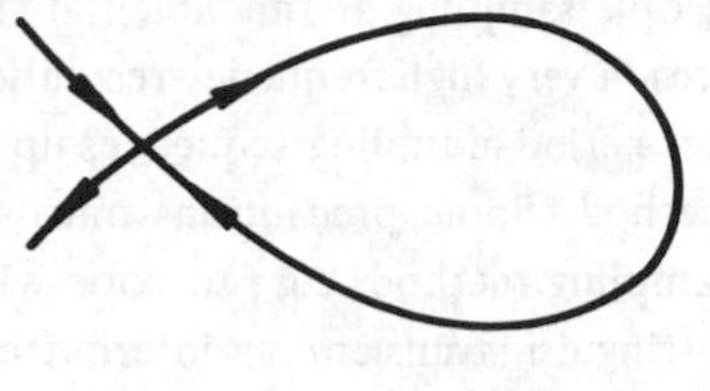

Fig. 5-1(a)
Unperturbed separatrix.

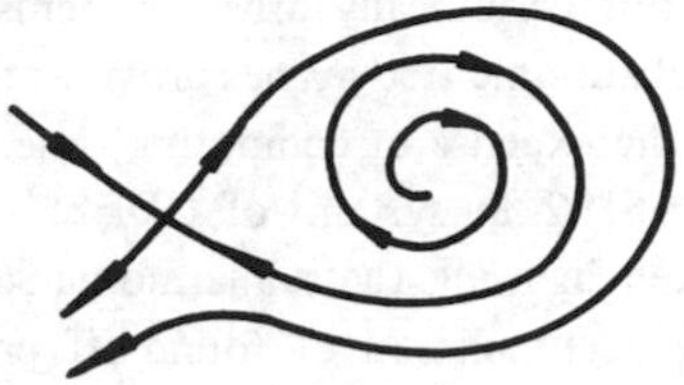

(b)

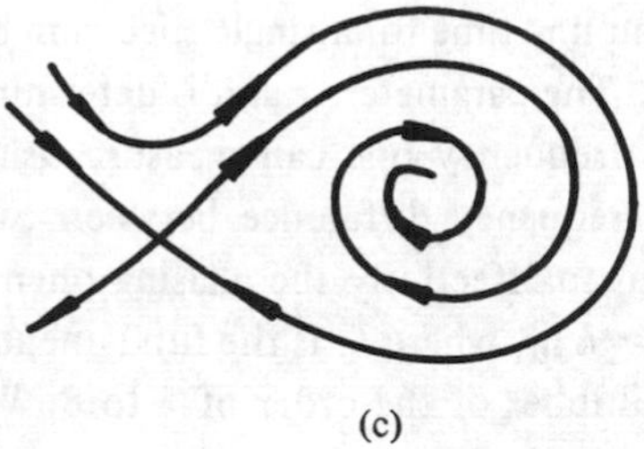

(c)

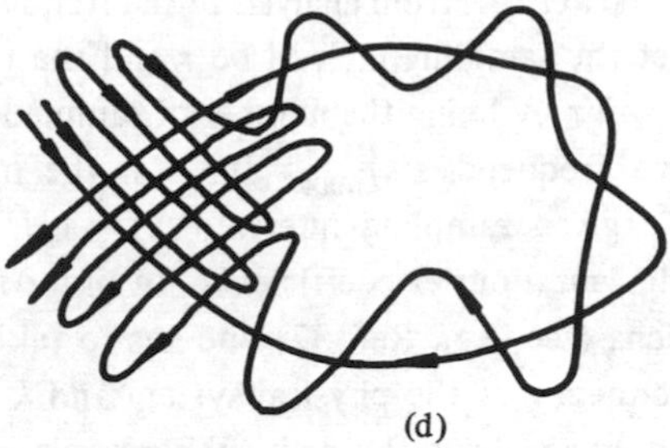

(d)

Fig. 5-1(b), (c) & (d)
Perturbed separatrices in dissipative system.

Irrational numbers can be approximated by rationals, chaotic regions are surrounded by periodic regimes. Our strategy in computer study consists in identifying the periodic orbits with confidence and then characterizing the "aperiodic" motion. We postpone the problem of characterizing the attractors to Chapter 8 and concentrate on how to reach high frequency resolution in numerical experiments. We shall compare the resolving power in terms of the recognizable order p of subharmonics with respect to the natural fundamental frequency of the system under study: $p = 2^n$ for period-doubling sequences, etc.

Direct observation of the trajectories is a method with the least resolution, but it enjoys the merit of having more physical intuition. In this way it is difficult to resolve subharmonics higher than $p = 32$. Sometimes it is useful to draw the stereoprojections of a trajectory in some three-dimensional subspaces of the phase space.

The plotting of Poincaré maps, i.e. the intersection points of a given trajectory with a fixed surface in the phase space, offers an effective means to explore the nature of the motion. A practical problem consists in matching the precision of the interpolation scheme used to locate the intersection point with that of the integration algorithm. In this respect a clever suggestion by Hénon[30] deserves to be mentioned.

For periodically driven systems the stroboscopic sampling at fundamental and subharmonic frequencies provides the means to reach very high frequency resolution at the expense of computing time. We identified period-doubling sequences up to $p = 8192$ in systems of ODE's[31] using this method. Some precautions must be taken in using the subharmonic stroboscopic sampling method, e.g., to cope with the accumulation of round-off errors or to distinguish transient or intermittent behaviour from chaotic behaviour. The interested reader may consult the original papers[31] for technical details.

Power spectrum analysis in the frequency domain deserves a few more comments. Let the sampling interval be τ and the total sampling time for a single spectrum be $L = N\tau$, N being the number of sampled points. The parameters τ and L determine two frequencies: $f_{\max} = 0.5/\tau$ is the maximal frequency one can measure using the given sampling interval; $\Delta f = 1/L$ is the frequency difference between two adjacent Fourier coefficients. In order to eliminate effectively the aliasing phenomena (see, e.g., Ref. 32) one has to take $f_{\max} = kf_0$, where f_0 is the fundamental frequency of the physical system and k is a multiplier of the order of 4 to 8. We aim at resolving the p-th subharmonic of f_0 and wish the subharmonic peak to be formed by s points in the spectrum, i.e., $f_0/p = s\Delta f$. Putting together all the above-mentioned relations, we get

$$p = N/(2ks) \quad . \tag{5-14}$$

Note that this is a relation independent on τ and f_0. Taking $N = 8192$, $k = 4$, $s = 8$, we have $p = 128$. This is the limit of resolution in power spectrum analysis on most medium-size computers. Nevertheless, power spectrum analysis remains the most useful means to tell different periods embedded in the chaotic bands and the presence of broad-band noise in the spectra is still the most practical symptom for chaos in computer and laboratory experiments.

We close this chapter by summarizing the main findings in computer experiments on various models in the following table.

Table Brief summary of computer experiments on differential equations

Model	Method	Findings	Reference
Lorenz (5-4)	Poincaré map Power spectra Lyapunov exponents Dimension	Period-doubling Intermittency Transient chaos Strange attractor	7
Forced Brusselator (5-12)	Stroboscopic sampling Power spectra Lyapunov exponents Dimension	Period-doubling Intermittency Quasiperiodic to chaos Hierarchy of chaotic bands *U*-sequence and attractor	21
Rossler (5-6)	Power spectra Dimension	Period-doubling	10
Coupled Brusselator	Poincaré map Lyapunov exponents Dimension	Period-doubling Intermittency Metastable chaos Quasiperiodic to chaos Strange attractor	15
Double-diffusive convection	Trajectory	Period-doubling	12
3-wave coupling	Poincaré map Power spectra Lyapunov exponents Dimension	Period-doubling Strange attractor	11
5-mode truncat. Navier-Stokes	Poincaré map Power spectra	Period-doubling	8
6-mode truncat. Navier-Stokes	Poincaré map Power spectra	Quasiperiodic to chaos Hysteresis, Intermittency	8
7-mode truncat. Navier-Stokes	Poincaré map	Quasiperiodic to chaos Hysteresis	8
Duffing (5-9)	Poincaré map Power spectra	Period-doubling Quasiperiodic to chaos Strange attractor	18
Parametric pendulum	Stroboscopic sampling Power spectra Lynapunov exponents	Period-doubling Strange attractor	20
PDE's of thermo-solutal convection	Trajectory Time evolution of Nusselt numbers	Period-doubling Chaos	24

References and guide to the bibliography

1. The Hénon mapping: Hénon and Pomeau (1977); Hénon (1976), Paper 13 in this volume; Feit (1978), Lozi (1978), Marotto (1979), Simo (1979), McLaughlin (1979b), Curry (1979a, 1981), Daido (1980), Misiurewicz and Szewc (1980), Tresser *et al.* (1980c), Hitzl (1981), Franceshini and Russo (1981), Hu (1981), Stefanski (1982), Grassberger (1983b).
2. The standard mapping: Chirikov (1979), Zaslavsky and Chirikov (1972), Bak (1982).
3. Other 2-dimensional and higher dimensional mappings: Marotto (1978), Curry and Yorke (1978), Aronson *et al.* (1982), Hamilton and Brumer (1982), Zisook (1981, 1982), Lanford (1983), Sun (1983a, b), Tel (1983a, b, c), Meiss and Cary (1983).
4. Collet *et al.* (1981a).
5. For $\delta = 8.7210\ldots$ in area-preserving mappings see Benettin *et al.* (1980), Derrida and Pomeau (1980), Collet and Eckmann (1981), Zisook (1981).
6. Complex mappings: Greene and Percival (1981), Cvitanovic and Myrheim (1983), Lee (1983), Manton and Nauenberg (1983), Widom *et al.* (1983), and the book by B. B. Mandelbrot, *Fractal Geometry in Nature*, Freeman, 1982.
7. Saltzman (1962); Lorenz (1963), Paper 14 in this volume; and book B32 (Sparrow, 1982). From numerous work on Lorenz model we indicate: Lorenz (1964, 1979, 1980a), Haken (1975), Guckenheimer (1976), Ruelle (1976, 1978b), Hénon and Pomeau (1977), Lanford (1977), Parry (1977), Williams (1977, 1979a, b), Nagashima and Shimada (1977), Rossler (1977b), Lucke (1976), Ibanez and Pomeau (1978), Rand (1978), Curry (1978, 1980), Marioka and Shimizu (1978), Shimizu and Marioka (1978a, b, c), Takeyama (1978, 1980), Guckenheimer and Williams (1979), Shimada (1979), Kaplan and Yorke (1979a, b), Knobloch (1979), Manneville and Pomeau (1979), Robbins (1979), Yorke and Yorke (1979), Brindley and Moroz (1980), Franceschini (1980), Gibbon and McGuinness (1980, 1982), Grab and Scholz (1980), Tomita and Tsuda (1980b), Tabor and Weiss (1981), Sinai and Vul (1981), Zippilius and Lucke (1981), Booty *et al*. (1982), Fowler and McGuinness (1982), Fowler *et al.* (1982, 1983), McGuinness (1983), Aizawa (1982).
8. Truncated Navier-Stokes equations with
   4 modes: Riela (1982a);
   5 modes: Baldrighini and Franceschini (1979), Franceschini and Tebaldi (1979), Paper 15 in this volume, Baive and Franceschini (1981), Riela (1982a), Gregorio *et al.* (1983);

6 modes: Angelo and Riela (1981), Riela (1982a, b, c);
7 modes: Franceschini and Tebaldi (1981), Tedeschini-Lalli (1982), Franceschini (1983);
14 modes: Curry (1978).

9. Tedeschini-Lalli (1982).
10. The Rössler models: Rössler (1976a, d, 1979a, b, c), Crutchfield *et al.* (1980), Fraser and Kapral (1982b), Gaspard and Niclis (1983).
11. The 3-wave coupling models: Wersinger *et al.* (1980a, b), Russell and Ott (1981).
12. The double-diffusive convection model or two-component Lorenz model: Siegman and Rubenfeld (1975), Huppert and Moore (1976), Rubenfeld and Siegman (1977), Knobloch and Weiss (1981), Knobloch and Proctor (1981), Da Costa *et al.* (1981), Verlarde (1981), Verlarde and Antoranz (1981).
13. Yahata (1978).
14. Nakamura (1977, 1978, 1979).
15. The coupled Brusselator: Schreiber and Marek (1982a, b), Sano and Sawada (1983).
16. G. Nicolis and I. Prigogine (1977). *Self-organization in Nonequilibrium Systems*, Wiley.
17. The forced van der Pol oscillator: Cartwright and Littlewood (1945), Cartwright (1948), Littlewood (1957), Grassman *et al.* (1976), Flaherty *et al.* (1978), Holmes and Rand (1978), Rand and Holmes (1980), Guckenheimer (1980b), Levi (1980, 1981).
18. Duffing's equation: Ueda (1979, 1980a, b), Holmes (1979a), Holmes *et al.* (1981), Ogura *et al.* (1981), Chui and Ma (1982), Elgin and Forster (1983), Sato *et al.* (1983), Holmes and Whitley (1983a, b), Hu and Mao (1983), and book B36.
19. Huberman and Crutchfield (1979), Herring and Huberman (1980).
20. Parametrically excited pendulum: McLaughlin (1981), Leven and Koch (1981), D'Humieres *et al.* (1982); Arneodo *et al.* (1983), Paper 35 in this volume.
21. The forced Brusselator. Review: Tomita (1982a), Hao (1984) to appear in *Chinese Science Reviews — Physics*, Science Press, Beijing. Papers: Tomita *et al.* (1977), Tomita and Kai (1978a, b, 1979a, b), Kai and Tomita (1979), Daido and Tomita (1979), Kai (1981), Hao (1982), Hao and Zhang (1982a, b, c), Paper 16 in this volume, and (1983), Hao *et al.* (1983), Paper 17 in this volume, Wang (1983), Wang *et al.* (1983); see also Broomhead *et al.* (1981), Ito (1979a, b, c).
22. The forced Lorenz model: Aizawa and Uezu (1982b), Uezu and Aizawa (1982a).

23. Gonzalez and Piro (1983), Paper 18 in this volume.
24. Moore *et al*. (1983), Paper 19 in this volume.
25. Perturbed sine-Gordon equation: Holmes (1981), Bennett *et al*. (1982), Nozaki (1982), Bishop *et al*. (1983).
26. Perturbed nonlinear Schrodinger equation: Nozaki and Bekki (1983a).
27. Ginsburg-Landau equation: Moon *et al*. (1982, 1983), Nozaki and Bekki (1983b).
28. Melnikov (1963), Broomhead and Rowlands (1982), Holmes and Marsden (1982a), Sanders (1982), and in books B31 and B36.
29. Silnikov (1965, 1969), Arneodo *et al*. (1982a), Gaspard (1983), Gaspard and Nicolis (1983).
30. Hénon (1982).
31. Hao and Zhang (1982a, c), Paper 16 in this volume, and (1983).
32. J. N. Rayner (1971). *An Introduction to Spectral Analysis*, Pion Ltd.

Chapter 6
THE INTERMITTENT TRANSITION

The term intermittency has been used in the hydrodynamical theory of turbulence to denote random burst of turbulent motion on the background of laminar flow[1]. However, its mechanism slightly differs from what is called intermittency in the context of chaos. In an open fluid the laminar and turbulent regions are divided by irregular and moving boundaries. The flow velocity sampled at a given point near such boundary will show alternating laminar or turbulent behaviour when the point enters or leaves the laminar region as the boundary moves. In one word, temporal intermittency is caused by spatially changing boundaries separating different flow regimes. However, in this volume we use the term intermittency exclusively to refer to random alternations of chaotic and regular behaviour in time evolution without involving any spatial degrees of freedom.

Among the various routes to chaos mentioned in Chapter 2, period-doubling and intermittency are in fact twin phenomena. Inspecting the bifurcation diagram of a one-dimensional mapping or the "phase diagram" showing the bifurcation and chaos structure of a system of ordinary differential equations, e.g., Fig. 1 in Paper 17, one sees many chaotic regions contained in between pairs of type RL^{n-1} and RL^n periods. (For the meaning of these letters see Chapter 4.) If one enters a given chaotic region from the RL^{n-1} side, then the transition takes place via period-doubling bifurcations. If one approaches the same chaotic region from the RL^n boundary, the intermittent transition shows itself. Both period-doubling and intermittency are described by the same renormalization group equation (4-24) with different boundary conditions[2] and are well understood by now. Recommending the reader to consult the reprinted papers and reviews for details, we only touch briefly a few points related to the intermittent transition.

1. The mechanism of intermittency in one-dimensional mappings has been fully understood. It occurs in the neighbourhood of a tangent bifurcation, whereas period-doubling is related to pitchfork bifurcations. (For figures and explanation of these bifurcations see May (1976), Paper 6 in this volume.) Take for example the logistic mapping (4-6). When the parameter μ is kept very close to, but still less than $\mu_c = 1.75$, its third iterate $F(3, \mu, x)$ (we use definition (4-4) for iterates) has three humps or valleys that come very close to the bisector in the F versus x diagram, but does not touch it, leaving thereby very narrow corridors between the curve $F(3, \mu, x)$ and the bisector (see Fig. 4 in Paper 20). Every time when one iterate point falls near the entrance to one of these corridors, it will take many iterations to pass through the corridor. When μ is very close to μ_c, this looks much like a segment of convergent iteration to the would-be fixed point at μ_c and leads to the laminar phase of iteration. Having passed one corridor, the point makes a number of large-step jumps before falling again to the entry of one or another corridor. These jumps constitute the random bursts of the turbulent phase. The nearer μ is to μ_c, the longer the averaged laminar time. Simple mean-field arguments give the passage time t [3]

$$t \propto (\mu_c - \mu)^{-(1-1/z)} \quad , \qquad (6\text{-}1)$$

where z denotes the order of maximum in the mapping (4-5). For the usual case $z = 2$ we have $t \propto (\mu_c - \mu)^{-1/2}$ which is to be compared with the divergence rate of correlation length ξ near a continuous phase transition point T_c

$$\xi \propto |T_c - T|^{-\nu} \qquad (6\text{-}2)$$

with its mean-field value $\nu = \frac{1}{2}$.

2. Intermittent transitions had been observed in computer experiments on ODE's before its mechanism was explained by Pomeau and Manneville[4]. In periodically forced systems, intermittency associated with tangent bifurcations of not very long period can be easily distinguished from chaotic or transient behaviour by their specific subharmonic stroboscopic sampling diagrams[5]. Another way to explore the intermittent transitions consists in constructing the (P_i, P_{i+n}) map near a tangent bifurcation of period n, where P_i may be any of the variables under study. If there were an exact period n, one would get n points on the bisector of the (P_i, P_{i+n}) diagram, but at the intermittent transition we can only see the "spirit" of the would-be fixed points, i.e., n clusters along the bisector together with a few points scattered away from it. This is a very effective method to recognize intermittency in practice.

Intermittent transitions have been reported in many laboratory experiments, e.g., in chemical turbulence, thermoconvective instability and nonlinear electronic circuits[6]. We would like to point out that since there are infinitely many periodic windows born from tangent bifurcations embedded in the chaotic regions of the parameter space, the intermittent transitions are typical phenomena to be observed in numerical and laboratory experiments.

3. A few words about the renormalization group equation for intermittent transitions. We know from Chapter 4 that a periodic window in one-dimensional mapping remains stable whenever the absolute value of its derivative does not exceed one (see (4-11)). In fact, the limit $F' = -1$ corresponds to period-doubling bifurcations, whereas $F' = +1$ indicates an intermittent transition. Therefore, it is not surprising that intermittent transitions can be described by the same renormalization group equation of Feigenbaum

$$g(x) = -\alpha g(g(-x/\alpha)) \tag{6-3}$$

with changed boundary conditions[2]

$$g(0) = 0, \qquad g'(0) = 1 \quad , \tag{6-4}$$

(cf. (4-25)). However, it is remarkable that Hirsch, Nauenberg and Scalapino[2] found an exact solution to (6-3) and (6-4). Their solution has been extended to give all the intermittent exponents and universal functions $g(x)$ for general z [7].

4. Our last remark concerns the effect of external noise on intermittency. Contrary to what one might expect by intuition, a small amount of noise increases the averaged passage time[8].

References and guide to the bibliography

1. For the hydrodynamical meaning of intermittency see, e.g., D. J. Tritton (1977), *Physical Fluid Mechanics*, Van-Nostrand-Reinhold, p. 265.
2. Renormalization group equation for intermittency was introduced by Hirsch *et al*. (1982b), Paper 22 in this volume.
3. The following papers provide a good review of the intermittent transition to chaos: Pomeau and Manneville (1980), Paper 20 in this volume; Pomeau (1980); Hirsch *et al*. (1982a), Paper 21 in this volume; and Scalapino *et al*. (1982).
4. Intermittent transitions were first observed by Pomeau in analog computer study of the Lorenz model and explained later in Manneville and Pomeau (1979). Actually, Fig. 4b in Morioka and Shimizu (1978), described there as diffusion around the formerly stable fixed points, and Fig. 4 in Franceschini and Tebaldi (1979) described as a nonzero segment in the Poincaré map, were all examples of intermittent transitions.

5. Wang *et al*. (1983).
6. Laboratory observation of intermittency in chemical turbulence: Pomeau *et al*. (1981); in hydrodynamical instabilities: Berge *et al*. (1980), Dubois *et al*. (1983); in nonlinear electronic circuits: Jeffries and Perez (1982). We mentioned also theoretical discussion of intermittency in Josephson junctions: Ben-Jacob *et al*. (1982), Seifert (1983); in lasers: Scholz *et al*. (1981).
7. For generalizations of the exact solution to the intermittency renormalization group equation see Hu and Rudnick (1982a), Paper 23 in this volume, and (1982b), and Hu (1982b).
8. On the effect of external noise on intermittency see Hirsch *et al*. (1982a, b), Papers 21 and 22 in this volume; Eckmann *et al*. (1981).
9. We list some other papers related to intermittent transitions: Manneville (1980a, b), Mayer-Kress and Haken (1981a), Bussac and Meunier (1982), Zisook (1982), Zisook and Shenker (1982), Aizawa (1983), Caroli *et al*. (1983), Daido (1983b), Fujisaka *et al*. (1983), Pikovsky (1983).

Chapter 7

THE TRANSITION FROM QUASIPERIODICITY TO CHAOS

The transition from quasiperiodicity to chaos has become a hot subject since 1982. This is provoked by the interest to elucidate whether there are universal aspects in the Ruelle-Takens route to chaos, which remains less understood theoretically in comparison with the period-doubling and intermittent routes.

As mentioned in Chapter 3, Ruelle and Takens[1] proposed an alternative to the Landau-Hopf picture of infinitely increasing number of incommensurable frequencies for the onset mechanism of turbulence. They showed that quasiperiodic motion on a 4-torus, i.e., with 4 incommensurable frequencies, is in general unstable and can be perturbed into a strange attractor corresponding to turbulent motion. A few years later, in collaboration with Newhouse they sharpened the above result to that of a 3-torus, i.e., quasiperiodic motion with three incommensurable frequencies is in general unstable and can be perturbed into chaotic motion[2].

Let us denote quasiperiodic motion with n incommensurable frequencies by Q_n. The Landau-Hopf route to chaos can be expressed as $Q_n \xrightarrow{n \to \infty}$ chaos. Ruelle and Takens first suggested the possibility of $Q_3 \to$ chaos, and then estimated it as nongeneric and replaced it by $Q_2 \to$ chaos.

In Ref. 2 the perturbation required to destroy Q_3 was rather specific, so in recent years the question has been raised concerning the existence of stable Q_3 in dynamical systems as a generic, i.e. typical, possibility. Indeed, stable Q_3 has been shown to exist in several mathematical models[3] and in hydrodynamical experiments[4]. Therefore, what was considered by Ruelle and Takens as nongeneric cases may well happen to be allowable in nature and there occurs the necessity to understand the mechanism of $Q_3 \to$ chaos transition.

Anyway, the transition $Q_2 \to$ chaos is generic and it does take place in many models and laboratory experiments. Two groups[5] studied the $Q_2 \to$ chaos transition by using renormalization group idea and discovered universal characteristics for this transition. Since it is impossible to have quasiperiodic motion in purely one-dimensional mappings, both groups turned to the next simplest model — the mapping of a circle onto itself[6] :

$$\theta_{n+1} = \theta_n + \Omega - \frac{\mu}{2\pi} \sin 2\pi\theta_n \qquad (\text{mod } 1) \tag{7-1}$$

which is a limiting case of the two-dimensional annular mapping[5].

In the linear case $\mu = 0$ the mapping

$$\theta_{n+1} = \theta_n + \Omega \qquad (\text{mod } 1) \tag{7-2}$$

has only two regimes: periodic for Ω rational and quasiperiodic for Ω irrational. (It is sufficient to restrict Ω to the interval (0, 1) due to taking (mod 1).) Nonzero μ introduces nonlinearity into the system and causes the periodic motion to persist for wider range of Ω values. This is the simplest occurrence of frequency-locking in nonlinear oscillator systems (see Fig. 7-1).

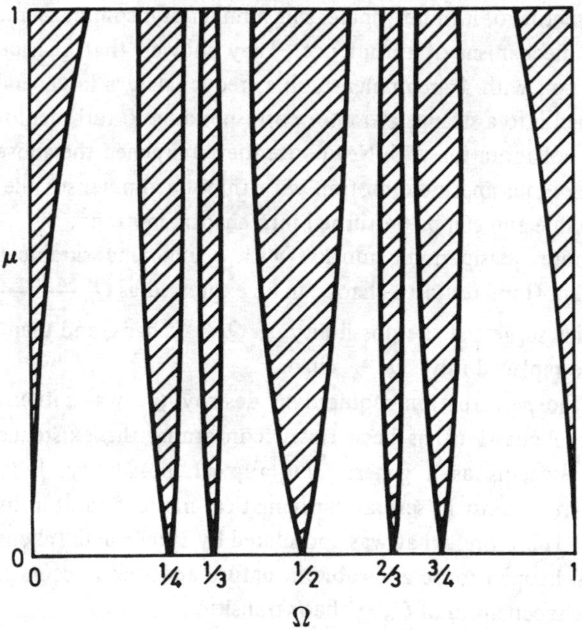

Fig. 7-1 Frequency-locking zones in the circle mapping (schematic).

The order of occurrence of the frequency-looking regions is given by the *Faray sequences* known in number theory[7]. An n-Faray sequence is the increasing succession of rational numbers p/q whose denominators are less than or equal to n, e.g., for $n = 5$ we have

$$0/1,\ 1/5,\ 1/4,\ 1/3,\ 2/5,\ 1/2,\ 3/5,\ 2/3,\ 3/4,\ 4/5,\ 1/1$$

(The case $n = 4$ is shown schematically in Fig. 7-1.)

Since for $\Omega = p/q$, p and q being relative prime, the mapping (7-2) has period q, it is the denominators in the Faray sequences which correspond to the observed periods. For $\mu = 0$ it is clear that periodic regimes can in no way fill up the whole Ω interval, because rational numbers have measure zero on the real axis. With increasing μ the frequency-locking "tongues" in Fig. 7.1 get wider and wider. They will contribute a positive measure on Ω. This picture persists for all μ less than one. In other words, when $\mu < 1$ we have periodic regimes immersed in the sea of quasiperiodicity. Chaotic motion can occur only for $\mu > 1$. Even for μ approaching 1 from below the periodic tongues do not approach each other closely and quasiperiodic gaps are left where the $Q_2 \rightarrow$ chaos transitions can take place.

However, it is impossible to stay within the quasiperiodic regime all the time when the parameters are varied. Sooner or later frequency-locking is destined to get in. Since the more irrational the Ω value is, the longer the quasiperiodic regime persists, as we have learned from the destruction of invariant KAM curves in conservative systems (cf. Chapter 2), it was suggested in Ref. 5 to stick to a sequence of rational numbers p_i/q_i which converges to the "most irrational" number the Golden Mean $(\sqrt{5} - 1)/2 = 0.61803\ldots$.

There are various ways to compare whether one irrational number is more irrational than another. Here the continuous fraction representation of numbers is used. Every irrational number can be represented by a unique infinite continuous fraction

$$(a_1, a_2, a_3, \ldots) \equiv \cfrac{1}{a_1 + \cfrac{1}{a_2 + \cfrac{1}{a_3 + \cfrac{1}{\ldots}}}} \tag{7-3}$$

If one asks how close a truncation of (7-4) approximates the irrational number it represents, then the worst fraction must be that with all a_i's equal to 1. We take this fact as the definition of the most irrational number. This number is nothing but the Golden Mean. It is known from elementary number theory that the truncations of the continuous fraction $(1, 1, 1, 1, \ldots)$ are given by the ratios of Fibonacci numbers:

$$\Omega_i = \frac{F_i}{F_{i+1}} \,, \qquad i = 1, 2, 3, \ldots \tag{7-4}$$

where $F_0 = 0$, $F_1 = 1$, and

$$F_{i+1} = F_i + F_{i-1} \,. \tag{7-5}$$

Instead of following a curve of quasiperiodic regime in the (μ, Ω) plane, it was suggested in Ref. 5 to stick to the sequence (7-4) up to the limit $i \to \infty$ where a transition from quasiperiodicity to chaos takes place. This can be done only when two parameters μ and Ω are both varied. In the language of dynamical system theory this is called a *codimension* two bifurcation sequence.

Using the recurrence relation (7-5) for Fibonacci numbers a pair of renormalization group equations similar to the Feigenbaum equation (4-24) was derived[5]:

$$\begin{aligned} f(x) &= \alpha f(\alpha(f(\alpha^{-2} x))) \\ f(x) &= \alpha^2 f(\alpha^{-1} f(\alpha^{-1} x)) \end{aligned} \tag{7-6}$$

and numerical study was carried out. We mention in passing that these two equations are not independent[8].

Being unable to present all the results obtained so far[9] on the $Q_2 \to$ chaos transition in this short chapter, we point out only that the transition is accomplished via loss of differentiability of certain functions which are analytic for $\mu < 1$. If one goes back to the case of two-dimensional annular mapping, the loss of differentiability can be visualized as progressively nested kinking of the invariant curve at all length scales.

Due to the presence of "rotational symmetry" in the annular mapping the destruction process takes place everywhere along the invariant curve which cannot be the general rule, say, in systems of ODE's. Indeed, there have appeared a few reports[10,11] on the $Q_2 \to$ chaos transitions in ODE's. A common feature of these numerical observations is the kinking and folding of the originally smooth Poincaré map at one or more points, but not everywhere simultaneously. These transitions are almost imperceptible at the beginning, but are reflected more clearly in the power spectra[11].

Besides the "local" universality studied in Ref. 5, there seems to be also similarity in the global bifurcation structure leading to the $Q_2 \to$ chaos transition. At least in some idealized models[7,11] the transition should be elucidated in the light of the relation between the Farray sequence immersed in the quasiperiodic sea and the *U*-sequence (see Chapter 4) immersed in chaotic regimes. This question is far from being solved at present.

References and guide to the bibliography

1. Ruelle and Takens (1971), Paper 5 in this volume.
2. Newhouse *et al*. (1978).
3. Mathematical models with stable Q_3: Grebogi *et al.* (1983b); R. K. Takavol, and A. S. Tworkowski, *Phys. Lett.*, **100A**, 65 (1984).
4. Experimental observation of stable Q_3: Gollub and Benson (1980), Paper 34 in this volume; Fauve and Libchaber (1981).
5. Renormalization-group theory of $Q_2 \to$ chaos transitions was developed by: Shenker (1982); Feigenbaum *et al.* (1982), Paper 25 in this volume; Rand *et al.* (1982), Paper 24 in this volume; Ostlund *et al.* (1983).
6. The circle mapping: Arnold (1965), Glass and Perez (1982), Perez and Glass (1982), Jensen *et al.* (1983), Kohyama and Aizawa (1983).
7. For Faray sequence see reference in Gonzalez and Piro (1983), Paper 18 in this volume.
8. Nauenberg (1982).
9. Among other papers on the $Q_2 \to$ chaos transition we mention: Kaneko (1982, 1983a, b), Hu (1983), Franceschini (1983), Arneodo, Coullet and Tresser (1983), Sano and Sawada (1983).
10. $Q_2 \to$ chaos transition in ODE's: Riela (1982c), Schreiber and Marek (1982b).
11. $Q_2 \to$ chaos transition in ODE's: S.-G. Chen, B.-L. Hao, G.-G. Wang, and S.-Y. Zhang, paper presented to the *15th International Conference on Thermo-dynamics and Statistical Physics*, July 1983, Edinburgh; and *Acta Physica Sinica*, to appear.

Chapter 8
CHARACTERIZING THE ATTRACTORS: DIMENSION AND LYAPUNOV EXPONENTS

We mentioned in Chapter 5 that the strategy in studying the bifurcation and chaos "spectrum" of a physical system consists in first identifying various periodic regimes with confidence and then trying to characterize the chaotic motion on the attractors. Both in the laboratory and on computers it is very difficult to distinguish a purely chaotic motion from a quasiperiodic one or from a motion perturbed by external noise by simply looking at the finite sequence of data. Moreover, although mathematicians have prepared such nice notions as hyperbolic attractor, Axiom A system or Smale's horseshoe, they are not directly applicable to systems of physical interest. Consequently, the term "strange attractor" is now being used in physical literature very liberally. Anyway, we need some quantitative means to recognize, characterize and classify attractors. It is also desirable to measure the strangeness of strange attractors from the data available, in particular, from a single time series of sampled points.

Power spectrum analysis is useful in telling quasiperiodic motion from periodic, and in identifying high order periodicities embedded in chaotic bands by the fine structures in spectra. Power spectra of chaotic attractors associated with period-doubling distinguish themselves by sharp peaks on the background of broad-band noise, but it makes little difference between attractors of other types. We indicate a few papers on scaling property of power spectra[1] and turn to more sophisticated ways of characterizing attractors, among which Lyapunov exponents and various definitions for dimension are widely used.

Lyapunov exponents, dimension, measure and entropy (in a mathematical sense) are closely related notions. They are backed up by a few mathematical theorems, complemented by plausible conjectures and numerical experience. We devote this chapter to a brief summary of the most important concepts and techniques.

The phase space volume of a dissipative system contracts in the process of evolution and the motion is confined to a certain attractor in the long time limit $t \to \infty$. The dimension of the attractor D is lower than the dimension n of the original phase space. (No Liouville's theorem holds for dissipative systems.) Using any reasonable definition of dimension, the dimension of trivial attractors can easily be calculated to be integers: $D = 0$ for fixed points, $D = 1$ for limit cycles, $D = 2$ for 2-torus, etc. However, the dimension of strange attractors often turns out to have noninteger value, yet what is an object with noninteger dimension?

Hausdorff introduced a generalized definition of dimension as early as in 1919. It can be explained very simply in the following way. Take usual "regular" geometrical object, say, a cube, and double its linear size in each spatial directions. We get a cube whose volume is eight times larger than the original one, because $2^3 = 8$. In general, taking an object of dimension D and increasing its linear size in each spatial directions ℓ times, one would have its volume increased to $k = \ell^D$ times of the original. Inverting this simple relation, we get a new definition for dimension:

$$D = \frac{\ln k}{\ln \ell} \qquad (8\text{-}1)$$

Now, we have freed ourselves from the restriction of D being an integer.

A precise definition of ***Hausdorff dimension*** requires the notion of Hausdorff measure (see the books by Mandelbrot[2]), but formula (8.1) is enough for our practical needs. We begin with the simplest example of a geometrical object having noninteger dimension — the Cantor set.

Take the interval (0, 1) and delete the central one third (1/3, 2/3), then repeat this operation with respect to the remaining segments again and yet again, ***ad infinitum.*** What is the dimension of the limiting set of points thus obtained? Let points left in the interval (0, 1/3) constitute our geometrical object under consideration. Increasing its linear size by a factor $\ell = 3$ would yield two copies ($k = 2$) of the same object, therefore

$$D = \frac{\ln 2}{\ln 3} = 0.6309\ldots \qquad (8\text{-}2)$$

Actually, many strange attractors have a Cantor set like structure (see, e.g., the frequently cited Figs. 3-6 in Hénon (1976), Paper 13 in this volume). To calculate the dimension numerically, a box-counting algorithm may be used. One divides that part of the phase space where the attractor lies into small cells of linear size ϵ, and counts the number $N(\epsilon)$ of such cells that contain at least one point of the orbit. Then, the dimension is calculated by taking the limit

$$D_C = \lim_{\epsilon \to 0} \frac{\ln N(\epsilon)}{\ln(1/\epsilon)} \tag{8-3}$$

To be precise, (8-3) is called the *Kolmogorov capacity* of the limiting set of points, hence the subscript C to D_C. In practice, one counts $N(\epsilon)$ for a series of finite ϵ's and obtains D_C from the slope of the $\ln N(\epsilon)$ versus $\ln(1/\epsilon)$ plot.

Thus far in counting $N(\epsilon)$ no attention has been paid to the possible inhomogeneity of the attractor, i.e., no matter how many times the orbit travels through a given cell, the cell is counted only once. To correct this inexactitude one introduces a weight according to how frequently a cell is visited. If the i-th cell is visited with a probability P_i, then the definition (8-3) is replaced by

$$D_I = \lim_{\epsilon \to 0} \frac{I(\epsilon)}{\ln(1/\epsilon)} \tag{8-4}$$

where

$$I(\epsilon) = -\sum_{i=1}^{N(\epsilon)} P_i \ln P_i \quad , \tag{8-5}$$

$N(\epsilon)$ being the total number of cells visited. If every cell is visited with equal probability, i.e., $P_i = 1/N(\epsilon)$, then (8-5) reduces to $I(\epsilon) = \ln N(\epsilon)$ and $D_I = D_C$. In general $D_I \leqslant D_C$. D_I is called the *information dimension.*

There exists a number of different ways to define dimension[3]; see especially the review by Farmer *et al.* (1983). It is conjectured that the Hausdorff dimension is equal to D_C and some other definitions are equivalent to D_I. Numerical calculations seem to support this conjecture.

The box-counting algorithm to compute the dimension is very time-consuming and becomes impractical when the dimension of the original phase space $n \geqslant 2$. Moreover, the dimension reflects only one essential aspect of dissipative dynamics, namely, the contraction of phase volume. However, a strange attractor cannot be generated merely by contraction. It must be stretched and folded in some directions as well. To describe these more subtle features we need the notion of *Lyapunov exponent* and we shall see that the dimension can be readily calculated from these exponents. In addition, Lyapunov exponents furnish us with a classification scheme of attractors.

Consider an autonomous systems of ODE's

$$\dot{x}_i = f_i(x_1, x_2, \ldots, x_n) \quad , \qquad i = 1, 2, \ldots, n \quad . \tag{8-6}$$

It can be linearized near a given point $\mathbf{x}_0$

$$x_i = x_{0i} + \delta x_i \qquad i = 1, 2, \ldots, n$$

to yield a linear, but nonautonomous (depending on $\mathbf{x}_0$) system of equations

$$\frac{d}{dt}\delta x_i = \sum_j^n U_{ij}(x_0)\,\delta x_j \qquad i = 1, 2, \ldots, n \tag{8-7}$$

where

$$U_{ij}(x_0) = \left.\frac{\partial f_i}{\partial x_j}\right|_{\mathbf{x}=\mathbf{x}_0} . \tag{8-8}$$

The eigenvalues of matrix (8-8) determine the local stability of Eq. (8-6) in the vacinity of $\mathbf{x}_0$: if the real parts of some eigenvalues happen to be positive numbers, then nearby trajectories will run away at an exponential rate. In principle, one can average the real parts of all eigenvalues to get a set of global characteristics — the Lyapunov exponents. However, there is another more intuitive way to calculate n real numbers playing the same role[4].

Let us build a local coordinate system with origin at $\mathbf{x}_0$ by choosing at random n small vectors coming out from $\mathbf{x}_0$ and then orthogonalizing them. Denote these vectors by $\left\{\mathbf{e}_0^{(1)}, \mathbf{e}_0^{(2)}, \ldots, \mathbf{e}_0^{(n)}\right\}$. Then integrate the Eq. (8-6) one time step forwards to reach a new position $\mathbf{x}_1$, and at the same time let $\left\{\mathbf{e}_0^{(i)}\right\}$ evolve according to the linearized Eq. (8-7) to form n new vectors with origin at $\mathbf{x}_1$. Now we can calculate a number of goemetrical ratios:

$$C_1^n = n \text{ length ratios } \left|\mathbf{e}_1^{(i)}\right| \Big/ \left|\mathbf{e}_0^{(i)}\right| \quad ;$$

$$C_2^n = n(n-1)/2 \text{ parallelogram area ratios } \left|\mathbf{e}_1^{(i)} \wedge \mathbf{e}_1^{(j)}\right| \Big/ \left|\mathbf{e}_0^{(i)} \wedge \mathbf{e}_0^{(j)}\right| \quad ;$$

... ...

$$C_n^n = 1 \;\; n\text{-dimensional hyperparallelepiped volume ratio}$$

$$\frac{\left|\mathbf{e}_1^{(1)} \wedge \mathbf{e}_1^{(2)} \wedge \ldots \wedge \mathbf{e}_1^{(n)}\right|}{\left|\mathbf{e}_0^{(1)} \wedge \mathbf{e}_0^{(2)} \wedge \quad \wedge \mathbf{e}_0^{(n)}\right|} \;;$$

(The exterior product $\wedge$ generalizes the vector product $\times$ to higher dimensional spaces.) Being geometrical quantities, all these ratios are real numbers. Repeating this process step by step along the trajectory and taking the long time average of these numbers, we get a set of global characteristics of the attractor. It was proved by Benettin *et al.* (1976) that provided the vectors $\{\mathbf{e}_0^{(i)}\}$ are taken at random, the length ratios converge to the maximal Lyapunov exponent λ_1, the parallelogram area ratios converge to the sum of the two largest Lyapunov exponents $\lambda_1 + \lambda_2, \ldots$ the n-dimensional hyperparallelepiped volume ratio converges to the sum of all Lyapunov exponents $\lambda_1 + \lambda_2 + \ldots + \lambda_n$. The last quantity must be equal to the contraction rate of a volume element along the trajectory, i.e.,

$$\lambda_1 + \lambda_2 + \ldots + \lambda_n = -\operatorname{div} \mathbf{f} \tag{8-9}$$

where $\mathbf{f} = (f_1, f_2, \ldots, f_n)$ is the vector field on the right hand side of Eq. (8-6). In practice, Eq. (8-9) serves as a numerical check for the correctness of computation. In stating the results above, we have introduced an ordering for all Lyapunov exponents:

$$\lambda_1 \geqslant \lambda_2 \geqslant \ldots \geqslant \lambda_n \tag{8-10}$$

We mention in passing that the step by step orthogonalization appears to be crucial in numerical calculation, since the initially independent vectors tend to evolve into a subspace of lower dimensionality due to phase volume contraction. For details see Shimada and Nagashima (1979), Paper 27 in this volume.

Kaplan and Yorke (1979) conjectured that the Kolmogorov capacity D_C is related to the Lyapunov exponents by

$$D_C = j + \left(\sum_{i=1}^{j} \lambda_i\right) \Big/ |\lambda_{j+1}| \tag{8-11}$$

where j is the number of Lyapunov exponents which assures the non-negativeness of the sum in (8-11), i.e.

$$\sum_{i=1}^{j} \lambda_i \geqslant 0.$$

In the extreme cases $j = 0$ or n, define $D_C = 0$ or n respectively.

Mori (1980) "derived" another relation between dimension and Lyapunov exponents

$$D_C = m_0 + m_+ (1 - \lambda_+/\lambda_-) \quad , \tag{8-12}$$

where

$$\lambda_+ = \frac{1}{m_+} \sum_{\lambda_i > 0} \lambda_i \ , \quad \lambda_- = \frac{1}{m_-} \sum_{\lambda_i < 0} \lambda_i \ ,$$

and m_-, m_0, m_+ denote the number of negative, zero, and positive exponents respectively. For sufficiently large n (8-11) differs from (8-12) significantly, because all negative exponents contribute to (8-12), whereas only a finite number of them enters (8-11). Numerical experiments seem to be in favour of (8-11). In general, the Lyapunov exponents converge fairly rapidly and provide an effective way to calculate the dimension D_C in comparison to the box-counting algorithm.

Yet the largest Lyapunov exponent alone is of much help to numerical work, since the appearance of a positive exponent signals the onset of chaos. It was suggested by many authors to put forward a qualitative classification of attractors according to the signature of the Lyapunov exponents. For example (–, –, –) corresponds to fixed point with $D = 0$, (0, –, –) to limit cycle with $D = 1$, (0, 0, –) to 2-tori with $D = 2$, and (+, 0, –) to strange attractor with $D = D_C$. In this scheme a signature (+, +, 0, –) would correspond to strange attractor displaying *superchaos*. Since the signature changes with control parameters, they provide also a means to classify transitions between various regimes.

Dimension and Lyapunov exponent for one-dimensional mappings have some peculiarities worth mentioning. First, there is only one Lyapunov exponent, sometimes called also Lyapunov number

$$\lambda(\mu) = \lim_{k \to \infty} \frac{1}{k} \sum_{i=0}^{k-1} \ln |f'(\mu, x_i)| \ . \tag{8-13}$$

Periodic orbits correspond to $\lambda < 0$, chaotic orbits to $\lambda > 0$.

Second, $\lambda = 0$ at every bifurcation point, at the accumulation point μ_∞ of the period-doubling cascade, and at tangent bifurcations. Grassberger (1981) showed that at μ_∞, where the attractor has a Contor set structure, the Hausdorff dimension is universal, with value $D_C = 0.538\ldots$. On the other hand, due to critical slowing down one always gets a cluster of points near every finite bifurcation point μ_k, no matter how long the iteration is being carried on. Wang and Chen[6] proved that this cluster acquires an "operational" dimension $D = 2/3$, if measured by a box-counting algorithm, although the true dimension of a periodic orbit must be zero. There exists a scaling function relating these two values $D = 2/3$ and $D = 0.538\ldots$[6].

Another remark concerns the reconstruction of strange attractors and the measurement of their dimension from experimental data[7]. For a real physical system neither the dimension of the phase space, nor that of the attractor are known in advance, therefore it has been suggested to extract information on dimensionality from single time series of the experimental data by linear regression and other methods of statistical estimation. For the time being only the integer part of the dimension can be evaluated. A few "real" strange attractors have been reconstructed in this way, e.g., the dimension of strange attractor in turbulent Couette flow was estimated to be 5 for not very high Reynolds number (Brandstäter) *et al.* 1983).

We conclude this chapter by discussing briefly the relation of Lyapunov exponents to measure and entropy defined on strange attractors[8]. Chaotic motion on strange attractors can be characterized by their ergodic properties. In particular, certain continuous measure exists on the attractor and a metric entropy (the K-entropy) invariant under measure-preserving transformations may be defined. K-entropy is essentially the sum of all positive Lyapunov exponents. Therefore, it is closely related to the Hausdorff dimension. Being metric properties, dimension, K-entropy, and Lyapunov exponents may change under continuous transformation of the phase space, e.g., when going from one coordinate system to certain other systems. In this context they are not as simple as topological characteristics such as the topological dimension or the topological entropy. However, it is just this flexibility which makes these metric properties more relevant to physics. In view of their possibly increasing role in physics we give some references to this more or less mathematical topic[8].

References and guide to bibliography

1. Power spectra of attractors: Feigenbaum (1980a, b, c, d), Farmer *et al.* (1980), Farmer (1981), Collet *et al.* (1981c), Huberman and Zisook (1981), Mori (1981), Nauenberg and Rudnick (1981), Thomae and Grossmann (1981a, b), Wolf and Swift (1981), Beloshapkin and Zaslavsky (1983).
2. B. B. Mandelbrot, *Fractals, Forms, Chance and Dimension*, Freeman, 1977; *Fractal Geometry in Nature*, Freeman, 1982.
3. Dimension of chaotic attractors: Russell *et al.* (1980), Paper 28 in this volume; Velarde (1981), Grassberger (1981, 1983b, c), Aizawa (1982), Greenside *et al.* (1982), Farmer (1982b, 1983), Farmer *et al.* (1983), Hentschel and Procaccia (1983), McGuinness (1983), Shtern (1983), Tel (1983c), Termonica and Alexandrowics (1983).
4. Lyapunov exponents: Oseledes (1968), Pesin (1976, 1977), Feit (1978), Benettin *et al.* (1976), Paper 26 in this volume, and (1978a, 1980), Shimada and Nagashima (1979), Paper 27 in this volume; Katok (1980), Geisel *et al.* (1981), Fraser and Kapral (1982a), Kai (1982), McCreadie and Rowlands (1982), Young (1982), Fujisaka (1982, 1983), Froyland (1983), Haken (1983).

5. Relation between dimension and Lyapunov exponents: Kaplan and Yorke (1979c), Mori (1980), Mori and Fujisaka (1980), Ledrappier (1981), Farmer *et al.* (1983).
6. Dimension of attractors in one-dimensional mappings: Grassberger (1981), Hu and Hao (1983); Y.-Q. Wang and S.-G. Chen (1984), *Acta Physica Sinica*, **33**, 351.
7. Reconstruction of strange attractors from experimental data: Packard *et al.* (1980), Froehling *et al.* (1981), Paper 29 in this volume; Takens (1981), Roux and Swinney (1981), Hudson and Mankin (1981); Roux *et al.* (1983), Paper 36 in this volume; Brandstäter *et al.* (1983).
8. Entropy and measure defined on strange attractors: Oseledes (1968), Pesin (1976, 1977), Casartelli *et al.* (1976); Benettin *et al.* (1976), Paper 26 in this volume; Benettin (1978, 1979), Ruelle (1978a, c, 1979a, 1980a, 1981a, b, c, 1983), Oono (1978a), Katok (1980), Sun and Froeschle (1981, 1982), Curry (1981), Dias De Deus (1982, 1983); Grassberger and Procaccia (1983a, c) and (1983b), Paper 30 in this volume.

Chapter 9
SCALING FOR EXTERNAL NOISE

We have emphasized the intrinsic nature of chaos in Chapter 1 and throughout the chapters thus far. However, external noise is an unavoidable factor in computer experiments (round-off errors) and laboratory experiments (thermal fluctuations of the ambience). A fuller understanding of chaos requires the inclusion of external noise into the theoretical framework.

Actually, in the theory of chaotic phenomena external noise plays a more constructive role than just to be an undesirable but inescapable disturbance from the outside. Only with the external noise taken into account shall we have a full parallel with the theory of continuous phase transitions which has enriched physics with so many new concepts since the mid-60's.

Phase transition occurs at certain critical value of control parameter such as the temperature T or the pressure p when the latter is subjected to a slow and progressive variation. Many thermodynamical quantities show singular behavior when $|T-T_c|\to 0$, where T_c denotes the critical point. Chaotic transition takes place at critical value μ_c of the control parameter such as the Reynold's number and shows up as singularity in $|\mu-\mu_c|$. The formation of the new phase may be described by the appearance of nonzero "order parameter", say, macroscopic magnetization in ferromagnetic phase transitions. Chaotic state can be characterized by certain "disorder parameter" as well. It was thought by some authors that the Lyapunov exponent may be taken as disorder parameter[1]. In fact, the Lyapunov exponent corresponds to the reciprocal of correlation length ξ in phase transitions as we shall see soon. One must take some other statistical characteristic of the chaotic motion on the strange attractor (entropy or the like) to be the disorder parameter.

Moreover, a new ordered phase may appear either spontaneously at the critical point T_c, or may be induced by suitable external fields. For example, the external magnetic field plays the role of "ordering field" in magnetic phase transitions. In phase transition theory the ordering field coupled to the order parameter is sometimes called a dual field. Does there exist a "disordering field" for chaotic transition? Chaotic states are characterized by spontaneous appearance of seemingly random motion showing noisy broad bands in the power spectra. Therefore, external noise may well play the role of the disordering field. To include the external noise, one adds a noisy source term to the nonlinear mapping (4-2) and thereby changes it into a discrete Langevin equation:

$$x_{n+1} = f(\mu, x_n) + \sigma \xi_n \quad , \tag{9-1}$$

where ξ_n are random numbers obeying certain statistical distribution with the constraints

$$\overline{\xi_n} = 0, \quad \overline{\xi_n \xi_m} = \delta_{nm}$$

and the coefficient σ, being a measure of the noise strength, can be compared with the external magnetic field in magnetic phase transitions.

An important notion in phase transition theory is the correlation length ξ which diverges at the critical point as

$$\xi \propto |T - T_c|^{-\nu} \tag{9-2}$$

where ν is a positive universal exponent. So far in this volume, chaotic behaviour has been considered only in connection with time evolution of nonlinear systems (except for a few words in Chapter 4 on PDE's). From the discussion in the last chapter we know that when the Lyapunov exponent λ is negative, the influence of initial conditions decays quickly and the system approaches an asymptotic state independent of the initial conditions, whereas when λ is positive, nearby orbits will run away exponentially and any minor change in initial conditions will be amplified with time. Therefore, λ^{-1} can be taken as a measure of time correlation. Since λ passes through zero at bifurcation points, in particular, at the limiting point μ_∞ of period-doubling sequence, one expects a relation similar to (9-2) to hold:

$$\lambda^{-1} \propto |\mu - \mu_\infty|^{-t} \tag{9-3}$$

Starting from the discrete Langevin equation (9-1) the whole arsenal of critical dynamics in phase transition theory (see, e.g., Ref. 2 for a review) can be brought to bear on chaotic transitions. We have included three pioneering papers along this line into this volume[3-5] (see also papers[6] on Champman-Kolmogorov equation related to the discrete Langevin equation). The main result of Ref. 5 was the scaling relation for λ in the presence of external noise:

$$\lambda(\mu_\infty - \mu, \sigma) = (\mu_\infty - \mu)^t \, \Phi\left(\frac{\sigma^\theta}{(\mu_\infty - \mu)^t}\right) \tag{9-4}$$

where Φ is a universal function and

$$t = \frac{\ln 2}{\ln \sigma} = 0.4498069\ldots$$

$$\theta = \frac{\ln 2}{\ln k} = 0.366739\ldots \tag{9-5}$$

$k = 6.6190\ldots$ being a new exponent related to the scaling of external noise. Putting $\sigma = 0$ in (9-4), we are led to

$$\lambda(\mu_\infty - \mu, 0) \propto (\mu_\infty - \mu)^t \tag{9-6}$$

which has the desired form of Eq. (9-3) and was obtained without involving external noise[7].

The scaling relation derived in Ref. 4 looks slightly different:

$$\lambda(\mu_\infty - \mu, \sigma) = \sigma^\theta \, L\left(\frac{\mu_\infty - \mu}{\sigma^{\theta/t}}\right) \tag{9-7}$$

It suffices to redefine the scaling function by

$$x L(1/x) \equiv \Phi(x) \tag{9-8}$$

with $x = \sigma^\theta / (\mu_\infty - \mu)^t$, to see that Eq. (9-7) is identical to (9-4).

Now we are prepared to summarize the role of external noise in chaotic dynamics as follows. First, external noise will wash out the details in high order bifurcation sequences including both the period-doubling cascades and the inverse sequences of chaotic bands[3]. In order to see one more bifurcation one must decrease the noise level by a factor of k. This explains the meaning of the noise exponent k. Second, the presence of external noise makes the bifurcation diagram fuzzy and lowers the accumulation point μ_∞ slightly as it can be expected from the role of noise as disordering field. Third, the Lyapunov exponent passes through zero at various bifurcation points, but this is no longer true when external noise is added. For more details see review papers[8].

The role of external noise was studied mainly on the logistic mapping[9]. There were a few reports on the effect of noise on strange attractors[10]. The influence of noise on the $Q_2 \to$ chaos transition and the intermittent transition has been studied in Ref. 11 and Ref. 12 respectively.

References and guide to the bibliography

1. For discussion on disorder parameter for chaos see: Oono (1978a), Oono and Osikawa (1980), and the review by Crutchfield *et al.* (1982).
2. P. C. Hohenberg, and B. I. Hanperin, *Rev. Mod. Phys.*, **49**, 435 (1977).
3. Crutchfield and Huberman (1980), Paper 31 in this volume.
4. Crutchfield, Nauenberg and Rudnick (1981), Paper 32 in this volume.
5. Sharaiman, Wayne and Martin (1981), Paper 33 in this volume.
6. Champman-Kolmogorov equation related to the discrete Langevin equation: Haken and Mayer-Kress (1981a, b), Haken and Wunderlin (1982).
7. Huberman and Rudnick (1980).
8. Review papers on external noise: Crutchfield *et al.* (1982), Guckenheimer (1982), Kapral *et al.* (1982).
9. Effect of noise on the logistic mapping: Mayer-Kress and Haken (1981b), Thomas and Grossmann (1981b), Hirschman and Whiteson (1982), Kai (1982), Karney *et al.* (1982).
10. Effect of noise on strange attractors: Zipplius and Lucke (1981), Ott and Hanson (1981), Rechester and White (1983).
11. Role of noise in the $Q_2 \to$ chaos transitions: Feigenbaum and Hasslacher (1982).
12. Role of noise in the intermittent transitions: Eckmann *et al.* (1981), Hirsch *et al.* (1982a, b), Papers 21 and 22 in this volume.

Chapter 10

EXPERIMENTAL OBSERVATION OF CHAOTIC PHENOMENA

Chaotic phenomena are not just a motley collection from the mathematical labyrinth, but a wide class of natural events found in the physical world. Generally speaking, chaos happens more frequently than order, just as there are many more irrational numbers than the rationals. Chaos can be thought of as a new regime of nonlinear oscillations, as a compromise between competing periodicities, as overlap of resonances, as accumulation of many instabilities, as the prelude to turbulence. Therefore, it is not difficult to imagine various experimental situations generating chaotic behaviour.

Moreover, Nature is always richer than mathematical models. Experimental study of chaos has at least a twofold mission: to verify the theoretical understanding gained from model studies and to bring about new physics by challenging the existing theory with unexpected findings, not to mention the prospect for technological development at our present level of knowledge. From this view-point laboratory observations can be roughly divided into three groups: those almost coinciding with theoretical expectations, e.g., experiments using nonlinear circuits; those showing much more variations and complications, only some facets of which resemble one or another aspect of theoretical models, e.g., experiments on hydrodynamical instabilities; and those occupying an intermediate state, e.g., most of the experiments on acoustic, optical, chemical, and solid-state turbulence.

In what follows various kinds of experiments will be listed briefly along with references to the bibliography, including some theoretical analysis and proposals which come under the same heading. Whenever appropriate we shall make a few remarks on the experimental results.

1. Forced vibration of shallow water wave in a finite container[1]

Historically this was the first experiment where transition to chaos would have been discovered, if Faraday had the idea of period-doubling route to chaos (see Faraday in Ref. 1). In fact, Faraday observed subharmonic component $f_0/2$ in the shallow water wave in a container forced to vibrate vertically at frequency f_0. Lord Rayleigh noticed, repeated and wrote about this experiment in his *Theory of Sound,* § 68b. Modern data acquisition techniques helped to reveal much longer bifurcation sequences and the onset of chaos[1]. However, the authors of Ref. 1 described their results as departures from period-doubling. For example, taking the driving period to be 1, they observed a sequence

1, 2, 4, 12, 14, 16, 18, 20, 22, 24, 28, 35,

but the power spectrum given for period 14 showed clearly a 2*7 fine structure, testifying to its being a secondary period 7 embedded in a chaotic band of period 2. Since the bifurcation and chaos "phase diagram" of a real physical system may be very intricate due to bending, folding and interpenetration of various sequences of periods (cf. Fig. 1 in Paper 17, or Fig. 1 in Paper 18 in this volume), it is very easy to mix up periods coming from different sequences or from different parts of the same sequence.

2. *Bifurcation and chaos in nonlinear circuits*

The most complete and beautiful results on chaos have been obtained on nonlinear circuits, because the experimental conditions can be precisely controlled and the circuits can be well represented by ODE's with only a few variables.

Earlier work on nonlinear circuits[2] were carried out on systems of many coupled nonlinear oscillators. Complicated combinations of periods and chaos were indeed observed, but could not be fitted into simple systematics as has now been done.

Most of recent experiments[3] deal with a single nonlinear oscillator driven by a periodical signal. Period-doubling bifurcations, tangent bifurcations, intermittency, crisis of the attractors, and the effect of external noise have been studied. In particular, the Feigenbaum constants δ and α, the noise scaling exponent k, and the ratio of successive period-doubling peaks in the power spectra were measured to be in good accord with that of the one-dimensional unimodal mappings. It seems desirable to carry out a detailed study on transitions from quasiperiodicity to chaos using nonlinear circuits, but probably more than one nonlinear oscillators should be used for this purpose.

3. *Birfurcation and chaos in mechanical or electromechanical systems*[4]

This is another group of experiments where results in keeping with theory can be expected. Some of these systems may serve as classroom demonstration of chaotic phenomena. We add also that both the Lorenz and Duffing equations can be modelled by simple mechanical systems (see book B32 and B36 respectively).

4. Onset of hydrodynamical instabilities[5,6]

Low temperature technique, laser Doppler velocimetry, and modern data acquisition systems have brought the study of turbulence back to physical laboratories. However, at the present time most of these studies still concentrate on instabilities in finite containers where the decay of many irrelevant modes of motion makes the results closer to theoretical predictions. As regards the more important shear flow instabilities, little progress has really been achieved with the help of the philosophy of chaos, because infinite degrees of freedom must be involved to interpret the experiments.

Thermoconvective instability of fluid in a container heated from below with free surface (the Bénard instability) or without space in between the fluid and the upper plate of the container (the Rayleigh-Bénard instability) has now become a classical setup to study the onset mechanism of turbulence. Ahlers first conducted the experiment at low temperatures on liquid helium in 1974. Libchaber and Maurer (1980) observed the successive appearance of quasiperiodic regime with two incommensurable frequencies, frequency locking, and period-doubling. They resolved 2^n ($n=0$ to 4) periods and chaotic bands with periods 2^n ($n=1$ to 4) from the power spectra at small Prandtl numbers. Three-dimensional instabilities and intermittency were observed at larger Prandtl numbers. Rayleigh-Bénard experiment using Hg as working substance revealed transition from quasiperiodicity with two or three incommensurable frequencies to chaos (Fauve and Libchaber, 1981).

Instabilities of Couette flow between rotating cylinders[6] show much more variations, some but not all of which may be put into the framework of the Ruelle-Takens route. Especially when the outer cylinder also rotates or stops suddenly, many more very complicated patterns have been recorded. Most of these experimental findings still await a better theoretical understanding. The same may be said as regards experiments using rotating spheres (Belyaev *et al.*, 1979).

5. Acoustic turbulence[7]

Subharmonic components have been known to exist in the spectra of cavitation (i.e., bubble formation in liquid subjected to high power supersonic irradiation) noise. In searching for the period-doubling route to chaos, these experiments have been improved to get 2^3 subharmonics, but higher bifurcations cannot be identified due to the high level of external noise in this kind of devices.

Subharmonic generation has been observed also in supersonic absorption in liquid helium. Above the superfluid transition temperature $T_\lambda = 2.17$ K, it does not differ from ordinary liquids. However, it is a little surprising that clear period-doubling bifurcation sequence and tangent bifurcation into periods 3, 5 and 7 show up at $T < T_\lambda$ when ordinary bubbles cannot form via the cavitation mechanism. Indeed, the first bifurcation occurs at the threshold of quantum vertex generation. Therefore, period-doubling in superfluid helium has a quantum origin, but enjoys the universality of other classical bifurcation phenomena. This is similar to universality in continuous phase transitions.

Other experiments on acoustic turbulence include simple acoustic systems in air, liquid nitrogen, or water.

6. *Optical turbulence*[8-10]

Lasers provide a typical class of nonlinear oscillator systems. The Maxwell-Bloch equations describing a unimodal laser in the semiclassical approach reduce to a system of three ODE's. Haken (1975) showed that this system can be transformed into the Lorenz model (5-4) under certain conditions, not realistic for a laser, but relevant to random spiking in the superradiance regime.

Ikeda (1979) predicted theoretically that bifurcation and chaos may appear in an optical bistability device using a ring cavity. This effect was first observed in so-called hybrid bistability devices by Gibbs *et al.* (1981), Paper 40 in this volume. Hybrid means that part of the optical output is delayed by an electronic circuit (or kept for a while by a microprocessor) and then fed back to the input signal. Recently, all-optical bistability devices have also been designed to show optical chaos, see, e.g., Harrison *et al.* (1983). The upper branch of the bistable states losses stability at certain threshold of the input power and gives rise to doubled periods $2t_R, 4t_R$, and chaos, where t_R is the delay time. This kind of devices with feedback corresponds to time-delayed ODE's, which in the simplest case may be reduced to a single equation for the phase:

$$\tau\dot{\varphi}(t) = -\varphi(t) + f(\mu, \varphi(t - t_R)) \quad , \tag{10-1}$$

where τ is the intrinsic relaxation time of the nonlinear medium causing bistability, and $f(\mu, \varphi)$ is a nonlinear function of φ characterizing the cavity with nonlinear medium, μ and t_R being adjustable parameters. In the limit of long delay $\tau \ll t_R$ the derivative term in (10-1) may be neglected and the equation reduces to a one-dimensional mapping by setting $\varphi_n \equiv \varphi(n\, t_R)$:

$$\varphi_n = f(\mu, \varphi_{n-1}) \tag{10-2}$$

Now there is nothing strange to have period-doubling and chaos in such systems. However, in bistability devices chaos may appear in other regimes too, e.g., when $\tau \gg t_R$ (Ikeda *et al.*, 1982).

In hybrid bistability devices usually the delay time t_R is of the order of milliseconds, so the frequency structure of the bifurcation and chaos has nothing to do with optics. Great effort has been made to search for purely optical chaos. Now we have quite a few reports on this subject[9], e.g., in Q-switched gas laser (Arechi *et al.*, 1982). We mention also some general or theoretical discussion on optical turbulence[10].

7. *Chemical turbulence*

The concentrations of intermediate products in some specially designed chemical reactions oscillate with time. The most thoroughly studied reaction of this type, the Belousov-Zhabotinskii reaction, consists of some 20 elementary reactions. During the last decade these systems served as prototype for self-organization phenomena, i.e., spontaneous occurrence of temporal and/or spatial structures in the system when energy and matter flows are supported from the outside. If well stirred, the reacting system may be considered as spatially homogeneous and described by autonomous nonlinear ODE's, which often turn out to be isomorphic to some coupled-oscillator systems[13].

Reports on experimental observation of chemical turbulence began to emerge since 1977[12] and chaotic solutions have been found in mathematical models describing such systems[13].

In order to keep the reaction in steady state, some reactants and products must be supplied or removed from the system constantly. The flow velocity of these chemicals can be taken as control parameter and the concentration of certain intermediate products, say, the concentration of Br^- ions in the BZ reaction, is to be monitored. At very small flow velocity the system approaches thermal equilibrium without showing complicated time behaviour. At very large flow velocity the chemicals rush through the container, having no time to get into reaction. Therefore, only at intermediate flow velocities does one observe complicated dynamics, including multisteady states, simple or relaxation oscillations, intermittency, period-doubling and chaos, hysteresis, etc. It has been proved experimentally that such dynamical behaviour was surely caused by chemical reactions, but not by the hydrodynamics of the flow.

It was thought at the beginning that chemical turbulence offers something quite different from other nonlinear systems. However, the strange attractors reconstructed from experimental data happened to have sufficient low dimensionality and the observed order of oscillation periods fits that of the *U*-sequence (q.v. Chapter 4). Therefore, people now tend to believe that in chemical turbulence there are more features in common with other nonlinear systems. Moreover, apart from nonlinear circuits, chemical systems are easy to control and the dynamics involves a quite different frequency range compared to that of nonlinear circuits. All these make chemical turbulence an interesting subject deserving more attention.

8. *Chaotic phenomena in solid state physics*[14-17]

Many models used in solid state physics consist of coupled or driven nonlinear oscillators whose time evolution must show chaotic behaviour within a suitable parameter range. On the other hand, competition of various commensurable or incommensurable periods makes chaos an intrinsic source for spatial disorder (or to be more precise, for order without periodicity). From numerous theoretical models and experimental proposals we mention only a few.

It has been known since 1977 that in rf driven Josephson junctions used as parametric amplifier noise grows anomalously with the gain level. The equivalent noise temperature reaches 50 000 K whereas the device is kept at 4 K. Since such a high noise level cannot be explained by any known noise source and its amplification, Huberman *et al.* (1980) attributed it to the intrinsic dynamics of the junction. In the resistively shunted junction model the current equation reads

$$C\dot{V} + V/R + I_c \sin\varphi = I_r \cos\omega t + I_0 \qquad (10\text{-}3)$$

where R, C, I_c, I_r, and I_0 denote the resistance, capacity, critical current, rf current and dc bias respectively. The phase difference across the junction is given by the Josephson equation

$$\dot{\varphi} = 2eV/\hbar \ . \qquad (10\text{-}4)$$

In terms of dimensionless quantities these two equations lead to the following equation of a forced damped pendulum

$$\ddot{\varphi} + a\dot{\varphi} + \sin\varphi = i_0 + i_r \cos\Omega t \qquad (10\text{-}5)$$

Both numerical study of (10-5) and experimental work revealed the appearance of chaos[14].

Another example of anomalous noise generation concerns the oscillatory circuit based on the Gunn diode. Nakamura[15] truncated the PDE describing the system into 40 coupled ODE's and found many attractors. He explained the anomalous fluctuation by the interweaving basins of attraction. The equation of motion for dislocation lines in the presence of a supersonic field reduces to that of a periodically forced anharmonic oscillator, i.e., Eq. (5-8). Consequently, Herring and Huberman suggested to observe chaos in this situation[16].

Now let us turn to spatial chaos as a source of disorder. In solid state physics there are many cases of competing periods leading to the formation of new periodic or disordered structures and phase transitions may take place in between these structures. For example, a monolayer of inert gas atoms adsorbed to a graphite substrate may form a two-dimensional lattice which does not necessarily match the lattice of the substrate. If the interaction between the monolayer and the substrate is weak enough, the adsorbed layer keeps its own periodicity. Conversely, when they interact strongly, the monolayer lattice is forced to match that of the substrate. In between these two extremes phase locking and spatial chaos may occur. This process resembles the behaviour of forced nonlinear oscillators with time evolution replaced by spatial ordering.

The simplest model for modulated structure was proposed by Frenckel and Kontorova in the 30's. Suppose a one-dimensional chain of atoms is put in a cosine potential of the substrate, the n-th atom being at x_n. The Hamilton function of the system may be written as

$$H = \sum_n \frac{1}{2}(x_{n+1} - x_n - a)^2 + \frac{\mu}{(2\pi)^2} \cos 2\pi x_n \qquad (10\text{-}6)$$

where the period of the substrate field has taken to be 1, and μ measures the coupling. The equilibrium positions of atoms are determined by solving $\frac{\partial H}{\partial x_n} = 0$, namely,

$$x_n - x_{n-1} - (x_{n+1} - x_n) - \frac{\mu}{2\pi} \sin 2\pi x_n = 0 \quad . \qquad (10\text{-}7)$$

Let $y_n = x_n - x_{n-1}$, we have

$$\begin{aligned} y_{n+1} &= y_n - \frac{\mu}{2\pi} \sin 2\pi x_n \\ x_{n+1} &= x_n + y_{n+1} \end{aligned} \qquad (10\text{-}8)$$

This is precisely the "standard mapping" (5-2) listed in Chapter 5. Now it is natural to have periodic or chaotic solutions depending on the parameter μ. When the characteristic wave number of the modulated structure varies continuously, commensurable phases may occur repeatedly at rational values, leading to the so-called devil's staircase. A similar situation appears in some statistical models and in spin glasses. Ruelle went even further to pose the question whether there exist turbulent crystals. We recommend the reader to consult the reviews by Bak, and by Fisher and Huse[17].

9. Chaos in biology[18,19] and in ecology[20]

We digress a little from physics to mention the impressive experiment with periodically stimulated beating of cultured chicken cardiac cells[18], displaying phase-locking, period-doubling, and chaos. It was modelled by a periodically forced oscillator and an attempt was made to connect the chaotic dynamics with cardiac arrhytmias.

There are many speculations on the role of chaos in biology[19]. Chaotic behaviour should exist in biochemical reactions. However, it is certainly too early to draw any far-reaching conclusion, at least in as much as chaos is a purely classical phenomenon.

In a sense the whole modern story of chaos has its origin in ecological models. Therefore, it is appropriate to conclude this chapter by listing a few more references[20].

References and guide to the bibliography

0. For recent reviews on experimental observation of chaos see Swinney (1983), and the conference proceedings B39.
1. Forced vibration of shallow water wave in finite container: Lord Rayleigh, *The Theory of Sound,* § 68B; Keolian *et al.* (1981), Paper 37 in this volume.
2. Earlier work on nonlinear circuits: Gollub *et al.* (1978, 1980).
3. Recent work on nonlinear circuits: Linsay (1981); Testa *et al.* (1982), Paper 39 in this volume; Perez and Jeffries (1982), Hunt (1982), Jeffries and Perez (1982, 1983), Rollins and Hunt (1982), Brorson *et al.* (1983), Cascais *et al.* (1983), Jeffries and Usher (1983).
4. Bifurcation and chaos in mechanical and electromechanical systems: Moon and Holmes (1979), Moon (1980), Dowel (1982), Ananthakrishnan and Valsakumar (1983), Viet *et al.* (1983), Shaw and Holmes (1983), Anishchenko *et al.* (1983).
5. Rayleigh-Bénard instabilities. Reviews: Normand *et al.* (1977), Velarde and Normand (1980), Busse (1981); Papers: Libchaber *et al.* (1978, 1980, 1982, 1983), Dubois and Berge (1978, 1979, 1980, 1981, 1983), Gollub and Benson (1978) and (1980), Paper 34 in this volume, Berge (1979), Berge *et al.* (1979, 1980), Maurer and Libchaber (1979, 1980), Fauve *et al.* (1981), Gilgio *et al.* (1981, 1983), Libchaber (1982), Manneville and Piquemal (1982),

Arneodo *et al.* (1983), Paper 35 in this volume.

6. Instabilities in Couette flow. Reviews: Swinney and Gollub (1978b), Prima and Swinney (1981); Papers: Gollub and Swinney (1975), Belyaev *et al.* (1979), Fenstermacher *et al.* (1979a, b), Walden and Donnelly (1979), Gorman *et al.* (1981), Park *et al.* (1981, 1983), Perrin (1982), Shaw *et al.* (1982), Lorenzen *et al.* (1983), Brandstäter *et al.* (1983); Related theory: Yahata (1977, 1978, 1979, 1980, 1983), Rand (1982).
7. Subharmonic generation in cavitation noise: Lauterborn and Cramer (1981a), Paper 38 in this volume, and (1981b, c), Lauterborn (1982), Cramer and Lauterborn (1982); Period-doubling in liquid helium: Carey *et al.* (1979), Smith *et al.* (1982, 1983); Subharmonic generation in other acoustic systems: Bindal *et al.* (1980), Kitano *et al.* (1983), Chrostowski *et al.* (1983), B.-R. Wang *et al.*, *Acta Physica Sinica*, **33**, 434 (1984).
8. Chaos in optical bistability: review by Englund *et al.* (1983); Ikeda *et al.* (1979, 1980, 1982a, b), Snapp *et al.* (1981); Gibbs *et al.* (1981), Paper 40 in this volume; Hopf *et al.* (1982), Savage *et al.* (1982), Zardecki (1982), Carmichael *et al.* (1983), Chrostowski *et al.* (1983), Gao *et al.* (1983a, b), Harrison *et al.* (1983), Mandel and Kapral (1983), McLaughlin *et al.* (1983), Singh and Agarwal (1983), X.-J. Zhang *et al.*, *Acta Physica Sinica*, to appear.
9. "Pure" optical chaos: E. Abraham *et al.* (1982), Weiss and King (1982), Arechi *et al.* (1982), Brunner and Paul (1983), Lugiato *et al.* (1983), Nakatsuka *et al.* (1983), Weiss *et al.* (1983).
10. General discussion and theory related to optical chaos: Haken (1975), Scholz *et al.* (1981), Verlarde and Antoranz (1981), N. Abraham (1983), Miloni *et al.* (1983), Otsuka and Iwamara (1983).
11. Reviews on chemical turbulence: Rössler (1981), Epstein (1983), Roux (1983).
12. Experiments on chemical chaos: Schmitz *et al.* (1977), Yamazaki *et al.* (1978), Hudson *et al.* (1979, 1981), Olsen and Degn (1979), Vidal *et al.* (1980, 1982), Roux *et al.* (1980, 1981, 1982, 1983), Turner *et al.* (1981), Stuchl and Marek (1982a, b), Simoyi *et al.* (1982), Lafon *et al.* (1983), Roux *et al.* (1983), Paper 36 in this volume.
13. Theoretical analysis of chemical chaos: Kuramoto and Yamada (1976a, b), Yamada and Kuramoto (1976), Rössler (1976b, c, 1977a), Bar-Eli and Noyes (1977), Fujisaka and Yamada (1977, 1980), Tyson (1978), Rössler and Wegmann (1978), Wegmann and Rössler (1978), Showalter *et al.* (1978), Tomita and Tsuda (1979, 1980a), Turner (1980), Tomita and Daido (1980), Tsuda (1981b, c), Kuramoto and Koga (1982), Y.-X. Li *et al.*, *Commun. Theor. Phys.*, **3** (1984), to appear.
14. Chaos in rf driven Josephson junction: Review by Beasley and Huberman

(1982); Huberman *et al.* (1980), Pederson and Davidson (1981), Kauta (1981), Ben-Jacob *et al.* (1981, 1982), Levinson (1982), Yeh and Kao (1982), Imry (1983), Miracky *et al.* (1983) , Seifert (1983).

15. Anomalous noise in Gunn oscillator: Nakamura (1977, 1978a, b, 1979).
16. Dislocation dynamics in supersonic field: Herring Huberman (1980).
17. Chaotic phase in modulated structures. Reviews: Bak (1982), Fisher and Huse (1982); Papers: Aubry (1978, 1979, 1983), Bak (1981, 1983), Bak and Bruinsma (1982), McKay *et al.* (1982), Ruelle (1983).
18. Periodically stimulated beating of cardiac cells: Guevara *et al.* (1981, 1982), Glass *et al.* (1983).
19. Chaos in biology: Mackey and Glass (1977), Bunov and Weiss (1979), Tomita and Daido (1980), Decroly and Goldbeter (1982), Tomita (1982b), Goldbeter and Decroly (1983).
20. Chaos in ecological models: May (1974) and (1976), Paper 6 in this volume; May and Oster (1976), Beddington *et al.* (1975), Guckenheimer *et al.* (1977), Mayer (1978), Picard and Johnston (1982).

Part Two

REPRINTED PAPERS

PRESERVATION OF CONDITIONALLY PERIODIC MOVEMENTS WITH SMALL CHANGE IN THE HAMILTON FUNCTION*

Academician A. N. Kolmogorov
Department of Mathematics
Moscow State University
117234 Moscow, B-234
U.S.S.R.

ABSTRACT

This paper is a translation of Kolmogorov's original article announcing the theorem now known as the KAM theorem.

THEOREM AND DISCUSSION OF PROOF

Let us consider in the 2s-dimensional phase space of a dynamic system with s degrees of freedom the region G, represented as the product of an s-dimensional torus, T, and a region S, of a Euclidian s-dimensional space. We will designate the points of the torus, T, by the circular coordinates $q_1, \ldots, q_s$ (replacing q_α with $q'_\alpha = q_\alpha + 2\pi$ does not change points q), and the coordinates of the points, p, of S we will designate as $p_1, \ldots, p_s$. We will assume that in region G, in the coordinates $(q_1, \ldots, q_s, p_1, \ldots, p_s)$ the equations of motion have the canonical form

$$\frac{dq_\alpha}{dt} = \frac{\partial}{\partial p_\alpha} H(q,p), \quad \frac{dp_\alpha}{dt} = -\frac{\partial}{\partial q_\alpha} H(q,p). \tag{1}$$

The Hamilton function, H, is further assumed as dependent on the parameter θ and determined for all $(q,p) \varepsilon G$, $\theta \varepsilon (-c; +c)$, but not time-dependent. Moreover, further considerations require fairly significant restrictions on the smoothness of the function $H(q, p, \theta)$, stronger than infinite differentiability. For simplicity, in the following it is assumed that the function $H(q, p, \theta)$ is analytic over the set of variables (q, p, θ).

Summation over the Greek indices is further assumed to be from 1 to s. The usual vector designations $(x,y) = \sum_\alpha x_\alpha y_\alpha$, $|x| = +\sqrt{(x,x)}$ are used. A whole number vector indicates a vector for which all the components are whole numbers. The set of points (q, p) of G with

* Los Alamos Scientific Laboratory translation LA-TR-71-67 by Helen Dahlby of Akad. Nauk. S.S.S.R., Doklady 98, 527 (1954).

$p = c$ are designated by T_c. In theorem 1 it is assumed that S contains the point $p = 0$, i.e., $T_0 \subseteq G$.

Theorem 1. Let

$$H(q,p,0) = m + \sum_{\alpha} \lambda_\alpha p_\alpha + \frac{1}{2} \sum_{\alpha\beta} \Phi_{\alpha\beta}(q) p_\alpha p_\beta + O(|p|^3), \qquad (2)$$

where m and λ_α are constants and for a certain choice of constants $c > 0$ and $\eta > 0$ for all whole-number vectors, n, the inequality

$$(n,\lambda) \geqslant \frac{c}{|n|^\eta} . \qquad (3)$$

is satisfied.

Let, moreover, the determinant composed of the average values

$$\varphi_{\alpha\beta}(0) = \frac{1}{(2\pi)^s} \int_0^{2\pi} \cdots \int_0^{2\pi} \Phi_{\alpha\beta}(q)\, dq_1 \ldots dq_s$$

of the functions

$$\Phi_{\alpha\beta}(q) = \frac{\partial^2}{\partial p_\alpha \partial p_\beta} H(q,0,0)$$

be different from zero:

$$|\varphi_{\alpha\beta}(0)| \neq 0. \qquad (4)$$

Then there exist analytical functions $F_\alpha(Q, P, \theta)$ and $G\,(Q, P, \theta)$ which are determined for all sufficiently small θ and for all points (Q, P) of some neighborhood, V, of the set T_0, which bring about a contact transformation

$$q_\alpha = Q_\alpha + \theta F_\alpha(Q,P,\theta), \quad p_\alpha = P_\alpha + \theta G\,(Q,P,\theta)$$

of V into $V' \subseteq G$, which reduced H to the form

$$H = M(\theta) + \sum_{\alpha} \lambda_\alpha P_\alpha + O(|P|^2) \qquad (5)$$

($M(\theta)$ does not depend on Q and P).

It is easy to grasp the meaning of Theorem 1 for mechanics. It indicates that an s-parametric family of conditionally periodic motions

$$q_\alpha = \lambda_\alpha t + q_\alpha^{(o)}, \quad p_\alpha = 0,$$

which exists at $\theta = 0$ cannot, under conditions (3) and (4), disappear as a result of a small change in the Hamilton function H: there occurs only a displacement of the s-dimensional torus, T_0, around which the trajectories of these motions run, into the torus $P = 0$, which remains filled by the trajectories of conditionally periodic motions with the same frequencies $\lambda_1, \ldots, \lambda_s$.

The transformation

$$(Q, P) = K_\theta(q, p),$$

the existence of which is confirmed in Theorem 1, can be constructed in the form of the limit of the transformations

$$(Q^{(k)}, P^{(k)}) = K_\theta^{(k)}(q, p),$$

where the transformations

$$(Q^{(1)}, P^{(1)}) = L_\theta^{(1)}(q, p), \quad (Q^{(k+1)}, P^{(k+1)}) = L_\theta^{(k+1)}(Q^{(k)}, P^{(k)}$$

are found by the "generalized Newton method" (see Ref. 1). In this note we confine ourselves to the construction of the transformation: $K_\theta^{(1)} = L_\theta^{(1)}$, which itself permits grasping the role of conditions (3) and (4) of Theorem 1. Let us apply the transformation $L_\theta^{(1)}$ to the equations

$$Q_\alpha^{(1)} = q_\alpha + \Theta Y_\alpha(q),$$

$$p_\alpha = P_\alpha^{(1)} = \theta \left\{ \sum_\beta P_\beta^{(1)} \frac{\partial Y_\beta}{\partial q_\alpha} + \xi_\alpha + \frac{\partial}{\partial q_\alpha} X(q) \right\} \tag{6}$$

(it is easy to verify that this is a contact transformation) and seek the constants ξ_α and ζ and the functions $X(q)$ and $Y_\beta(q)$, starting from the requirement that

$$H = m + \sum_\alpha \lambda_\alpha p_\alpha + \frac{1}{2} \sum_{\alpha\beta} \Phi_{\alpha\beta}(q) p_\alpha p_\beta + \theta \left\{ A(q) + \sum_\alpha B_\alpha(q) p_\alpha \right\} + O(|p|^3 + \theta|p|^2 + \theta^2) \tag{7}$$

take the form

$$H = m + \theta\zeta + \sum \lambda_\alpha P_\alpha^{(1)} + O(|P^{(1)}|^2 + \theta^2). \tag{8}$$

Substituting (6) into (7), we get

$$H = m + \sum_\alpha \lambda_\alpha P_\alpha^{(1)} + \theta \left\{ A + \sum_\alpha \lambda_\alpha \left(\xi_\alpha + \frac{\partial X}{\partial q_\alpha} \right) \right\} +$$

$$+ \theta \sum_\alpha P_\alpha^{(1)} \left\{ B_\alpha + \sum_\beta \Phi_{\alpha\beta}(q) \left(\xi_\beta + \frac{\partial X}{\partial q_\beta} \right) + \sum_\beta \lambda_\beta \frac{\partial Y_\beta}{\partial q_\beta} \right\} + O(|P^{(1)}|^2 + \theta^2).$$

Thus, our requirement (8) reduces to the equations

$$A + \sum_\alpha \lambda_\alpha \left(\xi_\alpha + \frac{\partial X}{\partial q_\alpha} \right) = \zeta \tag{9}$$

$$B_\alpha + \sum_\beta \Phi_{\alpha\beta} \left(\xi_\beta + \frac{\partial X}{\partial q_\beta} \right) + \sum_\beta \lambda_\beta \frac{\partial Y_\alpha}{\partial q_\beta} = 0. \tag{10}$$

being fulfilled.

Let us introduce the functions

$$Z_\alpha(q) = \sum_\beta \Phi_{\alpha\beta}(q) \frac{\partial}{\partial q_\beta} X(q). \tag{11}$$

Expanding the functions $\Phi_{\alpha\beta}$, A, B_α, X, Y_α, Z_α into a Fourier series of the type

$$X(q) = \sum x(n) e^{i(n,q)}$$

and assuming for definiteness that

$$x(0) = 0, \; y(0) = 0, \tag{12}$$

we get for the remaining Fourier coefficients $x(n)$, $y_\alpha(n)$, and $z_\alpha(n)$ and constants ξ_α and ζ of the equation which are relevant to the determination

$$a(0) + \sum \lambda_\alpha \xi_\alpha = \zeta, \tag{13}$$

$$a(n) + (n,\lambda)\, x(n) = 0 \quad \text{for } n \neq 0, \tag{14}$$

$$b_\alpha(0) + \sum_\beta \varphi_\alpha(0) \xi_\beta + z_\alpha(0) = 0, \tag{15}$$

$$b_\alpha(n) + \sum_\beta \varphi_{\alpha\beta}(n) \xi_\beta + z_\alpha(n) + (n,\lambda)\, y_\alpha(n) = 0 \quad \text{for } n \neq 0. \tag{16}$$

It is easy to see that the system (11) - (16) is unambiguously solved under conditions (3) and (4). Condition (3) is important in the determination of x(n) from (14), and in the determination of $y_\alpha(n)$ from (16). Condition (4) is important in the determination of ξ_β from (15). Since, as $|n|$ increases, the coefficients of the Fourier series of the analytical functions $\varphi_{\alpha\beta}$, A, and B_α have an order of decrease not less than $\rho^{|h|}$, $\rho < 1$, then from condition (3) there results not only the formal solvability of equations (13) - (16) but also the convergence of the Fourier series for the functions X, Y_α, and Z_α and the analyticity of these functions. The construction of further approximations is not associated with new difficulties. Only the use of condition (3) for proving the convergence of the recursions, $K_\theta^{(k)}$, to the analytical limit for the recursion K_θ is somewhat more subtle.

The condition of the absence of "small denominators" (3) should be considered, "generally speaking," as fulfilled, since for any $\eta > s - 1$ for all points of an s-dimensional space $\lambda = (\lambda_1, \ldots, \lambda_s)$ except the set of Lebesque measure zero it is possible to find $c(\lambda)$, for which

$$(n, \lambda) \geqslant \frac{c(\lambda)}{|n|^\eta}$$

whatever the integers $n_1, n_2, \ldots, n_s$ were[2]. It is also natural to consider condition (4) as, "generally speaking," fulfilled. Since

$$\varphi_{\alpha\beta}(0) = \frac{\partial}{\partial p_\alpha} \lambda_\beta(0),$$

where

$$\lambda_\beta(p) = \frac{1}{(2\pi)^s} \int_0^{2\pi} \cdots \int_0^{2\pi} \frac{dq_\beta}{dt}\, dq_1 \ldots dq_s$$

is the frequency averaged over the coordinate q_β with fixed momenta $p_1, \ldots, p_s$, condition (3) means that the Jacobian of the average frequencies over the momenta is different from zero.

Let us turn now to a consideration of the special case where H(q, p, 0) depends only on p, i.e., H(q, p, 0) = W(p). In this case, for $\theta = 0$ each torus, T_p, consists of the complete trajectories of the conditionally periodic movements with frequencies

$$\lambda_\alpha(p) = \frac{\partial W}{\partial p_\alpha}.$$

If the Jacobian

$$J = \left| \frac{\partial \lambda_\alpha}{\partial p_\beta} \right| = \left| \frac{\partial^2 W}{\partial p_\alpha \partial p_\beta} \right| \qquad (17)$$

is different from zero, then it is possible to apply Theorem 1 to almost all tori, T_p. There arises the natural hypothesis that at small θ the "displaced tori" obtained in accordance with Theorem 1 fill a larger part of region G. This is also confirmed by Theorem 2, pointed out later. In the formulation of this theorem we will consider the region S to be bounded and will introduce into the consideration the set, M_θ, of those points $(q^{(0)}, p^{(0)}) \in G$ for which the solution

$$q_\alpha(t) = f_\alpha(t; q^{(0)}, p^{(0)}, \theta), \quad p_\alpha(t) = G_\alpha(t; q^{(0)}, p^{(0)}, \theta)$$

of the system of equations (1) with initial conditions

$$q_\alpha(0) = q_\alpha^{(0)}, \quad p_\alpha(0) = p_\alpha^{(0)}$$

leads to trajectories not moving out of region G with change in t from $-\infty$ to $+\infty$, and conditionally periodic with periods $\lambda_\alpha = \lambda_\alpha(q^{(0)}, p^{(0)}, \theta)$, i.e., it has the form

$$f_\alpha(t) = \varphi_\alpha(e^{i\lambda_1 t}, \ldots, e^{i\lambda_s t}), \quad g_\alpha(t) = \psi_\alpha(e^{i\lambda_1 t}, \ldots, e^{i\lambda_s t}).$$

Theorem 2. If $H(q, p, 0) = W(p)$ and determinant (17) is not equal to zero in region S, then for $\theta \to 0$ the Lebesque degree of the set M_θ converges to the complete degree of region S.

Apparently, in the usual sense of the phrase, "general case" is when the set M_θ at all positive θ is everywhere dense. In such a case the complications arising in the theory of analytical dynamic systems are indicated more specifically in my note.[3]

REFERENCES

1. L. V. Kantorovich. Uspekhimatem. Nauk 3, 163 (1948).

2. J. F. Koksma. Diophantische Approximationen, Chelsea 1936. 157p.

3. A. N. Kolmogorov. Doklady Akad. Nauk 93, 763 (1953).

PHYSICAL REVIEW VOLUME 188, NUMBER 1 5 DECEMBER 1969

Amplitude Instability and Ergodic Behavior for Conservative Nonlinear Oscillator Systems*

Grayson H. Walker and Joseph Ford
School of Physics, Georgia Institute of Technology, Atlanta, Georgia 30332
(Received 27 March 1969)

Several earlier computer studies of nonlinear oscillator systems have revealed an amplitude instability marking a sharp transition from conditionally periodic to ergodic-type motion, and several authors have explained the observed instabilities in terms of a mathematical theorem due to Kolmogorov, Arnol'd, and Moser. In view of the significance of these results to several diverse fields, especially to statistical mechanics, this paper attempts to provide an elementary introduction to Kolmogorov-Arnol'd-Moser amplitude instability and to provide a verifiable scheme for predicting the onset of this instability. This goal is achieved by demonstrating that amplitude instability can occur even in simple oscillator systems which admit to a clear and detailed analysis. The analysis presented here is related to several earlier studies. Special attention is given to the relevance of amplitude instability for statistical mechanics.

I. INTRODUCTION

In attempting to determine the behavior of nonlinear oscillator systems governed by Hamiltonians of the form

$$H = H_0 + V , \tag{1}$$

where H_0 represents an integrable system of oscillators and where V represents a weak nonlinear and nonintegrable perturbation, most investigators have proceeded along one of two divergent paths. One approach assumes that the weak perturbation V changes the unperturbed motion only to the extent of slightly shifting the frequencies of the motion and introducing small nonlinear harmonics. This approach is used most frequently when the number of oscillators is relatively small, and it is exemplified in certain perturbation expansions due to Poincaré,[1] Birkhoff,[2] and Kryloff and

Bogoliubov.[3] The second approach assumes that V, even though weak, has a profound and pathological effect on the unperturbed motion, converting it into ergodic[4] motion. This latter approach uses the methods of statistical mechanics (presumed valid when the number of oscillators is large) and is exemplified in the work of Fermi[5] and Peierls.[6]

A brief paper by Kolmogorov[7] enunciated a theorem which can perhaps provide a cornerstone for linking the two aforementioned divergent views on the effects of the weak perturbation V in Eq. (1). Kolmogorov did not present a detailed proof of his theorem; the missing proof, which is quite long and mathematically sophisticated, was supplied almost a decade later by Arnol'd[8] and independently by Moser.[9] As a consequence perhaps, the physical scientist has largely remained unaware of Kolmogorov's theorem and its implications. For details of the theory with applications, the reader is referred to the review article by Arnol'd.[10] However, in order to make this paper self-contained, we briefly present here those details of the theory relevant to this paper; in particular we may restrict our attention to systems with two degrees of freedom without significant loss of generality.

Introducing action-angle type variables (J_i, φ_i) for a two-oscillator system, Hamiltonian (1) may be written

$$H = H_0(J_1, J_2) + V(J_1, J_2, \varphi_1, \varphi_2) . \tag{2}$$

If we set $V \equiv 0$, then Hamiltonian (2) generates motion for which the J's are constant and $\varphi_i = \omega_i(J_1, J_2)t + \varphi_{i0}$, where the unperturbed frequencies ω_i are given by $\omega_i = \partial H_0/\partial J_i$. Following Kolmogorov, we view the unperturbed system motion in phase space as lying on two-dimensional tori where (φ_1, φ_2) are the angle coordinates on the tori and (J_1, J_2) are the "radii" of the tori. By assuming that V is sufficiently small and by assuming that the Jacobian of the frequencies $\partial(\omega_1, \omega_2)/\partial(J_1, J_2) \not\equiv 0$, Kolmogorov-Arnol'd-Moser (hereafter referred to as KAM) are able to show that most of the unperturbed tori bearing conditionally periodic motion with incommensurate frequencies continue to exist, being only slightly distorted by the perturbation. On the other hand, the tori bearing periodic motion, or very nearly periodic motion, with commensurate frequencies, or with incommensurate frequencies whose ratio is approximated extremely well by (r/s) where r and s are relatively small integers, are grossly deformed by the perturbation and no longer remain close to the unperturbed tori. Since the unperturbed tori with commensurate frequencies which are destroyed by the perturbation are everywhere dense, it is remarkable indeed that KAM are able to show that the majority – in the sense of measure theory – of initial conditions for Hamiltonian (2) lie on the preserved tori bearing conditionally periodic motion when V is sufficiently small.

Thus for small V, KAM theory proves that for most initial conditions Hamiltonian (2) generates nonergodic motion thus justifying the view that the perturbation V largely serves only to slightly shift the frequencies and introduce small nonlinear harmonics into the motion.[11] Nonetheless, the relatively small set of initial conditions leading to motion not on preserved tori is, from a physical point of view, pathologically interspersed between the preserved tori. Moreover, Arnol'd[10] conjectures that the system phase-space trajectory in regions of the destroyed tori is quite complicated indeed, perhaps ergodically filling the destroyed region. Thus, if Hamiltonian (2) is ever to provide generally ergodic motion best described in terms of statistical mechanics, the source of such behavior must lie in the reasons for the very existence of this relatively small set of destroyed tori. Hence we now investigate the properties of V which lead to the destruction of tori.

To this end, we expand the V of Hamiltonian (2) in a Fourier series and write

$$H = H_0(J_1, J_2) + f_{mn}(J_1, J_2) \times \cos(m\varphi_1 + n\varphi_2) + \cdots, \tag{3}$$

where we have explicitly written only one term in the series. The KAM formalism seeks to eliminate the angle-dependent terms using a convergent sequence of canonical transformations, each of which is close to the identity transformation, thus obtaining a Hamiltonian which is a function of the transformed action variables alone and which is close to the original Hamiltonian. If this can be accomplished in some general sense, then one immediately finds that the perturbed motion, for the most part, lies on tori close to the unperturbed tori. As illustration, let us seek to eliminate the explicit angle-dependent term in Hamiltonian (3) by introducing the canonical transformation generated[12] by

$$F = \mathcal{J}_1\varphi_1 + \mathcal{J}_2\varphi_2 + B_{mn}(\mathcal{J}_1, \mathcal{J}_2)\sin(m\varphi_1 + n\varphi_2), \tag{4}$$

where $(\mathcal{J}_i, \theta_i)$ are the transformed action-angle variables and $B_{mn}(\mathcal{J}_1, \mathcal{J}_2)$ is to be determined. We note that if $B_{mn} = 0$, we have the identity transformation $J_i = \mathcal{J}_i$ and $\varphi_i = \theta_i$.

Introducing the canonical transformation generated by Eq. (4) into Hamiltonian (3), we obtain

$$H = H_0(\mathcal{J}_1, \mathcal{J}_2) + \{[m\omega_1(\mathcal{J}_1, \mathcal{J}_2) - n\omega_2(\mathcal{J}_1, \mathcal{J}_2)]$$

$$\times B_{mn}(\mathcal{J}_1,\mathcal{J}_2)+f_{mn}(\mathcal{J}_1,\mathcal{J}_2)\}$$
$$\cos(m\theta_1+n\theta_2)+\cdots, \tag{5}$$

where $\omega_i(\mathcal{J}_1,\mathcal{J}_2)=\partial H_0(\mathcal{J}_1,\mathcal{J}_2)/\partial\mathcal{J}_i$ and where we have explicitly retained only the lowest-order terms. We may now eliminate the given angle-dependent term, provided we set

$$B_{mn}(\mathcal{J}_1,\mathcal{J}_2)=-\frac{f_{mn}(\mathcal{J}_1,\mathcal{J}_2)}{m\omega_1(\mathcal{J}_1,\mathcal{J}_2)+n\omega_2(\mathcal{J}_1,\mathcal{J}_2)}, \tag{6}$$

and provided that the denominator in Eq. (6) is not very small (or zero) relative to f_{mn}. If the denominator in Eq. (6) is very small, then the coefficient B_{mn} is large, the transformation generated by Eq. (4) is not close to the identity transformation, and the transformed coordinate motion is not close to the unperturbed motion. As a consequence, if there exists a band of frequencies ω_i for Hamiltonian (3) satisfying

$$|m\omega_1(J_1,J_2)+n\omega_2(J_1,J_2)|\ll|f_{mn}(J_1,J_2)|, \tag{7}$$

then the angle-dependent term $\cos(m\varphi_1+n\varphi_2)$ grossly distorts an associated zone of unperturbed tori bearing the frequencies satisfying the inequality (7).

Moreover, when a zone of unperturbed tori is grossly distorted by a specified angle-dependent term $\cos(m\varphi_1+n\varphi_2)$, one must in general anticipate that there will be a host of angle-dependent terms $\cos(m'\varphi_1+n'\varphi_2)$ in Hamiltonian (3) whose (m'/n') ratios are sufficiently close to the specified ratio (m/n) that the analog of the inequality (7) is satisfied for them also. Hence the zone of unperturbed tori distorted by $\cos(m\varphi_1+n\varphi_2)$ will simultaneously be affected by a large number of other angle-dependent terms. Physically speaking, the inequality (7) is a resonance relationship which, if satisfied, asserts that $\cos(m\varphi_1+n\varphi_2)$ resonantly couples the unperturbed oscillators when their frequencies lie in the designated band. If a number of angle-dependent resonant terms couple the oscillators in this band, then one has the situation envisioned in the quantum-mechanical Golden Rule[13] in which an initial state is resonantly coupled to a density of final states leading to statistically irreversible behavior. In analogy, one would anticipate that the motion generated by Hamiltonian (3) in the overlapping resonant zones of destroyed tori is highly complicated, perhaps even ergodic.

When V is very small and hence all f_{mn} are small, the inequality (7) shows that the resonance zones are very narrow. Moreover, KAM show that the totality of all resonant destroyed zones is small[14] relative to the measure of the allowed phase space. However, as V and the f_{mn} increase, or equivalently as the total energy increases, one anticipates from the inequality (7) that the measure of the overlapping resonant zones may increase until most of phase space is filled with highly complicated trajectories moving under the influence of many resonances. In short, KAM theory indicates, but certainly does not prove, the existence of an amplitude instability for conservative nonlinear oscillator systems which permits a transition from motion which is predominantly conditionally periodic to motion which is predominantly ergodic. Since this transition lies outside the scope of KAM theory, we now review some of the computer generated evidence which supports the existence of an amplitude instability.

One of the first computer demonstrations of an amplitude instability was made in an investigation of unimolecular dissociation by Thiele and Wilson[15] and by Bunker.[16] These investigators noted that for small amplitude motion the harmonic modes of triatomic molecules exhibited very little energy exchange. As the energy of the molecule was increased, an amplitude instability occurred which allowed free and rapid interchange of energy between the harmonic modes. Consequently, as the energy was further increased to slightly above that needed to dissociate one atom from the molecule, almost all initial configurations led to dissociation. Thiele and Wilson[15] then used these results to argue that nonlinearity must be given a central role in developing a statistical theory of unimolecular dissociation. These molecular systems were not analyzed in terms of the KAM theory; however, using the techniques discussed in the following paragraphs, the present authors have shown that the observed amplitude instability of Wilson's oscillator model occurs concurrently with a large-scale disappearance of preserved tori.

The second major computer demonstration of amplitude instability occurred in an astronomical study. Henon and Heiles[17] studied the bounded motion of the system

$$H=\tfrac{1}{2}(p_1^2+p_2^2+q_1^2+q_2^2)+q_1^2q_2-\tfrac{1}{3}q_2^3 \tag{8}$$

in order to determine whether or not a well-behaved constant of the motion exists for Hamiltonian (8) in addition to H itself. This study was motivated by empirical and theoretical evidence[18] that a star moving in a cylindrically symmetric potential appeared to have a well-behaved constant of the motion in addition to the total energy and the z component of angular momentum. Since we intend to rely heavily in the main body of this paper on the techniques used by Henon and Heiles, we now outline their approach. The reader who finds the following discussion unclear is urged to read the extremely clear Henon and Heiles paper.

For fixed energy below the dissociation energy, all the phase-space trajectories generated by Hamiltonian (8) must lie on a bounded, three-dimensional energy surface. If an additional well-behaved constant of the motion $I(q_1, p_1, q_2, p_2)$ exists, then all system trajectories are further constrained to lie on bounded two-dimensional surfaces; if no well-behaved constant I exists, then the system trajectories will move randomly over some or all of the energy surface. If we now imagine a two-dimensional plane cutting through this three-dimensional energy surface, then the existence of a well-behaved constant I ensures that each system trajectory will intersect this plane along a curve, called a level curve; if a well-behaved I does not exist, the intersections of each trajectory will form a set of randomly scattered points. As a test then for the existence of a well-behaved constant I, Henon and Heiles integrated the equations of motion for Hamiltonian (8) and graphically plotted the intersection points of the system trajectory with the (q_2, p_2) plane determined[19] by the conditions $q_1 = 0$ and $p_1 \geq 0$. Using the Henon-Heiles method, we have integrated Hamiltonian (8) on a UNIVAC 1108 computer to obtain the level curves shown in Figs. 1–5.

Henon and Heiles analyzed these figures only in regard to constants of the motion. However, when the constants of the motion H and I restrict the system motion to lie on a two-dimensional surface in $(q_1 p_1, q_2, p_2)$ space, one may use canonical transformation theory[20] to show that this two-dimensional surface is topologically equivalent to a KAM torus. A level curve in the (q_2, p_2) plane is thus topologically equivalent to a cross section of the torus. Figures 1–5 may then be regarded as profiles of KAM tori.

Figure 1, for energy $E = \frac{1}{12}$, shows that, to computer accuracy,[21] all trajectories lie on tori. However, the inequality (7) makes it clear that there are zones of instability which could be observed with sufficient computer accuracy. Nonetheless, Fig. 1 demonstrates that $E = \frac{1}{12}$ is deep within the region of KAM stability. Figures 2 and 3 illustrate the characteristic behavior of the tori at the onset of macroscopic instability when the microscopic KAM zones of instability become large enough to be seen by the computer. Figure 2, at about $E = 0.106$, shows the appearance of a new type of torus which consists of a chain of eight islands surrounding the central invariant point on the upper p_2 axis. Figure 3 shows two zones on instability which appear simultaneously with the island chains; both instabilities characteristically first appear as a replacement for the self-intersecting curve of Fig. 1, called a separatrix. Jefferys[22] explains this latter fact in terms of the KAM theory and the examples studied later in this paper support his explanation. Figure 4, at $E = \frac{1}{8}$, depicts the intermediate situation in which preserved tori still cover about 70% of the available area while intersections of the single orbit shown rather uniformly cover the remaining 30%. Figure 5, at the dissociation energy $E = \frac{1}{6}$, demonstrates that almost all the system motion is now statistical in character.

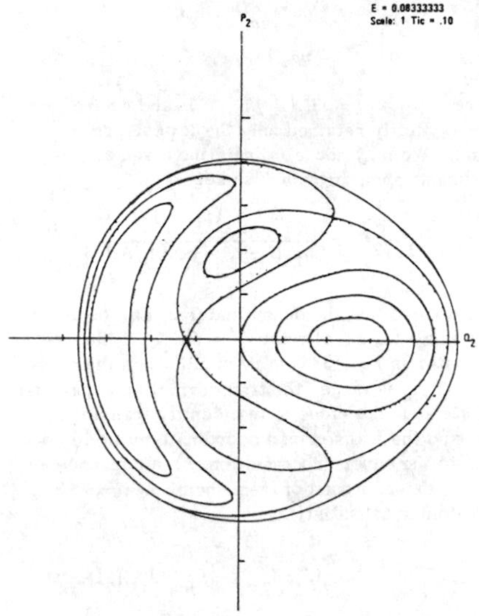

FIG. 1. Level curves for the Henon-Heiles system in the (q_2, p_2) plane. The microscopic zones of KAM instability lie below the computer integration accuracy. Here and in all level curve diagrams whenever a trajectory yields intersection points obviously lying on a smooth curve, the indicated curve has been drawn in freehand. The only exception occurs in Fig. 2 where the full level curves for the two chains of eight islands have not been freehanded in.

The final example of KAM instability is provided by the Hamiltonian system

$$H = \tfrac{1}{2} \sum_k (p_k^2 + \omega_k^2 q_k^2) + \alpha \sum_{i,j,k} A_{ijk} q_i q_j q_k + \lambda \sum_{h,i,j,k} B_{hijk} q_h q_i q_j q_k, \qquad (9)$$

originally proposed by Fermi, Pasta, and Ulam[23] as a model for the study of the approach to thermal equilibrium, although their study did not reveal an approach to equilibrium. Additional studies of this system have been made by Ford and Waters,[24] by Jackson,[25] and perhaps most interestingly by Kruskal and Zabusky.[26] However, Izrailev and Chirikov[27] were the first to suggest that Hamiltonian (9) should exhibit a KAM instability leading to statistical behavior. Zabusky and Deem[28] investi-

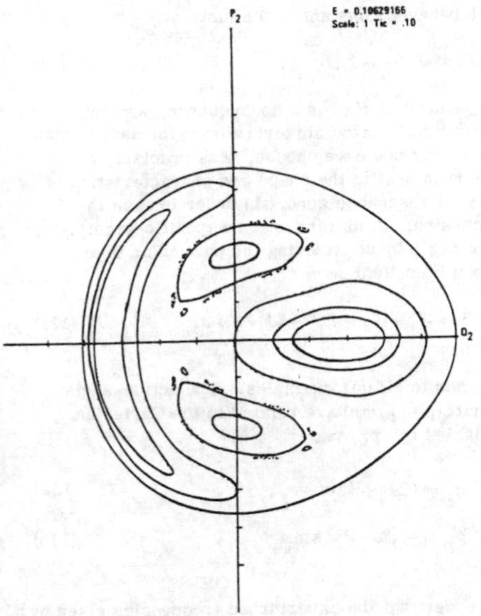

FIG. 2. Level curves for the Henon-Heiles system. The onset of macroscopic KAM instability is seen in the two chains of eight islands.

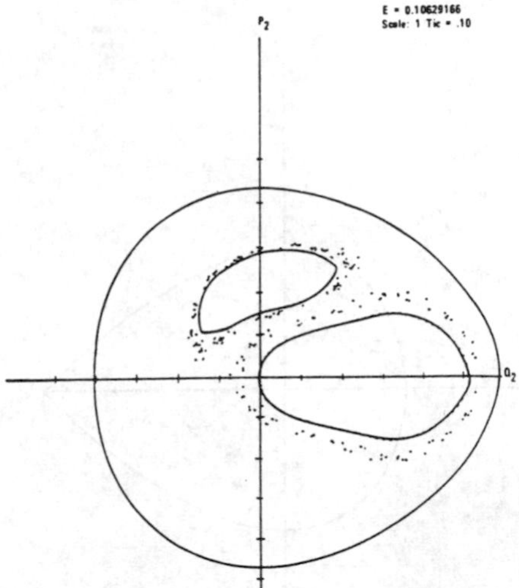

FIG. 3. Level curves for the Henon-Heiles system. This figure is a continuation of Fig. 2, using the same energy, and shows two macroscopic zones of KAM instability. The zone of instability centered on the p_2 axis encompasses the chain of eight islands shown in the upper part of Fig. 2.

gated this possibility and they demonstrated that the large amplitude motion, contrary to the small amplitude motion, does indeed exhibit widespread energy sharing among the harmonic modes. Nonetheless, complete equipartition of energy was not achieved and the motion exhibited correlations inconsistent with complete ergodicity or thermal equilibrium. At present, it is unclear whether Zabusky and Deem were observing incomplete KAM instability such as that observed in Fig. 4 or whether they were observing the constant high-order correlations of the type derived by Prigogine and coworkers.[29]

While there can be little doubt that KAM instability is the source of the amplitude instabilities observed in the above computer experiments, the theorist's ability to predict the onset and completion of macroscopic instability is less certain. Certainly, the three recent papers[27, 29, 30] attacking this problem have treated quite complicated systems for which prediction is especially difficult. Consequently, in this paper, we attempt to illustrate the origins and verifiably predict the onset of macroscopic amplitude instability using extremely simple oscillator models. Our intent is to provide an elementary introduction to KAM instability and to provide a reasonably accurate calculational scheme for predicting this instability

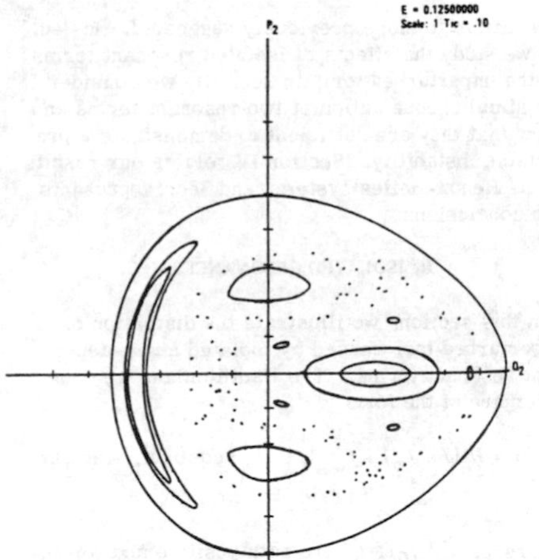

FIG. 4. Level curves for the Henon-Heiles system showing the increase in the zone of instability with increasing energy.

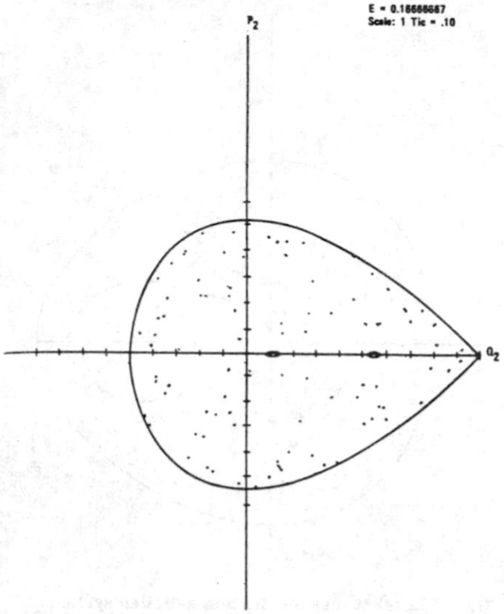

FIG. 5. Level curves for the Henon-Heiles system. The isolated dots represent the level curve for a single trajectory; however, integration accuracy for this highly unstable orbit is questionable.

alternative to those previously suggested. In Sec. II, we study the effects of isolated resonant terms on the unperturbed tori; in Sec. III, we consider the simultaneous action of two resonant terms and show that they are sufficient to demonstrate a predictable instability. Section IV relates our results to the Henon-Heiles system, and Sec. V presents our conclusions.

II. ISOLATED RESONANCES

In this section, we illustrate the distortion of unperturbed tori caused by isolated angle-dependent resonant terms. The Hamiltonians we consider are of the form

$$H = H_0(J_1, J_2) + f_{mn}(J_1, J_2)\cos(m\varphi_1 + n\varphi_2), \tag{10}$$

where $\omega_i = \partial H_0/\partial J_i$ are both positive and where m and n are integers such that the inequality (7) can be satisfied. For brevity, an isolated perturbation of this type is called an m–n resonance, and the associated zone of highly distorted tori, loosely specified by inequality (7), is called an m–n resonance zone. Such perturbations are especially easy to analyze since they give rise to a well-behaved constant of the motion

$$I = nJ_1 - mJ_2, \tag{11}$$

independent of H. As a consequence, we may algebraically calculate level curves for Hamiltonian (10) in any plane we choose, thus precisely determining in profile the shape and characteristics of the m-n resonance zone. In order to ease this discussion, let us turn to some specific examples.

We begin by considering the particular unperturbed Hamiltonian

$$H_0 = J_1 + J_2 - J_1^2 - 3J_1J_2 + J_2^2 \tag{12}$$

common to all our examples. The action-angle variables (J_i, φ_i) are related to the Cartesian variables (q_i, p_i) via

$$q_i = (2J_i)^{1/2}\cos\varphi_i, \tag{13a}$$

$$p_i = -(2J_i)^{1/2}\sin\varphi_i. \tag{13b}$$

In order that the unperturbed frequencies given by

$$\omega_1 = 1 - 2J_1 - 3J_2, \tag{14a}$$

$$\omega_2 = 1 - 3J_1 + 2J_2, \tag{14b}$$

be positive, we require that the energy E lie in the range $0 \leqslant E \leqslant \frac{3}{13}$ and that the values used for the J_i lie on the branch which goes to zero with E. Now Eq. (13) may be used to show that

$$J_i = \tfrac{1}{2}(p_i^2 + q_i^2). \tag{15}$$

Thus the unperturbed level curves in the (q_2, p_2) plane, hereafter called the J_2 plane, are concentric circles centered on the origin since J_2 is a constant. Similarly points on the level curves in the (q_1, p_1) plane or J_1 plane, defined by $q_2 = 0$, $p_2 \geqslant 0$ (or equivalently $\varphi_2 = \frac{1}{2}3\pi$), also lie on concentric circles. These circular level curves in either plane are enclosed by a bounding level curve representing the intersection of the energy surface with each plane.

We now introduce a 2–2 resonance and write

$$H = H_0(J_1, J_2) + \alpha J_1 J_2 \cos(2\varphi_1 - 2\varphi_2). \tag{16}$$

Now this system has the additional constant of the motion

$$I = J_1 + J_2. \tag{17}$$

If we now use Eq. (17) to eliminate J_2 from Eq. (16) and if we set $\varphi_2 = \frac{1}{2} 3\pi$, we obtain

$$(3 + \alpha \cos 2\varphi_1)J_1{}^2 - (5I + I \cos 2\varphi_1)J_1 + I + I^2 - E = 0 \quad (18)$$

as the algebraic equation for level curves in the J_1 plane. A typical level-curve diagram for Eq. (18) is shown in Fig. 6. The unperturbed circular level curves are only slightly distorted except in the 2 – 2 resonance zone enclosed by the self-intersecting separatrix level curve. The two self-intersection points represent distinct unstable periodic solutions while the two invariant points at the center of each crescent region represent distinct stable periodic solutions. Since the central point of each crescent represents a distinct periodic orbit, the two crescents are not a chain of two islands. The central points of an island chain represent a single periodic orbit.

For all four of the above periodic orbits, we have $\dot{J}_1 = \dot{J}_2 = (\dot{\varphi}_1 - \dot{\varphi}_2) = 0$, where a dot denotes time differentiation. For the stable periodic orbits we find

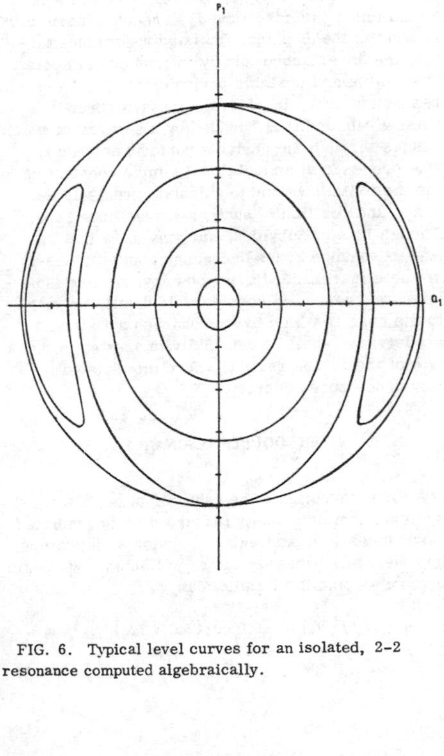

FIG. 6. Typical level curves for an isolated, 2–2 resonance computed algebraically.

$$J_1 = [(5 + \alpha)/(1 + \alpha)]J_2 , \quad (19a)$$

$$(\varphi_1 - \varphi_2) = \tfrac{1}{2}\pi \text{ or } \tfrac{3}{2}\pi . \quad (19b)$$

For the unstable periodic orbits, we have

$$J_1 = [(5 - \alpha)/(1 - \alpha)]J_2 , \quad (20a)$$

$$(\varphi_1 - \varphi_2) = 0 \text{ or } \pi . \quad (20b)$$

Using Eq. (14) we see that $2\omega_1 = 2\omega_2$ implies

$$J_1 = 5J_2 . \quad (21)$$

Thus as predicted by the inequality (7), the 2-2 resonance zone of highly distorted tori occurs in a neighborhood of the unperturbed torus bearing the frequencies $2\omega_1 = 2\omega_2$, designated as the 2-2 torus. Putting Eq. (21) into Eq. (12), we find

$$J_1 = \tfrac{5}{13}\left[1 - (1 - \tfrac{13}{3}E)^{1/2}\right] , \quad (22a)$$

$$J_2 = \tfrac{1}{13}\left[1 - (1 - \tfrac{13}{3}E)^{1/2}\right] , \quad (22b)$$

as the values of J_1 and J_2 on the unperturbed 2-2 torus. Consequently, the unperturbed 2-2 torus and the perturbed 2-2 resonance zone exist for all allowed energies $0 \leq E \leq \frac{3}{13}$. As the energy increases from zero, the 2-2 resonance zone moves out from the origin and increases in width.

The closest (low-order) resonance to the 2-2 is the 3-2 or the 2-3. We investigate each. First consider

$$H = H_0(J_1, J_2) + \beta J_1{}^{3/2} J_2 \cos(3\varphi_1 - 2\varphi_2) . \quad (23)$$

The additional constant of the motion is

$$I = 2J_1 + 3J_2 , \quad (24)$$

and the level curves in the J_1 plane are given by

$$E = \tfrac{1}{3}I + \tfrac{1}{9}I^2 + (\tfrac{1}{3} - \tfrac{13}{9}I)J_1 + \tfrac{13}{9}J_1{}^2 - (\tfrac{1}{3}\beta)(IJ_1{}^{3/2} - 2J_1{}^{5/2})\cos 3\varphi_1 . \quad (25)$$

Typical level curves for Eq. (25) are presented in Fig. 7. Here the points at the center of each of the three crescent regions do represent a single periodic solution, and thus the 3-2 resonance zone consists of a chain of three islands. Similarly, the three self-intersecting points on the separatrix represent a single unstable periodic solution.

Again setting $\dot{J}_1 = \dot{J}_2 = (3\dot{\varphi}_1 - 2\dot{\varphi}_2) = 0$ yields

$$J_2 = (1 + 2J_1{}^{3/2})/(13 + \tfrac{9}{12}J_1{}^{1/2}) , \quad (26a)$$

$$(3\varphi_1 - 2\varphi_2) = \pi, 3\pi, 5\pi , \quad (26b)$$

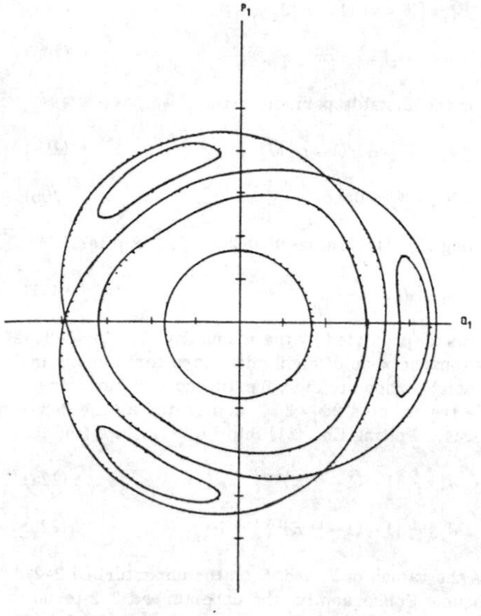

FIG. 7. Typical level curves for an isolated, 3-2 resonance computed algebraically. The dots represent points computed using Eq. (25); the curves were drawn in freehand. The chain of three islands first appears at the origin for $E=0.08$. All the widths of the islands including this one increase with increasing energy.

for the stable periodic orbit; while

$$J_2=(1-2J_1^{3/2})/(13+\tfrac{9}{2}J_1^{1/2}) \ , \qquad (27a)$$

$$(3\varphi_1-2\varphi_2)=0, 2\pi, 4\pi \ , \qquad (27b)$$

for the unstable periodic solution. As one expects, the 3-2 resonance zone lies near the unperturbed 3-2 torus. Indeed setting $3\omega_1=2\omega_2$ and using Eq. (12), we obtain

$$J_2=\tfrac{1}{13} \ , \qquad (28a)$$

$$J_1=\tfrac{5}{13}-(\tfrac{3}{13}-E)^{1/2} \ , \qquad (28b)$$

as the values of J_1 and J_2 on the unperturbed 3-2 torus where Eq. (28a) should be compared with Eqs. (26) and (27). Now Eq. (15) requires that $J_1 \geq 0$. Thus in Eq. (28b) we must have $E \geq \frac{14}{169} \approx 0.08$. At $E=\frac{14}{169}$, $J_1=0$. Hence the unperturbed 3-2 torus and the 3-2 resonance zone appear abruptly at the origin of the J_1 plane; they appear abruptly in the J_2 plane when the bounding level curve moves out to $J_2=\frac{1}{13}$, i.e., at $E=\frac{14}{169}$. The fact that resonance zones in the form of island chains may appear abruptly allows one to understand certain features of amplitude instability. For example, the onset of the instability observed by Thiele and Wilson[15] is evidently due to the sudden appearance of an island chain.

Next we briefly mention the 2-3 resonance. Here

$$H=H_0(J_1,J_2)+\beta J_1J_2^{3/2}\cos(2\varphi_1-3\varphi_2) \ . \qquad (29)$$

The additional constant is

$$I=3J_1+2J_2 \ . \qquad (30)$$

Level curves in the J_2 plane are found from

$$E=\tfrac{1}{3}I-\tfrac{1}{9}I^2+(\tfrac{1}{3}-\tfrac{1}{9}5I)J_2+\tfrac{23}{9}J_2^2 + \beta[\tfrac{2}{3}J_2^{5/2}-(\tfrac{1}{3}I)J_2^{3/2}]\cos 3\varphi_2 \ . \qquad (31)$$

As for the 3-2 resonance, the 2-3 resonance zone appears in the J_2 plane around the unperturbed 2-3 torus which can be shown to appear abruptly at $E=0.16$. The level curves for this resonance are quite similar to those of Fig. 7 except that the chain of three islands appears now in the J_2 plane.

These three examples suffice to give the general picture. An $m-n$ resonance for $m\neq n$ introduces a chain of m islands in the J_1 plane and a chain of n islands in the J_2 plane. Isolated resonances distort the unperturbed tori by introducing, in pairs, new stable and unstable periodic orbits. An $m-n$ resonance zone, in general, appears abruptly at some $E\geq 0$, and it is bounded by a separatrix which passes through the unstable periodic solutions. The presence of an additional simple constant of the motion allows one to calculate precisely the shape and position of each $m-n$ resonance zone. Though it is not obvious, one may show that the m-n resonance zones decrease rapidly in size as m and n increase. Having now investigated isolated resonances in considerable detail, we turn to the case in which two resonances act simultaneously; we shall be especially interested in the fate of the tori in regions where the isolated resonance zones overlap.

III. DOUBLE RESONANCE

We now investigate the behavior of oscillator systems simultaneously perturbed by two isolated resonances. In particular, we wish to determine how well this behavior can be predicted. We begin by considering the Hamiltonian

$$H=H_0(J_1,J_2)+\alpha J_1J_2\cos(2\varphi_1-2\varphi_2) + \beta J_1J_2^{3/2}\cos(2\varphi_1-3\varphi_2) \ , \qquad (32)$$

where $\alpha = \beta = 0.02$ in all calculations. Since the unperturbed 2-3 torus does not exist for energies $E \leqslant 0.16$, one may use the KAM technique indicated in Eqs. (3)–(6) to eliminate the 2-3 perturbation term. Consequently, one would expect the level curves for Hamiltonian (32) for $E \leqslant 0.16$ to be almost identical to those of Hamiltonian (16). This expectation is verified in Fig. 8 which presents the level curves for Hamiltonian (32), obtained by direct integration, for $E = 0.056$. Figure 8 should be compared with Fig. 6. In particular, Eq. (19a), valid for $\beta = 0$, predicts that a stable periodic orbit should occur at $q_2 = 0.142$, which to three-figure accuracy is exactly where it does occur in Fig. 8. In addition, the function $J_1 + J_2$, exactly constant for $\beta = 0$, is now constant to between four and six decimals while $3J_1 + 2J_2$ is constant only to two decimals.

Using data calculated from the equations of Sec. II, we determined that, for energies slightly greater than $E = 0.16$, the 2-2 and 2-3 resonance zones should be widely separated. Figure 9 shows the level curves for Hamiltonian (32) at $E = 0.18$. The details of Fig. 9 are accurately predicted from the data of Sec. II. Next we conjectured that a Henon-Heiles-type zone of instability might occur when the 2-2 and 2-3 resonance zones begin to overlap. At low energies, the 2-3 zone lies inside the 2-2 zone, thus the zones should first overlap when the outer edge of the 2-3 zone touches the inner edge of the 2-2 zone. In order to estimate the energy for the onset of overlap, we calculated the q_2-axis intercept of each zone boundary using the equations of Sec. II. In Fig. 10, we plot these q_2 intercepts as a function of energy. Overlap first occurs at energy $E = 0.2095$. In Fig. 11, we plot level curves for Hamiltonian (32), obtained by direct integration, at $E = 0.2095$. Here one observes that indeed a small zone of instability has rather abruptly appeared with the overlap of the 2-2 and 2-3 resonance zones as conjectured.

In order to illustrate the dynamics of breakdown, which is quite complicated, we show, in Fig. 12, the level curves for Hamiltonian (32) at $E = 0.20$, slightly below the predicted overlap energy. The 2-2 and 2-3 resonance zones occur in their predicted places; however, the computer also detects the chain of five islands shown in the figure but not previously predicted. There is also a detectable chain of seven islands near the chain of five which is not shown. We now discuss the origin of these higher-order resonances.

The hierarchy of resonances implicit in Hamiltonian (32) may be made explicit via the following

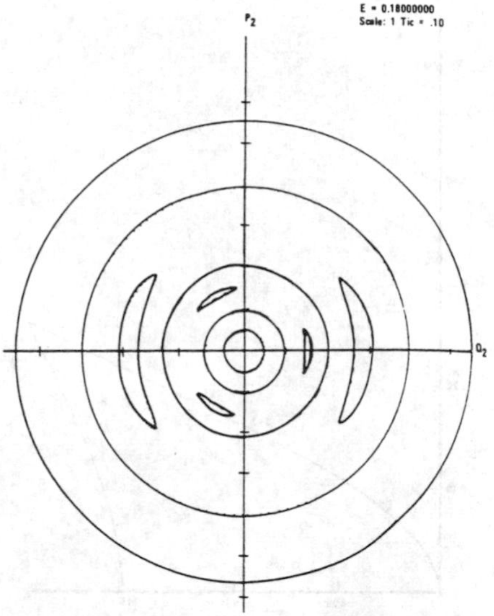

FIG. 9. Typical level curves for the 2-2, 2-3, doubly resonant Hamiltonian for energies yielding widely separated 2-2 and 2-3 resonances.

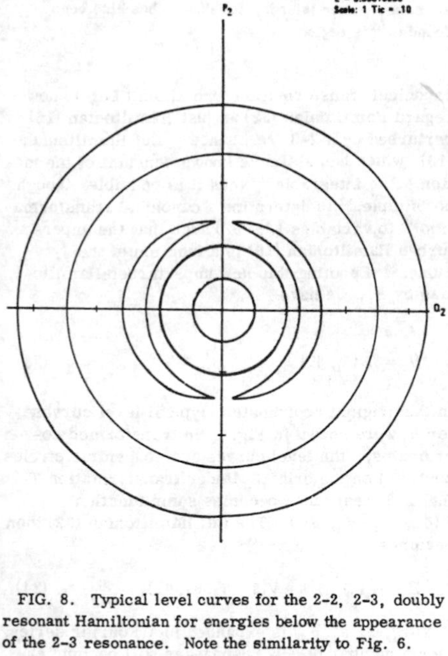

FIG. 8. Typical level curves for the 2-2, 2-3, doubly resonant Hamiltonian for energies below the appearance of the 2-3 resonance. Note the similarity to Fig. 6.

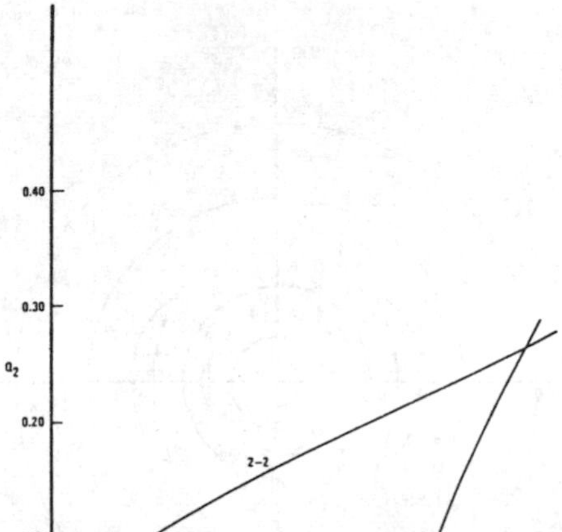

FIG. 10. q_2-axis intercepts of the inner 2-2 and the outer 2-3 separatrices are plotted as a function of total energy.

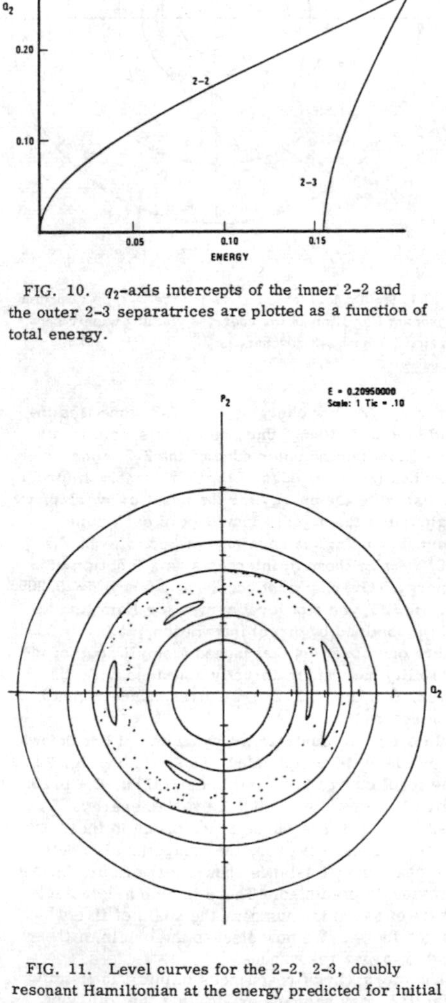

FIG. 11. Level curves for the 2-2, 2-3, doubly resonant Hamiltonian at the energy predicted for initial overlap of the 2-2 and 2-3 resonance zones.

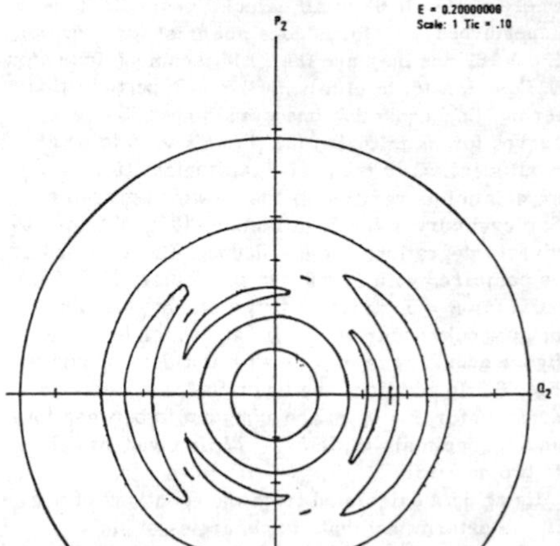

FIG. 12. Level curves for the 2-2, 2-3, doubly resonant Hamiltonian at $E = 0.20$, slightly below the predicted overlap energy. The dots between the 2-2 and 2-3 crescents are part of a chain of five islands. A chain of seven islands, not shown, has also been found in this region.

canonical transformation formalism. Let us now regard Hamiltonian (32) as just Hamiltonian (16) perturbed by a 2-3 resonance. But Hamiltonian (16), which has a simple known constant of the motion I, is integrable[20]; thus it is possible, though not simple, [22] to determine a canonical transformation T to variables $(\mathcal{J}_i, \theta_i)$ such that the unperturbed Hamiltonian (16) is a function of the $\mathcal{J}_i$ alone. [20] Denoting this new unperturbed Hamiltonian by H_1, we have

$$H_1 = H_1(\mathcal{J}_1, \mathcal{J}_2)\,. \tag{33}$$

In the original coordinates, typical level curves for H_1 were shown in Fig. 6; in transformed coordinates, the level curves are concentric circles centered on the origin. Under transformation T, the 2-3 resonance becomes some function $V(\mathcal{J}_1, \mathcal{J}_2, \theta_1, \theta_2)$. The full Hamiltonian (32) then becomes

$$H = H_1(\mathcal{J}_1, \mathcal{J}_2) + V(\mathcal{J}_1, \mathcal{J}_2, \theta_1, \theta_2). \tag{34}$$

If $V(\mathcal{J}_1, \mathcal{J}_2, \theta_1, \theta_2)$ is expanded in a Fourier series, then a number of new resonances will become explicit. These new resonances we shall call secondary resonances as opposed to the primary resonances explicitly appearing in Hamiltonian (32). Using the methods of Sec. II and treating each secondary resonance as if isolated, we could estimate the position and shape of each secondary-resonance-zone island chain. Transforming back to original coordinates would then reveal the position and shape of the secondary island chains. Let us note that any single resonance in Hamiltonian (34), when added to H_1, forms an integrable system. Thus, we could repeat the process, eliminate this resonance, and reveal tertiary resonances. Indeed repeating this process to arbitrary order would reveal, upon transformation to original coordinates, a complicated network of island chains, some nested within each other.[22, 30] The process outlined here is the germinal concept in Hamilton-Jacobi theory,[20] and all perturbation methods including the KAM method are various approximations which seek to reveal the nature of the Hamilton-Jacobi transformation.

It thus becomes clear that the chains of five and seven islands detected in Fig. 12 are due to secondary resonances. Rather than present the formidable calculations[31] necessary to verify this fact, let us observe only that the chains of five and seven islands occur at precisely the positions of the unperturbed $4\omega_1 = 5\omega_2$ and $6\omega_1 = 7\omega_2$ tori calculated using Eq. (14). The following picture thus emerges. As the primary 2-2 and 2-3 resonance zones approach overlap, certain higher-order resonances begin to macroscopically distort some of the intervening preserved tori. As a consequence the narrowing region between the 2-2 and the 2-3 resonance zones contains that host of overlapping resonances, anticipated in the paragraph following the inequality (7), which yields a macroscopic zone of instability. Clearly, instability is due to a host of stable and unstable periodic orbits which now lie in a narrow region. Evidence for this instability, again at $E = 0.20$, appears in Fig. 13 where we plot a ragged level curve for an orbit near the original 2-2 separatrix. In essence then by examining the overlap of macroscopic primary resonance zones, we have illustrated on a macroscopic scale the nature of the microscopic KAM zones of instability. Moreover, we have calculated with reasonable accuracy the energy of onset for this macroscopic instability. Rather than attempting to improve this estimate by including the secondary resonances, we now seek to demonstrate that the macroscopic zone of instability grows with increasing energy.

For energies much above $E = 0.20$, the unperturbed frequencies for Hamiltonian (32) become so small that obtaining level curves by direct integration requires prohibitively long integration times. Consequently, we now increase α and β to $\alpha = 0.95$ and $\beta = 0.25$ and consider the Hamil-

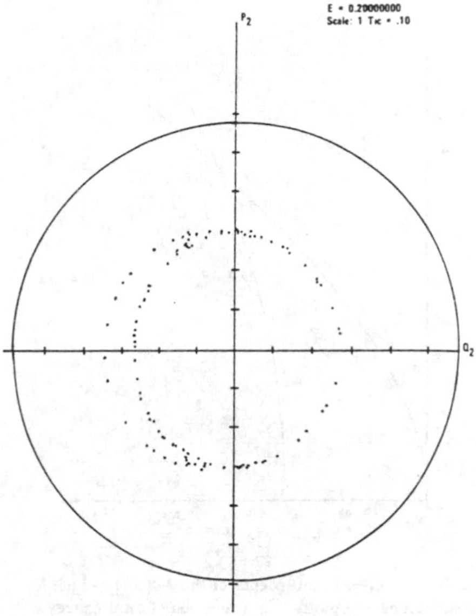

FIG. 13. This figure is a continuation of Fig. 12 and shows that a small zone of instability exists at energies below the predicted 2-2, 2-3 overlap.

tonian

$$H = H_0(J_1, J_2) + \alpha J_1 J_2 \cos(2\varphi_1 - 2\varphi_2) + \beta J_1^{3/2} J_2 \cos(3\varphi_1 - 2\varphi_2) , \qquad (35)$$

for which instability occurs at a lower energy. Here we choose to plot level curves in the J_1 plane, and Fig. 14 plots the q_1 intercept of the inner 2-2 and the outer 3-2 separatrix versus energy as calculated using the equations of Sec. II. Here the 3-2 resonance zone is seen to first appear at $E \cong 0.08$ while overlap is predicted at $E \cong 0.12$.

In Fig. 15, we plot level curves for energy $E = 0.05$. While these curves all appear regular, using high accuracy a chain of five islands has been detected at a radius of 0.08. The $5\omega_1 = 4\omega_2$ unperturbed torus lies at a radius of 0.098. In Fig. 16, we plot level curves for energy $E = 0.08$ The chain of five islands is now clearly visible and the 2-2 separatrix is now a small zone of instability. Instability thus occurs no later than $E = 0.08$ as compared with the predicted $E = 0.12$. Since the 3-2 resonance has not yet appeared at $E = 0.08$, it is especially clear that one cannot ignore secondary resonances here. Again, however, let us not pursue the details of secondary

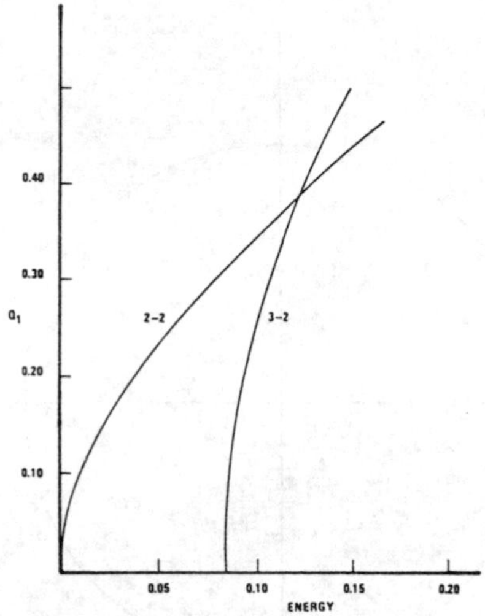

FIG. 14. q_1-axis intercepts of the 2-2 and 3-2 separatrices are plotted as a function of total energy.

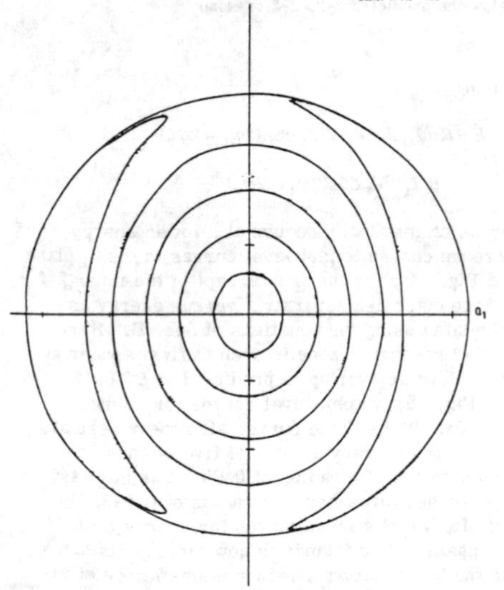

FIG. 15. Typical level curves for the 2-2, 3-2, doubly resonant Hamiltonian at low energy. Even at this low energy, a chain of five islands, not shown, has been detected.

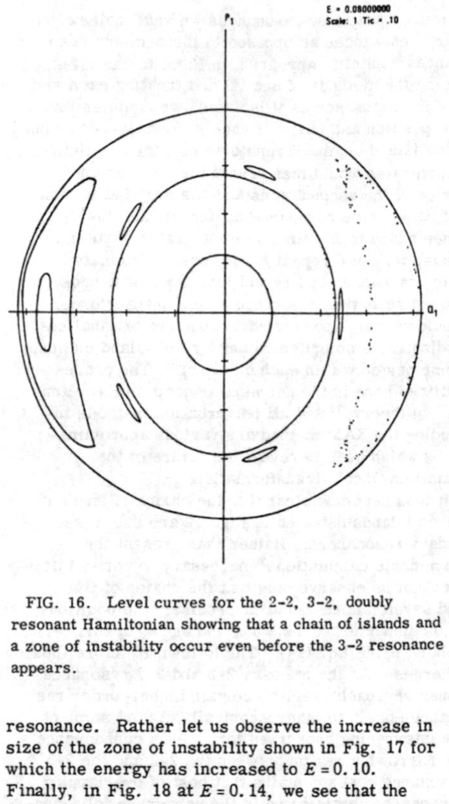

FIG. 16. Level curves for the 2-2, 3-2, doubly resonant Hamiltonian showing that a chain of islands and a zone of instability occur even before the 3-2 resonance appears.

resonance. Rather let us observe the increase in size of the zone of instability shown in Fig. 17 for which the energy has increased to $E = 0.10$. Finally, in Fig. 18 at $E = 0.14$, we see that the 3-2 resonance has at last moved into the ever enlarging zone of instability. It is now clear that our model Hamiltonians exhibit many of the characteristics of the more complicated systems previously studied. As illustration, we now turn to a discussion of the Henon-Heiles system in terms of our model Hamiltonians.

IV. RESONANCE IN TRANSFORMED COORDINATES

The isolated resonances of Sec. II were easily handled because one could without difficulty determine a well-behaved constant of the motion. Though it is less obvious, the small amplitude motion of the Henon-Heiles Hamiltonian (8) is dominated by an isolated resonance. Using a modified Birkhoff canonical transformation, Gustavson[32] and independently Walker[31] show that the Henon-Heiles Hamiltonian may be written

$$H = H_0[J_1, J_2, (\varphi_1 - \varphi_2)] + V(J_1, J_2, \varphi_1, \varphi_2) \; , \quad (36)$$

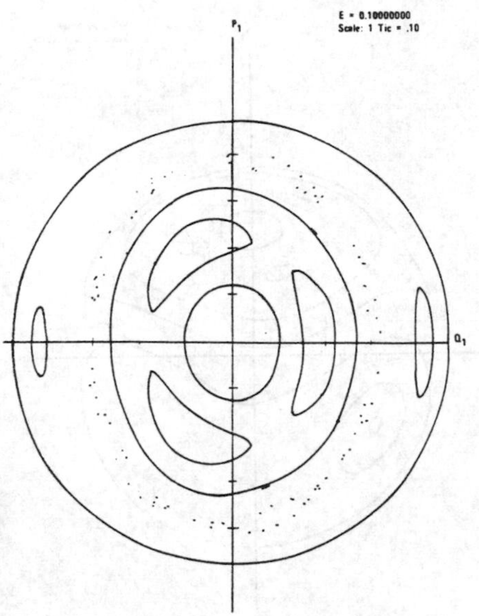

FIG. 17. Level curves for the 2-2, 3-2, doubly resonant Hamiltonian showing the increase of the instability zone with energy.

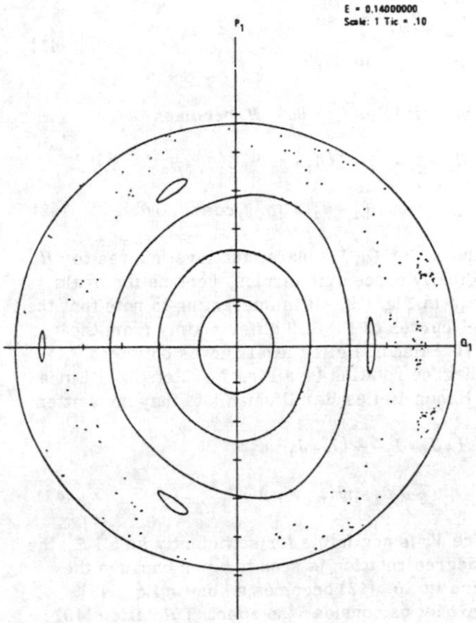

FIG. 18. Level curves for the 2-2, 3-2, doubly resonant Hamiltonian showing the 3-2 resonance as it moves into the ever-increasing zone of instability. The ragged looking chain of three islands in the right of the diagram represents a single level curve.

where V is very small compared to H_0 for sufficiently small-amplitude motion. Neglecting V, Hamiltonian (36) has the additional constant

$$I = J_1 + J_2 \quad . \tag{37}$$

One may then canonically transform Eq. (37) back to original coordinates obtaining the algebraic equation $I = I(q_1, p_1, q_2, p_2)$. This approximate constant of the motion and Hamiltonian (8) may be combined to obtain an analytic expression for level curves. For $E = \frac{1}{12}$ this expression gives level curves congruent with the directly integrated level curves of Fig. 1.

As the energy increases, the resonances in V begin to grossly distort the tori of Fig. 1. In order to determine the onset of macroscopic instability, we note that Hamiltonian (36), neglecting V, is integrable[20]; thus there exists a canonical transformation to coordinates $(\mathcal{J}_i, \theta_i)$ such that the full Hamiltonian (36) becomes

$$H = H_0(\mathcal{J}_1, \mathcal{J}_2) + V(\mathcal{J}_1, \mathcal{J}_2, \theta_1, \theta_2) \quad , \tag{38}$$

where H_0 is now angle-independent. Since V has a Fourier expansion, we may determine the behavior of this multiply resonant Hamiltonian in exactly the same way as we analyzed the doubly resonant Hamiltonians of Sec. III. In essence then, the Henon-Heiles Hamiltonian is only one (compounded) canonical transformation away from the analysis of Secs. II and III. In view of the complexity of actually reducing the Henon-Heiles problem to manageable form, however, it is perhaps worthwhile illustrating by example the effects that even a simple canonical transformation can introduce.

We begin with the unperturbed Hamiltonian (12),

$$H_0 = J_1 + J_2 - J_1^2 - 3J_1J_2 + J_2^2 \quad . \tag{39}$$

For Hamiltonian (39) we recall that the level curves are concentric circles in either the J_1 or the J_2 planes. We now introduce the coordinate rotation

$$Q_1 = (\tfrac{1}{2})^{1/2}(q_1 + q_2), \quad P_1 = (\tfrac{1}{2})^{1/2}(p_1 + p_2) \ ,$$
$$Q_2 = (\tfrac{1}{2})^{1/2}(-q_1 + q_2), \quad P_2 = (\tfrac{1}{2})^{1/2}(-p_1 + p_2) \ , \tag{40}$$

where (q_i, p_i) are related to the (J_i, φ_i) by Eq. (13), and the (Q_i, P_i) are related to the $(\mathcal{J}_i, \theta_i)$ via

$$Q_i = (2\mathcal{J}_i)^{1/2}\cos\theta_i \ ,$$
$$P_i = -(2\mathcal{J}_i)^{1/2}\sin\theta_i \ . \qquad (41)$$

In the variables $(\mathcal{J}_i, \theta_i)$, H_0 becomes

$$H_0 = \mathcal{J}_1 + \mathcal{J}_2 - \tfrac{3}{4}(\mathcal{J}_1 + \mathcal{J}_2)^2 + 2(\mathcal{J}_1 + \mathcal{J}_2)(\mathcal{J}_1\mathcal{J}_2)^{1/2}$$
$$\times \cos(\theta_1 - \theta_2) + 3\mathcal{J}_1\mathcal{J}_2\cos^2(\theta_1 - \theta_2) \ . \qquad (42)$$

In the $\mathcal{J}_1$ (or Q_1, P_1) plane, the level curves for H_0, originally concentric circles, become the ovals shown in Fig. 19. It is interesting to note that the level curves of Fig. 19 differ mainly from the $E = 10^{-4}$ Henon-Heiles level curves only by a 90-degree rotation (see Fig. 7 of Ref. 32). Since the Henon-Heiles Hamiltonian (36) may be written

$$H = J_1 + J_2 - \tfrac{5}{12}(J_1 + J_2)^2$$
$$+ \tfrac{7}{3} J_1 J_2 \sin^2(\varphi_1 - \varphi_2) + V_2 \ , \qquad (43)$$

where V_2 is negligible for sufficiently small E, the 90-degree rotation is seen to arise because the cosine in Eq. (42) becomes a sine in Eq. (43).

Next let us consider the effect of Rotation (40) on the level curves of Hamiltonian (35), where

$$H = H_0(J_1, J_2) + \alpha J_1 J_2 \cos(2\varphi_1 - 2\varphi_2)$$
$$+ \beta J_1^{3/2} J_2 \cos(3\varphi_1 - 2\varphi_2) \ , \qquad (44)$$

and where H_0 is given by Eq. (39). Changing variables according to Eq. (40), we may write Hamiltonian (44) as

$$H = H_0[\mathcal{J}_1, \mathcal{J}_2, (\theta_1 - \theta_2)] + V(\mathcal{J}_1, \mathcal{J}_2, \theta_1, \theta_2) \ , \qquad (45)$$

where H_0 is now given by Eq. (42) and V is a complicated but calculable function. Figure 20 shows the directly integrated, $\mathcal{J}_1$ plane level curves for Hamiltonian (45) at $E = 0.09$. First, comparing Fig. 17 with Fig. 20, we see the rather dramatic distortion in level curves produced by a simple rotation of coordinates. Next, comparing Fig. 2 with Fig. 20, we see that similarity which is to be expected from the similar forms of Hamiltonians (36) and (45). In short, we have made it quite plausible that the Henon-Heiles Hamiltonian is only a coordinate transformation away from the analysis of Secs. II and III.

In concluding this section, we note that the level curves of Fig. 20 were not obtained by integrating Hamiltonian (45). Rather we integrated Eq. (44)

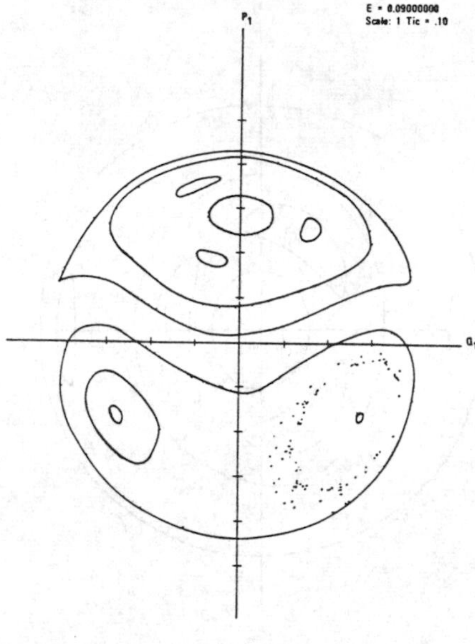

FIG. 20. Level curves for the 2-2, 3-2, doubly resonant Hamiltonian in the rotated coordinate system. These curves are close cousins of those appearing in Fig. 4.

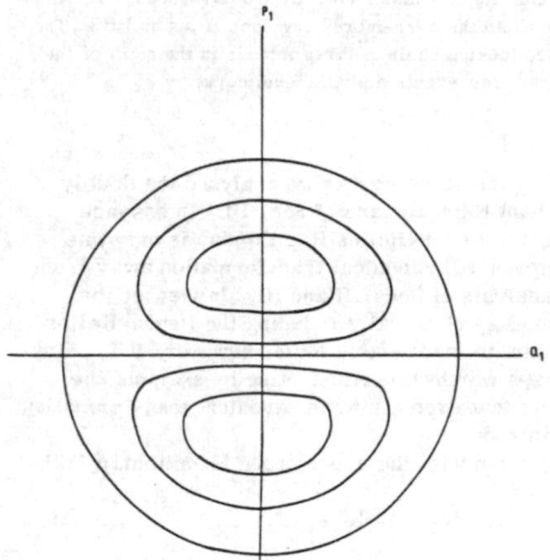

FIG. 19. Level curves for $H = J_1 + J_2 - J_1^2 - 3J_1J_2 + J_2^2$ at $E = 0.09$ in the rotated coordinate system. If this picture is rotated by 90 degrees, one very nearly obtains the low-energy level curves for the Henon-Heiles system.

expressed in (q_i, p_i) coordinates and then determined the trajectory intersections with the rotated (Q_1, P_1) plane defined by $q_1 = q_2$ and $p_1 \geqslant p_2$. Thus geometrically speaking Figs. 17 and 20 differ only in that they represent the intersections of a given set of trajectories with two different planes. As a consequence we see that the shape of the level curves for a given Hamiltonian can dramatically depend on the intersection plane used. In particular, in order to obtain level curves for the Henon-Heiles problem similar to those of Fig. 9, instead of those shown in Fig. 1, one would have not only to rotate the intersection plane but also distort it into a curved surface. Therein lies the complexity of the Henon-Heiles problem.

V. CONCLUSIONS

For over fifty years, Poincaré's theorem on the nonexistence of well-behaved constants of the motion (other than the total energy) has stood as a pillar in the foundations of equilibrium statistical mechanics.[5] It provides a central argument supporting the view that states of equal energy are equally likely for an isolated system. KAM theory on the other hand proves that the nonexistence of well-behaved constants of the motion is insufficient to ensure that states of equal energy are equally likely. For sufficiently small-amplitude motion, most initial conditions lead either to motion uninfluenced by any resonances, in which each action variable is a constant, or to motion influenced by isolated resonances, in which linear combinations of the action variables are constant. The existence of these constants of the motion do not violate Poincaré's theorem since the minority of trajectories moving under the influence of many resonances are densely woven between the well-behaved majority. In addition to clarifying and perhaps reducing the significance of Poincaré's theorem, however, the KAM theory points to an amplitude instability beyond which the irregular trajectories begin to dominate and the system motion perhaps becomes statistical.[33] It is thus quite possible that KAM instability can be made a cornerstone for statistical mechanics.

In this paper, we have attempted to illustrate the origin and nature of KAM instability using simple examples. For these simple systems, we have demonstrated not only that the large-amplitude motion does indeed become quite erratic but also that the onset of the instability can, in principle, be predicted. Moreover, several computer studies (in addition to ours) show that for sufficiently large amplitudes almost all trajectories are highly erratic. In this paper, we have considered systems with only two degrees of freedom; however, the procedures used can, with considerable labor, be extended to more general systems. In order for the KAM instability to be shown to be universally relevant to statistical mechanics, however, several rather serious questions must be answered.

All the computer studies thus far discussed indicate that widespread KAM instability occurs only after the amplitude of the motion becomes quite large indeed. Physical systems, on the other hand, apparently obey the laws of statistical mechanics even at cryogenic temperatures. One must therefore establish that physically realistic models exhibit KAM instability even for small-amplitude motion. One might suggest that physical potential energies are more complicated than the cubic terms of the Henon-Heiles potential or that physical systems have an enormous number of particles. However, Thiele and Wilson[15] used a Morse potential in their calculations and observed instability only for energies greater than one-half the dissociation energy. Equally, Zabusky and Deem[28] studied Hamiltonian (9) for a 200-particle system and they observed an instability only for relatively large energies. However, the work of Northcote and Potts[34] suggests a possible mechanism for introducing a small-amplitude instability. They used an infinitely steep hard-core repulsion superimposed on an otherwise harmonic potential between particles and obtained ergodic-type behavior for almost all amplitudes. It is thus quite possible that an extremely steep repulsive hard-core potential could be responsible for small-amplitude instability in physical systems. Such a nonlinearity would, in addition, cause those rather sharply defined "collisions" between harmonic normal modes usually assumed in statistical mechanics. Nonetheless, whatever the conjecture, the general existence of a small-amplitude instability for physically realistic models has not been proved.

Moreover, erratic or statistical behavior of the system trajectory in the full phase space is not really required for statistical mechanics.[29, 35, 36] Most physically measured quantities require only that projections of most trajectories be ergodic on some subspace of the full phase space. In short, statistical mechanics may follow from some type of "coarse graining." Consequently, even if KAM instability is applicable to a wide class of physical systems, there still might exist an equally wide class for which it is irrelevant.

Finally, there is an open question concerning the extent to which the zones of instability are in fact ergodic. In Fig. 4, for example, the unstable level curve shown does rather uniformly cover a certain zone of instability; however, successive intersection points for this orbit do not randomly fill the zone. Indeed, extremely accurate integration shows that they circle the stable invariant points in an orderly, though not regular, fashion. This rather low degree of order which persists even in zones of instability also appears in the work of Zabusky and Deem.[28] Successive points in Fig. 5 do jump randomly, as pointed out by Henon and Heiles[17]; however, this randomness might disappear with improved integration accuracy.[21] In short, the following situation may prevail. For small amplitudes only, a few low-order resonances influence the motion.[24] Here only simple measurable quantities would be correctly predicted by statistical mechanics. As the amplitudes increase a larger number of resonances become significant, allowing a wider variety of physical measurements (but not all) to be correctly predicted. In this picture the sudden onset of KAM instability would mark only a sudden increase in the predictive capacity of statistical mechanics rather than the issuance of an unrestricted license for its use.

Whatever the final resolution of these questions, we suggest that their study is fascinating. Answers would provide not only an increased understanding of irreversibility but perhaps also even a practical device for violating the second law of thermodynamics. In any event this paper represents at best only a very modest contribution toward their resolution, and much of the material in this paper will be quite familiar to that small group of astronomers, mathematicians, and physicists who have made similar or vastly superior contributions. Nonetheless, by couching the discussions in language familiar to physicists, it is our hope that these questions can attract the attention of a broader audience.

---

*Work supported in part by the National Science Foundation.

[1]H. Poincaré, Méthodes Nouvelles de la Mécanique Celeste (Dover Publications, Inc., New York, 1957).

[2]G. D. Birkhoff, Dynamical Systems (American Mathematical Society Colloquium Publications, New York, 1927), p. 82.

[3]N. Kryloff and N. Bogoliuboff, Introduction to Nonlinear Mechanics (Princeton University Press, Princeton, N.J., 1947).

[4]In this paper, we define ergodic motion as motion for which the time average of functions of the system variables is equal to the statistical mechanical phase-space average for all functions of physical interest and for most initial conditions. Since functions of physical interest are usually simple ones, this definition places less stringent requirements on the motion than do most earlier definitions.

[5]E. Fermi, Z. Physik 24, 261 (1923).

[6]R. E. Peierls, Ann. Physik 3, 1055 (1929).

[7]A. N. Kolmogorov, Dokl. Akad. Nauk. SSSR 98, 527 (1954).

[8]V. I. Arnol'd, Russian Math. Surveys 18, 9 (1963).

[9]J. Moser, Nachr. Akad. Wiss. Göttingen, II Math. Physik Kl. 1 (1962).

[10]V. I. Arnol'd, Russian Math. Surveys 18, 85 (1963).

[11]For a proof of the convergence of the formal series expansions used in Refs. 1–3, see J. Moser, Math. Annalen 169, 136 (1967).

[12]H. Goldstein, Classical Mechanics (Addison-Wesley Press, Inc., Cambridge, Mass., 1951), p. 241.

[13]E. Merzbacher, Quantum Mechanics (John Wiley & Sons, Inc., New York, 1964).

[14]Since the resonant zones have width and are everywhere dense, even the casual reader will ask why all the unperturbed tori are not destroyed. In every neighborhood of a preserved torus with incommensurate frequencies there exist zones of destroyed tori loosely specified by the inequality (7). However, as m and n satisfying the inequality (7) increase, specifying successive regions destroyed by the perturbations $\cos(m\varphi_1 + n\varphi_2)$ which are ever closer to the preserved tori, the width of these destroyed regions decreases since the f_{mn} decrease with increasing m and n. Thus, in order to demonstrate that a preserved torus exists, one need show only that the width of the destroyed regions are always smaller than the distance to the preserved torus. These notions are made quantitative in a rather simple fashion in a footnote on p. 299 of the paper by M. N. Rosenbluth, R. Z. Sagdeev, J. B. Taylor, and G. M. Zaslavski, Nucl. Fusion 6, 297 (1966).

[15]E. Thiele and D. J. Wilson, J. Chem. Phys. 35, 1256 (1961).

[16]D. Bunker, J. Chem. Phys. 37, 393 (1962).

[17]M. Henon and C. Heiles, Astron. J. 69, 73 (1964). Similar results have also been reported by G. Contopoulos and L. Woltjer, Astrophys. J. 140, 1106 (1964); W. H. Jefferys, Astron. J. 71, 306 (1966); B. Barbanis, *ibid.* 71, 415 (1966); G. Contopoulos and J. D. Hadjidemetriou, *ibid.* 73, 86 (1968).

[18]G. Contopoulos, Astron. J. 68, 1 (1963).

[19]In order to verify that the system trajectory should intersect the (q_2, p_2) plane in a curve when a well-behaved $I(q_1, p_1, q_2, p_2)$ exists, algebraically solve Eq. (8) for p_1, taking the positive square root. Putting this expression for p_1 into the equation $I(q_1, p_1, q_2, p_2) = I_0$, we may invert to obtain $p_2 = p_2(q_1, E, q_2, I_0)$, where E and I_0 are the constant values of H and I. Setting $q_1 = 0$, now yields the (q_2, p_2) plane level-curve equation $p_2 = p_2\,(0, E, q_2, I_0)$. This last equation gives those (q_2, p_2) points on a system trajectory for which $q_1 = 0$ and $p_1 \geq 0$.

[20]E. T. Whittaker, Analytical Dynamics (Cambridge University Press, London, 1965), p. 280 and Chap. 11. Also see Chap. 9 of Ref. 12.

[21]For the total integration times used, up to $t = 1500$ for the stable orbits, the energy was constant through six decimals. Moreover, at $t = 1500$, the velocities could be reversed and the motion integrated back to $t = 0$ maintaining four decimal accuracy. However, for the highly unstable orbits, the reversed integrations showed that the solutions were accurate only for $t < 200$ even though the energy continued to be constant to high

accuracy. Contopoulos (Ref. 30) obtains similar results.
[22]W. H. Jefferys, Astron. J. 71, 306 (1966).
[23]See Enrico Fermi: Collected Papers, Vol. II (University of Chicago Press, Chicago, 1965), p. 978.
[24]J. Ford and J. Waters, J. Math. Phys. 4, 1293 (1963).
[25]E. A. Jackson, J. Math. Phys. 4, 551, (1963); 4, 686 (1963).
[26]See the review article by N. J. Zabusky, in Proceedings of the Symposium of Nonlinear Partial Differential Equations (Academic Press Inc., New York, 1967). Also see M. Toda, J. Phys. Soc. Japan 22, 431 (1967); 26 Suppl., 235 (1969).
[27]F. M. Izrailev and B. V. Chirikov, Dokl. Akad. Nauk SSSR 166, 57 (1966) [English transl.: Soviet Phys. – Doklady 11, 30 (1966)]. This paper discusses undriven oscillator systems. For a discussion of the statistical behavior of driven oscillator systems, see V. A. Alekseev, Math. Sbornik 119, 545 (1968).
[28]N. J. Zabusky and G. J. Deem, J. Computational Phys. 2, 126 (1967). Also see H. Hirooka and N. Saito, J. Phys. Soc. Japan 26, 624 (1969) and N. Saito, H. Hirooka, and N. Ooyama, *ibid.* 26 Suppl., 223 (1969).
[29]G. M. Zaslavskii and R. Z. Sagdeev, Zh. Eksperim. i Teor. Fiz. 52, 1081 (1967) [English transl.: Soviet Phys. – JETP 25, 718 (1967)]. This paper extends the general theory discussed in I. Prigogine, Non-Equilibrium Statistical Mechanics (Wiley-Interscience, Inc., New York, 1962).
[30]G. Contopoulos, Bull. Astron. 2, 223 (1967).
[31]G. H. Walker, thesis, Georgia Institute of Technology, Atlanta, Ga., 1968 (unpublished).
[32]F. G. Gustavson, Astron. J. 71, 670 (1966).
[33]Physicists have long sought an underlying physical explanation for the purely mathematical probability arguments used to derive statistical mechanics from classical mechanics. The KAM zones of instability have an obvious random element in the positions of the unstable periodic orbits. The erratic path followed by a system trajectory may be attributed to "collisions" with these randomly positioned unstable periodic orbits.
[34]R. A. Northcote and R. B. Potts, J. Math. Phys. 5, 383 (1964).
[35]P. Mazur and E. Montroll, J. Math. Phys. 1, 70 (1960).
[36]J. H. Weiner and W. F. Adler, Phys. Rev. 144, 511 (1966).

3. The Problem of Onset of Turbulence

52. ON THE PROBLEM OF TURBULENCE

ALTHOUGH the turbulent motion has been extensively discussed in literature from different points of view, the very essence of this phenomenon is still lacking sufficient clearness. To the author's opinion, the problem may appear in a new light if the process of initiation of turbulence is examined thoroughly.

In the case of incompressible fluids the unsteadiness of the laminar motion is known to be determined as follows. Upon the principal motion with a velocity distribution $\boldsymbol{v}_0(x, y, z)$ there is superimposed a small disturbance $\boldsymbol{v}_1(x, y, z, t)$; the substitution of $\boldsymbol{v} = \boldsymbol{v}_0 + \boldsymbol{v}_1$ in the equation of motion of a viscous fluid and the neglect of terms oȷ the second order of smallness lead to a linear differential equation for the perturbation $\boldsymbol{v}_1$. Further, $\boldsymbol{v}_1$ is sought in the form

$$\boldsymbol{v}_1 = A(t)\, \boldsymbol{f}(x, y, z), \tag{1}$$

where the time function $A(t)$ may be represented as

$$A(t) = \text{const} \cdot e^{-i\Omega t}. \tag{2}$$

The problem of determining the possible values of the "frequencies" Ω where the boundary conditions of motion are given, is an "Eigenwert" problem. By solving it one will obtain a spectrum of proper frequencies Ω (which are complex values in the general case). This spectrum, generally speaking, contains separate, isolated, values ("discrete spectrum") and also contains frequencies continuously filling whole intervals of values ("continuous spectrum"). It may be supposed that the frequencies of the continuous spectrum correspond to such motions $\boldsymbol{v}_1$ as are not damped at infinity, while the frequencies of the discrete spectrum correspond to motions which are damped at infinity rather rapidly (as is the case in many other Eigenwert problems).

For the problem of steadiness of the principal motion those of the frequencies $\Omega = \omega + i\gamma$ (ω, γ are real) are relevant in which the imaginary part is negative ($\gamma < 0$). The presence of such proper frequencies in the spectrum indicates the unsteadiness of the principal motion with respect to infinitely small perturbations. Such values of Ω are only possible among the frequencies of the discrete spectrum. In fact, the principal motion presents at infinity a plane-parallel homogeneous flow (we mean a flow past a body of finite dimensions. In so far as a plane–parallel flow is in no case steady, it will be evident that any perturbation that fails to disappear at infinity must necessarily be damped in time, or, in other words, correspond to frequencies Ω with $\gamma > 0$ Accordingly, only the Ω frequencies of the discrete spectrum can be considered below.

Л. Ландау, К проблеме турбулентности, *Доклады Академии Наук СССР*, **44**, 339 (1944).
L. Landau, On the problem of turbulence, *C. R. Acad. Sci. URSS*, **44**, 311 (1944).

In the case of sufficiently small velocities the principal motion is a steady one (inasmuch as a resting fluid is in any case steady). On the other hand, with sufficiently large Reynolds numbers the laminar flow past a body is unsteady at any rate. In fact, with large Reynolds numbers the motion far away from the body is not appreciably different from a plane-parallel flow unless in the region of the narrow "track". Now it follows from Lord Rayleigh's work that no motion with a two-dimensional velocity distribution of such a type is steady, and one may expect that the same is true of the three-dimensional track.

If the values of the proper frequencies Ω are taken to be functions of the Reynolds number of the principal motion, then the critical value Re_{cr} is determined by the fact that for $\mathrm{Re} = \mathrm{Re}_{cr}$ the imaginary part of one of the frequencies Ω will vanish; suppose this frequency to be $\Omega_1 = \omega_1 + i\gamma_1$. For $\mathrm{Re} > \mathrm{Re}_{cr}$ we have $\gamma_1 > 0$; for such Reynolds numbers as are near to the critical value Re_{cr}, γ_1 is small in comparison with ω_1. However, the expression (1–2) for the respective function $\boldsymbol{v}_1(x, y, z, t)$ (with $\Omega = \Omega_1$) is only true for a very brief interval of time, as measured from the instant at which the stationary regime is broken. This is owing to the fact that the factor $e^{\gamma_1 t}$ grows rapidly with time. As a matter of fact, the modulus $|A|$ of the amplitude of non-stationary motion does not increase infinitely, but rather tends to a certain limit. With Re near to Re_{cr} (Re is always supposed to be greater than Re_{cr}), this limit is yet very small, too, and for determining it one may proceed as follows.

For very small times, when (2) is still applicable, we have

$$\frac{d|A|^2}{dt} = 2\gamma_1 |A|^2.$$

In substance this expression is but the first term of a series of powers of A and A^*. With the increase of the modulus $|A|$ the subsequent three terms of this series must be taken into account. The next terms are terms of the third order. We, however, are interested not in the exact value of the differential quotient $d|A|^2/dt$, but in its mean value with respect to time, the averaging being made over time intervals that are large in comparison with the period $2\pi/\omega_1$ of the periodic spectrum $e^{-i\omega_1 t}$ (as $\omega_1 \gg \gamma_1$, this period is small compared to the time $1/\gamma_1$ during which the modulus $|A|$ changes appreciably). But the terms of the third order involve a periodic spectrum, and so they are eliminated upon averaging. (Strictly speaking, they do not vanish altogether, but yield quantities of order four; these quantities are supposed to be included into the terms of the fourth order). The terms of the fourth order include a term proportional to $A^2 A^{*2} = |A|^4$; this term is not eliminated by averaging. Thus, up to terms of the fourth order we have

$$\overline{\frac{d|A|^2}{dt}} = 2\gamma_1 |A|^2 - \alpha |A|^4. \tag{3}$$

Here α is a positive constant (the case of negative α is considered below).

There are no signs of averaging over $|A|^2$ and $|A|^4$, because this operation is carried out over such time intervals as are small in comparison with $1/\gamma_1$. For the same reason in solving this equation we must disregard the bar over the derivative in the left hand member. The solution of the equation (3) has the form

$$\frac{1}{|A|^2} = \frac{\alpha}{2\gamma_1} + \text{const} \cdot e^{-2\gamma_1 t},$$

i.e. $|A|^2$ tends asymptotically to a limit

$$|A|^2_{\max} = 2\gamma_1/\alpha; \tag{4}$$

γ_1 is a function of Reynolds' number; it vanishes with $\mathrm{Re} = \mathrm{Re}_{\mathrm{cr}}$. Therefore for small $\mathrm{Re} - \mathrm{Re}_{\mathrm{cr}}$ we have $\gamma_1 = \text{const} \cdot (\mathrm{Re} - \mathrm{Re}_{\mathrm{cr}})$. Substituting this in (4) we shall see that

$$|A|_{\max} \sim \sqrt{\mathrm{Re} - \mathrm{Re}_{\mathrm{cr}}}. \tag{5}$$

Thus, the unsteadiness of the laminar motion for $\mathrm{Re} > \mathrm{Re}_{\mathrm{cr}}$ leads to the appearance of a non-stationary periodic motion. When Re is close to $\mathrm{Re}_{\mathrm{cr}}$, this motion can be represented as a superposition of a stationary motion $\boldsymbol{v}_0(x, y, z)$ over a periodic motion $\boldsymbol{v}_1(x, y, z, t)$, having a small but finite amplitude which varies with Re directly as $\sqrt{\mathrm{Re} - \mathrm{Re}_{\mathrm{cr}}}$. The velocity distribution in this motion has therefore the form

$$\boldsymbol{v}_1 = f(x, y\ z)\, e^{-i(\omega_1 t + \beta_1)}, \tag{6}$$

where β_1 is a constant initial phase. When the differences $\mathrm{Re} - \mathrm{Re}_{\mathrm{cr}}$ are large, there is no longer any sense in separating the velocities into two parts $\boldsymbol{v}_0$ and $\boldsymbol{v}_1$. Here we have to deal simply with a periodic motion of frequency ω_1. If instead of the time the phase $\varphi_1 \equiv \omega_1 t + \beta_1$ is used as the independent variable, the function $\boldsymbol{v}(x, y, z, \varphi_1)$ may be said to be a periodic function of φ_1 with a period 2π, but no simple trigonometric function. It can be represented as a Fourier series

$$\boldsymbol{v} = \sum A_p(x, y, z)\, e^{-i\varphi_1 p} \tag{7}$$

(the summation is carried out over all positive and negative integers p).

The essential fact is that only the absolute value of the factor, but not its phase are determined by the equation (3). The phase φ_1 remains in substance indefinite and depends upon the initial conditions which are a matter of change and may cause β_1 to take any value. It will be obvious that the periodic motion under consideration is not determined uniquely by the given stationary boundary conditions of motion; one quantity, the phase, remains arbitrary. This motion may be said to have one degree of freedom, whereas stationary motion is completely determined by the given boundary conditions, and enjoys not a single degree of freedom.

As Re is further increased, this periodic motion, too, eventually becomes unsteady. The investigation of its unsteadiness should be conducted in a manner analogous to that described above. The role of the principal motion is now played by the periodic motion $\boldsymbol{v}_0(x, y, z, t)$ of frequency ω_1. Substituting $\boldsymbol{v} = \boldsymbol{v}_0 + \boldsymbol{v}_2$ with small $\boldsymbol{v}_2$ into the equation of motion, we shall again obtain for $\boldsymbol{v}_2$ a linear equation, but this time the coefficients of this equation are not only functions of the co-ordinates, but of time also; with respect to time, they are periodic functions with a period $2\pi/\omega_1$. The solution of such an equation should be sought in the form $\boldsymbol{v}_2 = \boldsymbol{\Pi}(x, y, z, t)e^{-i\Omega t}$ where $\boldsymbol{\Pi}(x, y, z, t)$ is a periodical function of time (with a period $2\pi/\omega_1$). Unsteadiness sets in again when the frequency $\Omega_2 = \omega_2 + i\gamma_2$ turns up whose imaginary part γ_2 is positive and the corresponding real part ω_2 determines then the newly appearing frequency.

The result is a quasi-periodic motion characterised by two different periods. It involves two arbitrary quantities (phases), i.e. has two degrees of freedom.

In the course of a further increase of the Reynolds number more and more new periods appear in succession, and the motion assumes an involved character typical of a developed turbulence. For every value of Re the motion has a definite number of degrees of freedom; in the limit as Re tends to infinity, the number of degrees of freedom becomes likewise infinitely large. With n degrees of freedom the velocity distribution is described by an expression of the type

$$\boldsymbol{v}(x, y, z, t) = \sum_{p_1, p_2, \ldots, p_n} \boldsymbol{A}_{p_1, \ldots, p_n}(x, y, z)\, e^{-i\sum_{i=1}^{n} p_i \varphi_i} \tag{8}$$

(summation over all integral numbers $p_1, p_2, \ldots, p_n$) where the phases are $\varphi_i = \omega_i t + \beta_i$; it contains n arbitrary initial phases β_i. The frequencies ω_i being incommensurable, it will be apparent that during a sufficiently long interval of time the fluid will pass through the states which are as close as we will it to a state set beforehand by choosing freely a set of simultaneous values for the phases φ_i. It should, of course, be borne in mind that the states whose phases differ only by a multiple of 2π are identical physically. So a turbulent motion is to a certain extent a quasi-periodical motion.

The setting-up of a turbulent regime has a somewhat different character in those exceptional cases (the Poiseuille motion and others) where the laminar motion remains stable with respect to infinitesimal perturbations, no matter how large are the Reynolds numbers. If the latter are sufficiently small, no non-stationary motion is possible here at all; a steady non-stationary motion becomes possible only after a certain value of Re is reached, which is here in the nature of a critical value. With very large Reynolds numbers, the stationary motion may, notoriously, be materialised only if one is careful enough in eliminating the perturbations superimposed upon the motion. Contrary to this, if Re is close to Re_{cr}, the non-stationary motion is diffcult to materialise. It may be thought therefore that the true value of Re_{cr}, say, in the case of Poiseuille motion, lies in any case below the value generally adopted at present. As for the properties of the turbulent motion that appears here with $\mathrm{Re} > \mathrm{Re}_{cr}$, it should, contrary to the preceding case, enjoy from the outset a large number of degrees of freedom.

Finally, in principle, there is one more possible type of the loss of steadiness by the laminar motion; this corresponds to the case where the coefficient before $|A|^4$ in (3) is positive, so that

$$\frac{d|A|^2}{dt} = 2\gamma\,|A|^2 + \alpha\,|A|^4$$

with positive α. If Re is somewhat smaller than Re_{cr}, the term of the second order is negative (since $\gamma_1 < 0$ for $\mathrm{Re} < \mathrm{Re}_{cr}$). But, the term of the fourth order being positive, the derivative $d|A|^2/dt$ will become positive when the amplitude of $|A|$ is sufficiently large. This means that the motion becomes steady with respect to sufficiently large perturbations even for $\mathrm{Re} > \mathrm{Re}_{cr}$. Thus, this type of unsteadiness is characterised by the fact that for a certain value, Re_{cr}, of the Reynolds number the motion becomes unsteady with respect to infinitesimal disturbances, but even with $\mathrm{Re} > \mathrm{Re}_{cr}$ there is unsteadiness in response to perturbations of a finite magnitude. In this case along with the above-mentioned critical Reynolds number there should exist another, "lower" number which determines the instant of appearance of stable non-stationary solutions of the equations of motion.

A Mathematical Example Displaying Features of Turbulence

By EBERHARD HOPF

Introduction

Before entering upon the study of the example in question we want to make some introductory remarks about the actual hydrodynamic problems, in particular, about what is known and what is conjectured concerning the future behavior of the solutions. Consider an incompressible and homogeneous viscous fluid within given material boundaries under given exterior forces. The boundary conditions and the outside forces are assumed to be stationary, i.e. independent of time. For that, it is not necessary that the walls be at rest themselves. Parts of the material walls may move in a stationary movement provided that the geometrical boundary as a whole stays at rest. An instance is a fluid between two concentric cylinders rotating with prescribed constant velocities or a fluid between two parallel planes which are translated within themselves with given constant velocities. As to the stationarity of the exterior forces we may cite the case of a flow through an infinitely long pipe with a pressure drop (regarded as an outside force). In this case the pressure drop is required to be a given constant independent of time.

Each motion of the fluid that is theoretically possible under these conditions satisfies the Navier-Stokes equations ($\rho = 1$)

$$\frac{\partial u_i}{\partial t} = -\sum_{\nu} u_\nu \frac{\partial u_i}{\partial x_\nu} - \frac{\partial p}{\partial x_i} + \mu \Delta u_i \tag{0.1}$$

together with the incompressibility condition

$$\sum \frac{\partial u_\nu}{\partial x_\nu} = 0 \tag{0.2}$$

and the given stationary boundary conditions. To an arbitrarily prescribed initial velocity field $u(x, 0)$ satisfying (0.2) and the boundary conditions there is expected to belong a unique solution $u(x, t; \mu)$ ($t \geq 0$) of (0.1) and (0.2) that fulfills these boundary conditions. The pressure $p(x, t; \mu)$ may be considered as an auxiliary variable which, at every moment t is (up to an additive constant) perfectly well determined by the instantaneous velocity field $u(x, t)$ (solution of a Neumann problem of potential theory). If p is eliminated in this manner the Navier-Stokes equations appear in the form of an integrodifferential space-time system for the u_i alone where the right hand sides consist of first and second degree terms in the u_ν .

It is convenient to visualize the solutions in the phase space Ω of the problem. A phase or state of the fluid is a vector field $u(x)$ in the fluid space that satisfies (0.2) and the boundary conditions. The totality Ω of these phases is therefore a functional space with infinitely many dimensions. A flow of the fluid represents a point motion in Ω and the totality of these phase motions forms a stationary flow in the phase space Ω, which, of course, is to be distinguished from the fluid flow itself. What is the asymptotic future behavior of the solutions, how does the phase flow behave for $t \to \infty$? And how does this behavior change as μ decreases more and more? How do the solutions which represent the observed turbulent motions fit into the phase picture? The great mathematical difficulties of these important problems are well known and at present the way to a successful attack on them seems hopelessly barred. There is no doubt, however, that many characteristic features of the hydrodynamical phase flow occur in a much larger class of similar problems governed by nonlinear space-time systems. In order to gain insight into the nature of hydrodynamical phase flows we are, at present, forced to find and to treat simplified examples within that class. The study of such models has been originated by J. M. Burgers in a well known memoir.[1] His principal example is essentially

$$\frac{\partial v}{\partial t} = -v\frac{\partial v}{\partial x} + w\frac{\partial w}{\partial x} + v - w + \mu\frac{\partial^2 v}{\partial x^2}$$

$$\frac{\partial w}{\partial t} = w\frac{\partial v}{\partial x} + v\frac{\partial w}{\partial x} + v + w + \mu\frac{\partial^2 w}{\partial x^2}$$

where $0 \leq x \leq 1$ and where the boundary conditions are $v = w = 0$ at $x = 0$ and $x = 1$. Though simpler in form than the hydrodynamic equations this example presents essentially the same difficulties and the future behavior of the solutions for small values of μ still is an unsolved problem.

In this paper another nonlinear example is presented and studied that differs from Burgers' model in that the future behavior of its solutions can be completely determined. In this respect our example seems to us to be the first of its kind. The detailed study of this space-time system reveals geometrical features of the phase flow which come close to the qualitative picture we believe to prevail in the hydrodynamic cases. It must, however, be said that, for reasons to appear later in the paper, the analogy does not extend to the quantitative relations found to hold in turbulent fluid flow.

The observational facts about hydrodynamic flow reduced to the case of fixed side conditions and with μ as the only variable parameter are essentially these: For μ sufficiently large, $\mu > \mu_0$, the only flow observed in the long run is a stationary one (laminar flow). This flow is stable against arbitrary initial

[1] J. M. Burgers, *Mathematical examples illustrating relations occurring in the theory of turbulent fluid motion*. Akademie van Wetenschappen, Amsterdam, Eerste Sectie, Deel XVII, No. 2, pp. 1–53, 1939.

disturbances. Theoretically, the corresponding exact solution is known to exist for every value of $\mu > 0$ and its stability in the large can be rigorously proved, though only for sufficiently large values of μ. The corresponding phase flow in phase space Ω thus possesses an extremely simple structure. The laminar solution represents a single point in Ω invariant under the phase flow. For $\mu > \mu_0$, every phase motion tends, as $t \to \infty$, toward this laminar point. For sufficiently small values of μ, however, the laminar solution is never observed. The turbulent flow observed instead displays a complicated pattern of apparently irregularly moving "eddies" of varying sizes. The view widely held at present is that, for $\mu > 0$ having a fixed value, there is a "smallest size" of eddies present in the fluid depending on μ and tending to zero as $\mu \to 0$. Thus, macroscopically, the flow has the appearance of an intricate chance movement whereas, if observed with sufficient magnifying power, the regularity of the flow would never be doubted.

The qualitative mathematical picture which the author conjectures to correspond to the known facts about hydrodynamic flow is this: To the flows observed in the long run after the influence of the initial conditions has died down there correspond certain solutions of the Navier-Stokes equations. These solutions constitute a certain manifold $\mathfrak{M} = \mathfrak{M}(\mu)$ in phase space invariant under the phase flow. Presumably owing to viscosity $\mathfrak{M}$ has a finite number $N = N(\mu)$ of dimensions. This effect of viscosity is most evident in the simplest case of μ sufficiently large. In this case $\mathfrak{M}$ is simply a single point, $N = 0$. Also the complete stability of $\mathfrak{M}$ is in this simplest case obviously due to viscosity. On the other hand, for smaller and smaller values of μ, the increasing chance character of the observed flow suggests that $N(\mu) \to \infty$ monotonically as $\mu \to 0$. This can happen only if at certain "critical" values

$$\mu_0 > \mu_1 > \mu_2 > \cdots \to 0$$

the number $N(\mu)$ jumps. The manifold $\mathfrak{M}(\mu)$ itself presumably changes analytically as long as no critical value is passed. Now we believe that when μ decreases through such a value μ_k a continuous branching phenomenon occurs. The manifold $\mathfrak{M}(\mu)$ of motions observed in the long run (more precisely its analytical continuation for $\mu < \mu_k$) loses its stability. The notion of stability here refers to the whole manifold and not to the single motions contained in it. The loss of stability implies that the motions on the analytically continued $\mathfrak{M}$ are no longer observed. What we observe after passing μ_k is not the analytical continuation of the previous $\mathfrak{M}$ but a new manifold $\mathfrak{M}(\mu)$ continuously branching away from $\mathfrak{M}(\mu_k)$ and slightly swelling in a new dimension. This new $\mathfrak{M}(\mu)$ takes over stability from the old one. Stability here means that the "majority" of phase motions tends for $t \to \infty$ toward $\mathfrak{M}(\mu)$. We must expect that there is a "minority" of exceptional motions that do not converge toward $\mathfrak{M}$ (for instance the motions on the analytical continuation of the old $\mathfrak{M}$ and of all the other manifolds left over from all the previous branchings). The simplest case of such a bifurcation with corresponding change of stability is the branching of a periodic motion from a stationary one. This case is clearly observed in the flow around an obstacle (transition from the laminar flow to a periodic one with periodic discharges of eddies from the boundary). The next simplest case is the branching of a one-parameter family of almost periodic solutions from a periodic one. The new solutions are expressed by functions

$$u(\phi_1 , \phi_2 ; \mu)$$

periodic in each ϕ with period 2π where

$$\phi_i = a_i t + c_i , \qquad a_i = a_i(\mu),$$

and where the c_i are arbitrary constants (we can without loss of generality assume $c_1 = 0$). The functions f with ϕ_i arbitrary describe the manifold $\mathfrak{M}(\mu)$ which, in our case, is of the type of a torus. If $\mathfrak{M}$, quite generally, continuously develops out of the laminar point there is a reasonable expectation that $\mathfrak{M}$ is a multidimensional torus-manifold described by functions

$$u(\phi_1 , \cdots , \phi_N ; \mu)$$

with period 2π in each of the ϕ and that the turbulent solutions are given by linear functions $\phi_i = a_i(\mu)t + c_i$ as before. This is what happens in our example which precisely exhibits this phenomenon of continuous growth of almost periodic solutions out of the laminar one with an infinite succession of branchings of the type described above.

The geometrical picture of the phase flow is, however, not the most important problem of the theory of turbulence. Of greater importance is the determination of the probability distributions associated with the phase flow, particularly of their asymptotic limiting forms for small μ. In the case of our example these distributions have limiting forms (normal distribution). Recent investigations, however, suggest that there are essential deviations from normality in the hydrodynamic case. It seems that the influence of the second degree terms is in this case essentially different and much more complicated than in the case of our over-simplified model.

Another observation on our model case is this: If we proceed to the limit $\mu \to 0$ within the "observed," i.e. the turbulent solutions the turbulent fluctuations are found to disappear and we obtain a special stationary solution in the "ideal case" (equations with $\mu = 0$). This shows, by way of analogy, how important a role viscosity plays in turbulence.

Formulation of the Problem

The space of our model is a one-dimensional circular line and our space variable is an angular variable x mod 2π. All space functions are thus periodic functions of x with period 2π. For two arbitrary space functions f, g we denote

by
$$f \circ g = \frac{1}{2\pi} \int_0^{2\pi} f(x + y)g(y)\, dy$$

their convolution product which is again a space function. $f \circ 1$ is a constant, the mean value of $f(x)$ over a period. Throughout this paper z^* denotes the conjugate of the complex number z. Our integrodifferential system written in complex form is

$$\begin{aligned} \partial u/\partial t &= -z \circ z^* - u \circ 1 + \mu\, (\partial^2 u/\partial x^2), \\ \partial z/\partial t &= z \circ u^* + z \circ F^* + \mu\, (\partial^2 z/\partial x^2), \end{aligned} \tag{1.0}$$

where $\mu > 0$ is a parameter and where

$$F(x) = a(x) + ib(x) \tag{1.1}$$

is an arbitrarily given complex-valued space function. $F(x)$ is supposed to be an even and absolutely integrable function of x,

$$F(-x) = F(x). \tag{1.2}$$

Further conditions upon F will be stated when they are needed. The unknowns are the two complex-valued functions $u(x, t)$ and $z(x, t)$. The real equations into which (1.0) splits up are four in number.

In what follows we confine ourselves to those solutions of (1.0) for which u, z are even functions of x and for which u is real. It will be proved, by using (1.2), that any solution u, z which is even for $t = 0$ must be even for all $t > 0$ and that, for such a solution, u is always real if it is real for $t = 0$. If we confine ourselves to the even solutions with u real, (1.0) splits upon setting

$$z = v + iw$$

into three real equations for u, v, w

$$\begin{aligned} \partial u/\partial t &= -v \circ v - w \circ w - u \circ 1 + \mu\, (\partial^2 u/\partial x^2), \\ \partial v/\partial t &= v \circ u + v \circ a + w \circ b + \mu\, (\partial^2 v/\partial x^2), \\ \partial w/\partial t &= w \circ u - v \circ b + w \circ a + \mu\, (\partial^2 w/\partial x^2), \end{aligned} \tag{2.0}$$

where $F(x) = a(x) + ib(x)$. Our problem is to study the real solutions of (2.0) which are even functions of x with period 2π.

Another equivalent but in some respects more straightforward formulation of our problem is obtained if we confine ourselves to the interval

$$0 \le x \le \pi.$$

We look for the real solutions of (2.0), where

$$f \circ g = \frac{1}{2\pi} \int_0^{\pi} f(|\, x - y\,|)g(y)\, dy + \frac{1}{2\pi} \int_0^{\pi} f(\pi - |\, x - y\,|)g(\pi - y)\, dy,$$

satisfying the boundary conditions

(3.0) $\partial u/\partial x = \partial v/\partial x = \partial w/\partial x = 0$ at $x = 0$ and $x = \pi$.

The equivalence of this formulation is a consequence of the following facts. An even function of period 2π may be arbitrarily prescribed in the interval $(0, \pi)$. If the first derivative exists for all x, (3.0) must be satisfied. On the other hand, if a function in $[0, \pi]$ has a second derivative in this closed interval and if (3.0) is satisfied the corresponding even and periodic function has a second derivative at every x (in particular, the first derivative is continuous throughout). That the convolution of two even space functions reduces, for $0 \leq x \leq \pi$, to the expression mentioned above follows from a simple calculation.

In what follows we use the handier complex form (1.0) of the problem with restriction to the even solutions whether u is real or not. Our second degree terms share an important property with those in the hydrodynamic case. When the time derivative of the kinetic energy

$$\frac{1}{2}\int_0^{2\pi} (uu^* + zz^*)\, dx, \qquad \frac{1}{2}\int_0^{2\pi} (u^2 + v^2 + w^2)\, dx$$

is computed from the equations the third degree terms obtained on the right hand side are found to cancel. In our case this follows from the identity

$$\int_0^{2\pi} (f \circ g)h dx = \int_0^{2\pi} (f \circ h) g dx.$$

Our second degree terms, however, strikingly differ in nature from the hydrodynamic ones in that they are pure integrals. The fact that they are convolutions enables one to calculate the solutions by spatial Fourier analysis.

Properties of the Even Solutions of (1.0) and (2.0)

There is an infinite number of critical values $\mu_1 > \mu_2 > \mu_3 > \cdots \to 0$ for μ with the following properties. For $\mu > \mu_1$ there is a stationary "laminar solution" which is stable in the large for $t \to +\infty$, i.e. which will, as $t \to +\infty$, be approached by any other solution for the same value of μ. For $\mu_n > \mu > \mu_{n+1}$ there is an n-dimensional manifold of "turbulent solutions" essentially stable in the large as $t \to +\infty$, i.e. "almost" all other solutions will approach some of the turbulent solutions with the same value of μ. These turbulent solutions are represented by almost periodic functions of t of very simple type. These solutions persist for $\mu < \mu_{n+1}$ but they are no longer stable. For any given value of μ there is a definite statistical distribution within the totality of the "velocity fields," i.e. in the function space of the sets of three arbitrary functions $u(x)$, $v(x)$, $w(x)$ satisfying the given boundary conditions. These statistics are simply defined by time averages: The statistical average $\overline{\mathfrak{F}}$ of an arbitrary functional $\mathfrak{F}[u(x), v(x), w(x)]$ of the three functions is

$$\overline{\mathfrak{F}} = \lim_{T \to +\infty} \frac{1}{T} \int_0^T \mathfrak{F}[u(x, t), v(x, t), w(x, t)]\, dt,$$

where $u(x, t), \cdots$ is the solution of our equations with $u(x), \cdots$ as initial values for $t = 0$. If the function $F(x)$ given in our model satisfies certain requirements (which will be fulfilled "in general") the average $\overline{\mathfrak{F}}$ turns out to be essentially independent of the initial values $u(x)$, $v(x)$, $w(x)$ (property of ergodicity). Of course it will depend on μ. The probability that the point $[u(x), \cdots]$ of our function space falls into a given subset of this space is defined by the average of the "characteristic functional" of this set (1 inside, 0 outside). The statistics defined in this natural manner varies with μ. For $\mu > \mu_1$ it is a trivial one while, for μ decreasing, it will be more complex. The following fact must, however, be noted. If the real part $a(x)$ of our given function $F(x)$ is not too smooth a function our probability distribution becomes, in the limit $\mu \to 0$, more and more a Gaussian or normal one. The values u, v, w of the solutions at arbitrarily given fixed points $x_1, x_2, \cdots, x_j$ may be regarded as chance variables. If these chance variables are denoted by $u_1, u_2, \cdots, u_k$ respectively then the probability that

$$u_1 > a_1,\ u_2 > a_2,\ \cdots,\ u_k > a_k$$

differs, as $\mu \to 0$, less and less (uniformly with respect to the a's) from

$$K \int_{a_1}^{\infty} \cdots\cdots \int_{a_k}^{\infty} \exp\left\{-\frac{1}{2} \sum A_{\nu\mu}(\xi_\nu - m_\nu)(\xi_\mu - m_\mu)\right\} d\xi_1 \cdots d\xi_k\,,$$

where K is a constant chosen in such a way that the expression equals 1 if each $a_\nu = -\infty$. We have $m_\nu = \overline{u}_\nu$ and the matrix (A_{ik}) is the inverse of the correlation matrix $\overline{(u_i - m_i)(u_k - m_k)}$. For very small values of μ the distribution is therefore nearly determined by its moments of first and second degree. In our case these moments, i.e. the mean values $\overline{u(x)}$, $\overline{v(x)}$, $\overline{w(x)}$ and the correlation functions $\overline{u(x)u(x')}$ etc., are easily evaluated. Their asymptotic forms for $\mu \to 0$ will be investigated. Appreciable statistical interdependence is found only in points x, x' sufficiently close to each other. Approximately the correlations depend only on the mutual differences $|x' - x|$. This is analogous to the tendency of turbulence toward spatial isotropy and homogeneity in certain hydrodynamic cases.

The asymptotic evaluation of these moments is the cardinal problem in the theory of turbulent flow in hydrodynamics. Its great mathematical difficulties apparently arise from the fact that the spatial Fourier components of the motion are interrelated with each other, in contrast to our simple model where there is no interaction between the different frequencies of the spatial Fourier pattern. The mathematical nature and the formulation of this problem will be the subject of a later paper on the foundations of statistical hydrodynamics. Still, the continued study of models of our particular kind seems not without interest to the author. There might perhaps be a starting point in devising and discussing simple models with slight interaction.

Commun. math. Phys. 20, 167—192 (1971)

On the Nature of Turbulence

DAVID RUELLE and FLORIS TAKENS*

I.H.E.S., Bures-sur-Yvette, France

Received October 5, 1970

Abstract. A mechanism for the generation of turbulence and related phenomena in dissipative systems is proposed.

§ 1. Introduction

If a physical system consisting of a viscous fluid (and rigid bodies) is not subjected to any external action, it will tend to a state of rest (equilibrium). We submit now the system to a steady action (pumping, heating, etc.) measured by a parameter μ[1]. When $\mu = 0$ the fluid is at rest. For $\mu > 0$ we obtain first a *steady state*, i.e., the physical parameters describing the fluid at any point (velocity, temperature, etc.) are constant in time, but the fluid is no longer in equilibrium. This steady situation prevails for small values of μ. When μ is increased various new phenomena occur; (a) the fluid motion may remain steady but change its symmetry pattern; (b) the fluid motion may become periodic in time; (c) for sufficiently large μ, the fluid motion becomes very complicated, irregular and chaotic, we have *turbulence*.

The physical phenomenon of turbulent fluid motion has received various mathematical interpretations. It has been argued by Leray [9] that it leads to a breakdown of the validity of the equations (Navier-Stokes) used to describe the system. While such a breakdown may happen we think that it does not necessarily accompany turbulence. Landau and Lifschitz [8] propose that the physical parameters x describing a fluid in turbulent motion are quasi-periodic functions of time:

$$x(t) = f(\omega_1 t, \ldots, \omega_k t)$$

where f has period 1 in each of its arguments separately and the frequences $\omega_1, \ldots, \omega_k$ are not rationally related[2]. It is expected that k becomes large for large μ, and that this leads to the complicated and irregular behaviour

* The research was supported by the Netherlands Organisation for the Advancement of Pure Research (Z.W.O.).

[1] Depending upon the situation, μ will be the Reynolds number, Rayleigh number, etc.

[2] This behaviour is actually found and discussed by E. Hopf in a model of turbulence [A mathematical example displaying features of turbulence. Commun. Pure Appl. Math. 1, 303–322 (1948)].

characteristic of turbulent motion. We shall see however that a dissipative system like a viscous fluid will not in general have quasi-periodic motions[3]. The idea of Landau and Lifschitz must therefore be modified.

Consider for definiteness a viscous incompressible fluid occupying a region D of $\mathbb{R}^3$. If thermal effects can be ignored, the fluid is described by its velocity at every point of D. Let H be the space of velocity fields v over D; H is an infinite dimensional vector space. The time evolution of a velocity field is given by the Navier-Stokes equations

$$\frac{dv}{dt} = X_\mu(v) \tag{1}$$

where X_μ is a vector field over H. For our present purposes it is not necessary to specify further H or X_μ[4].

In what follows we shall investigate the nature of the solutions of (1), making only assumptions of a very general nature on X_μ. It will turn out that the fluid motion is *expected* to become chaotic when μ increases. This gives a justification for turbulence and some insight into its meaning. To study (1) we shall replace H by a finite-dimensional manifold[5] and use the qualitative theory of differential equations.

For $\mu = 0$, every solution $v(\cdot)$ of (1) tends to the solution $v_0 = 0$ as the time tends to $+\infty$. For $\mu > 0$ we know very little about the vector field X_μ. Therefore it is reasonable to study *generic* deformations from the situation at $\mu = 0$. In other words we shall ignore possibilities of deformation which are in some sense exceptional. This point of view could lead to serious error if, by some law of nature which we have overlooked, X_μ happens to be in a special class with exceptional properties[6]. It appears however that a three-dimensional viscous fluid conforms to the pattern of generic behaviour which we discuss below. Our discussion should in fact apply to very general dissipative systems[7].

The present paper is divided into two chapters. Chapter I is oriented towards physics and is relatively untechnical. In Section 2 we review

[3] Quasi-periodic motions occur for other systems, see Moser [10].

[4] A general existence and uniqueness theorem has not been proved for solutions of the Navier-Stokes equations. We assume however that we have existence and uniqueness locally, i.e., in a neighbourhood of some $v_0 \in H$ and of some time t_0.

[5] This replacement can in several cases be justified, see § 5.

[6] For instance the differential equations describing a Hamiltonian (conservative) system, have very special properties. The properties of a conservative system are indeed very different from the properties of a dissipative system (like a viscous fluid). If a viscous fluid is observed in an experimental setup which has a certain symmetry, it is important to take into account the invariance of X_μ under the corresponding symmetry group. This problem will be considered elsewhere.

[7] In the discussion of more specific properties, the behaviour of a viscous fluid may turn out to be nongeneric, due for instance to the local nature of the differential operator in the Navier-Stokes equations.

some results on differential equations; in Section 3–4 we apply these results to the study of the solutions of (1). Chapter II contains the proofs of several theorems used in Chapter I. In Section 5, center-manifold theory is used to replace H by a finite-dimensional manifold. In Sections 6–8 the theory of Hopf bifurcation is presented both for vector fields and for diffeomorphisms. In Section 9 an example of "turbulent" attractor is presented.

Acknowledgements. The authors take pleasure in thanking R. Thom for valuable discussion, in particular introducing one of us (F. T.) to the Hopf bifurcation. Some inspiration for the present paper was derived from Thom's forthcoming book [12].

Chapter I

§ 2. Qualitative Theory of Differential Equations

Let $B = \{x : |x| < R\}$ be an open ball in the finite dimensional euclidean space H[8]. Let X be a vector field with continuous derivatives up to order r on $\bar{B} = \{x : |x| \leqq R\}$, r fixed $\geqq 1$. These vector fields form a Banach space $\mathscr{B}$ with the norm

$$\|X\| = \sup_{1 \leqq i \leqq \nu} \sup_{|\varrho| \leqq r} \sup_{x \in B} \left| \frac{\partial^{|\varrho|}}{\partial x^{\varrho}} X^i(x) \right|$$

where

$$\frac{\partial^{|\varrho|}}{\partial x^{\varrho}} = \left(\frac{\partial}{\partial x^1} \right)^{\varrho_1} \cdots \left(\frac{\partial}{\partial x^{\nu}} \right)^{\varrho_{\nu}}$$

and $|\varrho| = \varrho_1 + \cdots + \varrho_\nu$. A subset E of $\mathscr{B}$ is called *residual* if it contains a countable intersection of open sets which are dense in $\mathscr{B}$. Baire's theorem implies that a residual set is again dense in $\mathscr{B}$; therefore a residual set E may be considered in some sense as a "large" subset of $\mathscr{B}$. A property of a vector field $X \in \mathscr{B}$ which holds on a residual set of $\mathscr{B}$ is called *generic.*

The *integral curve* $x(\cdot)$ through $x_0 \in B$ satisfies $x(0) = x_0$ and $dx(t)/dt = X(x(t))$; it is defined at least for sufficiently small $|t|$. The dependence of $x(\cdot)$ on x_0 is expressed by writing $x(t) = \mathscr{D}_{X,t}(x_0)$; $\mathscr{D}_{X,\cdot}$ is called *integral* of the vector field X; $\mathscr{D}_{X,1}$ is the time one integral. If $x(t) \equiv x_0$, i.e. $X(x_0) = 0$, we have a *fixed point* of X. If $x(\tau) = x_0$ and $x(t) \neq x_0$ for $0 < t < \tau$ we have a closed orbit of period τ. A natural generalization of the idea of *closed orbit* is that of *quasi-periodic* motion:

$$x(t) = f(\omega_1 t, \ldots, \omega_k t)$$

where f is periodic of period 1 in each of its arguments separately and the frequencies $\omega_1, \ldots, \omega_k$ are not rationally related. We assume that f is

[8] More generally we could use a manifold H of class C^r.

a C^k-function and its image a k-dimensional torus T^k imbedded in B. Then however we find that a quasi-periodic motion is non-generic. In particular for $k=2$, Peixoto's theorem[9] shows that quasi-periodic motions on a torus are in the complement of a dense open subset Σ of the Banach space of C^r vector fields on the torus: Σ consists of vector fields for which the non wandering set Ω[10] is composed of a finite number of fixed points and closed orbits only.

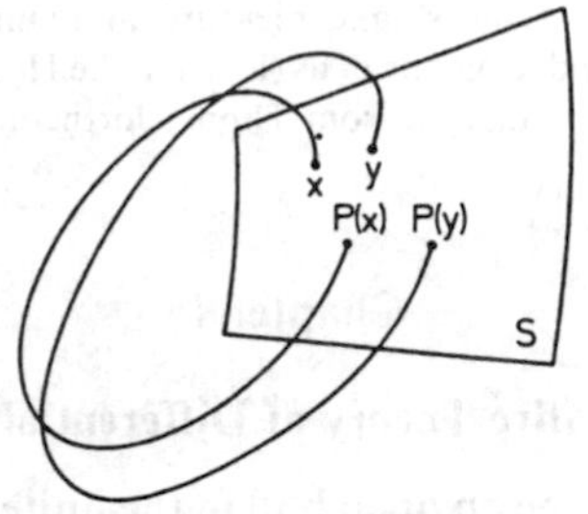

Fig. 1

As $t \to +\infty$, an integral curve $x(t)$ of the vector field X may be attracted by a fixed point or a closed orbit of the vector field, or by a more general attractor[11]. It will probably not be attracted by a quasi-periodic motion because these are rare. It is however possible that the orbit be attracted by a set which is not a manifold. To visualize such a situation in n dimensions, imagine that the integral curves of the vector field go roughly parallel and intersect transversally some piece of $n-1$-dimensional surface S (Fig. 1). We let $P(x)$ be the first intersection of the integral curve through x with S (P is a Poincaré map).

Take now $n-1=3$, and assume that P maps the solid torus Π_0 into itself as shown in Fig. 2,

$$P\Pi_0 = \Pi_1 \subset \Pi_0 .$$

The set $\bigcap_{n>0} P^n \Pi_0$ is an attractor; it is locally the product of a Cantor set and a line interval (see Smale [11], Section I.9). Going back to the vector field X, we have thus a "strange" attractor which is locally the product of a Cantor set and a piece of two-dimensional manifold. Notice that we

[9] See Abraham [1].

[10] A point x belongs to Ω (i.e. is non wandering) if for every neighbourhood U of x and every $T>0$ one can find $t>T$ such that $\mathcal{D}_{X,t}(U) \cap U \neq \emptyset$. For a quasi-periodic motion on T^k we have $\Omega = T^k$.

[11] A closed subset A of the non wandering set Ω is an attractor if it has a neighbourhood U such that $\bigcap_{t>0} \mathcal{D}_{X,t}(U) = A$. For more attractors than those described here see Williams [13].

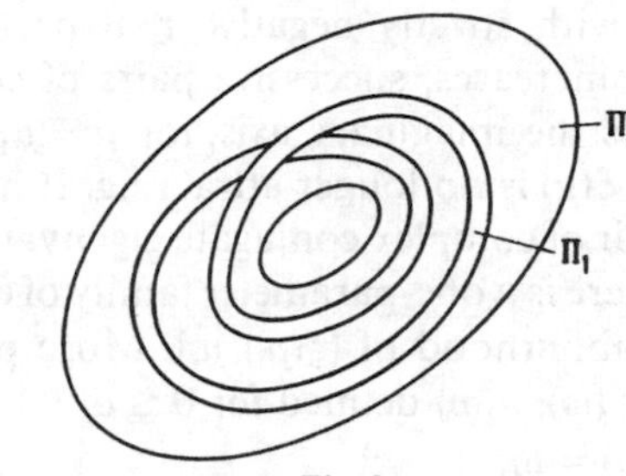

Fig. 2

keep the same picture if X is replaced by a vector field Y which is sufficiently close to X in the appropriate Banach space. An attractor of the type just described can therefore not be thrown away as non-generic pathology.

§ 3. A Mathematical Mechanism for Turbulence

Let X_μ be a vector field depending on a parameter μ[12]. The assumptions are the same as in Section 2, but the interpretation we have in mind is that X_μ is the right-hand side of the Navier-Stokes equations. When μ varies the vector field X_μ may change in a number of manners. Here we shall describe a pattern of changes which is physically acceptable, and show that it leads to something like turbulence.

For $\mu = 0$, the equation

$$\frac{dx}{dt} = X_\mu(x)$$

has the solution $x = 0$. We assume that the eigenvalues of the Jacobian matrix A^j_k defined by

$$A^j_k = \frac{\partial X^j_0}{\partial x^k}(0)$$

have all strictly negative real parts; this corresponds to the fact that the fixed point 0 is attracting. The Jacobian determinant is not zero and therefore there exists (by the implicit function theorem) $\xi(\mu)$ depending continuously on μ and such that

$$X_\mu(\xi(\mu)) = 0\,.$$

In the hydrodynamical picture, $\xi(\mu)$ describes a steady state.

We follow now $\xi(\mu)$ as μ increases. For sufficiently small μ the Jacobian matrix $A^j_k(\mu)$ defined by

$$A^j_k(\mu) = \frac{\partial X^j_\mu}{\partial x^k}(\xi(\mu)) \qquad (2)$$

[12] To be definite, let $(x, \mu) \to X_\mu(x)$ be of class C^r.

has only eigenvalues with strictly negative real parts (by continuity). We assume that, as μ increases, successive pairs of complex conjugate eigenvalues of (2) cross the imaginary axis, for $\mu = \mu_1, \mu_2, \mu_3, \ldots$[13]. For $\mu > \mu_1$, the fixed point $\xi(\mu)$ is no longer attracting. It has been shown by Hopf[14] that when a pair of complex conjugate eigenvalues of (2) cross the imaginary axis at μ_i, there is a one-parameter family of closed orbits of the vector field in a neighbourhood of $(\xi(\mu_i), \mu_i)$. More precisely there are continuous functions $y(\omega)$, $\mu(\omega)$ defined for $0 \leqq \omega < 1$ such that

(a) $y(0) = \xi(\mu_i)$, $\mu(0) = \mu_i$,

(b) the integral curve of $X_{\mu(\omega)}$ through $y(\omega)$ is a closed orbit for $\omega > 0$.

Generically $\mu(\omega) > \mu_i$ or $\mu(\omega) < \mu_i$ for $\omega \neq 0$. To see how the closed orbits are obtained we look at the two-dimensional situation in a neighbourhood of $\xi(\mu_1)$ for $\mu < \mu_1$ (Fig. 3) and $\mu > \mu_1$ (Fig. 4). Suppose that when μ crosses μ_1 the vector field remains like that of Fig. 3 at large distances of $\xi(\mu)$; we get a closed orbit as shown in Fig. 5. Notice that Fig. 4 corresponds to $\mu > \mu_1$ and that the closed orbit is attracting. Generally we shall assume that the closed orbits appear for $\mu > \mu_i$ so that the vector field at large distances of $\xi(\mu)$ remains attracting in accordance with physics. As μ crosses we have then replacement of an attracting fixed point by an attracting closed orbit. The closed orbit is physically interpreted as a periodic motion, its amplitude increases with μ.

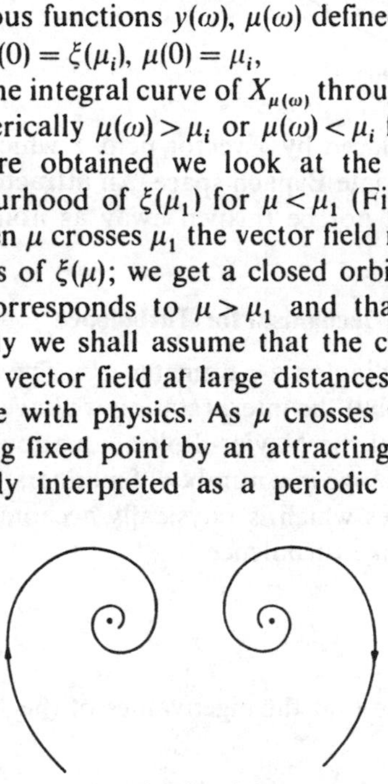

Figs. 3 and 4

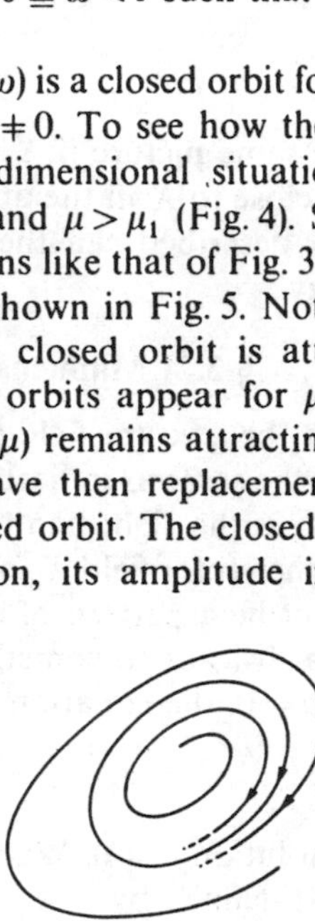

Fig. 5

§ 3a) Study of a Nearly Split Situation

To see what happens when μ crosses the successive μ_i, we let E_i be the two-dimensional linear space associated with the i-th pair of eigenvalues of the Jacobian matrix. In first approximation the vector field X_μ is, near $\xi(\mu)$, of the form

$$\tilde{X}_\mu(x) = \tilde{X}_{\mu 1}(x_1) + \tilde{X}_{\mu 2}(x_2) + \cdots \tag{3}$$

[13] Another less interesting possibility is that a real eigenvalue vanishes. When this happens the fixed point $\xi(\lambda)$ generically coalesces with another fixed point and disappears (this generic behaviour is changed if some symmetry is imposed to the vector field X_μ).

[14] Hopf [6] assumes that X is real-analytic; the differentiable case is treated in Section 6 of the present paper.

where $\tilde{X}_{\mu i}$, x_i are the components of $\tilde{X}_\mu$ and x in E_i. If μ is in the interval (μ_k, μ_{k+1}), the vector field $\tilde{X}_\mu$ leaves invariant a set $\tilde{T}^k$ which is the cartesian product of k attracting closed orbits $\Gamma_1, \ldots, \Gamma_k$ in the spaces $E_1, \ldots, E_k$. By suitable choice of coordinates on $\tilde{T}^k$ we find that the motion defined by the vector field on $\tilde{T}^k$ is quasi-periodic (the frequencies $\tilde{\omega}_1, \ldots, \tilde{\omega}_k$ of the closed orbits in $E_1, \ldots, E_k$ are in general not rationally related).

Replacing $\tilde{X}_\mu$ by X_μ is a perturbation. We assume that this perturbation is small, i.e. we assume that X_μ nearly splits according to (3). In this case there exists a C^r manifold (torus) T^k close to $\tilde{T}^k$ which is invariant for X_μ and attracting[15]. The condition that $X_\mu - \tilde{X}_\mu$ be small depends on how attracting the closed orbits $\Gamma_1, \ldots, \Gamma_k$ are for the vector field $\tilde{X}_{\mu 1}, \ldots, \tilde{X}_{\mu k}$; therefore the condition is violated if μ becomes too close to one of the μ_i.

We consider now the vector field X_μ restricted to T^k. For reasons already discussed, we do not expect that the motion will remain quasi-periodic. If $k = 2$, Peixoto's theorem implies that generically the non-wandering set of T^2 consists of a finite number of fixed points and closed orbits. What will happen in the case which we consider is that there will be one (or a few) attracting closed orbits with frequencies ω_1, ω_2 such that ω_1/ω_2 goes continuously through rational values.

Let $k > 2$. In that case, the vector fields on T^k for which the non-wandering set consists of a finite number of fixed points and closed orbits are no longer dense in the appropriate Banach space. Other possibilities are realized which correspond to a more complicated orbit structure; "strange" attractors appear like the one presented at the end of Section 2. Taking the case of T^4 and the C^3-topology we shall show in Section 9 that in any neighbourhood of a quasi-periodic $\tilde{X}$ there is an open set of vector fields with a strange attractor.

We propose to say that the motion of a fluid system is turbulent when this motion is described by an integral curve of the vector field X_μ which tends to a set A[16], and A is neither empty nor a fixed point nor a closed orbit. In this definition we disregard nongeneric possibilities (like A having the shape of the figure 8, etc.). This proposal is based on two things:

(a) We have shown that, when μ increases, it is not unlikely that an attractor A will appear which is neither a point nor a closed orbit.

[15] This follows from Kelley [7], Theorem 4 and Theorem 5, and also from recent work of Pugh (unpublished). That T^k is attracting means that it has a neighbourhood U such that $\bigcap_{t>0} \mathcal{D}_{X,t}(U) = T^k$. We cannot call T^k an attractor because it need not consist of non-wandering points.

[16] More precisely A is the ω^+ limit set of the integral curve $x(\cdot)$, i.e., the set of points ξ such that there exists a sequence (t_n) and $t_n \to \infty$, $x(t_n) \to \xi$.

(b) In the known generic examples where A is not a point or a closed orbit, the structure of the integral curves on or near A is complicated and erratic (see Smale [11] and Williams [13]).

We shall further discuss the above definition of turbulent motion in Section 4.

§ 3b) Bifurcations of a Closed Orbit

We have seen above how an attracting fixed point of X_μ may be replaced by an attracting closed orbit γ_μ when the parameter crosses the value μ_1 (Hopf bifurcation). We consider now in some detail the next bifurcation; we assume that it occurs at the value μ' of the parameter[17] and that $\lim_{\mu \to \mu'} \gamma_\mu$ is a closed orbit $\gamma_{\mu'}$ of $X_{\mu'}$[18].

Let Φ_μ be the Poincaré map associated with a piece of hypersurface S transversal to γ_μ, for $\mu \in (\mu_1, \mu']$. Since γ_μ is attracting, $p_\mu = S \cap \gamma_\mu$ is an attracting fixed point of Φ_μ for $\mu \in (\mu_1, \mu')$. The derivative $d\Phi_\mu(p_\mu)$ of Φ_μ at the point p_μ is a linear map of the tangent hyperplane to S at p_μ to itself.

We assume that the spectrum of $d\Phi_{\mu'}(p_{\mu'})$ consists of a finite number of isolated eigenvalues of absolute value 1, and a part which is contained in the open unit disc $\{z \in \mathbb{C} \mid |z| < 1\}$[19]. According to § 5, Remark (5.6), we may assume that S is finite dimensional. With this assumption one can say rather precisely what kind of generic bifurcations are possible for $\mu = \mu'$. We shall describe these bifurcations by indicating what kind of attracting subsets for X_μ (or Φ_μ) there are near $\gamma_{\mu'}$ (or $p_{\mu'}$) when $\mu > \mu'$.

Generically, the set E of eigenvalues of $d\Phi_{\mu'}(p_{\mu'})$, with absolute value 1, is of one of the following types:

1. $E = \{+1\}$,
2. $E = \{-1\}$,
3. $E = \{\alpha, \bar{\alpha}\}$ where α, $\bar{\alpha}$ are distinct.

For the cases 1 and 2 we can refer to Brunovsky [3]. In fact in case 1 the attracting closed orbit disappears (together with a hyperbolic closed orbit); for $\mu > \mu'$ there is no attractor of X_μ near $\gamma_{\mu'}$. In case 2 there is for $\mu > \mu'$ (or $\mu < \mu'$) an attracting (resp. hyperbolic) closed orbit near $\gamma_{\mu'}$, but the period is doubled.

If we have case 3 then Φ_μ has also for μ slightly bigger than μ' a fixed point p_μ; generically the conditions (a)', ..., (e) in Theorem (7.2) are

[17] In general μ' will differ from the value μ_2 introduced in § 3a).

[18] There are also other possibilities: If γ_μ tends to a point we have a Hopf bifurcation with parameter reversed. The cases where $\lim_{\mu \to \mu'} \gamma_\mu$ is not compact or where the period of γ_μ tends to ∞ are not well understood; they may or may not give rise to turbulence.

[19] If the spectrum of $d\Phi_{\mu'}(p_{\mu'})$ is discrete, this is a reasonable assumption, because for $\mu_1 < \mu < \mu'$ the spectrum is contained in the open unit disc.

satisfied. One then concludes that when $\gamma_{\mu'}$ is a "vague attractor" (i.e. when the condition (f) is satisfied) then, for $\mu > \mu'$, there is an attracting circle for Φ_μ; this amounts to the existence of an invariant and attracting torus T^2 for X_μ. If $\gamma_{\mu'}$ is not a "vague attractor" then, generically, X_μ has no attracting set near $\gamma_{\mu'}$ for $\mu > \mu'$.

§ 4. Some Remarks on the Definition of Turbulence

We conclude this discussion by a number of remarks:

1. The concept of genericity based on residual sets may not be the appropriate one from the physical view point. In fact the complement of a residual set of the μ-axis need not have Lebesgue measure zero. In particular the quasi-periodic motions which we had eliminated may in fact occupy a part of the μ-axis with non vanishing Lebesgue measure[20]. These quasi-periodic motions would be considered turbulent by our definition, but the "turbulence" would be weak for small k. There are arguments to define the quasi-periodic motions, along with the periodic ones, as non turbulent (see (4) below).

2. By our definition, a periodic motion (= closed orbit of X_μ) is not turbulent. It may however be very complicated and appear turbulent (think of a periodic motion closely approximating a quasi-periodic one, see § 3b) second footnote).

3. We have shown that, under suitable conditions, there is an attracting torus T^k for X_μ if μ is between μ_k and μ_{k+1}. We assumed in the proof that μ was not too close to μ_k or μ_{k+1}. In fact the transition from T^1 to T^2 is described in Section 3b, but the transition from T^k to T^{k+1} appears to be a complicated affair when $k > 1$. In general, one gets the impression that the situations not covered by our description are more complicated, hard to describe, and probably turbulent.

4. An interesting situation arises when statistical properties of the motion can be obtained, via the pointwise ergodic theorem, from an ergodic measure m supported by the attracting set A. An observable quantity for the physical system at a time t is given by a function x_t on H, and its expectation value is $m(x_t) = m(x_0)$. If m is "mixing" the time correlation functions $m(x_t y_0) - m(x_0)\, m(y_0)$ tend to zero as $t \to \infty$. This situation appears to prevail in turbulence, and "pseudo random" variables with correlation functions tending to zero at infinity have been studied by Bass[21]. With respect to this property of time correlation functions the quasi-periodic motions should be classified as non turbulent.

[20] On the torus T^2, the rotation number ω is a continuous function of μ. Suppose one could prove that, on some μ-interval, ω is non constant and is absolutely continuous with respect to Lebesgue measure; then ω would take irrational values on a set of non zero Lebesgue measure.

[21] See for instance [2].

5. In the above analysis the detailed structure of the equations describing a viscous fluid has been totally disregarded. Of course something is known of this structure, and also of the experimental conditions under which turbulence develops, and a theory should be obtained in which these things are taken into account.

6. Besides viscous fluids, other dissipative systems may exhibit time-periodicity and possibly more complicated time dependence; this appears to be the case for some chemical systems[22].

Chapter II

§ 5. Reduction to Two Dimensions

Definition (5.1). Let $\Phi: H \to H$ be a C^1 map with fixed point $p \in H$, where H is a Hilbert space. The spectrum of Φ at p is the spectrum of the induced map $(d\Phi)_p: T_p(H) \to T_p(H)$.

Let X be a C^1 vectorfield on H which is zero in $p \in H$. For each t we then have $d(\mathscr{D}_{X,t})_p: T_p(H) \to T_p(H)$, induced by the time t integral of X. Let $L(X): T_p(H) \to T_p(H)$ be the unique continuous linear map such that $d(\mathscr{D}_{X,t})_p = e^{t \cdot L(X)}$.

We define the spectrum of X at p to be the spectrum of $L(X)$, (note that $L(X)$ also can be obtained by linearizing X).

Proposition (5.2). *Let X_μ be a one-parameter family of C^k vectorfields on a Hilbert space H such that also X, defined by $X(h, \mu) = (X_\mu(h), 0)$, on $H \times \mathbb{R}$ is C^k. Suppose:*

(a) *X_μ is zero in the origin of H.*

(b) *For $\mu < 0$ the spectrum of X_μ in the origin is contained in* $\{z \in \mathbb{C} \mid \mathrm{Re}(z) < 0\}$.

(c) *For $\mu = 0$, resp. $\mu > 0$, the spectrum of X_μ at the origin has two isolated eigenvalues $\lambda(\mu)$ and $\overline{\lambda(\mu)}$ with multiplicity one and* $\mathrm{Re}(\lambda(\mu)) = 0$, *resp.* $\mathrm{Re}(\lambda(\mu)) > 0$. *The remaining part of the spectrum is contained in* $\{z \in \mathbb{C} \mid \mathrm{Re}(z) < 0\}$.

Then there is a (small) 3-dimensional C^k-manifold $\tilde{V}^c$ of $H \times \mathbb{R}$ containing $(0, 0)$ *such that:*

1. *$\tilde{V}^c$ is locally invariant under the action of the vectorfield X (X is defined by $X(h, \mu) = (X_\mu(h), 0)$); locally invariant means that there is a neighbourhood U of $(0, 0)$ such that for $|t| \leqq 1$, $\tilde{V}^c \cap U = \mathscr{D}_{X,t}(\tilde{V}^c) \cap U$.*

2. *There is a neighbourhood U' of $(0, 0)$ such that if $p \in U'$, is recurrent, and has the property that $\mathscr{D}_{X,t}(p) \in U'$ for all t, then $p \in \tilde{V}^c$*

3. *in $(0, 0)$ $\tilde{V}^c$ is tangent to the μ axis and to the eigenspace of $\lambda(0)$, $\overline{\lambda(0)}$.*

---

[22] See Pye, K., Chance, B.: Sustained sinusoidal oscillations of reduced pyridine nucleotide in a cell-free extract of Saccharomyces carlbergensis. Proc. Nat. Acad. Sci. U.S.A. 55, 888–894 (1966).

Proof. We construct the following splitting $T_{(0,0)}(H \times \mathbb{R}) = V^c \oplus V^s$: V^c is tangent to the μ axis and contains the eigenspace of $\lambda(\mu)$, $\overline{\lambda(\mu)}$; V^s is the eigenspace corresponding to the remaining (compact) part of the spectrum of $L(X)$. Because this remaining part is compact there is a $\delta > 0$ such that it is contained in $\{z \in \mathbb{C} \mid \mathrm{Re}(z) < -\delta\}$. We can now apply the centermanifold theorem [5], the proof of which generalizes to the case of a Hilbert space, to obtain $\tilde{V}^c$ as the centermanifold of X at $(0,0)$ [by assumption X is C^k, so $\tilde{V}^c$ is C^k; if we would assume only that, for each μ, X_μ is C^k (and X only C^1), then $\tilde{V}^c$ would be C^1 but, for each μ_0, $\tilde{V}^c \cap \{\mu = \mu_0\}$ would be C^k].

For positive t, $d(\mathscr{D}_{X,t})_{0,0}$ induces a contraction on V^s (the spectrum is contained in $\{z \in \mathbb{C} \mid |z| < e^{-\delta t}\}$). Hence there is a neighbourhood U' of $(0,0)$ such that

$$U' \cap \left[\bigcap_{t=1}^{\infty} \mathscr{D}_{X,t}(U')\right] \subset (U' \cap \tilde{V}^c).$$

Now suppose that $p \in U'$ is recurrent and that $\mathscr{D}_{X,t}(p) \in U'$ for all t. Then given $\varepsilon > 0$ and $N > 0$ there is a $t > N$ such that the distance between p and $\mathscr{D}_{X,t}(p)$ is $< \varepsilon$. It then follows that $p \in (U' \cap \tilde{V}^c) \subset \tilde{V}^c$ for U' small enough. This proves the proposition.

Remark (5.3). The analogous proposition for a one parameter set of diffeomorphisms Φ_μ is proved in the same way. The assumptions are then:

(a)′ The origin is a fixed point of Φ_μ.

(b)′ For $\mu < 0$ the spectrum of Φ_μ at the origin is contained in $\{z \in \mathbb{C} \mid |z| < 1\}$.

(c)′ For $\mu = 0$ resp. $\mu > 0$ the spectrum of Φ_μ at the origin has two isolated eigenvalues $\lambda(\mu)$ and $\overline{\lambda(\mu)}$ with multiplicity one and $|\lambda(\mu)| = 1$ resp. $|\lambda(\mu)| > 1$. The remaining part of the spectrum is contained in $\{z \in \mathbb{C} \mid |z| < 1\}$.

One obtains just as in Proposition (5.2) a 3-dimensional center manifold which contains all the local recurrence.

Remark (5.4). If we restrict the vectorfield X, or the diffeomorphism Φ [defined by $\Phi(h, \mu) = (\Phi_\mu(h), \mu)$], to the 3-dimensional manifold $\tilde{V}^c$ we have locally the same as in the assumptions (a), (b), (c), or (a)′, (b)′, (c)′ where now the Hilbert space has dimension 2. So if we want to prove a property of the local recurrent points for a one parameter family of vectorfield, or diffeomorphisms, satisfying (a) (b) and (c), or (a)′, (b)′ and (c)′, it is enough to prove it for the case where $\dim(H) = 2$.

Remark (5.5). Everything in this section holds also if we replace our Hilbert space by a Banach space with C^k-norm; a Banach space B has C^k-norm if the map $x \to \|x\|$, $x \in B$ is C^k except at the origin. This C^k-norm is needed in the proof of the center manifold theorem.

Remark (5.6). The Propositions (5.2) and (5.3) remain true if
1. we drop the assumptions on the spectrum of X_μ resp. Φ_μ for $\mu>0$.
2. we allow the spectrum of X_ϱ resp. Φ_ϱ to have an arbitrary but finite number of isolated eigenvalues on the real axis resp. the unit circle. The dimension of the invariant manifold $\tilde{V}^c$ is then equal to that number of eigenvalues plus one.

§ 6. The Hopf Bifurcation

We consider a one parameter family X_μ of C^k-vectorfields on $\mathbb{R}^2$, $k \geq 5$, as in the assumption of proposition (5.2) (with $\mathbb{R}^2$ instead of H); $\lambda(\mu)$ and $\overline{\lambda(\mu)}$ are the eigenvalues of X_μ in $(0,0)$. Notice that with a suitable change of coordinates we can achieve $X_\mu = (\mathrm{Re}\lambda(\mu)x_1 + \mathrm{Im}\lambda(\mu)x_2)\dfrac{\partial}{\partial x_1}$ $+(-\mathrm{Im}\lambda(\mu)x_1 + \mathrm{Re}\lambda(\mu)x_2)\dfrac{\partial}{\partial x_2}$ + terms of higher order.

Theorem (6.1).(Hopf [6]). *If* $\left(\dfrac{d(\lambda(\mu))}{d\mu}\right)_{\mu=0}$ *has a positive real part, and if* $\lambda(0) \neq 0$, *then there is a one-parameter family of closed orbits of* $X(=(X_\mu, 0))$ *on* $\mathbb{R}^3 = \mathbb{R}^2 \times \mathbb{R}^1$ *near* $(0,0,0)$ *with period near* $\dfrac{2\pi}{|\lambda(0)|}$; *there is a neighbourhood* U *of* $(0,0,0)$ *in* $\mathbb{R}^3$ *such that each closed orbit of* X, *which is contained in* U, *is a member of the above family.*

If $(0,0)$ *is a "vague attractor" (to be defined later) for* X_0, *then this one-parameter family is contained in* $\{\mu>0\}$ *and the orbits are of attracting type.*

Proof. We first have to state and prove a lemma on polar-coordinates:

Lemma (6.2). *Let* X *be a* C^k *vectorfield on* $\mathbb{R}^2$ *and let* $X(0,0)=0$. *Define polar coordinates by the map* $\Psi: \mathbb{R}^2 \to \mathbb{R}^2$, *with* $\Psi(r,\varphi) = (r\cos\varphi, r\sin\varphi)$. *Then there is a unique* C^{k-2}*-vectorfield* $\tilde{X}$ *on* $\mathbb{R}^2$, *such that* $\Psi_*(\tilde{X}) = X$ *(i.e. for each* (r,φ) $d\Psi(\tilde{X}(r,\varphi)) = X(r\cos\varphi, r\sin\varphi)$*).*

Proof of Lemma (6.2). We can write

$$\begin{aligned} X &= X_1 \frac{\partial}{\partial x_1} + X_2 \frac{\partial}{\partial x_2} \\ &= \frac{x_1 X_1 + x_2 X_2}{\sqrt{x_1^2 + x_2^2}} \left(\frac{1}{\sqrt{x_1^2+x_2^2}} \left(x_1 \frac{\partial}{\partial x_1} + x_2 \frac{\partial}{\partial x_2} \right)\right) \\ &\quad + \frac{(-x_2 X_1 + x_1 X_2)}{(x_1^2 + x_2^2)} \left(-x_2 \frac{\partial}{\partial x_1} + x_1 \frac{\partial}{\partial x_2} \right) \\ &= \frac{f_r(x_1,x_2)}{r} \cdot \Psi_*(\tilde{Z}_r) + \frac{f_\varphi(x_1,x_2)}{r^2} \Psi_*(\tilde{Z}_\varphi). \end{aligned}$$

Where $\tilde{Z}_r\left(=\dfrac{\partial}{\partial r}\right)$ and $\tilde{Z}_\varphi\left(=\dfrac{\partial}{\partial \varphi}\right)$ are the "coordinate vectorfields" with respect to (r, φ) and $r = \pm\sqrt{x_1^2 + x_2^2}$. (Note that r and $\Psi_*(\tilde{Z}_r)$ are bi-valued.)

Now we consider the functions $\Psi^*(f_r) = f_r \circ \Psi$ and $\Psi^*(f_\varphi)$. They are zero along $\{r = 0\}$; this also holds for $\dfrac{\partial}{\partial r}(\Psi^*(f_r))$ and $\dfrac{\partial}{\partial r}(\Psi^*(f_\varphi))$. By the division theorem $\dfrac{\Psi^*(f_r)}{r}$, resp. $\dfrac{\Psi^*(f_\varphi)}{r^2}$, are C^{k-1} resp. C^{k-2}.

We can now take $\tilde{X} = \dfrac{\Psi^*(f_r)}{r}\tilde{Z}_r + \dfrac{\Psi^*(f_\varphi)}{r^2}\tilde{Z}_\varphi$; the uniqueness is evident.

Definition (6.3). We define a Poincaré map P_X for a vectorfield X as in the assumptions of Theorem (6.1):

P_X is a map from $\{(x_1, x_2, \mu) \mid |x_1| < \varepsilon,\ x_2 = 0,\ |\mu| \leq \mu_0\}$ to the (x_1, μ) plane; μ_0 is such that $\operatorname{Im}(\lambda(\mu)) \neq 0$ for $|\mu| \leq \mu_0$; ε is sufficiently small. P_X maps (x_1, x_2, μ) to the first intersection point of $\mathscr{D}_{X,t}(x_1, x_2, \mu)$, $t > 0$, with the (x_1, μ) plane, for which the sign of x_1 and the x_1 coordinate of $\mathscr{D}_{X,t}(x_1, x_2, \mu)$ are the same.

Remark (6.4). P_X preserves the μ coordinate. In a plane $\mu =$ constant the map P_X is illustrated in the following figure $\operatorname{Im}(\lambda(u)) \neq 0$ means that

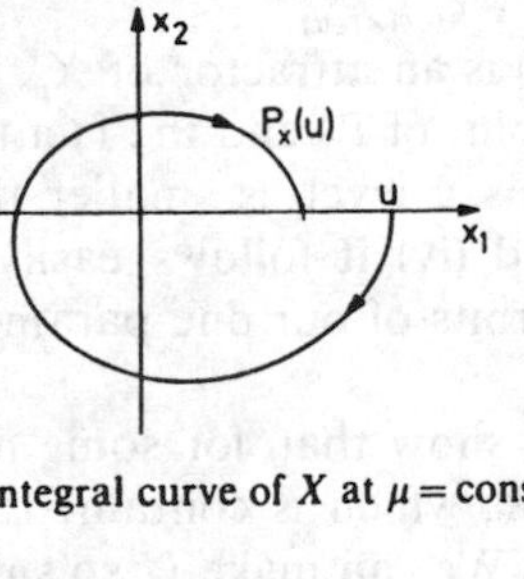

Fig. 6. Integral curve of X at $\mu =$ constant

X has a "non vanishing rotation"; it is then clear that P_X is defined for ε small enough.

Remark (6.5). It follows easily from Lemma (6.2) that P_X is C^{k-2}. We define a *displacement function* $V(x_1, \mu)$ on the domain of P_X as follows:

$$P_X(x_1, 0, \mu) = (x_1 + V(x_1, \mu), 0, \mu); \qquad V \text{ is } C^{k-2}.$$

This displacement function has the following properties:

(i) V is zero on $\{x_1 = 0\}$; the other zeroes of V occur in pairs (of opposite sign), each pair corresponds to a closed orbit of X. If a closed orbit γ of X is contained in a sufficiently small neighbourhood of $(0, 0)$, and intersects $\{x_1 = 0\}$ only twice then V has a corresponding pair of zeroes (namely the two points $\gamma \cap$(domain of P_X)).

(ii) For $\mu < 0$ and $x_1 = 0$, $\frac{\partial V}{\partial x_1} < 0$; for $\mu > 0$ and $x_1 = 0$, $\frac{\partial V}{\partial x_1} > 0$ and for $\mu = 0$ and $x = 0$, $\frac{\partial^2 V}{\partial \mu \, \partial x_1} > 0$. This follows from the assumptions on $\lambda(\mu)$. Hence, again by the division theorem, $\tilde{V} = \frac{V}{x_1}$ is C^{k-3}. $\tilde{V}(0,0)$ is zero, $\frac{\partial \tilde{V}}{\partial \mu} > 0$, so there is locally a unique C^{k-3}-curve l of zeroes of $\tilde{V}$ passing through $(0,0)$. Locally the set of zeroes of V is the union of l and $\{x_1 = 0\}$. l induces the one-parameter family of closed orbits.

(iii) Let us say that $(0,0)$ is a "vague attractor" for X_0 if $V(x_1, 0) = -Ax_1^3 +$ terms of order > 3 with $A > 0$. This means that the 3rd order terms of X_0 make the flow attract to $(0,0)$. In that case $\tilde{V} = \alpha_1 \mu - Ax_1^2 +$ terms of higher order, with α_1 and $A > 0$, so $\tilde{V}(x_1, \mu)$ vanishes only if $x_1 = 0$ or $\mu > 0$. This proves that the one-parameter family is contained in $\{\mu > 0\}$.

(iv) The following holds in a neighbourhood of $(0,0,0)$ where $\frac{\partial V}{\partial x_1} > -1$.

If $V(x_1, \mu) = 0$ and $\left(\frac{\partial V}{\partial x_1}\right)_{(x_1, \mu)} < 0$, then the closed orbit which cuts the domain of P_X in (x_1, μ) is an attractor of X_μ. This follows from the fact that (x_1, μ) is a fixed point of P_X and the fact that the derivative of P_X in (x_1, μ), restricted to this μ level, is smaller than 1 (in absolute value).

Combining (iii) and (iv) it follows easily that, if $(0,0)$ is a vague attractor, the closed orbits of our one parameter family are, near $(0,0)$, of the attracting type.

Finally we have to show that, for some neighbourhood U of $(0,0)$, every closed orbit of X, which is contained in U, is a member of our family of closed orbits. We can make U so small that every closed orbit γ of X, which is contained in U, intersects the domain of P_X.

Let $p = (x_1(\gamma), 0, \mu(\gamma))$ be an intersection point of a closed orbit γ with the domain of P_X. We may also assume that U is so small that $P_X[U \cap (\text{domain of } P_X)] \subset (\text{domain of } P_X)$. Then $P_X(p)$ is in the domain of P_X but also $P_X(p) \subset U$ so $(P_X)^2(p)$ is defined etc.; so $P_X^i(p)$ is defined.

Restricted to $\{\mu = \mu(\gamma)\}$, P_X is a local diffeomorphism of a segment of the half line $(x_1 \geqq 0$ or $x_1 \leqq 0$, $x_2 = 0$, $\mu = \mu(\gamma))$ into that half line.

If the x_1 coordinate of $P_X^i(p)$ is $<$ (resp. $>$) than $x_1(\gamma)$ then the x_1 coordinate of $P_X^{i+1}(p)$ is $<$ (resp. $>$) than the x_1 coordinate of $P_X^i(p)$, so p does not lie on a closed orbit. Hence we must assume that the x_1 coordinate of $P_X(p)$ is $x_1(\gamma)$, hence p is a fixed point of P_X, hence p is a zero of V, so, by property (ii), γ is a member of our one parameter family of closed orbits.

§ 7. Hopf Bifurcation for Diffeomorphisms*

We consider now a one parameter family $\Phi_\mu: \mathbb{R}^2 \to \mathbb{R}^2$ of diffeomorphisms satisfying (a)′, (b)′ and (c)′ (Remark (5.3)) and such that:

(d) $\frac{d}{d\mu}(|\lambda(\mu)|)_{\mu=0} > 0.$

Such a diffeomorphism can for example occur as the time one integral of a vectorfield X_μ as we studied in Section 2. In this diffeomorphism case we shall of course not find any closed (circular) orbit (the orbits are not continuous) but nevertheless we shall, under rather general conditions, find, near (0, 0) and for μ small, a one parameter family of invariant circles.

We first bring Φ_μ, by coordinate transformations, into a simple form:

We change the μ coordinate in order to obtain

(d)′ $|\lambda(\mu)| = 1 + \mu$.

After an appropriate (μ dependent) coordinate change of $\mathbb{R}^2$ we then have $\Phi(r, \varphi, \mu) = ((1+\mu)r, \varphi + f(\mu), \mu) +$ terms of order r^2, where $x_1 = r\cos\varphi$ and $x_2 = r\sin\varphi$; "$\Phi = \Phi' +$ terms of order r^l" means that the derivatives of Φ and Φ' up to order $l-1$ with respect to (x_1, x_2) agree for $(x_1, x_2) = (0, 0)$.

We now put in one extra condition:

(e) $f(0) \neq \frac{k}{l} \cdot 2\pi$ for all $k, l \leqq 5$.

Proposition (7.1). *Suppose Φ_μ satisfies* (a)′, (b)′, (c)′, (d)′ *and* (e) *and is C^k, $k \geqq 5$. Then for μ near 0, by a μ dependent coordinate change in $\mathbb{R}^2$, one can bring Φ_μ in the following form:*

$$\Phi_\mu(r, \varphi) = ((1+\mu)r - f_1(\mu)\cdot r^3, \varphi + f_2(\mu) + f_3(\mu)\cdot r^2) + \text{terms of order } r^5 .$$

For each μ, the coordinate transformation of $\mathbb{R}^2$ is C^∞; the induced coordinate transformation on $\mathbb{R}^2 \times \mathbb{R}$ is only C^{k-4}.

The next paragraph is devoted to the proof of this proposition**. Our last condition on Φ_μ is:

* *Note added in proof.* J. Moser kindly informed us that the Hopf bifurcation for diffeomorphisms had been worked out by Neumark (reference not available) and R. Sacker (Thesis, unpublished). An example of "decay" (loss of differentiability) of T^2 under perturbations has been studied by N. Levinson [a second order differentiable equation with singular solutions. Ann. of Math. **50**, 127–153 (1949)].

** *Note added in proof.* The desired normal form can also be obtained from § 21 of C. L. Siegel, Vorlesungen über Himmelsmechanik, Springer, Berlin, 1956 (we thank R. Jost for emphasizing this point).

(f) $f_1(0) \neq 0$. We assume even that $f_1(0) > 0$ (this corresponds to the case of a vague attractor for $\mu = 0$, see Section 6); the case $f_1(0) < 0$ can be treated in the same way (by considering $\Phi_{-\mu}^{-1}$ instead of Φ_μ).

Notation. We shall use $N\Phi_\mu$ to denote the map

$$(r, \varphi) \to ((1+\mu)r - f_1(\mu) \cdot r^3, \varphi + f_2(\mu) + f_3(\mu) \cdot r^2)$$

and call this "the simplified Φ_μ".

Theorem (7.2). *Suppose Φ_μ is at least C^5 and satisfies* (a)′, (b)′, (c)′, (d)′ *and* (e) *and $N\Phi_\mu$, the simplified Φ_μ, satisfies* (f). *Then there is a continuous one parameter family of invariant attracting circles of Φ_μ, one for each $\mu \in (0, \varepsilon)$, for ε small enough.*

Proof. The idea of the proof is as follows: the set $\Sigma = \{\mu = f_1(\mu) \cdot r^2\}$ in (r, q, μ)-space is invariant under $N\Phi$; $N\Phi$ even "attracts to this set". This attraction makes Σ stable in the following sense: $\{\Phi^n(\Sigma)\}_{n=0}^\infty$ is a sequence of manifolds which converges (for μ small) to an invariant manifold (this is actually what we have to prove); the method of the proof is similar to the methods used in [4, 5].

First we define $U_\delta = \left\{(r, \varphi, \mu) \mid r \neq 0 \text{ and } \frac{\mu}{r^2} \in [f_1(\mu) - \delta, f_1(\mu) + \delta]\right\}$, $\delta \ll f_1(\mu)$, and show that $N\Phi(U_\delta) \subset U_\delta$ and also, in a neighbourhood of $(0, 0, 0)$, $\Phi(U_\delta) \subset U_\delta$. This goes as follows:

If $p \in \partial U_\delta$, and $r(p)$ is the r-coordinate of p, then the r-coordinate of $N\Phi(p)$ is $r(p) \pm \delta \cdot (r(p))^3$ and p goes towards the interior of U_δ. Because Φ equals $N\Phi$, modulo terms of order r^5, also, locally, $\Phi(U_\delta) \subset U_\delta$. From this it follows that, for ε small enough and all $n \geqq 0$ $\Phi^n(\Sigma_\varepsilon) \subset U_\delta$; $\Sigma_\varepsilon = \Sigma \cap \{0 < \mu < \varepsilon\}$.

Next we define, for vectors tangent to a μ level of U_δ, the slope by the following formula: for X tangent to $U_\delta \cap \{\mu = \mu_0\}$ and $X = X_r \frac{\partial}{\partial r} + X_\varphi \frac{\partial}{\partial \varphi}$ the slope of X is $\left|\frac{X_r}{\mu_0 \cdot X_\varphi}\right|$; for $X_\varphi = 0$ the slope is not defined.

By direct calculations it follows that if X is a tangent vector of $U_\delta \cap \{\mu = \mu_0\}$ with slope $\leqq 1$, and μ_0 is small enough, then the slope of $d(N\Phi)(X)$ is $\leqq (1 - K\mu_0)$ for some positive K. Using this, the fact that $\frac{\mu}{r^2} \sim$ constant on U_δ and the fact that Φ and $N\Phi$ only differ by terms of order r^5 one can verify that for ε small enough and X a tangent vector of $U_\delta \cap \{\mu = \mu_0\}$, $\mu_0 \leqq \varepsilon$, with slope $\leqq 1$, $d\Phi(X)$ has slope < 1.

From this it follows that for ε small enough and any $n \geqq 0$,

1. $\Phi^n(\Sigma_\varepsilon) \subset U_\delta$ and
2. the tangent vectors of $\Phi^n(\Sigma_\varepsilon) \cap \{\mu = \mu_0\}$, for $\mu_0 \leqq \varepsilon$ have slope < 1.

This means that for any $\mu_0 \leqq \varepsilon$ and $n \geqq 0$

$$\Phi^n(\Sigma_\varepsilon) \cap \{\mu = \mu_0\} = \{(f_{n,\mu_0}(\varphi), \varphi, \mu_0)\}$$

where f_{n,μ_0} is a unique smooth function satisfying:

1′. $f_{n,\mu_0}(\varphi) \in \left[\sqrt{\frac{\mu_0}{f_1(\mu_0)+\delta}}, \sqrt{\frac{\mu_0}{f_1(\mu_0)-\delta}}\right]$ for all φ

2′. $\frac{d}{d\varphi}(f_{n,\mu_0}(\varphi)) \leqq \mu_0$ for all φ.

We now have to show that, for μ_0 small enough, $\{f_{n,\mu_0}\}_{n=0}^{\infty}$ converges. We first fix a φ_0 and define

$$p_1 = (f_n(\varphi_0), \varphi_0, \mu_0), \qquad p_1' = \Phi(p_1) = (r_1', \varphi_1', \mu),$$

$$p_2 = (f_{n+1}(\varphi_0), \varphi_0, \mu_0), \qquad p_2' = \Phi(p_2) = (r_2', \varphi_2', \mu).$$

Using again the fact that $(f_{n,\mu_0}(\varphi))^2/\mu_0 \sim$ constant (independent of μ_0), one obtains:

$$|r_1' - r_2'| \leqq (1 - K_1\mu_0)\,|f_{n,\mu_0}(\varphi_0) - f_{n+1,\mu_0}(\varphi_0)|$$

and

$$|\varphi_1' - \varphi_2'| \leqq K_2\sqrt{\mu_0}\cdot|f_{n,\mu_0}(\varphi_0) - f_{n+1,\mu_0}(\varphi_0)| \quad \text{where} \quad K_1, K_2 > 0$$

and independent of μ_0.

By definition we have $f_{n+1,\mu_0}(\varphi_1') = r_1'$ and $f_{n+2,\mu_0}(\varphi_2') = r_2'$. We want however to get an estimate for the difference between $f_{n+1,\mu_0}(\varphi_1')$ and $f_{n+2,\mu_0}(\varphi_1')$. Because

$$\frac{d}{d\varphi}(f_{n+2,\mu_0}(\varphi)) \leqq \mu_0,$$

$$|f_{n+2,\mu_0}(\varphi_2') - f_{n+2,\mu_0}(\varphi_1')| \leqq \mu_0|\varphi_2' - \varphi_1'| \leqq K_2\cdot\mu_0^{\frac{3}{2}}|f_{n,\mu_0}(\varphi_0) - f_{n+1,\mu_0}(\varphi_0)|.$$

We have seen that

$$|f_{n+1,\mu_0}(\varphi_1') - f_{n+2,\mu_0}(\varphi_2')| = |r_1' - r_2'|$$
$$\leqq (1 - K_1\mu_0)\,|f_{n,\mu_0}(\varphi_0) - f_{n+1,\mu_0}(\varphi_0)|.$$

So

$$|f_{n+1,\mu_0}(\varphi_1') - f_{n+2,\mu_0}(\varphi_1')| \leqq (1 + K_2\mu_0^{\frac{3}{2}} - K_1\mu_0)\,|f_{n,\mu_0}(\varphi_0) - f_{n+1,\mu_0}(\varphi_0)|.$$

We shall now assume that μ_0 is so small that $(1 + K_2\mu_0^{\frac{3}{2}} - K_1\mu_0) = K_3(\mu_0) < 1$, and write $\varrho(f_{n,\mu_0}, f_{n+1,\mu_0}) = \max_{\varphi}(|f_{n,\mu_0}(\varphi) - f_{n+1,\mu_0}(\varphi)|)$.

It follows that

$$\varrho(f_{m,\mu_0}, f_{m+1,\mu_0}) \leqq (K_3(\mu_0))^m \cdot \varrho(f_{0,\mu_0}, f_{1,\mu_0}).$$

This proves convergence, and gives for each small $\mu_0 > 0$ an invariant and attracting circle. This family of circles is continuous because the limit functions f_{∞,μ_0} depend continuously on μ_0, because of uniform convergence.

Remark (7.3). For a given μ_0, f_{∞,μ_0} is not only continuous but even Lipschitz, because it is the limit of functions with derivative $\leqq \mu_0$. Now we can apply the results on invariant manifolds in [4, 5] and obtain the following:

If Φ_μ is C^r for each μ then there is an $\varepsilon_r > 0$ such that the circles of our family which are in $\{0 < \mu < \varepsilon_r\}$ are C^r. This comes from the fact that near $\mu = 0$ in U_δ the contraction in the r-direction dominates sufficiently the maximal possible contraction in the φ-direction.

§ 8. Normal Forms (the Proof of Proposition (7.1))

First we have to give some definitions. Let $\underline{V}_r$ be the vectorspace of r-jets of vectorfields on $\mathbb{R}^2$ in 0, whose $(r-1)$-jet is zero (i.e. the elements of $\underline{V}_r$ can be uniquely represented by a vectorfield whose component functions are homogeneous polynomials of degree r). V_r is the set of r-jets of diffeomorphisms $(\mathbb{R}^2, 0) \to (\mathbb{R}^2, 0)$, whose $(r-1)$-jet is "the identity". $\mathrm{Exp}: \underline{V}_r \to V_r$ is defined by: for $\alpha \in \underline{V}_r$, $\mathrm{Exp}(\alpha)$ is the (r-jet of) the diffeomorphism obtained by integrating α over time 1.

Remark (8.1). For $r \geqq 2$, Exp is a diffeomorphism onto and $\mathrm{Exp}(\alpha) \circ \mathrm{Exp}(\beta) = \mathrm{Exp}(\alpha + \beta)$. The proof is straightforward and left to the reader.

Let now $A: (\mathbb{R}^2, 0) \to (\mathbb{R}^2, 0)$ be a linear map. The induced transformations $A_r: \underline{V}_r \to \underline{V}_r$ are defined by $A_r(\alpha) = A_* \alpha$, or, equivalently, $\mathrm{Exp}(A_r(\alpha)) = A \circ \mathrm{Exp}\,\alpha \circ A^{-1}$.

Remark (8.2). If $[\Psi]_r$ is the r-jet of $\Psi: (\mathbb{R}^2, 0) \to (\mathbb{R}^2, 0)$ and $d\Psi$ is A, then, for every $\alpha \in \underline{V}_r$, the r-jets $[\Psi]_r \circ \mathrm{Exp}(\alpha)$ and $\mathrm{Exp}(A_r \alpha) \circ [\Psi]_r$ are equal. The proof is left to the reader.

A splitting $\underline{V}_r = \underline{V}_r' \oplus \underline{V}_r''$ of $\underline{V}_r$ is called an *A-splitting*, $A: (\mathbb{R}^2, 0) \to (\mathbb{R}^2, 0)$ linear, if

1. $\underline{V}_r'$ and $\underline{V}_r''$ are invariant under the action of A_r.
2. $A_r \mid V_r''$ has no eigenvalue one.

Example (8.3). We take A with eigenvalues $\lambda, \bar{\lambda}$ and such that $|\lambda| \neq 1$ or such that $|\lambda| = 1$ but $\lambda \neq e^{k/l\, 2\pi i}$ with $k, l \leqq 5$. We may assume that A

is of the form

$$|\lambda|\begin{pmatrix}\cos\alpha & \sin\alpha\\ -\sin\alpha & \cos\alpha\end{pmatrix}.$$

For $2 \leqq i \leqq 4$ we can obtain a A-splitting of $\underline{V}_i$ as follows:

$\underline{V}_i'$ is the set of those (i-jets of) vectorfields which are, in polar coordinates of the form $\alpha_1 r^i \frac{\partial}{\partial r} + \alpha_2 r^{i-1}\frac{\partial}{\partial\varphi}$. More precisely $\underline{V}_2' = 0$, $\underline{V}_3'$ is generated by $r^3\frac{\partial}{\partial r}$ and $r^2\frac{\partial}{\partial\varphi}$ and $\underline{V}_4' = 0$ (the other cases give rise to vectorfields which are not differentiable, in ordinary coordinates).

$\underline{V}_i''$ is the set of (i-jets of) vectorfields of the form

$$g_1(\varphi)\, r^i\frac{\partial}{\partial r} + g_2(\varphi)\, r^{i-1}\frac{\partial}{\partial\varphi} \quad \text{with} \quad \int_0^{2\pi} g_1(\varphi) = \int_0^{2\pi} g_2(\varphi) = 0\,.$$

$g_1(\varphi)$ and $g_2(\varphi)$ have to be linear combinations of $\sin(j\cdot\varphi)$ and $\cos(j\cdot\varphi)$, $j \leqq 5$, because otherwise the vectorfield will not be differentiable in ordinary coordinates (not all these linear combinations are possible).

Proposition (8.4). *For a given diffeomorphism* $\Phi: (\mathbb{R}^2, 0) \to (\mathbb{R}^2, 0)$ *with* $(d\Phi)_0 = A$ *and a given A-splitting* $\underline{V}_i = \underline{V}_i' \oplus \underline{V}_i''$ *for* $2 \leqq i \leqq i_0$, *there is a coordinate transformation* $\varkappa: (\mathbb{R}^2, 0) \to (\mathbb{R}^2, 0)$ *such that:*

1. $(d\varkappa)_0 =$ *identity.*
2. *For each* $z \leqq i \leqq i_0$ *the i-jet of* $\Phi' = \varkappa\circ\Phi\circ\varkappa^{-1}$ *is related to its* $(i-1)$*-jet as follows: Let* $[\Phi']_{i-1}$ *be the polynomial map of degree* $\leqq i-1$ *which has the same* $(i-1)$*-jet. The i-jet of* Φ' *is related to its* $(i-1)$*-jet if there is an element* $\alpha \in \underline{V}_i'$ *such that* $\operatorname{Exp}\alpha \circ [\Phi']_{i-1}$ *has the same i-jet as* Φ'.

Proof. We use induction: Suppose we have a map $\varkappa$ such that 1 and 2 hold for $i < i_1 \leqq i_0$. Consider the i_1 jet of $\varkappa\circ\Phi\circ\varkappa^{-1}$. We now replace $\varkappa$ by $\operatorname{Exp}\alpha\circ\varkappa$ for some $\alpha \in \underline{V}_{i_1}''$. $\varkappa\circ\Phi\circ\varkappa^{-1}$ is then replaced by $\operatorname{Exp}(\alpha)\circ\varkappa \circ\Phi\circ\varkappa^{-1}\circ\operatorname{Exp}(-\alpha)$, according to remark (8.2) this equal to $\operatorname{Exp}(-A_{i_1}\alpha) \circ\operatorname{Exp}(\alpha)\circ\varkappa\circ\Phi\circ\varkappa^{-1} = \operatorname{Exp}(\alpha - A_{i_1}\alpha)\circ\varkappa\circ\Phi\circ\varkappa^{-1}$.

$A_{i_1} \mid \underline{V}_{i_1}''$ has no eigenvalue one, so for each $\beta \in \underline{V}_{i_1}''$ there is a unique $\alpha \in \underline{V}_{i_1}''$ such that if we replace $\varkappa$ by $\operatorname{Exp}\alpha\circ\varkappa$, $\varkappa\circ\Phi\circ\varkappa^{-1}$ is replaced by $\operatorname{Exp}\beta\circ\varkappa\circ\Phi\circ\varkappa^{-1}$. It now follows easily that there is a unique $\alpha \in \underline{V}_{i_1}''$ such that $\operatorname{Exp}\alpha\circ\varkappa$ satisfies condition 2 for $i \leqq i_1$. This proves the proposition.

Proof of Proposition (7.1). For μ near 0, $d\Phi_\mu$ is a linear map of the type we considered in example (8.3). So the splitting given there is a $d\Phi_\mu$-splitting of $\underline{V}_i$, $i = 2, 3, 4$, for μ near zero. We now apply Proposition (8.4) for each μ and obtain a coordinate transformation $\varkappa_\mu$ for each μ which brings Φ_μ in the required form. The induction step then becomes:

Given $\varkappa_\mu$, satisfying 1 and 2 for $i<i_1$ there is for each μ a unique $\alpha_\mu \in \underline{V}''_{i_1}$ such that $\operatorname{Exp}\alpha_\mu \circ \varkappa_\mu$ satisfies 1 and 2 for $i \leqq i_1$. α_μ depends then C^r on μ if the i_1-jet of Φ depends C^r on μ; this gives the loss of differentiability in the μ direction.

§ 9. Some Examples

In this section we show how a small perturbation of a quasi-periodic flow on a torus gives flows with strange attractors (Proposition (9.2)) and, more generally, flows which are not Morse-Smale (Proposition (9.1)).

Proposition (9.1). *Let ω be a constant vector field on $T^k=(\mathbb{R}/\mathbb{Z})^k$, $k \geqq 3$. In every C^{k-1}-small neighbourhood of ω there exists an open set of vector fields which are not Morse-Smale.*

We consider the case $k=3$. We let $\omega=(\omega_1,\omega_2,\omega_3)$ and we may suppose $0 \leqq \omega_1 \leqq \omega_2 \leqq \omega_3$. Given $\varepsilon>0$ we may choose a constant vector field ω' such that

$$\|\omega'-\omega\|_2 = \|\omega'-\omega\|_0 < \varepsilon/2\,,$$

$$\omega_3'>0, \quad 0<\frac{\omega_1'}{\omega_3'}=\frac{p_1}{q_1}<1, \quad 0<\frac{\omega_2'}{\omega_3'}=\frac{p_2}{q_2}<1\,,$$

where p_1, p_2, q_1, q_2 are integers, and $p_1 q_2$ and $p_2 q_1$ have no common divisor. We shall also need that q_1, q_2 are sufficiently large and satisfy

$$\tfrac{1}{2} < q_1/q_2 < 2\,.$$

All these properties can be satisfied with $q_1=2^{m_1}$, $q_2=3^{m_2}$.

Let $I=\{x \in \mathbb{R} : 0 \leqq x \leqq 1\}$ and define $g, h : I^3 \to T^3$ by

$$\begin{aligned} g(x_1,x_2,x_3) &= (x_1(\operatorname{mod}1), x_2(\operatorname{mod}1), x_3(\operatorname{mod}1)) \\ h(x_1,x_2,x_3) &= (q_1^{-1}x_1+p_1q_2x_3(\operatorname{mod}1), q_2^{-1}x_2+p_2q_1x_3(\operatorname{mod}1), \\ &\qquad q_1q_2x_3(\operatorname{mod}1))\,. \end{aligned}$$

We have $gI^3 = hI^3 = T^3$ and g (resp. h) has a unique inverse on points gx (resp. hx) with $x \in \mathring{I}^3$.

We consider the map f of a disc into itself (see [11] Section I.5, Fig. 7) used by Smale to define the horseshoe diffeomorphism. Imbedding Δ in T^2:

$$\Delta \subset \{(x_1,x_2) : \tfrac{1}{3}<x_1<\tfrac{2}{3}, \tfrac{1}{3}<x_2<\tfrac{2}{3}\} \subset T^2$$

we can arrange that f appears as Poincaré map in $T^3 = T^2 \times T^1$. More precisely, it is easy to define a vector field $X=(\tilde{X},1)$ on $T^2 \times T^1$ such that if $\xi \in \Delta$, we have

$$(f(\xi), 0) = \mathscr{D}_{X,1}(\xi, 0)$$

where $\mathscr{D}_{X,1}$ is the time one integral of X (see Fig. 7). Finally we choose the restriction of X to a neighbourhood of $g(\partial I^2 \times I)$ to be $(0, 1)$ (i.e. $\tilde{X} = 0$).

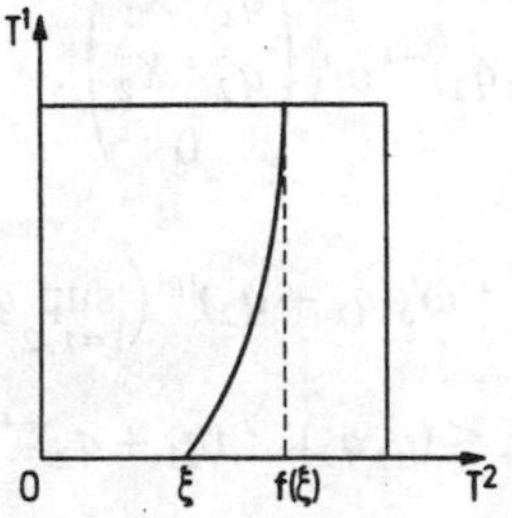

Fig. 7

If $x \in g\mathring{I}^3$, then $\Phi x = h \circ g^{-1}$ is uniquely defined and the tangent mapping to Φ applied to X gives a vector field Y:

$$Y(\Phi(x)) = [d\Phi(x)]\, X(x)$$

where

$$[d\Phi(x)] = \begin{pmatrix} q_1^{-1} & & p_1 q_2 \\ & q_2^{-1} & p_2 q_1 \\ & & q_1 q_2 \end{pmatrix}.$$

Y has a unique smooth extension to T^3, again called Y. Let now $Z = (q_1 q_2)^{-1} \omega_3' Y$. We want to estimate

$$\|Z - \omega'\|_r = \sup_{\varrho;\, |\varrho| \le r} N^\varrho$$

where

$$N^\varrho = \sup_{y \in T^3} \sup_{i=1,2,3} |D^\varrho Z_i(y) - D^\varrho \omega_i'| \qquad (*)$$

and D^ϱ denotes a partial differentiation of order $|\varrho|$. Notice that it suffices to take the first supremum in (*) over $y \in h\mathring{I}^3$, i.e. $y = \Phi x$ where $x \in g\mathring{I}^3$. We have

$$\frac{\partial}{\partial y} = \begin{pmatrix} q_1 & & \\ & q_2 & \\ -p_1 & -p_2 & (q_1 q_2)^{-1} \end{pmatrix} \frac{\partial}{\partial x}$$

so that

$$\sup_i \left| \frac{\partial}{\partial y_i} \right| < (q_1 + q_2) \sup_i \left| \frac{\partial}{\partial x_i} \right|.$$

Notice also that

$$Z_i(y) - \omega_i' = (q_1 q_2)^{-1} \omega_3' \begin{pmatrix} q_1^{-1} X_1 + p_1 q_2 \\ q_2^{-1} X_2 + p_2 q_1 \\ q_1 q_2 \end{pmatrix} - \omega_3' \begin{pmatrix} p_1 q_1^{-1} \\ p_2 q_2^{-1} \\ 1 \end{pmatrix}$$

$$= (q_1 q_2)^{-1} \omega_3' \begin{pmatrix} q_1^{-1} X_1 \\ q_2^{-1} X_2 \\ 0 \end{pmatrix}.$$

Therefore

$$N^{\varrho} \leqq (q_1 q_2)^{-1} \, \omega_3' (q_1 + q_2)^{|\varrho|} \Big(\sup_{i=1,2} q_i^{-1} \Big) \sup_{i=1,2} \|X_i\|_{|\varrho|}$$

$$\|Z - \omega'\|_r \leqq (q_1 q_2)^{-2} \, (q_1 + q_2)^{r+1} \, [\omega_3' \, \|\tilde{X}\|_r] .$$

If we have chosen q_1, q_2 sufficiently large, we have

$$\|Z - \omega'\|_2 < \varepsilon/2$$

and therefore

$$\|Z - \omega\|_2 < \varepsilon .$$

Consider the Poincaré map $P: T^2 \to T^2$ defined by the vector field Z on $T^3 = T^2 \times T^1$. By construction the non wandering set of P contains a Cantor set, and the same is true if Z is replaced by a sufficiently close vector field Z'. This concludes the proof for $k = 3$.

In the general case $k \geqq 3$ we approximate again ω by ω' rational and let

$$0 < \frac{\omega_i'}{\omega_k'} = \frac{p_i}{q_i} < 1 \quad \text{for} \quad i = 1, \dots, k-1 .$$

We assume that the integers $p_1 \prod_{i \neq 1} q_i, \dots, p_{k-1} \prod_{i \neq k-i} q_i$ have no common divisor. Furthermore $q_1, \dots, q_{k-1}$ are chosen sufficiently large and such that

$$(\max_i q_i)/(\min_i q_i) < C$$

where C is a constant depending on k only.

The rest of the proof goes as for $k = 2$, with the horseshoe diffeomorphism replaced by a suitable $k-1$-diffeomorphism. In particular, using the diffeomorphism of Fig. 2 (end of § 2) we obtain the following result.

Proposition (9.2). *Let ω be a constant vector field on T^k, $k \geqq 4$. In every C^{k-1}-small neighbourhood of ω there exists an open set of vector fields with a strange attractor.*

Appendix

Bifurcation of Stationary Solutions of Hydrodynamical Equations

In this appendix we present a bifurcation theorem for fixed points of a non linear map in a Banach space. Our result is of a known type[23], but has the special interest that the fixed points are shown to depend differentiably on the bifurcation parameter. The theorem may be used to study the bifurcation of stationary solutions in the Taylor and Bénard[24] problems for instance. By reference to Brunovský (cf. § 3 b) we see that the bifurcation discussed below is *nongeneric*. The bifurcation of stationary solutions in the Taylor and Bénard problems is indeed nongeneric, due to the presence of an invariance group.

Theorem. *Let H be a Banach space with* C^k *norm*, $1 \leqq k < \infty$, *and* $\Phi_\mu : H \to H$ *a differentiable map such that* $\Phi_\mu(0) = 0$ *and* $(x, \mu) \to \Phi_\mu x$ *is* C^k *from* $H \times \mathbb{R}$ *to* H. *Let*

$$L_\mu = [d\Phi_\mu]_0, \qquad N_\mu = \Phi_\mu - L_\mu . \tag{1}$$

We assume that L_μ *has a real simple isolated eigenvalue* $\lambda(\mu)$ *depending continuously on* μ *such that* $\lambda(0) = 1$ *and* $(d\lambda/d\mu)(0) > 0$; *we assume that the rest of the spectrum is in* $\{\mathfrak{z} \in \mathbb{C} : |\mathfrak{z}| < 1\}$.

(a) *There is a one parameter family (a* C^{k-1} *curve l) of fixed points of* $\Phi : (x, \mu) \to (\Phi_\mu x, \mu)$ *near* $(0, 0) \in H \times \mathbb{R}$. *These points and the points* $(0, \mu)$ *are the only fixed points of* Φ *in some neighbourhood of* $(0, 0)$.

(b) *Let* $\mathfrak{z}$ *(resp.* $\mathfrak{z}^*$*) be an eigenvector of* L_0 *(resp. its adjoint* L_0^* *in the dual* H^* *of H) to the eigenvalue* 1, *such that* $(\mathfrak{z}^*, \mathfrak{z}) = 1$. *Suppose that for all* $\alpha \in \mathbb{R}$

$$(\mathfrak{z}^*, N_0 \alpha \mathfrak{z}) = 0 . \tag{2}$$

Then the curve l of (a) *is tangent to* $(\mathfrak{z}, 0)$ *at* $(0, 0)$.

From the center-manifold theorem of Hirsch, Pugh, and Shub[25] it follows that there is a 2-dimensional C^k-manifold V^c, tangent to the vectors $(\mathfrak{z}, 0)$ and $(0, 1)$ at $(0, 0) \in H \times \mathbb{R}$ and locally invariant under Φ. Furthermore there is a neighbourhood U of $(0, 0)$ such that every fixed point of Φ in U is contained in $V^c \cap U$[26].

We choose coordinates (α, μ) on V^c so that

$$\Phi(\alpha, \mu) = (f(\alpha, \mu), \mu)$$

---

[23] See for instance [16] and [14].

[24] The bifurcation of the Taylor problem has been studied by Velte [18] and Yudovich [19]. For the Benard problem see Rabinowitz [17], Fife and Joseph [15].

[25] Usually the center manifold theorem is only formulated for diffeomorphisms; C. C. Pugh pointed out to us that his methods in [5], giving the center manifold, also work for differentiable maps which are not diffeomorphisms.

[26] See § 5.

with $f(0,\mu)=0$ and $\frac{\partial f}{\partial \alpha}(0,\mu)=\lambda(\mu)$. The fixed points of Φ in V^c are given by $\alpha=f(\alpha,\mu)$, they consist of points $(0,\mu)$ and of solutions of

$$g(\alpha,\mu)=0$$

where, by the division theorem, $g(\alpha,\mu)=\frac{f(\alpha,\mu)}{\alpha}-1$ is C^{k-1}. Since $\frac{\partial g}{\partial \mu}(0,0)=\frac{d\lambda}{d\mu}(0)>0$, the implicit function theorem gives (a).

Let $(x,0)\in V^c$. We may write

$$x=\alpha\mathfrak{z}+Z \tag{3}$$

where $(\mathfrak{z}^*,Z)=0$, $Z=0(\alpha^2)$. Since $(\mathfrak{z}^*,\mathfrak{z})=1$ we have

$$\begin{aligned} f(\alpha,0)&=(\mathfrak{z}^*,\Phi_0(\alpha\mathfrak{z}+Z))\\ &=\alpha+(\mathfrak{z}^*,L_0Z+N_0(\alpha\mathfrak{z}+Z))\\ &=\alpha+(\mathfrak{z}^*,L_0Z+N_0\alpha\mathfrak{z})+0(\alpha^3)\,. \end{aligned} \tag{4}$$

Notice that

$$(\mathfrak{z}^*,L_0Z)=(L_0^*\mathfrak{z}^*,Z)=(\mathfrak{z}^*,Z)=0\,.$$

We assume also that (2) holds:

$$(\mathfrak{z}^*,N_0\alpha\mathfrak{z})=0\,.$$

Then

$$\begin{aligned} f(\alpha,0)&=\alpha+0(\alpha^3)\,,\\ f(\alpha,\mu)&=\alpha(\lambda(\mu)+0^2)\,, \end{aligned} \tag{5}$$

where 0^2 represents terms of order 2 and higher in α and μ. The curve l of fixed points of Φ introduced in (a) is given by

$$\lambda(\mu)-1+0^2=0 \tag{6}$$

and (b) follows from $\lambda(0)=0$, $\frac{d\lambda}{d\mu}(0)>0$.

Remark 1. From (2) and the local invariance of V^c we have

$$\Phi_0(\alpha\mathfrak{z}+Z)=\alpha\mathfrak{z}+Z+0(\alpha^3)$$

hence

$$\begin{aligned} Z&=L_0Z+N_0\alpha\mathfrak{z}+0(\alpha^3)\,,\\ Z&=(1-L_0)^{-1}N_0\alpha\mathfrak{z}+0(\alpha^3)\,, \end{aligned}$$

and (5) can be replaced by the more precise

$$f(\alpha,0)=\alpha+(\mathfrak{z}^*,N_0[\alpha\mathfrak{z}+(1-L_0)^{-1}N\alpha\mathfrak{z}])+0(\alpha^4) \tag{7}$$

from which one can compute the coefficient A of α^3 in $f(\alpha, 0)$. Then (6) is, up to higher order terms

$$\left[\frac{d\lambda}{d\mu}(0)\right]\mu + A\alpha^2 = 0\,.$$

Depending on whether $A<0$ or $A>0$, this curve lies in the region $\mu>0$ or $\mu<0$, and consists of attracting or non-attracting fixed points. This is seen by discussing the sign of $f(\alpha, \mu) - \alpha$ (see Fig. 8b, c); the fixed points $(0, \mu)$ are always attracting for $\mu<0$, non attracting for $\mu>0$.

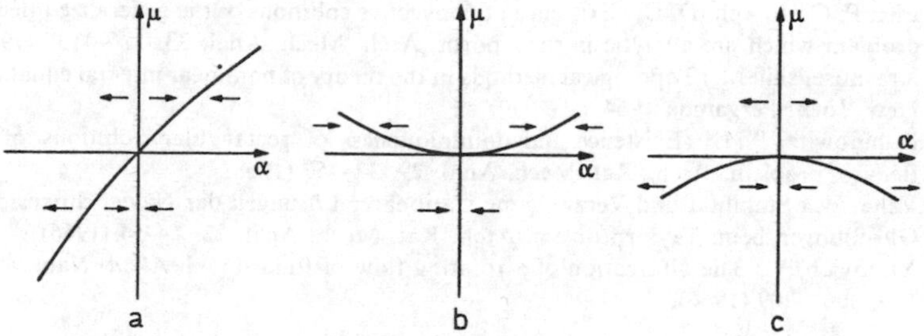

Fig. 8

Remark 2. If the curve l of fixed points is not tangent to $(\mathfrak{z}, 0)$ at $(0, 0)$, then the points of l are attracting for $\mu>0$, non-attracting for $\mu<0$ (see Fig. 8a).

Remark 3. If it is assumed that L_μ has the real simple isolated eigenvalue $\lambda(\mu)$ as in the theorem and that the rest of the spectrum lies in $\{\mathfrak{z}\in\mathbb{C}: |\mathfrak{z}| \neq 1\}$ (rather than $\{\mathfrak{z}\in\mathbb{C}: |\mathfrak{z}|<1\}$, the theorem continues to hold but the results on the attractive character of fixed points are lost.

References

1. Abraham, R., Marsden, J.: Foundations of mechanics. New York: Benjamin 1967.
2. Bass, J.: Fonctions stationnaires. Fonctions de corrélation. Application à la représentation spatio-temporelle de la turbulence. Ann. Inst. Henri Poincaré. Section B **5**, 135—193 (1969).
3. Brunovsky, P.: One-parameter families of diffeomorphisms. Symposium on Differential Equations and Dynamical Systems. Warwick 1968—69.
4. Hirsch, M., Pugh, C. C., Shub, M.: Invariant manifolds. Bull. A.M.S. **76**, 1015—1019 (1970).
5. — — — Invariant manifolds. To appear.
6. Hopf, E.: Abzweigung einer periodischen Lösung von einer stationären Lösung eines Differentialsystems. Ber. Math.-Phys. Kl. Sächs. Akad. Wiss. Leipzig **94**, 1—22 (1942).
7. Kelley, A.: The stable, center-stable, center, center-unstable, and unstable manifolds. Published as Appendix C of R. Abraham and J. Robbin: Transversal mappings and flows. New York: Benjamin 1967.

8. Landau, L. D., Lifshitz, E. M.: Fluid mechanics. Oxford: Pergamon 1959.
9. Leray, J.: Sur le mouvement d'un liquide visqueux emplissant l'espace. Acta Math. **63**, 193—248 (1934).
10. Moser, J.: Perturbation theory of quasiperiodic solutions of differential equations. Published in J. B. Keller and S. Antman: Bifurcation theory and nonlinear eigenvalue problems. New York: Benjamin 1969.
11. Smale, S.: Differentiable dynamical systems. Bull. Am. Math. Soc. **73**, 747—817 (1967).
12. Thom, R.: Stabilité structurelle et morphogénèse. New York: Benjamin 1967.
13. Williams, R. F.: One-dimensional non-wandering sets. Topology **6**, 473—487 (1967).
14. Berger, M.: A bifurcation theory for nonlinear elliptic partial differential equations and related systems. In: Bifurcation theory and nonlinear eigenvalue problems. New York: Benjamin 1969.
15. Fife, P. C., Joseph, D. D.: Existence of convective solutions of the generalized Bénard problem which are analytic in their norm. Arch. Mech. Anal. **33**, 116—138 (1969).
16. Krasnosel'skii, M.: Topological methods in the theory of nonlinear integral equations. New York: Pergamon 1964.
17. Rabinowitz, P. H.: Existence and nonuniqueness of rectangular solutions of the Bénard problem. Arch. Rat. Mech. Anal. **29**, 32—57 (1968).
18. Velte, W.: Stabilität und Verzweigung stationärer Lösungen der Navier-Stokesschen Gleichungen beim Taylorproblem. Arch. Rat. Mech. Anal. **22**, 1—14 (1966).
19. Yudovich, V.: The bifurcation of a rotating flow of fluid. Dokl. Akad. Nauk SSSR **169**, 306—309 (1966).

D. Ruelle
The Institute for Advanced Study
Princeton, New Jersey 08540, USA

F. Takens
Universiteit van Amsterdam
Roetersstr. 15
Amsterdam, The Netherlands

Commun. math. Phys. 23, 343—344 (1971)

Note Concerning our Paper "On the Nature of Turbulence"

D. Ruelle and F. Takens

Commun. math. Phys. **20**, 167—192 (1971)

Received November 2, 1971

After the paper referred to in the title was published, references to the relevant Russian literature were made available to us by Ya. G. Sinai and V. I. Arnold. From these it appears that much of the work on bifurcation described in our paper duplicates results already published by Russian authors. On the other hand the mathematical interpretation which we give of turbulence seems to remain our own responsability!

We thank Sinai and Arnold for informing us of the Russian references. The following list has been compiled from their indications.

Brušlinskaja, N. N.: Qualitative integration of a system of n differential equations in a region containing a singular point and a limit cycle. Dokl. Akad. Nauk SSSR **139** N1, 9—12 (1961); — Soviet Math. Dokl. **2**, 845—848 (1961). MR 26, No. 5212; Errata MR 30, p. 1203.

— Limit cycles for equations of motion of a rigid body and Galerkin's equations for hydrodynamics. Dokl. Akad. Nauk SSSR **157**, N5, 1017—1020 (1964); — Soviet Math. Dokl. **5**, 1051—1054 (1964). MR 29, No. 6133.

— The behavior of solutions of the equations of hydrodynamics when the Reynolds number passes through a critical value. Dokl. Akad. Nauk SSSR **162** N4, 731—734 (1965);— Soviet Math. Dokl. **6**, 724—728 (1965). MR 31, No. 6460.

— On the generation of periodic flows and tori from laminar flows, p. 57—79 in *Nekotorje voprosy mehaniki gornych porod* (some problems of mechanics of minerals), Trudy gornogo instituta, Moscow, 1968.

Gurtovnik, A. S., Neĭmark, Ju. I.: On the question of the stability of quasi-periodic motions. Diff. Uravn. **5** N5, 824—832 (1969).

Neĭmark, Ju. I.: On some cases of dependence of periodic solutions on parameters. Dokl. Akad. Nauk SSSR **129** N4, 736—739 (1959).

— Motions close to doubly-asymptotic motion. Dokl. Akad. Nauk SSSR **172** N5, 1021—1024 (1967); — Soviet Math. Dokl. **8**, 228—231 (1967). MR 38, No. 669.

— Izv. Vysš. Učebn. Zav., Radiofizika **1** N5—6 (1958); **2** N3 (1959); **8** N3 (1965); **10** N3 (1967).

Judovič, V. I.: An example of the generation of a secondary or periodic flow due to the loss of stability of a laminar flow of a viscous incompressible fluid. Prikl. Mat. Meh. **29** N3, 453—467 (1965); — J. Appl. Math. Mech. **29**, 527—544 (1965).

— On the bifurcation of rotationary flows of fluids. Dokl. Akad. Nauk SSSR **169** N2, 306—309 (1966).

Judovič, V. I.: An example of loss of stability and generation of a secondary flow in a closed volume. Mat. Sbornik **74** (116) N 4, 565—579 (1967).

— Generation of secondary stationary and periodic solutions by destabilization of a stable stationary flow of fluid. Abstracts of short scientific communications, Sec. 12. Internat. Congress of Math. (Moscow, 1966). Mir, Moscow, 1968.

— Questions of the mathematical theory of stability of flows of fluids. Vsesojusnyi siesd po teoretičeskoi i prikladnoi mehanike. Moscow: Annotacii Dokladov 1968.

— On the stability of forced oscillations of fluid. Dokl. Akad. Nauk SSSR **195** N 2, 292—295 (1970); — Soviet Math. Dokl. **11**, 1473—1477 (1970).

— Appearance of self-oscillations in fluids, I. Prikl. Mat. Meh., 1971.

D. Ruelle
Institut des Hautes Etudes
Scientifiques
F-91 Bures-sur Yvette, France

F. Takens
Dept. of Mathematics
University of Groningen
Groningen. The Netherlands

Nature Vol. 261 June 10 1976 459

Simple mathematical models with very complicated dynamics

Robert M. May*

First-order difference equations arise in many contexts in the biological, economic and social sciences. Such equations, even though simple and deterministic, can exhibit a surprising array of dynamical behaviour, from stable points, to a bifurcating hierarchy of stable cycles, to apparently random fluctuations. There are consequently many fascinating problems, some concerned with delicate mathematical aspects of the fine structure of the trajectories, and some concerned with the practical implications and applications. This is an interpretive review of them.

THERE are many situations, in many disciplines, which can be described, at least to a crude first approximation, by a simple first-order difference equation. Studies of the dynamical properties of such models usually consist of finding constant equilibrium solutions, and then conducting a linearised analysis to determine their stability with respect to small disturbances: explicitly nonlinear dynamical features are usually not considered.

Recent studies have, however, shown that the very simplest nonlinear difference equations can possess an extraordinarily rich spectrum of dynamical behaviour, from stable points, through cascades of stable cycles, to a regime in which the behaviour (although fully deterministic) is in many respects "chaotic", or indistinguishable from the sample function of a random process.

This review article has several aims.

First, although the main features of these nonlinear phenomena have been discovered and independently rediscovered by several people, I know of no source where all the main results are collected together. I have therefore tried to give such a synoptic account. This is done in a brief and descriptive way, and includes some new material: the detailed mathematical proofs are to be found in the technical literature, to which signposts are given.

Second, I indicate some of the interesting mathematical questions which do not seem to be fully resolved. Some of these problems are of a practical kind, to do with providing a probabilistic description for trajectories which seem random, even though their underlying structure is deterministic. Other problems are of intrinsic mathematical interest, and treat such things as the pathology of the bifurcation structure, or the truly random behaviour, that can arise when the nonlinear function $F(X)$ of equation (1) is not analytical. One aim here is to stimulate research on these questions, particularly on the empirical questions which relate to processing data.

Third, consideration is given to some fields where these notions may find practical application. Such applications range from the abstractly metaphorical (where, for example, the transition from a stable point to "chaos" serves as a metaphor for the onset of turbulence in a fluid), to models for the dynamic behaviour of biological populations (where one can seek to use field or laboratory data to estimate the values of the parameters in the difference equation).

*King's College Research Centre, Cambridge CB2 1ST; on leave from Biology Department, Princeton University, Princeton 08540.

Fourth, there is a very brief review of the literature pertaining to the way this spectrum of behaviour—stable points, stable cycles, chaos—can arise in second or higher order difference equations (that is, two or more dimensions; two or more interacting species), where the onset of chaos usually requires less severe nonlinearities. Differential equations are also surveyed in this light; it seems that a three-dimensional system of first-order ordinary differential equations is required for the manifestation of chaotic behaviour.

The review ends with an evangelical plea for the introduction of these difference equations into elementary mathematics courses, so that students' intuition may be enriched by seeing the wild things that simple nonlinear equations can do.

First-order difference equations

One of the simplest systems an ecologist can study is a seasonally breeding population in which generations do not overlap[1–4]. Many natural populations, particularly among temperate zone insects (including many economically important crop and orchard pests), are of this kind. In this situation, the observational data will usually consist of information about the maximum, or the average, or the total population in each generation. The theoretician seeks to understand how the magnitude of the population in generation $t+1$, X_{t+1}, is related to the magnitude of the population in the preceding generation t, X_t: such a relationship may be expressed in the general form

$$X_{t+1} = F(X_t) \tag{1}$$

The function $F(X)$ will usually be what a biologist calls "density dependent", and a mathematician calls nonlinear; equation (1) is then a first-order, nonlinear difference equation.

Although I shall henceforth adopt the habit of referring to the variable X as "the population", there are countless situations outside population biology where the basic equation (1) applies. There are other examples in biology, as, for example in genetics[5,6] (where the equation describes the change in gene frequency in time) or in epidemiology[7] (with X the fraction of the population infected at time t). Examples in economics include models for the relationship between commodity quantity and price[8], for the theory of business cycles[9], and for the temporal sequences generated by various other economic quantities[10]. The general equation (1) also is germane to the social sciences[11], where it arises, for example, in theories of learning (where X may be the number of bits of information that can be remembered after an interval t), or in the propagation of rumours in variously structured societies (where X is the number of pêople to have heard the rumour after time t). The imaginative reader will be able to invent other contexts for equation (1).

In many of these contexts, and for biological populations in particular, there is a tendency for the variable X to increase from one generation to the next when it is small, and for it to decrease when it is large. That is, the nonlinear function $F(X)$ often has the following properties: $F(0)=0$; $F(X)$ increases monotonically as X increases through the range $0<X<A$ (with $F(X)$ attaining its maximum value at $X=A$); and $F(X)$ decreases monotonically as X increases beyond $X=A$. Moreover, $F(X)$ will usually contain one or more parameters which "tune" the severity of this nonlinear behaviour; parameters which tune the steepness of the hump in the $F(X)$ curve. These parameters will typically have some biological or economic or sociological significance.

A specific example is afforded by the equation[1,4,12–23]

$$N_{t+1} = N_t(a-bN_t) \tag{2}$$

This is sometimes called the "logistic" difference equation. In the limit $b=0$, it describes a population growing purely exponentially (for $a>1$); for $b\neq 0$, the quadratic nonlinearity produces a growth curve with a hump, the steepness of which is tuned by the parameter a. By writing $X=bN/a$, the equation may be brought into canonical form[1,4,12–23]

$$X_{t+1} = aX_t(1-X_t) \tag{3}$$

In this form, which is illustrated in Fig. 1, it is arguably the simplest nonlinear difference equation. I shall use equation (3) for most of the numerical examples and illustrations in this article. Although attractive to mathematicians by virtue of its extreme simplicity, in practical applications equation (3) has the disadvantage that it requires X to remain on the interval $0<X<1$; if X ever exceeds unity, subsequent iterations diverge towards $-\infty$ (which means the population becomes extinct). Furthermore, $F(X)$ in equation (3) attains a maximum value of $a/4$ (at $X=\frac{1}{2}$); the equation therefore possesses non-trivial dynamical behaviour only if $a<4$. On the other hand, all trajectories are attracted to $X=0$ if $a<1$. Thus for non-trivial dynamical behaviour we require $1<a<4$; failing this, the population becomes extinct.

Fig. 1 A typical form for the relationship between X_{t+1} and X_t described by equation (1). The curves are for equation (3), with $a = 2.707$ (a); and $a = 3.414$ (b). The dashed lines indicate the slope at the "fixed points" where $F(X)$ intersects the 45° line: for the case a this slope is less steep than $-45°$ and the fixed point is stable; for b the slope is steeper than $-45°$, and the point is unstable.

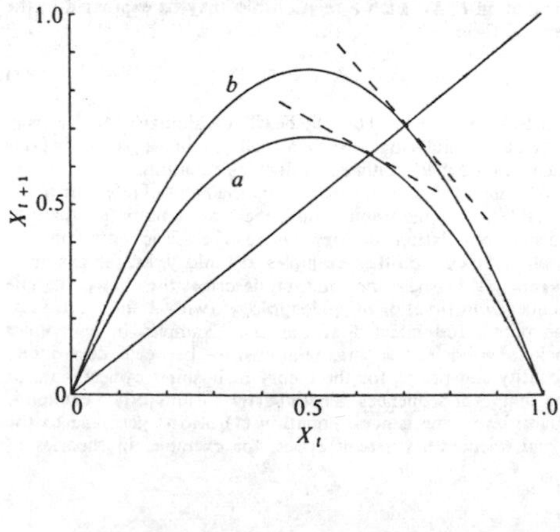

Another example, with a more secure provenance in the biological literature[1,23–27], is the equation

$$X_{t+1} = X_t \exp[r(1-X_t)] \tag{4}$$

This again describes a population with a propensity to simple exponential growth at low densities, and a tendency to decrease at high densities. The steepness of this nonlinear behaviour is tuned by the parameter r. The model is plausible for a single species population which is regulated by an epidemic disease at high density[28]. The function $F(X)$ of equation (4) is slightly more complicated than that of equation (3), but has the compensating advantage that local stability implies global stability[1] for all $X>0$.

The forms (3) and (4) by no means exhaust the list of single-humped functions $F(X)$ for equation (1) which can be culled from the ecological literature. A fairly full such catalogue is given, complete with references, by May and Oster[1]. Other similar mathematical functions are given by Metropolis *et al.*[16]. Yet other forms for $F(X)$ are discussed under the heading of "mathematical curiosities" below.

Dynamic properties of equation (1)

Possible constant, equilibrium values (or "fixed points") of X in equation (1) may be found algebraically by putting $X_{t+1}=X_t=X^*$, and solving the resulting equation

$$X^* = F(X^*) \tag{5}$$

An equivalent graphical method is to find the points where the curve $F(X)$ that maps X_t into X_{t+1} intersects the 45° line, $X_{t+1}=X_t$, which corresponds to the ideal nirvana of zero population growth; see Fig. 1. For the single-hump curves discussed above, and exemplified by equations (3) and (4), there are two such points: the trivial solution $X=0$, and a non-trivial solution X^* (which for equation (3) is $X^*=1-[1/a]$).

The next question concerns the stability of the equilibrium point X^*. This can be seen[24,25,19–21,1,4] to depend on the slope of the $F(X)$ curve at X^*. This slope, which is illustrated by the dashed lines in Fig. 1, can be designated

$$\lambda^{(1)}(X^*) = [dF/dX]_{X=X^*} \tag{6}$$

So long as this slope lies between 45° and $-45°$ (that is, $\lambda^{(1)}$ between $+1$ and -1), making an acute angle with the 45° ZPG line, the equilibrium point X^* will be at least locally stable, attracting all trajectories in its neighbourhood. In equation (3), for example, this slope is $\lambda^{(1)}=2-a$: the equilibrium point is therefore stable, and attracts all trajectories originating in the interval $0<X<1$, if and only if $1<a<3$.

As the relevant parameters are tuned so that the curve $F(X)$ becomes more and more steeply humped, this stability-determining slope at X^* may eventually steepen beyond $-45°$ (that is, $\lambda^{(1)}<-1$), whereupon the equilibrium point X^* is no longer stable.

What happens next? What happens, for example, for $a>3$ in equation (3)?

To answer this question, it is helpful to look at the map which relates the populations at successive intervals 2 generations apart; that is, to look at the function which relates X_{t+2} to X_t. This second iterate of equation (1) can be written

$$X_{t+2} = F[F(X_t)] \tag{7}$$

or, introducing an obvious piece of notation,

$$X_{t+2} = F^{(2)}(X_t) \tag{8}$$

The map so derived from equation (3) is illustrated in Figs 2 and 3.

Population values which recur every second generation (that

Nature Vol. 261 June 10 1976

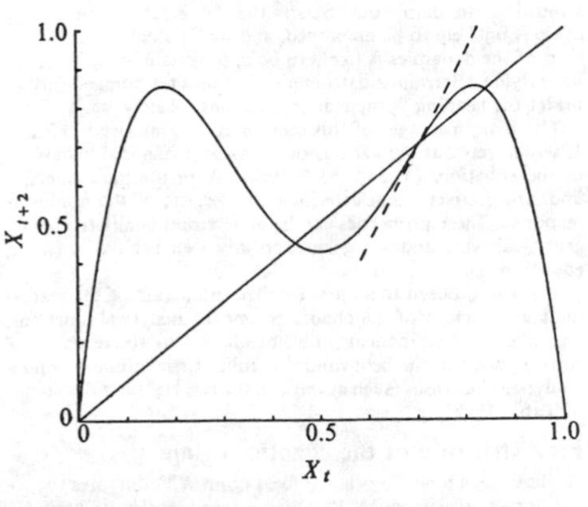

Fig. 2 The map relating X_{t+2} to X_t, obtained by two iterations of equation (3). This figure is for the case (*a*) of Fig. 1, $a = 2.707$: the basic fixed point is stable, and it is the only point at which $F^{(2)}(X)$ intersects the 45° line (where its slope, shown by the dashed line, is less steep than 45°).

is, fixed points with period 2) may now be written as X^*_2, and found either algebraically from

$$X^*_2 = F^{(2)}(X^*_2) \tag{9}$$

or graphically from the intersection between the map $F^{(2)}(X)$ and the 45° line, as shown in Figs 2 and 3. Clearly the equilibrium point X^* of equation (5) is a solution of equation (9); the basic fixed point of period 1 is a degenerate case of a period 2 solution. We now make a simple, but crucial, observation[1]: the slope of the curve $F^{(2)}(X)$ at the point X^*, defined as $\lambda^{(2)}(X^*)$ and illustrated by the dashed lines in Figs 2 and 3, is the square of the corresponding slope of $F(X)$

$$\lambda^{(2)}(X^*) = [\lambda^{(1)}(X^*)]^2 \tag{10}$$

This fact can now be used to make plain what happens when the fixed point X^* becomes unstable. If the slope of $F(X)$ is less than $-45°$ (that is, $|\lambda^{(1)}| < 1$), as illustrated by curve *a* in Fig. 1, then X^* is stable. Also, from equation (10), this implies $0 < \lambda^{(2)} < 1$ corresponding to the slope of $F^{(2)}$ at X^* lying between 0° and 45°, as shown in Fig. 2. As long as the fixed point X^* is stable, it provides the only non-trivial solution to equation (9). On the other hand, when $\lambda^{(1)}$ steepens beyond $-45°$ (that is, $|\lambda^{(1)}| > 1$), as illustrated by curve *b* in Fig 1, X^* becomes unstable. At the same time, from equation (10) this implies $\lambda^{(2)} > 1$, corresponding to the slope of $F^{(2)}$ at X^* steepening beyond 45°, as shown in Fig. 3. As this happens, the curve $F^{(2)}(X)$ must develop a "loop", and two new fixed points of period 2 appear, as illustrated in Fig. 3.

In short, as the nonlinear function $F(X)$ in equation (1) becomes more steeply humped, the basic fixed point X^* may become unstable. At exactly the stage when this occurs, there are born two new and initially stable fixed points of period 2, between which the system alternates in a stable cycle of period 2. The sort of graphical analysis indicated by Figs 1, 2 and 3, along with the equation (10), is all that is needed to establish this generic result[1,4].

As before, the stability of this period 2 cycle depends on the slope of the curve $F^{(2)}(X)$ at the 2 points. (This slope is easily shown to be the same at both points[1,20], and more generally to be the same at all *k* points on a period *k* cycle.) Furthermore, as is clear by imagining the intermediate stages between Figs 2 and 3, this stability-determining slope has the value $\lambda = +1$ at the birth of the 2-point cycle, and then decreases through zero towards $\lambda = -1$ as the hump in $F(X)$ continues to steepen. Beyond this point the period 2 points will in turn become unstable, and bifurcate to give an initially stable cycle of period 4. This in turn gives way to a cycle of period 8, and thence to a hierarchy of bifurcating stable cycles of periods 16, 32, 64, . . ., 2^n. In each case, the way in which a stable cycle of period *k* becomes unstable, simultaneously bifurcating to produce a new and initially stable cycle of period $2k$, is basically similar to the process just adumbrated for $k=1$. A more full and rigorous account of the material covered so far is in ref. 1.

This "very beautiful bifurcation phenomenon"[22] is depicted in Fig. 4, for the example equation (3). It cannot be too strongly emphasised that the process is generic to most functions $F(X)$ with a hump of tunable steepness. Metropolis *et al.*[16] refer to this hierarchy of cycles of periods 2^n as the harmonics of the fixed point X^*.

Although this process produces an infinite sequence of cycles with periods 2^n ($n \to \infty$), the "window" of parameter values wherein any one cycle is stable progressively diminishes, so that the entire process is a convergent one, being bounded above by some critical parameter value. (This is true for most, but not all, functions $F(X)$: see equation (17) below.) This critical parameter value is a point of accumulation of period 2^n cycles. For equation (3) it is denoted a_c: $a_c = 3.5700\ldots$

Beyond this point of accumulation (for example, for $a > a_c$ in equation (3)) there are an infinite number of fixed points with different periodicities, and an infinite number of different periodic cycles. There are also an uncountable number of initial points X_0 which give totally aperiodic (although bounded) trajectories; no matter how long the time series generated by $F(X)$ is run out, the pattern never repeats. These facts may be established by a variety of methods[1,4,20,22,29]. Such a situation, where an infinite number of different orbits can occur, has been christened "chaotic" by Li and Yorke[29].

As the parameter increases beyond the critical value, at first all these cycles have even periods, with X_t alternating up and down between values above, and values below, the fixed point X^*. Although these cycles may in fact be very complicated (having a non-degenerate period of, say, 5,726 points before repeating), they will seem to the casual observer to be rather like a somewhat "noisy" cycle of period 2. As the parameter value continues to increase, there comes a stage (at $a = 3.6786$ . . for equation (3)) at which the first odd period cycle appears. At first these odd cycles have very long periods, but as the parameter value continues to increase cycles with smaller and smaller odd periods are picked up, until at last the three-point

Fig. 3 As for Fig. 2, except that here $a = 3.414$, as in Fig. 1*b*. The basic fixed point is now unstable: the slope of $F^{(2)}(X)$ at this point steepens beyond 45°, leading to the appearance of two new solutions of period 2.

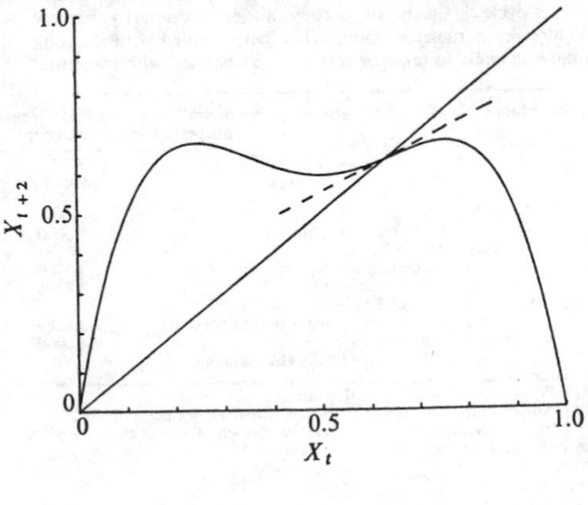

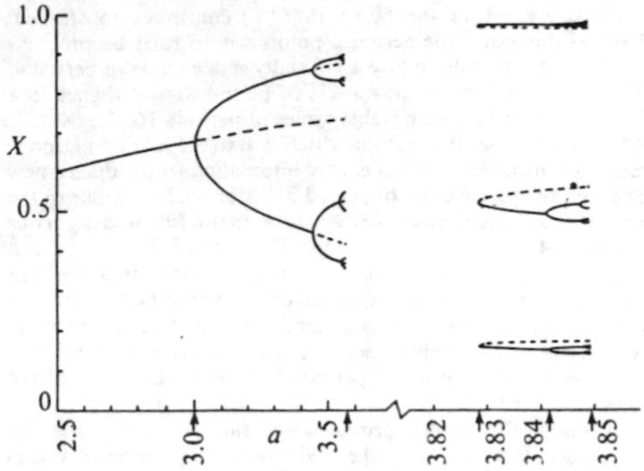

Fig. 4 This figure illustrates some of the stable (———) and unstable (— — — —) fixed points of various periods that can arise by bifurcation processes in equation (1) in general, and equation (3) in particular. To the left, the basic stable fixed point becomes unstable and gives rise by a succession of pitchfork bifurcations to stable harmonics of period 2^n; none of these cycles is stable beyond $a = 3.5700$. To the right, the two period 3 cycles appear by tangent bifurcation: one is initially unstable; the other is initially stable, but becomes unstable and gives way to stable harmonics of period 3×2^n, which have a point of accumulation at $a = 3.8495$. Note the change in scale on the a axis, needed to put both examples on the same figure. There are infinitely many other such windows, based on cycles of higher periods.

cycle appears (at a=3.8284 . . for equation (3)). Beyond this point, there are cycles with every integer period, as well as an uncountable number of asymptotically aperiodic trajectories: Li and Yorke[20] entitle their original proof of this result "Period Three Implies Chaos".

The term "chaos" evokes an image of dynamical trajectories which are indistinguishable from some stochastic process. Numerical simulations[12,15,21,23,25] of the dynamics of equation (3), (4) and other similar equations tend to confirm this impression. But, for smooth and "sensible" functions $F(X)$ such as in equations (3) and (4), the underlying mathematical fact is that for any specified parameter value there is one unique cycle that is stable, and that attracts essentially all initial points[22,29] (see ref. 4, appendix A, for a simple and lucid exposition). That is, there is one cycle that "owns" almost all initial points; the remaining infinite number of other cycles, along with the asymptotically aperiodic trajectories, own a set of points which, although uncountable, have measure zero.

As is made clear by Tables 3 and 4 below, any one particular stable cycle is likely to occupy an extraordinarily narrow window of parameter values. This fact, coupled with the long time it is likely to take for transients associated with the initial conditions to damp out, means that in practice the unique cycle is unlikely to be unmasked, and that a stochastic description of the dynamics is likely to be appropriate, in spite of the underlying deterministic structure. This point is pursued further under the heading "practical applications", below.

The main messages of this section are summarised in Table 1, which sets out the various domains of dynamical behaviour of the equations (3) and (4) as functions of the parameters, a and r respectively, that determine the severity of the nonlinear response. These properties can be understood qualitatively in a graphical way, and are generic to any well behaved $F(X)$ in equation (1).

We now proceed to a more detailed discussion of the mathematical structure of the chaotic regime for analytical functions, and then to the practical problems alluded to above and to a consideration of the behavioural peculiarities exhibited by non-analytical functions (such as those in the two right hand columns of Table 1).

Fine structure of the chaotic regime

We have seen how the original fixed point X^* bifurcates to give harmonics of period 2^n. But how do new cycles of period k arise?

The general process is illustrated in Fig. 5, which shows how period 3 cycles originate. By an obvious extension of the notation introduced in equation (8), populations three generations apart are related by

$$X_{t+3} = F^{(3)}(X_t) \tag{11}$$

If the hump in $F(X)$ is sufficiently steep, the threefold iteration will produce a function $F^{(3)}(X)$ with 4 humps, as shown in Fig. 5 for the $F(X)$ of equation (3). At first (for $a<3.8284$ . . in equation 3) the 45° line intersects this curve only at the single point X^* (and at X=0), as shown by the solid curve in Fig. 5. As the hump in $F(X)$ steepens, the hills and valleys in $F^{(3)}(X)$ become more pronounced, until simultaneously the first two valleys sink and the final hill rises to touch the 45° line, and then to intercept it at 6 new points, as shown by the dashed curve in Fig. 5. These 6 points divide into two distinct three-point cycles. As can be made plausible by imagining the intermediate stages in Fig. 5, it can be shown that the stability-determining slope of $F^{(3)}(X)$ at three of these points has a common value, which is $\lambda^{(3)} = +1$ at their birth, and thereafter steepens beyond +1: this period 3 cycle is never stable. The slope of $F^{(3)}(X)$ at the other three points begins at $\lambda^{(3)} = +1$, and then decreases towards zero, resulting in a stable cycle of period 3. As $F(X)$ continues to steepen, the slope $\lambda^{(3)}$ for this intially stable three-point cycle decreases beyond −1; the cycle becomes unstable, and gives rise by the bifurcation process discussed in the previous section to stable cycles of period 6, 12, 24, . . ., 3×2^n. This birth of a stable and unstable pair of period 3 cycles, and the subsequent harmonics which arise as the initially stable cycle becomes unstable, are illustrated to the right of Fig. 4.

Table 1 Summary of the way various "single-hump" functions $F(X)$, from equation (1), behave in the chaotic region, distinguishing the dynamical properties which are generic from those which are not

The function $F(X)$ of equation (1)	$aX(1-X)$	$X\exp[r(1-X)]$	aX; if $X<\frac{1}{2}$ / $a(1-X)$; if $X>\frac{1}{2}$	λX; if $X<1$ / λX^{1-b}; if $X>1$
Tunable parameter	a	r	a	b
Fixed point becomes unstable	3.0000	2.0000	1.0000*	2.0000
"Chaotic" region begins [point of accumulation of cycles of period 2^n]	3.5700	2.6924	1.0000	2.0000
First odd-period cycle appears	3.6786	2.8332	1.4142	2.6180
Cycle with period 3 appears [and therefore every integer period present]	3.8284	3.1024	1.6180	3.0000
"Chaotic" region ends	4.0000†	∞‡	2.000†	∞‡
Are there stable cycles in the chaotic region?	Yes	Yes	No	No

* Below this a value, X 0 is stable.
† All solutions are attracted to $-\infty$ for a values beyond this.
‡ In practice, as r or b becomes large enough, X will eventually be carried so low as to be effectively zero, thus producing extinction in models of biological populations.

Table 2 Catalogue of the number of periodic points, and of the various cycles (with periods k = 1 up to 12), arising from equation (1) with a single-humped function $F(X)$

k	1	2	3	4	5	6	7	8	9	10	11	12
Possible total number of points with period k	2	4	8	16	32	64	128	256	512	1,024	2,048	4,096
Possible total number of points with non-degenerate period k	2	2	6	12	30	54	126	240	504	990	2,046	4,020
Total number of cycles of period k, including those which are degenerate and/or harmonics and/or never locally stable	2	3	4	6	8	14	20	36	60	108	188	352
Total number of non-degenerate cycles (including harmonics and unstable cycles)	2	1	2	3	6	9	18	30	56	99	186	335
Total number of non-degenerate, stable cycles (including harmonics)	1	1	1	2	3	5	9	16	28	51	93	170
Total number of non-degenerate, stable cycles whose basic period is k (that is, excluding harmonics)	1	–	1	1	3	4	9	14	28	48	93	165

There are, therefore, two basic kinds of bifurcation processes[1,4] for first order difference equations. Truly new cycles of period k arise in pairs (one stable, one unstable) as the hills and valleys of higher iterates of $F(X)$ move, respectively, up and down to intercept the 45° line, as typified by Fig. 5. Such cycles are born at the moment when the hills and valleys become tangent to the 45° line, and the initial slope of the curve $F^{(k)}$ at the points is thus $\lambda^{(k)}=+1$: this type of bifurcation may be called[1,4] a tangent bifurcation or a $\lambda=+1$ bifurcation. Conversely, an originally stable cycle of period k may become unstable as $F(X)$ steepens. This happens when the slope of $F^{(k)}$ at these period k points steepens beyond $\lambda^{(k)}=-1$, whereupon a new and initially stable cycle of period $2k$ is born in the way typified by Figs 2 and 3. This type of bifurcation may be called a pitchfork bifurcation (borrowing an image from the left hand side of Fig. 4) or a $\lambda=-1$ bifurcation[1,4].

Putting all this together, we conclude that as the parameters in $F(X)$ are varied the fundamental, stable dynamical units are cycles of basic period k, which arise by tangent bifurcation, along with their associated cascade of harmonics of periods $k2^n$, which arise by pitchfork bifurcation. On this basis, the constant equilibrium solution X^* and the subsequent hierarchy of stable cycles of periods 2^n is merely a special case, albeit a conspicuously important one (namely $k=1$), of a general phenomenon. In addition, remember[1,4,22,29] that for sensible, analytical functions (such as, for example, those in equations (3) and (4)) there is a unique stable cycle for each value of the parameter in $F(X)$. The entire range of parameter values ($1<a<4$ in equation (3), $0<r$ in equation (4)) may thus be regarded as made up of infinitely many windows of parameter

Fig. 5 The relationship between X_{t+3} and X_t, obtained by three iterations of equation (3). The solid curve is for a = 3.7, and only intersects the 45° line once. As a increases, the hills and valleys become more pronounced. The dashed curve is for a = 3.9, and six new period 3 points have appeared (arranged as two cycles, each of period 3).

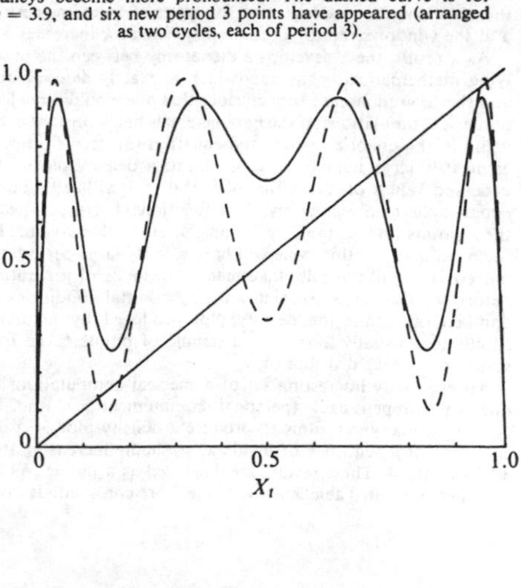

values—some large, some unimaginably small—each corresponding to a single one of these basic dynamical units. Tables 3 and 4, below, illustrate this notion. These windows are divided from each other by points (the points of accumulation of the harmonics of period $k2^n$) at which the system is truly chaotic, with no attractive cycle: although there are infinitely many such special parameter values, they have measure zero on the interval of all values.

How are these various cycles arranged along the interval of relevant parameter values? This question has to my knowledge been answered independently by at least 6 groups of people, who have seen the problem in the context of combinatorial theory[16,30], numerical analysis[13,14], population biology[1], and dynamical systems theory[22,31] (broadly defined).

A simple-minded approach (which has the advantage of requiring little technical apparatus, and the disadvantage of being rather clumsy) consists of first answering the question, how many period k points can there be? That is, how many distinct solutions can there be to the equation

$$X^*_k = F^{(k)}(X^*_k)? \qquad (12)$$

If the function $F(X)$ is sufficiently steeply humped, as it will be once the parameter values are sufficiently large, each successive iteration doubles the number of humps, so that $F^{(k)}(X)$ has 2^{k-1} humps. For large enough parameter values, all these hills and valleys will intersect the 45° line, producing 2^k fixed points of period k. These are listed for $k\leq 12$ in the top row of Table 2. Such a list includes degenerate points of period k, whose period is a submultiple of k; in particular, the two period 1 points ($X=0$ and X^*) are degenerate solutions of equation (12) for all k. By working from left to right across Table 2, these degenerate points can be subtracted out, to leave the total number of non-degenerate points of basic period k, as listed in the second row of Table 2. More sophisticated ways of arriving at this result are given elsewhere[13,14,16,22,30,31].

For example, there eventually are $2^6=64$ points with period 6. These include the two points of period 1, the period 2 "harmonic" cycle, and the stable and unstable pair of triplets of points with period 3, for a total of 10 points whose basic period is a submultiple of 6; this leaves 54 points whose basic period is 6.

The 2^k period k points are arranged into various cycles of period k, or submultiples thereof, which appear in succession by either tangent or pitchfork bifurcation as the parameters in $F(X)$ are varied. The third row in Table 2 catalogues the total number of distinct cycles of period k which so appear. In the fourth row[14], the degenerate cycles are subtracted out, to give the total number of non-degenerate cycles of period k: these numbers must equal those of the second row divided by k. This fourth row includes the (stable) harmonics which arise by pitchfork bifurcation, and the pairs of stable–unstable cycles arising by tangent bifurcation. By subtracting out the cycles which are unstable from birth, the total number of possible stable cycles is given in row five; these figures can also be obtained by less pedestrian methods[13,16,30]. Finally we may subtract out the stable cycles which arise by pitchfork bifurcation, as harmonics of some simpler cycle, to arrive at the final

Table 3 A catalogue of the stable cycles (with basic periods up to 6) for the equation $X_{t+1} = aX_t(1-X_t)$

Period of basic cycle	*a* value at which: Basic cycle first appears	Basic cycle becomes unstable	Subsequent cascade of "harmonics" with period $k2^n$ all become unstable	Width of the range of *a* values over which the basic cycle, or one of its harmonics, is attractive
1	1.0000	3.0000	3.5700	2.5700
3	3.8284	3.8415	3.8495	0.0211
4	3.9601	3.9608	3.9612	0.0011
5(*a*)	3.7382	3.7411	3.7430	0.0048
5(*b*)	3.9056	3.9061	3.9065	0.0009
5(*c*)	3.99026	3.99030	3.99032	0.00006
6(*a*)	3.6265	3.6304	3.6327	0.0062
6(*b*)	3.937516	3.937596	3.937649	0.000133
6(*c*)	3.977760	3.977784	3.977800	0.000040
6(*d*)	3.997583	3.997585	3.997586	0.000003

row in Table 2, which lists the number of stable cycles whose basic period is k.

Returning to the example of period 6, we have already noted the five degenerate cycles whose periods are submultiples of 6. The remaining 54 points are parcelled out into one cycle of period 6 which arises as the harmonic of the only stable three-point cycle, and four distinct pairs of period 6 cycles (that is, four initially stable ones and four unstable ones) which arise by successive tangent bifurcations. Thus, reading from the foot of the column for period 6 in Table 2, we get the numbers 4, 5, 9, 14.

Using various labelling tricks, or techniques from combinatorial theory, it is also possible to give a generic list of the order in which the various cycles appear[1,13,16,22]. For example, the basic stable cycles of periods 3, 5, 6 (of which there are respectively 1, 3, 4) must appear in the order 6, 5, 3, 5, 6, 6, 5, 6: compare Tables 3 and 4. Metropolis *et al.*[16] give the explicit such generic list for all cycles of period $k \leq 11$.

As a corollary it follows that, given the most recent cycle to appear, it is possible (at least in principle) to catalogue all the cycles which have appeared up to this point. An especially elegant way of doing this is given by Smale and Williams[22], who show, for example, that when the stable cycle of period 3 first originates, the total number of other points with periods k, N_k, which have appeared by this stage satisfy the Fibonacci series, $N_k = 2, 4, 5, 8, 12, 19, 30, 48, 77, 124, 200, 323$ for $k = 1, 2, \ldots, 12$: this is to be contrasted with the total number of points of period k which will eventually appear (the top row of Table 2) as $F(X)$ continues to steepen.

Such catalogues of the total number of fixed points, and of their order of appearance, are relatively easy to construct. For any particular function $F(X)$, the numerical task of finding the windows of parameter values wherein any one cycle or its harmonics is stable is, in contrast, relatively tedious and inelegant. Before giving such results, two critical parameter values of special significance should be mentioned.

Hoppensteadt and Hyman[21] have given a simple graphical method for locating the parameter value in the chaotic regime at which the first odd period cycle appears. Their analytic recipe is as follows. Let α be the parameter which tunes the steepness of $F(X)$ (for example, $\alpha = a$ for equation (3), $\alpha = r$ for equation (4)), $X^*(\alpha)$ be the fixed point of period 1 (the non-trivial solution of equation (5)), and $X_{max}(\alpha)$ the maximum value attainable from iterations of equation (1) (that is, the value of $F(X)$ at its hump or stationary point). The first odd period cycle appears for that value of α which satisfies[21,31]

$$X^*(\alpha) = F^{(2)}(X_{max}(\alpha)) \qquad (13)$$

As mentioned above, another critical value is that where the period 3 cycle first appears. This parameter value may be found numerically from the solutions of the third iterate of equation (1): for equation (3) it is[14] $a = 1 + \sqrt{8}$.

Myrberg[13] (for all $k \leq 10$) and Metropolis *et al.*[16]. (for all $k \leq 7$) have given numerical information about the stable cycles in equation (3). They do not give the windows of parameter values, but only the single value at which a given cycle is maximally stable; that is, the value of a for which the stability-determining slope of $F^{(k)}(X)$ is zero, $\lambda^{(k)} = 0$. Since the slope of the k-times iterated map $F^{(k)}$ at any point on a period k cycle is simply equal to the product of the slopes of $F(X)$ at each of the points X^*_k on this cycle[1,8,20], the requirement $\lambda^{(k)} = 0$ implies that $X = A$ (the stationary point of $F(X)$, where $\lambda^{(1)} = 0$) is one of the periodic points in question, which considerably simplifies the numerical calculations.

For each basic cycle of period k (as catalogued in the last row of Table 2), it is more interesting to know the parameter values at which: (1) the cycle first appears (by tangent bifurcation); (2) the basic cycle becomes unstable (giving rise by successive pitchfork bifurcations to a cascade of harmonics of periods $k2^n$); (3) all the harmonics become unstable (the point of accumulation of the period $k2^n$ cycles). Tables 3 and 4 extend the work of May and Oster[1], to give this numerical information for equations (3) and (4), respectively. (The points of accumulation are not ground out mindlessly, but are calculated by a rapidly convergent iterative procedure, see ref. 1, appendix A.) Some of these results have also been obtained by Gumowski and Mira[32].

Practical problems

Referring to the paradigmatic example of equation (3), we can now see that the parameter interval $1 < a < 4$ is made up of a one-dimensional mosaic of infinitely many windows of a-values, in each of which a unique cycle of period k, or one of its harmonics, attracts essentially all initial points. Of these windows, that for $1 < a < 3.5700\ldots$ corresponding to $k = 1$ and its harmonics is by far the widest and most conspicuous. Beyond the first point of accumulation, it can be seen from Table 3 that these windows are narrow, even for cycles of quite low periods, and the windows rapidly become very tiny as k increases.

As a result, there develops a dichotomy between the underlying mathematical behaviour (which is exactly determinable) and the "commonsense" conclusions that one would draw from numerical simulations. If the parameter a is held constant at one value in the chaotic region, and equation (3) iterated for an arbitrarily large number of generations, a density plot of the observed values of X_t on the interval 0 to 1 will settle into k equal spikes (more precisely, delta functions) corresponding to the k points on the stable cycle appropriate to this a-value. But for most a-values this cycle will have a fairly large period, and moreover it will typically take many thousands of generations before the transients associated with the initial conditions are damped out: thus the density plot produced by numerical simulations usually looks like a sample of points taken from some continuous distribution.

An especially interesting set of numerical computations are due to Hoppensteadt (personal communication) who has combined many iterations to produce a density plot of X_t for each one of a sequence of a-values, gradually increasing from $3.5700\ldots$ to 4. These results are displayed as a movie. As can be expected from Table 3, some of the more conspicuous cycles do show up as sets of delta functions: the 3-cycle and its first few harmonics; the first 5-cycle; the first 6-cycle. But for most values of a the density plot looks like the sample function of a random process. This is particularly true in the neighbourhood of the a-value where the first odd cycle appears ($a=3.6786\ldots$), and again in the neighbourhood of $a=4$: this is not surprising, because each of these locations is a point of accumulation of points of accumulation. Despite the underlying discontinuous changes in the periodicities of the stable cycles, the observed density pattern tends to vary smoothly. For example, as a increases toward the value at which the 3-cycle appears, the density plot tends to concentrate around three points, and it smoothly diffuses away from these three points after the 3-cycle and all its harmonics become unstable.

I think the most interesting mathematical problem lies in designing a way to construct some approximate and "effectively continuous" density spectrum, despite the fact that the exact density function is determinable and is always a set of delta functions. Perhaps such techniques have already been developed in ergodic theory[33] (which lies at the foundations of statistical mechanics), as for example in the use of "coarse-grained observers". I do not know.

Such an effectively stochastic description of the dynamical properties of equation (4) for large r has been provided[28], albeit by tactical tricks peculiar to that equation rather than by any general method. As r increases beyond about 3, the trajectories generated by this equation are, to an increasingly good approximation, almost periodic with period $(1/r)\exp(r-1)$.

The opinion I am airing in this section is that although the exquisite fine structure of the chaotic regime is mathematically fascinating, it is irrelevant for most practical purposes. What seems called for is some effectively stochastic description of the deterministic dynamics. Whereas the various statements about the different cycles and their order of appearance can be made in generic fashion, such stochastic description of the actual dynamics will be quite different for different $F(X)$: witness the difference between the behaviour of equation (4), which for large r is almost periodic "outbreaks" spaced many generations apart, versus the behaviour of equation (3), which for $a\to 4$ is not very different from a series of Bernoulli coin flips.

Mathematical curiosities

As discussed above, the essential reason for the existence of a succession of stable cycles throughout the "chaotic" regime is that as each new pair of cycles is born by tangent bifurcation (see Fig. 5), one of them is at first stable, by virtue of the way the smoothly rounded hills and valleys intercept the 45° line. For analytical functions $F(X)$, the only parameter values for which the density plot or "invariant measure" is continuous and truly ergodic are at the points of accumulation of harmonics, which divide one stable cycle from the next. Such exceptional parameter values have found applications, for example, in the use of equation (3) with $a=4$ as a random number generator[34,35]: it has a continuous density function proportional to $[X(1-X)]^{-\frac{1}{2}}$ in the interval $0<X<1$.

Non-analytical functions $F(X)$ in which the hump is in fact a spike provide an interesting special case. Here we may imagine spikey hills and valleys moving to intercept the 45° line in Fig. 5, and it may be that both the cycles born by tangent bifurcation are unstable from the outset (one having $\lambda^{(k)}>1$, the other $\lambda^{(k)}<-1$), for all $k>1$. There are then no stable cycles in the chaotic regime, which is therefore literally chaotic with a continuous and truly ergodic density distribution function.

One simple example is provided by

$$X_{t+1} = aX_t;\ \text{if } X_t<\tfrac{1}{2} \qquad\qquad (14)$$
$$X_{t+1} = a(1-X_t);\ \text{if } X_t>\tfrac{1}{2}$$

defined on the interval $0<X<1$. For $0<a<1$, all trajectories are attracted to $X=0$; for $1<a<2$, there are infinitely many periodic orbits, along with an uncountable number of aperiodic trajectories, none of which are locally stable. The first odd period cycle appears at $a=\sqrt{2}$, and all integer periods are represented beyond $a=(1+\sqrt{5})/2$. Kac[36] has given a careful discussion of the case $a=2$. Another example, this time with an extensive biological pedigree[1–3], is the equation

$$X_{t+1} = \lambda X_t;\ \text{if } X_t<1 \qquad\qquad (15)$$
$$X_{t+1} = \lambda X_t^{1-b};\ \text{if } X_t>1$$

If $\lambda>1$ this possesses a globally stable equilibrium point for $b<2$. For $b>2$ there is again true chaos, with no stable cycles: the first odd cycle appears at $b=(3+\sqrt{5})/2$, and all integer periods are present beyond $b=3$. The dynamical properties of equations (14) and (15) are summarised to the right of Table 2.

The absence of analyticity is a necessary, but not a sufficient, condition for truly random behaviour[31]. Consider, for example,

$$X_{t+1} = (a/2)X_t;\ \text{if } X_t<\tfrac{1}{2}$$
$$X_{t+1} = aX_t(1-X_t)\ ;\ \text{if } X_t>\tfrac{1}{2} \qquad\qquad (16)$$

This is the parabola of equation (3) and Fig. 1, but with the left hand half of $F(X)$ flattened into a straight line. This equation does possess windows of a values, each with its own stable cycle, as described generically above. The stability-determining slopes $\lambda^{(k)}$ vary, however, discontinuously with the parameter a, and the widths of the simpler stable regions are narrower than for equation (3): the fixed point becomes unstable at $a=3$; the point of accumulation of the subsequent harmonics is at $a=3.27\ldots$; the first odd cycle appears at $a=3.44\ldots$; the 3-point cycle at $a=3.67\ldots$ (compare the first column in Table 1).

These eccentricities of behaviour manifested by non-analytical functions may be of interest for exploring formal questions in ergodic theory. I think, however, that they have no relevance to models in the biological and social sciences, where functions such as $F(X)$ should be analytical. This view is elaborated elsewhere[37].

As a final curiosity, consider the equation

$$X_{t+1} = \lambda X_t[1+X_t]^{-\beta} \qquad\qquad (17)$$

Table 4 Catalogue of the stable cycles (with basic periods up to 6) for the equation $X_{t+1} = X_t\exp[r(1-X_t)]$

Period of basic cycle	r value at which: Basic cycle first appears	Basic cycle becomes unstable	Subsequent cascade of "harmonics" with period $k2^n$ *all* become unstable	Width of the range of r values over with the basic cycle, or one of its harmonics, is attractive
1	0.0000	2.0000	2.6924	2.6924
3	3.1024	3.1596	3.1957	0.0933
4	3.5855	3.6043	3.6153	0.0298
5(*a*)	2.9161	2.9222	2.9256	0.0095
5(*b*)	3.3632	3.3664	3.3682	0.0050
5(*c*)	3.9206	3.9295	3.9347	0.0141
6(*a*)	2.7714	2.7761	2.7789	0.0075
6(*b*)	3.4558	3.4563	3.4567	0.0009
6(*c*)	3.7736	3.7745	3.7750	0.0014
6(*d*)	4.1797	4.1848	4.1880	0.0083

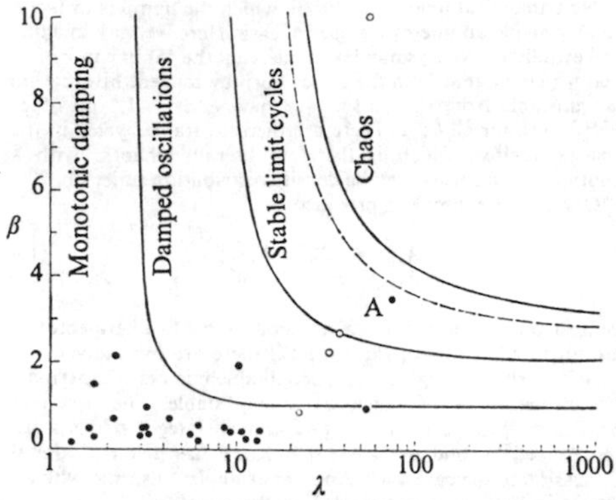

Fig. 6 The solid lines demarcate the stability domains for the density dependence parameter, β, and the population growth rate, λ, in equation (17); the dashed line shows where 2-point cycles give way to higher cycles of period 2^n. The solid circles come from analyses of life table data on field populations, and the open circles from laboratory populations (from ref. 3, after ref. 39).

This has been used to fit a considerable amount of data on insect populations[38,39]. Its stability behaviour, as a function of the two parameters λ and β, is illustrated in Fig. 6. Notice that for $\lambda < 7.39 \ldots$ there is a globally stable equilibrium point for all β; for $7.39 \ldots < \lambda < 12.50 \ldots$ this fixed point becomes unstable for sufficiently large β, bifurcating to a stable 2-point cycle which is the solution for all larger β; as λ increases through the range $12.50 \ldots < \lambda < 14.77 \ldots$ various other harmonics of period 2^n appear in turn. The hierarchy of bifurcating cycles of period 2^n is thus truncated, and the point of accumulation and subsequent regime of chaos is not achieved (even for arbitrarily large β) until $\lambda > 14.77 \ldots$.

Applications

The fact that the simple and deterministic equation (1) can possess dynamical trajectories which look like some sort of random noise has disturbing practical implications. It means, for example, that apparently erratic fluctuations in the census data for an animal population need not necessarily betoken either the vagaries of an unpredictable environment or sampling errors: they may simply derive from a rigidly deterministic population growth relationship such as equation (1). This point is discussed more fully and carefully elsewhere[1].

Alternatively, it may be observed that in the chaotic regime arbitrarily close initial conditions can lead to trajectories which, after a sufficiently long time, diverge widely. This means that, even if we have a simple model in which all the parameters are determined exactly, long term prediction is nevertheless impossible. In a meteorological context, Lorenz[15] has called this general phenomenon the "butterfly effect": even if the atmosphere could be described by a deterministic model in which all parameters were known, the fluttering of a butterfly's wings could alter the initial conditions, and thus (in the chaotic regime) alter the long term prediction.

Fluid turbulence provides a classic example where, as a parameter (the Reynolds number) is tuned in a set of deterministic equations (the Navier–Stokes equations), the motion can undergo an abrupt transition from some stable configuration (for example, laminar flow) into an apparently stochastic, chaotic regime. Various models, based on the Navier–Stokes differential equations, have been proposed as mathematical metaphors for this process[15,40,41]. In a recent review of the theory of turbulence, Martin[42] has observed that the one-dimensional difference equation (1) may be useful in this context. Compared with the earlier models[15,40,41], it has the disadvantage of being even more abstractly metaphorical, and the advantage of having a spectrum of dynamical behaviour which is more richly complicated yet more amenable to analytical investigation.

A more down-to-earth application is possible in the use of equation (1) to fit data[1,2,3,38,39,43] on biological populations with discrete, non-overlapping generations, as is the case for many temperate zone arthropods. Figure 6 shows the parameter values λ and β that are estimated[39] for 24 natural populations and 4 laboratory populations when equation (17) is fitted to the available data. The figure also shows the theoretical stability domains: a stable point; its stable harmonics (stable cycles of period 2^n); chaos. The natural populations tend to have stable equilibrium point behaviour. The laboratory populations tend to show oscillatory or chaotic behaviour; their behaviour may be exaggeratedly nonlinear because of the absence, in a laboratory setting, of many natural mortality factors. It is perhaps suggestive that the most oscillatory natural population (labelled *A* in Fig. 6) is the Colorado potato beetle, whose present relationship with its host plant lacks an evolutionary pedigree. These remarks are only tentative, and must be treated with caution for several reasons. Two of the main caveats are that there are technical difficulties in selecting and reducing the data, and that there are no single species populations in the natural world: to obtain a one-dimensional difference equation by replacing a population's interactions with its biological and physical environment by passive parameters (such as λ and β) may do great violence to the reality.

Some of the many other areas where these ideas have found applications were alluded to in the second section, above[5–11]. One aim of this review article is to provoke applications in yet other fields.

Related phenomena in higher dimensions

Pairs of coupled, first-order difference equations (equivalent to a single second-order equation) have been investigated in several contexts[4,44–46], particularly in the study of temperate zone arthropod prey–predator systems[2–4,23,47]. In these two-dimensional systems, the complications in the dynamical behaviour are further compounded by such facts as: (1) even for analytical functions, there can be truly chaotic behaviour (as for equations (14) and (15)), corresponding to so-called "strange attractors"; and (2) two or more different stable states (for example, a stable point and a stable cycle of period 3) can occur together for the same parameter values[4]. In addition, the manifestation of these phenomena usually requires less severe nonlinearities (less steeply humped $F(X)$) than for the one-dimensional case.

Similar systems of first-order ordinary differential equations, or two coupled first-order differential equations, have much simpler dynamical behaviour, made up of stable and unstable points and limit cycles[48]. This is basically because in continuous two-dimensional systems the inside and outside of closed curves can be distinguished; dynamic trajectories cannot cross each other. The situation becomes qualitatively more complicated, and in many ways analogous to first-order difference equations, when one moves to systems of three or more coupled, first-order ordinary differential equations (that is, three-dimensional systems of ordinary differential equations). Scanlon (personal communication) has argued that chaotic behaviour and "strange attractors", that is solutions which are neither points nor periodic orbits[48], are typical of such systems. Some well studied examples arise in models for reaction–diffusion systems in chemistry and biology[49], and in the models of Lorenz[15] (three dimensions) and Ruelle and Takens[40] (four dimensions) referred to above. The analysis of these systems is, by virtue of their higher dimensionality, much less transparent than for equation (1).

An explicit and rather surprising example of a system which

Nature Vol. 261 June 10 1976 **467**

has recently been studied from this viewpoint is the ordinary differential equations used in ecology to describe competing species. For one or two species these systems are very tame: dynamic trajectories will converge on some stable equilibrium point (which may represent coexistence, or one or both species becoming extinct). As Smale[50] has recently shown, however, for 3 or more species these general equations can, in a certain reasonable and well-defined sense, be compatible with any dynamical behaviour. Smale's[50] discussion is generic and abstract: a specific study of the very peculiar dynamics which can be exhibited by the familiar Lotka-Volterra equations once there are 3 competitors is given by May and Leonard[51].

Conclusion

In spite of the practical problems which remain to be solved, the ideas developed in this review have obvious applications in many areas.

The most important applications, however, may be pedagogical.

The elegant body of mathematical theory pertaining to linear systems (Fourier analysis, orthogonal functions, and so on), and its successful application to many fundamentally linear problems in the physical sciences, tends to dominate even moderately advanced University courses in mathematics and theoretical physics. The mathematical intuition so developed ill equips the student to confront the bizarre behaviour exhibited by the simplest of discrete nonlinear systems, such as equation (3). Yet such nonlinear systems are surely the rule, not the exception, outside the physical sciences.

I would therefore urge that people be introduced to, say, equation (3) early in their mathematical education. This equation can be studied phenomenologically by iterating it on a calculator, or even by hand. Its study does not involve as much conceptual sophistication as does elementary calculus. Such study would greatly enrich the student's intuition about nonlinear systems.

Not only in research, but also in the everyday world of politics and economics, we would all be better off if more people realised that simple nonlinear systems do not necessarily possess simple dynamical properties.

I have received much help from F. C. Hoppensteadt, H. E. Huppert, A. I. Mees, C. J. Preston, S. Smale, J. A. Yorke, and particularly from G. F. Oster. This work was supported in part by the NSF.

1 May, R. M., and Oster, G. F., *Am. Nat.*, 110 (in the press).
2 Varley, G. C., Gradwell, G. R., and Hassell, M. P., *Insect Population Ecology* (Blackwell, Oxford, 1973).
3 May, R. M. (ed.), *Theoretical Ecology: Principles and Applications* (Blackwell, Oxford, 1976).
4 Guckenheimer, J., Oster, G. F., and Ipaktchi, A., *Theor. Pop. Biol.* (in the press).
5 Oster, G. F., Ipaktchi, A., and Rocklin, I., *Theor. Pop. Biol.* (in the press).
6 Asmussen, M. A., and Feldman, M. W., *J. theor. Biol.* (in the press).
7 Hoppensteadt, F. C., *Mathematical Theories of Populations: Demographics, Genetics and Epidemics* (SIAM, Philadelphia, 1975).
8 Samuelson, P. A., *Foundations of Economic Analysis* (Harvard University Press, Cambridge, Massachusetts, 1947).
9 Goodwin, R. E., *Econometrica*, 19, 1–17 (1951).
10 Baumol, W. J., *Economic Dynamics*, 3rd ed. (Macmillan, New York, 1970).
11 See, for example, Kemeny, J., and Snell, J. L., *Mathematical Models in the Social Sciences* (MIT Press, Cambridge, Massachusetts, 1972).
12 Chaundy, T. W., and Phillips, E., *Q. Jl Math. Oxford*, 7, 74–80 (1936).
13 Myrberg, P. J., *Ann. Akad. Sc. Fennicae, A*, I, No. 336/3 (1963).
14 Myrberg, P. J., *Ann. Akad. Sc. Fennicae, A*, I, No. 259 (1958).
15 Lorenz, E. N., *J. Atmos. Sci.*, 20, 130–141 (1963); *Tellus*, 16, 1–11 (1964).
16 Metropolis, N., Stein, M. L., and Stein, P. R., *J. Combinatorial Theory*, 15(A), 25–44 (1973).
17 Maynard Smith, J., *Mathematical Ideas in Biology* (Cambridge University Press, Cambridge, 1968).
18 Krebs, C. J., *Ecology* (Harper and Row, New York, 1972).
19 May, R. M., *Am. Nat.*, 107, 46–57 (1972).
20 Li, T-Y., and Yorke, J. A., *Am. Math. Monthly*, 82, 985–992 (1975).
21 Hoppensteadt, F. C., and Hyman, J. M. (Courant Institute, New York University: preprint, 1975).
22 Smale, S., and Williams, R. (Department of Mathematics, Berkeley: preprint, 1976).
23 May, R. M., *Science*, 186, 645–647 (1974).
24 Moran, P. A. P., *Biometrics*, 6, 250–258 (1950).
25 Ricker, W. E., *J. Fish. Res. Bd. Can.*, 11, 559–623 (1954).
26 Cook, L. M., *Nature*, 207, 316 (1965).
27 Macfadyen, A., *Animal Ecology: Aims and Methods* (Pitman, London, 1963).
28 May, R. M., *J. theor. Biol.*, 51, 511–524 (1975).
29 Guckenheimer, J., *Proc. AMS Symposia in Pure Math.*, *XIV*, 95–124 (1970).
30 Gilbert, E. N., and Riordan, J., *Illinois J. Math.*, 5, 657–667 (1961).
31 Preston, C. J. (King's College, Cambridge: preprint, 1976).
32 Gumowski, I., and Mira, C., *C. r. hebd. Séanc. Acad. Sci., Paris*, 281a, 45–48 (1975); 282a, 219–222 (1976).
33 Layzer, D., *Sci. Am.*, 233(6), 56–69 (1975).
34 Ulam, S. M., *Proc. Int. Congr. Math. 1950, Cambridge, Mass.; Vol. II*, pp. 264–273 (AMS, Providence R.I., 1950).
35 Ulam, S. M., and von Neumann, J., *Bull. Am. math. Soc.* (abstr.), 53, 1120 (1947).
36 Kac, M., *Ann. Math.*, 47, 33–49 (1946).
37 May, R. M., *Science*, 181, 1074 (1973).
38 Hassell, M. P., *J. Anim. Ecol.*, 44, 283–296 (1974).
39 Hassell, M. P., Lawton, J. H., and May, R. M., *J. Anim. Ecol.* (in the press).
40 Ruelle, D., and Takens, F., *Comm. math. Phys.*, 20, 167–192 (1971).
41 Landau, L. D., and Lifshitz, E. M., *Fluid Mechanics* (Pergamon, London, 1959).
42 Martin, P. C., *Proc. Int. Conf. on Statistical Physics, 1975, Budapest* (Hungarian Acad. Sci., Budapest, in the press).
43 Southwood, T. R. E., in *Insects, Science and Society* (edit. by Pimentel, D.), 151–199 (Academic, New York, 1975).
44 Metropolis, N., Stein, M. L., and Stein, P. R., *Numer. Math.*, 10, 1–19 (1967).
45 Gumowski, I., and Mira, C., *Automatica*, 5, 303–317 (1969).
46 Stein, P. R., and Ulam, S. M., *Rosprawy Mat.*, 39, 1–66 (1964).
47 Beddington, J. R., Free, C. A., and Lawton, J. H., *Nature*, 255, 58–60 (1975).
48 Hirsch, M. W., and Smale, S., *Differential Equations, Dynamical Systems and Linear Algebra* (Academic, New York, 1974).
49 Kolata, G. B., *Science*, 189, 984–985 (1975).
50 Smale, S. (Department of Mathematics, Berkeley: preprint, 1976).
51 May, R. M., and Leonard, W. J., *SIAM J. Appl. Math.*, 29, 243–253 (1975).

Journal of Statistical Physics, Vol. 19, No. 1, 1978

Quantitative Universality for a Class of Nonlinear Transformations

Mitchell J. Feigenbaum[1]

Received October 31, 1977

A large class of recursion relations $x_{n+1} = \lambda f(x_n)$ exhibiting infinite bifurcation is shown to possess a rich quantitative structure essentially independent of the recursion function. The functions considered all have a unique differentiable maximum $\bar{x}$. With $f(\bar{x}) - f(x) \sim |x - \bar{x}|^z$ (for $|x - \bar{x}|$ sufficiently small), $z > 1$, the universal details depend only upon z. In particular, the local structure of high-order stability sets is shown to approach universality, rescaling in successive bifurcations, asymptotically by the ratio α ($\alpha = 2.5029078750957...$ for $z = 2$). This structure is determined by a universal function $g^*(x)$, where the 2^nth iterate of f, $f^{(n)}$, converges locally to $\alpha^{-n}g^*(\alpha^n x)$ for large n. For the class of f's considered, there exists a λ_n such that a 2^n-point stable limit cycle including $\bar{x}$ exists; $\lambda_\infty - \lambda_n \sim \delta^{-n}$ ($\delta = 4.669201609103...$ for $z = 2$). The numbers α and δ have been computationally determined for a range of z through their definitions, for a variety of f's for each z. We present a recursive mechanism that explains these results by determining g^* as the fixed-point (function) of a transformation on the class of f's. At present our treatment is heuristic. In a sequel, an exact theory is formulated and specific problems of rigor isolated.

KEY WORDS: Recurrence; bifurcation; limit cycles; attractor; universality; scaling; population dynamics.

1. INTRODUCTION

Recursion equations $x_{n+1} = f(x_n)$ provide a description for a variety of problems. For example, a numerical computation of a zero of $h(x)$ is obtained recursively according to

$$x_{n+1} = x_n + \frac{\epsilon h(x_n)}{h(x_n - \epsilon) - h(x_n)} \equiv f(x_n)$$

Research performed under the auspices of the U.S. Energy Research and Development Administration.

[1] Theoretical Division, Los Alamos Scientific Laboratory, Los Alamos, New Mexico.

If $\bar{x} = \lim_{n\to\infty} x_n$ exists, then $h(\bar{x}) = 0$. As $\bar{x}$ satisfies

$$\bar{x} = f(\bar{x})$$

the desired zero of h is obtained as the "fixed point" of the transformation f. In a natural context, a (possibly fictitious) discrete population satisfies the formula $p_{n+1} = f(p_n)$, determining the population at one time in terms of its previous value. We mention these two examples purely for illustrative purposes. The results of this paper, of course, apply to any situation modeled by such a recursion equation. Nevertheless, we shall focus attention throughout this section on the population example, both for the intuitive appeal of so tangible a realization as well as for a definite viewpoint, rather different from the usual one toward this situation, that shall emerge in the discussion. It is to be emphasized, though, that our results are generally applicable.

If the population referred to is that of a dilute group of organisms, then

$$p_{n+1} = bp_n \tag{1}$$

accurately describes the population growth so long as it remains dilute, with the solution $p_n = p_0 b^n$. For a given species of organism in a fixed milieu, b is a constant—the static birth rate for the configuration. As the population grows, the dilute approximation will ultimately fail: sufficient organisms are present and mutually interfere (e.g., competition for nutrient supply). At this point, the next value of the population will be determined by a *dynamic* or effective birth rate:

$$p_{n+1} = b_{\text{eff}} p_n$$

with $b_{\text{eff}} < b$. Clearly b_{eff} is a function of p, with

$$\lim_{p\to 0} b_{\text{eff}}(p) = b$$

the only model-independent quantitative feature of b_{eff}. Since the volume and nutrient available to a population are limited, it is clear that $b_{\text{eff}} \simeq 0$ for p sufficiently large. Accordingly, the simplest form of $b_{\text{eff}}(p)$ to reproduce the qualitative dynamics of such a population should resemble Fig. 1, where $b_{\text{eff}}(0) = b$ is an adjustable parameter [say, the nutrient level of the milieu held fixed independent of $p(t)$, and measurable by observing very dilute populations in that milieu]. A simple specific form of b_{eff} is

$$b_{\text{eff}} = b - ap$$

so that

$$p_{n+1} = bp_n - ap_n^2$$

By defining $p_n \equiv (b/a)x_n$, we obtain the standard form

$$x_{n+1} = bx_n(1 - x_n) \tag{2}$$

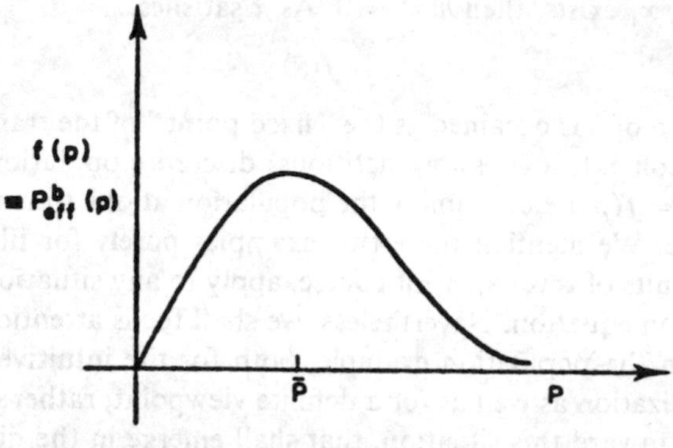

Fig. 1

In (2) the adjustable parameter b is purely multiplicative. With a different choice of b_{eff}, x_{n+1} would not in general depend upon b in so simple a fashion. Nevertheless, the internal b dependence may be (and often is) sufficiently mild in comparison to the multiplicative dependence that at least for qualitative purposes the internal dependence can be ignored. Thus, with $f(p) = pb_{\text{eff}}(p)$ any function like Fig. 1,

$$p_{n+1} = bf(p_n) \tag{3}$$

is compatible and representative of the population discussed. So long as $f'(0) = 1$ (so that the static birth rate is b and the dilute regime is correctly modeled) and f goes to zero for large p with a single central maximum, relation (3) correctly (at least qualitatively) models the situation. However, $f_2(p) = \sin(ap)$ affords an (a priori) equally good modeling as $f_1(p) = p - ap^2$. Thus only detailed quantitative results of (3) could determine which (if either) is empirically correct. One should then ask what the dynamical behavior of (3) is with f as in Fig. 1. It turns out that (3) enjoys a rich spectrum of excitations, with a universal behavior that would frustrate any attempt to discriminate among possible f's qualitatively. That is, providing (3) affords an honest model of a population's dynamics, so far as qualitative aspects are concerned, f is sufficiently specified by Fig. 1: the data could not qualitatively determine any more specific form [such as (2), say]. Conversely, *any* such choice of f—say Eq. (2)—is fully sufficient for study to comprehend all qualitative aspects of the dynamics. If the data should in any way disagree qualitatively with the predictions of (2), then (3) for any believable f must be an incorrect model.

The qualitative information available pertaining to (3) for any f of the form considered (see Appendix A for the exact requirements on f) is quite specific and detailed. In discussing the numerical solution to $h(x) = 0$ a fixed point was considered. In a population context, a fixed point

$$p^* = bf(p^*)$$

signifies zero population growth: $p_n = p^*$ for all n. However, p^* is "interesting" only so long as it is stable: if p fluctuates away from p^*, it should return to p^* in successive generations. For example, if $g(\bar{x})$ is finite, then

$$x_{n+1} = x_n + h(x_n)g(x_n) \tag{4}$$

will possess $\bar{x}$ as a fixed point if $h(\bar{x}) = 0$. However, unless

$$x_n \to \bar{x}$$

(4) is of no value to obtain $\bar{x}$; indeed, g is *chosen* so as to maximize the stability of $\bar{x}$. A stable fixed point is termed an "attractor," since points in its neighborhood approach it when iterated. An attractor is "global" if almost all points are eventually attracted to it. It is not necessary that an attractor be a unique isolated point. Thus, there might be n points $\bar{x}_1, \bar{x}_2, \ldots, \bar{x}_n$ such that

$$\bar{x}_{i+1} = f(\bar{x}_i), \quad i = 1, \ldots, n-1; \quad \bar{x}_1 = f(\bar{x}_n)$$

Such a set is called an "n-point limit cycle." Every n applications of f return an $\bar{x}_i$ to itself: each $\bar{x}_i$ is a fixed point of the nth iterate of f, $f^{(n)}$:

$$f^{(n)}(\bar{x}_i) = \bar{x}_i, \qquad i = 1, \ldots, n$$

Accordingly, $\{\bar{x}_1, \ldots, \bar{x}_n\}$ is a stable n-point limit cycle if each $\bar{x}_i$ is a stable fixed point of $f^{(n)}$. If it is a global attractor, then for almost every x_0, the sequence $x_n = f^{(n)}(x_0)$, $n = 1, 2, \ldots$, approaches the sequence

$$\bar{x}_1, \ldots, \bar{x}_n, \bar{x}_1, \ldots, \bar{x}_n, \ldots$$

Finally, there can be infinite stability sets $\{\bar{x}_i\}$ with

$$\bar{x}_{i+1} = f(\bar{x}_i)$$

such that the sequence $x_n = f^{(n)}(x_0)$ eventually becomes the sequence $\{\bar{x}_i\}$.

With this terminology, some of the detailed qualitative features of (3) can be stated as follows. (See Appendix A for more precise statements.) Depending upon the parameter value b, (3) possesses stable attractors of every order, with one attractor present and global for each fixed choice of b. As b is increased from a sufficiently small positive value, a fixed point $p^* > 0$ is stable until a value B_0 is reached when it becomes unstable. As b increases above B_0, a two-point cycle is stable, until at B_1 it becomes unstable, giving rise to a stable four-point cycle. As b is increased, this phenomenon recurs, with a 2^n-point cycle stable for

$$B_{n-1} < b < B_n$$

giving rise to a 2^{n+1}-point cycle above B_n until B_{n+1}, etc. The sequence of B_n is *bounded above* converging to a finite B_∞. This set of cycles (of order 2^n, $n = 1, 2,...$) is termed the set of "harmonics" of the two-point cycle. For any value of $b > B_\infty$ (but not too large) some particular stable n-point cycle will be present. As b is increased, it becomes unstable, and is replaced with a stable $2n$-point cycle. Until the cycle has doubled ad infinitum, no new stability sets save for these appear. Moreover, the ordering (with respect to b) of the onset of new size stability sets (e.g., seven-point before five-point) is also independent of f. Thus, if b is the unique parameter governing a population, any deviation of the ordering of stability sets upon increase of b from that determined by (2), say, constitutes empirical proof that (3) for any believable f incorrectly models the population. On the other hand, if (3) is appropriate for some f, then (2), for all qualitative purposes, comprises the full theory of the population's evolution. The exact quantitative theory reduces to the problem of determining the particular f. Unfortunately, even if (3) might be applicable, the data of biological populations are too crude at present to significantly discriminate among f's.

With so much specific qualitative information about (3) independent of f available, we may ask if the form of (3) might not also imply some *quantitative* information independent of f. It is the content of the following to answer this inquiry in the affirmative. Thus, the local structure of high-order stability sets (the quantitative locations of all elements of a stability set nearby one another) is independent of f. The role of a specific f is to set a local scale size for each cluster of stability points and to set the spacing between them. If one plots the points of, say, a 2^8-point limit cycle of (2) (or any cycle highly bifurcated from some low-order one), then by unevenly stretching the axis, the same 2^8-point cycle of (3) for another f is produced. The points are distributed unevenly in clusters sufficiently small that the stretching is essentially a pure magnification over the scale of a cluster. Moreover, for a fixed f, if b is increased to produce a 2^9-point cycle, that cluster about $\bar{x}$ (the maximum point) reproduces itself on a scale approximately α times smaller, where

$$\alpha = 2.5029078750957...$$

when f has a normal (i.e., quadratic) maximum. (This shall be assumed unless specifically stated otherwise.) The presence of the number α is a binding test on whether or not (3) is a correct model. α is a reflection of the infinitely bifurcative structure of (3), independent of any particular f. That is, the great bulk of the detailed quantitative aspect of solutions to (3) is independent of a specific choice of f: Eq. (3) and Fig. 1 comprise the bulk of the quantitative theory of such a population. Indeed, it is very difficult to extract the exact form of f from data, as so much quantitative information is determined purely by (3). In addition to α, another universal number determined by (3) should leave its mark on the data of a system described by (3). Thus, let b_0 be the value of b such that $\bar{x}$ (the abscissa of the maximum) is an element of a stable r-point cycle, and generally b_n the value of b such that $\bar{x}$ is an element of the stable $(r \times 2^n)$-point cycle n times bifurcated from the original. Then

$$\delta = \lim_{n \to \infty} \frac{b_{n+1} - b_n}{b_{n+2} - b_{n+1}}$$

is universal, with

$$\delta = 4.6692016091029...$$

It must be stressed that the numbers α and δ are *not* determined by, say, the set of all derivatives of (an analytic) f at same point. (Indeed, f need not be analytic.) Rather, universal functions exist that describe the local structure of stability sets, and these functions obey functional equations [independent of the f of (3)] implicating α and δ in a fundamental way.

2. QUALITATIVE ASPECTS OF BIFURCATION AND UNIVERSALITY

For definiteness (with no loss of generality), f is taken to map [0, 1] *onto* itself. At the unique differentiable maximum $\bar{x}$, $f(\bar{x}) = 1$,

$$x_{n+1} = \lambda f(x_n)$$

and λ lies in the interval [0, 1] to guarantee that if $x_0 \in [0, 1]$ then so, too, will all its iterates. When $\lambda = \bar{x}$,

$$\lambda f(\bar{x}) = \bar{x} f(\bar{x}) = \bar{x}$$

and $\bar{x}$ is a fixed point (Fig. 2). There is a simple graphical technique to determine the successive iterates of an initial point x_0:

(a) Draw a vertical segment along $x = x_0$ up to $\lambda f(x)$, intersecting at P.
(b) Draw a horizontal segment from P to $y = x$. The abscissa of the point of intersection is x_1.
(c) Repeat (a) and (b) to obtain x_{n+1} from x_n.

It is obvious from Fig. 2 that $\bar{x}$ is stable. Stability is locally analyzed by linear approximation about a fixed point. Setting

$$x_n = \bar{x} + \xi_n, \qquad \bar{x} f(x) \equiv g(x), \qquad g(\bar{x}) = \bar{x}$$

$$x_{n+1} = g(x_n) \Rightarrow \bar{x} + \xi_{n+1} = g(\bar{x} + \xi_n) = g(\bar{x}) + \xi_n g'(\bar{x}) + \cdots$$

$$\Rightarrow \xi_{n+1} = g'(\bar{x})\xi_n + O(\xi_n^2)$$

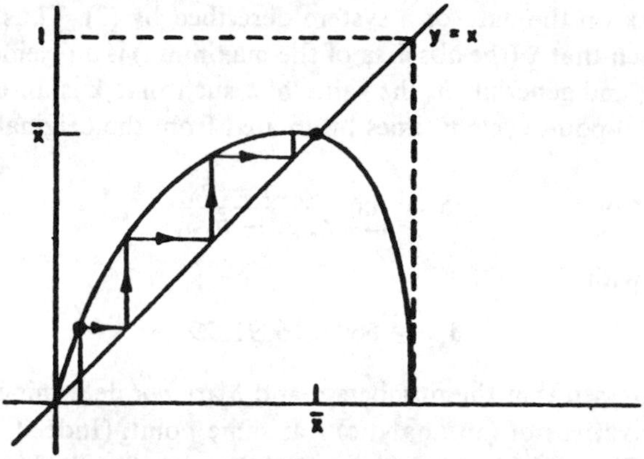

Fig. 2

Clearly $\xi_n \to 0$ if $|g'(\bar{x})| < 1$, the criterion for local stability. But $g'(\bar{x}) = \bar{x}f'(\bar{x}) = 0$, so that $\bar{x}$ is stable. With $r \equiv |g'(\bar{x})| < 1$,

$$\xi_n \propto r^n$$

so that convergence is geometric for $r \neq 0$. For $r = 0$, convergence is faster than geometric, and $\lambda = \bar{x}$ is that value of λ determining the most stable fixed point. We denote this value of λ by λ_0. Increasing λ just above λ_0 causes the fixed point x^* to move to the right with $g'(x^*) < 0$. At $\lambda = \Lambda_0$, $g'(x^*) = -1$ and x^* is marginally stable; for $\lambda > \Lambda_0$ it is unstable. According to Metropolis *et al.*,[1] a two-point cycle should now become stable. Stability of either of these points, say $x_1{}^*$, is determined by $|g^{(2)}(x_1{}^*)|$, where

$$g^{(2)}(x) = g(g(x)); \qquad g^{(n+1)}(x) = g(g^{(n)}(x)) = g^{(n)}(g(x))$$

Accordingly, consider $g^{(2)}(x)$ when $g'(x^*) < -1$ (Fig. 3). Several details of Fig. 3 are especially important. First, $g^{(2)}$ has two maxima: this because $\bar{x}$ has two inverses for $\lambda > \lambda_0$. Each maximum is of identical character to that of g: a neighborhood of $x_m^{(1)}$ is mapped into a neighborhood about $\bar{x}$ by g; g has a nonvanishing derivative at $x_m^{(1)}$, so that the imaged neighborhood is the original simply translated and stretched; accordingly, g applied to this new neighborhood is simply a magnification of g about $\bar{x}$. Thus, if $g(x) \propto |x - \bar{x}|^z + g(\bar{x})$, $z > 1$ for $|x - \bar{x}|$ small, then $g^{(2)} \propto |x - x_m^{(1)}|^z + g(\bar{x})$ for $|x - x_m^{(1)}|$ small. Similarly, the minimum (located at $\bar{x}$) is of order z. This is, of course, the content of the chain rule: $g^{(n)\prime}(x_0) = \prod_{i=0}^{n-1} g'(x_i)$ with $x_i = g^{(i)}(x_0)$ [$g^{(0)}(x) \equiv x$]. In particular, observe that $\bar{x}$ is a point of extremum of $g^{(n)}$ for all n. Also, if $g(x^*) = x^*$, then $g^{(n)\prime}(x^*) = [g'(x^*)]^n$. With $g'(x^*) < -1$,

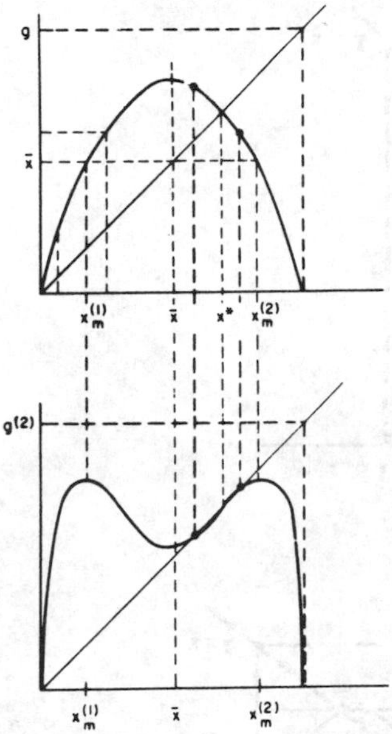

Fig. 3

$g^{(2)\prime}(x^*) > 1$, so that $g^{(2)}$ must develop two fixed points besides x^*: these two new fixed points are a two-point cycle of g itself, and for $\lambda - \Lambda_0$ sufficiently small, $0 < g^{(2)\prime} < 1$ at these points. Moreover, since $g(x_1^*) = x_2^*$ and $g(x_2^*) = x_1^*$, the chain rule implies that $g^{(2)\prime}(x_1^*) = g^{(2)\prime}(x_2^*)$, so that each element of the cycle enjoys identical stability. As λ is increased, the maxima of $g^{(2)}$ ($g^{(2)} = \lambda$ at maximum) also increase until a value λ_1 is reached when the abscissa of the rightmost maximum $x_m^{(2)} = \lambda_1$. By the chain rule, the other fixed point is now also at an extremum, and must be at $\bar{x}$ (Fig. 4).

As λ increases above λ_1, $g^{(2)}(\bar{x})$ decreases below $\bar{x}$, so that $g^{(2)\prime} < 0$ for the leftmost fixed point, and so, for the rightmost one. At $\lambda = \Lambda_1$, $g^{(2)\prime} = -1$ for both: otherwise the two-point cycle would always remain stable, in violation of the results of Metropolis *et al.* Thus, $g^{(2)\prime} < -1$ for $\lambda > \Lambda_1$, the two-point cycle is unstable, and we are now motivated to consider $g^{(4)}$, as a four-point cycle should now be stable. Alternatively, the region "a" of $g^{(2)}$ of Fig. 4 bears a distinct resemblance to g of Fig. 2 turned upside down and reduced in scale: the transition that led from Fig. 2 to Fig. 4 is now being reexperienced, with $g^{(2)}$ replacing g and $g^{(4)}$ replacing $g^{(2)}$. In particular, at $\lambda = \lambda_2 > \Lambda_1$ the fixed points of $g^{(4)}$ beyond those of $g^{(2)}$ will occur at extrema (Fig. 5). The region "a" of $g^{(4)}$ is again an upside-down, reduced version of that of $g^{(2)}$ in Fig. 4; the square box construction including $\bar{x}$ for $g^{(2)}$ of Fig. 5 is an upside-down, reduced version of that of g in Fig. 4. Since the boxes are *squares*, the Fig. 5 box is reduced by the *same* scale on both height

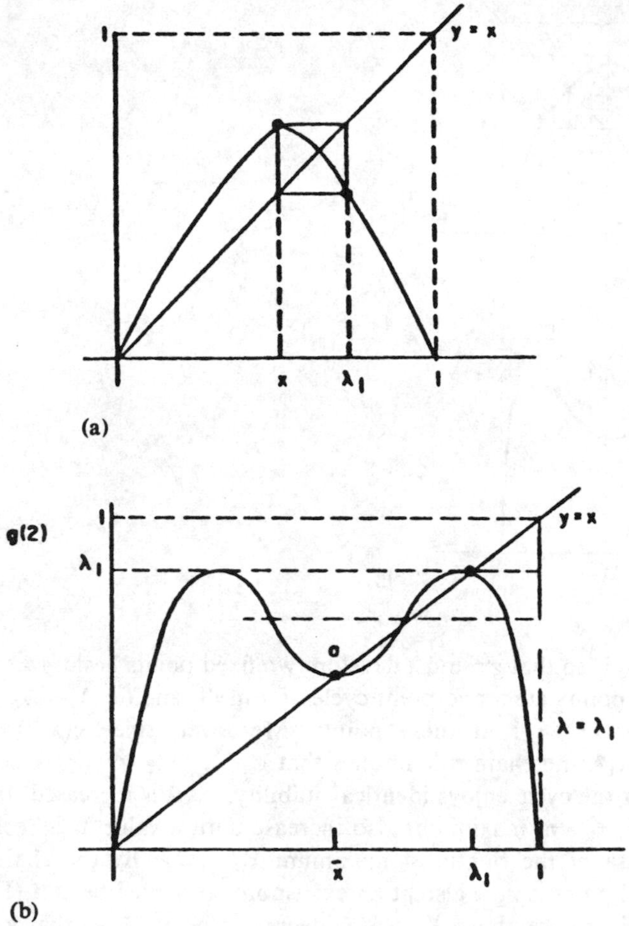

Fig. 4

and width from Fig. 4. Accordingly, the regions "a" are also rescaled identically on height and width.

It is very important to realize that in Fig. 5, g itself was not drawn since it is unnecessary: $g^{(2)}$ is sufficient to determine $g^{(4)}$:

$$g^{(4)}(x) = g(g(g(g(x)))) = g(g(g^{(2)}(x))) = g^{(2)}(g^{(2)}(x))$$

[and similarly, $g^{(2^{n+1})}(x) = g^{(2^n)}(g^{(2^n)}(x))$]. At the level of discussion of Fig. 5, $g^{(2)}$ has effectively replaced g as the fundamental function considered. $g^{(2)}$, though, is not simply proportional to λ, possessing internal λ dependence: the underlying role of g is exposed by $g^{(2)}$ in the simultaneous occurrence of the two box constructions. Similarly, by the nth bifurcation, only $g^{(2^{n-1})}$ and $g^{(2^n)}$ are important. If at λ_{n+1} (at $\lambda = \lambda_n$, $\bar{x}$ is an element of a 2^n-point cycle)

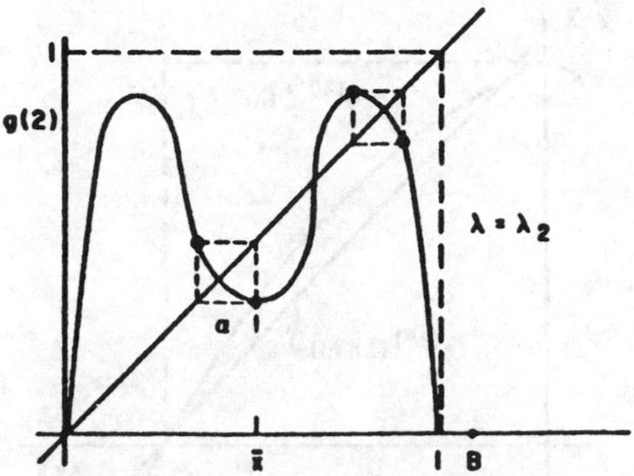

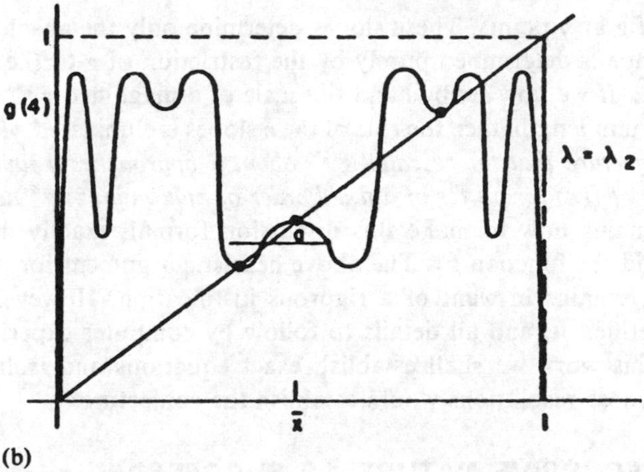

Fig. 5

we magnify the box containing $\bar{x}$ of $g^{(2^n)}$ and invert it to overlay that of $g^{(2^{n-1})}$ at $\lambda = \lambda_n$ (Fig. 6), we have two curves of identical order of maximum z, of identical height with identical zeros. Through a set of operations, $g^{(2^{n-1})}$ determines $g^{(2^n)}$, just as will $g^{(2^n)}$ determine $g^{(2^{n+1})}$. Referring back to Fig. 4, observe that the restriction of $g^{(2)}$ to the interval between maxima is determined entirely by the restriction of g itself to this same interval. The region "a" of $g^{(2)}$ is determined by g restricted a smaller interval plus essentially just the slope of g at λ_1 if g is sufficiently smooth. Analogously, the restriction of $g^{(2^n)}$ to its box part is determined through a similar restriction of $g^{(2^{n-1})}$. With the n scale reductions that have taken place by this level of iteration, $g^{(2^n)}$ is determined by g restricted to an increasingly small interval about $\bar{x}$ together with

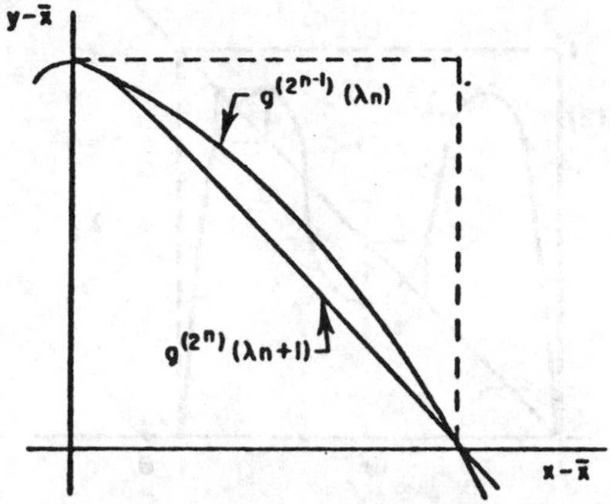

Fig. 6

the slope of g at n points. These slopes determine only the absolute scale of $g^{(2^n)}$: its shape is determined purely by the restriction of g to the immediate vicinity of $\bar{x}$. If we now set by hand the scale of a magnified $g^{(2^n)}$ so that the square is of unit length, then the role of the n slopes is eliminated. *Accordingly, we now conjecture that the rescaled $g^{(2^n)}$ about $\bar{x}$ approaches a function $g^*(x)$ independent of $f(x)$ for all f's of a fixed order of maximum z: g^* depends only on z.* It remains now to make this discussion formal, exactly defining the rescaling and the function g^*. The above heuristic argument for universality regrettably remains in want of a rigorous justification. However, we have carefully verified it, and all details to follow by computer experiment. In a sequel to this work we shall establish exact equations and isolate specific questions whose resolutions would establish the conjecture.

3. THE RECURSIVE NATURE OF SUCCESSIVE BIFURCATION

We have described a process that can be summarized as follows.

(0) We start at $\lambda = \lambda_n$, and look at $g^{(2^n)}$ near $x = \bar{x}$. Alternatively, we might look at $g^{(2^{n-1})}$ for the same λ and range of x, as depicted in Fig. 7.

(i) Form $g^{(2^n)}(x) = g^{(2^{n-1})}(g^{(2^{n-1})}(x))$, depicted in Fig. 8.

(ii) Increase λ from λ_n to λ_{n+1}, depicted in Fig. 9.

(iii) Rescale: $g^{(2^n)}(x) \to \alpha_n g^{(2^n)}(x/\alpha_n)$, depicted in Fig. 10 ($|\alpha| > 1$).

Calling the operations (i)–(iii) B_{n-1}, we have

$$\tilde{g}_n(x) = B_{n-1}[\tilde{g}_{n-1}(x)], \qquad n = 2, 3, \ldots$$

and are claiming $\tilde{g}_n(x) \to g^*(x)$ locally about $\bar{x}$.

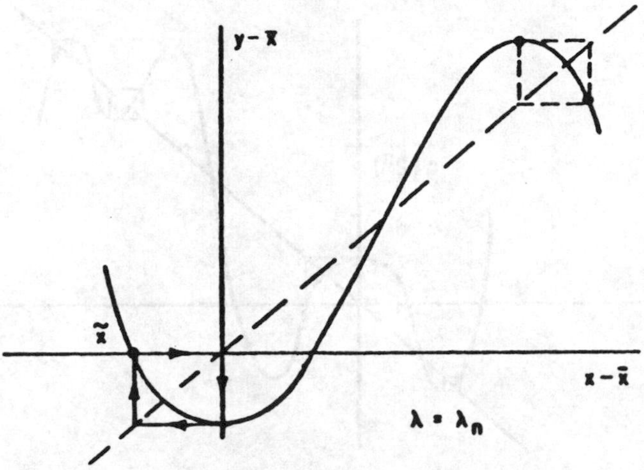

Fig. 7

Clearly (i) of B_n is recursive and n-independent; we call this part of B_n "doubling." We will motivate that (ii) becomes asymptotically n-independent; we term this part of B_n "λ-shifting." Also, with $\alpha_n \to \alpha$ essentially by (i), part (iii) of B_n becomes asymptotically n-independent; we term this part (obviously) "rescaling." Thus, $B_n \to B$. That is,

$$\lim_{r \to \infty} B^r[\tilde{g}_n(x)] = g^*(x)$$

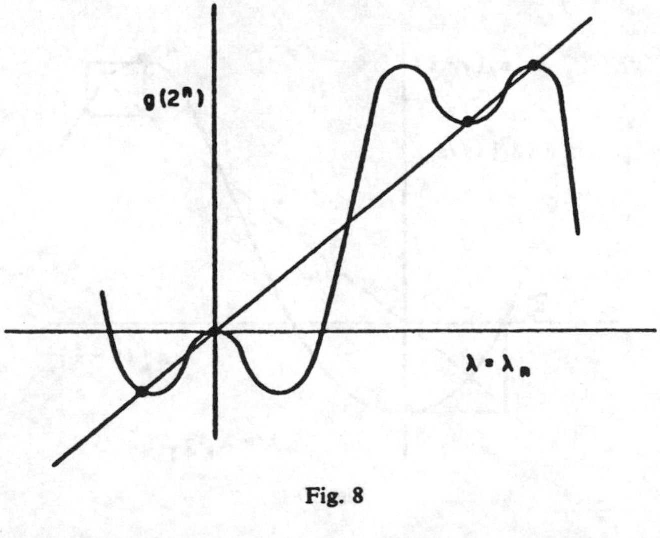

Fig. 8

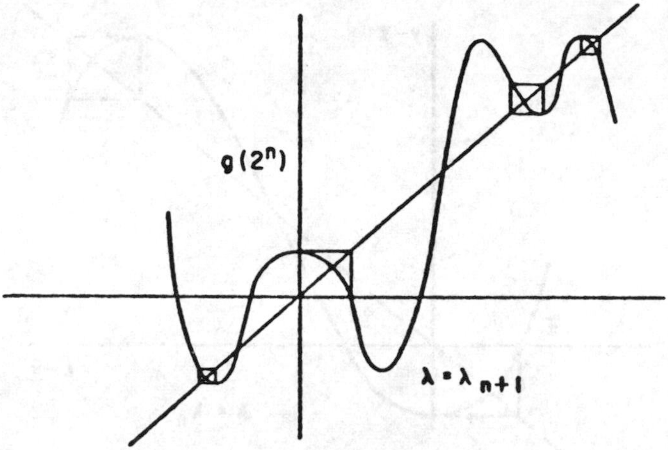

Fig. 9

Accordingly, g^* satisfies the equation

$$g^* = B[g^*] \tag{5}$$

Universality, thus, is the consequence of a recursion on the class of functions $f(x)$ considered. Under high-order bifurcation, the fixed point of B is approached—that fixed point being, within a certain domain, a property of B itself and not of the starting $f(x)$. Evidently, domains of the various fixed points of B are disjoint for different z. Also, each fixed-z domain clearly exceeds the class of f's specified by properties 1–4 of Appendix A, since $(f)^{(2^n)}$ for each n is also in the domain. At present we cannot specify just how

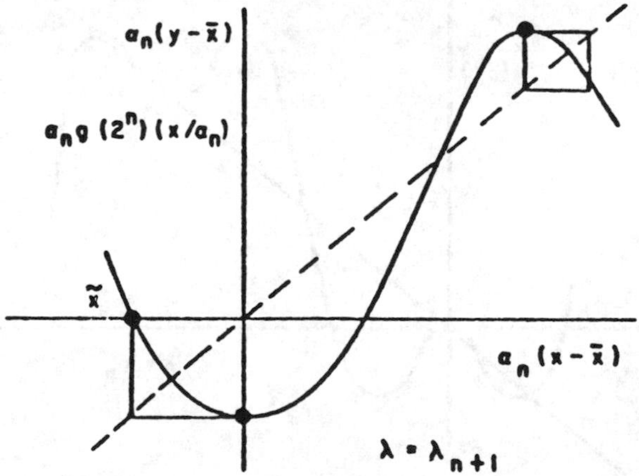

Fig. 10

large this domain is. The fixed-point equation (5) will certainly, for a given z, determine the rescaling ratio α as well as g^*. [For a variety of functions $f(x)$ with $z = 2$, we have determined g^*, with $\bar{x}$ of Fig. 10 set to unit length.]

4. DETAILED FEATURES OF THE BIFURCATION RECURSION

We first indicate roughly how the parameters α and δ are interrelated and determined by g^*. At $\lambda = \lambda_n$, $\tilde{g}_{n-1}$ and $\tilde{g}_{n-1} \circ \tilde{g}_{n-1}$ appear as in Fig. 11. Increasing λ has $\tilde{g}_{n-1}(0)$ increase above 1, producing Fig. 12, where $\tilde{g}_{n-1}$ and $\tilde{g}_{n-1} \circ \tilde{g}_{n-1}$ at λ_n are shown dashed. By the definition of α_n, h_n of Fig. 12 satisfies

$$h_n = \alpha_n^{-1}$$

Clearly, though, in some rough sense

$$h_n \simeq (h_{n-1} - 1)|\tilde{g}'_{n-1}(1)| \equiv \delta h_{n-1}|\tilde{g}'_{n-1}(1)|$$

i.e.,

$$\delta h_{n-1} \simeq |\alpha_n \tilde{g}'_{n-1}(1)|^{-1} \tag{6}$$

Also,

$$\delta h_{n-2} \simeq |\tilde{g}'_{n-2}(1)|^{-1} \delta h_{n-1} \tag{7}$$

This is more nearly accurate than (6), since $\tilde{g}_{n-2}$ shifts less than $\tilde{g}_{n-1}$ for the same λ increase. Thus,

$$\delta h_{n-1} \simeq \prod_2^n |\tilde{g}'_{n-i}(1)| \, \delta h_0 \tag{8}$$

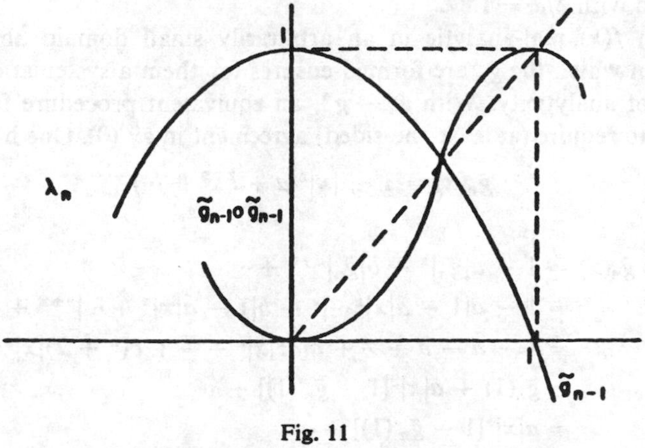

Fig. 11

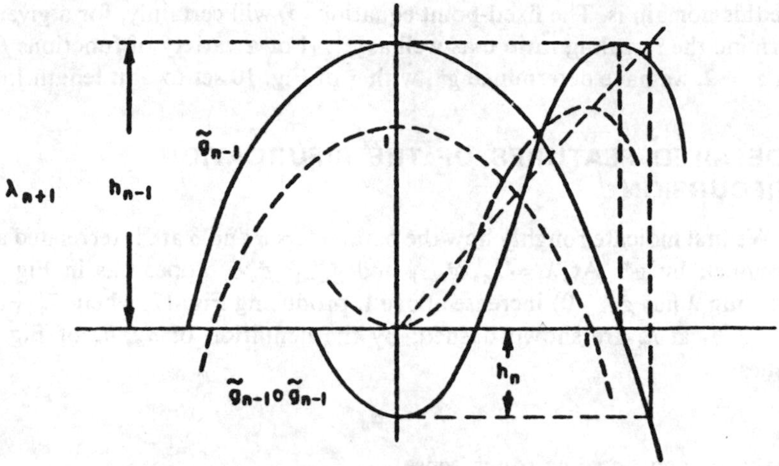

Fig. 12

However, $\delta h_0 = \delta\lambda_n = \lambda_{n+1} - \lambda_n$. Assuming $\tilde{g}_n \to g^*$ (this is not quite correct; see Section 5) one has, so far as n dependence is concerned,

$$\delta h_{n-1} \simeq \mu |g^{*\prime}(1)|^{n-1}\, \delta\lambda_n \tag{9}$$

with $\mu \sim 1$ an asymptotically n-independent factor. Substituting in (6),

$$\delta\lambda_n \simeq \mu^{-1}/|\alpha g^{*\prime}(1)|^n$$

with $\alpha = \lim \alpha_n$.

Accordingly, $\delta\lambda_n \propto \delta^{-n}$, with

$$\delta \simeq \alpha |g^{*\prime}(1)| \tag{10}$$

For $z = 2$, the computer-experimental value for $|g^{*\prime}(1)|$ is $+1.89$, to be compared with $\delta/\alpha = 1.87$.

With $f(x)$ real-analytic in an arbitrarily small domain about $\bar{x}$, the manner in which the $\tilde{g}_n$ are formed ensures for them a systematically larger domain of analyticity. With $\tilde{g}_n \to g^*$, an equivalent procedure for defining the α_n is to require (at least one-sided) agreement in $\tilde{g}_n^{(2)}(0)$. One has

$$\tilde{g}_n(x) = 1 - |x|^z(a + bx^2 + \cdots) \tag{11}$$

Then,

$$\begin{aligned}
\tilde{g}_n \circ \tilde{g}_n(x) &= 1 - a|\tilde{g}_n|^z - b|\tilde{g}_n|^{z+2} + \cdots \\
&= 1 - a|1 - a|x|^z \cdots|^z - b|1 - a|x|^z + \cdots|^{z+2} + \cdots \\
&= 1 - a - b + \cdots + a[az|x|^z + \cdots + b(z+2)|x|^z + \cdots] \\
&= \tilde{g}_n(1) + a|x|^z[1 - \tilde{g}_n'(1)] + \cdots \\
&= a|x|^z[1 - \tilde{g}_n'(1)] + \cdots
\end{aligned}$$

Next, the λ shift is performed:

$$\tilde{g}_n \to \tilde{g}_n \circ \tilde{g}_n \to -\nu + \mu a|x|^z[1 - \tilde{g}_n'(1)] + \cdots$$

and finally, α rescaling:

$$\tilde{g}_n \to -\{1 - \mu(a/\alpha^{z-1})[1 - \tilde{g}^{*\prime}(1)]|x|^z + \cdots\} \tag{12}$$

For (11) and (12) to agree, one has

$$\alpha^{z-1} \sim 1 - g^{*\prime}(1) \tag{13}$$

where $\mu \lesssim 1$ corresponds to λ-shifting being mostly a displacement in the immediate environs of $\bar{x}$. Again, for $z = 2$, one compares $\alpha = 2.50$ with $1 - g^{*\prime}(1) = 2.87$. Combining (10) and (13), one has

$$\delta \simeq |g^{*\prime}(1)|[1 - g^{*\prime}(1)]^{1/z-1} \tag{14}$$

While (13) and (14) are crude, they are roughly correct for $z \gtrsim 2$, but more important, indicate that g^* ultimately determines everything.

We now proceed to describe the situation more carefully, tacitly assuming convergence, and successively illustrating its details through consistency arguments.

By definition

$$g^*(x) = \lim(-1)^n\alpha^n g^{(2^n)}(x/\alpha^n, \lambda_{n+1}) \equiv \lim \tilde{g}_n(x) \tag{15}$$

where α^n is symbolic for α_n which becomes asymptotically a multiple of α^n: the multiple has been absorbed in $g^{(2^n)}$. For all n, $\tilde{g}_n$ satisfies

$$\tilde{g}_n(1) = 0, \qquad \tilde{g}_n(0) = 1, \qquad \tilde{g}_n'(0) = 0 \tag{16}$$

and near $x = 0$, $1 - \tilde{g}_n(x) \sim |x|^z$.

We now furnish an approximate equation for g^*:

$$\begin{aligned}(-1)^n\alpha^{n-1}g^{(2^n)}(x, \lambda_n) &= (-1)^n\alpha^{n-1}g^{(2^{n-1})}(g^{(2^{n-1})}(x, \lambda_n), \lambda_n)\\ &= (-1)^n\alpha^{n-1}g^{(2^{n-1})}\left(\frac{1}{\alpha^{n-1}}\alpha^{n-1}g^{(2^{n-1})}(x, \lambda_n), \lambda_n\right)\\ &= -\tilde{g}_n \circ \tilde{g}_n(x\alpha^{n-1}) \end{aligned}\tag{17}$$

or

$$(-1)^n\alpha^n g^{(2^n)}(x/\alpha^n, \lambda_n) = -\alpha\tilde{g}_n \circ \tilde{g}_n(x/\alpha) \tag{18}$$

or

$$-\alpha\tilde{g}_n \circ \tilde{g}_n(x/\alpha) = \tilde{g}_{n+1}(x) - (-1)^n\alpha^n(g^{(2^n)}(x/\alpha_n, \lambda_{n+1}) - g^{(2^n)}(x/\alpha_n, \lambda_n))$$

or

$$-\alpha\tilde{g}_n \circ \tilde{g}_n(x/\alpha) \simeq \tilde{g}_{n+1}(x) - (-1)^n\alpha^n(\lambda_{n+1} - \lambda_n)\,\partial_\lambda g^{(2^n)}(x/\alpha_n, \lambda_n) \tag{19}$$

assuming a "mild" λ-shifting.

Clearly, $\alpha^n\,\partial_\lambda g^{(2^n)}(x/\alpha_n, \lambda_n)$ diverges with n since $\lambda_{n+1} - \lambda_n \to 0$. Thus, a more careful analysis, like that used to treat Eq. (10), needs to be done. By (17),

$$\begin{aligned}\partial_\lambda g^{(2^n)}(x, \lambda_n) &= \partial_\lambda g^{(2^{n-1})}(g^{(2^{n-1})}(x, \lambda_n), \lambda_n)\\ &\quad + \partial_x g^{(2^{n-1})}(g^{(2^{n-1})}(x, \lambda_n), \lambda_n)\,\partial_\lambda g^{(2^{n-1})}(x, \lambda_n)\\ &= \partial_\lambda g^{(2^{n-1})}(g^{(2^{n-1})}(x, \lambda_n), \lambda_n)\\ &\quad + \partial_x \tilde{g}_{n-1}(\tilde{g}_{n-1}(x/\alpha^{n-1}))\partial_\lambda g^{(2^{n-1})}(x, \lambda_n)\end{aligned}$$

So,

$$\begin{aligned}\alpha^n\,\partial_\lambda g^{(2^n)}\left(\frac{x}{\alpha^n}, \lambda_n\right) &= \alpha^n\,\partial_\lambda g^{(2^{n-1})}\left(\frac{1}{\alpha^{n-1}}\,\tilde{g}_{n-1}\left(\frac{x}{\alpha}\right), \lambda_n\right)\\ &\quad + \tilde{g}'_{n-1}\circ\tilde{g}_{n-1}\left(\frac{x}{\alpha}\right)\alpha^n\,\partial_\lambda g^{(2^{n-1})}\left(\frac{x}{\alpha^n}, \lambda_n\right)\end{aligned} \tag{20}$$

At $x = 0$,

$$\alpha^n\,\partial_\lambda g^{(2^n)}(0, \lambda_n) = \alpha^n\,\partial_\lambda g^{(2^{n-1})}(1/\alpha^{n-1}, \lambda_n) + \tilde{g}'_{n-1}(1)\alpha^n\,\partial_\lambda g^{(2^{n-1})}(0, \lambda_n) \tag{21}$$

With

$$\alpha^n\,\partial_\lambda g^{(2^{n-1})}(1/\alpha^{n-1}, \lambda_n) = \mu\alpha^n\,\partial_\lambda g^{(2^{n-1})}(0, \lambda_n)$$

(such a μ exists if λ-shifting becomes n-independent), (21) becomes

$$\begin{aligned}\alpha^n\,\partial_\lambda g^{(2^n)}(0, \lambda_n) &= [\mu + \tilde{g}'_{n-1}(1)]\alpha^n\,\partial_\lambda g^{(2^{n-1})}(0, \lambda_n)\\ &\simeq [\mu + \tilde{g}'(1)]\alpha^n\,\partial_\lambda g^{(2^{n-1})}(0, \lambda_{n-1})\end{aligned} \tag{22}$$

($g^{(2^{n-1})}$ shifts more slowly than $g^{(2^n)}$: higher order λ derivatives have been neglected). Iterating (22), one has

$$\partial_\lambda g^{(2^n)}(0, \lambda_n) \simeq \rho[\mu + \tilde{g}'(1)]^n \tag{23}$$

with $\rho \sim 1$, n-independent. So,

$$(-1)^n(\lambda_{n+1} - \lambda_n)\alpha^n\,\partial_\lambda g^{(2^n)}(0, \lambda_n) \simeq \rho[\alpha(-\tilde{g}'(1) - \mu)]^n(\lambda_{n+1} - \lambda_n)$$

By (19) this is n-independent, and so,

$$\lambda_{n+1} - \lambda_n \sim \delta^{-n}$$

with

$$\delta \simeq \alpha(-\tilde{g}'(1) - \mu) \tag{24}$$

Defining $\tilde{h}_n(x) = (-1)^n\alpha^n(\lambda_{n+1} - \lambda_n)\,\partial_\lambda g^{(2^n)}(x/\alpha_n, \lambda_n)$, (19) reads

$$\tilde{g}_{n+1}(x) = \tilde{h}_n(x) - \alpha\tilde{g}_n\circ\tilde{g}_n(x/\alpha) \tag{25}$$

or

$$g^*(x) = h^*(x) - \alpha g^*\circ g^*(x/\alpha) \tag{26}$$

Returning to (20), multiplied by $\lambda_{n+1} - \lambda_n$, neglecting higher order derivatives,

$$\tilde{h}_n(x) \simeq -\omega(\tilde{h}_{n-1} \circ \tilde{g}_{n-1}(x/\alpha) + \tilde{h}_{n-1}(x/\alpha)\tilde{g}'_{n-1} \circ \tilde{g}_{n-1}(x/\alpha)) \tag{27}$$

with some $\omega \sim 1$, or, as $n \to \infty$, and repeating (26),

$$h^*(x) = -\omega(h^* \circ g^*(x/\alpha) + h^*(x/\alpha)g^{*\prime} \circ g^*(x/\alpha))$$

and

$$g^*(x) = h^*(x) - \alpha g^* \circ g^*(x/\alpha) \tag{28}$$

These constitute first-order (approximate) fixed-point equations, satisfying the boundary conditions

$$g^*(0) = 1, \quad g^{*\prime}(0) = 0, \quad g^*(1) = 0, \quad h^*(0) = 1 \tag{29}$$

[We comment that (28) is recursively stable, and for $z = 2$ affords a 10% approximate solution.]

At this point, some remarks concerning convergence (say of $\tilde{g}_n \to g^*$) are in order. The function $g^*(x)$ describes the stability set for large n in the vicinity of $\bar{x}$: those x_i such that

$$g^* \circ g^*(x_i) = x_i$$

[and, of course, $g^{*\prime}(g^*(x_i)) \cdot g^{*\prime}(x_i) = 0$] are the stability set points near $\bar{x}$. Accordingly, all such x_i scale with α upon bifurcation: $|x_i - x_j| \to (1/\alpha)|x_i - x_j|$. For example, the distance between $\bar{x}$ and the nearest element to it of the stability set of order 2^n is α times greater than that distance in the stability set of order 2^{n+1}. (Also, if x_1 is the nearest point to $\bar{x}$ and x_2 the next nearest, then for all n large enough, $|\bar{x} - x_1|/|\bar{x} - x_2| \equiv \gamma$ is fixed.) This immediately leads to a difficulty: distances near $\bar{x}$ and those near λ_n (the furthest right element of a stability set) cannot possibly scale identically.

As is obvious from Fig. 13, with Δ_n the distance from $\bar{x}$ to x_1, and d_n the distance from λ_n to $\tilde{x}$ (the next to rightmost point), $d_n \sim \Delta_n{}^z$, so that with $\Delta_n \propto \alpha^{-n}$,

$$d_n \propto (\alpha^z)^{-n} \neq \alpha^{-n} \tag{30}$$

Thus, convergence of $\tilde{g}_n(x)$ to $g^*(x)$ must be local in nature. The scale for which $g^*(0) = 1$ and $g^*(1) = 0$ is, of course, α^{-n} finer than usual measure on $[0, 1]$: for large n, $\sup|\tilde{g}_n - g^*| < \epsilon_{N_n}$ for $|x| < N_n$ is uniform convergence in "real" x of $|x| < N/\alpha^n$. To allow for a shifting rescaling of parts of g^*, $N_n \ll \alpha^n$. Thus, one anticipates that $\tilde{g}_n \to g^*$ (say in sup-norm) over *any* bounded part of R *but* with the $g^*(1) = 0$ measure. In any (small) interval about a given point in the stability set of order n, one sets the origin of $g^{(2^n)}$ at the point in question and forms $\tilde{g}_n$ with an appropriate (local) scale factor. As n increases, in the $\tilde{g}_n(1) = 0$ measure, any other point a finite distance away

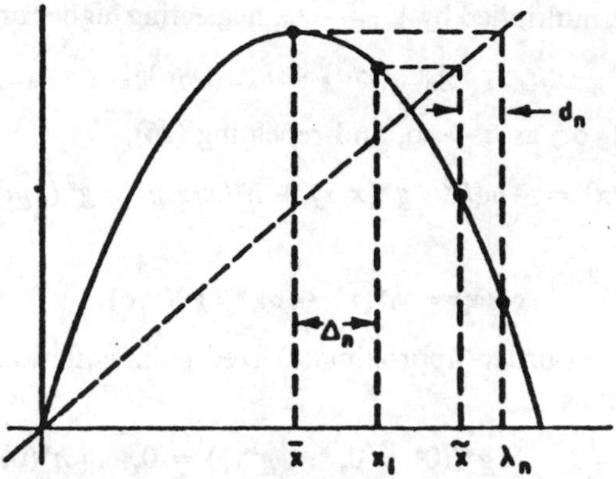

Fig. 13

in *usual* [0, 1] measure grows far remote from the chosen point: so far, that the local $\tilde{g}_n$ never converge to it. Thus, in effect, a class of g^* exist, each determining the large-n limiting stability set about a point.

For example, defining $\tilde{f}_n(x)$ about λ_{n+1} by Fig. 14 [$x_n \equiv g^{(2^n)}(\lambda_{n+1})$, $\lambda_{n+1} = g^{(2^n)}(x_n)$], we have

$$\tilde{f}_n(x) = [g^{(2^n)}((\lambda_{n+1} - x_n)x + x_n) - x_n]/(\lambda_{n+1} - x_n) \to f^*(x) \tag{31}$$

[so that $\tilde{f}_n(0) = 1, \tilde{f}_n'(0) = 0, \tilde{f}_n(1) = 0$]. In the notation of Fig. 13,

$$\tilde{f}_n(x) = [g^{(2^n)}(xd_n + x_n) - x_n]/d_n$$

and so $\tilde{f}_n$ scales by d^z rather than d. It is straightforward to relate f^* to g^*:

$$g^{(2^n)}(\lambda_{n+1}f(x)) = \lambda_{n+1}f(g^{(2^n)}(x)) \tag{32}$$

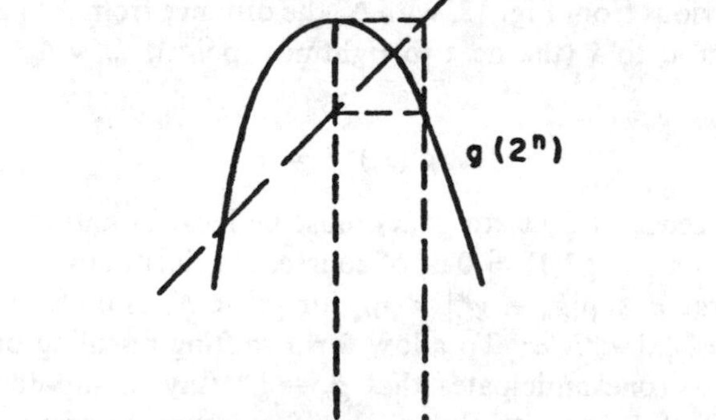

Fig. 14

so that for $x \sim 0$ (we have conveniently set $\bar{x} = 0$), $\lambda_{n+1}f(x)$ is near λ_{n+1} and (32) relates $g^{(2^n)}$ about λ_{n+1} to $g^{(2^n)}$ about 0. Thus

$$x \text{ small} \Rightarrow \lambda_{n+1}f(x) = \lambda_{n+1} - a\lambda_{n+1}|x|^z + O(|x|^{z+1})$$

$$\begin{aligned} g^{(2^n)}(\lambda_{n+1}f(x)) &= g^{(2^n)}((\lambda_{n+1} - x_n)(1 - a|x|^z) - ax_n|x|^z + x_n) \\ &= x_n + (\lambda_{n+1} - x_n)\tilde{f}_n\left(1 - a|x|^z - \frac{ax_n}{\lambda_{n+1} - x_n}|x|^z\right) \end{aligned}$$

Also,

$$\begin{aligned} \lambda_{n+1}f(g^{(2^n)}(x)) &= \lambda_{n+1} - a\lambda_{n+1}|g^{(2^n)}(x)|^z + \cdots \\ &= \lambda_{n+1} - a\lambda_{n+1}\Delta_n^z|\tilde{g}_n(x/\Delta_n)|^z \end{aligned}$$

Accordingly, (32) implies for small x that

$$(\lambda_{n+1} - x_n)\left[1 - \tilde{f}_n\left(1 - a|x|^z - \frac{ax_n}{\lambda_{n+1} - x_n}|x|^z\right)\right] = a\lambda_{n+1}\Delta_n^z\left|\tilde{g}_n\left(\frac{x}{\Delta_n}\right)\right|^z$$

By Fig. 14, $\lambda_{n+1} - x_n = d_n = a\Delta_n^z\lambda_{n+1}$, so that

$$1 - \tilde{f}_n\left(1 - \frac{\lambda_{n+1}a|x|^z}{\lambda_{n+1} - x_n}\right) = \left|\tilde{g}_n\left(\frac{x}{\Delta_n}\right)\right|^z$$

or

$$1 - \tilde{f}_n(1 - |x|^z/\Delta_n^z) = |\tilde{g}_n(x/\Delta_n)|^z$$

or

$$\tilde{f}_n(1 - |\xi|^z) = -|\tilde{g}_n(\xi)|^z + 1$$

or

$$\tilde{f}_n(x) = -|\tilde{g}_n((1 - x)^{1/z})|^z + 1 + \cdots \tag{33}$$

For large n, neglected terms are powers of $\Delta_n \to 0$. So,

$$f^*(x) = -|g^*((1 - x)^{1/z})|^z + 1 \tag{34}$$

and g^* determines f^*. (This has, of course, been computationally verified to full precision.) We are unsure of the size of the set of rescalings: clearly α and α^z belong to the set. However, about any point a fixed, finite number of iterates prior to $\bar{x}$, scaling goes by α, whereas any region a finite number of iterates after λ_{n+1} scales with α^z. On the other hand, points 2^{n-1} iterates from $\bar{x}$, as well as $2^{n-2},\ldots, 2^{n-N}$ from $\bar{x}$, are just the N points nearest $\bar{x}$ that scale with α. That is, we are uncertain how to define a region which possesses a scaling intermediate between α and α^z; possibly the situation is simply an interspersal of regions scaling by either α or α^z. What is missing is some notion of ordered measure along a stability set.

At this point it is perhaps illuminating to indicate just what g^* looks like. Evidently Fig. 5 is a somewhat distorted version of g^* very near $x = 0$.

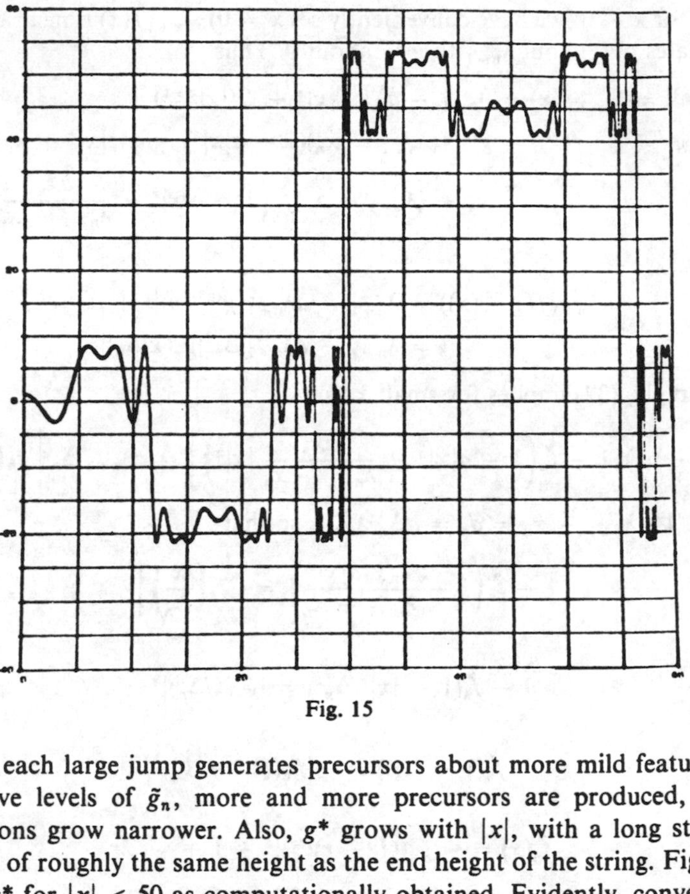

Fig. 15

Indeed, each large jump generates precursors about more mild features. At successive levels of $\tilde{g}_n$, more and more precursors are produced, whose oscillations grow narrower. Also, g^* grows with $|x|$, with a long string of features of roughly the same height as the end height of the string. Figure 15 shows g^* for $|x| < 50$ as computationally obtained. Evidently, convergence to such a function worsens with increasing $|x|$.

5. CONCLUSIONS AND A BRIEF SUMMARY OF THE EXACT THEORY

In this paper we have attempted to heuristically motivate our conjecture of universality, and indicate the form of an exact theory of highly bifurcated attractor sets. Our conclusion is that both of the numbers α and δ as well as the local structure of highly bifurcated attractors as determined by the universal function g^* are determined by functional equations. We have provided approximate such equations, but failed in establishing exact equations for an inability to exactly reflect an increase in the parameter λ. As described, the local structure determined by g^* pertains to values of λ asymptotically near λ_∞ at the specific values λ_n: we have not described that structure for values of

λ between λ_n and λ_{n+1}. However, at these choices of λ, the theory we describe holds regardless of the attractor from which the bifurcated attractors arise. (That is, λ_n refers to that attractor of order $m \cdot 2^n$, which includes the critical point $\bar{x}$ for any m.) Or, the local structure of all *infinite* attractors of $x_{n+1} = \lambda f(x_n)$ is described by g^*. Also, the rescaling parameter α and the convergence rate δ are common to all highly bifurcated attractors.

At this point we should like to briefly summarize some results of the exact theory; a full treatment of these results will shortly appear in a sequel.[(2)]

Figure 5a shows, near $\bar{x}$, what shall evolve into g^*, and represents the local structure at a "two-point" level. Figure 5b represents the graph of a function that shall evolve into $g^* \circ g^*$, and accordingly must also be universal: it represents the identical local structure as does g^*, but now at a "one-point" level of description. Evidently, one can view this local structure at a "2^r-point" level from the function

$$g_r(x) = \lim_{n\to\infty} (-\alpha)^n \tilde{g}^{(2^n)}(\lambda_{n+r}, x/(-\alpha)^n), \qquad r = 0, 1, 2, \ldots \tag{35}$$

where $g_1(x)$ is exactly $g^*(x)$. All g_r are of identical qualitative shape: each bump of g_1 aligned along $y = x$ contains two points of an attractor set, whereas each bump of g_r similarly situated now contains 2^r points of that set. Evidently, the lower r, the more magnified the local structure. Following immediately from the definition,

$$g_{r-1}(x) = -\alpha g_r(g_r(x/\alpha)) \tag{36}$$

(all the functions g_r are symmetric). The content of this equation is essentially the Cantor-set-like nature of highly bifurcated attractors: at each bifurcation, the rough locations of attractor points are unchanged, with a "microscopic" splitting of each such point; the scale of splitting is α below the previous level, so that rescaling by α after a bifurcation reveals the set at a next level of magnification. In fact, (36) provides the entire exact description, as we now synoptically elucidate. The central bump of g_r is effectively a $\lambda f(x)$ containing a 2^r-point cycle, which, as r increases, quickly approaches a λf containing the infinite attractor. That is, one expects

$$g(x) \equiv \lim_{r\to\infty} g_r(x) \tag{37}$$

to exist; g no longer affords the same description of attractor points as does g_r. Rather, g is the description at the level of infinite clusters of points, which is again a universal property. But g defined by (37) is simply a fixed point of (36). Accordingly,[2]

$$g(x) = -\alpha g(g(x/\alpha)) \tag{38}$$

[2] This exact equation was discovered by P. Cvitanović during discussion and in collaboration with the author.

The great virtue of g is that λ has been set at λ_∞ at the outset, and so the difficulty of modeling λ-shifting is totally bypassed. The price paid for this is that (38) defines no recursively stable equation like

$$\bar{g}_{n+1}(x) = -\alpha_n \bar{g}_n(\bar{g}_n(x/\alpha_n)) \tag{39}$$

[By (36), iterating produces g_r's for smaller values of r and hence diverging from g.] There are a variety of ways to solve (38). A method based on the fact that $\bar{g}_0(x) = f(\lambda_\infty(f)x)$ *must* cause (39) to converge [by (35), $g = \lim(-\alpha)^n \bar{g}^{(2^n)}(\lambda_\infty, x/(-\alpha)^n)$] together with the general recursive instability of (39) allows very fast, high-accuracy estimates of all λ_∞ for any chosen f. Alternatively, one can simply solve (38) by a numerical functional-Newton's method. (The result of the latter method is a 20-place determination of both α and g for $z = 2$.)

The g of (38) is a fixed point of (36). By setting

$$g = g_r + y_r$$

in (38), employing (36), and expanding to first order in y, one obtains

$$y_{r-1}(x) = -\alpha[y_r(g(x/\alpha)) + g'(g(x/\alpha))y_r(x/\alpha)] \tag{40}$$

(40) simply separates with the substitution

$$y_r = \lambda^{-r}\psi(x) \tag{41}$$

where ψ obeys

$$\mathscr{L}[\psi(x)] \equiv -\alpha[\psi(g(x/\alpha)) + g'(g(x/\alpha))\psi(x/\alpha)] = \lambda\psi(x) \tag{42}$$

The eigenvalue λ can clearly attain the value +1 corresponding to

$$\psi = g - xg'$$

reflecting the dilatation invariance of (38). In addition to a spectrum $|\lambda| \leqslant 1$, computationally there exists a unique alternate value δ strictly greater than +1.

It is possible to show that the eigenvalue δ is exactly that convergence rate discussed in this paper. Heuristically, if λ is held fixed at λ_n for $n \gg 1$, and $\lambda_n f$ iterated, it is indistinguishable from the iterates of $\lambda_\infty f$ that approximate g after an initial transient, until roughly n iterations have been performed to magnify the deviation of λ_n from λ_∞. Thus the argument about Eq. (8) can be made exact where the function $\tilde{g}_{n-i}$ there is essentially g for (logarithmically) all iterations.

One can next begin to investigate the nature of the n limit of (35). Defining

$$g_{r,n}(x) \equiv (-\alpha)^n \tilde{g}^{(2^n)}(\lambda_{n+r}, x/(-\alpha)^n)$$

it is immediate to verify that

$$g_{r-1,n+1}(x) = -\alpha g_{r,n}(g_{r,n}(-x/\alpha)) \tag{43}$$

But (36) is the large-n fixed point of (43), and so one can discuss in linear approximation the stability of (36). (We mention at this point that the antisymmetric parts of $g_{r,n}$ vanish in the large-n limit at the rate of $-\alpha$; this result is exactly observed computationally.)

With α and g obtained from (38), (42) determines both h and δ. (Again, both have been obtained to 20-place accuracy.) We stress that (38) and (42) are totally free of any reference to (1) and do produce the same α and δ to the 14-place accuracy of our best recursion data. Next,

$$g_r \sim g - \delta^{-r}h \tag{44}$$

so that a g_r for $r \gg 1$ is available. By successive application of (36) to this asymptotic g_r, g_1 can be obtained. (Regarding $r = 3$ as asymptotic produces a g_1 to six-place accuracy, to give an idea of the speed of onset of the asymptotic regime.) The approximate equations (28) are a high-z approximation to (38) and (42).

Since $\delta^{-r} \sim \lambda_\infty - \lambda_r$, (44) has an immediate continuation:

$$g_\lambda(x) \sim g(x) - (\lambda_\infty - \lambda)h(x)$$

which allows the determination of local structure for λ between λ_n and λ_{n+1} as well. Thus, the bifurcation points Λ_n also geometrically converge to λ_∞ at the rate δ, and logarithmically the behavior of bifurcation is periodic with period $\log \delta$. A demonstration that (36) is in fact a stable fixed point of (43) would constitute a proof of our universality conjecture: with exact (functional) equations at hand, it is possible to focus on the exact details requiring proof.[3]

APPENDIX A

In the formula $x_{n+1} = \lambda f(x_n)$ with $0 < \lambda < 1$, $f(x)$ satisfies the following conditions:

1. $f(x)$ is continuous, single-valued, piecewise $C^{(1)}$ on $[0, 1]$ possessing a unique, differentiable maximum at $\bar{x}$ with $f(\bar{x}) = 1$.
2. $f(x) > 0$ on $(0, 1)$, $f(0) = f(1) = 0$, and f is strictly decreasing on $(\bar{x}, 1)$ and strictly increasing on $(0, \bar{x})$.
3. For $\Lambda_0 < \lambda < 1$, $\lambda f(x)$ has two fixed points [$x^* = 0$, and some other $x^* \in (x, 1)$] both of which are repellant (i.e., $|f'(x)| > 1/\lambda$).
4. In the interval N about $\bar{x}$ such that $|f'(x)| < 1$, f is concave downward.

[3] We are in possession of extensive high-precision data pertaining to all details discussed in this paper, as well as for the solutions to the functional equations discussed in the last sections. We will consider reasonable requests from individuals for copies of specific parts of this library.

Given these conditions, Metropolis *et al.*[1] have established among others the following universal, qualitative features:

(a) For $\Lambda_0 < \lambda < \lambda_\infty$, there exists stability sets of order 2^n, $n = 1, 2, \ldots$ only, with n increasing with λ.

(b) For $\Lambda_0 < \Lambda_{n-1} < \lambda < \Lambda_n < \lambda_\infty$ only 2^n-order stability sets exist. In particular, at λ_n, with $\Lambda_{n-1} < \lambda_n < \Lambda_n$, the 2^n-order stability set contains $\bar{x}$ as an element.

(c1) For $\lambda = \lambda_1$, under repeated application of λf, one has $\bar{x} \to x' \to \bar{x} \to \cdots$, where $x' > \bar{x}$. Calling an x "R" if $x > \bar{x}$ and "L" if $x < \bar{x}$, the "pattern" of motion through the stability set is abbreviated as R—meaning $\bar{x} \to R \to \bar{x}$.

(c2) The "harmonic" of a pattern P is a stability set of twice the order of P, with pattern PLP if P contains an odd number of Rs and PRP otherwise. The 2^n-order stability sets are exactly the successive harmonics of R (e.g., for λ_2, RLR; for λ_3, RLRRRLR; etc.).

(d) If P is a basic pattern (say, for an r-point cycle), then (a)–(c) hold with 2^n replaced by $r \cdot 2^n$.

APPENDIX B. COMPUTATIONAL RESULTS

The parameter values λ_n for a given recurrence function f are obtained by definition from

$$(\lambda_n f)^{(2^n m)}(\bar{x}) - \bar{x} = 0 \tag{B1}$$

(B1) possesses in general many roots. Accordingly, λ_0 is first obtained for a given fundamental pattern. λ is slowly increased to find the first new zero for $n = 1$; this λ by definition is λ_1. Next, λ_2 is similarly found as the next largest zero of (B1) for $n = 2$. At this point δ_1 is calculated

$$\delta_1 = (\lambda_1 - \lambda_0)/(\lambda_2 - \lambda_1)$$

and used to estimate–predict λ_3

$$\lambda_3 \simeq \lambda_2 + \delta_1^{-1}(\lambda_2 - \lambda_1) \tag{B2}$$

As n increases, $\delta_n \to \delta$ and the predicted value increases in precision, so that for large n, several Newton's-method iterations suffice to locate λ_n to full precision. Since the number of iterations increases geometrically and the number of zeros of (B1) similarly increases with collateral decrease of spacing between them, the prediction method is essential to locate high-n λ's. (For example, the set of all λ_n up to $n = 20$ for $f = x - x^2$ to 29-place precision requires just a few minutes of CDC 6600 time.)

Analogous to δ_n, one can also compute the rate of convergence of δ_n to δ through δ_n':

$$\delta_n' \equiv (\delta_{n+1} - \delta_n)/(\delta_{n+2} - \delta_{n+1})$$

With λ_n of 29-place accuracy, δ_n converges to δ to 13 places by $n = 20$ and δ_n' converges to three or four places. We quote some typical results in Tables I–III. Observe that for an f symmetric about its maximum, $\delta' \simeq \delta$ for $z \leqslant 2$, whereas $\delta' < \delta$ for $z > 2$. As related in the text, we shall explain these results in the sequel to this work.

With the λ_n determined, the parameter α is next obtained. We transform to variables in which $\bar{x} = 0$. The element of the limit cycle nearest to $\bar{x}$ is obtained as the 2^{n-1} iterate of $\bar{x}$,

$$z_n \equiv (\lambda_n f)^{(2^{n-1})}(\bar{x})$$

and the nth rescaling α_n, defined by

$$\alpha_n = -z_n/z_{n+1}$$

These α_n converge to α, also typically to 13 places.

Table I.[a] Two-Cycle Data for f = 1 − 2x²

N	λ	δ	δ'
1	0.7071067811865475244008443621	4.74430946893705	—
2	0.8095377203493463168459541018	4.67444782765301	2.7504
3	0.8311279938830304702482833891	4.67079115022921	6.7888
4	0.8357467797438888850823009395	4.66946164833746	3.7990
5	0.8367356455938705846037094966	4.66926580979910	5.2553
6	0.8369474185828047108022721846	4.66921427043589	4.3595
7	0.8369927732483047323090713162	4.66920445137251	4.8560
8	0.8370024868024425943459682976	4.66920220132661	4.5641
9	0.8370045671470149993313732630	4.66920173797283	4.7307
10	0.8370050126930596349457550266	4.66920163645133	4.6340
11	0.8370051081153758334851887620	4.66920161499127	4.6896
12	0.8370051285519137318702724660	4.66920161036023	4.6575
13	0.8370051329287943173583990344	4.66920160937272	4.6759
14	0.8370051338661881055761765511	4.66920160916069	4.6684
15	0.8370051340669491492492744646	4.66920160911533	
16	0.8370051341099460169105929249	4.66920160910564	
17	0.8370051341191546292732244007		
18	0.8370051341211268320465365015		

[a] In this and the following tables, the cycle of size 2^N for two-cycle data and $3 \times 2^{N-1}$ for three-cycle data is referenced by N. The parameter is denoted by λ; $\delta_N = (\lambda_{N+1} - \lambda_N)/(\lambda_{N+2} - \lambda_{N-1})$ and $\delta_N' = (\delta_{N+1} - \delta_N)/(\delta_{N+2} - \delta_{N+1})$.

Table II.[a] Three-Cycle Data for f = 1 − 2x²

N	λ	δ	δ'
1	0.9367170507273508470331116496	3.36345171892599	5.6876
2	0.9415128526423905912356034646	4.42501749338226	4.0902
3	0.9429387098616436488660308130	4.61166338330503	4.9053
4	0.9432609362079542493235619775	4.65729621052512	4.5349
5	0.9433308082518413013515037559	4.66659897033153	4.7439
6	0.9433458109574302537416522425	4.66865036240409	4.6264
7	0.9433490258695502652989696376	4.66908278613357	4.6938
8	0.9433497144865745558168755447	4.66917625427465	4.6550
9	0.9433498619710069020581737454	4.66919616732989	4.6774
10	0.9433498935578274276740167548	4.66920044506702	4.6774
11	0.9433499003227649872965866091	4.66920135962593	
12	0.9433499017716078115042431354		
13	0.9433499020819055832053356825		

[a] See footnote to Table I.

Table III.[a] Two-Cycle Data for f = x(1 − x²)

N	λ	δ	δ'
1	2.1213203435596425732025330870	4.59165349403582	4.7073
2	2.2629896545363471897845547810	4.65266937815338	4.6198
3	2.2938433134983575307521580080	4.66563137254676	4.6704
4	2.3004747021602907471027097710	4.66843710204515	4.6686
5	2.3018960293360623306427444230	4.66903785250679	4.6680
6	2.3022004839414695502983032340	4.66916653116789	4.6700
7	2.3022656910815752484787734520	4.66919409734649	4.6687
8	2.3022796565590690203670882640	4.66920000018325	4.6695
9	2.3022826475414960966948147840	4.66920126453840	4.6690
10	2.3022832881185600388981314270	4.66920153530566	4.6693
11	2.3022834253105609079220759030	4.66920159329811	4.6691
12	2.3022834546928867363664720370	3.66920160571803	4.6692
13	2.3022834609856811769873007170	4.66920160837802	4.6831
14	2.3022834623334050432933708930	4.66920160894770	4.2652
15	2.3022834626220462242369870730	4.66920160906935	
16	2.3022834626838643261562712190	4.66920160909788	
17	2.3022834626971036705355950340	4.66920160909268	
18	2.3022834626999393755079069360		
19	2.3022834627005466538414854460		

[a] See footnote to Table I.

Finally, one computes the functions (where $\bar{x} = 0$)

$$g_n^*(x) \equiv (-1)^n \alpha_n^{-1} (\lambda_n f)^{(2^n - 1)} (\alpha_n x)$$

so normalized that

$$g_n^*(0) = 1, \qquad g_n^*(1) = 0$$

and observes convergence to g^* (in the interval [0, 1] also to 13 places).

ACKNOWLEDGMENTS

Initial thoughts on this work occurred during the author's stay as Aspen; accordingly, he thanks the Aspen Institute for Physics for their hospitality, and especially E. Kerner for early discussions. A crucial hint at the existence of δ followed from conversation with P. Stein. The author's rapid acquaintance with computational technique was strongly abetted by M. Bolsterli and L. Carruthers. Discussion with E. Larsen and R. Haymaker was especially useful during initial attempts at a comprehension of the existence of δ. To E. Lieb we owe strong thanks for his most profitable, critical remarks as well as for his high enthusiasm. Throughout this research, D. Campbell, F. Cooper, R. Menikoff, M. Nieto, D. Sharp, and P. Stein have offered continued critical interest. Finally, we thank P. Stein and P. Cvitanović for criticism of the manuscript. Beyond these acknowledgments, the author feels the strongest gratitude to Peter Carruthers, whose unflagging support of the author's research career has made this work possible.

REFERENCES

1. N. Metropolis, M. L. Stein, and P. R. Stein, On Finite Limit Sets for Transformations on the Unit Interval, *J. Combinatorial Theory* **15**(1): 25 (1973).
2. M. Feigenbaum, The Formal Development of Recursive Universality, Los Alamos preprint LA-UR-78-1155.
3. B. Derrida, A. Gervois, Y. Pomeau, Iterations of Endomorphisms on the Real Axis and Representations of Numbers, Saclay preprint (1977).
4. J. Guckenheimer, *Inventiones Math.* **39**: 165 (1977).
5. R. H. May, *Nature* **261**: 459 (1976).
6. J. Milnor, W. Thurston, Warwick Dynamical Systems Conference, *Lecture Notes in Mathematics*, Springer Verlag (1974).
7. P. Stefan, *Comm. Math. Phys.* **54**: 237 (1977).

Journal of Statistical Physics, Vol. 21, No. 6, 1979

The Universal Metric Properties of Nonlinear Transformations

Mitchell J. Feigenbaum[1]

Received May 29, 1979

The role of functional equations to describe the exact local structure of highly bifurcated attractors of $x_{n+1} = \lambda f(x_n)$ independent of a specific f is formally developed. A hierarchy of universal functions $g_r(x)$ exists, each descriptive of the same local structure but at levels of a cluster of 2^r points. The hierarchy obeys $g_{r-1}(x) = -\alpha g_r(g_r(x/\alpha))$, with $g = \lim_{r\to\infty} g_r$ existing and obeying $g(x) = -\alpha g(g(x/\alpha))$, an equation whose solution determines both g and α. For r asymptotic

$$g_r \sim g - \delta^{-r}h \qquad (*)$$

where $\delta > 1$ and h are determined as the associated eigenvalue and eigenvector of the operator $\mathscr{L}$:

$$\mathscr{L}[\psi] = -\alpha[\psi(g(x/\alpha)) + g'(g(x/\alpha))\psi(-x/\alpha)]$$

We conjecture that $\mathscr{L}$ possesses a unique eigenvalue in excess of 1, and show that this δ is the λ-convergence rate. The form (*) is then continued to all λ rather than just discrete λ_r and bifurcation values Λ_r and dynamics at such λ is determined. These results hold for the high bifurcations of any fundamental cycle. We proceed to analyze the approach to the asymptotic regime and show, granted $\mathscr{L}$'s spectral conjecture, the stability of the g_r limit of highly iterated λf's, thus establishing our theory in a local sense. We show in the course of this that highly iterated λf's are conjugate to g_r's, thereby providing some elementary approximation schemes for obtaining λ_r for a chosen f.

KEY WORDS: Recurrence; bifurcation; attractor; universal; functional equations; scaling; conjugacy; spectrum of linearized operator.

Work performed under the auspices of the U.S. Energy Research and Development Administration.

[1] Theoretical Division, Los Alamos Scientific Laboratory, University of California, Los Alamos, New Mexico.

1. INTRODUCTION

In a previous paper[1] (hereafter referred to as I), a viewpoint was advanced that detailed information about large stability sets of a recursion relation

$$x_{n+1} = \lambda f(x_n) \tag{1}$$

is available independent of the exact form of f for a wide class of functions. A heuristic argument (corroborated by computer computation) was offered to the effect that appropriate functional equations, free of reference to the recursion equation, furnish all this detailed quantitative information. Specifically, the exact distribution of points of large limit cycles of the recursion equation within local clusters is determined by a certain universal function $g^*(x)$ obeying a functional equation we conjectured to exist, but only approximately could specify. A parameter α implicated in that equation, and presumably determined by it collaterally with $g^*(x)$, plays the role of a fundamental scale factor: upon bifurcation of a high-order cycle, the points of the bifurcated cycle are identically distributed, save for a reduction in scale by the factor α. Another fundamental parameter δ, the convergence rate of a variety of universal details, was crudely determined from $g^*(x)$.

In the present paper we shall vindicate these conjectures in exhibiting an exact equation determining α and a universal function g closely related to g^* of I. Indeed, two functions $g^*(x)$ and $-\alpha g^*(g^*(x/\alpha))$ were discussed in I; these are the first two (g_1 and g_0, respectively) of an infinite sequence of functions $g_r(x)$ linked by the shift operation

$$g_{r-1}(x) = -\alpha g_r(g_r(x/\alpha)) \tag{2}$$

The equation

$$g(x) = -\alpha g(g(x/\alpha)) \tag{3}$$

is obeyed by $g(x) = \lim_{r\to\infty} g_r(x)$, a function determining the local distribution of infinite clusters of elements of *all* the infinite attractors of (1). We then proceed to determine $g_r(x)$ for large r in terms of an auxiliary function $h(x)$ obeying a functional equation implicating δ and determining both $h(x)$ and δ. Utilizing (2), one can then step down to determine from a g_r for $r \gg 1$ the g^* of I. Thus, as conjectured in I, the entire local structure of high-order stability sets of (1) is determined in a framework liberated from (1).

With the structure of the infinite limiting attractors laid out in Sections 2 and 3, we investigate in Section 4 the asymptotic approach to this structure. The equation obeyed by h is a linear functional eigenvalue equation, whose eigenvalues in excess of 1 lead to convergence of g_r to g; the eigenvalues bounded by 1 in absolute value represent potential instabilities.

However, in analyzing the large-n approach to a g_r, we discover that exactly these eigenvalues lead to convergence, so that in the infinite-n limit, g_r possesses no unstable components. In this fashion, though, the large eigenvalues destroy convergence to g_r. We discover, however, that the choice of the λ_r dictated by the recursion equation exactly suppresses this instability *providing* there is a *unique* eigenvalue in excess of 1. This eigenvalue is δ and we conjecture its uniqueness. Proof of this conjecture would constitute a local proof of universality. (At present we have only computer corroboration.)

Finally, in Section 5 we discuss techniques for the solution of the fundamental functional equations and various approximation schemes.

2. THE SEQUENCE $\{g_r\}$ OF UNIVERSAL FUNCTIONS AND THE BASIC FUNCTIONAL EQUATION

As heuristically argued in I, defining

$$g(\lambda, x) \equiv \lambda f(x)$$

where f possesses a differentiable zth-order maximum at $x = 0$,

$$f(0) - f(x) \propto |x|^z \qquad z > 1 \text{ for } |x| \text{ small}$$

we have

$$(-\alpha)^n g^{(2^n)}(\lambda_{n+1}, x/\alpha^n) \sim \mu g^*(x/\mu) \tag{4}$$

for large n, where $g^{(n)}(x)$ is the nth iterate of g:

$$g^{(2)}(x) \equiv g(g(x)); \qquad g^{(n+1)}(x) \equiv g(g^{(n)}(x))$$

and μ depends upon the specific form of f. Rescaling g^* on both height and width by μ removes all vestige of the specific form of f, and is accomplished through the definition of an absolute scale:

$$g^*(0) = 1$$

We understand this absolute rescaling implicitly, and simply write (4) as

$$(-\alpha)^n g^{(2^n)}(\lambda_{n+1}, x/\alpha^n) \sim g^*(x) \tag{5}$$

The value of λ, λ_{n+1} is determined by the condition

$$g^{(2^n)}(\lambda_n, 0) = 0, \qquad g^{(2^{n'})}(\lambda_n, 0) \neq 0 \qquad \text{for } n' < n$$

g^* accordingly describes a two-point cycle near $x = 0$, since $g^*(0) = 1$ and $g^*(1) = 0$ (see Fig. 1). With g^* the limit of (5) as $n \to \infty$, and universal for

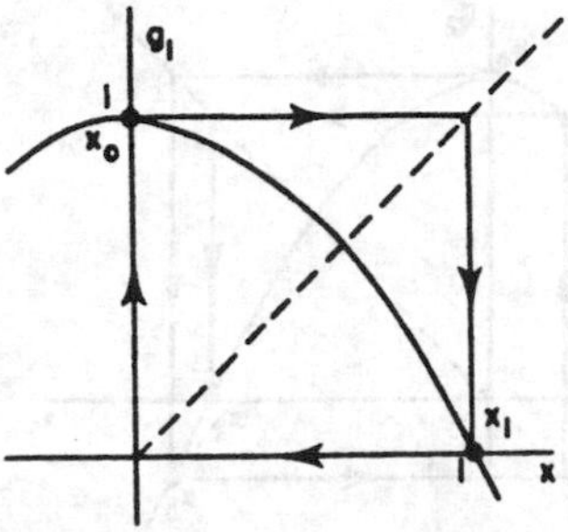

Fig. 1. The function g_1 normalized to $g_1(0) = 1$; the 2-cycle $x_0 \to x_1 \to x_0 \to x$, etc., is indicated.

all f of fixed z, $g^*(g^*(x))$ is itself universal, possessing 0 and 1 as fixed points. $g^*(g^*)$ is also a limit of highly iterated g's:

$$\begin{aligned} g^*(g^*(x)) &\sim (-\alpha)^n g^{(2^n)}(\lambda_{n+1}, g^{(2^n)}(\lambda_{n+1}, x/\alpha_n)) \\ &= (-\alpha)^n g^{(2^{n+1})}(\lambda_{n+1}, x/\alpha^n) \end{aligned}$$

or,

$$\begin{aligned} g_0 \equiv -\alpha g^*(g^*(x/\alpha)) &\sim (-\alpha)^{n+1} g^{(2^{n+1})}(\lambda_{n+1}, x/\alpha^{n+1}) \\ &\sim (-\alpha)^n g^{(2^n)}(\lambda_n, x/\alpha^n) \end{aligned} \tag{6}$$

Thus, $g^*(x)$ is obtained from $g_0(x)$ by increasing λ into the next bifurcation; conversely, g_0 describes a one-cycle near $x = 0$ (Fig. 2). Both g_0 and $g^* \equiv g_1$ describe the identical local structure of the elements of a large 2^n-cycle near $x = 0$, but at different "magnifications." Generalizing,

$$g_r(x) \equiv \lim_{n \to \infty} (-\alpha)^n g^{(2^n)}(\lambda_{n+r}, x/\alpha^n) \tag{7}$$

again describes the identical local structure, but now at a level of magnification such that 2^r elements of the cycle are clustered about the central bump (Fig. 3). From the definition (7),

$$g_{r-1}(x) = -\alpha g_r(g_r(x/\alpha)) \tag{8}$$

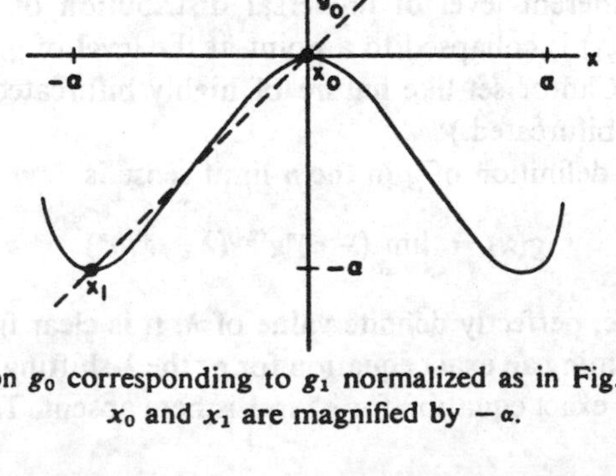

Fig. 2. The function g_0 corresponding to g_1 normalized as in Fig. 1; the locations of x_0 and x_1 are magnified by $-\alpha$.

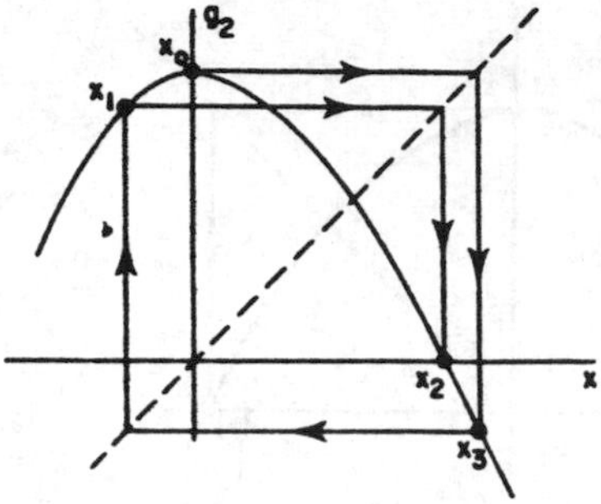

Fig. 3. The function g_2; the 4-cycle $x_0 \to x_3 \to x_1 \to x_2 \to x_0$, etc., is indicated. x_0 and x_1 are as in Figs. 1 and 2, but now *reduced* from Fig. 1 by $-\alpha$.

(Universality implies that all g_r are symmetric functions.) Given any g_r, simply iterating produces all other g_m for $m < n$, and each contains the identical information.

When r is so large that n can become large and yet much smaller than r, then

$$g^{(2^n)}(\lambda_{n+r+1}) \approx g^{(2^n)}(\lambda_{n+r})$$

since $\lambda_m \to \lambda_\infty$ and $\lambda_{n+r+1} \approx \lambda_{n+r}$: n must be of the order of r before any error in λ_{n+r} can become significant. Alternatively, the central bump suffers very slight distortion to accommodate the infinite attractor when it already accommodates a very large attractor. That is, we intuitively conjecture that the limit

$$\lim_{r\to\infty} g_r(x) \equiv g(x)$$

exists. This granted, (8) implies that g satisfies

$$g(x) = -\alpha g(g(x/\alpha)) \tag{9}$$

Qualitatively, $g(x)$ looks like the curve of Fig. 3. Yet, $g(x)$ contains different information from $g_r(x)$ for any finite r: quite simply, any two stability points located by g_r possess a minimum separation, which is not true for g. Rather, g represents a different level of universal distribution of stability points: the entirety of $g_1(x)$ is collapsed to a point at the level of g. (This is again a reflection of the Cantor set-like nature of highly bifurcated stability sets—indeed, infinitely bifurcated.)

An alternate definition of g in the n-limit sense is

$$g(x) = \lim_{n\to\infty} (-\alpha)^n g^{(2^n)}(\lambda_\infty, x/\alpha^n) \tag{10}$$

since λ_∞ is a finite, perfectly definite value of λ. It is clear from (10) why we succeeded in obtaining an exact equation for g: the λ-shifting which frustrated our attempt at an exact equation for g^* in I is here absent. There is, however, a strong price to be exacted for this grace: unlike the hoped-for equation in I, (9) must be recursively unstable. To understand this, let us rederive (9) from (10). Define

$$\tilde{g}_n(x) \equiv (-1)^n \beta_n g^{(2^n)}(\lambda_\infty, x/\beta_n) \tag{11}$$

or

$$(1/\beta_n)\tilde{g}_n(\beta_n x) = (-1)^n g^{(2^n)}(\lambda_\infty, x)$$

Then

$$\begin{aligned}(-1)^n g^{(2^{n+1})}(\lambda_\infty, x) &= (-1)^n g^{(2^n)}(\lambda_\infty, (-1)^n g^{(2^n)}(\lambda_\infty, x)) \\ &= \frac{1}{\beta_n}\tilde{g}_n((-1)^n \beta_n g^{(2^n)}(\lambda_\infty, x)) \\ &= \frac{1}{\beta_n}\tilde{g}_n(\tilde{g}_n(\beta_n x))\end{aligned}$$

or,

$$-\frac{1}{\beta_{n+1}}\tilde{g}_{n+1}(\beta_{n+1}x) = \frac{1}{\beta_n}\tilde{g}_n(\tilde{g}_n(\beta_n x))$$

or, with $\beta_{n+1}/\beta_n \equiv \alpha_n$,

$$\tilde{g}_{n+1}(x) = -\alpha_n \tilde{g}_n(\tilde{g}_n(x/\alpha_n)) \tag{12}$$

Setting an absolute scale

$$\tilde{g}_n(0) = 1 \qquad \text{for all } n$$

(12) implies that $g_n(x)$ determines α_n:

$$\tilde{g}_{n+1}(0) = -\alpha_n \tilde{g}_n(\tilde{g}_n(0))$$

or

$$1 = -\alpha_n \tilde{g}_n(1) \tag{13}$$

Accordingly, choosing a $\tilde{g}_0(x)$ satisfying $\tilde{g}_0(0) = 1$ and possessing a zth order maximum at $x = 0$, we can use (12) and (13) to recursively generate $\tilde{g}_n(x)$. Should $\tilde{g}_n(x) \to g(x)$, then

$$g(x) = -\alpha g(g(x/\alpha))$$

with $\alpha = \lim_{n\to\infty} \alpha_n$.

Apart from manipulations, the regimen of (12) and (13) is simply a machine to perform the (attempted) limit of (10) starting with a $\tilde{g}_0$ that is essentially $\lambda_\infty f(x)$, or more exactly

$$\tilde{g}_0(x) = f(\lambda_\infty x) \tag{14}$$

[So that $\tilde{g}_0(0) = 1$, we have rescaled on height and width by λ_∞: (9) is invariant to such rescaling.] Since λ_∞ depends upon f,

$$\tilde{g}_0(x) = f(\lambda_\infty(f)x) \tag{15}$$

Indeed, for any f of our class, $\tilde{g}_0$ as given by (15) must result in $\tilde{g}_n \to g$. *However*, for

$$\tilde{g}_0(x) = f(ax), \qquad a \neq \lambda_\infty$$

it must be *impossible* for $\tilde{g}_n$ to converge: unless $a = \lambda_\infty$ for some harmonic sequence, there is no infinite attractor and no sequence of g_r's converging to g. For example, if we choose

$$\tilde{g}_0(x) = 1 - ax^2$$

then unless a is chosen at special isolated values, the $\tilde{g}_n$ will not converge. Rather, a could in general be a value λ_m; after a number of iterations, $\tilde{g}_n$ would, by definition, be a g_r [Eq. (7)] approximately. Since (12) is (8) (with some rescaling), successive iterations would move toward g_0 rather than g and then divergently away. Figure 4 represents a suggestive picture of the situation. That is, the fixed point g is repellent, and unless $\tilde{g}_0$ is correctly chosen so that the $\tilde{g}_n$ will "aim" into g, they will at first approach $g(\lambda_r \simeq \lambda_\infty$ so far as $g^{(2^n)}$ is concerned until $n \sim r$), but then diverge away from it along the "path" of decreasing g_r's. That is, (9) in general defines a recursively unstable problem.

This instability can, however, be turned to excellent advantage. Since an arbitrary $\tilde{g}_0$ will lead to divergence, a "good" $\tilde{g}_0$ must already be a good approximation to g. With $g(0) = 1$, (9) implies that

$$g(1) = -1/\alpha$$

By (14) one should then estimate

$$f(\lambda_\infty(f)) = \tilde{g}_0(1) \simeq -1/\alpha \tag{16}$$

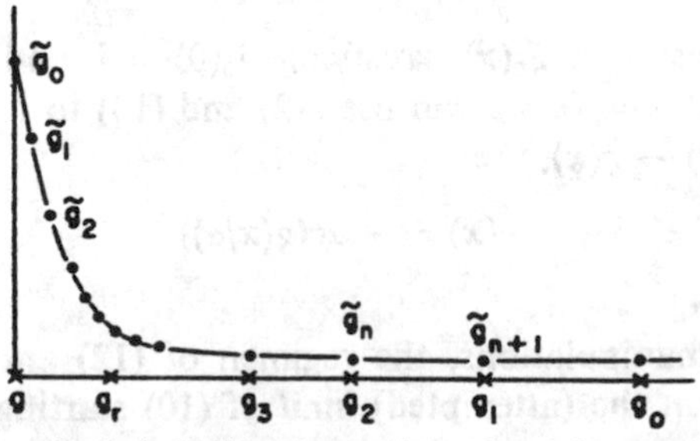

Fig. 4. The sequence of g_r's is indicated as points along the x axis. The iterates of $\tilde{g}_0$ are shown at first approaching g, the trajectory spending many iterations in the vicinity of g (or g_r for large r), but ultimately diverging away near low-lying g_r's.

that is, the instability of (9) provides an estimation of $\lambda_\infty(f)$ for any suitable f. The closer an f is to g (in some sense), the better the estimate. For example, consider

$$x_{n+1} = \lambda(1 - 2x_n^2)$$

or

$$f = 1 - 2x^2$$

and

$$f(\lambda_\infty x) = 1 - 2\lambda_\infty{}^2 x^2$$

With $\alpha = 2.5029\cdots$ for $z = 2$, we obtain from $1 - 2\lambda_\infty{}^2 \simeq -1/\alpha$ a value of $\lambda_\infty \simeq 0.8365$, to be compared to $\lambda_\infty = 0.8370$ for the limit of 2^n-cycles. In Section 5 we shall pursue this idea to obtain a technique for solving (9).

3. THE INFINITESIMAL λ SHIFT AND CONVERGENCE

Increasing λ from λ_n to λ_{n+1} maps g_r into g_{r+1} for all r. Calling this operation R, the λ shift, we write

$$R(g_r) = g_{r+1} \tag{17}$$

In I, R was applied to g_0 to produce g_1. Equation (8), written as

$$g_{r-1} = -\alpha g_r(g_r(x/\alpha)) \equiv L(g_r)$$

accomplishes the operation inverse to R. The combined operation

$$B \equiv L \cdot R$$

is the bifurcation transformation of I, which serves as an identity on the sequence $\{g_r\}$:

$$B(g_r) = g_r, \qquad r = 0, 1, \ldots \tag{18}$$

or, each g_r is a fixed point of the transformation B. We select g_1 by imposing the conditions

$$g_1(0) = 1, \qquad g_1(1) = 0 \tag{19}$$

and our universality conjecture is phrased in this language by saying the fixed point g_1 of B is *stable*, so that if any $\tilde{g}_0$ satisfying (19) with a zth-order maximum is chosen,

$$\tilde{g}_{n+1} = B[\tilde{g}_n]$$

will result in

$$\tilde{g}_n \rightarrow g_1$$

for the zth-order maximum universal g_1. The empirical computer evidence for universality, together with the instability of (12), means that R stabilizes B, as R reduces trivially to the identity only for the fixed-point g. Indeed, our approximate modelings of R in I resulted in recursively stable functional equations. We now determine R restricted to operation on g_r's in the limit of infinite r. The interchangeability of r and n in (7)

$$(-\alpha)^n g^{(2^n)}(\lambda_{n+r}, x/\alpha_n) \sim (-\alpha)^{n-s} g^{(2^{n-s})}(\lambda_{n+(r-s)}, x/\alpha^{n-s}), \qquad n - s \gg 1$$

together with the shifting operators implies that our study shall provide information about large-n convergence properties, and so determine δ as well.

We want to compute

$$\delta g^{(2^n)}(\lambda_{n+r}, x) \equiv g^{(2^n)}(\lambda_{n+r+1}, x) - g^{(2^n)}(\lambda_{n+r}, x) \tag{20}$$

Defining

$$\delta g_r(x) \equiv (-\alpha)^n \, \delta g^{(2^n)}(\lambda_{n+r}, x/\alpha^n) \tag{21}$$

(20) becomes

$$\delta g_r(x) = g_{r+1}(x) - g_r(x) = (R - 1)(g_r) \tag{22}$$

Substituting (22) in (8),

$$\begin{aligned} g_r(x) &= -\alpha(g_r + \delta g_r)[g_r(x/\alpha) + \delta g_r(x/\alpha)] \\ &= -\alpha g_r(g_r(x/\alpha)) - \alpha g_r'(g_r(x/\alpha)) \, \delta g_r(x/\alpha) \\ &\quad - \alpha \, \delta g_r(g_r(x/\alpha)) + O((\delta g_r)^2) \\ &= g_{r-1}(x) - \alpha[\delta g_r(g_r(x/\alpha)) + g_r'(g_r(x/\alpha)) \, \delta g_r(x/\alpha)] + O((\delta g_r)^2) \end{aligned} \tag{23}$$

or,

$$\delta g_{r-1}(x) = -\alpha[\delta g_r(g_r(x/\alpha)) + g_r'(g_r(x/\alpha)) \, \delta g_r(x/\alpha)] + O((\delta g_r)^2) \tag{24}$$

Since $g_r \to g$, $\delta g_r \to 0$, and in the limit of infinite r

$$\begin{aligned} \delta g_{r-1}(x) &\sim -\alpha[\delta g_r(g_r(x/\alpha)) + g_r'(g_r(x/\alpha)) \, \delta g_r(x/\alpha)] \\ &\sim -\alpha[\delta g_r(g(x/\alpha)) + g'(g(x/\alpha)) \, \delta g_r(x/\alpha)] \end{aligned} \tag{25}$$

Separating $\delta g_r(x)$ as

$$\delta g_r(x) = \eta_r h(x) \tag{26}$$

with $\eta_r \to 0$ as $r \to \infty$, we obtained a closed equation for $h(x)$—the generator of infinitesimal λ shifts—and an equation for η_r:

$$\eta_{r-1} = \delta \eta_r \tag{27}$$

and

$$h(x) = -(\alpha/\delta)[h(g(x/\alpha)) + g'(g(x/\alpha))h(x/\alpha)] \tag{28}$$

In fact, (28) represents a rederivation of (28) of I, where (28) of I was an approximate realization of R applied to g_0, the approximation consisting of "mild" λ shifting, which becomes rigorous in the present context, and in this context, involving g and not g_1. Given $g(x)$ and α obeying (9), (28) determines both $h(x)$ and δ and defines a recursively stable equation. We return to this in the last section.

Equation (27) is trivially solved:

$$\eta_r = \delta^{-r} \tag{29}$$

so that

$$\delta g_r(x) = \delta^{-r} h(x) \tag{30}$$

or,

$$g_{r+1}(x) - g_r(x) = \delta^{-r} h(x), \qquad r \gg 1 \tag{31}$$

Summing (31) from $r = r_0$ to ∞, we obtain

$$g_{r_0}(x) = g(x) - \frac{\delta^{-r_0}}{1 - \delta^{-1}} h(x), \qquad r_0 \gg 1 \tag{32}$$

so that $g_r \to g$ (asymptotically) geometrically at the rate δ.

We now show that the δ determined by (28) and (9) is the δ of I: the argument is that of I resulting in (13) made exact. By (30),

$$\delta g_r(x) = \delta^{-1} \, \delta g_{r-1}(x)$$

which, by (21) reads for $x = 0$

$$(-\alpha)^n \, \delta g^{(2^n)}(\lambda_{n+r}, 0) = \delta^{-1}(-\alpha)^{n+1} \, \delta g^{(2^{n+1})}(\lambda_{n+r}, 0)$$

or

$$\delta g^{(2^{n-r})}(\lambda_n, 0) = -(\alpha/\delta) \, \delta g^{(2^{n-(r-1)})}(\lambda_n, 0)$$

or,

$$\delta g^{(2^s)}(\lambda_n, 0) = -(\alpha/\delta) \, \delta g^{(2^{s+1})}(\lambda_n, 0) \qquad \text{for } 1 \ll s \ll n$$

Thus, for n very large, the change in $g^{(2^r)}$ is the constant multiple $-\delta/\alpha$ of the change in $g^{(2^{r-1})}$ induced by increasing λ_n to λ_{n+1} for all r except for the very small (initial transient) and very large (the bottom of the g_r sequence). Accordingly,

$$(-1)^n \, \delta g^{(2^n)}(\lambda_n, 0) \sim (\delta/\alpha)^n \, \delta g^{(1)}(\lambda_n, 0) = (\lambda_{n+1} - \lambda_n)(\delta/\alpha)^n$$

in the sense of logarithms. Since

$$\begin{aligned}(-\alpha)^n \, \delta g^{(2^n)}(\lambda_n, 0) &= (-\alpha)^n g^{(2^n)}(\lambda_{n+1}, 0) - (-\alpha)^n g^{(2^n)}(\lambda_n, 0) \\ &= g_1(0) - g_0(0) \sim 1\end{aligned}$$

we have

$$\lambda_{n+1} - \lambda_n \sim \delta^{-n}$$

logarithmically, or $(\lambda_{n+1} - \lambda_n)/(\lambda_{n+2} - \lambda_{n+1}) \to \delta$ as $n \to \infty$, which is the original definition of δ in I.

Combining (9) and (28), we can obtain $g(x)$, $h(x)$, α, and δ. By (32) we next obtain g_r for large r, and then by repeated application of (8) obtain low-lying g_r's. We have thus succeeded in determining all local quantitative properties of all highly bifurcated (and infinite) attractors of (1) in a framework independent of (1), and its unspecified $f(x)$. *The theory of high-order attractors is fully posed in a functional equation framework*, and represents the common residue of all specifically posed recursion equations $x_{n+1} = \lambda f(x_n)$.

It is important to make two observations pertaining to Eq. (28) at this point: one concerning the uniqueness of δ and the other concerning the linearity of (28) to any scaling of h.

Equation (32) can be derived by setting

$$g_r(x) = g(x) + \eta_r(x)$$

substituting in (8), and expanding to first order in η: we are simply analyzing the manner of approach of g_r to g. The separation of (26), namely,

$$\eta_r(x) = \eta_r h(x) = \delta^{-r} h(x)$$

demonstrates that

$$\lim_{r \to \infty} g_r = g$$

provided the eigenvalue δ of (28) is strictly greater than 1. In fact, it is easy to see that $\delta = 1$ with $h(x) = g(x) - xg'(x)$ exactly satisfies (28). To see the significance of this solution, observe that

$$g(x) - xg'(x) = (1 - x\, d/dx)g(x)$$

is exactly the generator of infinitesimal magnifications:

$$\begin{aligned}(1 + \mu)g(x/1 + \mu) &= (1 + \mu)g(x - x\mu) + O(\mu^2)\\ &= g(x) + \mu(g(x) - xg'(x)) + O(\mu^2)\end{aligned}$$

However, the magnifications comprise a degeneracy group of (9): if $g(x)$ obeys (9), then so too does $\mu g(x/\mu) \equiv g_\mu(x)$:

$$g_\mu\left(g_\mu\left(\frac{x}{\alpha}\right)\right) = \mu g\left(\frac{1}{\mu}\mu g\left(\frac{x}{\alpha\mu}\right)\right) = \mu g\left(g\left(\frac{x/\mu}{\alpha}\right)\right) = -\frac{1}{\alpha}\mu g\left(\frac{x}{\mu}\right) = -\frac{1}{\alpha}g_\mu(x)$$

Thus, the r-independent piece of λ_r corresponding to the eigenvalue $+1$ simply represents a convergence of g_r to a suitably magnified g; by *choosing*

$g(0) = 1$, this freedom is eliminated, and the r-independent piece of η_r set to 0.

We shall see in the next section that a spectrum of eigenvalues bounded by 1 in absolute value also exists. Anticipating some of that discussion, it turns out that g_r is orthogonal to the span of this part of the spectrum, so that only the large (convergence-producing) eigenvalues matter here.

Observe that (28) is linear in h, so that if $h(x)$ is a solution, so too is $\mu h(x)$ for any μ. That is, $h(0)$, say, is free. By (32), with $g(0) = 1$ by convention, this leaves $g_r(0)$ free in the asymptotic-r regime. However, a definite choice of $h(0)$ is necessary to ensure that $g_0(0) = 0$. A different choice of $h(0)$ would, for no number of iterations of (8), result in a g_r satisfying $g_r(0) = 0$.

It is easy to comprehend the meaning of other choices of $h(0)$. Since

$$g_r(x) = g(x) - \delta^{-r}h(x)$$

if $\bar{h}(x)$ [for a definite $\bar{h}(0)$] guarantees that $g_0(0) = 0$, then

$$\bar{h}_1(x) \equiv \delta\bar{h}(x)$$

produces a $g_1(x)$ such that $g_1(0) = 0$: that is, by increasing $h(0)$ we need perform fewer iterations to obtain a g_r satisfying $g_r(0) = 0$. Differently put, the absolute size of $h(x)$ is logarithmically periodic with period $\log \delta$: if

$$\bar{h}(x) \Rightarrow g_0(0) = 0$$

then

$$\delta^n\bar{h}(x) \Rightarrow g_n(0) = 0$$

All this means is that as $\log \bar{h}(0)$ is increased by $\log \delta$, one has moved through an entire bifurcation. Choices of $h(0) \neq \delta^n\bar{h}(0)$ determine a sequence of g_r's whose λ's are chosen *not* at λ_n's, but rather at intermediate values of λ between λ_n and λ_{n+1}. In particular, there is choice of $h(0) \equiv H(0)$ such that the λ's are the *bifurcation* values Λ_n. That is, our results determine the behavior of stability points not just at those values of λ_n such that $\bar{x}$ is an element, but indeed the entire behavior as λ is *continually* increased to λ_∞. The reason is simple: since $\lambda_n \sim \lambda_\infty - \delta^{-n}$, $\delta^{-n} \sim \lambda_\infty - \lambda_n$, and $g_r(x)$ is

$$g_r(x) = g(x) - (\lambda_\infty - \lambda_n)h(x)$$

and so,

$$g_\lambda(x) \equiv g(x) - (\lambda_\infty - \lambda)h(x)$$

is the continuation from discrete λ to continuous λ. Deviations of λ from

λ_∞ are most naturally measured logarithmically to the base δ: the bifurcation values Λ_n obey

$$\lambda_\infty - \Lambda_n \sim \delta^{-n}$$

as do the λ widths of a given harmonic

$$\Lambda_{n+1} - \Lambda_n \sim \delta^{-n}$$

A given "kind" of cycle recurs in the next harmonic periodic logarithmically. Moreover, had we started with, say, a stable three-cycle, bifurcated to the six-cycle, and considered $\lambda_n \equiv \lambda$ such that the 3×2^n cycle is stable and includes $\bar{x}$, then $g^{(3)}(\lambda, x)$ about $\bar{x}$ describes a two-cycle and

$$g_r(x) = (-\alpha)^n g^{(3\times 2^n)}(\lambda_{n+r}, x/(-\alpha)^n)$$

more generally are of the same character as the g_r's obtained from the harmonics of the two-cycle. Clearly (8) is again obeyed, leading to (9). With α and g unique solutions to (9) for a fixed z, we now realize that the entirety of the above treatment carries over *un*changed in every way to the structure of *every* highly bifurcated cycle of (1) no matter from which fundamental the bifurcations are obtained. That is, the local description of stability points at both the isolated-point and infinite-cluster level as well as α and δ are unique for every highly bifurcated cycle of (1) independent of f for any fixed z. Thus, in the so-called "chaotic" regime of (1) where most values of λ correspond to high bifurcation of a high-order fundamental, the local description of the attractor is essentially that of the g's.

We mention in passing that once g_r has been continued to a continuous index, Eq. (8) in the form

$$g_{r-s}(x) = (-\alpha)^s g_r^{(2^s)}(x/\alpha^s)$$

defines the notion of a continuous interaction, since every ingredient of the equation has received a natural continuation, save for the 2^s iterations.

4. THE APPROACH TO THE $\{g_r\}$ FIXED POINT

At this point we return to the fundamental question of the large-n limit of (7). In general f is not symmetric about $\bar{x}$ (although universality implies that even for an asymmetric f the g_r's must be symmetric) and the correct form of (7) is

$$g_r(x) = \lim(-\alpha)^n g^{(2^n)}(\lambda_{n+r}, x/(-\alpha)^n) \tag{33}$$

Defining

$$g_{r,n}(x) \equiv (-\alpha)^n g^{(2^n)}(\lambda_{n+r}, x/(-\alpha)^n) \tag{34}$$

we can immediately verify that

$$g_{r-1,n+1}(x) = -\alpha g_{r,n}(g_{r,n}(-x/\alpha)) \tag{35}$$

Viewing (8) as a fixed point (in n) of (35), we could establish our theory in at least a local sense by ascertaining the stability of (8). Thus, we are led to consider

$$g_{r,n}(x) = g_r(x) + \eta_{r,n}(x) \tag{36}$$

and attempt to show that

$$\lim_{n\to\infty} \eta_{r,n}(x) = 0$$

Substituting (36) in (35), with (8) valid at the fixed point, we have in linear approximation (in η)

$$\eta_{r-1,n+1}(x) = -\alpha[\eta_{r,n}(g_r(x/\alpha)) + g_r'(g_r(x/\alpha))\eta_{r,n}(-x/\alpha)] \tag{37}$$

[Neglecting the n index, and replacing $g_r \to g$, we find that (37) reduces to (25): we shall have more to say about the eigenvalues of (28).] Since the $g_r(x)$ have been normalized to $g(0) = 1$, in general

$$\lim_{n\to\infty} g_{r,n} \neq g_r$$

but rather $g_{r,n}$ will approach a suitably magnified g_r. [We could alternatively have defined $\eta_{r,n} = g_{r,n} - \mu_{r+n}g_r(x/\mu_{r+n})$ with $\lim_{t\to\infty} \mu_t = \mu \neq 1$.] We shall account for this with $\eta^*_{r,n} \sim \mu(g_r - xg_r')$ a piece of $\eta_{r,n}$ to be determined from (37). Also, by defining

$$\eta_{r,n} \equiv \psi_{r,n+r}; \qquad n + r \equiv t$$

we find that (37) becomes

$$\psi_{r-1,t}(x) = -\alpha[\psi_{r,t}(g_r(x/\alpha)) + g_r'(g_r(x/\alpha))\psi_{r,t}(-x/\alpha)]$$

so that (37) itself is insufficient to determine any $n + r$ dependence; rather,

$$\eta_{r,0} = g_{r,0} - g_r = \lambda_r f - g_r \tag{38}$$

shall serve as initial data to fix $\eta_{r,n}$ uniquely.

We proceed to solve (37) by an artificial quadrature. Setting

$$\eta_{r,n} \equiv \eta^1_{r,n} + \eta^2_{r,n} \tag{39}$$

we find that (37) becomes

$$\begin{aligned}\eta^1_{r-1,n+1}(x) &+ \alpha[\eta^2_{r,n}(g_r(x/\alpha)) + g_r'(g_r(x/\alpha))\eta^1_{r,n}(-x/\alpha) + \eta^2_{r,n}(g_r(x/\alpha))] \\ &= -\{\eta^2_{r-1,n+1}(x) + \alpha g_r'(g_r(x/\alpha))\eta^2_{r,n}(-x/\alpha)\} \equiv 0\end{aligned}$$

thereby defining η^2. That is (38) is replaced with the pair of equations

$$\eta^1_{r-1,n+1}(x) = -\alpha[\eta^1_{r,n}(g_r(x/\alpha)) + g_r'(g_r(x/\alpha))\eta^1_{r,n}(-x/\alpha) + \eta^2_{r,n}(g_r(x/\alpha))] \tag{40}$$

and

$$\eta^2_{r-1,n+1}(x) = -\alpha g_r'(g_r(x/\alpha))\eta^2_{r,n}(-x/\alpha) \tag{41}$$

However, by (8)

$$g'_{r-1}(x) = -g_r'(g_r(x/\alpha))g_r'(x/\alpha)$$

so that (41) becomes

$$\frac{\eta^2_{r-1,n+1}(x)}{g'_{r-1}(x)} = \alpha\,\frac{\eta^2_{r,n}(-x/\alpha)}{g_r'(x/\alpha)}$$

Since $g_r(x)$ is symmetric,

$$g_r'(-x/\alpha) = -g_r'(x/\alpha)$$

so that

$$\frac{\eta^2_{r-1,n+1}(x)}{g'_{r-1}(x)} = -\alpha\,\frac{\eta^2_{r,n}(-x/\alpha)}{g_r'(-x/\alpha)} \tag{42}$$

Defining

$$f_{r,n}(x) \equiv \frac{\eta^2_{r,n}(x)}{g_r'(x)} \tag{43}$$

we find that (39) reads

$$f_{r-1,n+1}(x) = -\alpha f_{r,n}(-x/\alpha)$$

with solution

$$f_{r,n}(x) = (-\alpha)^n f_{r+n,0}(x/(-\alpha)^n) \equiv (-\alpha)^n F_{r+n}(x/(-\alpha)^n)$$

so that we have for η^2

$$\eta^2_{r,n}(x) = (-\alpha)^n F_{r+n}(x/(-\alpha)^n)g_r'(x) \tag{44}$$

With (44), (40) now reads

$$\begin{aligned}\eta^1_{r-1,n+1} = {}& -\alpha\{\eta^1_{r,n}(g_r'(x/\alpha)) + g_r(g_r'(x/\alpha))\\ &\times [\eta^1_{r,n}(-x/\alpha) + (-\alpha)^n F_{r+n}(g_r(x/\alpha)/(-\alpha)^n)]\}\end{aligned} \tag{45}$$

Setting

$$\eta^1_{r,n}(x) \equiv \eta^0_{r,n}(x) - (-\alpha)^n F_{r+n}(g_r(x)/(-\alpha)^n)$$

we can immediately verify from (45) that

$$\eta^0_{r-1,n+1}(x) = -\alpha[\eta^0_{r,n}(g_r(x/\alpha)) + g_r(g_r'(x/\alpha))\eta^0_{r,n}(-x/\alpha)]$$

that is, if $\eta^0_{r,n}$ obeys (37), then for any F_{r+n}, so too does

$$\eta_{r,n}(x) = \eta^0_{r,n}(x) + [(-\alpha)^n F_{r+n}(x/(-\alpha)^n)g_r'(x) - (-\alpha)^n F_{r+n}(g_r(x)/(-\alpha)^n)] \tag{46}$$

so that we have obtained a particular solution of (37). We now regard $\eta^0_{r,n}(x)$ to be a homogeneous transient to the F solution which we utilize to meet initial data. Specifically, we absorb all antisymmetric parts of $f(x)$ into F, leaving $\eta^0_{r,0}(x)$ and hence all $\eta^0_{r,n}(x)$ symmetric, and so obeying the "intrinsic" equation

$$\eta^0_{r-1,n+1}(x) = -\alpha[\eta^0_{r,n}(g_r(x/\alpha)) + g_r'(g_r(x/\alpha))\eta^0_{r,n}(x/\alpha)] \tag{47}$$

η^0 is viewed as built exclusively of g_r's with minimal dependence on the initial $f(x)$. [It will be seen that the decomposition of (39)–(41) is determined and not at all artificial if η^1 and η^2 are respectively taken to be the symmetric and antisymmetric parts of η.]

We now utilize the initial data (38):

$$\lambda_r f(x) - g_r(x) = \eta_{r,0}(x) = \eta^0_{r,0}(x) + F_r(x)g_r'(x) - F_r(g_r(x)) \tag{48}$$

With an overbar denoting symmetry and a circumflex denoting antisymmetry, (48) reads

$$\lambda_r \bar{f}(x) = \bar{F}_r(x)g_r'(x) \tag{49}$$

and

$$\lambda_r f(x) - g_r(x) = \eta^0_{r,0}(x) + \hat{F}_r(x)g_r'(x) - \hat{F}_r(g_r(x)) - \bar{F}_r(g_r(x)) \tag{50}$$

By (49), $\bar{F}_r(x)$ is nonvanishing only when f is asymmetric, in which case it can absorb all the antisymmetry.

Let us specialize to the case $\bar{f} \neq 0$. We then have from (49)

$$\bar{F}_r(x) = \frac{\lambda_r}{g_r'(x)}\bar{f}(x)$$

and so, by (46), a piece of $\eta_{r,n}$ of the form

$$\eta^*_{r,n}(x) = \lambda_{r+n}\left\{(-\alpha)^n \frac{\bar{f}(x/(-\alpha)^n)}{g'(x/(-\alpha)^n)} g_r'(x) - (-\alpha)^n \frac{\bar{f}(g_r(x)/(-\alpha)^n)}{g'_{r+n}(g_r(x)/(-\alpha)^n)}\right\}$$

$$\underset{n\to\infty}{\sim} \lambda_\infty\left\{(-\alpha)^n \frac{\bar{f}(x/(-\alpha)^n)}{g'(x/(-\alpha)^n)} g_r'(x) - (-\alpha)^n \frac{\bar{f}(g_r(x)/(-\alpha)_n)}{g'(g_r(x)/(-\alpha)^n)}\right\}$$

With $1 - \bar{f}(x) \propto |x|^z$, $\bar{f}(x) \propto |x|^{z+\epsilon}\operatorname{sgn} x$ ($\epsilon > 0$: otherwise f not extreme at $x = 0$), and $g'(x) \propto |x|^{z-1}\operatorname{sgn} x$,

$$(\bar{f}/g')(x) \propto |x|^{1+\epsilon}$$

and so

$$\eta^*_{r,n}(x) \underset{n\to\infty}{\sim} (-1)^n\alpha^{-n\epsilon}(|x|^{1+\epsilon}g_r'(x) - |g_r(x)|^{1+\epsilon}) \to 0$$

Thus, the g_r fixed point is stable against antisymmetric perturbations. As an example, if

$$f(x) = 1 - ax^2 - bx^3$$

then $\epsilon = 1$ and $\eta_{r,n}$ converge to zero geometrically at the rate $-\alpha^{-1}$, in perfect agreement with the computer data for this f, since the $\eta_{r,n}$ convergence rate is exactly the α_n convergence rate:

$$\alpha_n^{(r)} \equiv -\frac{g^{(2^n)}(\lambda_{n+r,0})}{g^{(2^{n+1})}(\lambda_{n+1,r+0})} = \alpha \frac{g_{r,n}(0)}{g_{r,n+1}(0)}$$

$$\approx \alpha\left\{1 + \frac{1}{g_r(0)}\left[\eta_{r,n}(0) - \eta_{r,n+1}(0)\right]\right\}$$

Accordingly, we consider now only symmetric f's so that

$$F_r(x) = \hat{F}_r(x)$$

and

$$\lambda_r f(x) - g_r(x) = \eta^0_{r,0}(x) + \hat{F}_r(x) g_r'(x) - \hat{F}_r(g_r(x)) \tag{51}$$

As a first observation, following the parenthetic remark below Eq. (37),

$$\eta_r(x) \equiv \hat{F}_r(x) g'(x) - \hat{F}_r(g(x))$$

with

$$\hat{F}_r(x) = \alpha^{-r}\hat{F}_0(\alpha^r x)$$

must obey

$$\eta_{r-1}(x) = -\alpha[\eta_r(g(x/\alpha)) + g'(g(x/\alpha))\eta_r(x/\alpha)]$$

For monomials,

$$\hat{F}_0(x) = |x|^z \operatorname{sgn} x, \qquad \hat{F}_{r-1}(x) = \alpha^{-z+1}\hat{F}_r(x)$$

and so

$$-\alpha[h^{(z)}(g(x/\alpha)) + g'(g(x/\alpha))h^{(z)}(x/\alpha)] = \alpha^{-z+1}h^{(z)}(x)$$

with

$$h^{(z)}(x) = |g(x)|^z \operatorname{sgn}(g(x)) - g'(x)|x|^z \operatorname{sgn} x$$

That is, the eigenvalue of (28) can assume any positive value less than or equal to 1 in addition to the value $\delta > 1$. These eigenvalues represent potential instabilities of the convergence of g_r to g. However, they are unexcited in every g_r exactly because they provide stable convergence of $g_{r,n}$ to g_r: The g_r meet no initial conditions save for their convergence to g. The potentially hazardous parts $h^{(z)}$ of a g_r are all shed in the approach of $g_{r,n}$ (for any suitable f) to g_r. That is,

$$g = \lim_{r\to\infty} g_r$$

must exist since all unstable eigenvalues are exactly those that vanish in the formation of g_r from $g_{r,n}$.

We return to (51) and now ask whether the $\hat{F}$'s can span the initial data, in which case $\eta^0 \equiv 0$ and our theory is complete. In fact, an $\eta_{r,n}$ built wholly from $\hat{F}$ must produce convergence at a rate $(-\alpha)^{1-z}$ for

$$\hat{F}_r(x) = \mu_r x + a_r x^z + \cdots \text{ higher order}$$

For example, if $f = 1 - zx^2 + \cdots$, the smallest value of z would be 3, providing a convergence rate α^{-2}. However, $\alpha_n^{(1)} \to \alpha$ at the rate δ^{-1} in this case. Since $\delta < \alpha^2$, only the δ rate survives asymptotically, and leaves the question as to how it enters $\eta_{r,n}$. This suggests that η^0 might not vanish in general, and so we examine the situation more carefully.

Returning to (35), define

$$r + n \equiv t, \qquad g_{r,n} \equiv \psi_{r+n,n} = \psi_{t,n}$$

so that

$$\psi_{t,n+1}(x) = -\alpha\psi_{t,n}(\psi_{t,n}(-x/\alpha)) \tag{52}$$

We next expand ψ about g:

$$\psi_{t,n} \equiv g + \omega_{t,n} \tag{53}$$

so that

$$\omega_{t,n+1}(x) = -\alpha[\omega_{t,n}(g(x/\alpha)) + g'(g(x/\alpha))\omega_{t,n}(-x/\alpha)] \equiv \mathscr{L}[\omega_{t,n}] \tag{54}$$

in linear approximation. Equation (54) is our familiar shift equation written in the form of (12). We already know a variety of eigenvalues of $\mathscr{L}$:

$$\psi_\rho(x) \equiv g^\rho(x) - x^\rho g'(x) \Rightarrow \mathscr{L}[\psi_\rho] = (-\alpha)^{1-\rho}\psi_\rho$$

and

$$h(x): \quad \mathscr{L}[h] = \delta h$$

Expanding $\omega_{t,n}$ along these eigenvectors, we have

$$\omega_{t,n} = \sum_\rho c_t^\rho(-\alpha)^{n(1-\rho)}(g^\rho - x^\rho g') + c_t\,\delta^n h + \omega_{t,n}^0 \tag{55}$$

Observe at this point a strong similarity to (46), and yet with the difference that (55) possesses an isolated h piece plus

$$\sum_\rho c_t^\rho(-\alpha)^{n(1-\rho)}(g^\rho - x^\rho g') = (-\alpha)^n F_t(g/(-\alpha)^n) - (-\alpha)^n F_t(x/(-\alpha)^n)g'$$

This is the same form of particular solution as in (46) but constructed from g rather than g_r. Now, if $c_t\delta^n = c_{r+n}\delta^n \sim \delta^{-r}$, then the piece $c_t\delta^n h$ is a fixed (in n) perturbation about g. Now, $\mathscr{L}$ accounts for only first-order perturbative effects. One would obtain a second-order correction by expanding about $g + c_t\delta^n h$. However, this is the form of expansion about g_r that would modify the $\psi_\rho(x)$ to be

$$\psi_\rho \to g_r{}^\rho - x^\rho g_r{}' = (g^\rho - x^\rho g') - \delta^{-r}(\rho g^{\rho-1}h - x^\rho h')$$

Thus, assuming $c_t\delta^n \sim \delta^{-r}$, the correct form of (55) including all first-order $c_t{}^\rho$ dependence is

$$\omega_{t,n} = \sum_\rho c_t{}^\rho(-\alpha)^{n(1-\rho)}(g^\rho - x^\rho g') + c_t\delta^n\left(h + \sum_\rho c_t{}^\rho(-\alpha)^{n(1-\rho)}(\rho g^{\rho-1}h - x^\rho h')\right) + \omega^0_{t,n} \qquad (56)$$

which possesses extra h dependence beyond the $F_t(g_r)$ terms than does (46): even with $\omega^0_{t,n} = 0$, (56) has already included a transient piece $\eta^{0'}_{r,n}$, which is still constructed from the span of $h \oplus \{\psi\}$. We now pose the (strong) conjecture that $\omega^0_{t,n} = 0$:

Conjecture. The spectrum of the operator $\mathscr{L}$ is δ and $(-\alpha)^{1-\rho}$, $\rho \leqslant 1$, and, moreover, the spectrum is complete.

(We possess computational evidence for this conjecture at least for $z = 2$ and 4, which we discuss in Section 5.) Accordingly, we have

$$\omega_{t,n} = \sum_\rho c_t{}^\rho(-\alpha)^{n(1-\rho)}(g^\rho - x^\rho g') + c_t\delta^n\left(h + \sum_\rho c_t{}^\rho(-\alpha)^{n(1-\rho)} \times (\rho g^{\rho-1}h - x^\rho h')\right) \qquad (57)$$

or

$$g_{r,n} = g + \sum_\rho c^\rho_{t+n}(-\alpha)^{n(1-\rho)}(g^\rho - x^\rho g') + c_{r+n}\delta^n\left(h + \sum_\rho c^\rho_{r+n}(-\alpha)^{n(1-\rho)}(\rho g^{\rho-1}h - x^\rho h')\right) \qquad (58)$$

in linear approximation.

It is easy to extend (58) to an exact solution of (35), since the first-order terms are exactly the generators of conjugacy transformations connected to the identity. Defining

$$F_t(x) = \sum_\rho c_t{}^\rho x^\rho$$

we find that (58) becomes

$$g_{r,n} = (g + c_{r+n}\delta^n h) + (-\alpha)^n F_{r+n}((-\alpha)^{-n}(g + c_{r+n}\delta^n h)) - (-\alpha)^n(g + c_{r+n}\delta^n h)' F_{r+n}(x/(-\alpha)^n) \qquad (59)$$

Defining

$$S_{t,n}(x) = x + (-\alpha)^n F_t(x/(-\alpha)^n)$$

we have that (59) constitutes the leading approximation to

$$g_{r,n} = S_{r+n,n} \circ (g + c_{r+n}\delta^n h) \circ S_{r+n,n}^{-1} \tag{60}$$

However, defining $c_t = \delta^{-t}d_t$, we have

$$g + c_{r+n}\delta^n h = g + d_{r+n}\delta^{-r}h \equiv \tilde{g}_{r,r+n} + O(\delta^{-2r})$$

which obeys

$$\tilde{g}_{r-1,t}(x) = -\alpha\tilde{g}_{r,t}(\tilde{g}_{r,t}(x/\alpha))$$

to first order for any choice of d_t, converging to g as $r \to \infty$ for t fixed. That is, (58) is the leading approximation to

$$g_{r,n} = S_{r+n,n} \circ \tilde{g}_{r,r+n} \circ S_{r+n,n}^{-1} \tag{61}$$

which is easily seen to exactly satisfy (35). However, while (58) is *compatible* with the solution (61), (61) is not the *general* solution to (35) containing (58) as its linear approximation. That is, if (35) is stable about the $\tilde{g}_{r,t}$ fixed point, the linear approximation becomes a conjugacy transformation upon $\tilde{g}_{r,t}$, deviations from conjugacy vanishing in the higher order transient.

According to the discussion on p. 680, the functions $\tilde{g}_{r,t}$ all converge to g as $r \to \infty$, but differ in the small-r regime: only $d_t = -1$ (for the properly normalized h) will lead to $\tilde{g}_{0,t}(0) = 0$. Equation (61) is correct for large n; setting $r = 0$, it reads

$$g_{0,n}(x) = S_{n,n} \circ g_{0,n} \circ S_{n,n}^{-1}$$

Since $S_{t,n}$ is connected to the identity,

$$g_{0,n}(0) = 0 \Rightarrow \tilde{g}_{0,n}(0) = 0$$

However, $g_{0,n}(0)$ *must* vanish for all n:

$$g_{0,n}(0) = (-\alpha)^n g^{(2^n)}(\lambda_n, 0) = 0$$

by the recursion-equation definition of λ_n. But, if $\tilde{g}_{0,t}(0) = 0$ for all t, then $d_t = -1$ for all t. Thus, the recursion-defined values of λ_r determine $c_t = \delta^{-t}d_t = -\delta^{-t}$, which, when entered in (58), yields

$$g_{r,n} = g + \sum_\rho c_{r+n}^\rho(-\alpha)^{n(1-\rho)}(g^\rho - x^\rho g')$$
$$- \delta^{-r}\left(h + \sum_\rho c_{r+n}^\rho(-\alpha)^{n(1-\rho)}(\rho g^{\rho-1}h - x^\rho h')\right) \tag{62}$$

so that the potentially divergent δ^n terms have been stabilized. Moreover, for n large, all terms decay with powers of $-\alpha$ save for $\rho = 1$:

$$g_{r,n} \underset{n\to\infty}{\sim} g + c^1_{r+n}(g - xg') - \delta^{-r}(h + c^1_{r+n}(h - xh'))$$

or, with $\mu_t \equiv 1 + c_t^1$,

$$g_{r,n} \sim \mu_t(g - \delta^{-r}h)(x/\mu_t)$$

or

$$g_{r,n} \sim \mu_{r+n} g_r(x/\mu_{r+n}) \tag{63}$$

a magnification of g_r. Thus, our conjecture implies the local stability of the g_r fixed point of (35). [Conversely, the one parameter λ could be adjusted to cancel the potentially growing δ mode; had $\mathscr{L}$ possessed several growing eigenvalues, it is difficult to see how this cancellation could be arranged. Also, although the conjugacy generators produce convergence at rates $(-\alpha)^{1-\rho}$, we can see from (63) how $\mu_t \to \mu_\infty$ can produce a different convergence scheme for α_n.] We have not investigated any higher order stability questions, and apart from some approximation schemes and computational methods which we shall discuss in the next section, have nothing further to say about the ingredients of a nonlocal proof.

5. APPROXIMATIONS AND METHODS OF SOLUTION

All infinite attracters are locally determined by the hierarchy (8),

$$g_{r-1}(x) = -\alpha g_r(g_r(x/\alpha))$$

As previously described, (8) is solved by first computing g and α through (9),

$$g(x) = -\alpha g(g(x/\alpha))$$

with g_r for asymptotic r given by (32),

$$g_r \sim g - \delta^{-r} h$$

where h and δ are obtained through (28),

$$-\alpha[h(g(x/\alpha)) + g'(g(x/\alpha))h(x/\alpha)] = \delta h(x), \qquad \delta > 1$$

To any desired accuracy, an r_0 is chosen such that (32) provides g_r for all $r \geqslant r_0$, and g_r for $r < r_0$ determined from g_{r_0} through (8). In particular $h(0)$ is fixed through the requirement that $g_0(0) = 0$.

We now seek an approximate equation for g_r for a fixed r that bypasses the above asymptotic ansatz. The virtue of such an equation is that it must define a recursively stable scheme for obtaining g_r. As we shall see, the approximate formula of I shall appear to linear approximation.

By (8),

$$g_1(x) = -\alpha g_2(g_2(x/\alpha)) \tag{64}$$

Relation (32) provides g_r up to first order in δ^{-r}. Since $\delta \to \infty$ as $z \to \infty$, for large enough z

$$g_1 \simeq g - \delta^{-1}h \quad \text{and} \quad g_2 \simeq g - \delta^{-2}h \tag{65}$$

will be arbitrarily accurate. Accordingly,

$$g_2 - g_1 \simeq \delta^{-1}(1 - \delta^{-1})h$$

or

$$g_2 \simeq g_1 + \delta^{-1}(1 - \delta^{-1})h \tag{66}$$

Substituting (66) in (64), we find

$$g_1(x) = -\alpha\{g_1(g_1(x/\alpha)) + (1 - \delta^{-1})\delta^{-1}[h(g_1(x/\alpha)) + g_1'(g_1(x/\alpha))h(x/\alpha)] + O(\delta^{-2})\}$$

or, by (65),

$$g_1(x) = -\alpha g_1(g_1(x/\alpha)) + (1 - \delta^{-1})(-\alpha\delta^{-1})[h(g(x/\alpha)) + g'(g(x/\alpha))h(x/\alpha)] + O(\delta^{-2})$$

which, by (28), is

$$g_1(x) = -\alpha g_1(g_1(x/\alpha)) + (1 - \delta^{-1})h(x) + O(\delta^{-2}) \tag{67}$$

Thus, to leading order in δ^{-1},

$$g_0(x) = -\alpha g_1(g_1(x/\alpha)) \simeq g_1(x) - (1 - \delta^{-1})h(x) = g(x) - h(x) \tag{68}$$

That is, as $z \to \infty$, the asymptotic form (32) becomes arbitrarily accurate for all $r \geqslant 0$. Since $g_0(0) = 0$, (68) produces

$$h(0) \simeq g(0) = 1$$

in this limit, a

$$g_1(0) \simeq 1 - \delta^{-1} \tag{69}$$

[To appreciate this estimate, for $z = 2$, $g_1(0) \doteq 0.733$, in comparison with $1 - \delta^{-1} \doteq 0.786$.] Defining

$$g^*(x) \equiv (1 - \delta^{-1})^{-1}g_1((1 - \delta^{-1})x)$$

and

$$h^*(x) \equiv h((1 - \delta^{-1})x)$$

we can write (67) as

$$g^*(x) = -\alpha g^*(g^*(x/\alpha)) + h^*(x) + O(\delta^{-2}) \tag{70}$$

Since

$$-\alpha[h(g_1(x/\alpha)) + g_1'(g_1(x/\alpha))h(x/\alpha)] = \delta h(x) + O(\delta^{-1})$$

we also have

$$h^*(x) = -(\alpha/\delta)[h^*(g^*(x/\alpha)) + g^{*\prime}(g^*(x/\alpha))h^*(x/\alpha)] + O(\delta^{-2}) \quad (71)$$

Equations (70) and (71) are exactly the approximate equations of I, since

$$h^*(0) = 1 \Rightarrow g^*(0) = 1 \qquad \text{by (70)}$$

while the definition of g_1 implies that

$$g^*(1) = 0$$

Since (70) and (71) constitute equations for g_1, their natural recursion forms ($g^*, h^* \to g_n^*, h_n^*$ on the right-hand sides and g^*_{n+1}, h^*_{n+1} on the left-hand sides) accomplish the recursion

$$g_{1,n+1} = f(g_{1,n})$$

which is stable.

We now exhibit a computational technique for solving (9) based on the observation about Eq. (16). The recursion form of (9) given by (12),

$$\tilde{g}_{n+1}(x) = -\alpha_n\tilde{g}_n(\tilde{g}_n(x/\alpha_n))$$
$$\tilde{g}_n(0) \equiv 1 \qquad \text{for all } n \Rightarrow \alpha_n = -[\tilde{g}_n(1)]^{-1}$$

must be convergent to g if $\tilde{g}_0$ is appropriately chosen. Thus, if f is any function of our class, there is a value λ_∞ such that

$$x_{n+1} = \lambda_\infty f(x_n)$$

will determine the infinite attractor bifurcated from the 2-cycle, in which case Eq. (15),

$$\tilde{g}_0(x) = f(\lambda_\infty x)$$

will lead to convergence. However, the strong instability of (12) requires that λ_∞ be known to very high precision in order that its high iterates will be accurate approximates of g. As a rough estimate, $\tilde{g}_0(x)$ should approximate g, and so (16),

$$f(\lambda_\infty) = \tilde{g}_0(1) \simeq g(1) = -\alpha^{-1}$$

provides an estimate for λ_∞. It is elementary to obtain better estimates. Clearly,

$$\alpha_0 = -[\tilde{g}_0(1)]^{-1} = -[f(\lambda_\infty)]^{-1}$$

and so

$$\tilde{g}_1(x) = [f(\lambda_\infty)]^{-1}f\{\lambda_\infty f(\lambda_\infty x f(\lambda_\infty))\} \quad (72)$$

But now, $\tilde{g}_1(x)$ is a better estimate of g and so

$$\tilde{g}_1(1) \simeq -\alpha^{-1}$$

will provide a better estimate of λ_∞. Accordingly, with α known accurately, we could, for any f of our class, determine high-accuracy estimates of $\lambda_\infty(f)$. With α unknown, we could seek to collaterally determine it with λ_∞ by setting

$$\tilde{g}_1(1) \simeq \tilde{g}_0(1) \tag{73}$$

which by (16) and (72) provides an equation purely for λ_∞. Evidently, by successively setting $\tilde{g}_{n+1}(1) \simeq \tilde{g}_n(1)$ more accurate estimations are obtained. We now show that this can be turned into a highly convergent scheme for λ_∞. It is immediate to see that (12) can be "solved" as

$$\tilde{g}_n(x) = (-1)^n \alpha_{n-1}\alpha_{n-2} \cdots \alpha_0 \tilde{g}_0^{(2n)}(x/\alpha_{n-1} \cdots \alpha_0)$$

so that

$$\tilde{g}_n(0) = (-1)^n \alpha_{n-1} \cdots \alpha_0 \tilde{g}_0^{(2n)}(0) \tag{74}$$

Since $\alpha_n \to \alpha$ if $\tilde{g}_0(x) = f(\lambda_\infty x)$ for the exact λ_∞,

$$\tilde{g}_0^{(2n)}(0) \sim (-\alpha)^{-n} \tag{75}$$

Also, by (74)

$$\alpha_n = -\tilde{g}_0^{(2n)}(0)/\tilde{g}_0^{(2n+1)}(0) \tag{76}$$

so that, with the definition

$$\xi_n \equiv \tilde{g}_0^{(2n-1)}(0)\tilde{g}_0^{(2n+1)}(0) - [\tilde{g}_0^{(2n)}(0)]^2 \tag{77}$$

one has

$$\frac{\xi_{n+1}}{\xi_n} = \frac{1}{\alpha_{n+1}\alpha_n} \frac{\alpha_{n+1} - \alpha_n}{\alpha_n - \alpha_{n-1}} \equiv \frac{1}{\alpha_{n+1}\alpha_n} \rho_n \tag{78}$$

with ρ_n the α convergence rate. With

$$\tilde{g}_0(x) = f(ax) \tag{79}$$

and a chosen exactly at $\lambda_\infty(f)$, $\alpha_n \to \alpha$ and so $\rho_n < 1$. Since $\alpha > 1$ for all $z > 1$ ($\alpha \to \infty$ as $z \to 1$ and $\alpha \to 1$ as $z \to \infty$), ξ_n converges to zero α^2 times faster than $\alpha_n \to \alpha$. Accordingly, if one sets $\xi_n = 0$ for each n, an equation for a results (whose solution is a λ_∞ of f) that is exact to linear order in the error of the estimation. (For $f = 1 - 2x^2$, $\xi_n = 0$ yields λ_∞ for the 2-cycle fundamental to $2n$ significant figures and α_n to n significant figures.) Had we wanted λ_∞ for a 3-cycle, $\tilde{g}_0(x) = f^{(3)}(\lambda_\infty x)$ provides the starting $\tilde{g}$, and similarly *all* λ_∞ of a chosen f are rapidly determined (with of course the same α resulting). Once λ_∞ is determined, the iterates of $f(\lambda_\infty x)$ converge toward g until the error in λ_∞ is sufficiently magnified to cause divergence. (A 25-significant-figure estimate of λ_∞ provides a g obeying (9) to one part in 10^{-14} on [0, 1]. As shall follow, one can do significantly better far more quickly.)

We now consider a Newton's-method scheme of solution of (9) which shall lead into deeper considerations of the spectral problem of $\mathscr{L}$, and simple hand computations of λ_r's to several significant figures.

Regarding $g(x)$ on a compact interval—say [0, 1]—as a matrix of its values at N points x_i *together* with an interpolation scheme (of at least zth order to protect a zth-order g), (9) evaluated at the N points x_i becomes a set of N coupled nonlinear equations for the N quantities $g(x_i)$. Accordingly, one can perform an N-dimensional Newton's-method recursion to obtain the $g(x_i)$ from an initial estimate. However, high-precision estimates of g require a high-order interpolation scheme upon a large-N matrix, leading to an inaccurate inversion. Schematically, one writes

$$\bar{g}(x) = g(x) + \delta g(x), \qquad \bar{\alpha} = \alpha + \delta\alpha \tag{80}$$

where $\bar{g}$ and $\bar{\alpha}$ satisfy (9) and g and α serve as an approximate solution. We insert (80) in (9) and expand about the approximate values to first order in δg and $\delta\alpha$. By setting

$$g(0) \equiv \bar{g}(0) = 1$$

in the approximate solution, we have $\delta g(0) = 0$, which determines $\delta\alpha$ in terms of the $\delta g(x)$'s, so that we have a linear equation for $\delta g(x)$ alone. Expressions like $\delta g(g(x/\alpha))$ appear: the equation is evaluated at each x_i and $\delta g(g(x_i/\alpha))$ is expressed through the interpolation procedure in terms of linear combinations of $\delta g(x_i)$. The equations are then inverted to obtain $\delta g(x_i)$ and the procedure iterated. Convergence is slow and precision-limited.

However, the matrix of $g(x_i)$ and interpolation scheme simply constitute a certain parametrization of $g(x)$. Accordingly, one can perform the method with far simpler parametrizations. In particular, setting

$$g_N(x) = 1 + \sum_{i=1}^{N} G_i x^{zi}, \qquad \delta g_N(x) = \sum_{i=1}^{N} \delta G_i x^{zi}$$

and evaluating the linear approximate equation at N points (say $x_m = m/N$) produces an N-dimensional linear system again to be inverted. {In fact, for $z = 2$, precision limitations occurred for $N = 14$, determining α and g consistent with (9) to within 10^{-20} on [0, 1] in 10 sec of CDC 6600 time.} The solution obtained (the G_i) of course provides a very rapidly computable g for any further usage.

With α and g determined, we now face the determination of δ and h from the solution of (28),

$$\mathscr{L}[\psi] = -\alpha[\psi(g(x/\alpha)) + g'(g(x/\alpha)\psi(-x/\alpha)] = \lambda\psi(x)$$

In light of the previous discussion, this is an infinite-dimensional linear eigenvalue problem, which we shall study in a finite-dimensional approximation. That is, set

$$\psi(x) = \sum_{n=0}^{N-1} \psi_n x^{zn} \qquad (\psi_0 \equiv 1) \tag{81}$$

so that (28) evaluated at N points x_i becomes

$$\sum_{n=0}^{N-1} \{\lambda x_i^{zn} + \alpha[g(x_i/\alpha)]^{zn} + \alpha g'(g(x_i/\alpha))(-x_i/\alpha)^{zn}\}\psi_n = 0 \tag{82}$$

or

$$(\lambda X_{in} - \tilde{L}_{in})\psi_n = 0 \qquad \text{(summation convention)} \tag{83}$$

with (83) defining from (82) the $N \times N$ matrices X and $\tilde{L}$. The matrix X is invertible, and so

$$(X^{-1}\tilde{L})_{in}\psi_n \equiv L_{in}\psi_n = \lambda\psi_n \tag{84}$$

Accordingly, the eigenvalues λ are determined from

$$\det(\lambda X - \tilde{L}) = 0 \tag{85}$$

producing N eigenvalues in the N-dimensional approximating space.

The computational results are highly interesting. Starting with $N = 1$ and $x_1 = 0$ there simply results

$$\lambda = -\alpha - \alpha g'(1) = \alpha^z - \alpha$$

which is an approximate formula for δ with $h(1) = h(0) = 1$, asymptotically accurate as $z \to \infty$. Setting $N = 2$ with $x_1 = 0$ and $x_2 = 1$ results in a larger eigenvalue more nearly δ and a smaller one quite close to $\lambda = 1$, with corresponding eigenvectors approximating h and $\psi_1 = g - xg'$. Increasing N and evaluating at equally spaced points in [0, 1] produces more accurate determinations of δ and various $(-\alpha)^{1-\rho}$, $\rho = 1, 2,\ldots$: *δ is the solitary eigenvalue of L greater than* 1, at least in the two cases we studied, $z = 2$ and $z = 4$. {At $N = 14$ we obtain δ and h to 20 places consistent with (28) on [0, 1] and agreeing to the 14 places of our best recursion data.} That is, the spectrum of $\mathscr{L}$ restricted to these discrete linear systems is comprised of the conjugacy eigenvalues smaller than or equal to 1 in absolute value, plus a solitary larger one equal to δ. The eigenvalues are always *nondegenerate*, so that *L is complete* despite its nonsymmetric form. That is, defining L^*, the adjoint of L (the transpose in the present context), and adjoint eigenvectors

$$L^*\psi_\lambda^* \equiv \psi_\lambda^* L = \lambda\psi_\lambda^* \tag{86}$$

normalized to

$$(\psi_{\lambda'}, \psi_\lambda) \equiv \sum_n \psi^*_{\lambda',n}\psi_{\lambda,n} = \delta_{\lambda\lambda'} \tag{87}$$

L can be spectrally decomposed:

$$L_{mn} = \sum_\lambda \lambda\psi_{\lambda,m}\psi^*_{\lambda,n} \tag{88}$$

or symbolically,

$$L = \sum_\lambda \lambda\psi_\lambda\psi_\lambda^*$$

so that

$$L^p = \sum_\lambda \lambda^p\psi_\lambda\psi_\lambda^*$$

The condition to be met for $L^p\phi$ to not contain any eigenvector $\psi_{\bar\lambda}$ is then

$$(\psi_{\bar\lambda}^*, \phi) = 0 \tag{89}$$

We now consider universality in the light of this framework. With $\tilde g_0(x) = f(\lambda_\infty x)$,

$$\tilde g_0^{(n)}(x) = \lambda_\infty^{-1}(\lambda_\infty f)^{(n)}(\lambda_\infty x)$$

or

$$(\lambda_\infty f)^{(n)}(x) = \lambda_\infty \tilde g_0^{(n)}(x/\lambda_\infty) \tag{90}$$

which is simply a magnification by λ_∞ of $\tilde g_0^{(n)}(x)$. Defining the recursion

$$\tilde g_{n+1}(x) = -\alpha\tilde g_n(\tilde g_n(x/\alpha)), \qquad \tilde g_0(x) = \lambda_\infty f(x) \tag{91}$$

will cause $\tilde g_n \to g$, where

$$\tilde g_n(x) = (-\alpha)^n \tilde g_0^{(2^n)}(x/\alpha^n)$$

Defining $\eta_n \equiv \tilde g_n - g$, then

$$\eta_{n+1} = \mathscr{L}[\eta_n]$$

in linear approximation. Expanding η_n along eigenvectors of $\mathscr{L}$, we have by the orthogonality (77)

$$\eta_n(x) = \sum_\lambda \lambda^n(\psi_\lambda^*, \eta_0)\psi_\lambda(x) \tag{92}$$

Since $\eta_n \to 0$, the solitary growing mode corresponding to $\lambda = \delta$ must be unexcited. Calling h^* the adjoint eigenvector of eigenvalue δ, this means

$$(h^*, \eta_0) = 0 \tag{93}$$

However,

$$\eta_0 = \tilde{g}_0 - g = \lambda_\infty f - g$$

so that (93) becomes

$$\lambda_\infty(f) = (h^*, g)/(h^*, f) \tag{94}$$

That is, the λ_∞ of the 2-cycle receives the interpretation as the unique value of λ to extinguish the diverging mode of $\mathscr{L}$. [For a fundamental cycle of order s, $\tilde{g}_0$ must be $(\lambda_\infty f)^{(s)}$, so that λ_∞ is no longer multiplicative, although (93) will still provide an equation to determine λ_∞.] For example, for $z = 2$ and $N = 2$

$$g \simeq 1 + g_1 x^2, \qquad h^* \simeq 1 + h_1^* x^2, \qquad f \simeq 1 + f_1 x^2$$

and (94) becomes

$$\lambda_\infty(f) \simeq (1 + g_1 h_1^*)/(1 + f_1 h_1^*) \tag{95}$$

To be a good estimate, η_0 must be in the linear domain, so that f should be "nice." Thus $f = 2x^2$ determines through (95) a 0.1% estimate of λ_∞. Provided f is nice, once g_n and h_n^* are determined, (95) allows for 5-sec estimates of $\lambda_\infty(f)$.

In view of the computer spectral evidence, h^* is the unique eigenvector to all conjugacy-generator eigenvectors. This is important in the application of (94): iterates of $\lambda_\infty f$ converge not to g, but to $\mu g(x/\mu)$. In writing $\eta_0 = \lambda_\infty f - g$, we never specified the normalization of g. Indeed, it is irrelevant: h^* is computed from the g normalized to $g(0) = 1$. Since h^* is orthogonal to all conjugacy generators of g,

$$(h^*, \mu g(x/\mu)) = (h^*, g)$$

for all μ (in linear approximation) and so (94) is correct for g with fixed normalization. Moreover, the conjugacy problem of g is solved: if

$$f = \psi \circ g \circ \psi^{-1}$$

for some ψ connected to the identity, then it must follow that

$$(h^*, f) = (h^*, g)$$

and conversely. (Clearly our spectral conjecture is quite strong.) This leads to another method of estimating λ_∞:

$$(\lambda_\infty f - g, h^*) = 0 \Rightarrow \lambda_\infty f \sim g \qquad \text{(conjugacy)}$$

Thus, should λ_∞ satisfy a *necessary* condition for conjugacy, $\lambda_\infty f$ must be conjugate to g. The condition is elementary: if

$$g(x^*) = x^* \qquad \text{and} \qquad \lambda f(\xi_\lambda^*) = \xi_\lambda^*$$

then

$$g \sim \lambda_\infty f \Rightarrow g'(x^*) = \lambda_\infty f(\xi_{\lambda_\infty}^*)$$

But $g'(x^*)$ is a fixed value for fixed z, and so upon calculation of the fixed point of λf an estimate of λ_∞ is had:

$$\lambda_\infty(f) = g'(x^*)/f'(\xi^*_{\lambda_\infty})$$

Again for $f = 1 - 2x^2$, $f = \sin \pi x$, $f = x - x^3$, and other "nice" f's, a 0.1% estimate is obtained for λ_∞. "Nice" here means that f is "close" to conjugate to g.

We now extend these ideas to the $g_{r,n}$ recursion to provide another proof that

$$\lambda_r \sim \lambda_\infty - \mu\delta^{-r}$$

with μ now determined by the same simple kind of estimates (and to equal precision) as was λ_∞. Moreover, we shall demonstrate how the convergence rate of α_n is computable and equal to δ. Repeating Eq. (62),

$$g_{r,n} = g + \sum_\rho c^\rho_{r+n}(-\alpha)^{n(1-\rho)}(g^\rho - x^\rho g') - \delta^{-r}\left(h + \sum_\rho c^\rho_{r+n}(-\alpha)^{n(1-\rho)}(\rho g^{\rho-1}h - x^\rho h')\right)$$

it is clear that the $g_{r,n}$ for large n are fixed by determining the $c_t{}^\rho$ from initial data. Thus, for $n = 0$

$$g_{r,0} = \lambda_r f(x) = g + \sum_\rho c_r{}^\rho(g^\rho - x^\rho g') - \delta^{-r}\left(h + \sum_\rho c_r{}^\rho(\rho g^{\rho-1}h - x^\rho h')\right) \tag{96}$$

Recall that

$$\psi_\rho \equiv g^\rho - x^\rho g'$$

is the eigenvector of $\mathscr{L}$ corresponding to $\lambda = (-\alpha)^{1-\rho}$ and so orthogonal to h^*. Defining

$$\rho g^{\rho-1}h - x^\rho h' \equiv h_\rho \tag{97}$$

and projecting (96) on h^*, we have

$$\lambda_r(h^*, f) = (h^*, g) - \delta^{-r}\left(1 + \sum_\rho c_r{}^\rho(h^*, h_\rho)\right) \tag{98}$$

or

$$\lambda_r = \frac{(h^*, g)}{(h^*, f)} - \frac{1 + \sum_\rho c_r{}^\rho(h^*, h_\rho)}{(h^*, f)}\,\delta^{-r} \equiv \lambda_\infty - \mu_r\delta^{-r}$$

Thus,

$$\lambda_\infty = (h^*, g)/(h^*, f)$$

as before and

$$\mu_r/\lambda_\infty = \left[1 + \sum_\rho c_r{}^\rho(h^*, h_\rho)\right]\Big/(h^*, g) \tag{99}$$

Projecting next upon $\psi_{\bar{\rho}}^*$, we have

$$\lambda_r(\psi_{\bar{\rho}}^*, f) = (\psi_{\bar{\rho}}^*, g) + c_r^{\bar{\rho}} - \delta^{-r} \sum_\rho c_r^\rho(\psi_{\bar{\rho}}^*, h_\rho) \tag{100}$$

In particular, as $r \to \infty$,

$$\lambda_\infty(\psi_{\bar{\rho}}^*, f) = (\psi_{\bar{\rho}}^*, g) + c_\infty^{\bar{\rho}}$$

so that

$$\lim_{r\to\infty} c_r^\rho = c_\infty^{\bar{\rho}} = \lambda_\infty(\psi_{\bar{\rho}}^*, f) - (\psi_{\bar{\rho}}^*, g) \tag{101}$$

exists and is finite to meet initial data. Accordingly, for large r we have

$$\lambda_r \sim \lambda_\infty - \mu\delta^{-r}$$

with

$$\frac{\mu}{\lambda_\infty} = \frac{1 + \sum_\rho c_\infty^\rho(h^*, h_\rho)}{(h^*, g)}$$

$$= \frac{1 - \sum_\rho (\psi_\rho^*, g)(h^*, h_\rho) + \lambda_\infty \sum_\rho (\psi_\rho^*, f)(h^*, h_\rho)}{(h^*, g)} \tag{102}$$

Accordingly, μ/λ_∞ is also available and easily computed for small N quite accurately.

We next obtain a sum rule for c_r^ρ. Setting $x = 0$ in (96), we have

$$\lambda_r = 1 + \sum_\rho c_r^\rho - \delta^{-r} h(0)\left(1 + \sum_\rho c_r^\rho\right) \tag{103}$$

which as $r \to \infty$ becomes

$$\lambda_\infty = 1 + \sum_\rho c_\infty^\rho \tag{104}$$

Since

$$\sum_\rho \rho c_r^\rho < \infty$$

by (103),

$$c_r^\rho < 1/\rho^{2+\epsilon}$$

for large ρ. Accordingly, truncation of the ρ sum allows high-accuracy estimates. Setting $N = 2$ so that only $\lambda = \delta$ and $\lambda = 1$ contribute, one has the rough result

$$\lambda_\infty \simeq 1 + c_\infty^1 \tag{105}$$

By (62),

$$g_{r,n} \underset{n\to\infty}{\sim} g + c^1_{r+n}(g - xg') - \delta^{-r}(h + c^1_{r+n}(h - xh'))$$

$$\simeq (1 + c^1_{r+n})(g - \delta^{-r}h)(x/1 + c^1_{r+n}) \tag{106}$$

or,

$$g_{r,n} \underset{n\to\infty}{\longrightarrow} (1 + c_\infty{}^1)g_r(x/1 + c_\infty{}^1) \tag{107}$$

With the rough estimate (105), this reads

$$g_{r,n} \underset{n\to\infty}{\longrightarrow} \lambda_\infty g_r(x/\lambda_\infty)$$

Also, assuming a rapid n approach, and setting $r \to \infty$, we have

$$g_{\infty,n} \simeq \lambda_\infty g(x/\lambda_\infty)$$

so that $g_{\infty,n} \simeq g_{\infty,0}$ is the estimate

$$\lambda_\infty f(x) \simeq \lambda_\infty g(x/\lambda_\infty)$$

or

$$g(x) \simeq f(\lambda_\infty x)$$

Thus we realize that all our approximation schemes produce estimates of the same accuracy. Next, Eqs. (98) and (100) for $N = 2$ are

$$\lambda_r(h^*, f) \simeq (h^*, g) - \delta^{-r}(1 + c_r{}^1(h^*, h_1))$$

and

$$\lambda_r(\psi_1{}^*, f) \simeq (\psi_1{}^*, g) + c_r{}^1(1 - \delta^{-r}(\psi_1{}^*, h_1))$$

The ratio of these equations produces

$$c_r{}^1 - c_\infty{}^1 \propto \delta^{-r} \tag{108}$$

Together with (106), we then have

$$g_{r,n} - g_{r,\infty} \sim \delta^{-n}$$

in fixed r, providing the mechanism for $\alpha_n \to \alpha$ at the rate δ. We are unsure as to why $\alpha_n \to \alpha$ at a rate $\delta' \neq \delta$ for $z > 2$, especially since the spectrum of L for $z = 4$ possesses δ as the unique growing eigenvalue. Presumably, higher order transients can here decay at a rate below that of the "aymptotic" features discussed here. But for this one defect, the above techniques explain to good accuracy every detail of all our recursion data.

6. AFTERWORD

The preceding parts of this paper were contained in a preprint first circulated in November 1976. This paper is incomplete insofar as the uniqueness of an appropriate solution to (9) as well as the basic spectral conjecture remain unproven. Failing to publish it immediately (because it was not self-contained), I allowed it to hover in a limbo while I anticipated some measure of success at a proof, foreshadowing its content in the final section of its predecessor.[1]

Early in 1979, I was informed that an effort by Collet *et al.*[2],[2] has succeeded in this task. These authors have proven existence and uniqueness of the appropriate solution to the fundamental equation (9) and verified the spectral conjecture of $\mathscr{L}$. (This demonstration is, so far, restricted to $z = 1 + \epsilon$ with ϵ small.) Accordingly, the theory presented here is now well-founded, although no extension beyond the local stability of the fixed point is expected in the immediate future.

At this time, I should like to mention another effort. In the special case $z = 1 + \epsilon$, ϵ small, it is easy to approximately solve (9) and the spectral problem of $\mathscr{L}$ since α^{-1} is perturbatively small. This result first appeared in a work by Derrida *et al.*[4] (DGP). In this and another interesting paper by these authors,[5] the work of Metropolis *et al.*[6] (MSS) has been significantly elaborated upon through the discovery of an "internal symmetry" of the MSS sequences which allows organizations of these sequences in manners approaching λ_∞ from *above* rather than from below along the harmonics. I will here briefly explore the connection of one aspect of their work with the present work.

There exists a unique fundamental 4-cycle above the λ_∞ of the 2-cycle. Related to the pattern of this 4-cycle by the operation of DGP is a fundamental 8-cycle, below the 4-cycle and closest to λ_∞. Similarly, for each n there is a fundamental 2^n-cycle below the 2^{n-1}-cycle and closest to and above λ_∞. Denoting the parameter value of these cycles that are superstable by $\hat{\lambda}_n$, we have

$$\hat{\lambda}_2 > \hat{\lambda}_3 > \cdots > \hat{\lambda}_n > \cdots > \lambda_\infty \tag{109}$$

DGP observe that

$$\hat{\lambda}_n - \lambda_\infty \propto \delta^{-n}$$

with the same δ as for the harmonics of the 2-cycle. It is easy to see why this can be so. For the harmonics, the functions g_r were constructed, with

$$g_r \sim g - \delta^{-r} h$$

In this form, the coefficient of h is *negative*. Indeed, this is required for the harmonics to guarantee that $g_0(0) = 0$. However, the term in h is perturbative about the fixed point g, so that nothing in the local analysis requires this negative coefficient. Indeed, for an appropriate *positive* coefficient the

[2] See Ref. 3 for a preview.

phenomena described above are explained, with the g's constructed determining the elements of these cycles.

To see how these phenomena are described, write

$$\lambda f = \lambda_\infty f + (\lambda - \lambda_\infty) f \equiv G_0 + (\lambda - \lambda_\infty) H_0 \tag{110}$$

Iterating 2^n times, and keeping terms to order $\lambda - \lambda_\infty$, we obtain

$$(\lambda f)^{2^n} = G_n + (\lambda - \lambda_\infty) H_n + O((\lambda - \lambda_\infty)^2) \tag{111}$$

where

$$G_n = (\lambda_\infty f)^{2^n}$$

Defining

$$\begin{aligned} (-\alpha)^n (\lambda f)^{2^n}(x/(-\alpha)^n) &\equiv f_n \\ (-\alpha)^n G_n(x/(-\alpha)^n) &\equiv g_n \\ (-\alpha)^n H_n(x/(-\alpha)^n) &\equiv h_n \end{aligned} \tag{112}$$

we can write (111) as

$$f_n(x) = g_n(x) + (x - \lambda_\infty) h_n(x) \tag{113}$$

where

$$\begin{aligned} h_{n+1}(x) &= -\alpha [h_n(g_n(-x/\alpha)) + g_n'(g_n(-x/\alpha)) h_n(-x/\alpha)] \\ &\equiv \mathscr{L}_n[h_n(x)] \end{aligned}$$

and

$$h_n = \mathscr{L}_{n-1} \mathscr{L}_{n-2} \cdots \mathscr{L}_0 f \tag{114}$$

By the definition of λ_∞ and g,

$$g_n \to g, \qquad \mathscr{L}_n \to \mathscr{L} \qquad \text{as } n \to \infty$$

Accordingly, (114) becomes

$$h_n \sim c(f) \delta^n h$$

and (113) reads

$$f_n \sim g + c(f)(\lambda - \lambda_\infty) \delta^n h \tag{115}$$

Equation (115) is approximately correct so long as $(\lambda - \lambda_\infty)^n$ is small, which is the case when

$$|\lambda - \lambda_\infty| \delta^n \leqslant \text{small constant}$$

With λ_n chosen as usual to determine the superstable 2^n-cycle harmonic of the 2-cycle,

$$f_n \to g_0$$

Since $g_0(0) = 0$ and $g(0)$, $h(0) \neq 0$ evidently

$$(\lambda_n - \lambda_\infty) \delta^n \sim 1$$

again establishing δ as the λ convergence rate. If an n-independent *finite* condition on the f_n can more generally be maintained, δ will again be the convergence rate and the corresponding limit of the f_n will be given by (115).

Accordingly, consider determining λ_n by the condition that

$$(\lambda_{nf})^{2^n}(\xi_n) = \xi_n$$
$$D(\lambda_{nf})^{2^n}(\xi_n) = \mu \qquad \text{(independent of } n\text{)}$$

where ξ_n is the fixed point closest to $x = 0$. By (112) these conditions transcribe to

$$f_n(x_n) = x_n, \qquad f_n'(x_n) = \mu, \qquad x_n = (-\alpha)^n \xi_n$$

As $n \to \infty, f_n \to f_\mu, x_n \to x_\mu$, by (115)

$$x_\mu \sim g(x_\mu) + c(f)(\lambda_n - \lambda_\infty)\delta^n h(x_\mu) \tag{116}$$
$$\mu \sim g'(x_\mu) + c(f)(\lambda_n - \lambda_\infty)\delta^n h'(x_\mu) \tag{117}$$

Denoting the fixed point of g by $\hat{x}$, and the slope of g at $\hat{x}$ by $\hat{\mu}$, then from (116) and (117) we immediately obtain the approximation

$$\lambda_n \sim \lambda_\infty + \delta^{-n}(\hat{\mu} - \mu)/c|h'(\hat{x})| \tag{118}$$

for $\mu \simeq \hat{\mu}$.

Thus, so long as $\mu \neq \hat{\mu}$, the corresponding λ_n converge to λ_∞ at the rate δ. (At $\mu = \hat{\mu}$, $\lambda_n \to \lambda_\infty$ faster than geometric at the rate δ.) Also by (118), the coefficient of h in f_μ, by (115), changes sign at $\mu = \hat{\mu}$: for $|\mu| < |\hat{\mu}|$, f_μ is a g_r or its continuous analog (for example, at bifurcation values) as described at the end of Section 3; for $|\mu| > |\hat{\mu}|$, $\lambda_n \to \lambda_\infty$ from *above*, and evidently the harmonics are not under consideration. For example, with $\bar{\lambda}_n$ of (109)

$$f_n = (-\alpha)^n(\bar{\lambda}_{n+1}f)^{2^n}(x/(-\alpha)^n)$$

corresponds to a limiting value of $|\mu| > |\hat{\mu}|$ and $\bar{\lambda}_n \to \lambda_\infty$ from above at rate δ, as was to be demonstrated. Accordingly, the fixed point g is the "organizing center" for all attractors with $\lambda \to \lambda_\infty$ whether from above or below λ_∞.

As a final comment, it is perhaps worthy to point out the resemblance of the theory presented to the renormalization-group notions of Wilson.[7] Essentially, the function g_r determine elements of infinitely bifurcated attractors at various levels of magnification, belying a self-similarity of their distribution; this structure is precisely reproduced through the operations of composition and rescaling $\mathscr{T}$, resulting in the next lower g_r. The function g itself is the fixed point of $\mathscr{T}$, while the g_r lie on the one-dimensional unstable manifold through g along h; δ and h indeed were determined by linearizing $\mathscr{T}$ about g. More generally, applied to any f, $\mathscr{T}$ can be viewed as a renormalization-group transformation with self-similarity (critical behavior) determined by the fixed point g. Viewing the parameter λ as temperature, λ_∞ is the critical point and δ emerges as a critical exponent. More intuitively, an analog of Kadanoff's block-spin notion is also available. Thus, consider the superstable 2^n cycles starting at $n = 0$, for which there is a single point at $x = 0$. For $n = 1$, this point is split into one at $x = 0$ again, and another point x_1 to the right. For $n = 2$, $x = 0$ again splits, with x_2 nearest to $x = 0$ and to the left, while x_1 splits into a more closely spaced pair with centroid roughly at x_1. By the definition of α, $x_1 \simeq -\alpha x_2$. As n increases, each point splits into a pair with the element nearest to $x = 0$ located $-\alpha$ times nearer to $x = 0$ than its predecessor. It is thus clear that if each closely spaced pair is replaced by a point at its centroid (viewing at lower resolution), then the same set of points about $x = 0$ is reproduced, but with all distances $-\alpha$ times larger. Accordingly, spin-blocking has here the analog of functional composition, while the following volume rescaling is here, rather than a geometrical factor of 2, now a dynamically determined factor of α. In this way, the theory presented in this work may be viewed as an instance arising mathematically of the renormalization-group notions of statistical mechanics.

APPENDIX

We include here some numerical results, useful for normal ($z = 2$) recursive calculations.

A1. $g(x) = 1 + \sum_{i=1}^{7} g_i x^{2i}$, determining g to ten significant figures as [0, 1]:

$$g_1 = 1.527632997$$
$$g_2 = 1.048151943 \times 10^{-1}$$
$$g_3 = 2.670567349 \times 10^{-2}$$
$$g_4 = -3.527413864 \times 10^{-3}$$
$$g_5 = 8.158191343 \times 10^{-5}$$
$$g_6 = 2.536842339 \times 10^{-5}$$
$$g_7 = -2.687772769 \times 10^{-6}$$
$$\Rightarrow -g'(1) = \alpha = 2.502907876$$

A2. From the above, $g(x^*) = x^*$ for $x^* = 0.5493052461$ and $g'(x^*) = -1.601191328$, which is required for the estimate

$$\lambda_\infty(f) \simeq g'(x^*)/f'(\xi^*)$$

where ξ^* satisfies

$$\lambda_\infty f(\xi^*) = \xi^*$$

For example, with $f = x(1 - x)$,

$$\xi^* = 1 - \lambda_\infty^{-1} \quad \text{and} \quad \lambda_\infty f'(\xi^*) = -\lambda_\infty + 2$$

so that

$$-\lambda_\infty + 2 \simeq g'(x^*)$$

or $\lambda_\infty \simeq 3.60119$, to be compared with the correct result $\lambda_\infty = 3.56995$.

A3. In order for g_r to be computed, one needs $h(x)$ normalized to $h(0) = 1$ together with the correct $h(0)$ to ensure that $g_0(0) = 0$. Regarding $r = 6$ as asymptotic, we have

$$h(0) = 1.318707$$

and a parametrization of similar accuracy to Section A1 is

$$h(x) = h(0)\left(1 + \sum_{i=1}^{6} h_i x^{2i}\right)$$

with

$$\begin{aligned}
h_1 &= -3.256513712 \times 10^{-1} \\
h_2 &= -5.055393508 \times 10^{-2} \\
h_3 &= 1.455982806 \times 10^{-2} \\
h_4 &= -8.810422078 \times 10^{-4} \\
h_5 &= -1.062170276 \times 10^{-4} \\
h_6 &= 1.983988805 \times 10^{-5}
\end{aligned}$$

Iterating (8), g_1 or g_0 is obtained for estimates of the locations of elements of a highly bifurcated cycle near $x = 0$. Observe that since

$$g_{r-s}(x) = (-\alpha)^s g_r^{(2^s)}(x/(-\alpha)^s)$$

with $s = r = 6$,

$$g_0(x) = (-\alpha)^6 g_6^{(2^6)}(x/\alpha^6) \simeq (-\alpha)^6(g - \delta^{-6}h)^{(2^6)}(x/\alpha^6)$$

so that g and h restricted to $[0, 1]$ provide g_0 or $[0, \alpha^6]$, thereby determining many elements near $x = 0$.

A4. Solving the eigenvalue problem of L for $N = 2$, we have

$$\delta \simeq 4.6736, \qquad \lambda_1 \simeq 0.9880$$

to be compared with

$$\delta = 4.6692, \qquad \lambda_1 = 1.0000$$

The corresponding eigenvectors and adjoint eigenvectors (unnormalized) are $\left[\psi = \begin{pmatrix} a \\ b \end{pmatrix} \Rightarrow \psi(x) = a + bx^2\right]$

$$h = \begin{pmatrix} 1 \\ -0.3644 \end{pmatrix}, \qquad \psi_1 = \begin{pmatrix} 1 \\ 1.1082 \end{pmatrix}$$

$$h^* = \begin{pmatrix} 1 \\ -0.9024 \end{pmatrix}, \qquad \psi_1^* = \begin{pmatrix} 1 \\ 2.7444 \end{pmatrix}$$

Writing g as

$$g \simeq \begin{pmatrix} 1 \\ g_1 + \frac{1}{3}g_2 \end{pmatrix} = \begin{pmatrix} 1 \\ -1.4927 \end{pmatrix}$$

it is trivial to estimate $\lambda_\infty \simeq (h^*, g)/(h^*, f)$. For example, with $f(x) = 1 - 2x^2$,

$f = \begin{pmatrix} 1 \\ -2 \end{pmatrix}$ and $\lambda_\infty(f) \simeq 0.8368$. Also, $h_1 = \begin{pmatrix} 1 \\ 0.3644 \end{pmatrix}$ and $(h^*, h_1) \simeq 0.5051$

(properly normalized), so that by (102)

$$\frac{\mu}{\lambda_\infty} \simeq \frac{1 - (\psi_1^*, g)(h^*, h_1) + \lambda_\infty(\psi_1^*, f)(h^*, h_1)}{(h^*, g)} \simeq 0.6851$$

or $\mu \simeq 0.5733$, to be compared with $\mu(f) = 0.5981$. While it is true that $N = 3$ significantly improves this result, this is already quite accurate and trivial to obtain.

ACKNOWLEDGMENTS

The first exact equation, Eq. (9) of the text, together with the scheme of solution incorporating Eq. (91), was obtained by Predrag Cvitanović in discussion and collaboration with the author. This result proved to be seminal in the construction of the theory presented. The author has profited from discussion with R. Menikoff, and thanks his colleagues, especially D. Campbell and F. Cooper, for critical interest.

REFERENCES

1. Mitchell J. Feigenbaum, *J. Stat. Phys.* **19**:25 (1978).
2. P. Collet, J.-P. Eckmann, and O. E. Lanford III, Universal Properties of Maps on an Interval, in draft.
3. P. Collet and J.-P. Eckmann, Bifurcations et Groupe de Renormalisation, IHES/P/78/250.

4. B. Derrida, A. Gervois, and Y. Pomeau, Universal Metric Properties of Bifurcations of Endomorphisms, Saclay preprint (1977).
5. B. Derrida, A. Gervois, and Y. Pomeau, Iterations of Endomorphisms on the Real Axis and Representation of Numbers, Saclay preprint (1977).
6. N. Metropolis, M. L. Stein, and P. R. Stein, *J. Combinatorial Theory* **15**:25 (1973).
7. K. Wilson and J. Kogut, *Phys. Rep.* **12C**:75 (1974).

Reprinted from JOURNAL OF COMBINATORIAL THEORY Vol. 15, No. 1, July 1973
All Rights Reserved by Academic Press, New York and London *Printed in Belgium*

On Finite Limit Sets for Transformations on the Unit Interval

N. METROPOLIS, M. L. STEIN, AND P. R. STEIN*

University of California, Los Alamos Scientific Laboratory, Los Alamos, New Mexico 87544

Communicated by G.-C. Rota

Received June 8, 1971

An infinite sequence of finite or denumerable limit sets is found for a class of many-to-one transformations of the unit interval into itself. Examples of four different types are studied in some detail; tables of numerical results are included. The limit sets are characterized by certain patterns; an algorithm for their generation is described and established. The structure and order of occurrence of these patterns is universal for the class.

1. Introduction. The iterative properties of 1-1 transformations of the unit interval into itself have received considerable study, and the general features are reasonably well understood. For many-to-one transformations, however. the situation is less satisfactory, only special and fragmentary results having been obtained to date [1, 2]. In the present paper we attempt to bring some coherence to the problem by exhibiting an infinite sequence of finite limit sets whose structure is common to a broad class of non 1-1 transformations of [0, 1] into itself. Generally speaking, the limit sets we shall construct are not the only possible ones belonging to an arbitrary transformation in the underlying class. Nevertheless, our sequence—which we shall call the "*U*-sequence"—constitutes perhaps the most interesting family of finite limit sets in virtue of the universality of their structure and of their order of occurrence. With regard to infinite limit sets we shall have little to say. There is reason to believe, however, that for a non-vacuous (in fact, infinite) subset of the class of transformations considered here, our construction—suitably extended to the limit of "periods of infinite length"—is exhaustive in the

* Work performed under the auspices of the U.S. Atomic Energy Commission under contract W-7405-ENG-36.

probabilistic sense, namely, that with "probability 1" every limit set belongs to the U-sequence.

2. We begin by describing the class of transformations to which our construction applies, but make no claim that the conditions imposed are strictly necessary. The description is complicated by our attempt to exclude, insofar as is possible, certain finite limit sets, not belonging to the U-sequence, whose existence and structure depend on detailed properties of the particular transformation in question.

All our transformations will be of the form

$$T_\lambda(x) : x' = \lambda f(x),$$

where x' denotes the first iterate of x (*not* the derivative!) and λ varies in a certain open interval to be specified below. The fundamental properties of $f(x)$ will be:

A.1. $f(x)$ is continuous, single-valued, and piece-wise $C^{(1)}$ on $[0, 1]$, and strictly positive on the open interval, with $f(0) = f(1) = 0$.

A.2. $f(x)$ has a unique maximum, $f_{\max} \leqslant 1$, assumed either at a point or in an interval. To the left or right of this point (or interval) $f(x)$ is strictly increasing or strictly decreasing, respectively.

A.3. At any x such that $f(x) = f_{\max}$, the derivative exists and is equal to zero.

We allow the possibility that $f(x)$ assumes its maximum in an interval so as to include certain broken-linear functions with a "flat top" (cf. example (3.4) below).

In addition to the properties (A) we need some further conditions which will serve to define the range of the parameter λ:

B. Let $\lambda_{\max} = 1/f_{\max}$. Then there exists a λ_0 such that, for $\lambda_0 < \lambda < \lambda_{\max}$, $\lambda f(x)$ has only two fixed points, the origin and $x_F(\lambda)$, say, both of which are repellent. For functions of class $C^{(1)}$ this means simply that

$$\left.\lambda \frac{df}{dx}\right|_{x=0} > 1 \qquad\qquad (\lambda_0 < \lambda \leqslant \lambda_{\max}).$$
$$\left.\lambda \frac{df}{dx}\right|_{x=x_F(\lambda)} < -1$$

For piece-wise $C^{(1)}$ functions the generalization of these conditions is obvious.

The above conditions are sufficient to guarantee the existence of the

U-sequence; that they are not necessary can be shown by various examples, but we shall not pursue this matter here.

$f(x)$ as defined above clearly has the property that its piecewise derivative is less than 1 in absolute value in some interval N which includes that for which $f(x) = f_{\max}$. In order to exclude certain unwanted finite limit sets, we append the following condition:

C. $f(x)$ is convex in the interval N; at every point $x \notin N$, the piece-wise derivative of $f(x)$ is greater than 1 in absolute value.

Unfortunately, property (C) is not sufficient to exclude all finite limit sets not given by our construction; to achieve this end it might be necessary to restrict the underlying class of transformations rather drastically. We shall return to this point in Section 4 below.

It will simplify the subsequent discussion to make the non-essential assumption that $f(x)$ assumes its maximum at the point $x = \frac{1}{2}$ (or, if the function assumes its maximum in an interval, that the interval includes $x = \frac{1}{2}$). In the sequel we shall make this assumption without further comment. A particular iterate x' will then be said to be of "type L" or of "type R" according as $x' < \frac{1}{2}$ or $x' > \frac{1}{2}$, respectively. Given an initial point x_0, the minimum distinguishing information about the sequence of iterates $T_\lambda^{(k)}(x_0)$, $k = 1, 2, \ldots$, will consist in a "pattern" of R's and L's, the k-th letter giving the relative position of the k-th iterate of x_0 with respect to the point $x = \frac{1}{2}$. The patterns turn out to play a fundamental role in our construction; they will be discussed in detail in the following sections.

3. Let us give some simple examples of the class of transformations we are considering:

$$Q_\lambda(x): \quad x' = \lambda x(1-x)$$

$$3 < \lambda < 4 \tag{3.1}$$

$$S_\lambda(x): \quad x' = \lambda \sin \pi x$$

$$\lambda_0 < \lambda < 1 \qquad (\text{with } .71 < \lambda_0 < .72) \tag{3.2}$$

$$C_\lambda(x): \quad x' = \lambda W(3 - 3W + W^2), \qquad W \equiv 3x(1-x)$$

$$\lambda_0 < \lambda < \frac{64}{63} \qquad (\text{with } .872 < \lambda_0 < .873) \tag{3.3}$$

In the last two examples, more precise limits for λ_0 are available, but they are not important for our discussion. All these examples are convex functions of class $C^{(\infty)}$ which are, moreover, symmetric about $x = \frac{1}{2}$.

With regard to the existence of the U-sequence, these restrictions are in no way essential. As will be remarked in Section 4, however, these examples happen to belong to the subclass for which our construction does exhaust all finite limit sets.

As a further example, consider the broken-linear mapping:

$$
\begin{aligned}
L_\lambda(x; e):\quad & x' = \frac{\lambda}{e}\, x, && 0 \leqslant x \leqslant e, \\
& x' = \lambda, && e \leqslant x \leqslant 1 - e, \qquad (3.4)\\
& x' = \frac{\lambda}{e}(1 - x), && 1 - e \leqslant x \leqslant 1, \\
& && \text{with}\quad 1 - e < \lambda < 1.
\end{aligned}
$$

Here e is a parameter characterizing the width 1-2e of the maximum, and may be chosen to have any value in the range $0 < e < \frac{1}{2}$. It remains fixed as λ is varied, and different choices of e yield distinct transformations.

4. The finite limit sets of our class of transformations—and, in particular, of the four special transformations given above—are attractive periods of order $k = 2, 3, \ldots$. (We exclude the case $k = 1$ by invoking property (B).) The reader will recall that an "attractive period of order k" is a set of k periodic points x_i , $i = 1, 2, \ldots, k$, with $T_\lambda(x_i) = x_{i+1}$ (in some order). Each of these is a fixed point of the k-th power of $T_\lambda : T_\lambda^{(k)}(x_i) = x_i$, for which, moreover, the (piece-wise) derivative satisfies

$$\left| \frac{dT_\lambda^{(k)}}{dx} \right|_{x=x_i} < 1.$$

(By the chain rule, the slope is the same at all points in the period.) As a consequence of this slope condition, there exists for each x_i an attractive neighborhood $n(x_i)$ such that for any $x^* \in n(x_i)$ the sequence of iterates $T_\lambda^{(jk)}(x^*), j = 1, 2, \ldots,$ will converge to x_i . Periodic points which do not satisfy this slope condition (more precisely, for which the absolute value of the derivative is greater than 1) have no attractive neighborhood; they are consequently termed repellent (or unstable). These points belong to what is sometimes called "the set of exceptional points," a set of measure zero in the interval which plays no role in a discussion of limit sets.

The sequence of finite periods which we shall exhibit will be characterized *inter alia* by the following property:

J. For every period belonging to the U-sequence there is a period point whose attractive neighborhood includes the point $x = \frac{1}{2}$.

Now it follows from a theorem of G. Julia [3] that, if $T_\lambda(x)$ is the restriction to [0, 1] of some function analytic in the complex plane whose derivative vanishes at a single point in the interval, then the only possible finite limit sets ($k > 1$) are those with the property (J). The transformations (3.1) through (3.3) are clearly of this type, so that, with respect to finite limit sets, the *U*-sequence will exhaust all possibilities for them. That Julia's criterion is not necessary is shown by example (3.4); in this case there cannot be any attractive periods which do not have a period point lying in the region $e \leqslant x \leqslant 1 - e$. Such a period, however, clearly is of the type described by property (J), and hence belongs to the *U*-sequence.

Taking our clue from property (J), we now investigate the solutions λ of the equation:

$$T_\lambda^{(k)}(\tfrac{1}{2}) = \tfrac{1}{2}. \tag{4.1}$$

The corresponding periodic limit sets will be attractive, since the slope of $\lambda f(x)$ at $x = \frac{1}{2}$ is zero by hypothesis (property (A.3)). By way of example, we choose $k = 5$. Then for each of the four transformations of Section 3 there are precisely three distinct solutions of equation (4.1). The three patterns—common to all four transformations—are:

$$\tfrac{1}{2} \to R \to L \to R \to R \to \tfrac{1}{2},$$
$$\tfrac{1}{2} \to R \to L \to L \to R \to \tfrac{1}{2},$$
$$\tfrac{1}{2} \to R \to L \to L \to L \to \tfrac{1}{2}.$$

Omitting the initial and final points as understood, we write these patterns in the simplified form:

$$\begin{aligned} &\mathrm{RLR}^2, \\ &\mathrm{RL}^2\mathrm{R}, \\ &\mathrm{RL}^3. \end{aligned} \tag{4.2}$$

In accordance with this convention, a pattern with $k - 1$ letters R or L will be said to be of "length k."

These solutions are clearly ordered on the parameter λ. In Table I we give the full set of solutions of (4.1), through $k = 7$, for all four special transformations; in the broken-linear case we choose $e = .45$. The numerical values of λ were found by a simple iterative technique (the "binary chopping process"); although they are given to only seven decimal digits, they are actually known to approximately twice that precision. Of course, once a particular λ has been found, the corresponding pattern can be generated by direct iteration.

TABLE I

			Values of λ_i			
i	k_i	P_i	$Q_\lambda(x)$	$S_\lambda(x)$	$C_\lambda(x)$	$L_\lambda(x; .45)$
1	2	R	3.2360680	.7777338	.9325336	.6581139
2	4	RLR	3.4985617	.8463822	.9764613	.7457329
3	6	RLR^3	3.6275575	.8811406	.9895107	.7806832
4	7	RLR^4	3.7017692	.9004906	.9955132	.8031673
5	5	RLR^2	3.7389149	.9109230	.9990381	.8180892
6	7	RLR^2LR	3.7742142	.9213346	1.0024311	.8318799
7	3	RL	3.8318741	.9390431	1.0073533	.8645337
8	6	RL^2RL	3.8445688	.9435875	1.0083134	.8858150
9	7	RL^2RLR	3.8860459	.9568445	1.0111617	.8977794
10	5	RL^2R	3.9057065	.9633656	1.0123766	.9085993
11	7	RL^2R^3	3.9221934	.9687826	1.0132699	.9187692
12	6	RL^2R^2	3.9375364	.9735656	1.0140237	.9278274
13	7	RL^2R^2L	3.9510322	.9782512	1.0146450	.9361518
14	4	RL^2	3.9602701	.9820353	1.0149542	.9462185
15	7	RL^3RL	3.9689769	.9857811	1.0152122	.9564172
16	6	RL^3R	3.9777664	.9892022	1.0154974	.9635343
17	7	RL^3R^2	3.9847476	.9919145	1.0156711	.9702076
18	5	RL^3	3.9902670	.9944717	1.0157727	.9775473
19	7	RL^4R	3.9945378	.9966609	1.0158320	.9846165
20	6	RL^4	3.9975831	.9982647	1.0158621	.9903134
21	7	RL^5	3.9993971	.9994507	1.0158718	.9957404

We note that the set of 21 patterns and its λ-ordering is common to all four transformations. This remains true when we extend our calculations through $k = 15$. As k increases, the total number of solutions of (4.1) becomes large, as indicated in Table II. Thus for $k \leqslant 15$ there is a total of 2370 distinct solutions of equation (4.1). In the appendix we give a complete list of ordered patterns for $k \leqslant 11$.

The fact that these patterns and their λ-ordering are a common property of four apparently unrelated transformations (note that they are not connected by ordinary conjugacy, a relation which will be discussed in Section (6) suggests that the pattern sequence is a general property of a wide class of mappings. For this reason we have called this sequence of patterns the U-sequence where "U" stands (with some exaggeration) for

TABLE II

k ...	2	3	4	5	6	7	8	9	10	11	12	13	14	15
Number of solutions ...	1	1	2	3	5	9	16	28	51	93	170	315	585	1091

"universal." In the next section we shall state and prove a logical algorithm which generates the U-sequence for any transformation having the properties (A) and (B) of Section 2. In the present section we confine ourselves to describing what might be called the "λ-structure" of the limit sets associated with the patterns of the U-sequence. No proofs are included, since the results given here will not be used in the proof of our main theorem.

As constructed, the patterns of the U-sequence correspond to distinct solutions of equation (4.1); they are attractive k-periods containing the point $x = \frac{1}{2}$ and possessing the property (J). It is clear by continuity that, given any solution λ (with finite k) and its associated pattern $P_k(\lambda)$, then for sufficiently small $\epsilon > 0$ there will exist periodic limit sets with the same pattern for all $\bar{\lambda}$ in the interval $\lambda - \epsilon \leqslant \bar{\lambda} \leqslant \lambda + \epsilon$. In other words, each period has a finite "λ-width." It is also clear that there exist critical values $m_1(\lambda)$ and $m_2(\lambda)$ such that, for $\bar{\lambda} < \lambda - m_1$ and $\bar{\lambda} > \lambda + m_2$, the pattern $P_k(\lambda)$ does *not* correspond to an attractive period of $T_{\bar{\lambda}}(x)$.

Consider now for simplicity the case in which the transformation is $C^{(1)}$, and take m_1 and m_2 to be boundary values such that for $\lambda - m_1 < \bar{\lambda} < \lambda + m_2$ the periodic limit set with pattern $P_k(\lambda)$ is attractive. As $\bar{\lambda}$ varies in this interval, the slope $dT_{\bar{\lambda}}^{(k)}/dx$ at a period point varies continuously from $+1$ to -1, the values ± 1 being assumed at the boundary points. It is natural to ask: what happens if $\bar{\lambda}$ lies just to the left or just to the right of the above interval? The question as to what the limit sets look like if $\bar{\lambda} = \lambda - m_1 - \delta$ (δ small) is a difficult one; the conjectured behavior will be described in Section 6, but rigorous proof is lacking. For $\bar{\lambda} = \lambda + m_2 + \delta$ we are in better case. As shown in Section 5, corresponding to any solution λ of (4.1) and its associated pattern there exists an infinite sequence of solutions $\lambda < \lambda^{(1)} < \lambda^{(2)} < \cdots < \lambda^{(\infty)}$ with associated patterns $H^{(1)}(\lambda^{(1)}), H^{(2)}(\lambda^{(2)}),\ldots,$ called "harmonics," with the property that they exhaust all possible solutions λ^* in the interval $\lambda \leqslant \lambda^* \leqslant \lambda^{(\infty)}$. The sequence of harmonics of a given solution is a set of periods of order $2^m k$, $m = 1, 2,\ldots,$ with contiguous λ-widths and well-defined pattern structure; no other periods of the U-sequence can exist for any λ^* in the given interval (harmonics have been encountered before in a more restrictive context: cf. reference 4). From the constructior Section 5 it will be obvious that $\lambda^{(\infty)}$ exists as a right-hand limit; question as to the nature of the limit sets for $\lambda^* = \lambda^{(\infty)} + \epsilon$ remains o (but cf. Conjecture 2 of Section 6).

In order to prepare the ground for the discussion in the next section, give here the following formal

DEFINITION 1. Let $P = RL^{\alpha_1}R^{\alpha_2}L^{\alpha_3}\cdots$ be a pattern correspondin; some solution of (4.1). Then the (first) harmonic of P is the patt $H = P\mu P$, where $\mu = L$ if P contains an odd number of R's, and μ = otherwise.

For example, the pattern RLR^2 has the harmonic $H = RLR^2LRL$ while for RL^2R we have $H = RL^2R^3L^2R$.

Naturally, the construction of the harmonic can be iterated, so that may speak of the second, third,..., m-th harmonic, etc. When necess; we shall write $H^{(j)}$ to denote the j-th harmonic.

In addition to the harmonic H of a pattern P, there is another for construct which will be used in the sequel:

DEFINITION 2. The "antiharmonic" A of a pattern P is construc analogously to the harmonic H except that $\mu = L$ when P contains even number of R's, while $\mu = R$ otherwise.

Thus in passing from a pattern to its harmonic the "R-parity" chan while for the antiharmonic the parity remains the same. The antiharm is a purely formal construct and never corresponds to any periodic li set; the reason for this will become clear in the next section. Note t like that of the harmonic, the antiharmonic construction can be iter to any desired order.

5. We begin by defining the "extension" of a pattern:

DEFINITION 3. The H-extension of a pattern P is the pattern gener by iterating the harmonic construction applied j times to P, whe increases indefinitely.

DEFINITION 4. The A-extension of P is the pattern $A^{(j)}(P)$, whe increases indefinitely. Here $A^{(j)}(P)$ denotes the j-th iterate of the a harmonic.

In these definitions we avoid writing $j \to \infty$, in order to avoid rai questions concerning the structure of the limiting pattern. In practice that will be required is that j is "sufficiently large."

We are now in a position to state

THEOREM 1. *Let K be an integer. Consider the complete ordered sequence of solutions of* (4.1) *and their associated patterns for* $2 \leqslant k \leqslant K$. *Let λ_1 be any such solution with pattern P_1 and length k_1, and let $\lambda_2 > \lambda_1$ be the "adjacent" solution (i.e., the next in order) with pattern P_2 and length k_2.*

Form the H-extension of P_1 and the A-extension of P_2. Reading from left to right, the two extensions $H(P_1)$ and $A(P_2)$ will have a maximal common leading subpattern P^* of length k^*, so that we may write

$$H(P_1) = P^*\mu_1 \ldots, \quad A(P_2) = P^*\mu_2 \ldots, \quad \mu_1 \neq \mu_2,$$

where μ_i stands for one of the letters L *or* R.

Case 1. $k^* \geqslant 2k_1$. *Then the solution λ^* of lowest order such that $\lambda_1 < \lambda^* < \lambda_2$ is the harmonic of P_1.*

Case 2. $k^* < 2k_1$. *Then the solution λ^* of lowest order such that $\lambda_1 < \lambda^* < \lambda_2$ corresponds to the pattern P^* of length k^* ($> K$ necessarily).*

A simple consequence of this theorem is the following:

COROLLARY. *Let $| k_1 - k_2 | = 1$ in Theorem* 1. *Then the lowest order solution λ^* with $\lambda_1 < \lambda^* < \lambda_2$ has length $k^* = 1 + \max(k_1, k_2)$.*

This follows from the theorem on noting that all patterns have the common leading subpatterns (not maximal!) RL; therefore, in forming the extensions, the first disagreement will indeed come at the indicated value of k^*.

We give some examples of the application of Theorem 1.

Example 1. Take $K = 9$. Reference to the table in the appendix shows that patterns #12 and #14 are adjacent. We have

$$P_1 = \mathrm{RLR}^4, \ k_1 = 7, \ H(P_1) = \mathrm{RLR}^4\mathrm{LRLR}^4\ldots,$$
$$P_2 = \mathrm{RLR}^4\mathrm{LR}, \ k_2 = 9, \ A(P_2) = \mathrm{RLR}^4\mathrm{LRLRL}\ldots,$$

so $P^* = \mathrm{RLR}^4\mathrm{LRLR}$, with $k^* = 11$, as verified by the table.

Example 2. Again take $K = 9$. Patterns #16 and #19 are adjacent and $P_1 = \mathrm{RLR}^2$ with $k_1 = 5$; here $k^* \geqslant 2k_1$. Therefore, by Case 1, the lowest order solution between the two patterns is the harmonic of P_1, namely, $\mathrm{RLR}^2\mathrm{LRLR}^2$, as given in the table (pattern #17).

To prove Theorem 1 we must first introduce some new concepts. Consider the transformation:

$$T_m(x)\colon x' = \lambda_{\max} f(x) \tag{5.1}$$

This transformation maps [0, 1] *onto* itself; hence, for any point in the interval, the inverses of all orders exist. Let us restrict ourselves for the moment to the point $x = \frac{1}{2}$ and its set (2^k in number) of k-th order inverses. At each step in constructing a k-th order inverse we have the free choice of taking a point on the right or on the left. For example, designating the point $x = \frac{1}{2}$ by the letter O, a possible inverse of order 5 would be represented by the sequence of letters

$$\mathrm{RLR^2O}, \tag{5.2}$$

which is to be read from *right to left*. Let us call a sequence like (5.2), when read from right to left, a "5-th order inverse path of the point $x = \frac{1}{2}$." Note that (5.2) is precisely the pattern associated with the first solution of equation (4.1) for $k = 5$. Another possible inverse path of the same order would be $\mathrm{L^2R^2O}$, but this clearly does not correspond to any solution of (4.1).

Choosing a particular k-th order inverse path of $x = \frac{1}{2}$, let us call the numerical value of the corresponding k-th inverse the "coordinate" of the path. Obviously, no path whose coordinate is less than $\frac{1}{2}$ can correspond to a pattern associated with a solution of (4.1). In order to achieve a 1-1 correspondence between a subclass of inverse paths and our periodic patterns we introduce the concept of a "legal inverse path," which we abbreviate as "l.i.p."

Definition 5. For the transformation $T_m(x)$ (cf. (5.1)), an l.i.p. of order k is a k-th order inverse path of $x = \frac{1}{2}$ whose coordinate x_k has the greatest numerical value of any point on the path.

In other words, of all the inverses constituting the path, the coordinate (i.e., the k-th inverse) lies farthest to the right. Note that any inverse path of $x = \frac{1}{2}$ can be inversely extended to an l.i.p. by appending on the left some suitable sequence, e.g., the sequence RL^α with α sufficiently large. Now consider the transformation $T_\lambda(x)$ corresponding to $T_m(x)$ with $\lambda < \lambda_{\max}$. As λ decreases, the original l.i.p. is deformed into an inverse path with varying coordinate $x_k(\lambda)$, but with the same pattern. By continuity, there clearly exists a λ^* for which

$$T_{\lambda^*}(\tfrac{1}{2}) = x_k(\lambda^*);$$

this in turn implies that for $\lambda = \lambda^*$ there exists a solution of equation (4.1) with the same pattern as that of the original l.i.p. On the other hand, for an inverse path (with, say, an R-type coordinate) which is *not* an l.i.p. the cycle will close on some intermediate point of the path (farther to the right than $x_k(\lambda)$), so that the path cannot be further inverted; this means that the original pattern cannot correspond to a solution of (4.1). Thus we have proved

LEMMA 1. *There is a* 1-1 *correspondence between the set of l.i.p.'s and the patterns associated with the solutions of equation* (4.1).

We note that the l.i.p.'s are naturally ordered on the values of their coordinates. By Lemma 1, any true statement about the pattern structure and coordinate ordering of the set of l.i.p.'s corresponds to a true statement about the pattern structure and λ-ordering of the set of solutions of (4.1).

Given some l.i.p. of order k, we construct an *inverse extension* $I(P)$ of the path according to the prescription $I(P) = P\mu PO$, where μ is R or L. Obviously, one choice corresponds to the harmonic, the other to the antiharmonic (Definitions 1 and 2). We can therefore speak of the harmonic or antiharmonic of an l.i.p. as well as of a pattern. Now, because of the monotonicity property (A.2) it follows that, given any two points, taking the left-hand inverse of both points preserves their relative order, while taking the right-hand inverse reverses it. A simple argument shows that $x_A < x < x_H$, where x is the coordinate of some l.i.p. and x_A, x_H are the coordinates of its antiharmonic and harmonic, respectively. This explains why the harmonic of an l.i.p. is again an l.i.p.. while the antiharmonic is not (and hence can never correspond to an attractive period of the U-sequence).

One final concept, the "projection" of an interval, will be of value in the subsequent discussion.

DEFINITION 6. Choose any two points x_1, x_2 in (0, 1); they define some interval I. Let $\bar{P}$ be an arbitrary sequence of R's and L's with $k - 1$ letters in all. Now, for some $T_m(x)$, construct the inverse paths $\bar{P}x_1$ and $\bar{P}x_2$. The coordinate $x_1{}^*$ and $x_2{}^*$ of these two paths define a new interval I^*, called the "projection under $\bar{P}$ of I." Because the defining inverse paths are of length k, we refer to it as a k-th order projection. (If we wish to exhibit explicitly the end-points x_1, x_2 of the interval I, we write I(x_1, x_2); in contrast to the usual notation for an interval, no ordering is implied.)

Proof of Theorem 1. It is clear that, if two intervals I, I^* are related by a k-th order projection, then for any point $x \in I^*$ we have $T_m^{(k)}(x) \in I$.

Consider now any l.i.p. with pattern PO and coordinate x_1. Its harmonic is again an l.i.p., with pattern PμPO and coordinate x_H, μ being either R or L depending on the R-parity of P. If x_μ is the point corresponding to the choice μ, then this construction shows that the interval $I^*(x_1, x_H)$ is the (k-th order) projection under P of the interval $I(\frac{1}{2}, x_\mu)$. Now any point x in the interior of I^* must map into the interior of I, and the end-points must map into end-points. Thus no inverse path of $x = \frac{1}{2}$—which is one of the end-points of I—can have a coordinate x^* satisfying $x_1 < x^* < x_H$. Precisely the same argument can be made for the antiharmonic. This proves

LEMMA 2. *Let PO be some l.i.p. with coordinate x_1. Form the $H^{(j)}$-extension of P, with coordinate $x_H^{(j)}$, and the $A^{(j)}$-extension of P with coordinate $x_A^{(j)}$. We then have $x_A^{(j)} < x_1 < x_H^{(j)}$. The intervals $I^*(x_1, x_H^{(j)})$ and $I(x_1, x_A^{(j)})$ do not contain the coordinate of any inverse path of $x = \frac{1}{2}$.*

The left-hand interval $I^*(x_1, x_A^{(j)})$ is of no significance for the limit sets of $T_\lambda(x)$; in fact, for λ a solution of (4.1) this interval shrinks to zero (and for $\lambda^* < \lambda$, neither the harmonic nor the antiharmonic exists). The right-hand, interval, however, is important. Using Lemma 2 and Lemma 1 we immediately derive

LEMMA 3. *If λ_1 is a solution of equation* (4.1) *and λ_H is the solution corresponding to its harmonic, then there does not exist any solution λ^* of* (4.1) *with the property $\lambda_1 < \lambda^* < \lambda_H$.*

Iterating this argument, we verify the statement of Section 4 that the sequence of harmonics is contiguous, i.e., that harmonics are always adjacent.

The adjacency property of harmonics serves to prove Case 1 of Theorem 1. We now proceed to Case 2.

Given some K, let (P_1, x_1, k_1) and (P_2, x_2, k_2) be two adjacent l.i.p.'s with $x_1 < x_2$ and $K + 1 < 2k_1$. Form the H-extension of P_1 and the A-extension of P_2; these can be written in the form

$$\begin{aligned} H(P_1) &= P^* \mu_1 \cdots \\ A(P_2) &= P^* \mu_2 \cdots \end{aligned} \qquad (\mu_1 \neq \mu_2).$$

The coordinates x_H and x_A define an interval I^* which is a projection of $I(x_{\mu_1}, x_{\mu_2})$; clearly, I^* is contained in the original interval $I(x_1, x_2)$. Since I contains the point $x = \frac{1}{2}$, there must exist an inverse path of $\frac{1}{2}$, P*O, with coordinate x^* satisfying $x_1 < x^* < x_2$. But P*O must be an l.i.p. since it is a leading subpattern of the interated harmonic of P_1. Moreover, by the adjacency assumption, its length k^* must necessarily be greater than k_1 or k_2. On the other hand, P*O is the shortest pattern for which an interval with non-zero content exists. Invoking Lemma 1, we see that the proof of the theorem is complete.

The formulation of a practical algorithm, using the results of Theorem 1, needs little comment. Given the complete U-sequence for $k \leqslant K$, one generates the sequence for $K+1$ by inserting the appropriate pattern of length $K+1$ between every two (non-harmonic) adjacent patterns whenever the theorem permits it. The pattern $R(k = 2)$ remains the lowest pattern; as is easily shown, for any k the last pattern is always of the form RL^{k-2}, and this is simply appended to the list. As previously mentioned, the algorithm has been checked (to $k \leqslant 15$) for the four special transformations of Section 3 by actually finding the corresponding solutions of equation (4.1)—a simple process in which there are no serious accuracy limitations.

We remark here that the combinatorial problem of enumerating all l.i.p.'s of a given length k has been solved [5]; the number of patterns turns out to be just the number of symmetry types of primitive periodic sequences (with two "values" or letters allowed) under the cyclic group C_k (so that the full symmetry group is $C_k \times S_2$, where S_2 is the symmetric group on two letters). For k a prime, this number is simple, and turns out to be given by the expression

$$\frac{1}{k}(2^{k-1} - 1).$$

We encountered this enumeration problem previously (cf. reference 4, Table 1); at that time we were not aware of the work of Gilbert and Riordan [5].

6. In this final section we collect some observations and conjectures concerning the nature of limit sets not belonging to the U-sequence, ending with a few remarks on the relation of conjugacy.

(1) *Other finite limit sets.* As remarked in the introduction, it does not seem possible to exclude "anomalous" limit sets without seriously restricting the underlying class of transformations. To convince the reader that such anomalous periods can in fact exist we give a simple example:

Let us alter the special transformation (3.4) in the following way (we take $e = .45$):

$$
\begin{aligned}
x' &= 4.5\lambda x, && 0 \leqslant x \leqslant .2 \\
x' &= \lambda(.4x + .82), && .2 \leqslant x \leqslant .45 \\
x' &= \lambda, && .45 \leqslant x \leqslant .55 \\
x' &= \frac{\lambda}{.45}(1 - x), && .55 \leqslant x \leqslant 1
\end{aligned}
\qquad (.55 < \lambda < 1). \qquad (6.1)
$$

Then, in addition to the U-sequence (with λ values different from those of the original transformation), there exists an attractive 2-period in the range $\lambda_1 < \lambda \leqslant 1$ with

$$\lambda_1 = \tfrac{1}{2} + \tfrac{1}{2}\sqrt{.19}.$$

Note that the 2-period remains attractive even for $\lambda = 1$. While the anomalous periods do not affect the existence of the U-sequence, they do cause additional partitioning of the unit interval because their existence implies that there is a set of points (with non-zero measure) whose sequence of iterated images will converge to the periods in question.

These anomalous periods, however, differ radically from those belonging to the U-sequence in that they do not possess the property (J). This in turn means that the slope at a period point is strongly bounded away from zero. Thus, at least for transformations with the property (C), it seems reasonable to conjecture that such periods cannot have arbitrary length and still remain attractive. Hence we make

CONJECTURE 1. *For transformations with properties* (A), (B) *and* (C), *the anomalous attractive periods constitute at most a finite sequence.*

(2) *Infinite limit sets.* For simplicity we consider the case in which there are no anomalous periods, e.g., functions covered by Julia's theorem (or some valid extension thereof). We assign to each period of the U-sequence a λ-measure equal to its λ-width. The question is then: is the λ-measure of the full U-sequence equal to $\lambda_{max} - \lambda_0$? Or, put otherwise, is there a set of non-zero measure in the interval $(\lambda_0, \lambda_{max})$ such that the sequence of iterates of $x = \frac{1}{2}$ does *not* converge to a member of the U-sequence? Numerical experiments with the four special transformations of Section 3 together with some heuristic arguments based on the iteration of the algorithm of Theorem 1 leads us to make the modest

CONJECTURE 2. *For an infinite subclass of transformations with properties* (A) *and* (B), *the* λ*-measure of the U-sequence is the whole* λ *interval.*

(3) *A limiting case*. Take the transformation $L_\lambda(x; e)$ of Section 3 and set $e = \frac{1}{2}$. We then have

$$L_\lambda(x; \tfrac{1}{2}): \begin{cases} x' = 2\lambda x, & 0 \leqslant x \leqslant \frac{1}{2} \\ x' = 2\lambda(1 - x), & \frac{1}{2} \leqslant x \leqslant 1 \end{cases} \qquad (\tfrac{1}{2} < \lambda < 1). \tag{6.2}$$

Although we cannot speak of attractive periods in this case (since the slope of the function is nowhere less than 1 in absolute value), it is still of interest to investigate the corresponding solutions of equation (4.1). These turn out to be a subset of the U-sequence in which the 2-period, all harmonics, and all patterns algorithmically generated from the harmonics and adjacent nonharmonics, are absent. The count through $k \leqslant 15$ is given in Table III, which may be compared with Table II.

TABLE III

k ...	3	4	5	6	7	8	9	10	11	12	13	14	15
Number of solutions	1	1	3	4	9	14	27	48	93	163	315	576	1085

One can explain this behavior by saying that, as the width 1-2e of the flat-top shrinks to zero, the harmonics and harmonic-generated periods "coalesce" in structure with their fundamentals. This provides another illustration of the nature of the harmonics outlined in Section 4.

(4) *Conjugacy*. Two transformations $f(x)$, $g(x)$ on [0, 1] are said to be conjugate to each other if there exists a continuous, 1-1 mapping $h(x)$ of [0, 1] onto itself such that

$$g(x) = hf[h^{-1}(x)], \qquad x \in [0, 1]. \tag{6.3}$$

If $f(x)$ and $g(x)$ are themselves 1-1, the question of the existence of an $h(x)$ satisfying (6.3) is settled by a theorem of Schreier and Ulam [6]. When $f(x)$, $g(x)$ are not homeomorphisms, very little is known about the existence or nonexistence of a conjugating function $h(x)$.

The importance of (6.3) for our purpose is that the attractive nature of limit sets is preserved under conjugacy; in particular, if $T_\lambda(x)$ possess the U-sequence, then so does every conjugate of it. Clearly, our class of transformations must be invariant under conjugation by the set of all continuous, 1-1 functions $h(x)$ on [0, 1]. (Incidentally, we now see why our special choice of the point $x = \frac{1}{2}$ is no restriction, since it can be shifted by conjugation with an appropriate $h(x)$.)

It has long been known [7] that the parabolic transformation (3.1) with $\lambda = \lambda_{max} = 4$ is conjugate to the broken-linear transformation (6.2) with $\lambda = 1$, the conjugating function being

$$h(x) = \frac{2}{\pi} \sin^{-1} (\sqrt{x}).$$

In general, no such pairwise conjugacy exists for the four special transformations of Section 3. For particular choices of the parameters this can be shown by making the following simple test. If $f(x_0) = x_0$ and $g(x_1) = x_1 (x_0, x_1 \neq 0)$, then a short calculation shows that

$$\left.\frac{df(x)}{dx}\right|_{x=x_0} = \left.\frac{dg(x)}{dx}\right|_{x=x_1}. \tag{6.4}$$

It is easily established that (6.4) does not hold in general for any pair of our special transformations.

In view of the existence of the U-sequence, it is of interest to speculate whether there is not some well-defined but less restrictive equivalence relation that will serve to replace conjugacy (for one such suggestion-due to S. Ulam—see the remarks in reference 1, p. 49). Of course, Theorem 1 itself provides such an equivalence relation, albeit not a very useful one:

Let $T_{1\lambda}(x)$, $T_{2\mu}(x)$ be two transformations with properties (A) and (B). Then there exists a mapping function M_{12} such that $M_{12}(\lambda) = \mu$, the domain of M being the union of the λ-widths of the U-sequence for T_1 and the range being the union of the μ-widths of the U-sequence for T_2.

Since at present nothing whatsoever is known about these mappings M_{ij}, the above correspondence amounts to nothing more than a restatement of the existence of the U-sequence itself.

APPENDIX

The following table gives the complete ordered set of patterns associated with the U-sequence for $K \leqslant 11$; i is a running index, K gives the pattern length, and $I(K)$ indicates the relative order of periods of given length K. The ordering corresponds to the λ-ordering of solutions of equation (4.1).

i	K	I(K)	Pattern	i	K	I(K)	Pattern
1	2	1	R	41	10	9	RL^2RLR^3L
2	4	1	RLR	42	7	3	RL^2RLR
3	8	1	RLR^3LR	43	10	10	$RL^2RLRLRL$
4	10	1	RLR^3LRLR	44	11	15	$RL^2RLRLRLR$
5	6	1	RLR^3	45	9	7	RL^2RLRLR
6	10	2	RLR^5LR	46	11	16	RL^2RLRLR^3
7	8	2	RLR^5	47	10	11	RL^2RLRLR^2
8	10	3	RLR^7	48	11	17	RL^2RLRLR^2L
9	11	1	RLR^8	49	8	5	RL^2RLRL
10	9	1	RLR^6	50	11	18	RL^2RLRL^2RL
11	11	2	RLR^6LR	51	5	2	RL^2R
12	7	1	RLR^4	52	10	12	$RL^2R^3L^2R$
13	11	3	RLR^4LRLR	53	11	19	$RL^2R^3L^2RL$
14	9	2	RLR^4LR	54	8	6	RL^2R^3L
15	11	4	RLR^4LR^3	55	11	20	$RL^2R^3LR^2L$
16	5	1	RLR^2	56	10	13	$RL^2R^3LR^2$
17	10	4	RLR^2LRLR^2	57	11	21	$RL^2R^3LR^3$
18	11	5	RLR^2LRLR^3	58	9	8	RL^2R^3LR
19	9	3	RLR^2LRLR	59	11	22	RL^2R^3LRLR
20	11	6	$RLR^2LRLRLR$	60	10	14	RL^2R^3LRL
21	7	2	RLR^2LR	61	7	4	RL^2R^3
22	11	7	RLR^2LR^3LR	62	10	15	RL^2R^5L
23	9	4	RLR^2LR^3	63	11	23	RL^2R^5LR
24	11	8	RLR^2LR^5	64	9	9	RL^2R^5
25	10	5	RLR^2LR^4	65	11	24	RL^2R^7
26	8	3	RLR^2LR^2	66	10	16	RL^2R^6
27	10	6	RLR^2LR^2LR	67	11	25	RL^2R^6L
28	11	9	$RLR^2LR^2LR^2$	68	8	7	RL^2R^4
29	3	1	RL	69	11	26	RL^2R^4LRL
30	6	2	RL^2RL	70	10	17	RL^2R^4LR
31	9	5	RL^2RLR^2L	71	11	27	$RL^2R^4LR^2$
32	11	10	$RL^2RLR^2LR^2$	72	9	10	RL^2R^4L
33	10	7	RL^2RLR^2LR	73	11	28	$RL^2R^4L^2R$
34	11	11	RL^2RLR^2LRL	74	6	3	RL^2R^2
35	8	4	RL^2RLR^2	75	11	29	$RL^2R^2LRL^2R$
36	11	12	RL^2RLR^4L	76	9	11	RL^2R^2LRL
37	10	8	RL^2RLR^4	77	11	30	$RL^2R^2LRLR^2$
38	11	13	RL^2RLR^5	78	10	18	RL^2R^2LRLR
39	9	6	RL^2RLR^3	79	11	31	RL^2R^2LRLRL
40	11	14	RL^2RLR^3LR	80	8	8	RL^2R^2LR

i	K	I(K)	Pattern	i	K	I(K)	Pattern
81	11	32	$RL^2R^2LR^3L$	121	9	17	RL^3R^3L
82	10	19	$RL^2R^2LR^3$	122	11	50	$RL^3R^3LR^2$
83	11	33	$RL^2R^2LR^4$	123	10	30	RL^3R^3LR
84	9	12	$RL^2R^2LR^2$	124	11	51	RL^3R^3LRL
85	11	34	$RL^2R^2LR^2LR$	125	8	11	RL^3R^3
86	10	20	$RL^2R^2LR^2L$	126	11	52	RL^3R^5L
87	7	5	RL^2R^2L	127	10	31	RL^3R^5
88	10	21	$RL^2R^2L^2RL$	128	11	53	RL^3R^6
89	11	35	$RL^2R^2L^2RLR$	129	9	18	RL^3R^4
90	9	13	$RL^2R^2L^2R$	130	11	54	RL^3R^4LR
91	11	36	$RL^2R^2L^2R^3$	131	10	32	RL^3R^4L
92	10	22	$RL^2R^2L^2R^2$	132	11	55	$RL^3R^4L^2$
93	11	37	$RL^2R^2L^2R^2L$	133	7	7	RL^3R^2
94	4	2	RL^2	134	11	56	$RL^3R^2LRL^2$
95	8	9	RL^3RL^2	135	10	33	RL^3R^2LRL
96	11	38	$RL^3RL^2R^2L$	136	11	57	RL^3R^2LRLR
97	10	23	$RL^3RL^2R^2$	137	9	19	RL^3R^2LR
98	11	39	$RL^3RL^2R^3$	138	11	58	$RL^3R^2LR^3$
99	9	14	RL^3RL^2R	139	10	34	$RL^3R^2LR^2$
100	11	40	RL^3RL^2RLR	140	11	59	$RL^3R^2LR^2L$
101	10	24	RL^3RL^2RL	141	8	12	RL^3R^2L
102	11	41	$RL^3RL^2RL^2$	142	11	60	$RL^3R^2L^2RL$
103	7	6	RL^3RL	143	10	35	$RL^3R^2L^2R$
104	11	42	$RL^3RLR^2L^2$	144	11	61	$RL^3R^2L^2R^2$
105	10	25	RL^3RLR^2L	145	9	20	$RL^3R^2L^2$
106	11	43	RL^3RLR^2LR	146	11	62	$RL^3R^2L^3R$
107	9	15	RL^3RLR^2	147	5	3	RL^3
108	11	44	RL^3RLR^4	148	10	36	RL^4RL^3
109	10	26	RL^3RLR^3	149	11	63	RL^4RL^3R
110	11	45	RL^3RLR^3L	150	9	21	RL^4RL^2
111	8	10	RL^3RLR	151	11	64	$RL^4RL^2R^2$
112	11	46	$RL^3RLRLRL$	152	10	37	RL^4RL^2R
113	10	27	RL^3RLRLR	153	11	65	RL^4RL^2RL
114	11	47	RL^3RLRLR^2	154	8	13	RL^4RL
115	9	16	RL^3RLRL	155	11	66	RL^4RLR^2L
116	11	48	RL^3RLRL^2R	156	10	38	RL^4RLR^2
117	10	28	RL^3RLRL^2	157	11	67	RL^4RLR^3
118	6	4	RL^3R	158	9	22	RL^4RLR
119	10	29	$RL^3R^3L^2$	159	11	68	RL^4RLRLR
120	11	49	$RL^3R^3L^2R$	160	10	39	RL^4RLRL

i	K	I(K)	Pattern	i	K	I(K)	Pattern
161	11	69	RL^4RLRL^2	201	11	89	RL^6R^2L
162	7	8	RL^4R	202	8	16	RL^6
163	11	70	$RL^4R^3L^2$	203	11	90	RL^7RL
164	10	40	RL^4R^3L	204	10	50	RL^7R
165	11	71	RL^4R^3LR	205	11	91	RL^7R^2
166	9	23	RL^4R^3	206	9	28	RL^7
167	11	72	RL^4R^5	207	11	92	RL^8R
168	10	41	RL^4R^4	208	10	51	RL^8
169	11	73	RL^4R^4L	209	11	93	RL^9
170	8	14	RL^4R^2				
171	11	74	RL^4R^2LRL				
172	10	42	RL^4R^2LR				
173	11	75	$RL^4R^2LR^2$				
174	9	24	RL^4R^2L				
175	11	76	$RL^4R^2L^2R$				
176	10	43	$RL^4R^2L^2$				
177	11	77	$RL^4R^2L^3$				
178	6	5	RL^4				
179	11	78	RL^5RL^3				
180	10	44	RL^5RL^2				
181	11	79	RL^5RL^2R				
182	9	25	RL^5RL				
183	11	80	RL^5RLR^2				
184	10	45	RL^5RLR				
185	11	81	RL^5RLRL				
186	8	15	RL^5R				
187	11	82	RL^5R^3L				
188	10	46	RL^5R^3				
189	11	83	RL^5R^4				
190	9	26	RL^5R^2				
191	11	84	RL^5R^2LR				
192	10	47	RL^5R^2L				
193	11	85	$RL^5R^2L^2$				
194	7	9	RL^5				
195	11	86	RL^6RL^2				
196	10	48	RL^6RL				
197	11	87	RL^6RLR				
198	9	27	RL^6R				
199	11	88	RL^6R^3				
200	10	49	RL^6R^2				

References

1. P. R. Stein and S. M. Ulam, Non-linear transformation studies on electronic computers, *Rozprawy Mat.* **39** (1964), 1–66.
2. O. W. Rechard, Invariant measures for many-one transformations, *Duke Math. J.* **23** (1956), 477.
3. G. Julia, Mémoire sur l'itération des functions rationelles, *J. de Math. Ser.* 7, **4** (1918), 47–245. The relevant theorem appears on p. 129 ff.
4. N. Metropolis, M. L. Stein, and P. R. Stein, Stable states of a non-linear transformation, *Numer. Math.* **10** (1967), 1–19.
5. E. N. Gilbert and J. Riordan, Symmetry types of periodic sequences, *Illinois J. Math.* **5** (1961), 657.
6. J. Schreier and S. Ulam, Eine Bemerkung über die Gruppe der topologischen Abbildung der Kreislinie auf sich selbst, *Studia Math.* **5** (1935), 155–159.
7. See reference 1, p. 52. The result is due to S. Ulam and J. von Neumann.

Reprinted from the AMERICAN MATHEMATICAL MONTHLY
Vol. 82, No. 10, December 1975
pp. 985–992

PERIOD THREE IMPLIES CHAOS

TIEN-YIEN LI AND JAMES A. YORKE

1. Introduction. The way phenomena or processes evolve or change in time is often described by differential equations or difference equations. One of the simplest mathematical situations occurs when the phenomenon can be described by a single number as, for example, when the number of children susceptible to some disease at the beginning of a school year can be estimated purely as a function of the number for the previous year. That is, when the number x_{n+1} at the beginning of the $n+1$st year (or time period) can be written

$$x_{n+1} = F(x_n), \tag{1.1}$$

where F maps an interval J into itself. Of course such a model for the year by year progress of the disease would be very simplistic and would contain only a shadow of the more complicated phenomena. For other phenomena this model might be more accurate. This equation has been used successfully to model the distribution of points of impact on a spinning bit for oil well drilling, as mentioned in **[8, 11]**, knowing this distribution is helpful in predicting uneven wear of the bit. For another example, if a population of insects has discrete generations, the size of the $n+1$st generation will be a function of the nth. A reasonable model would then be a generalized logistic equation

$$x_{n+1} = rx_n[1 - x_n/K]. \tag{1.2}$$

A related model for insect populations was discussed by Utida in **[10]**. See also Oster *et al* **[14, 15]**.

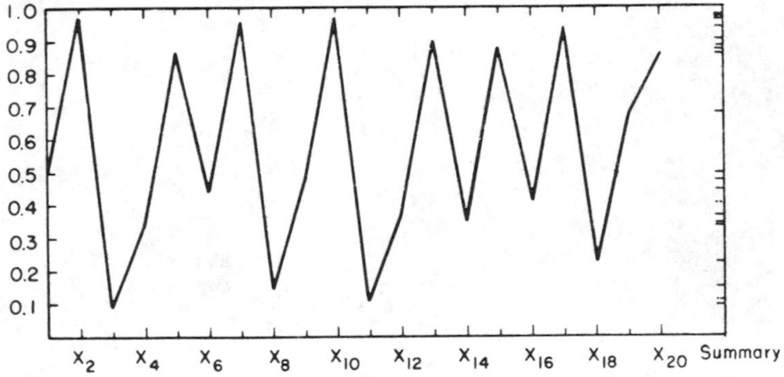

FIG. 1. For $K = 1$, $r = 3.9$, with $x_1 = .5$, the above graph is obtained by iterating Eq. (1.2) 19 times. At right the 20 values are repeated in summary. No value occurs twice. While $x_2 = .975$ and $x_{10} = .973$ are close together, the behavior is not periodic with period 8 since $x_{18} = .222$.

These models are highly simplified, yet even this apparently simple equation (1.2) may have surprisingly complicated dynamic behavior. See Figure 1. We approach these equations with the viewpoint that irregularities and chaotic oscillations of complicated phenomena may sometimes be understood in terms of the simple model, even if that model is not sufficiently sophisticated to allow accurate numerical predictions. Lorenz **[1–4]** took this point of view in studying turbulent behavior in a fascinating series of papers. He showed that a certain complicated fluid flow could be modelled by such a sequence $x, F(x), F^2(x), \cdots$, which retained some of the chaotic aspects of the original flow. See Figure 2. In this paper we analyze a situation in which the sequence $\{F^n(x)\}$ is non-periodic and might be called "chaotic." Theorem 1 shows that chaotic behavior for (1.1) will result in any situation in which a "population" of size x can grow for two or more successive generations and then having reached an unsustainable height, a population bust follows to the level x or below.

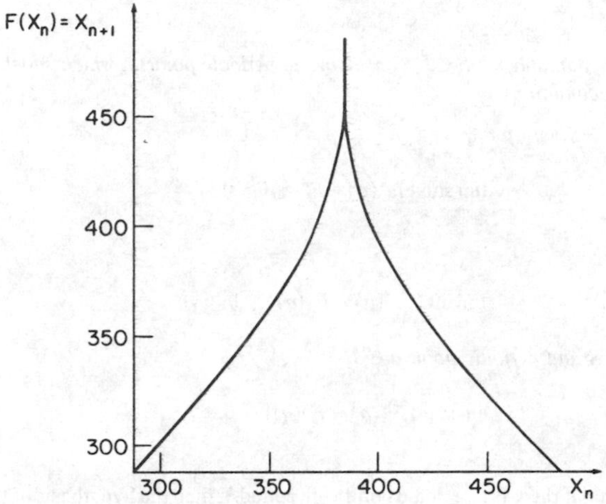

FIG. 2. Lorenz [1] studied the equations for a rotating water-filled vessel which is circularly symmetric about its vertical axis. The vessel is heated near the rim and cooled near its center. When the vessel is annular in shape and the rotation rate high, waves develop and alter their shape irregularly. From a simplified set of equations solved numerically, Lorenz let X_n be in essence the maximum kinetic energy of successive waves. Plotting X_{n+1} against X_n, and connecting the points, the above graph is obtained.

In section 3 we give a well-known simple condition which guarantees that a periodic point is stable and then in section 4 we quote a result applicable when F is like the one in Figure 2. It implies that there is an interval $J_\infty \subset J$ such that for almost every $x \in J$, the set of limit points of the sequence $\{F^n(x)\}$ is J_∞.

A number of questions remain unanswered. For example, is the closure of the periodic points an interval or at least a finite union of intervals? Other questions are mentioned later.

Added in proof. May has recently discovered other strong properties of these maps in his independent study of how the behavior changes as a parameter is varied [17].

2. The main theorem. Let $F: J \to J$. For $x \in J$, $F^0(x)$ denotes x and $F^{n+1}(x)$ denotes $F(F^n(x))$ for $n = 0, 1, \cdots$. We will say p is a **periodic point with period** n if $p \in J$ and $p = F^n(p)$ and $p \neq F^k(p)$ for $1 \leqq k < n$. We say p is **periodic** or is a **periodic point** if p is periodic for some $n \geqq 1$. We say q is **eventually periodic** if for some positive integer m, $p = F^m(q)$ is periodic. Since F need not be one-to-one, there may be points which are eventually periodic but are not periodic. Our objective is to understand the situations in which iterates of a point are very irregular. A special case of our main result says that if there is a periodic point with period 3, then for each integer $n = 1, 2, 3, \cdots$, there is a periodic point with period n. Furthermore, there is an uncountable subset of points x in J which are not even "asymptotically periodic."

THEOREM 1. *Let J be an interval and let $F: J \to J$ be continuous. Assume there is a point $a \in J$ for which the points $b = F(a)$, $c = F^2(a)$ and $d = F^3(a)$, satisfy*

$$d \leqq a < b < c \text{ (or } d \geqq a > b > c).$$

Then

T1: *for every $k = 1, 2, \cdots$ there is a periodic point in J having period k.*

Furthermore,

T2: *there is an uncountable set $S \subset J$ (containing no periodic points), which satisfies the following conditions:*

(A) *For every $p, q \in S$ with $p \neq q$,*

$$\limsup_{n \to \infty} |F^n(p) - F^n(q)| > 0 \tag{2.1}$$

and

$$\liminf_{n \to \infty} |F^n(p) - F^n(q)| = 0. \tag{2.2}$$

(B) *For every $p \in S$ and periodic point $q \in J$,*

$$\limsup_{n \to \infty} |F^n(p) - F^n(q)| > 0.$$

REMARKS. Notice that if there is a periodic point with period 3, then the hypothesis of the theorem will be satisfied.

An example of a function satisfying the hypotheses of the theorem is $F(x) = rx[1 - x/K]$ as in (1.2) for $r \in (3.84, 4]$ with $J = [0, K]$ and for $r > 4$, $F(x) = \max\{0, rx[1 - x/K]\}$ with $J = [0, K]$. See **[2]** for a detailed description of iterates of this function for $r \in [0, 4)$. The case $r = 4$ is discussed in **[6, 7, 12]**.

While the existence of a point of period 3 implies the existence of one of period 5, the converse is false. (See Appendix 1).

We say $x \in J$ is **asymptotically periodic** if there is a periodic point p for which

$$F^n(x) - F^n(p) \to 0 \quad \text{as} \quad n \to \infty. \tag{2.3}$$

It follows from (B) that the set S contains no asymptotically periodic points. We remark that it is unknown what the infimum of r is for which the equation (1.2) has points which are not asymptotically periodic.

Proof of Theorem 1. The proof of T1 introduces the main ideas for both T1 and T2. We now give the proof of T1 with necessary lemmas and relegate the tedious proof of T2 to Appendix 2.

LEMMA 0. *Let $G: I \to R$ be continuous, where I is an interval. For any compact interval $I_1 \subset G(I)$ there is a compact interval $Q \subset I$ such that $G(Q) = I_1$.*

Proof. Let $I_1 = [G(p), G(q)]$, where $p, q \in I$. If $p < q$, let r be the last point of $[p, q]$ where $G(r) = G(p)$ and let s be the first point after r where $G(s) = G(q)$. Then $G([r, s]) = I_1$. Similar reasoning applies when $p > q$.

LEMMA 1. *Let $F: J \to J$ be continuous and let $\{I_n\}_{n=0}^{\infty}$ be a sequence of compact intervals with $I_n \subset J$ and $I_{n+1} \subset F(I_n)$ for all n. Then there is a sequence of compact intervals Q_n such that $Q_{n+1} \subset Q_n \subset I_0$ and $F^n(Q_n) = I_n$ for $n \geqq 0$. For any $x \in Q = \cap Q_n$ we have $F^n(x) \in I_n$ for all n.*

Proof. Define $Q_0 = I_0$. Then $F^0(Q_0) = I_0$. If Q_{n-1} has been defined so that $F^{n-1}(Q_{n-1}) = I_{n-1}$, then $I_n \subset F(I_{n-1}) = F^n(Q_{n-1})$. By Lemma 0 applied to $G = F^n$ on Q_{n-1} there is a compact interval $Q_n \subset Q_{n-1}$ such that $F^n(Q_n) = I_n$. This completes the induction.

The technique of studying how certain sequences of sets are mapped into or onto each other is often used in studying dynamical systems. For instance, Smale uses this method in his famous "horseshoe example" in which he shows how a homeomorphism on the plane can have infinitely many periodic points [13].

LEMMA 2. *Let $G: J \to R$ be continuous. Let $I \subset J$ be a compact interval. Assume $I \subset G(I)$. Then there is a point $p \in I$ such that $G(p) = p$.*

Proof. Let $I = [\beta_0, \beta_1]$. Choose $\alpha_i (i = 0, 1)$ in I such that $G(\alpha_i) = \beta_i$. It follows $\alpha_0 - G(\alpha_0) \geqq 0$ and $\alpha_1 - G(\alpha_1) \leqq 0$ and so continuity implies $G(\beta) - \beta$ must be 0 for some β in I.

Assume $d \leqq a < b < c$ as in the theorem. The proof for the case $d \geqq a > b > c$ is similar and so is omitted. Write $K = [a, b]$ and $L = [b, c]$.

Proof of T1: Let k be a positive integer. For $k > 1$ let $\{I_n\}$ be the sequence of intervals $I_n = L$ for $n = 0, \cdots, k-2$ and $I_{k-1} = K$, and define I_n to be periodic inductively, $I_{n+k} = I_n$ for $n = 0, 1, 2, \cdots$. If $k = 1$, let $I_n = L$ for all n.

Let Q_n be the sets in the proof of Lemma 1. Then notice that $Q_k \subset Q_0$ and $F^k(Q_k) = Q_0$ and so by Lemma 2, $G = F^k$ has a fixed point p_k in Q_k. It is clear that p_k cannot have period less than k for F; otherwise we would need to have $F^{k-1}(p_k) = b$, contrary to $F^{k+1}(p_k) \in L$. The point p_k is a periodic point of period k for F.

3. Behavior near a periodic point. For some functions F, the asymptotic behavior of iterates of a point can be understood simply by studying the periodic points. For

$$F(x) = ax(1-x) \tag{3.1}$$

a detailed discussion of the points of period 1 and 2 may be found in [1] for $a \in [0, 4]$ and we now summarize some of those results. For $a \in [0, 4]$, F: $[0, 1] \to [0, 1]$.

For $a \in [0, 1]$, $x = 0$ is the only point of period 1; in fact, for $x \in [0, 1]$, the sequence $F^n(x) \to 0$ as $n \to \infty$.

For $a \in (1, 3]$, there are two points of period 1, namely 0 and $1 - a^{-1}$, and for $x \in (0, 1)$, $F^n(x) \to 1 - a^{-1}$ as $n \to \infty$.

For $a > 3$ there are also two points of period 2 which we may call p and q and of course $F(p) = q$ and $F(q) = p$. For $a \in (3, 1 + \sqrt{6} \approx 3.449)$ and $x \in (0, 1)$, $F^{2n}(x)$ converges to either p or q while $F^{2n+1}(x)$ converges to the other, except for those x for which there is an n for which $F^n(x)$ equals the point $1 - a^{-1}$ of period 1. There are only a countable number of such points so that the behavior of $\{F^n(x)\}$ can be understood by studying the periodic points.

For $a > 1 + \sqrt{6}$, there are 4 points of period 4 and for a slightly greater than $1 + \sqrt{6}$, $F^{4n}(x)$ tends to one of these 4 unless for some n, $F^n(x)$ equals one of the points of period 1 or 2. Therefore we may summarize this situation by saying that each point in [0, 1] is asymptotically periodic.

For those values of a for which each point is asymptotically periodic, it is sufficient to study only the periodic points and their "stability properties." For any function F a point $y \in J$ with period k is said to be **asymptotically stable** if for some interval $I = (y - \delta, y + \delta)$ we have

$$|F^k(x) - y| < |x - y| \qquad \text{for all} \quad x \in I.$$

If F is differentiable at the points $y, F(y), \cdots, F^{k-1}(y)$, there is a simple condition that will guarantee this behavior, namely

$$\left| \frac{d}{dx} F^k(x) \right| < 1.$$

By the chain rule

$$\frac{d}{dx}F^k(y)=\frac{d}{dx}F(F^{k-1}(y))\cdot\frac{d}{dx}F^{k-1}(y)$$

$$=\frac{d}{dx}F(F^{k-1}(y))\times\frac{d}{dx}F(F^{k-2}(y))\times\cdots\times\frac{d}{dx}F(y) \tag{3.2}$$

$$=\prod_{n=0}^{k-1}\frac{d}{dx}F(y_n),$$

where y_n is the nth iterate, $F^n(y)$. Therefore y is asymptotically stable if

$$\left|\prod_{i=0}^{k-1}\frac{d}{dx}F(y_i)\right|<1,\qquad\text{where}\qquad y_i=F^i(y).$$

This condition of course guarantees nothing about the limiting behavior of points which do not start "near" the periodic point or one of its iterates. The function in Figure 2 which was studied by Lorenz has the opposite behavior, namely, where the derivative exists we have

$$\left|\frac{d}{dx}F(x)\right|>1.$$

For such a function every periodic point is "unstable" since for x near a periodic point y of period k, the kth iterate $F^k(x)$ is further from y than x is. To see this, approximate $F^k(x)$ by

$$F^k(y)+\frac{d}{dx}F^k(y)[y-x]=y+\frac{d}{dx}F^k(y)[y-x].$$

Thus for x near y, $|F^k(x)-y|$ is approximately $|x-y|\,|(d/dx)F^k(y)|$. From (3.2) $|(d/dx)F^k(y)|$ is greater than 1. Therefore $F^n(x)$ is further from y than x is.

We do not know when values of a begin to occur for which F in (3.1) has points which are not asymptotically periodic. For $a=3.627$, F has a periodic point (which is asymptotically stable) of period 6 (approx. $x=.498$). This x is therefore a point of period 3 for F^2 and so Theorem 1 may be applied to F^2. Since F^2 has points which are not asymptotically periodic, the same is true of F.

In order to contrast the situations in this section with other possible situations discussed in the next section, we define the limit set of a point x. The point y is a **limit point** of a sequence $\{x_n\}\subset J$ if there is a subsequence x_{n_i} converging to y. The **limit set** $L(x)$ is defined to be the set of limit points of $\{F^n(x)\}$. If x is asymptotically periodic, then $L(x)$ is the set $\{y, F(y),\cdots,F^{k-1}(y)\}$ for some periodic point y of period k.

4. Statistical properties of $\{F^n(x)\}$. Theorem 1 establishes the irregularity of the behavior of iterates of points. What is also needed is a description of the regular behavior of the sequence $\{F^n(x)\}$ when F is piecewise continuously differentiable (as is Lorenz's function in Figure 2) and

$$\inf_{x\in J_1}\left|\frac{dF}{dx}\right|>1\qquad\text{where}\qquad J_1=\left\{x:\frac{dF}{dx}\text{ exists}\right\}. \tag{4.1}$$

One approach to describing the asymptotic behavior for such functions is to describe $L(x)$, if possible. A second approach, which turns out to be related, is to examine the average behavior of $\{F^n(x)\}$. The fraction of the iterates $\{x,\cdots,F^{N-1}(x)\}$ of x that are in $[a_1,a_2]$ will be denoted by $\phi(x,N,[a_1,a_2])$. The limiting fraction will be denoted

$$\phi(x,[a_1,a_2])=\lim_{N\to\infty}\phi(x,N,[a_1,a_2])$$

when the limit exists. The subject of ergodic theory, which studies transformation on general spaces, motivates the following definition. We say g is the **density of** x (for F) if the limiting fraction satisfies

$$\phi(x,[a_1,a_2]) = \int_{a_1}^{a_2} g(x)dx \quad \text{for all} \quad a_1, a_2 \in J; \quad a_1 < a_2.$$

The techniques for the study of densities use non-elementary techniques of measure theory and functional analysis, so that we shall only summarize the results. But their value lies in the fact that for certain F almost all $x \in J$ have the same density. Until recently the existence of such densities had not been proved, except for the simplest of functions F. The following result has recently been proved:

THEOREM 2. **[5]**. *Let $F: J \to J$ satisfy the following conditions:*
1) *F is continuous.*
2) *Except at one point $t \in J$, F is twice continuously differentiable.*
3) *F satisfies* (4.1).

Then there exists a function $g: J \to [0,\infty)$, such that for almost all $x \in J$, g is the density of x. Also for almost all $x \in J$, $L(x) = \{y: g(y) > 0\}$ which is an interval. Moreover, the set $J_\infty = \{y: g(y) > 0\}$ is an interval, and $L(x) = J_\infty$ for almost all x.

The proof makes use of results in **[8]**. The problem of computationally finding the density is solved in **[9]**.

A detailed discussion of (3.1) is given in **[16]**, describing how $L(x)$ varies as the parameter a in (3.1) varies between 3.0 and 4.0.

A major question left unsolved is whether (for some nice class of functions F) the existence of a stable periodic point implies that almost every point is asymptotically periodic.

Appendix 1: Period 5 does not imply period 3. In this Appendix we give an example which has a fixed point of period 5 but no fixed point of period 3.

Let $F: [1,5] \to [1,5]$, be defined such that $F(1) = 3$, $F(2) = 5$, $F(3) = 4$, $F(4) = 2$, $F(5) = 1$ and on each interval $[n, n+1]$, $1 \leqq n \leqq 4$, assume F is linear. Then

$$F^3([1,2]) = F^2([3,5]) = F([1,4]) = [2,5].$$

Hence, F^3 has no fixed points in $[1,2]$. Similarly, $F^3([2,3]) = [3,5]$ and $F^3([4,5]) = [1,4]$, so neither of these intervals contains a fixed point of F^3. On the other hand,

$$F^3([3,4]) = F^2([2,4]) = F([2,5]) = [1,5] \supset [3,4].$$

Hence, F^3 must have a fixed point in $[3,4]$. We shall now demonstrate that the fixed point of F^3 is unique and is also a fixed point of F.

Let $p \in [3,4]$ be a fixed point of F^3. Then $F(p) \in [2,4]$. If $F(p) \in [2,3]$, then $F^3(p)$ would be in $[1,2]$ which is impossible since then p could not be a fixed point. Hence $F(p) \in [3,4]$ and $F^2(p) \in [2,4]$. If $F^2(p) \in [2,3]$ we would have $F^3(p) \in [4,5]$, an impossibility. Hence p, $F(p)$, $F^2(p)$ are all in $[3,4]$. On the interval $[3,4]$, F is defined linearly and so $F(x) = 10 - 2x$. It has a fixed point 10/3 and it is easy to see that F^3 has a unique fixed point, which must be 10/3. Hence there is no point of period 3.

Appendix 2. Proof of T2 of Theorem 1. Let $\mathcal{M}$ be the set of sequences $M = \{M_n\}_{n=1}^\infty$ of intervals with

(A.1) $$M_n = K \quad \text{or} \quad M_n \subset L, \quad \text{and} \quad F(M_n) \supset M_{n+1}$$

$$\text{if} \quad M_n = K \quad \text{then}$$

(A.2) $$n \text{ is the square of an integer and } M_{n+1}, M_{n+2} \subset L,$$

where $K = [a, b]$ and $L = [b, c]$. Of course if n is the square of an integer, then $n + 1$ and $n + 2$ are not, so the last requirement in (A.2) is redundant. For $M \in \mathcal{M}$, let $P(M, n)$ denote the number of i's in $\{1, \cdots, n\}$ for which $M_i = K$. For each $r \in (3/4, 1)$ choose $M^r = \{M_n^r\}_{n=1}^{\infty}$ to be a sequence in $\mathcal{M}$ such that

(A.3) $$\lim_{n\to\infty} P(M^r, n^2)/n = r.$$

Let $M_0 = \{M^r : r \in (3/4, 1)\} \subset M$. Then $\mathcal{M}_0$ is uncountable since $M^{r_1} \neq M^{r_2}$ for $r_1 \neq r_2$. For each $M^r \in \mathcal{M}_0$, by Lemma 1, there exists a point x_r with $F^n(x_r) \in M_n^r$ for all n. Let $S = \{x_r : r \in (3/4, 1)\}$. Then S is also uncountable. For $x \in S$, let $P(x, n)$ denote the number of i's in $\{1, \cdots, n\}$ for which $F^i(x) \in K$. We can never have $F^k(x_r) = b$, because then x_r would eventually have period 3, contrary to (A.2). Consequently $P(x_r, n) = P(M^r, n)$ for all n, and so

$$\rho(x_r) = \lim_{n\to\infty} P(X_r, n^2) = r$$

for all r. We claim that

(A.4) for $p, q \in S$, with $p \neq q$, there exist infinitely many n's such that $F^n(p) \in K$ and $F^n(q) \in L$ or vice versa.

We may assume $\rho(p) > \rho(q)$. Then $P(p, n) - P(q, n) \to \infty$, and so there must be infintely many n's such that $F^n(p) \in K$ and $F^n(q) \in L$.

Since $F^2(b) = d \leqq a$ and F^2 is continuous, there exists $\delta > 0$ such that $F^2(x) < (b + d)/2$ for all $x \in [b - \delta, b] \subset K$. If $p \in S$ and $F^n(p) \in K$, then (A.2) implies $F^{n+1}(p) \in L$ and $F^{n+2}(p) \in L$. Therefore $F^n(p) < b - \delta$. If $F^n(q) \in L$, then $F^n(q) \geqq b$ so

$$|F^n(p) - F^n(q)| > \delta.$$

By claim (A.4), for any $p, q \in S$, $p \neq q$, it follows

$$\limsup_{n\to\infty} |F^n(p) - F^n(q)| \geqq \delta > 0.$$

Hence (2.1) is proved. This technique may be similarly used to prove (B) is satisfied.

Proof of 2.2. Since $F(b) = c$, $F(c) = d \leqq a$, we may choose intervals $[b^n, c^n]$, $n = 0, 1, 2, \cdots$, such that

(a) $[b, c] = [b^0, c^0] \supset [b^1, c^1] \supset \cdots \supset [b^n, c^n] \supset \cdots$,
(b) $F(x) \in (b^n, c^n)$ for all $x \in (b^{n+1}, c^{n+1})$,
(c) $F(b^{n+1}) = c^n$, $F(c^{n+1}) = b^n$.

Let $A = \bigcap_{n=0}^{\infty} [b^n, c^n]$, $b^* = \inf A$ and $c^* = \sup A$, then $F(b^*) = c^*$ and $F(c^*) = b^*$, because of (c).

In order to prove (2.2) we must be more specific in our choice of the sequences M^r. In addition to our previous requirements on $M \in M$, we will assume that if $M_k = K$ for both $k = n^2$ and $(n + 1)^2$ then $M_k = [b^{2n-(2j-1)}, b^*]$ for $k = n^2 + (2j - 1)$, $M_k = [c^*, c^{2n-2j}]$ for $k = n^2 + 2j$ where $j = 1, \cdots, n$. For the remaining k's which are not squares of integers, we assume $M_k = L$.

It is easy to check that these requirements are consistent with (A.1) and (A.2), and that we can still choose M^r so as to satisfy (A.3). From the fact that $\rho(x)$ may be thought of as the limit of the fraction of n's for which $F^{n^2}(x) \in K$, it follows that for any $r^*, r \in (3/4, 1)$ there exist infinitely many n such that $M_k^r = M_k^{r^*} = K$ for both $k = n^2$ and $(n + 1)^2$. To show (2.2), let $x_r \in S$ and $x_{r^*} \in S$. Since $b^n \to b^*$, $c^n \to c^*$ as $n \to \infty$, for any $\varepsilon > 0$ there exists N with $|b^n - b^*| < \varepsilon/2$, $|c^n - c^*| < \varepsilon/2$ for all $n > N$. Then, for any n with $n > N$ and $M_k^r = M_k^{r^*} = K$ for both $k = n^2$ and $(n + 1)^2$, we have

$$F^{n^2+1}(x_r) \in M_k^r = [b^{2n-1}, b^*]$$

with $k = n^2 + 1$ and $F^{n^2+1}(x_r)$ and $F^{n^2+1}(x_{r*})$ both belong to $[b^{2n-1}, b^*]$. Therefore, $|F^{n^2+1}(x_r) - F^{n^2+1}(x_{r*})| < \varepsilon$. Since there are infinitely many n with this property, $\liminf_{n\to\infty} |F^n(x_r) - F^n(x_{r*})| = 0$. □

REMARK. The theorem can be generalized by assuming that $F: J \to R$ without assuming that $F(J) \subset J$ and we leave this proof to the reader. Of course $F(J) \cap J$ would be nonempty since it would contain the points a, b, and c, assuming that b, c, and d, are defined.

Research partially supported by National Science Foundation grant GP-31386X.

References

1. E. N. Lorenz, The problem of deducing the climate from the governing equations, Tellus, 16 (1964) 1–11.
2. ———, Deterministic nonperiodic flows, J. Atmospheric Sci., 20 (1963) 130–141.
3. ———, The mechanics of vacillation, J. Atmospheric Sci., 20 (1963) 448–464.
4. ———, The predictability of hydrodynamic flow, Trans. N.Y. Acad. Sci., Ser. II, 25 (1963) 409–432.
5. T. Y. Li and J. A. Yorke, Ergodic transformations from an interval into itself, (submitted for publication).
6. P. R. Stein and S. M. Ulam, Nonlinear transformation studies on electronic computers, Los Alamos Sci. Lab., Los Alamos, New Mexico, 1963.
7. S. M. Ulam, A Collection of Mathematical Problems, Interscience, New York, 1960, p. 150.
8. A. Lasota and J. A. Yorke, On the existence of invariant measures for piecewise monotonic transformations, Trans. Amer. Math. Soc., 186 (1973) 481–488.
9. T. Y. Li, Finite approximation for the Frobenius-Perron operator — A Solution to Ulam's conjecture, (submitted for publication).
10. Syunro Utida, Population fluctuation, an experimental and theoretical approach, Cold Spring Harbor Symposia on Quantitative Biology, 22 (1957) 139–151.
11. A. Lasota and P. Rusek, Problems of the stability of the motion in the process of rotary drilling with cogged bits, Archium Gornictua, 15 (1970) 205–216 (Polish with Russian and German Summary).
12. N. Metropolis, M. L. Stein and P. R. Stein, On infinite limit sets for transformations on the unit interval, J. Combinatorial Theory Ser. A 15, (1973) 25–44.
13. S. Smale, Differentiable dynamical systems, Bull. A.M.S., 73 (1967) 747–817 (see §1.5).
14. G. Oster and Y. Takahashi, Models for age specific interactions in a periodic environment, Ecology, in press.
15. D. Auslander, G. Oster and C. Huffaker, Dynamics of interacting populations, a preprint.
16. T. Y. Li and James A. Yorke, The "simplest" dynamics system, (to appear).
17. R. M. May, Biological populations obeying difference equations, stable cycles, and chaos, J. Theor. Biol. (to appear).

DEPARTMENT OF MATHEMATICS, UNIVERSITY OF UTAH, SALT LAKE CITY, UT 84112.
INSTITUTE FOR FLUID DYNAMICS AND APPLIED MATHEMATICS, UNIVERSITY OF MARYLAND, COLLEGE PARK, MD 20742.

Ann. Inst. Henri Poincaré,
Vol. XXIX, n° 3, 1978, p. 305-356.

Section A :
Physique théorique.

Iteration of endomorphisms on the real axis and representation of numbers

by

B. DERRIDA (*), A. GERVOIS, Y. POMEAU
C. E. N. Saclay, B. P. 2, 91190 Gif-sur-Yvette, France

ABSTRACT. — We study a class of endomorphisms of the set of real numbers x of the form: $x \to \lambda f(x)$, $x \in [0, 2]$. The function f is continuous, convex with a single maximum but otherwise arbitrary; λ is a real parameter.

We focus our attention on periodic points: $x \in [0, 2]$ is periodic if there exists an integer n such that the nth iterate of x by λf coincides with x. Because of their special importance, we restrict ourselves to periods involving the maximum.

As shown by Metropolis *et al.*, for each mapping, one may represent in a non ambiguous way these periods by finite sequences of symbols R and L [the ith iterate of the maximum is represented by R (right) or L (left) depending on its position relatively to maximum] and these sequences have many universal properties. For example they can be ordered in a way which does not depend on details of the mapping.

In this paper, we prove two points:

i) the ordered set of all the symbolic sequences possesses a property of internal similarity: it is possible to find a monotonous application of the whole set into one of its subsets;

ii) we give a simple criterion for recognizing whether a sequence is allowed or not and to know in which order two given sequences appear.

For reasons of universality property, it will be sufficient to derive these results for the simplest case, namely the « linear » transform, i. e.

$$\begin{aligned} g(x) &= x, & 0 < x < 1 \\ &= 2 - x, & 1 < x < 2. \end{aligned}$$

We are led to define an expansion of real numbers analogous to β-expansion of Renyi. A few other properties of this peculiar case are briefly discussed.

(*) Institut von Laue-Langevin, B. P. 156, 38042 Grenoble Cedex

RÉSUMÉ. — On étudie une classe d'endomorphismes de l'ensemble des nombres réels x, de la forme

$$x \to \lambda f(x), x \in [0, 2].$$

La fonction f est continue, convexe avec un seul maximum mais par ailleurs arbitraire; λ est un paramètre réel.

Nous nous intéressons surtout aux points périodiques: $x \in [0,2]$ est périodique s'il existe un entier n tel que le $n^{\text{ième}}$ itéré de x par λf coïncide avec x. En raison de leur importance particulière, nous nous limitons aux périodes comprenant le maximum.

Comme l'ont montré Métropolis et ses collaborateurs, ces périodes peuvent être représentées de manière non ambiguë par des séquences finies de symboles R et L [le $i^{\text{ème}}$ itéré du sommet est représenté par R (« right ») ou L (« left ») suivant sa position relative par rapport au maximum] et ces séquences ont des propriétés universelles. Par exemple, on peut les ordonner d'une manière qui ne dépend pas de l'application.

Dans cet article, on clarifie 2 points :

i) l'ensemble ordonné de toutes ces séquences symboliques présente une propriété d'homothétie interne : on peut trouver une application monotone de tout l'ensemble dans l'une de ses parties,

ii) on donne un critère simple pour décider si une séquence est autorisée et pour savoir dans quel ordre relatif deux séquences apparaissent.

En raison de l'universalité, il suffit de montrer ces propriétés pour le cas le plus simple, l'application « linéaire »

$$\begin{aligned} g(x) &= x, & 0 < x < 1 \\ &= 2 - x, & 1 < x < 2. \end{aligned}$$

On est amené à définir un développement des nombres réels semblables au β-développement de Renyi. Quelques propriétés supplémentaires liées à ce cas particulier sont également discutées.

CONTENTS

1. INTRODUCTION

Let f be a real valued function defined for $0 \leq x \leq 2$ which satisfies the following requirements

i) f is continuous and differentiable in [0,2]
ii) $f(0) = f(2) = 0$ (1.1)
iii) f has a unique maximum at $x = c$, $0 < c < 2$. $f(x) < f(c)$, if $x \neq c$. $f'(x) > 0$ if $x < c$ and $f'(x) < 0$ if $x > c$. (see fig. 1).

Let λ be a real number $0 \leq \lambda \leq 2/f(c)$. Consider the transformation

$$T_\lambda(x) \equiv \lambda f(x) \tag{1.2}$$

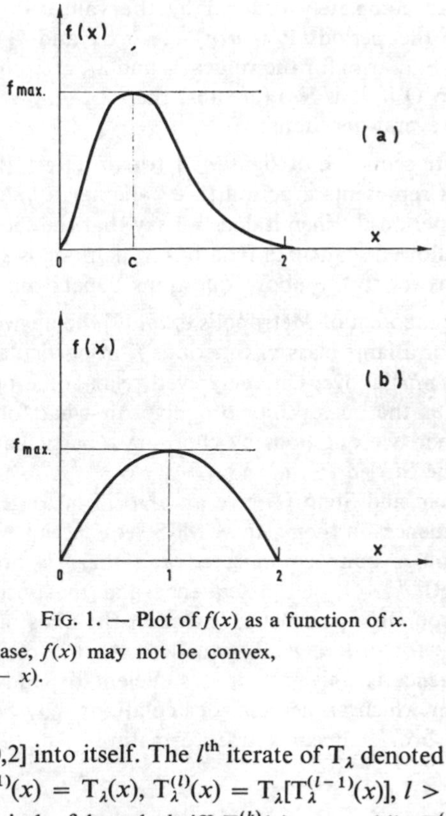

FIG. 1. — Plot of $f(x)$ as a function of x.
a) in the general case, $f(x)$ may not be convex,
b) for $f(x) = x(2 - x)$.

of the interval [0,2] into itself. The lth iterate of T_λ denoted as $T_\lambda^{(l)}$, is defined by recursion, $T_\lambda^{(1)}(x) = T_\lambda(x)$, $T_\lambda^{(l)}(x) = T_\lambda[T_\lambda^{(l-1)}(x)]$, $l > 1$. The number x belongs to a period of length k iff $T_\lambda^{(k)}(x) = x$ while $T_\lambda^{(l)}(x) \neq x$, $l < k$, and, given x, this will happen only for particular values of λ. The periods containing c and the corresponding values λ are of special importance and we will consider only these in what follows. Such a period may be represented by a finite sequence of signs ± 1 (or of letters R and L) as follows. Without loss of generality we start with c. If $T_\lambda(c) = c$, then c belongs to a period of length 1 and we denote this period by a blank (or 0 or the letter *b*). If $T_\lambda(c) \neq c$ and c belongs to a period of length $k > 1$, then $T_\lambda^{(k)}(c) - c = 0$, while $T_\lambda^{(l)}(c) - c \neq 0$ if $1 \leq l \leq k - 1$. The period (or the values of λ for which this period exists) is represented by the $k - 1$ signs of the differences $c - T_\lambda^{(l)}(c)$ (or replace each $+ 1$ by an L and each $- 1$ by an R). Thus for example, if $T_\lambda(c) > c$, $T_\lambda^{(2)}(c) < c$ and $T_\lambda^{(3)}(c) = c$, then c belongs to the period $(- 1, + 1)$ (or RL).

Note that if $T_\lambda(c) < c$, then $T_\lambda^{(l)}(c) = c$, $l \geq 1$ is impossible under the conditions on f. Thus every period starts with a $- 1$ (or the letter R). Also not every finite sequence of signs will represent a period.

The periods are completely ordered by the values of λ corresponding to them. Thus if the periods $P = \sigma_1\sigma_2 \ldots \sigma_{k-1}$ and $Q = \tau_1\tau_2 \ldots \tau_{j-1}$, $\sigma_i = \pm 1, \tau_l = \pm 1$, arise for the values λ_P and λ_Q of λ, then P is said to be less (greater) than Q if λ_P is less (greater) than λ_Q.

One may ask several questions:

1. Given a finite sequence of signs ± 1 (or of letters R and L), how to decide whether it represents a period (i. e. whether a value of λ exists for which this is the period)? When it does we say that the sequence is allowed.
2. Given two allowed sequences P and Q, which one is smaller?
3. Does the answer to the above questions depend on the details of f?

According to a theorem of Metropolis *et al.* [*1*] the answer to question (3) above is « no » for a large class of functions f, in particular for those satisfying conditions *i*)-*iii*) above. One may even relax some of them. We refer to this property as the universality property. In what follows we will try to answer the other two questions by choosing a particular f for which all calculations can be carried to the end.

Metropolis, Stein and Stein [*1*] give an algorithm to get all the allowed sequences (U sequences in their paper, MSS sequences here).

The prescription is quite complicated and there is no simple relation between the length $k - 1$ of the sequence and the position on the real axis of the corresponding parameter λ. One of the aims of this article is to give a simple rule for ordering all sequences. As the order of occurrence of the MSS sequence is universal, it is sufficient to study it on a special transformation for which all necessary calculations may be done. We have considered the « broken linear » transformation

$$L_\lambda(x) = \lambda g(x) \quad 1 < \lambda < 2 \tag{1.3}$$

where $g(x)$ is the function

$$\begin{aligned} g(x) &= x & 0 < x < 1 \\ &= 2 - x & 1 < x < 2. \end{aligned} \tag{1.4}$$

Obviously, $g(x)$ is not differentiable at $x = 1$ and condition *i*) is violated. Nevertheless, the order of the periods is the same as for functions f satisfying conditions (1.1) except that some of them (mostly « the harmonics ») are absent [*1*]. It is easy to see which ones and why and then to reconstruct the whole set of MSS sequences.

As shown in section 4, many properties of these sequences are connected with a representation of the real numbers, which we have called the λ-expansion by reference to the β-expansion of Renyi [*2*]-[*3*]. This λ-expansion is defined as follows. Given λ, $1 < \lambda < 2$, the λ-expansion of a number x $1 < x < 2$ in the basis λ is represented by the sequence $c_0, c_1, c_2, \ldots$ such that

i) $c_0 = 1, c_n = \pm 1$ or 0 for $n \geq 1$

ii) let $x_n = \sum_{k=0}^{n} \frac{c_k}{\lambda^k}$; then $x_n < x \Rightarrow c_{n+1} = +1$

$$x_n > x \Rightarrow c_{n+1} = -1$$

and $x_n = x \Rightarrow c_{n'} = 0$ for $n' > n$.

This is an expansion of the number x, as $|x - x_m| < 1/\lambda^m$. The digits used in a λ-expansion are $+1$ or -1, although the digits used in a β-expansion are 0 or 1 for $1 < \beta \leq 2$ (The digit 0 does not really exist in the λ-expansion, as it cannot be inserted between two non zero digits, it just marks the end of a λ-expansion when it is finite).

A particular case of λ-expansion is the so-called auto-expansion of the basis, that is the set $\{ c_n \}$ constructed by choosing for λ and x the same number. The « digits » of these auto-expansions are simply related with the symbolic sequences of Metropolis *et al.* This connection becomes clear when one considers the particular « broken linear » transform of eqs. (1.3)-(1.4).

More precisely, in sections 3 to 5, we derive:

i) a criterion indicating whether a given sequence of symbols R and L (in the sense of Metropolis *et al.*) does or does not correspond to a period (in other term, whether this sequence is « allowed » or not);

ii) the ordering criterion for the periods;

iii) the Sarkovskii theorem [*4-5-6-7*];

iv) some complementary properties of the λ-numbers related to the broken linear mapping L_λ.

In part 2, we show that the whole set of MSS sequences is similar to some parts of it. This property is called internal similarity and is proved just by using the algorithm of construction given by Metropolis *et al.* After recalling briefly the main results of ref. [*1*], we display the mapping which manifests the internal similarity property. The sequences given by the construction of Metropolis *et al.* is countable and fully ordered. Thus, one may define accumulation points for this set. Using the law of internal similarity we may show that the set of accumulation points has the power of the continuum.

2. INTERNAL SIMILARITY

In this section, we show that the family of MSS sequences is similar to some of its parts. This result does not depend on the particular expression of the function $f(x)$ of eqs. (1.1-1.2).

After recalling some notations and defining the ordering of the sequences, we build recursively all the MSS sequences (subsection 2.1).

It is then possible to define mappings of this set of sequences into itself which preserve the ordering relation. We exhibit a particular class of them which reduces the domain of values for the parameter λ (subsection 2.2); we get thus the principle of internal similarity.

At the end of the section, we study other properties of this mapping considered as a composition law; one consequence of its existence is that the accumulation points of the parameter set associated to the MSS family of sequences has the power of the continuum.

2.1. The construction of Metropolis *et al.*

We intend to order the values of λ for which MSS sequences exist i. e. values of λ for which

$$T_\lambda^{(k)}(c) = c \tag{2.1 a}$$

with

$$T_\lambda^{(i)}(c) \neq c \qquad \text{for} \qquad 1 \leq i \leq k-1. \tag{2.1 b}$$

The application $T_\lambda^{(i)}$ is defined recursively by

$$\begin{aligned} T_\lambda^{(o)}(x) &= x \\ T_\lambda^{(i)}(x) &= T_\lambda[T_\lambda^{(i-1)}(x)] \quad , \quad i \geq 1. \end{aligned} \tag{2.2}$$

Following Metropolis *et al.* [*1*] one may associate to this period a sequence of $(k-1)$ characters or symbols R (« right ») or L (« left ») in the following manner:

$$P = \sigma_1\sigma_2 \ldots \sigma_{k-1} \tag{2.3 a}$$

where

$$\sigma_i = \text{R} \quad \text{if} \quad T_\lambda^{(i)}(c) > c \quad , \quad \sigma_i = \text{L} \quad \text{if} \quad T_\lambda^{(i)}(c) < c.$$

This may be written also

$$P = R^{\nu_1}L^{\mu_1}R^{\nu_2}L^{\mu_2} \ldots \tag{2.3 b}$$

with ν_i, μ_i positive integers.

From now on, we shall denote by P or Q such sequences.

Several sequences may exist for a given k. For example, if $k = 5$, 3 sequences are found in the construction of Metropolis *et al.*:

$$RLR^2, \quad RL^2R, \; RL^3.$$

These « allowed » sequences correspond to three values λ_1, λ_2, λ_3 of the parameter λ.

In their paper, Metropolis *et al.* [*1*] show that the λ associated with the allowed sequences are ordered in a way which is independent of f provided f belongs to a large enough class of functions (roughly conditions 1.1). In the example above, one shows in this way that $\lambda_1 < \lambda_2 < \lambda_3$.

Remarks. — *a*) All sequences of R and L characters are not allowed. For example no allowed sequence begins with an L since in that case, we should have $T_\lambda(c) < c$, $T_\lambda^{(l)}(c) < T_\lambda^{(l-1)}(c)$ since $T_\lambda(x)$ is strictly increasing for $x < c$ and the equation $T_\lambda^k(c) = c$ is never satisfied.

b) The sequence P corresponding to $k = 1$ $[T_\lambda(c) = c]$ has length 0 and we shall denote it from now on by the symbol b (= blank)

$$P = b \tag{2.4}$$

Ordering relation on the sequences

Being given P and P′ associated with λ and λ', we shall write

$$P < P' \quad \text{whenever} \quad \lambda < \lambda'. \tag{2.5}$$

This orders totally the MSS-sequences. Note that there is no simple connection between the length $k - 1$ (or cardinality) of a sequence and its order. For example

$$b < RLR^r < RL < RL^m \text{ for every } m, n.$$

2.1.1. HARMONIC AND ANTI-HARMONIC MAPPING

Again let P be an allowed sequence of $(k - 1)$ characters R or L.

The harmonic of P is defined by the mapping $H : P \to H(P)$ where

$$H(P) = P\sigma P \tag{2.6}$$

with $\sigma = L$ (resp. $\sigma = R$) if P contains an odd (resp. even) number of R symbols. For example,

when $P_1 = RL^2R$ and $P_2 = RLR^2$
one gets $H(P_1) = RL^2R^3L^2R$ and $H(P_2) = RLR^2LRLR^2$.

Metropolis *et al.* [*1*] prove that, if P is allowed, H(P) is allowed too (their theorem 1); we have $P < H(P)$ and the harmonics are adjacent i. e. no allowed sequence exists between P and H(P). When we iterate the process

$$P < H(P) < H^2(P) \ldots < H^j(P).$$

In a similar way, the *anti-harmonic mapping* $A : P \to A(P)$ is defined as

$$A(P) = P\tau P \tag{2.7}$$

with $\tau = R$ (resp. $\tau = L$) if P has an odd (resp. even) number of R symbols.

For the sequences P_1 and P_2 of the above example:

$$A(P_1) = RL^2RLRL^2R \quad \text{and} \quad A(P_2) = RLR^4LR^2.$$

In general, P being an allowed sequence, A(P) *is not* allowed and must be considered only as a mathematical tool.

Let $\lambda_1 < \lambda_2$ be two values of λ corresponding to two allowed sequences P_1 and P_2. The theorems of Metropolis *et al.* give an iterative algorithm for constructing all the allowed sequences between P_1 and P_2, that is any sequence corresponding to a value of λ, say λ^*, such as $\lambda_1 < \lambda^* < \lambda_2$. For that purpose construct $H(P_1)$ and $A(P_2)$. If $H(P_1) = P^*\mu_1\mu_2 \ldots$ and $A(P_2) = P^*\nu_1\nu_2 \ldots, \mu_1 \neq \nu_1$, then P^* is an allowed sequence lying between P_1 and P_2 and it has the smallest possible length. One may replace P_1 or P_2 by P^* and start again.

Example. — Consider $P_1 = RLR^4$ and $P_2 = RLR^4LR$, then $P_2 > P_1$, $H(P_1) = RLR^4LRLR^4$, $A(P_2) = RLR^4LRLRLR^4LR$ and $P^* = RLR^4LRLR$.

2.1.2. Construction of all periods

The above prescription makes possible the construction of any allowed sequence between two given allowed sequences P_1 and P_2 corresponding to λ_1, λ_2 with $\lambda_1 < \lambda_2$.

If $\lambda < c/f(c)$, $T_\lambda^{(l)}c = c$ has no solution. The first period appears for $\lambda = c/f(c)$ and corresponds to a sequence of length zero,

$$P_1 = b.$$

When i increases the set of sequences RL^i increases (i. e. $RL^i > RL^j$ if $i > j$) it is easy to see that the greatest allowed value for λ, i. e. $\lambda = 2$ corresponds to the « limit » sequence when $i \to \infty$, which is written RL^∞ for obvious reasons (*).

(*) The notion of infinite sequence is intuitive in the case of RL^∞. More generally, as the set of all allowed sequences is countable, there are certainly accumulation points of the corresponding set of the parameter λ. These accumulation points correspond to infinite sequences that are defined as follows. One first considers an equivalence relation in the monotonous sets of sequences. Let $\{ P_i \}$ and $\{ P'_j \}$ be two increasing sets [$\{ S_i \}$ is increasing iff $S_i > S_k \Leftrightarrow i > k$], then $\{ P_i \} \sim \{ P'_j \}$ iff any Q that is larger than any P_i is also larger than any P'_j and conversely. If $\{ P_i \}$ is increasing and $\{ P'_j \}$ decreasing, then $\{ P_i \} \sim \{ P'_j \}$ iff, given a sequence Q larger than any P_i, a sequence P'_j exists that is smaller than Q although any P'_j is greater than any P_i. The quotient set of the monotonous sequences by this equivalence relation is by definition the set of the infinite sequences. They are totally ordered in an obvious way, finite allowed sequences being a subset of them. Furthermore, this set of infinite sequences is closed under the formation of accumulation points by increasing and decreasing sets. A simple example of such an infinite sequence is sequence RL^∞ above. Similarly, the set $\{ RLR^{2l}, l \in \mathbb{N} \}$ is increasing although $\{ RLR^{2l+1}, l \in \mathbb{N} \}$ is decreasing (fig. 2). It is not difficult to show that, in the

FIG. 2. — Values of the parameters defining the sequences R and RLR^n ($n = 0,4$); $\bar{\lambda}$ is the accumulation point.

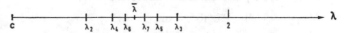

All sequences appear when λ increases from $cf(c)$ to 2 between the limitings equences

$$P_1 = b, \qquad P_2 = RL^{\infty}.$$

Using the method of Metropolis *et al.* which extends at once to the case of infinite sequences, one gets step by step all the allowed sequences which are between b and RL^{∞}. The first steps of the construction are shown on table I. As P_1 and P_2 do not depend on the exact form of the defining function $f(x)$, the ordering is universal.

TABLE I. — *Construction of all the first sequences by the MSS method.*

a)	$k' = 1$	$P_1 = b$
		$P_2 = RL^{\infty}$
b)	$k = 1$	b
	$k = 2$	R
		RL^{∞}
c)	$k = 1$	b
	$k = 2$	R
	$k = 3$	RL
		RL^{∞}
d)	$k = 1$	b
	$k = 2$	R
	$k = 4$	RLR
	$k = 3$	RL
	$k = 4$	RL^2
		RL^{∞}
e)	$k = 1$	b
	$k = 2$	R
	$k = 4$	RLR
	$k = 5$	RLR^2
	$k = 3$	RL
	$k = 5$	RL^2R
	$k = 4$	RL^2
	$k = 5$	RL^3
		RL^{∞}
f)	$k = 1$	b
	$k = 2$	R
	$k = 4$	RLR
	$k = 6$	RLR^3
	$k = 5$	RLR^2
	$k = 3$	RL
	$k = 6$	RL^2RL
	$k = 5$	RL^2R
	$k = 6$	RL^2R^2
	$k = 4$	RL^2
	$k = 6$	RL^3R
	$k = 5$	RL^3
	$k = 6$	RL^4
		RL^{∞}

Before ending this subsection let us point out some auxiliary results which are of some interest and will be recovered in a simple manner thereafter

above sense these two sets define a single accumulation point that is the infinite sequence RLR^{∞}. It must be noticed that it is yet not proved—although presumably true as judged from numerical calculations—that the upper bound of the increasing set $\{ \lambda_{2l} \}$ and the lower bound of the decreasing set $\{ \lambda_{2l+1} \}$ are the same. This common bound, if it exists should be the value of λ corresponding to the infinite sequence RLR^{∞}. For the broken linear transform to be studied below, any infinite sequence defined as above, does actually correspond to a single value of the parameter λ.

i) R is the smallest sequence (after b) as no sequence can be built between b and R. So $R < RL^n$ for every integer n.

ii) if $p < n$, $R < RL^p < RL^n$.

Between RL^{n-1}, and RL^n, all the sequences begin necessarily by the pattern $RL^nR^\alpha \ldots$ Conversely $RL^{n-1} < Q < RL^n$ whenever $Q = RL^nR^\alpha \ldots$, $\alpha \neq 0$.

2.2. The internal similarity

One can thus reconstruct the whole set of allowed MSS sequences starting with $P_1 = b$ and $P_2 = RL^N$, N arbitrarily large but finite. We will see in this section that this set is similar (in a sense to be made more precise) to some of its own parts; we will refer to this property as the law of internal similarity.

Starting with P'_1 and P'_2, $P_1 < P'_1 < P'_2 < P_2$, one can, by the algorithm of Metropolis *et al.* described above, construct all sequences P' such that $P'_1 < P' < P'_2$. With a convenient choice of P'_1 and P'_2 we will find a monotonous bijection between the sequences P, $P_1 < P < P_2$ and the sequences P', $P'_1 < P' < P'_2$.

2.2.1. THE * COMPOSITION LAW

Let $P = \sigma_1\sigma_2 \ldots \sigma_{p-1}$ with σ_i either R or L, be a sequence, allowed or not, of $p - 1$ symbols. Similarly let Q be a sequence of $n - 1$ symbols R or L. Define $Q * P_1 \equiv Q * b = Q$ for $p = 1$, while for $p > 1$

$$Q * P = Q\sigma_1Q\sigma_2Q \ldots Q\sigma_{p-1}Q \tag{2.8 a}$$

if the number of R symbols in Q is even, and

$$Q * P = Q\tau_1Q\tau_2Q \ldots Q\tau_{p-1}Q, \qquad \tau_i \neq \sigma_i \tag{2.8 b}$$

otherwise.

For example, $R * RL^n = RLR^{2n+1}$, while

$$RLR * RL^n = RLR^3(LR)^{2n+1}.$$

Some properties of the * mapping may be noted.

i) For two sequences Q_1 and Q_2 of symbols R or L, $Q_1 * Q_2 \neq Q_2 * Q_1$ in general. For example, $R * RL = RLR^3$, while $RL * R = RL^2RL$.

ii) The associative law holds, i. e.

$$Q_1 * (Q_2 * Q_3) = (Q_1 * Q_2) * Q_3.$$

This can be verified directly by first observing that the number of R symbols in Q_1 and Q_2 has the same parity as that number in $Q_1 * Q_2$.

iii) If H(P) is the harmonic and A(P) the anti-harmonic of P, then,

$$H(P) = P * R, \qquad A(P) = P * L. \tag{2.9}$$

So that for any Q,

$$Q * H(P) = H(Q * P), \qquad Q * A(P) = A(Q * P). \tag{2.10}$$

iv) If P and Q are allowed sequences, then so is $P * Q$. The period of $P * Q$ is the product of the periods of P and Q.

v) If P, Q_1 and Q_2 are allowed sequences and $Q_1 < Q_2$, then $P * Q_1 < P * Q_2$.

A direct proof of this point is rather tedious. It becomes obvious by using the criterion of classification and ordering of MSS sequences given in section 4.

vi) Let the allowed sequences P, Q_1, Q_2, Q_1', Q_2' satisfy

$$P * Q_1 = Q_1' < Q_2' = P * Q_2.$$

Then corresponding to any allowed Q' with $Q_1' < Q' < Q_2'$ one can always find an allowed sequence Q such that $Q' = P * Q$.

We show on Table II the first allowed sequences between Q_1' and Q_2' when $P = b$, R and RL respectively.

TABLE II. — *MSS method applied together to sequences* Q *of lower period* $k(\leq 4)$ *and to sequences between* $Q_1' = R$, $Q_2' = R * RL^\infty$ *(multiplication of periods by 2) and sequences between* $Q_1'' = RL$, $Q_2'' = RL * RL^\infty$.

	Q		R * Q		RL * Q
a) $k = 1$	$P_1 = b$	$k = 2$	$Q_1' = R$	$k = 3$	$Q_1'' = RL$
	$P_2 = RL^\infty$		$Q_2' = RLR^\infty$		$Q_2'' = RL^2(RLR)^\infty$
b) $k = 1$	b				
$k = 2$	R	$k = 2$	R	$k = 3$	RL
		$k = 4$	RLR	$k = 6$	RL^2RL
	RL^∞		RLR^∞		$RL^2(RLR)^\infty$
c) $k = 1$	b				
$k = 2$	R	$k = 2$	R	$k = 3$	RL
$k = 3$	RL	$k = 4$	RLR	$k = 6$	RL^2RL
		$k = 6$	RLR^3	$k = 9$	RL^2RLR^2L
	RL^∞		RLR^∞		$RL^2(RLR)^\infty$
d) $k = 1$	b				
$k = 2$	R	$k = 2$	R	$k = 3$	RL
$k = 4$	RLR	$k = 4$	RLR	$k = 6$	RL^2RL
$k = 3$	RL	$k = 8$	RLR^3LR	$k = 12$	$RL^2RLR^2L^2RL$
$k = 4$	RL^2	$k = 6$	RLR^3	$k = 9$	RL^2RLR^2L
		$k = 8$	RLR^5	$k = 12$	$RL^2RLR^2LR^2L$
	RL^∞		RLR^∞		$RL^2(RLR)^\infty$

vii) For any allowed sequence Q, one has

$$b < Q < Q * Q < Q * Q * Q < \dots$$

and

$$RL^n > Q * RL^n > Q * Q * RL^n > Q * Q * Q * RL^n > \dots$$

viii) Let $P_1 = b$, $P_2 = RL^n$ and P and Q any allowed sequences, $P_1 < P$, $Q < P_2$. Set $P_i' = Q * P_i$, $i = 1,2$; $P' = Q * P$. Then to every P corresponds a P′. Conversely for any P′ with $Q * P_1 = P_1' < P' < P_2' = Q * P_2$ corresponds a P with $P_1 < P < P_2$, $P' = Q * P$.

This is the law of internal similarity alluded to at the beginning of the section.

Remark 1. — Given any sequence Q, can it be written as $Q_1 * Q_2$ with $Q_1 \neq b$, $Q_2 \neq b$? As the number of symbols in $Q_1 * Q_2$ is $(q_1 + 1)(q_2 + 1) - 1$ where q_1 and q_2 are the numbers of symbols in Q_1 and Q_2 respectively, one sees that if the number q of symbols in Q is such that $q + 1$ is a prime integer, then one cannot factorize Q as $Q_1 * Q_2$. Even if $q + 1$ were not a prime integer, the factorization of Q may not exist. For example, the (allowed) sequences RL^2R^2, RL^3R and RL^4 cannot be so factorized. Such sequences are called primary sequences.

For a given q, the number of allowed sequences of length q is known [*1*] and is roughly $2^q/(q + 1)$. If $q + 1$ is prime, all of these are primary. If $q + 1$ is composite, the number of factorizable allowed sequences becomes rapidly negligible for large q. For $q = 14$, there are 1,091 allowed sequences ($2^{14}/15 \sim 1{,}092.3$), out of which 6 are non-primary.

Remark 2. — The mapping $P \to Q * P$, Q fixed, considered above defines the law of internal similarity, because of property (*iii*) above. The other mapping $P \to P * Q$ does not define any law of internal similarity.

2.2.2. Accumulation points

Let Q be an allowed sequence, so that

$$Q^{*j} = \underbrace{Q * \dots * Q}_{j\text{-times}}$$

is allowed. Let λ_j and μ_j be the values of the parameter λ for the sequences $Q^{*j} * b$ and $Q^{*j} * RL^n$ respectively. Then according to (*v*) and (*vii*) above $\lambda_{j-1} < \lambda_j < \mu_j < \mu_{j-1}$ for any j. The λ_j's and μ_j's tend to definite limits λ_∞ and μ_∞ with $\lambda_\infty \leq \mu_\infty$. It is conjectured that $\lambda_\infty = \mu_\infty$. If so, we may allow the infinite sequence $Q^{*\infty}$, since $Q^{*\infty} * P$ for any allowed P will be independent of P (fig. 3).

This infinite sequence may be also considered as an accumulation point, and we shall prove that the set of the accumulation points has the power of the continuum.

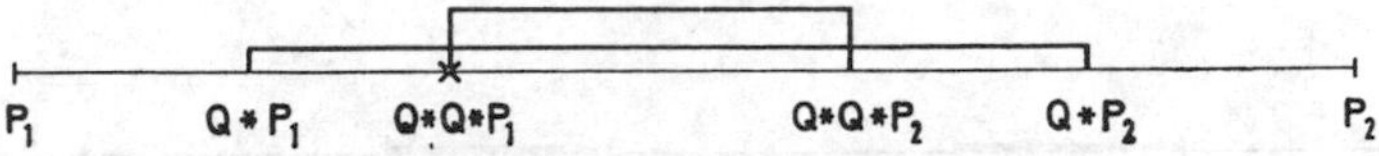

FIG. 3. — Nested intervals corresponding to sequences $Q^{*j} * P_1$ and $Q^{*j} * P_2$ ($P_1 = b$, $P_2 = RL^\infty$) when j increases.

Proof. — Let Q_1 and Q_2 be two allowed finite sequences and suppose there exists a finite sequence P′ such that $Q_1 * P < P' < Q_2$ whatever P is. This is the case, for instance, with $Q_1 = R$ and $Q_2 = RL$. We look to the sequences

$$Q_1^{*\alpha_1} * Q_2^{*\beta_1} * Q_1^{*\alpha_2} * Q_2^{*\beta_2} * \ldots * Q_1^{*\alpha_i} * Q_2^{*\beta_i} * P, \qquad (2.11)$$

$\alpha_k, \beta_k \in N_+$, $1 \leq k \leq i$.

The property of internal similarity shows that any sequence of the form (2.11) is included between the sequence corresponding to $P = b$ and to $P = RL^\infty$ respectively, and any sequence located between these two bounds is of the form (2.11). Let us call this subset an *interval* in the set of the allowed sequences.

We shall consider the structure of these intervals as

$$N = \sum_{j=1}^{i} \alpha_j + \sum_{m=1}^{i} \beta_m \quad \text{grows, with} \quad Q_1 = R \quad \text{and} \quad Q_2 = RL.$$

If $N = 1$, either $\alpha_1 = 1$, $\beta_1 = 0$ or $\alpha_1 = 0$ and $\beta_1 = 1$. The intervals generated by these two different choices are $[Q_1 * b, Q_1 * RL^\infty[$ and $[Q_2 * b, Q_2 * RL^\infty[$. These intervals are disconnected due to our particular choice of Q_1 and Q_2 (a limite sequence P′ exists such as $Q_1 * RL^\infty < P' < Q_2 * b$).

Let $N = 2$, then four intervals are found, corresponding to four different choices for $\{\alpha_i\}$ and $\{\beta_i\}$:

$$[Q_1^{*2} * b, Q_1^{*2} * RL^\infty[, \; [Q_1 * Q_2 * b, Q_1 * Q_2 * RL^\infty[,$$
$$[Q_2 * Q_1 * b, Q_2 * Q_1 * RL^\infty[\text{ and } [Q_2^{*2} * b, Q_2^{*2} * RL^\infty[.$$

As $Q_2 > Q_1 * P$ whatever P is, and as the mapping Q^* is strictly monotonous (if $P > S$ and $P \neq S$, then $Q * P > Q * S$ and $Q * P \neq Q * S$), the above four intervals are disconnected, the first two being included into $[Q_1 * b, Q_1 * RL^\infty[$ and the two others into $[Q_2 * b, Q_2 * RL^\infty[$. Iterating this construction, one finds at the order N, 2^N disconnected intervals, these intervals being included by pair into the intervals of the previous order. This construction is similar to the one of the triadic Cantor set, and generates a set of accumulation point with the power of the continuum. As the whole set of the *accumulation* points is a subset of the values of the real parameter λ, the whole set of the accumulation points has the power of the continuum (fig. 4).

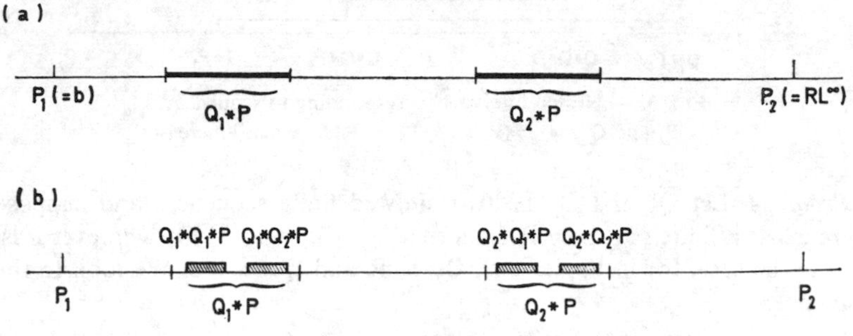

FIG. 4. — Construction of a Cantor ensemble of accumulation points by $Q_1^{*\alpha_1} * Q_2^{*\beta_1} * \ldots * P$, $N = \Sigma\alpha_i + \Sigma\beta_i$.

a) N = 1 *b*) N = 2.

REFERENCES

[1] N. METROPOLIS, M. L. STEIN and P. R. STEIN, *J. of Combinatorial theory*, t. A **15**, 1973, p. 25.
See also J. GUCKENHEIMER, *Inventiones Math.*, t. **39**, 1977, p. 165.

[2] A. RENYI, *Acta Math. Acad. Sci. Hung.*, t. **8**, 1957, p. 477.

[3] W. PARRY, *Acta Math. Acad. Sci. Hung,*, t. **11**, 1960, p. 401.

[4] A. N. ŠARKOVSKII, *Ukrainian Math. J.*, t. **16**, nº 1, 1964, p. 61.

[5] P. ŠTEFAN, *Comm. Math. Phys.*, t. **54**, 1977, p. 237.

[6] M. Y. COSNARD, A. EBERHARD, *Sur les cycles d'une application continue de la variable réelle*, Séminaire analyse numérique nº 274 Lab. Math. Appl., 1977. Université Scientifique et Mathématique de Grenoble, U. S. M.G.

[7] T. Y. LI and J. A. YORKE, *A. M. M.*, t. **82**, 10, 1975, p. 985.

[8] B. DERRIDA, Y. POMEAU, *in preparation.*

[9] J. MILNOR, W. THURSTON, « The kneading matrix ». *Preprint*, IHES (1977).

[10] B. DERRIDA, A. GERVOIS, Y. POMEAU, *in preparation.*

[11] R. L. ADLER, A. G. KONHEIM and M. H. MC ANDREW, *Trans. Amer. Math. Soc.*, t. **114**, 1965, p. 309.

[12] P. ŠTEFAN, *Private communication.*

[13] A. LASOTA, J. A. YORKE, *Trans. Amer. Math. Soc.*, t. **186**, 1973, p. 481.

(Manuscrit reçu le 15 juin 1978)

PHYSICAL REVIEW LETTERS

VOLUME 48 31 MAY 1982 NUMBER 22

Chaotic Attractors in Crisis

Celso Grebogi
Laboratory for Plasma and Fusion Energy Studies, University of Maryland, College Park, Maryland 20742

and

Edward Ott
Laboratory for Plasma and Fusion Energy Studies, Department of Electrical Engineering, and Department of Physics and Astronomy, University of Maryland, College Park, Maryland 20742

and

James A. Yorke
Institute for Physical Science and Technology, and Department of Mathematics, University of Maryland, College Park, Maryland 20742

(Received 16 April 1982)

The occurrence of sudden qualitative changes of chaotic (or "turbulent") dynamics is discussed and illustrated within the context of the one-dimensional quadratic map. For this case, the chaotic region can suddenly widen or disappear, and the cause and properties of these phenomena are investigated.

PACS numbers: 02.50.+s, 05.20.-y

.n dissipative physical systems, such as occur plasmas, fluids, acoustics, optical systems, lid-state devices, etc., it is often observed ıt the system settles into a state of sustained haotic" or "turbulent" motion (cf. Refs. 1 and or a partial listing of some recent relevant ysical examples). Furthermore, this chaotic havior is now understood to result from the esence of strange attractors. [A strange at-ıctor may be thought of as a complicatedly aped surface in the phase space of the dynam-.l variables, to which the system orbit is as-ptotic in time and on which it wanders in a ıotic fashion (cf. Ott[3] for a recent review).] e features of such states have recently been wn to be well described by surprisingly simple ılinear dynamical models (e.g., the one-dimen-nal quadratic map to be discussed below). In ny experiments, changes in the system behavior are studied as some parameter of the system is varied. Thus, much theoretical interest has focused on characterizing the evolution of the dynamics as a function of a system parameter.[4-9] In this paper we investigate sudden qualitative changes in chaotic dynamical behavior which occur at parameter values at which the attractor collides with an unstable periodic orbit. We call such events *crises*.

In order to fix ideas and provide a clear, simple illustration of the phenomenon in question, we first consider an elementary case involving the one-dimensional map given by

$$x_{n+1} = C - x_n^{\,2} = F(x_n, C). \tag{1}$$

For $C < -\frac{1}{4}$, no fixed point of the map exists, and all orbits are asymptotic to $x = -\infty$. At $C = -\frac{1}{4}$ a tangent bifurcation occurs at which a stable and an unstable fixed point are created. It is well known[7,10,11] that as C is increased past $-\frac{1}{4}$, the stable fixed point undergoes period doubling followed by chaos. [For $C > -\frac{1}{4}$, Eq. (1) can be transformed by a change of variables to the logistic map, $x_{n+1} = rx_n(1-x_n)$; note, however, that the logistic map does not possess a tangent bifurcation analogous to that of Eq. (1) at $C = -\frac{1}{4}$ due to its nongeneric behavior at $r = 1$.] As C is increased past $C = 2$, the chaotic attracting orbit is destroyed, and all initial conditions lead to orbits which approach $x = -\infty$ (corresponding to the logistic map with $r > 4$). Figure 1 gives a bifurcation diagram illustrating the above. In this figure we have plotted the position of the unstable fixed point created at $C = -\frac{1}{4}$, $x = -x_* = -\frac{1}{2} - [\frac{1}{4} + C]^{1/2}$, as a dashed curve. For $2 \geq C \geq -\frac{1}{4}$, and for almost any initial point in the range $|x| < x^*$ the orbit generated by (1) is asymptotic to the bounded orbits shown in Fig. 1. Conversely, any point in $|x| > x_*$ generates an orbit which is asymptotic to $x = -\infty$. Thus, for $-\frac{1}{4} \leq C \leq 2$, the range $|x| < x_*$ is the *basin of attraction* for bounded orbits, while $|x| > x_*$ is the basin of attraction for $x = -\infty$. Note from the figure that destruction of the chaotic orbit at $C = 2$ coincides with the *intersection of the chaotic band with the unstable fixed point* $x = -x_*$. To understand why this happens, consider C to be slightly larger than 2. In this case, a typical initial condition in the region which was chaotic for C slightly less than 2 will

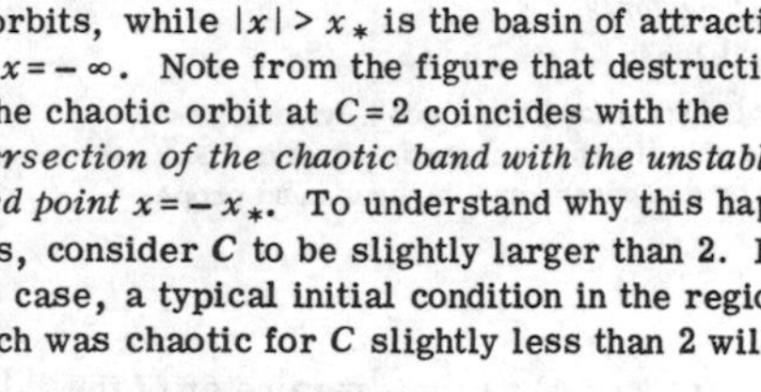

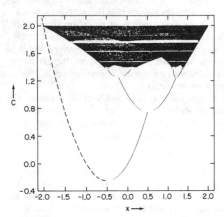

FIG. 1. Bifurcation diagram for the map Eq. (1). The dashed curve is the unstable fixed point. This figure is generated by first preiterating the orbit from an initial condition and then plotting the subsequent orbit in x for a given C, for many different values of C.

generate a chaotic-looking orbit (a chaotic tra sient[5]) until the orbit puts x below $-x_*$. After this happens, the orbit rapidly accelerates to large negative values of x.

One of the points which we wish to convey in this Letter is that such intersections of a chao region and a coexisting unstable orbit are prev lent in many circumstances and systems and lead to *discontinuous* qualitative changes in the character of the long-time behavior of the orbi For example, in the case of the two-dimension Henon map ($x_{n+1} = 1 - \alpha x_n^2 + y_n$, $y_{n+1} = 0.3x_n$), w find, in a certain range of the parameter, α, strange attractors, each with its own basin of attraction. However, as the parameter is rais a critical value is reached. At this critical va one of the attractors collides with an unstable (saddle) periodic point on the boundary separat ing the basins of attraction of the two strange attractors. This collision marks the death of that strange attractor and its basin, and, for values of the parameter immediately above thi critical value, that strange attractor is gone. Further discussion of this case will appear in future publication.[12] In addition, similar crisi induced deaths of strange attractors and their basins are probably present in several reporte ordinary differential-equation examples where hysteresis occurs (e.g., in the Lorenz system, as discussed by Kaplan and Yorke,[4] in a model of Josephson junctions given by Huberman and Crutchfield,[2] and in the nonlinear coupled-plas ma-wave problem of Russell and Ott[13]).

The example of Fig. 1 concerns a crisis in which the unstable orbit is on the boundary of t basin, and the crisis causes termination of the attractor and its basin (we call this a *boundary crisis*). When the collision occurs within the b sin of attraction (we call this an *interior crisi* a sudden expansion of the attractor almost alw occurs. Note that for a boundary crisis the ba of attraction disappears discontinuously, rathe than by shrinking continuously to zero (e.g., a the crisis point $C = 2$ of Fig. 1, the basin of att tion for the bounded chaotic orbit is $|x| < 2$). A an example of an interior crisis, consider Fig This figure is an enlargement of the bifurcatic diagram of Fig. 1 for C between 1.72 and 1.82. This range encompasses the region where sta period-three orbits appear by tangent bifurcat Also shown in Fig. 2 are dashed curves denoti the *unstable* period-three orbit created at the gent bifurcation. Note from Fig. 2 that for a range of C less than a certain critical value,

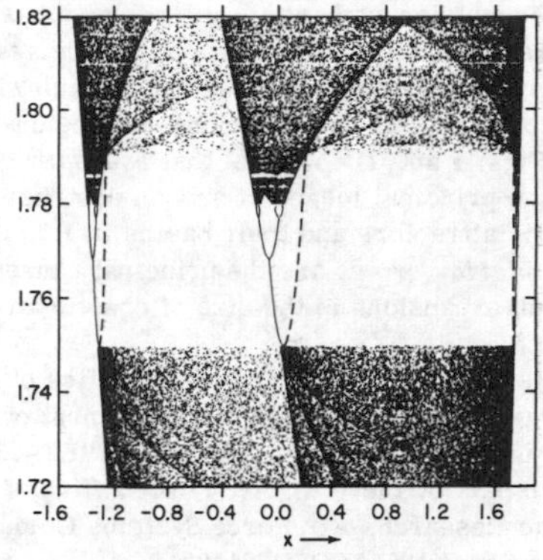

FIG. 2. Blowup of the bifurcation diagram of Fig. 1 the region of the period-three tangent bifurcation. he dashed curves denote the unstable period-three bit created at the tangent bifurcation.

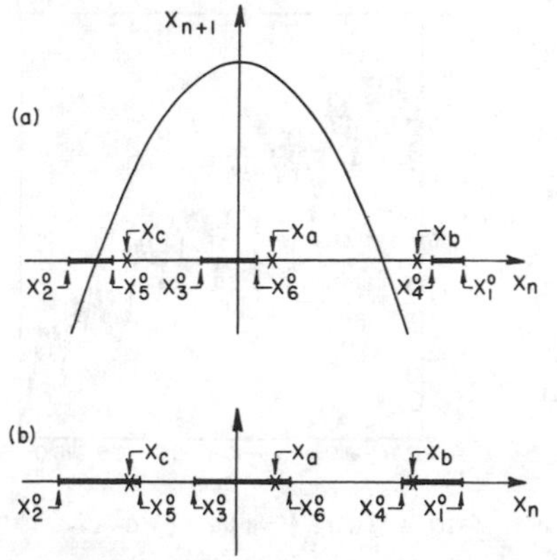

FIG. 3. (a) Schematic illustration of the quadratic map, Eq. (1), for a value of C slightly less than C_{*3}. The three chaotic bands are indicated on the x_n axis with boundary points x_1^0, x_2^0, x_3^0, x_4^0, x_5^0, and x_6^0. Also shown as crosses are the components of the unstable period-three orbit, x_a, x_b, and x_c. (b) Schematic illustration of the x_n axis for C slightly larger than C_{*3}.

aos occurs in three distinct bands, but that, hen C increases past $C_{*3} \simeq 1.79$, the three cha-ic regions suddenly widen to form a single band. urthermore, this coincides precisely with the tersection of the unstable period-three orbit eated at the original tangent bifurcation with e chaotic region. We have noted similar crisis-duced widenings associated with the other tan-nt bifurcations occurring in the chaotic range e., $C_\infty < C \leq 2$, where C_∞ is the accumulation int for period-doubling bifurcations of the orig-al stable fixed point).

Figure 3(a) shows the map (1) for a value of C ightly less than C_{*3}. The three chaotic bands e indicated on the x_n axis as the intervals $[x_3^0$, $^0]$, $[x_4^0, x_1^0]$, and $[x_2^0, x_5^0]$. Also, the unstable riod-three points, x_a, x_b, and x_c, are indi-ted as crosses. The rightmost boundary of the aotic region, x_1^0, is clearly the image of $x=0$, nce $F(x, C)$ is maximum at $x=0$. Thus, x_1^0 $F(0, C)$. We denote F composed with itself n nes by $F^{(n)}(x, C)$; i.e., $F^{(n)}(x, C) = F(F^{(n-1)}(x,$ $, C)$, and $F^{(1)}(x, C) \equiv F(x, C)$. Examination Fig. 3 then shows that $x_n^0 = F^{(n)}(0, C)$, $n = 1$, 2, 4, 5, and 6. Now consider x_4^0. At $C = C_{*3}$, $^0 = x_b$, and hence $x_4^0 = x_7^0$, or

$$F^{(4)}(0, C_{*3}) = F^{(7)}(0, C_{*3}). \tag{2}$$

uation (2) provides a means for the accurate merical determination of C_{*3}. We obtain C_{*3} .790 327 492. . . .

For C slightly larger than C_{*3}, the unstable orbit x_a, x_b, x_c will lie within the bands $[x_6^0, x_3^0]$, $[x_1^0, x_4^0]$, $[x_2^0, x_5^0]$ [cf. Fig. 3(b)]; x_a will be slightly less than x_6^0, x_b will be slightly greater than x_4^0, and x_c will be slightly less than x_5^0. An orbit started within one of the regions $[x_3^0, x_a]$, $[x_b, x_1^0]$, $[x_2^0, x_c]$ will typically *initially* move about in a chaotic way, cycling between the three regions, as in the case $C < C_{*3}$. After a while, the point will eventually fall within one of the small regions $[x_a, x_6^0]$, $[x_4^0, x_b]$, $[x_c, x_5^0]$. It will then be repelled by the unstable period-three orbit and be pushed into the formerly empty region.

Let f denote the fraction of time which an orbit spends in the formerly empty regions, $[x_5^0, x_3^0]$ and $[x_6^0, x_4^0]$. Consideration of the action of the map leads us to suspect that this fraction will have a functional dependence on $C - C_{*3} \equiv c$ which is approximately of the form $(0 < c \ll 1)$

$$f(c) = c^{1/2} P(\ln c) + k_1 c^{1/2} \ln(k_2/c), \tag{3}$$

where k_1 and k_2 are constants, P is a periodic function, $P(\zeta) = P(\zeta + \alpha)$, and the periodicity α is given by $\alpha \simeq \ln\beta$,

$$\beta = F'(x_a, C_{*3}) F'(x_b, C_{*3}) F'(x_c, C_{*3}),$$

with $F' \equiv dF/dx$. This yields $\alpha \simeq 1.312$. The origin of (3) will be discussed in a future publication.[12] Note that the first term in (3), denoted

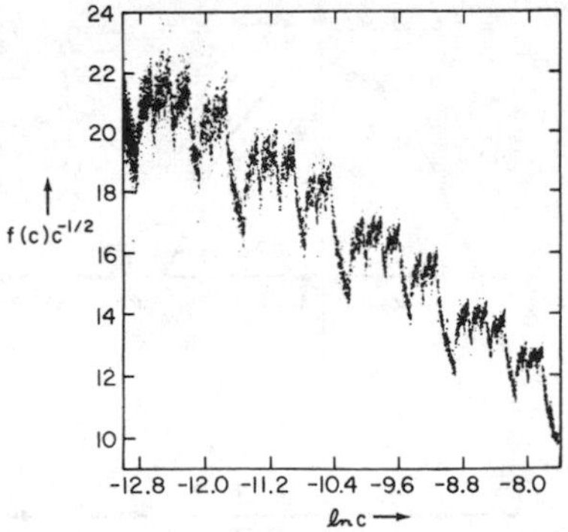

FIG. 4. $f(c)/c^{1/2}$ vs ln$c$. $c = C - C_{*3}$.

$f_1(c)$, is scale invariant, $f_1(\beta c) = \beta^{1/2} f_1(c)$. Figure 4 shows a plot of $c^{-1/2} f(c)$ vs ln$c$ obtained by numerical iteration of Eq. (1). It is seen that the result is in agreement with (3), including the predicted periodicity α.

Now we turn to a consideration of the Lyapunov number of the map at $C = C_{*3}$. For a $C = C_{*3}$ both $x_3{}^0$ and $x_6{}^0$ map to $x_4{}^0$ ($x_3{}^0 = -x_6{}^0$). In this case, after three iterations, the middle interval is symmetrically stretched, folded in two, and mapped back onto itself. To the extent that the map F has small curvature in the side intervals $[x_4{}^0, x_1{}^0]$, $[x_2{}^0, x_5{}^0]$, the map $F^{(3)}$ acting on one of the three intervals is approximately parabolic and stretches and folds the interval in two and then maps it onto itself. Thus, appropriate to this situation, we predict that the Liapunov exponent of $F^{(3)}$ is approximately ln2 and that the Liapunov exponent for F is approximately (ln2)/3. In fact, it can be shown that ln2 is also an exact upper bound for the Liapunov exponent of a map like $F^{(3)}$ (cf. Ref. 12 for a simple proof). Numerical calculation indeed reveals that lnλ, the Liapunov exponent for F, is very close to its upper bound,

$$\ln\lambda \simeq [(\ln 2)/3](1 - 4\times 10^{-4}).$$

Even more precise agreement is found for the case $C = C_{*5}$, corresponding to the crisis point following the tangent bifurcation to a period-five orbit,

$$\ln\lambda \simeq [(\ln 2)/5](1 - 4\times 10^{-6}).$$

In conclusion we have identified two types of crises, *boundary crises* and *interior crises*, ar we have illustrated and investigated each within the context of the quadratic one-dimensional ma (cf. Figs. 1 and 2). We feel that *boundary crise* are the principal means of sudden destruction o chaotic attractors and their basins in R^n, and that *interior crises* are the principal causes of sudden expansions in the size of chaotic attractors.[14]

This work was supported by the Office of Basi Energy Sciences of the U. S. Department of Energy under Contract No. DE-AC05-79ET-53044 and in part by the U. S. Air Force Office of Scientific Research, Air Force Systems Command under Grant No. AFOSR-81-0217.

---

[1]H. L. Swinney and J. P. Gollub, Phys. Today 31, No. 8, 41 (1978); J. M. Wersinger, J. M. Finn, and E. Ott, Phys. Rev. Lett. 44, 453 (1980); G. Ahlers and R. W. Walden, Phys. Rev. Lett. 44, 445 (1980); M. Giglio, S. Musazzi, and U. Perini, Phys. Rev. Le 47, 243 (1981); W. Lauterborn and E. Cramer, Phys Rev. Lett. 47, 1445 (1981); J. Testa, J. Perez, and C. Jeffries, Phys. Rev. Lett. 48, 714 (1982); K. Iked and D. Akimoto, Phys. Rev. Lett. 48, 617 (1982).

[2]B. A. Huberman and J. P. Crutchfield, Phys. Rev. Lett. 43, 1743 (1979).

[3]E. Ott, Rev. Mod. Phys. 53, 655 (1982).

[4]J. L. Kaplan and J. A. Yorke, Commun. Math. Ph 67, 93 (1979).

[5]J. A. Yorke and E. D. Yorke, J. Stat. Phys. 21, 2 (1979).

[6]D. Ruelle and F. Takens, Commun. Math. Phys. 2 167 (1971).

[7]M. J. Feigenbaum, Los Alamos Science 1, 4 (198 and J. Stat. Phys. 19, 25 (1978).

[8]P. Manneville and Y. Pomeau, Physica (Utrecht) 219 (1980).

[9]J. P. Eckmann, Rev. Mod. Phys. 53, 643 (1981).

[10]R. M. May, Nature (London) 261, 459 (1976).

[11]P. Collet and J. P. Eckmann, *Iterated Maps on th Interval as Dynamical Systems* (Birkhauser, Boston 1980).

[12]C. Grebogi, E. Ott, and J. A. Yorke, to be publis

[13]D. A. Russell and E. Ott, Phys. Fluids 24, 1976 (1981).

[14]Other means of sudden destruction of bounded ch attractors and their basins and other sudden expans of chaotic attractors do occur, but we conjecture th they are exceptional and depend on special symmet

5. Bifurcation and Chaos in Higher Dimensional Systems

Commun. math. Phys. 50, 69—77 (1976)

Communications in Mathematical Physics

A Two-dimensional Mapping with a Strange Attractor

M. Hénon

Observatoire de Nice, F-06300 Nice, France

Abstract. Lorenz (1963) has investigated a system of three first-order differential equations, whose solutions tend toward a "strange attractor". We show that the same properties can be observed in a simple mapping of the plane defined by: $x_{i+1}=y_i+1-ax_i^2$, $y_{i+1}=bx_i$. Numerical experiments are carried out for $a=1.4$, $b=0.3$. Depending on the initial point (x_0, y_0), the sequence of points obtained by iteration of the mapping either diverges to infinity or tends to a strange attractor, which appears to be the product of a one-dimensional manifold by a Cantor set.

1. Introduction

Lorenz (1963) proposed and studied a remarkable system of three coupled first-order differential equations, representing a flow in three-dimensional space. The divergence of the flow has a constant negative value, so that any volume shrinks exponentially with time. Moreover, there exists a bounded region R into which every trajectory becomes eventually trapped. Therefore, all trajectories tend to a set of measure zero, called *attractor*. In some cases the attractor is simply a point (which is then a stable equilibrium point) or a closed curve (known as a limit cycle). But in other cases the attractor has a much more complex structure; it appears to be locally the product of a two-dimensional manifold by a Cantor set. This is known as a *strange attractor*. Inside the attractor, trajectories wander in an apparently erratic manner. Moreover, they are highly sensitive to initial conditions. These phenomena are of interest for weather prediction (Lorenz, 1963) and more generally for turbulence theory (Ruelle and Takens, 1971; Ruelle, 1975). Further numerical explorations of the Lorenz system have been made by Lanford (1975) and Pomeau (1976).

We present her a "reductionist" approach in which we try to find a model problem which is as simple as possible, yet exhibits the same essential properties as the Lorenz system. Our aim is (i) to make the numerical exploration faster and more accurate, so that solutions can be followed for a longer time, more

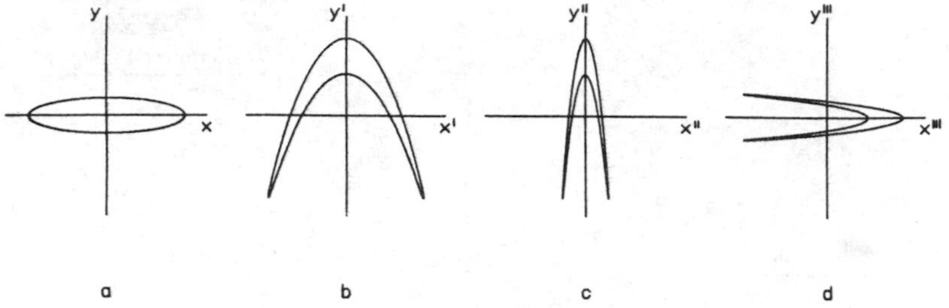

Fig. 1. The initial area a is mapped by T' into b, then by T'' into c, and finally by T''' into d

detailed explorations can be conducted, etc.; (ii) to provide a model which might lend itself more easily to mathematical analysis.

2. The Model

Our first step is classical (Birkhoff, 1917) and consists in considering not the whole trajectories in the three-dimensional space, but only their successive intersections with a two-dimensional *surface of section* S. We define a mapping T of S into itself as follows: given a point A of S, we follow the trajectory which originates from A until it intersects S again; this new point is $T(A)$. This mapping is sometimes called a *Poincaré map*. A trajectory is thus replaced by an infinite set of points in S, obtained by repeated application of the mapping T. The essential properties of the trajectory are reflected into corresponding properties of the set of points. We have thus formally reduced the problem to the study of a two-dimensional mapping.

At this point, however, the only advantage really gained is in clarity of presentation of the results; the actual computation of the mapping still requires the numerical integration of the differential equations. Now comes the second and decisive step: we forget about the differential system, and we define a mapping T by explicit equations, giving directly $T(A)$ when A is known. This of course simplifies the computation drastically. The new mapping T does not any more correspond to the Lorenz system; however, by choosing it carefully we may hope to retain the essential properties which we wish to study. Past experience in the measure-preserving case (see Hénon, 1969, and references therein) has shown indeed that the same features are found in dynamical systems defined by differential equations and in mappings defined as such.

The third step consists in specifying T. Here we have been inspired by the numerical results of Pomeau (1976) on the Lorenz system, which show clearly how a volume is stretched in one direction, and at the same time folded over itself, in the course of one revolution. This folding effect has been also described by Ruelle (1975, Fig. 5 and 6). We simulate it by the following chain of three mappings of the (x, y) plane onto itself. Consider a region elongated along the x axis (Fig. 1a). We begin the folding by

$$T' : x' = x, \quad y' = y + 1 - ax^2, \tag{1}$$

which produces Figure 1b; a is an adjustable parameter. We complete the folding by a contraction along the x axis:

$$T'': x'' = bx', \quad y'' = y', \tag{2}$$

which produces Figure 1c; b is another parameter, which should be less than 1 in absolute value. Finally we come back to the orientation along the x axis by

$$T''': x''' = y'', \quad y''' = x'', \tag{3}$$

which results in Figure 1d.

Our mapping will be defined as the product $T = T'''T''T'$. We write now (x_i, y_i) for (x, y) and (x_{i+1}, y_{i+1}) for (x''', y''') (as a reminder that the mapping will be iterated) and we have

$$T: x_{i+1} = y_i + 1 - ax_i^2, \quad y_{i+1} = bx_i. \tag{4}$$

This mapping has some interesting properties. Its Jacobian is a constant:

$$\frac{\partial(x_{i+1}, y_{i+1})}{\partial(x_i, y_i)} = -b. \tag{5}$$

The geometrical interpretation is quite simple: T' preserves areas; T''' also preserves areas but reverses the sign; and T'' contracts areas, multiplying them by the constant factor b. The property (5) is welcome because it is the natural counterpart of the constant negative divergence in the Lorenz system.

A polynomial mapping satisfying (5) is known as an *entire Cremona transformation*, and the inverse mapping is also given by polynomials (Engel, 1955, 1958). Indeed we have here

$$T^{-1}: x_i = b^{-1} y_{i+1}, \quad y_i = x_{i+1} - 1 + ab^{-2} y_{i+1}^2. \tag{6}$$

Thus T is a one-to-one mapping of the plane onto itself. This is also a welcome property, because it is the natural counterpart of the fact that in the Lorenz system there is a unique trajectory through any given point.

The selection of T could have been approached in a different way, by looking for the "simplest" non-trivial mapping. It is natural then to consider polynomial mappings of progressively increasing order. Linear mappings are trivial, so the polynomials must be at least of degree 2. The most general quadratic mapping is

$$x_{i+1} = f + ax_i + by_i + cx_i^2 + dx_iy_i + ey_i^2,$$
$$y_{i+1} = f' + a'x_i + b'y_i + c'x_i^2 + d'x_iy_i + e'y_i^2 \tag{7}$$

and depends on 12 parameters. But if we impose the condition that the Jacobian is a constant, some relations must be satisfied by these parameters. We can further reduce the number of parameters by an appropriate linear change of coordinates in the plane. In this way, by a slight extension of the results of Engel (1958), it can be shown that the general form (7) is reducible to a "canonical form" depending on two parameters only. This is a generalization of our earlier result (Hénon, 1969) that a quadratic *area-preserving* mapping can be brought into a form depending on one parameter only. The canonical form can be written in several different ways; and one of them turns out to be identical with (4), which is thus reached by an entirely different road! The mapping (4), which was initially constructed in empirical fashion, is in fact the most general quadratic mapping with constant Jacobian.

One difference with the Lorenz problem is that the successive points obtained by repeated application of T do not always converge towards an attractor; sometimes they "escape" to infinity. This is because the quadratic term in (4) dominates when the distance from the origin becomes large. However, for particular values of a and b it is still possible to prove the existence of a bounded "trapping region" R, from which the points can never escape once they have entered it (see below Section 5).

T has two invariant points, given by

$$x=(2a)^{-1}[-(1-b)\pm\sqrt{(1-b)^2+4a}], \quad y=bx. \tag{8}$$

These points are real for

$$a>a_0=(1-b)^2/4. \tag{9}$$

When this is the case, one of the points is always linearly unstable, while the other is unstable for

$$a>a_1=3(1-b)^2/4. \tag{10}$$

3. Choice of Parameters

We select now particular values of a and b for a numerical study. b should be small enough for the folding described by Figure 1 to occur really, yet not too small if one wishes to observe the fine structure of the attractor. The value $b=0.3$ was found to be adequate. A good value of a was found only after some experimenting. For $a<a_0$ or $a>a_3$, where a_0 is given by (9) and a_3 is of the order of 1.55 for $b=0.3$, the points always escape to infinity: apparently there exists no attractor in these cases. For $a_0<a<a_3$, depending on the initial values (x_0, y_0), either the points escape to infinity or they converge towards an attractor, which appears to be unique for a given value of a. We concentrate now on this attractor. For $a_0<a<a_1$, where a_1 is given by (10), the attractor is the stable invariant point. When a is increased over a_1, at first the attractor is still simple and consists of a periodic set of p points. (An equivalent attractor in the Lorenz problem would be a limit cycle intersecting the surface of section p times). The value of p increases through successive "bifurcations" as a increases, and appears to tend to infinity as a approaches a critical value a_2, of the order of 1.06 for $b=0.3$. For $a_2<a<a_3$, the attractor is no more simple, and the behaviour of the points becomes erratic. This is the case in which we are interested. We adopt the following values:

$$a=1.4, \quad b=0.3. \tag{11}$$

4. Numerical Results

Figure 2 shows the result of plotting 10000 successive points, obtained by iteration of T, starting from the arbitrarily chosen initial point $x_0=0$, $y_0=0$; the vertical scale is enlarged to give a better picture. Figure 3 shows the result of 10000

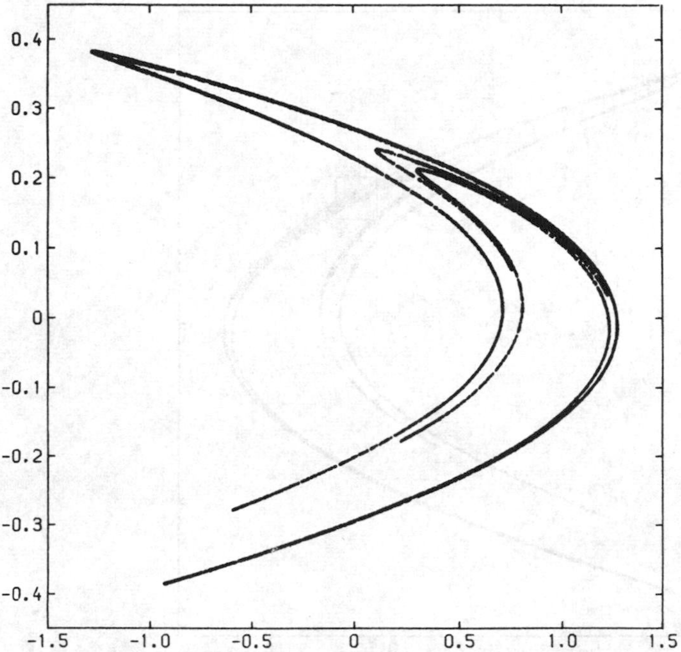

Fig. 2. 10000 successive points obtained by iteration of the mapping T starting from $x_0=0$, $y_0=0$

iterations of T again, starting from a different point: $x_0=0.63135448$, $y_0=0.18940634$ (this choice will be explained below). The two figures are seen to be almost identical. This suggests strongly that what we see in both figures is essentially the attractor itself: the successive points quickly approach the attractor and soon become undistinguishable from it at the scale of the figure. This is confirmed if one looks at the first few points on Figure 2. The initial point at $x_0=0$, $y_0=0$ and the first iterate at $x_1=1$, $y_1=0$ are clearly visible; the second iterate is still visible at $x_2=-0.4$, $y_2=0.3$; the third iterate can barely be distinguished at $x_3=1.076$, $y_3=-0.12$; and the fourth iterate at $x_4=-0.7408864$, $y_4=0.3228$ is already lost inside the attractor at the resolution of Figure 2. The following points then wander over the attractor in an apparently erratic manner.

One of the two unstable invariant points has the coordinates, given by (8):

$$x=0.63135448\ldots, \quad y=0.18940634\ldots . \tag{12}$$

This point appears to belong to the attractor. The two eigenvalues λ_1, λ_2 and the slopes p_1, p_2 of the corresponding eigenvectors are

$$\lambda_1=0.15594632\ldots, \quad p_1=1.92373886\ldots,$$

$$\lambda_2=-1.92373886\ldots, \quad p_2=-0.15594632\ldots . \tag{13}$$

The instability is due to λ_2. The corresponding slope p_2 appears to be tangent to the "curves" in Figure 2.

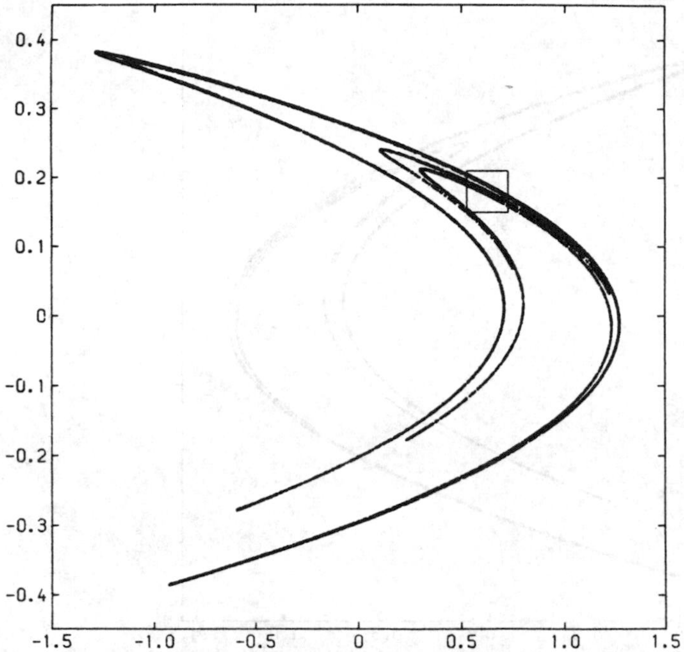

Fig. 3. Same as Figure 2, but starting from $x_0 = 0.63135448$, $y_0 = 0.18940634$

These properties allow us to eliminate the "transient regime" in which the points approach the attractor, and which is not of much interest: we simply start from the close vicinity of the unstable point (12), by rounding off its coordinates to 8 digits. This is done in Figure 3 and in the following figures. The points quickly move away along the line of slope p_2 since $|\lambda_2|$ is appreciably larger than 1.

The attractor appears to consist of a number of more or less parallel "curves"; the points tend to distribute themselves densely over these curves. The few gaps that can still be seen on Figures 2 and 3 have probably no particular significance. Their locations are not the same on the two figures. They are simply due to statistical fluctuations in the quasi-random distribution of points, and they would disappear if more moints were plotted. Thus, the *longitudinal structure* of the attractor (along the curves) appears to be simple, each curve being essentially a one-dimensional manifold.

The *transversal structure* (across the curves) appears to be entirely different, and much more complex. Already on Figures 2 and 3 a number of curves can be seen, and the visible thickness of some of them suggests that they have in fact an underlying structure. Figure 4 is a magnified view of the small square of Figure 3: some of the previous "curves" are indeed resolved now into two or more components. The number n of iterations has been increased to 10^5, in order to have a sufficient number of points in the small region examined. The small square in Figure 4 is again magnified to produce Figure 5, with n increased to 10^6: again the

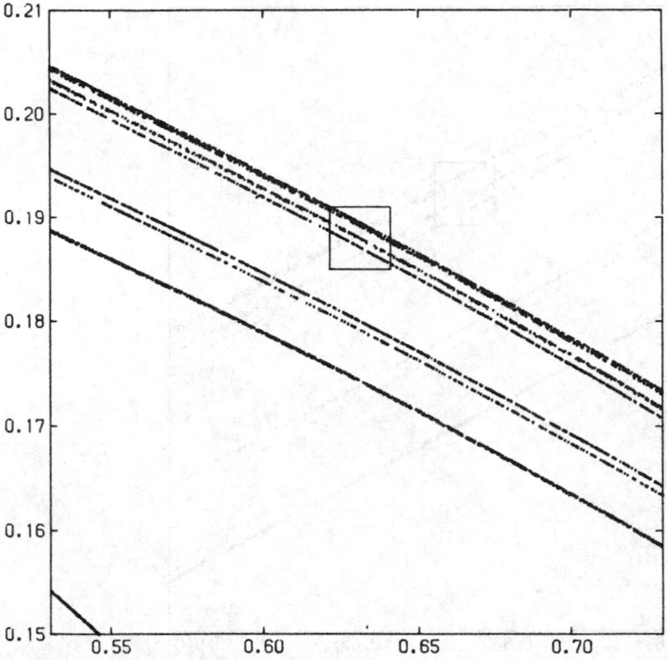

Fig. 4. Enlargement of the squared region of Figure 3. The number of computed points is increased to $n = 10^5$

number of visible "curves" increases. One more enlargement results in Fig. 6, with $n = 5 \times 10^6$: the points become sparse but new curves can still easily be traced.

These figures strongly suggest that the process of multiplication of "curves" will continue indefinitely, and that each apparent "curve" is in fact made of an infinity of quasi-parallel curves. Moreover, Figures 4 to 6 indicate the existence of a hierarchical sequence of "levels", the structure being practically identical at each level save for a scale factor. This is exactly the structure of a Cantor set.

The frames of Figures 4 to 6 have been chosen so as to contain the invariant point (12). This point appears to lie on the upper boundary of the attractor. Surprisingly, its presence is completely invisible on the figures; this contrasts with the area-preserving case, were stable and unstable invariant points play a very conspicuous role (see for instance Hénon, 1969). On the other hand, the presence of the invariant point explains, locally at least, the hierarchy of similar structures: at each application of the mapping, the scale of the transversal structure is multiplied by λ_1 given by (13). At the same time, the points spread out along the curves, as dictated by the value of λ_2.

5. A Trapping Region

The fact that even after 5×10^6 iterations the points have not diverged to infinity suggests that there is a region of the plane from which the points cannot escape.

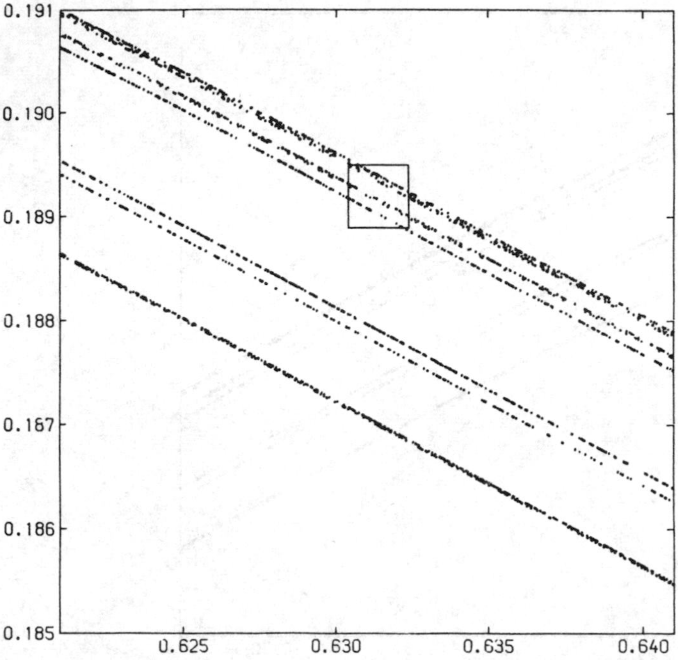

Fig. 5. Enlargement of the squared region of Figure 4; $n=10^6$

This can be actually proved by finding a region R which is mapped inside itself. An example of such a region is the quadrilateral $ABCD$ defined by

$$x_A = -1.33, \quad y_A = 0.42, \quad x_B = 1.32, \quad y_B = 0.133,$$

$$x_C = 1.245, \quad y_C = -0.14, \quad x_D = -1.06, \quad y_D = -0.5. \tag{14}$$

The image of $ABCD$ is a region bounded by four arcs of parabola, and it can be shown by elementary algebra that this image lies inside $ABCD$. Plotting the quadrilateral on Figure 2 or 3, one can verify that it encloses the observed attractor.

6. Conclusions

The simple mapping (4) appears to have the same basic properties as the Lorenz system. Its numerical exploration is much simpler: in fact most of the exploratory work for the present paper was carried out with a programmable pocket computer (HP-65). For the more extensive computations of Figures 2 to 6, we used a IBM 7040 computer, with 16-digit accuracy. The solutions can be followed over a much longer time than in the case of a system of differential equations. The accuracy is also increased since there are no integration errors.

Lorenz (1963) inferred the Cantor-set structure of the attractor from reasoning, but could not observe it directly because the contracting ratio after one "circuit"

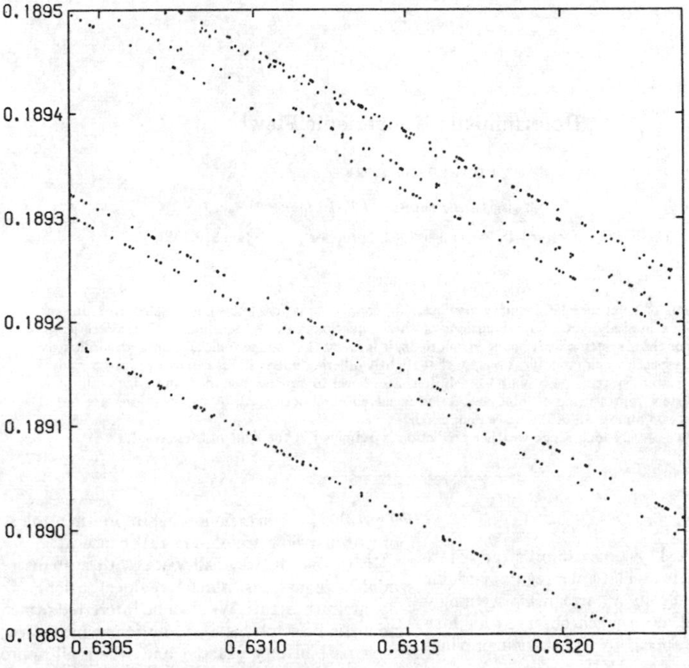

Fig. 6. Enlargement of the squared region of Figure 5; $n=5\times 10^6$

was too small: 7×10^{-5}. A similar experience was reported by Pomeau (1976). In the present mapping, the contracting ratio after one iteration is 0.3, and one can easily observe a number of successive levels in the hierarchy. This is also facilitated by the larger number of points.

Finally, for mathematical studies the mapping (4) might also be easier to handle than a system of differential equations.

References

Birkhoff, G. D.: Trans. Amer. Math. Soc. **18**, 199 (1917)
Engel, W.: Math. Annalen **130**, 11 (1955)
Engel, W.: Math. Annalen **136**, 319 (1958)
Hénon, M.: Quart. Appl. Math. **27**, 291 (1969)
Lanford, O.: Work cited by Ruelle, 1975
Lorenz, E. N.: J. atmos. Sci. **20**, 130 (1963)
Pomeau, Y.: to appear (1976)
Ruelle, D., Takens, F.: Comm. math. Phys. **20**, 167; **23**, 343 (1971)
Ruelle, D.: Report at the Conference on "Quantum Dynamics Models and Mathematics" in Bielefeld, September 1975

Communicated by K. Hepp

Received March 25, 1976

Reprinted from JOURNAL OF THE ATMOSPHERIC SCIENCES, Vol. 20, No. 2, March, 1963, pp. 130–141
Printed in U. S. A.

Deterministic Nonperiodic Flow[1]

EDWARD N. LORENZ

Massachusetts Institute of Technology

(Manuscript received 18 November 1962, in revised form 7 January 1963)

ABSTRACT

Finite systems of deterministic ordinary nonlinear differential equations may be designed to represent forced dissipative hydrodynamic flow. Solutions of these equations can be identified with trajectories in phase space. For those systems with bounded solutions, it is found that nonperiodic solutions are ordinarily unstable with respect to small modifications, so that slightly differing initial states can evolve into considerably different states. Systems with bounded solutions are shown to possess bounded numerical solutions.

A simple system representing cellular convection is solved numerically. All of the solutions are found to be unstable, and almost all of them are nonperiodic.

The feasibility of very-long-range weather prediction is examined in the light of these results.

1. Introduction

Certain hydrodynamical systems exhibit steady-state flow patterns, while others oscillate in a regular periodic fashion. Still others vary in an irregular, seemingly haphazard manner, and, even when observed for long periods of time, do not appear to repeat their previous history.

These modes of behavior may all be observed in the familiar rotating-basin experiments, described by Fultz, *et al.* (1959) and Hide (1958). In these experiments, a cylindrical vessel containing water is rotated about its axis, and is heated near its rim and cooled near its center in a steady symmetrical fashion. Under certain conditions the resulting flow is as symmetric and steady as the heating which gives rise to it. Under different conditions a system of regularly spaced waves develops, and progresses at a uniform speed without changing its shape. Under still different conditions an irregular flow pattern forms, and moves and changes its shape in an irregular nonperiodic manner.

Lack of periodicity is very common in natural systems, and is one of the distinguishing features of turbulent flow. Because instantaneous turbulent flow patterns are so irregular, attention is often confined to the statistics of turbulence, which, in contrast to the details of turbulence, often behave in a regular well-organized manner. The short-range weather forecaster, however, is forced willy-nilly to predict the details of the large-scale turbulent eddies—the cyclones and anticyclones—which continually arrange themselves into new patterns. Thus there are occasions when more than the statistics of irregular flow are of very real concern.

In this study we shall work with systems of deterministic equations which are idealizations of hydrodynamical systems. We shall be interested principally in nonperiodic solutions, i.e., solutions which never repeat their past history exactly, and where all approximate repetitions are of finite duration. Thus we shall be involved with the ultimate behavior of the solutions, as opposed to the transient behavior associated with arbitrary initial conditions.

A closed hydrodynamical system of finite mass may ostensibly be treated mathematically as a finite collection of molecules—usually a very large finite collection—in which case the governing laws are expressible as a finite set of ordinary differential equations. These equations are generally highly intractable, and the set of molecules is usually approximated by a continuous distribution of mass. The governing laws are then expressed as a set of partial differential equations, containing such quantities as velocity, density, and pressure as dependent variables.

It is sometimes possible to obtain particular solutions of these equations analytically, especially when the solutions are periodic or invariant with time, and, indeed, much work has been devoted to obtaining such solutions by one scheme or another. Ordinarily, however, nonperiodic solutions cannot readily be determined except by numerical procedures. Such procedures involve replacing the continuous variables by a new finite set of functions of time, which may perhaps be the values of the continuous variables at a chosen grid of points, or the coefficients in the expansions of these variables in series of orthogonal functions. The governing laws then become a finite set of ordinary differential

[1] The research reported in this work has been sponsored by the Geophysics Research Directorate of the Air Force Cambridge Research Center, under Contract No. AF 19(604)-4969.

equations again, although a far simpler set than the one which governs individual molecular motions.

In any real hydrodynamical system, viscous dissipation is always occurring, unless the system is moving as a solid, and thermal dissipation is always occurring, unless the system is at constant temperature. For certain purposes many systems may be treated as conservative systems, in which the total energy, or some other quantity, does not vary with time. In seeking the ultimate behavior of a system, the use of conservative equations is unsatisfactory, since the ultimate value of any conservative quantity would then have to equal the arbitrarily chosen initial value. This difficulty may be obviated by including the dissipative processes, thereby making the equations nonconservative, and also including external mechanical or thermal forcing, thus preventing the system from ultimately reaching a state of rest. If the system is to be deterministic, the forcing functions, if not constant with time, must themselves vary according to some deterministic rule.

In this work, then, we shall deal specifically with finite systems of deterministic ordinary differential equations, designed to represent forced dissipative hydrodynamical systems. We shall study the properties of nonperiodic solutions of these equations.

It is not obvious that such solutions can exist at all. Indeed, in dissipative systems governed by finite sets of *linear* equations, a constant forcing leads ultimately to a constant response, while a periodic forcing leads to a periodic response. Hence, nonperiodic flow has sometimes been regarded as the result of nonperiodic or random forcing.

The reasoning leading to these concludions is not applicable when the governing equations are nonlinear. If the equations contain terms representing advection—the transport of some property of a fluid by the motion of the fluid itself—a constant forcing can lead to a variable response. In the rotating-basin experiments already mentioned, both periodic and nonperiodic flow result from thermal forcing which, within the limits of experimental control, is constant. Exact periodic solutions of simplified systems of equations, representing dissipative flow with constant thermal forcing, have been obtained analytically by the writer (1962a). The writer (1962b) has also found nonperiodic solutions of similar systems of equations by numerical means.

2. Phase space

Consider a system whose state may be described by M variables $X_1, \cdots, X_M$. Let the system be governed by the set of equations

$$dX_i/dt = F_i(X_1, \cdots X_M), \quad i=1, \cdots, M, \tag{1}$$

where time t is the single independent variable, and the functions F_i possess continuous first partial derivatives. Such a system may be studied by means of *phase space*—an M-dimensional Euclidean space Γ whose coordinates are $X_1, \cdots, X_M$. Each *point* in phase space represents a possible instantaneous state of the system. A state which is varying in accordance with (1) is represented by a moving *particle* in phase space, traveling along a *trajectory* in phase space. For completeness, the position of a stationary particle, representing a steady state, is included as a trajectory.

Phase space has been a useful concept in treating finite systems, and has been used by such mathematicians as Gibbs (1902) in his development of statistical mechanics, Poincaré (1881) in his treatment of the solutions of differential equations, and Birkhoff (1927) in his treatise on dynamical systems.

From the theory of differential equations (e,g., Ford 1933, ch. 6), it follows, since the partial derivatives $\partial F_i/\partial X_j$ are continuous, that if t_0 is any time, and if $X_{10}, \cdots X_{M0}$ is any point in Γ, equations (1) possess a unique solution

$$X_i = f_i(X_{10}, \cdots, X_{M0}, t), \quad i=1, \cdots, M, \tag{2}$$

valid throughout some time interval containing t_0, and satisfying the condition

$$f_i(X_{10}, \cdots, X_{M0}, t_0) = X_{i0}, \quad i=1, \cdots, M. \tag{3}$$

The functions f_i are continuous in $X_{10}, \cdots, X_{M0}$ and t. Hence there is a unique trajectory through each point of Γ. Two or more trajectories may, however, approach the same point or the same curve asymptotically as $t \to \infty$ or as $t \to -\infty$. Moreover, since the functions f_i are continuous, the passage of time defines a continuous deformation of any region of Γ into another region.

In the familiar case of a conservative system, where some positive definite quantity Q, which may represent some form of energy, is invariant with time, each trajectory is confined to one or another of the surfaces of constant Q. These surfaces may take the form of closed concentric shells.

If, on the other hand, there is dissipation and forcing, and if, whenever Q equals or exceeds some fixed value Q_1, the dissipation acts to diminish Q more rapidly then the forcing can increase Q, then $(-dQ/dt)$ has a positive lower bound where $Q \geqq Q_1$, and each trajectory must ultimately become trapped in the region where $Q < Q_1$. Trajectories representing forced dissipative flow may therefore differ considerably from those representing conservative flow.

Forced dissipative systems of this sort are typified by the system

$$dX_i/dt = \sum_{j,k} a_{ijk} X_j X_k - \sum_j b_{ij} X_j + c_i, \tag{4}$$

where $\sum a_{ijk} X_i X_j X_k$ vanishes identically, $\sum b_{ij} X_i X_j$ is positive definite, and $c_1, \cdots, c_M$ are constants. If

$$Q = \tfrac{1}{2} \sum_i X_i^2, \tag{5}$$

and if $e_1, \cdots, e_M$ are the roots of the equations

$$\sum_j (b_{ij}+b_{ji})e_j=c_i, \qquad (6)$$

it follows from (4) that

$$dQ/dt=\sum_{i,j} b_{ij}e_ie_j-\sum_{i,j} b_{ij}(X_i-e_i)(X_j-e_j). \qquad (7)$$

The right side of (7) vanishes only on the surface of an ellipsoid E, and is positive only in the interior of E. The surfaces of constant Q are concentric spheres. If S denotes a particular one of these spheres whose interior R contains the ellipsoid E, it is evident that each trajectory eventually becomes trapped within R.

3. The instability of nonperiodic flow

In this section we shall establish one of the most important properties of deterministic nonperiodic flow, namely, its instability with respect to modifications of small amplitude. We shall find it convenient to do this by identifying the solutions of the governing equations with trajectories in phase space. We shall use such symbols as $P(t)$ (variable argument) to denote trajectories, and such symbols as P or $P(t_0)$ (no argument or constant argument) to denote points, the latter symbol denoting the specific point through which $P(t)$ passes at time t_0.

We shall deal with a phase space Γ in which a unique trajectory passes through each point, and where the passage of time defines a continuous deformation of any region of Γ into another region, so that if the points $P_1(t_0)$, $P_2(t_0)$, $\cdots$approach $P_0(t_0)$ as a limit, the points $P_1(t_0+\tau)$, $P_2(t_0+\tau)$, $\cdots$must approach $P_0(t_0+\tau)$ as a limit. We shall furthermore require that the trajectories be uniformly bounded as $t\rightarrow\infty$; that is, there must be a bounded region R, such that every trajectory ultimately remains with R. Our procedure is influenced by the work of Birkhoff (1927) on dynamical systems, but differs in that Birkhoff was concerned mainly with conservative systems. A rather detailed treatment of dynamical systems has been given by Nemytskii and Stepanov (1960), and rigorous proofs of some of the theorems which we shall present are to be found in that source.

We shall first classify the trajectories in three different manners, namely, according to the absence or presence of transient properties, according to the stability or instability of the trajectories with respect to small modifications, and according to the presence or absence of periodic behavior.

Since any trajectory $P(t)$ is bounded, it must possess at least one *limit point* P_0, a point which it approaches arbitrarily closely arbitrarily often. More precisely, P_0 is a limit point of $P(t)$ if for any $\epsilon>0$ and any time t_1 there exists a time $t_2(\epsilon,t_1)>t_1$ such that $|P(t_2)-P_0|<\epsilon$. Here absolute-value signs denote distance in phase space. Because Γ is continuously deformed as t varies, every point on the trajectory through P_0 is also a limit point of $P(t)$, and the set of limit points of $P(t)$ forms a trajectory, or a set of trajectories, called the *limiting trajectories* of $P(t)$. A limiting trajectory is obviously contained within R in its entirety.

If a trajectory is contained among its own limiting trajectories, it will be called *central*; otherwise it will be called *noncentral*. A central trajectory passes arbitrarily closely arbitrarily often to any point through which it has previously passed, and, in this sense at least, separate sufficiently long segments of a central trajectory are statistically similar. A noncentral trajectory remains a certain distance away from any point through which it has previously passed. It must approach its entire set of limit points asymptotically, although it need not approach any particular limiting trajectory asymptotically. Its instantaneous distance from its closest limit point is therefore a transient quantity, which becomes arbitrarily small as $t\rightarrow\infty$.

A trajectory $P(t)$ will be called *stable at a point* $P(t_1)$ if any other trajectory passing sufficiently close to $P(t_1)$ at time t_1 remains close to $P(t)$ as $t\rightarrow\infty$; i.e., $P(t)$ is stable at $P(t_1)$ if for any $\epsilon>0$ there exists a $\delta(\epsilon,t_1)>0$ such that if $|P_1(t_1)-P(t_1)|<\delta$ and $t_2>t_1$, $|P_1(t_2)-P(t_2)|<\epsilon$. Otherwise $P(t)$ will be called *unstable* at $P(t_1)$. Because Γ is continuously deformed as t varies, a trajectory which is stable at one point is stable at every point, and will be called a *stable* trajectory. A trajectory unstable at one point is unstable at every point, and will be called an *unstable* trajectory. In the special case that $P(t)$ is confined to one point, this definition of stability coincides with the familiar concept of stability of steady flow.

A stable trajectory $P(t)$ will be called uniformly stable if the distance within which a neighboring trajectory must approach a point $P(t_1)$, in order to be certain of remaining close to $P(t)$ as $t\rightarrow\infty$, itself possesses a positive lower bound as $t_1\rightarrow\infty$; i.e., $P(t)$ is uniformly stable if for any $\epsilon>0$ there exists a $\delta(\epsilon)>0$ and a time $t_0(\epsilon)$ such that if $t_1>t_0$ and $|P_1(t_1)-P(t_1)|<\delta$ and $t_2>t_1$, $|P_1(t_2)-P(t_2)|<\epsilon$. A limiting trajectory $P_0(t)$ of a uniformly stable trajectory $P(t)$ must be uniformly stable itself, since all trajectories passing sufficiently close to $P_0(t)$ must pass arbitrarily close to some point of $P(t)$ and so must remain close to $P(t)$, and hence to $P_0(t)$, as $t\rightarrow\infty$.

Since each point lies on a unique trajectory, any trajectory passing through a point through which it has previously passed must continue to repeat its past behavior, and so must be *periodic*. A trajectory $P(t)$ will be called *quasi-periodic* if for some arbitrarily large time interval τ, $P(t+\tau)$ ultimately remains arbitrarily close to $P(t)$, i.e., $P(t)$ is quasi-periodic if for any $\epsilon>0$ and for any time interval τ_0, there exists a $\tau(\epsilon,\tau_0)>\tau_0$ and a time $t_1(\epsilon,\tau_0)$ such that if $t_2>t_1$, $|P(t_2+\tau)-P(t_2)|$

$<\epsilon$. Periodic trajectories are special cases of quasi-periodic trajectories.

A trajectory which is not quasi-periodic will be called *nonperiodic*. If $P(t)$ is nonperiodic, $P(t_1+\tau)$ may be arbitrarily close to $P(t_1)$ for some time t_1 and some arbitrarily large time interval τ, but, if this is so, $P(t+\tau)$ cannot remain arbitrarily close to $P(t)$ as $t\rightarrow\infty$. Nonperiodic trajectories are of course representations of deterministic nonperiodic flow, and form the principal subject of this paper.

Periodic trajectories are obviously central. Quasi-periodic central trajectories include multiple periodic trajectories with incommensurable periods, while quasi-periodic noncentral trajectories include those which approach periodic trajectories asymptotically. Nonperiodic trajectories may be central or noncentral.

We can now establish the theorem that a trajectory with a stable limiting trajectory is quasi-periodic. For if $P_0(t)$ is a limiting trajectory of $P(t)$, two distinct points $P(t_1)$ and $P(t_1+\tau)$, with τ arbitrarily large, may be found arbitrary close to any point $P_0(t_0)$. Since $P_0(t)$ is stable, $P(t)$ and $P(t+\tau)$ must remain arbitrarily close to $P_0(t+t_0-t_1)$, and hence to each other, as $t\rightarrow\infty$, and $P(t)$ is quasi-periodic.

It follows immediately that a stable central trajectory is quasi-periodic, or, equivalently, that a nonperiodic central trajectory is unstable.

The result has far-reaching consequences when the system being considered is an observable nonperiodic system whose future state we may desire to predict. It implies that two states differing by imperceptible amounts may eventually evolve into two considerably different states. If, then, there is any error whatever in observing the present state—and in any real system such errors seem inevitable—an acceptable prediction of an instantaneous state in the distant future may well be impossible.

As for noncentral trajectories, it follows that a uniformly stable noncentral trajectory is quasi-periodic, or, equivalently, a nonperiodic noncentral trajectory is not uniformly stable. The possibility of a nonperiodic noncentral trajectory which is stable but not uniformly stable still exists. To the writer, at least, such trajectories, although possible on paper, do not seem characteristic of real hydrodynamical phenomena. Any claim that atmospheric flow, for example, is represented by a trajectory of this sort would lead to the improbable conclusion that we ought to master long-range forecasting as soon as possible, because, the longer we wait, the more difficult our task will become.

In summary, we have shown that, subject to the conditions of uniqueness, continuity, and boundedness prescribed at the beginning of this section, a central trajectory, which in a certain sense is free of transient properties, is unstable if it is nonperiodic. A noncentral trajectory, which is characterized by transient properties, is not uniformly stable if it is nonperiodic, and, if it is stable at all, its very stability is one of its transient properties, which tends to die out as time progresses. In view of the impossibility of measuring initial conditions precisely, and thereby distinguishing between a central trajectory and a nearby noncentral trajectory, all nonperiodic trajectories are effectively unstable from the point of view of practical prediction.

4. Numerical integration of nonconservative systems

The theorems of the last section can be of importance only if nonperiodic solutions of equations of the type considered actually exist. Since statistically stationary nonperiodic functions of time are not easily described analytically, particular nonperiodic solutions can probably be found most readily by numerical procedures. In this section we shall examine a numerical-integration procedure which is especially applicable to systems of equations of the form (4). In a later section we shall use this procedure to determine a nonperiodic solution of a simple set of equations.

To solve (1) numerically we may choose an initial time t_0 and a time increment Δt, and let

$$X_{i,n}=X_i(t_0+n\Delta t). \tag{8}$$

We then introduce the auxiliary approximations

$$X_{i(n+1)}=X_{i,n}+F_i(P_n)\Delta t, \tag{9}$$

$$X_{i((n+2))}=X_{i(n+1)}+F_i(P_{(n+1)})\Delta t, \tag{10}$$

where P_n and $P_{(n+1)}$ are the points whose coordinates are

$$(X_{1,n},\cdots,X_{M,n}) \quad \text{and} \quad (X_{1(n+1)},\cdots,X_{M(n+1)}).$$

The simplest numerical procedure for obtaining approximate solutions of (1) is the forward-difference procedure,

$$X_{i,n+1}=X_{i(n+1)}. \tag{11}$$

In many instances better approximations to the solutions of (1) may be obtained by a centered-difference procedure

$$X_{i,n+1}=X_{i,n-1}+2F_i(P_n)\Delta t. \tag{12}$$

This procedure is unsuitable, however, when the deterministic nature of (1) is a matter of concern, since the values of $X_{1,n},\cdots,X_{M,n}$ do not uniquely determine the values of $X_{1,n+1},\cdots,X_{M,n+1}$.

A procedure which largely overcomes the disadvantages of both the forward-difference and centered-difference procedures is the double-approximation procedure, defined by the relation

$$X_{i,n+1}=X_{i,n}+\tfrac{1}{2}[F_i(P_n)+F_i(P_{(n+1)})]\Delta t. \tag{13}$$

Here the coefficient of Δt is an approximation to the time derivative of X_i at time $t_0+(n+\frac{1}{2})\Delta t$. From (9) and (10), it follows that (13) may be rewritten

$$X_{i,n+1}=\tfrac{1}{2}(X_{i,n}+X_{i((n+2))}). \tag{14}$$

A convenient scheme for automatic computation is the successive evaluation of $X_{i(n+1)}$, $X_{i((n+2))}$, and $X_{i,n+1}$ according to (9), (10) and (14). We have used this procedure in all the computations described in this study.

In phase space a numerical solution of (1) must be represented by a jumping particle rather than a continuously moving particle. Moreover, if a digital computer is instructed to represent each number in its memory by a preassigned fixed number of bits, only certain discrete points in phase space will ever be occupied. If the numerical solution is bounded, repetitions must eventually occur, so that, strictly speaking, every numerical solution is periodic. In practice this consideration may be disregarded, if the number of different possible states is far greater than the number of iterations ever likely to be performed. The necessity for repetition could be avoided altogether by the somewhat uneconomical procedure of letting the precision of computation increase as n increases.

Consider now numerical solutions of equations (4), obtained by the forward-difference procedure (11). For such solutions,

$$Q_{n+1}=Q_n+(dQ/dt)_n\Delta t+\tfrac{1}{2}\sum_i F_i^2(P_n)\Delta t^2. \quad (15)$$

Let S' be any surface of constant Q whose interior R' contains the ellipsoid E where dQ/dt vanishes, and let S be any surface of constant Q whose interior R contains S'.

Since $\sum F_i^2$ and dQ/dt both possess upper bounds in R', we may choose Δt so small that P_{n+1} lies in R if P_n lies in R'. Likewise, since $\sum F_i^2$ possesses an upper bound and dQ/dt possesses a *negative* upper bound in $R-R'$, we may choose Δt so small that $Q_{n+1}<Q_n$ if P_n lies in $R-R'$. Hence Δt may be chosen so small that any jumping particle which has entered R remains trapped within R, and the numerical solution does not blow up. A blow-up may still occur, however, if initially the particle is exterior to R.

Consider now the double-approximation procedure (14). The previous arguments imply not only that $P_{(n+1)}$ lies within R if P_n lies within R, but also that $P_{((n+2))}$ lies within R if $P_{(n+1)}$ lies within R. Since the region R is convex, it follows that P_{n+1}, as given by (14), lies within R if P_n lies within R. Hence if Δt is chosen so small that the forward-difference procedure does not blow up, the double-approximation procedure also does not blow up.

We note in passing that if we apply the forward-difference procedure to a conservative system where $dQ/dt=0$ everywhere,

$$Q_{n+1}=Q_n+\tfrac{1}{2}\sum_i F_i^2(P_n)\Delta t^2. \quad (16)$$

In this case, for any fixed choice of Δt the numerical solution ultimately goes to infinity, unless it is asymptotically approaching a steady state. A similar result holds when the double-approximation procedure (14) is applied to a conservative system.

5. The convection equations of Saltzman

In this section we shall introduce a system of three ordinary differential equations whose solutions afford the simplest example of deterministic nonperiodic flow of which the writer is aware. The system is a simplification of one derived by Saltzman (1962) to study finite-amplitude convection. Although our present interest is in the nonperiodic nature of its solutions, rather than in its contributions to the convection problem, we shall describe its physical background briefly.

Rayleigh (1916) studied the flow occurring in a layer of fluid of uniform depth H, when the temperature difference between the upper and lower surfaces is maintained at a constant value ΔT. Such a system possesses a steady-state solution in which there is no motion, and the temperature varies linearly with depth, If this solution is unstable, convection should develop.

In the case where all motions are parallel to the $x-z$-plane, and no variations in the direction of the y-axis occur, the governing equations may be written (see Saltzman, 1962)

$$\frac{\partial}{\partial t}\nabla^2\psi=-\frac{\partial(\psi,\nabla^2\psi)}{\partial(x,z)}+\nu\nabla^4\psi+g\alpha\frac{\partial\theta}{\partial x}, \quad (17)$$

$$\frac{\partial}{\partial t}\theta=-\frac{\partial(\psi,\theta)}{\partial(x,z)}+\frac{\Delta T}{H}\frac{\partial\psi}{\partial x}+\kappa\nabla^2\theta. \quad (18)$$

Here ψ is a stream function for the two-dimensional motion, θ is the departure of temperature from that occurring in the state of no convection, and the constants g, α, ν, and κ denote, respectively, the acceleration of gravity, the coefficient of thermal expansion, the kinematic viscosity, and the thermal conductivity. The problem is most tractable when both the upper and lower boundaries are taken to be free, in which case ψ and $\nabla^2\psi$ vanish at both boundaries.

Rayleigh found that fields of motion of the form

$$\psi=\psi_0\sin(\pi aH^{-1}x)\sin(\pi H^{-1}z), \quad (19)$$

$$\theta=\theta_0\cos(\pi aH^{-1}x)\sin(\pi H^{-1}z), \quad (20)$$

would develop if the quantity

$$R_a=g\alpha H^3\Delta T\nu^{-1}\kappa^{-1}, \quad (21)$$

now called the *Rayleigh number*, exceeded a critical value

$$R_c=\pi^4a^{-2}(1+a^2)^3. \quad (22)$$

The minimum value of R_c, namely $27\pi^4/4$, occurs when $a^2=\tfrac{1}{2}$.

Saltzman (1962) derived a set of ordinary differential equations by expanding ψ and θ in double Fourier series in x and z, with functions of t alone for coefficients, and substituting these series into (17) and (18). He arranged the right-hand sides of the resulting equations in double-Fourier-series form, by replacing products of trigonometric functions of x (or z) by sums of trigonometric functions, and then equated coefficients of similar functions of x and z. He then reduced the resulting infinite system to a finite system by omitting reference to all but a specified finite set of functions of t, in the manner proposed by the writer (1960).

He then obtained time-dependent solutions by numerical integration. In certain cases all except three of the dependent variables eventually tended to zero, and these three variables underwent irregular, apparently nonperiodic fluctuations.

These same solutions would have been obtained if the series had at the start been truncated to include a total of three terms. Accordingly, in this study we shall let

$$a(1+a^2)^{-1}\kappa^{-1}\psi = X\sqrt{2}\sin(\pi a H^{-1}x)\sin(\pi H^{-1}z), \tag{23}$$

$$\pi R_c^{-1}R_a\Delta T^{-1}\theta = Y\sqrt{2}\cos(\pi a H^{-1}x)\sin(\pi H^{-1}z) - Z\sin(2\pi H^{-1}z), \tag{24}$$

where X, Y, and Z are functions of time alone. When expressions (23) and (24) are substituted into (17) and (18), and trigonometric terms other than those occurring in (23) and (24) are omitted, we obtain the equations

$$X^{\cdot} = -\sigma X + \sigma Y, \tag{25}$$

$$Y^{\cdot} = -XZ + rX - Y, \tag{26}$$

$$Z^{\cdot} = XY - bZ. \tag{27}$$

Here a dot denotes a derivative with respect to the dimensionless time $\tau = \pi^2 H^{-2}(1+a^2)\kappa t$, while $\sigma = \kappa^{-1}\nu$ is the *Prandtl number*, $r = R_c^{-1}R_a$, and $b = 4(1+a^2)^{-1}$. Except for multiplicative constants, our variables X, Y, and Z are the same as Saltzman's variables A, D, and G. Equations (25), (26), and (27) are the convection equations whose solutions we shall study.

In these equations X is proportional to the intensity of the convective motion, while Y is proportional to the temperature difference between the ascending and descending currents, similar signs of X and Y denoting that warm fluid is rising and cold fluid is descending. The variable Z is proportional to the distortion of the vertical temperature profile from linearity, a positive value indicating that the strongest gradients occur near the boundaries.

Equations (25)–(27) may give realistic results when the Rayleigh number is slightly supercritical, but their solutions cannot be expected to resemble those of (17) and (18) when strong convection occurs, in view of the extreme truncation.

6. Applications of linear theory

Although equations (25)–(27), as they stand, do not have the form of (4), a number of linear transformations will convert them to this form. One of the simplest of these is the transformation

$$X' = X, \quad Y' = Y, \quad Z' = Z - r - \sigma. \tag{28}$$

Solutions of (25)–(27) therefore remain bounded within a region R as $\tau \rightarrow \infty$, and the general results of Sections 2, 3 and 4 apply to these equations.

The stability of a solution $X(\tau)$, $Y(\tau)$, $Z(\tau)$ may be formally investigated by considering the behavior of small superposed perturbations $x_0(\tau)$, $y_0(\tau)$, $z_0(\tau)$. Such perturbations are temporarily governed by the linearized equations

$$\begin{bmatrix} x_0 \\ y_0 \\ z_0 \end{bmatrix}^{\cdot} = \begin{bmatrix} -\sigma & \sigma & 0 \\ (r-Z) & -1 & -X \\ Y & X & -b \end{bmatrix} \begin{bmatrix} x_0 \\ y_0 \\ z_0 \end{bmatrix}. \tag{29}$$

Since the coefficients in (29) vary with time, unless the basic state X, Y, Z is a steady-state solution of (25)–(27), a general solution of (29) is not feasible. However, the variation of the volume V_0 of a small region in phase space, as each point in the region is displaced in accordance with (25)–(27), is determined by the diagonal sum of the matrix of coefficients; specifically

$$V_0^{\cdot} = -(\sigma + b + 1)V_0. \tag{30}$$

This is perhaps most readily seen by visualizing the motion in phase space as the flow of a fluid, whose divergence is

$$\frac{\partial X^{\cdot}}{\partial X} + \frac{\partial Y^{\cdot}}{\partial Y} + \frac{\partial Z^{\cdot}}{\partial Z} = -(\sigma + b + 1). \tag{31}$$

Hence each small volume shrinks to zero as $\tau \rightarrow \infty$, at a rate independent of X, Y, and Z. This does not imply that each small volume shrinks to a point; it may simply become flattened into a surface. It follows that the volume of the region initially enclosed by the surface S shrinks to zero at this same rate, so that all trajectories ultimately become confined to a specific subspace having zero volume. This subspace contains all those trajectories which lie entirely within R, and so contains all central trajectories.

Equations (25)–(27) possess the steady-state solution $X = Y = Z = 0$, representing the state of no convection. With this basic solution, the characteristic equation of the matrix in (29) is

$$[\lambda + b][\lambda^2 + (\sigma+1)\lambda + \sigma(1-r)] = 0. \tag{32}$$

This equation has three real roots when $r > 0$; all are negative when $r < 1$, but one is positive when $r > 1$. The criterion for the onset of convection is therefore $r = 1$, or $R_a = R_c$, in agreement with Rayleigh's result.

When $r > 1$, equations (25)–(27) possess two additional steady-state solutions $X = Y = \pm\sqrt{b(r-1)}$, $Z = r - 1$.

For either of these solutions, the characteristic equation of the matrix in (29) is

$$\lambda^3+(\sigma+b+1)\lambda^2+(r+\sigma)b\lambda+2\sigma b(r-1)=0. \quad (33)$$

This equation possesses one real negative root and two complex conjugate roots when $r>1$; the complex conjugate roots are pure imaginary if the product of the coefficients of λ^2 and λ equals the constant term, or

$$r=\sigma(\sigma+b+3)(\sigma-b-1)^{-1}. \quad (34)$$

This is the critical value of r for the instability of steady convection. Thus if $\sigma<b+1$, no positive value of r satisfies (34), and steady convection is always stable, but if $\sigma>b+1$, steady convection is unstable for sufficiently high Rayleigh numbers. This result of course applies only to idealized convection governed by (25)–(27), and not to the solutions of the partial differential equations (17) and (18).

The presence of complex roots of (34) shows that if unstable steady convection is disturbed, the motion will oscillate in intensity. What happens when the disturbances become large is not revealed by linear theory. To investigate finite-amplitude convection, and to study the subspace to which trajectories are ultimately confined, we turn to numerical integration.

TABLE 1. Numerical solution of the convection equations. Values of X, Y, Z are given at every fifth iteration N, for the first 160 iterations.

N	X	Y	Z
0000	0000	0010	0000
0005	0004	0012	0000
0010	0009	0020	0000
0015	0016	0036	0002
0020	0030	0066	0007
0025	0054	0115	0024
0030	0093	0192	0074
0035	0150	0268	0201
0040	0195	0234	0397
0045	0174	0055	0483
0050	0097	−0067	0415
0055	0025	−0093	0340
0060	−0020	−0089	0298
0065	−0046	−0084	0275
0070	−0061	−0083	0262
0075	−0070	−0086	0256
0080	−0077	−0091	0255
0085	−0084	−0095	0258
0090	−0089	−0098	0266
0095	−0093	−0098	0275
0100	−0094	−0093	0283
0105	−0092	−0086	0297
0110	−0088	−0079	0286
0115	−0083	−0073	0281
0120	−0078	−0070	0273
0125	−0075	−0071	0264
0130	−0074	−0075	0257
0135	−0076	−0080	0252
0140	−0079	−0087	0251
0145	−0083	−0093	0254
0150	−0088	−0098	0262
0155	−0092	−0099	0271
0160	−0094	−0096	0281

7. Numerical integration of the convection equations

To obtain numerical solutions of the convection equations, we must choose numerical values for the constants. Following Saltzman (1962), we shall let $\sigma=10$ and $a^2=\frac{1}{2}$, so that $b=8/3$. The critical Rayleigh number for instability of steady convection then occurs when $r=470/19=24.74$.

We shall choose the slightly supercritical value $r=28$. The states of steady convection are then represented by the points $(6\sqrt{2}, 6\sqrt{2}, 27)$ and $(-6\sqrt{2}, -6\sqrt{2}, 27)$ in phase space, while the state of no convection corresponds to the origin (0,0,0).

We have used the double-approximation procedure for numerical integration, defined by (9), (10), and (14). The value $\Delta\tau=0.01$ has been chosen for the dimensionless time increment. The computations have been performed on a Royal McBee LGP-30 electronic com-

TABLE 2. Numerical solution of the convection equations. Values of X, Y, Z are given at every iteration N for which Z possesses a relative maximum, for the first 6000 iterations.

N	X	Y	Z	N	X	Y	Z
0045	0174	0055	0483	3029	0117	0075	0352
0107	−0091	−0083	0287	3098	0123	0076	0365
0168	−0092	−0084	0288	3171	0134	0082	0383
0230	−0092	−0084	0289	3268	0155	0069	0435
0292	−0092	−0083	0290	3333	−0114	−0079	0342
0354	−0093	−0083	0292	3400	−0117	−0077	0350
0416	−0093	−0083	0293	3468	−0125	−0083	0361
0478	−0094	−0082	0295	3541	−0129	−0073	0378
0540	−0094	−0082	0296	3625	−0146	−0074	0413
0602	−0095	−0082	0298	3695	0127	0079	0370
0664	−0096	−0083	0300	3772	0136	0072	0394
0726	−0097	−0083	0302	3853	−0144	−0077	0407
0789	−0097	−0081	0304	3926	0129	0072	0380
0851	−0099	−0083	0307	4014	0148	0068	0421
0914	−0100	−0081	0309	4082	−0120	−0074	0359
0977	−0100	−0080	0312	4153	−0129	−0078	0375
1040	−0102	−0080	0315	4233	−0144	−0082	0404
1103	−0104	−0081	0319	4307	0135	0081	0385
1167	−0105	−0079	0323	4417	−0162	−0069	0450
1231	−0107	−0079	0328	4480	0106	0081	0324
1295	−0111	−0082	0333	4544	0109	0082	0329
1361	−0111	−0077	0339	4609	0110	0080	0334
1427	−0116	−0079	0347	4675	0112	0076	0341
1495	−0120	−0077	0357	4741	0118	0081	0349
1566	−0125	−0072	0371	4810	0120	0074	0360
1643	−0139	−0077	0396	4881	0130	0081	0376
1722	0140	0075	0401	4963	0141	0068	0406
1798	−0135	−0072	0391	5035	−0133	−0081	0381
1882	0146	0074	0413	5124	−0151	−0076	0422
1952	−0127	−0078	0370	5192	0119	0075	0358
2029	−0135	−0070	0393	5262	0129	0083	0372
2110	0146	0083	0408	5340	0140	0079	0397
2183	−0128	−0070	0379	5419	−0137	−0067	0399
2268	−0144	−0066	0415	5495	0140	0081	0394
2337	0126	0079	0368	5576	−0141	−0072	0405
2412	0137	0081	0389	5649	0135	0082	0384
2501	−0153	−0080	0423	5752	0160	0074	0443
2569	0119	0076	0357	5816	−0110	−0081	0332
2639	0129	0082	0371	5881	−0113	−0082	0339
2717	0136	0070	0395	5948	−0114	−0075	0346
2796	−0143	−0079	0402				
2871	0134	0076	0388				
2962	−0152	−0072	0426				

›uting machine. Approximately one second per itera-ion, aside from output time, is required.

For initial conditions we have chosen a slight de-›arture from the state of no convection, namely (0,1,0). Cable 1 has been prepared by the computer. It gives the ·alues of N (the number of iterations), X, Y, and Z at ·very fifth iteration for the first 160 iterations. In the ›rinted output (but not in the computations) the values f X, Y, and Z are multiplied by ten, and then only hose figures to the left of the decimal point are printed. Chus the states of steady convection would appear as ›084, 0084, 0270 and −0084, −0084, 0270, while the tate of no convection would appear as 0000, 0000, 0000.

The initial instability of the state of rest is evident. All hree variables grow rapidly, as the sinking cold fluid s replaced by even colder fluid from above, and the ising warm fluid by warmer fluid from below, so that by tep 35 the strength of the convection far exceeds that f steady convection. Then Y diminishes as the warm luid is carried over the top of the convective cells, so hat by step 50, when X and Y have opposite signs, varm fluid is descending and cold fluid is ascending. The notion thereupon ceases and reverses its direction, as ndicated by the negative values of X following step 60. Sy step 85 the system has reached a state not far from hat of steady convection. Between steps 85 and 150 it xecutes a complete oscillation in its intensity, the light amplification being almost indetectable.

The subsequent behavior of the system is illustrated n Fig. 1, which shows the behavior of Y for the first 3000 iterations. After reaching its early peak near step 35 and then approaching equilibrium near step 85, it ndergoes systematic amplified oscillations until near tep 1650. At this point a critical state is reached, and hereafter Y changes sign at seemingly irregular inter-·als, reaching sometimes one, sometimes two, and some-imes three or more extremes of one sign before changing ign again.

Fig. 2 shows the projections on the X-Y- and Y-Z-›lanes in phase space of the portion of the trajectory orresponding to iterations 1400–1900. The states of teady convection are denoted by C and C'. The first ›ortion of the trajectory spirals outward from the ·icinity of C', as the oscillations about the state of teady convection, which have been occurring since step 5, continue to grow. Eventually, near step 1650, it rosses the X-Z-plane, and is then deflected toward the eighborhood of C. It temporarily spirals about C, but rosses the X-Z-plane after one circuit, and returns to he neighborhood of C', where it soon joins the spiral ›ver which it has previously traveled. Thereafter it rosses from one spiral to the other at irregular intervals.

Fig. 3, in which the coordinates are Y and Z, is based ıpon the printed values of X, Y, and Z at every fifth teration for the first 6000 iterations. These values deter-nine X as a smooth single-valued function of Y and Z ›ver much of the range of Y and Z; they determine X

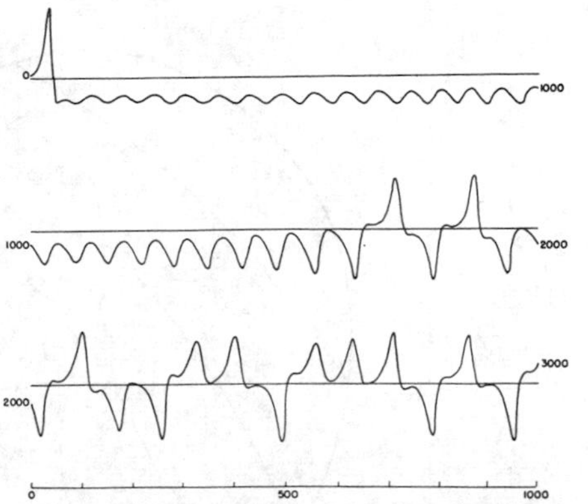

FIG. 1. Numerical solution of the convection equations. Graph of Y as a function of time for the first 1000 iterations (upper curve), second 1000 iterations (middle curve), and third 1000 iterations (lower curve).

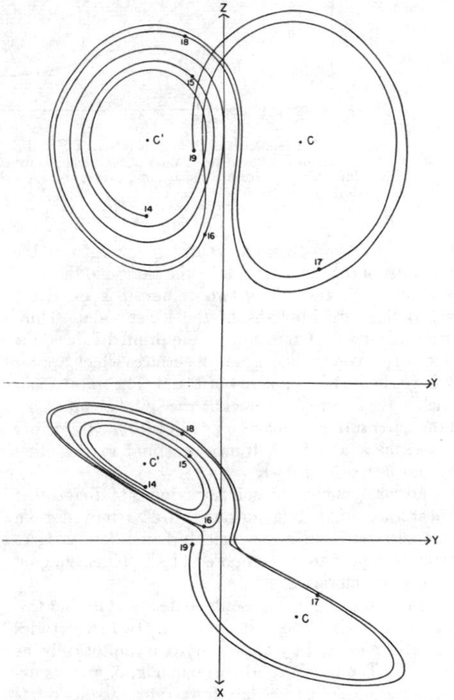

FIG. 2. Numerical solution of the convection equations. Projections on the X-Y-plane and the Y-Z-plane in phase space of the segment of the trajectory extending from iteration 1400 to iteration 1900. Numerals "14," "15," etc., denote positions at iterations 1400, 1500, etc. States of steady convection are denoted by C and C'.

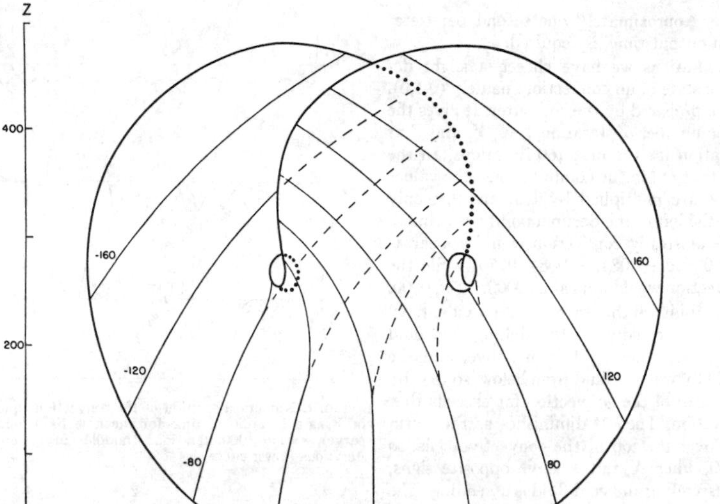

FIG. 3. Isopleths of X as a function of Y and Z (thin solid curves), and isopleths of the lower of two values of X, where two values occur (dashed curves), for approximate surfaces formed by all points on limiting trajectories. Heavy solid curve, and extensions as dotted curves, indicate natural boundaries of surfaces.

as one of two smooth single-valued functions over the remainder of the range. In Fig. 3 the thin solid lines are isopleths of X, and where two values of X exist, the dashed lines are isopleths of the lower value. Thus, within the limits of accuracy of the printed values, the trajectory is confined to a pair of surfaces which appear to merge in the lower portion of Fig. 3. The spiral about C lies in the upper surface, while the spiral about C' lies in the lower surface. Thus it is possible for the trajectory to pass back and forth from one spiral to the other without intersecting itself.

Additional numerical solutions indicate that other trajectories, originating at points well removed from these surfaces, soon meet these surfaces. The surfaces therefore appear to be composed of all points lying on limiting trajectories.

Because the origin represents a steady state, no trajectory can pass through it. However, two trajectories emanate from it, i.e., approach it asymptotically as $\tau \rightarrow -\infty$. The heavy solid curve in Fig. 3, and its extensions as dotted curves, are formed by these two trajectories. Trajectories passing close to the origin will tend to follow the heavy curve, but will not cross it, so that the heavy curve forms a natural boundary to the region which a trajectory can ultimately occupy. The holes near C and C' also represent regions which cannot be occupied after they have once been abandoned.

Returning to Fig. 2, we find that the trajectory apparently leaves one spiral only after exceeding some critical distance from the center. Moreover, the extent to which this distance is exceeded appears to determine the point at which the next spiral is entered; this in turn seems to determine the number of circuits to be executed before changing spirals again.

It therefore seems that some single feature of a given circuit should predict the same feature of the following circuit. A suitable feature of this sort is the maximum value of Z, which occurs when a circuit is nearly completed. Table 2 has again been prepared by the computer, and shows the values of X, Y, and Z at only those iterations N for which Z has a relative maximum. The succession of circuits about C and C' is indicated by the succession of positive and negative values of X and Y. Evidently X and Y change signs following a maximum which exceeds some critical value printed as about 385.

Fig. 4 has been prepared from Table 2. The abscissa is M_n, the value of the nth maximum of Z, while the ordinate is M_{n+1}, the value of the following maximum. Each point represents a pair of successive values of Z taken from Table 2. Within the limits of the round-off in tabulating Z, there is a precise two-to-one relation between M_n and M_{n+1}. The initial maximum $M_1=483$ is shown as if it had followed a maximum $M_0=385$, since maxima near 385 are followed by close approaches to the origin, and then by exceptionally large maxima.

It follows that an investigator, unaware of the nature of the governing equations, could formulate an empirical prediction scheme from the "data" pictured in Figs. 2 and 4. From the value of the most recent maximum of Z, values at future maxima may be obtained by repeated applications of Fig. 4. Values of X, Y, and Z between maxima of Z may be found from Fig. 2, by interpolating between neighboring curves. Of course, the accuracy of predictions made by this method is limited by the exactness of Figs. 2 and 4, and, as we shall see, by the accuracy with which the initial values of X, Y, and Z are observed.

Some of the implications of Fig. 4 are revealed by considering an idealized two-to-one correspondence between successive members of sequences $M_0, M_1, \cdots,$ consisting of numbers between zero and one. These sequences satisfy the relations

$$\begin{aligned} M_{n+1} &= 2M_n && \text{if } M_n<\tfrac{1}{2} \\ M_{n+1} &\text{ is undefined} && \text{if } M_n=\tfrac{1}{2} \\ M_{n+1} &= 2-2M_n && \text{if } M_n>\tfrac{1}{2}. \end{aligned} \quad (35)$$

The correspondence defined by (35) is shown in Fig. 5, which is an idealization of Fig. 4. It follows from repeated applications of (35) that in any particular sequence,

$$M_n = m_n \pm 2^n M_0, \quad (36)$$

where m_n is an even integer.

Consider first a sequence where $M_0=u/2^p$, where u is odd. In this case $M_{p-1}=\frac{1}{2}$, and the sequence terminates. These sequences form a denumerable set, and correspond to the trajectories which score direct hits upon the state of no convection.

Next consider a sequence where $M_0=u/2^p v$, where u and v are relatively prime odd numbers. Then if $k>0$, $M_{p+1+k}=u_k/v$, where u_k and v are relatively prime and u_k is even. Since for any v the number of proper fractions u_k/v is finite, repetitions must occur, and the sequence is periodic. These sequences also form a denumerable set, and correspond to periodic trajectories.

The periodic sequences having a given number of distinct values, or phases, are readily tabulated. In particular there are a single one-phase, a single two-phase, and two three-phase sequences, namely,

$2/3, \cdots,$

$2/5, 4/5, \cdots,$

$2/7, 4/7, 6/7, \cdots,$

$2/9, 4/9, 8/9, \cdots.$

The two three-phase sequences differ qualitatively in that the former possesses two numbers, and the latter only one number, exceeding $\frac{1}{2}$. Thus the trajectory corresponding to the former makes two circuits about C, followed by one about C' (or vice versa). The trajectory corresponding to the latter makes three circuits about C, followed by three about C', so that actually only Z varies in three phases, while X and Y vary in six.

Now consider a sequence where M_0 is not a rational fraction. In this case (36) shows that M_{n+k} cannot equal

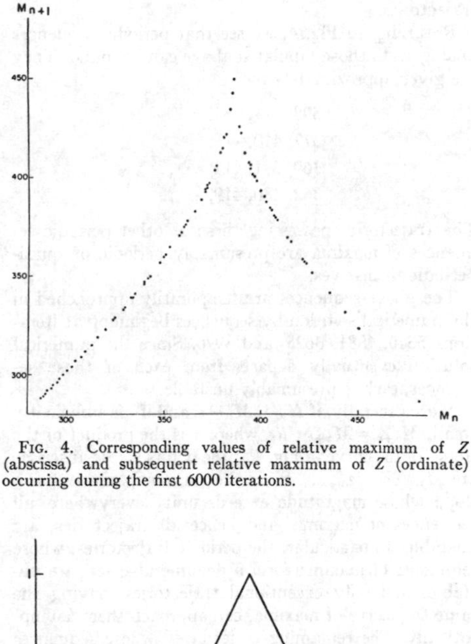

FIG. 4. Corresponding values of relative maximum of Z (abscissa) and subsequent relative maximum of Z (ordinate) occurring during the first 6000 iterations.

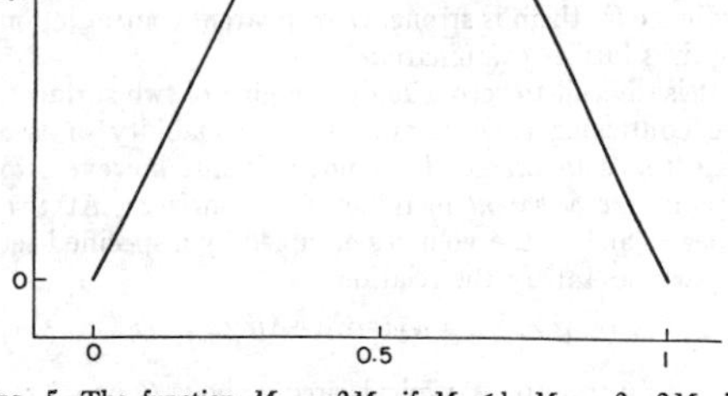

FIG. 5. The function $M_{n+1}=2M_n$ if $M_n<\frac{1}{2}$, $M_{n+1}=2-2M_n$ if $M_n>\frac{1}{2}$, serving as an idealization of the locus of points in Fig. 4.

M_n if $k>0$, so that no repetitions occur. These sequences, which form a nondenumerable set, may conceivably approach periodic sequences asymptotically and be quasi-periodic, or they may be nonperiodic.

Finally, consider two sequences M_0, M_1, $\cdots$ and M_0', M_1', $\cdots$, where $M_0'=M_0+\epsilon$. Then for a given k, if ϵ is sufficiently small, $M_k'=M_k\pm 2^k\epsilon$. All sequences are therefore unstable with respect to small modifications. In particular, all periodic sequences are unstable, and no other sequences can approach them asymptotically. All sequences except a set of measure zero are therefore nonperiodic, and correspond to nonperiodic trajectories.

Returning to Fig. 4, we see that periodic sequences analogous to those tabulated above can be found. They are given approximately by

$$\begin{array}{l} 398, \cdots, \\ 377, 410, \cdots, \\ 369, 391, 414, \cdots, \\ 362, 380, 419, \cdots. \end{array}$$

The trajectories possessing these or other periodic sequences of maxima are presumably periodic or quasi-periodic themselves.

The above sequences are temporarily approached in the numerical solution by sequences beginning at iterations 5340, 4881, 3625, and 3926. Since the numerical solution eventually departs from each of these sequences, each is presumably unstable.

More generally, if $M_n'=M_n+\epsilon$, and if ϵ is sufficiently small, $M_{n+k}'=M_{n+k}+\Lambda\epsilon$, where Λ is the product of the slopes of the curve in Fig. 4 at the points whose abscissas are M_n, $\cdots$, M_{n+k-1}. Since the curve apparently has a slope whose magnitude exceeds unity everywhere, all sequences of maxima, and hence all trajectories, are unstable. In particular, the periodic trajectories, whose sequences of maxima form a denumerable set, are unstable, and only exceptional trajectories, having the same sequences of maxima, can approach them asymptotically. The remaining trajectories, whose sequences of maxima form a nondenumerable set, therefore represent deterministic nonperiodic flow.

These conclusions have been based upon a finite segment of a numerically determined solution. They cannot be regarded as mathematically proven, even though the evidence for them is strong. One apparent contradiction requires further examination.

It is difficult to reconcile the merging of two surfaces, one containing each spiral, with the inability of two trajectories to merge. It is not difficult, however, to explain the *apparent* merging of the surfaces. At two times τ_0 and τ_1, the volumes occupied by a specified set of particles satisfy the relation

$$V_0(\tau_1)=e^{-(\sigma+b+1)(\tau_1-\tau_0)}V_0(\tau_0), \tag{37}$$

according to (30). A typical circuit about C or C' requires about 70 iterations, so that, for such a circuit, $\tau_2=\tau_1+0.7$, and, since $\sigma+b+1=41/3$,

$$V_0(\tau_1)=0.00007\,V_0(\tau_0). \tag{38}$$

Two particles separated from each other in a suitable direction can therefore come together very rapidly, and appear to merge.

It would seem, then, that the two surfaces merely appear to merge, and remain distinct surfaces. Following these surfaces along a path parallel to a trajectory, and circling C or C', we see that each surface is really a pair of surfaces, so that, where they appear to merge, there are really four surfaces. Continuing this process for another circuit, we see that there are really eight surfaces, etc., and we finally conclude that there is an infinite complex of surfaces, each extremely close to one or the other of two merging surfaces.

The infinite set of values at which a line parallel to the X-axis intersects these surfaces may be likened to the set of all numbers between zero and one whose decimal expansions (or some other expansions besides binary) contain only zeros and ones. This set is plainly nondenumerable, in view of its correspondence to the set of all numbers between zero and one, expressed in binary. Nevertheless it forms a set of measure zero. The sequence of ones and zeros corresponding to a particular surface contains a history of the trajectories lying in that surface, a one or zero immediately to the right of the decimal point indicating that the last circuit was about C or C', respectively, a one or zero in second place giving the same information about the next to the last circuit, etc. Repeating decimal expansions represent periodic or quasi-periodic trajectories, and, since they define rational fractions, they form a denumerable set.

If one first visualizes this infinite complex of surfaces, it should not be difficult to picture nonperiodic deterministic trajectories embedded in these surfaces.

8. Conclusion

Certain mechanically or thermally forced nonconservative hydrodynamical systems may exhibit either periodic or irregular behavior when there is no obviously related periodicity or irregularity in the forcing process. Both periodic and nonperiodic flow are observed in some experimental models when the forcing process is held constant, within the limits of experimental control. Some finite systems of ordinary differential equations designed to represent these hydrodynamical systems possess periodic analytic solutions when the forcing is strictly constant. Other such systems have yielded nonperiodic numerical solutions.

A finite system of ordinary differential equations representing forced dissipative flow often has the property that all of its solutions are ultimately confined within the same bounds. We have studied in detail the properties of solutions of systems of this sort. Our principal results concern the instability of nonperiodic solutions. A nonperiodic solution with no transient component must be unstable, in the sense that solutions temporarily approximating it do not continue to do so. A nonperiodic solution with a transient component is sometimes stable, but in this case its stability is one of its transient properties, which tends to die out.

To verify the existence of deterministic nonperiodic flow, we have obtained numerical solutions of a system of three ordinary differential equations designed to represent a convective process. These equations possess three steady-state solutions and a denumerably infinite set of periodic solutions. All solutions, and in particular the periodic solutions, are found to be unstable. The remaining solutions therefore cannot in general approach the periodic solutions asymptotically, and so are nonperiodic.

When our results concerning the instability of nonperiodic flow are applied to the atmosphere, which is ostensibly nonperiodic, they indicate that prediction of the sufficiently distant future is impossible by any method, unless the present conditions are known exactly. In view of the inevitable inaccuracy and incompleteness of weather observations, precise very-long-range forecasting would seem to be non-existent.

There remains the question as to whether our results really apply to the atmosphere. One does not usually regard the atmosphere as either deterministic or finite, and the lack of periodicity is not a mathematical certainty, since the atmosphere has not been observed forever.

The foundation of our principal result is the eventual necessity for any bounded system of finite dimensionality to come arbitrarily close to acquiring a state which it has previously assumed. If the system is stable, its future development will then remain arbitrarily close to its past history, and it will be quasi-periodic.

In the case of the atmosphere, the crucial point is then whether analogues must have occurred since the state of the atmosphere was first observed. By analogues, we mean specifically two or more states of the atmosphere, together with its environment, which resemble each other so closely that the differences may be ascribed to errors in observation. Thus, to be analogues, two states must be closely alike in regions where observations are accurate and plentiful, while they need not be at all alike in regions where there are no observations at all, whether these be regions of the atmosphere or the environment. If, however, some unobserved features are implicit in a succession of observed states, two successions of states must be nearly alike in order to be analogues.

If it is true that two analogues have occurred since atmospheric observation first began, it follows, since the atmosphere has not been observed to be periodic, that the successions of states following these analogues must eventually have differed, and no forecasting scheme could have given correct results both times. If, instead, analogues have not occurred during this period, some accurate very-long-range prediction scheme, using observations at present available, may exist. But, if it does exist, the atmosphere will acquire a quasi-periodic behavior, never to be lost, once an analogue occurs. This quasi-periodic behavior need not be established, though, even if very-long-range forecasting is feasible, if the variety of possible atmospheric states is so immense that analogues need never occur. It should be noted that these conclusions do not depend upon whether or not the atmosphere is deterministic.

There remains the very important question as to how long is "very-long-range." Our results do not give the answer for the atmosphere; conceivably it could be a few days or a few centuries. In an idealized system, whether it be the simple convective model described here, or a complicated system designed to resemble the atmosphere as closely as possible, the answer may be obtained by comparing pairs of numerical solutions having nearly identical initial conditions. In the case of the real atmosphere, if all other methods fail, we can wait for an analogue.

Acknowledgments. The writer is indebted to Dr. Barry Saltzman for bringing to his attention the existence of nonperiodic solutions of the convection equations. Special thanks are due to Miss Ellen Fetter for handling the many numerical computations and preparing the graphical presentations of the numerical material.

REFERENCES

Birkhoff, G. O., 1927: *Dynamical systems.* New York, Amer. Math. Soc., Colloq. Publ., 295 pp.

Ford, L. R., 1933: *Differential equations.* New York, McGraw-Hill, 264 pp.

Fultz, D., R. R. Long, G. V. Owens, W. Bohan, R. Kaylor and J. Weil, 1959: Studies of thermal convection in a rotating cylinder with some implications for large-scale atmospheric motions. *Meteor. Monog,* 4(21), Amer. Meteor. Soc., 104 pp.

Gibbs, J. W., 1902: *Elementary principles in statistical mechanics.* New York, Scribner, 207 pp.

Hide, R., 1958: An experimental study of thermal convection in a rotating liquid. *Phil. Trans. Roy. Soc. London,* (A), **250**, 441–478.

Lorenz, E. N., 1960: Maximum simplification of the dynamic equations. *Tellus,* **12**, 243–254.

——, 1962a: Simplified dynamic equations applied to the rotating-basin experiments. *J. atmos. Sci.,* **19**, 39–51.

——, 1962b: The statistical prediction of solutions of dynamic equations. *Proc. Internat. Symposium Numerical Weather Prediction,* Tokyo, 629–635.

Nemytskii, V. V., and V. V. Stepanov, 1960: *Qualitative theory of differential equations.* Princeton, Princeton Univ. Press, 523 pp.

Poincaré, H., 1881: Mémoire sur les courbes définies par une équation différentielle. *J. de Math.,* **7**, 375–442.

Rayleigh, Lord, 1916: On convective currents in a horizontal layer of fluid when the higher temperature is on the under side. *Phil. Mag.,* **32**, 529–546.

Saltzman, B., 1962: Finite amplitude free convection as an initial value problem—I. *J. atmos. Sci.,* **19**, 329–341.

Journal of Statistical Physics, Vol. 21, No. 6, 1979

Sequences of Infinite Bifurcations and Turbulence in a Five-Mode Truncation of the Navier–Stokes Equations

Valter Franceschini[1] **and Claudio Tebaldi**[2]

Received May 21, 1979

Two infinite sequences of orbits leading to turbulence in a five-mode truncation of the Navier–Stokes equations for a 2-dimensional incompressible fluid on a torus are studied in detail. Their compatibility with Feigenbaum's theory of universality in certain infinite sequences of bifurcations is verified and some considerations on their asymptotic behavior are inferred. An analysis of the Poincaré map is performed, showing how the turbulent behavior is approached gradually when, with increasing Reynolds number, no stable fixed point or periodic orbit is present and all the unstable ones become more and more unstable, in close analogy with the Lorenz model.

KEY WORDS: Navier–Stokes equations; turbulence; strange attractors; Poincarè map; infinite sequences of periodic orbits; stable and hyperbolic orbits collapse; universal properties in infinite sequences of bifurcations.

1. INTRODUCTION

A model obtained by a suitable five-mode truncation of the Navier–Stokes equations for a two-dimensional incompressible fluid on a torus has been presented in Ref. 1.

The system of nonlinear ordinary differential equations resulting from such a truncation is

$$\begin{aligned}
\dot{x}_1 &= -2x_1 + 4x_2x_3 + 4x_4x_5 \\
\dot{x}_2 &= -9x_2 + 3x_1x_3 \\
\dot{x}_3 &= -5x_3 - 7x_1x_2 + r \\
\dot{x}_4 &= -5x_4 - x_1x_5 \\
\dot{x}_5 &= -x_5 - 3x_1x_4
\end{aligned}$$

[1] Istituto Matematico, Università di Modena, Modena, Italy.
[2] Dipartimento di Matematica, Università di Ancona, Ancona, Italy and Istituto di Fisica, Università di Bologna, Bologna, Italy.

(where r is the Reynolds number), and exhibits an interesting variety of different behaviors for different ranges of r. Keeping the same symbols as in Ref. 1 for the critical values of r, the most interesting feature is the stochastic behavior observed when $R_{12} < r < R_{13}$, with $28.73 < R_{12} < 29.0$ and $R_{13} \approx 33.43$.

In recent years much attention has been devoted to the study of models exhibiting such a feature when one or more parameters increase beyond certain critical values. The best known models of this kind are certainly the ones by Lorenz[(2–4)] and Hénon.[(5,6)] Ruelle and Takens[(7)] explain this stochastic behavior as a consequence of the appearance of an attractor with a complicated nature ("strange attractor"), on which the motion seems completely chaotic ("turbulence"). In addition to detailed studies on the nature of these attractors (see, for example, Lanford,[(8)] and Hénon and Pomeau[(6)]), strong interest has been focused upon the study of the mechanism of their generation.

In Ref. 1 it is shown that turbulence is reached through a long and rather complicated sequence of bifurcations related to two sequences of orbits: the former consists of four orbits with periods T, $2T$, $4T$, and $8T$, respectively, and the latter of five orbits of a different type with periods T^*, $2T^*$, $4T^*$, $8T^*$, and $16T^*$. In the following, $\mathscr{C}_i$, $i = 0, 1, 2, 3$, will refer to the orbits of the former sequence and $\mathscr{C}_i^*$, $i = 0, 1,..., 4$, to the orbits of the latter.[3] The orbit $\mathscr{C}_0^*$ is found for a value of r larger than but very close to the largest value for which $\mathscr{C}_3$ is still found.[4] It is then suggested that the sequence $\mathscr{C}_i$ is finite, different from $\mathscr{C}_i^*$, and $\mathscr{C}_3$ bifurcates in $\mathscr{C}_0^*$. Since this transition remains an obscure point, because it does not fit very well with the ideas of bifurcation theory, it seems interesting to us to investigate more deeply the two sequences of orbits.

A further reason for this investigation is to verify if the behavior of the sequence $\mathscr{C}_i^*$ is compatible with the strongly suggestive idea of universality in certain infinite sequences of bifurcations developed by Feigenbaum.[(9)] This exhaustive study has been possible because we have been able to apply numerical schemes that are more efficient for studying the stability of orbits and especially in searching for new orbits, even unstable ones. With these new techniques, for a better understanding of the generation of

[3] More precisely, $\mathscr{C}_i$ must be regarded as one of four symmetrically placed, identical orbits going through identical behavior, and the same for $\mathscr{C}_i^*$. There are then four sequences $\mathscr{C}_i$ and four $\mathscr{C}_i^*$, although we will refer simply to "the" sequence $\mathscr{C}_i$ or $\mathscr{C}_i^*$, when not otherwise required. The presence of quadruples of periodic solutions is accounted for by the symmetries $(x_1, x_2, x_3, -x_4, -x_5) \leftrightarrow (x_1, x_2, x_3, x_4, x_5)$, $(-x_1, -x_2, x_3, -x_4, x_5) \leftrightarrow (x_1, x_2, x_3, x_4, x_5)$, $(-x_1, -x_2, x_3, x_4, -x_5) \leftrightarrow (x_1, x_2, x_3, x_4, x_5)$.

[4] $\mathscr{C}_3$ is observed up to $r = 28.6660$, while $\mathscr{C}_0^*$ is first found for $r = 28.6662$.

the "strange attractor" we also have been able to reconsider the transitions through R_{12}, for r increasing, and R_{13}, for r decreasing, with which the system goes over to turbulent behavior. The results of these investigations are described in the following.

In Section 2 a detailed analysis of the two sequences of orbits $\mathscr{C}_i$ and $\mathscr{C}_i^*$ is given, showing that both of them are very likely to be infinite, with a phenomenon of hysteresis because of the simultaneous presence of the orbits $\mathscr{C}_i$, $i \geqslant 3$, with $\mathscr{C}_0^*$.

In Section 3 it is shown that the appearance of the "strange attractor" for r decreasing to R_{13} follows the collapse of the stable orbit present for the high-r regime[5] with an unstable hyperbolic one and that analogous phenomenology is present in the appearance of $\mathscr{C}_0^*$. An interpretation using the theory of the bifurcation of periodic orbits in generic conditions[(7)] is tried.

In Section 4 the compatibility of the, now two, infinite sequences of bifurcations with the universality theory developed by Feigenbaum is verified and some considerations on their asymptotic behavior are inferred.

In Section 5 a detailed analysis of the Poincaré map shows how the turbulent behavior is approached gradually when the previously stable periodic orbits, now all unstable, become more and more unstable for r increasing.

Finally, a schematic picture of the features exhibited by the system is presented in Section 6, together with some concluding remarks.

2. TWO INFINITE SEQUENCES OF BIFURCATIONS

In Ref. 1 it is shown that for a certain value of the Reynolds number r ($r = R_3 = 22.85370163 \cdots$) four previously stable fixed points become unstable and four stable periodic orbits, referred to as $\mathscr{C}_0$ in Section 1, arise via a Hopf bifurcation[6] around each fixed point. With increasing r, the periodic orbits are shown to go through a number of bifurcations, doubling in period and winding up twice as many times around the fixed points from which they are generated. This is shown to happen up to $r = 28.6660$, when three successive bifurcations have taken place, giving rise to the orbits $\mathscr{C}_1$, $\mathscr{C}_2$, $\mathscr{C}_3$. For $r = 28.6662$ four new stable orbits $\mathscr{C}_0^*$ are found, with structure and period different from the previous ones of $\mathscr{C}_3$, each of them winding up around two of the fixed points. It is stressed that no definite statement can be made about the fact that no further similar bifurcation

[5] We recall that for $r \to R_{13}$ from above, this stable periodic orbit bifurcates with a real eigenvalue crossing the unit circle at $+1$.

[6] For a detailed theory concerning the Hopf bifurcation and its applications see Ref. 11.

takes place in the sequence $\mathscr{C}_i$; all the phenomenology observed, however, leads to the conjecture that $\mathscr{C}_3$ bifurcates in $\mathscr{C}_0^*$.

The difficulty in interpreting this point according to the bifurcation theory motivated us to reconsider it, applying Newton's method to obtain and analyze periodic solutions. Once the approximate initial point is close enough, the convergence of the method is fast, both for stable and unstable periodic orbits. The main purpose of such a method is in fact to be able to find unstable periodic solutions too.

Going back to the study of the first sequence of orbits, we have been able to determine the bifurcation points with a very good accuracy and, much more important, to find two more orbits in the sequence $\mathscr{C}_i$, i.e., $\mathscr{C}_4$ and $\mathscr{C}_5$ (see Table I). We have also verified that each of the orbits in the sequence $\mathscr{C}_i$ becomes unstable when an eigenvalue of the Liapunov matrix of the Poincaré map crosses the unit circle at the point -1.[7] The agreement with what is predicted by the bifurcation theory in this case is now complete, since upon bifurcation the previously stable orbit becomes unstable and a new stable orbit appears, with the period doubled (see Ref. 7). At this point it is reasonable to infer that the sequence $\mathscr{C}_i$ is infinite too. We have not carried out a further investigation for higher bifurcated orbits because of the high amount of computational time required even to consider only the next one.

Also for the sequence of orbits $\mathscr{C}_i^*$ the bifurcation points have been determined very accurately using the Newton method, up to the orbit $\mathscr{C}_4^*$ (see Table II). It has been verified that $\mathscr{C}_i^*$ is generated by a sequence of bifurcations of the same kind as those for $\mathscr{C}_i$, since also each orbit $\mathscr{C}_i^*$ becomes unstable with an eigenvalue of -1 in the Liapunov matrix of the Poincaré map.

[7] This was already seen in Ref. 1 for the orbit $\mathscr{C}_0$.

Table I. Bifurcation points ρ_i of the Periodic Orbits in the Sequence $\mathscr{C}_i$ and Relative Periods $T(\rho_i)$

i	ρ_i	$T(\rho_i)$
0	28.4105	0.81621
1	28.6399	1.64567
2	28.6641	3.29334
3	28.66776	6.58741
4	28.668463	13.17507
5	28.668611	26.35026

Table II. Bifurcation points ρ_i^* of the Periodic Orbits in the Sequence $\mathscr{C}_i^*$ and Relative Periods $T^*(\rho_i^*)$

i	ρ_i^*	$T^*(\rho_i^*)$
0	28.7013	3.80928
1	28.71606	7.62056
2	28.71926	15.24271
3	28.719947	30.48597
4	28.720103	60.97222

A question left open is how the orbit $\mathscr{C}_0^*$ appears. The answer is that an unstable orbit exists simultaneously with $\mathscr{C}_0^*$, very close to it, and the two collapse upon bifurcation for r decreasing. All the details will be given in the next section.

An important feature must be emphasized concerning the two sequences of orbits $\mathscr{C}_i$ and $\mathscr{C}_i^*$, i.e., the simultaneous presence of different stable orbits for the same value of the Reynolds number r in a certain range of r. For $r \geqslant 28.663$ in fact the stable orbit $\mathscr{C}_0^*$ is present together with one of the sequence $\mathscr{C}_i$. It is clear that $\mathscr{C}_0^*$ appears in the beginning with a very small basin of attraction; with increasing r, this becomes larger and larger, while that of the simultaneous orbit $\mathscr{C}_i$ gets smaller and smaller. $\mathscr{C}_4$ already appears with a very small basin of attraction and $\mathscr{C}_5$ even more so. The simultaneous presence of more than one attracting orbit is termed hysteresis and has the effect of causing a rather sensitive dependence of the asymptotic solution on the initial conditions. In this kind of model this was found by Curry[12] in a generalized Lorenz system and by us[13] as a very strong feature in a seven-mode truncation model of the two-dimensional Navier–Stokes equations.

3. COLLAPSE OF A STABLE ORBIT WITH AN UNSTABLE ONE

One of the interesting results of Ref. 1 is the observation that, after the second infinite sequence of orbits, the system shows the presence of two symmetric attractors, with all the characteristics of a "strange attractor," on which a random motion takes place. With increasing Reynolds number r, each attractor seems to shrink to a stable periodic orbit, present for all the high-r regime considered. Analysis of the stability of this orbit shows that, with decreasing r toward $R_{13} = 33.43$, an eigenvalue of the Poincaré map approaches the unit circle at the point $+1$. It seems of interest to us to attempt to obtain a better understanding of the way this bifurcation takes place, as it is connected with the transition to turbulence.

We have verified that after the bifurcation, i.e., for $r < R_{13}$, the orbit is no longer present. This fact suggests the hypothesis that the stable orbit could disappear by collapse with an unstable one present at the same time (see, for example, Brunowsky[14]). We have looked at how the fixed point of the Poincaré map for the stable orbit, using a fixed hyperplane, would move when r decreases toward R_{13}. Keeping in mind the hypothesis of collapse, by extrapolation we have been able to find the fixed point for an unstable orbit at a value of r close to the critical one. We have then followed the unstable orbit present together with the stable one, quite close to it, and disappearing for $r < R_{13}$. It has been verified that, with decreasing r toward R_{13}, the two orbits become closer and closer (see Fig. 1) and so do their periods, that for the stable orbit increasing, that of the unstable one decreasing. Figure 2, where the fixed points of the Poincaré map for the stable and unstable orbits are represented for different values of r approaching R_{13} from above, shows the phenomenon of collapse quite clearly.

The same detailed analysis performed on the stable orbit present in the high-r regime has been carried on for $\mathscr{C}_0{}^*$, since a study of its stability shows an eigenvalue of the Poincaré map approaching the unit circle at $+1$ for r

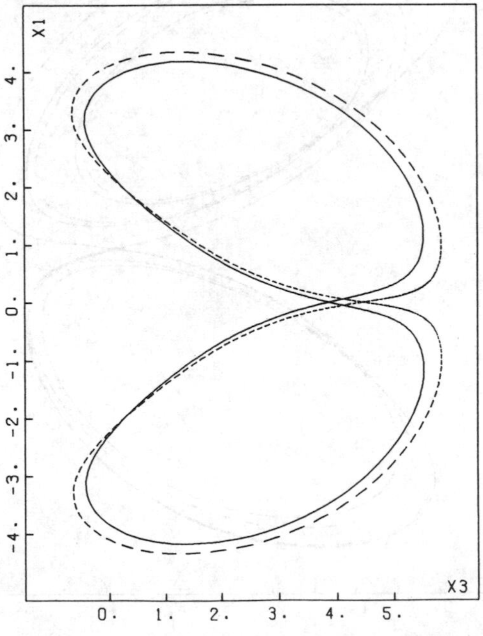

Fig. 1. Stable (——) and hyperbolic (- - -) orbits for $r = 33.60$.

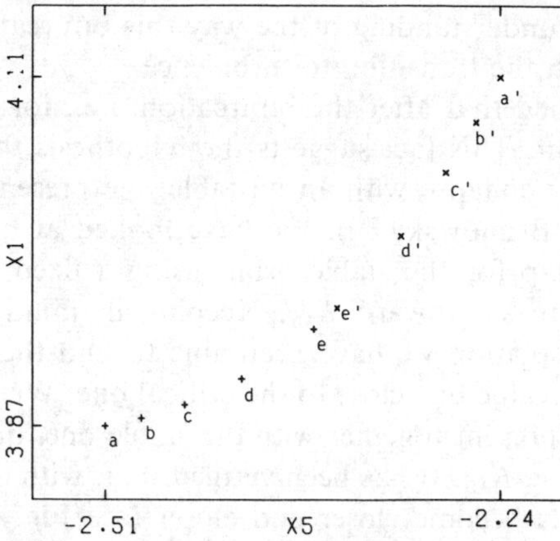

Fig. 2. Fixed points of the Poincaré map for the stable (+) and hyperbolic (×) orbits for r approaching R_{13} from above: (a, a') $r = 33.80$; (b, b') $r = 33.70$; (c, c') $r = 33.60$; (d, d') $r = 33.50$; (e, e') $r = 33.44$.

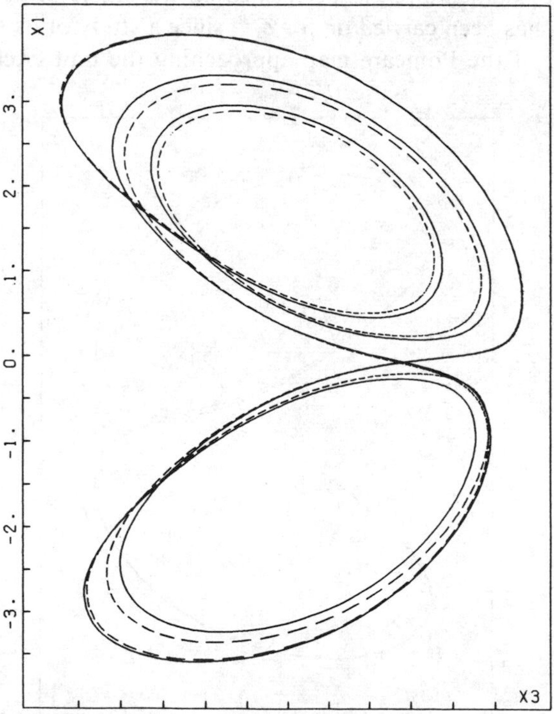

Fig. 3. Stable (——) and hyperbolic (- - -) orbits for $r = 28.695$.

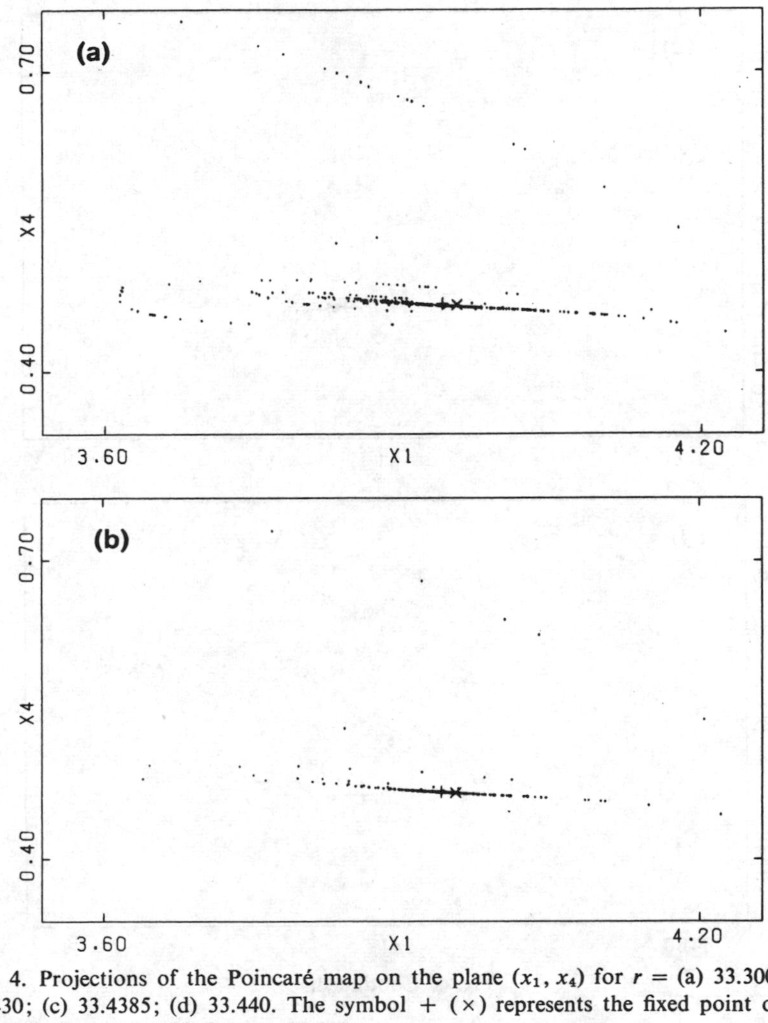

Fig. 4. Projections of the Poincaré map on the plane (x_1, x_4) for $r =$ (a) 33.300; (b) 33.430; (c) 33.4385; (d) 33.440. The symbol $+$ $(\times)$ represents the fixed point of the stable (hyperbolic) orbit for $r = 33.440$.

decreasing toward 28.663, the orbit disappearing below that value. For $\mathscr{C}_0^*$ too we have verified the identical phenomenon of collapse with an unstable orbit $\bar{\mathscr{C}}_0^*$ (see Fig. 3); the difference from the previous case is that no attractor close to the orbit is present after the bifurcation. The presence of the orbit $\bar{\mathscr{C}}_0^*$, besides explaining the bifurcation for $r = 28.663$, plays a role in the explanation of how the model goes over to "turbulent" behavior, as will be seen in Section 5.

In their fundamental paper, among other considerations, Ruelle and

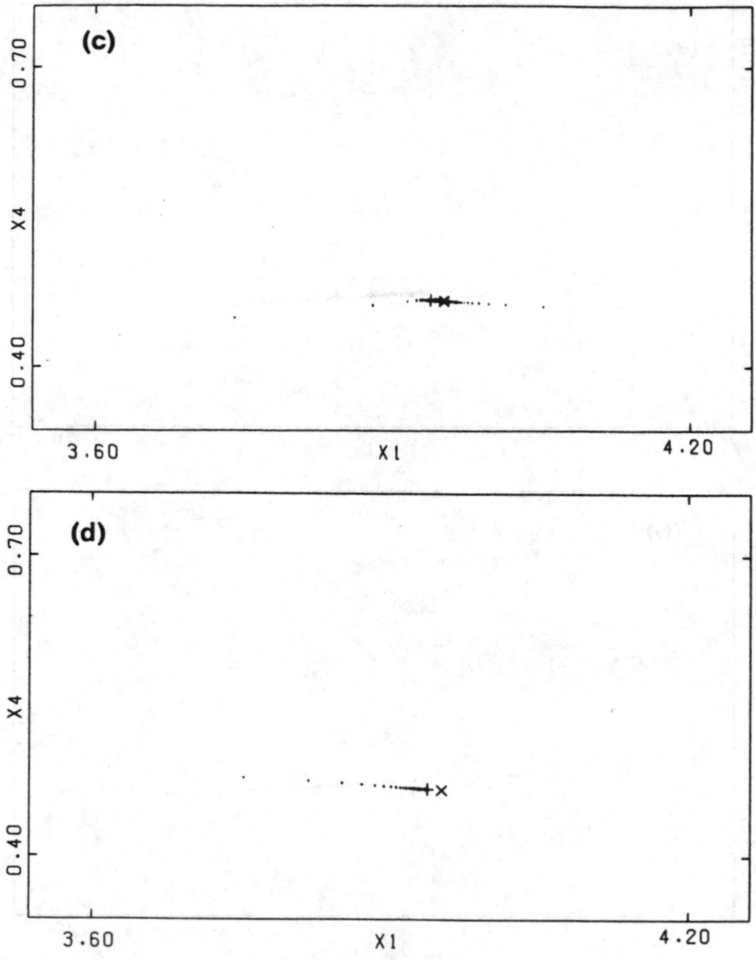

Fig. 4. Continued.

Takens[7] analyze the bifurcations of a stable periodic orbit in generic systems when the Poincaré map has only a finite number of isolated eigenvalues with modulus 1 and the others inside the open unit circle. They find that when only one eigenvalue crosses the unit circle at the point +1, one should expect the attracting closed orbit to disappear together with a hyperbolic closed one and no attractor close to the orbit to appear after the bifurcation has taken place. This is the exact phenomenology found for $\mathscr{C}_0^*$, but at first sight not that for the orbit in the high-r regime. In this last case in fact at the bifurcation point we observe the collapse of the two orbits, but after the bifurcation a strange attractor is present, apparently close to the orbits.

This phenomenology does not seem to be in agreement with Ref. 7, so we have tried to study in more detail the transition from the strange attractor to the periodic orbits.

For different values of r approaching R_{13} from below, we have considered the Poincaré map on the hyperplane $x_3 = 3.0$, $x_1 \geqslant 0$, plotting its projection on (x_1, x_4) together with, as a reference, the fixed points of the stable and hyperbolic orbits for r slightly greater than R_{13} ($r = 33.44$).

Figures 4a–4c clearly show how, as r approaches R_{13}, the points tend to dispose themselves along a line, getting denser and denser on a segment containing the fixed points of the two orbits. Looking at the projection of the intersection points on the plane (x_1, x_4), this segment is described from left to right, with a return mechanism which redescribes it always progressing in the same direction. Moreover, numerical evidence has been found for the two following facts: for $r \to R_{13}$, $33.4385 < R_{13} < 33.4390$, the length of the segment does not seem to tend to zero; at $r = R_{13}$ a stable fixed point for the Poincaré map appears on the segment.

A possible qualitative explanation for the phenomenology is the existence of some attracting variety, containing the "strange attractor" for $r < R_{13}$, and a stable periodic orbit, together with an unstable one, for $r > R_{13}$. The presence of such a manifold, attracting also for $r > R_{13}$, appears to be confirmed in Fig. 4d, where the approach to the fixed point of the stable orbit on the Poincaré map is shown for $r = 33.44$. Looking at the phenomenon for r increasing, we have then that the strange attractor does not "shrink" to the orbit, in agreement with the bifurcation theory.

For r decreasing, the stable orbit and the hyperbolic one collapse and, disappearing, are replaced by a "larger" attractor which occupies a portion of a manifold to which the orbits are always attracted for r near R_{13}, no matter whether larger or smaller; the diameter of the attractor does not tend to zero as $r \to R_{13}$.

4. COMPATIBILITY WITH A CONJECTURE OF UNIVERSALITY IN INFINITE SEQUENCES

In a recent paper, Feigenbaum[9] develops a very interesting theory concerning a large class of recursion relations $x_{n+1} = \lambda f(x_n)$ exhibiting infinite bifurcations, varying the parameter λ in the open interval (0, 1). They are shown to possess a structure essentially independent of the recursion function that, among other properties, is supposed to map the closed interval [0, 1] on itself and have a unique, twice differentiable maximum $\bar{x}$. For such a class of f, a λ_n exists such that a stable 2^n-point limit cycle including $\bar{x}$ exists. It is shown as numerical evidence that

$$\lim_{n\to\infty} \delta_n = \lim_{n\to\infty} \frac{\lambda_n - \lambda_{n-1}}{\lambda_{n+1} - \lambda_n} = \delta$$

i.e.,, the λ_n geometrically converge to a certain λ_∞ at the rate δ, independent of the specific function f.

The same asymptotic behavior is shown to take place for

$$\frac{\Lambda_n - \Lambda_{n-1}}{\Lambda_{n+1} - \Lambda_n}$$

where Λ_n is the nth bifurcation point. Moreover, when λ is increased in order to obtain the transition from a stable 2^n-point to a stable 2^{n+1}-point limit cycle, the local structure about $\bar{x}$ reproduces itself on a scale α_n times smaller. It is shown that

$$\lim_{n\to\infty} \alpha_n = \alpha$$

with α also f-independent. Both the numbers α and δ depend only on the order of the maximum $\bar{x}$ of f; for a normal (i.e., quadratic) maximum, it is found that $\delta = 4.6692 \cdots$ and $\alpha = 2.5029 \cdots$.

In a previous section we have described two sequences of bifurcations that are very likely to be infinite, even if for the reasons exposed there we could obtain only a limited number of terms. We have tried to verify numerically if the universal metric properties pointed out by Feigenbaum for one-dimensional mappings could hold in our dynamical system too, hoping for a convergence of α_n and δ_n as fast as that of one of the examples in Ref. 9. We have computed the ratios $\delta_i = (\rho_i - \rho_{i-1})/(\rho_{i+1} - \rho_i)$, $i = 1,..., 5$, for the sequence $\mathscr{C}_i$ and $\delta_i^* = (\rho_i^* - \rho_{i-1}^*)/(\rho_{i+1}^* - \rho_i^*)$ for $\mathscr{C}_i^*$, where ρ_i and ρ_i^* are the bifurcation points given in Tables I and II. Both sequences δ_i and δ_i^*, listed in Table III, seem to indicate a convergence, more rapid for δ_i^*, to numbers quite compatible with the one found by Feigenbaum. A comment is required by the last term δ_4^*. This has been computed knowing quite well that it might not have been completely reliable. In fact the numerical errors due to the large value of the period of $\mathscr{C}_4^*$ now become relevant compared with the very small variation of the parameter r. We have computed the term anyway because of the small number of terms available otherwise, to verify at least a persistence of the sequence around the value of δ.

Table III

i	δ_i	δ_i^*
1	24.22	2.57
2	9.48	4.63
3	6.54	4.64
4	5.29	4.42
5	4.73	—

An even more striking instance of the compatibility of our sequences with the one considered by Feigenbaum is found by looking at how the fixed points of the Poincaré map reproduce themselves upon transition from each orbit to the next one in the sequence. Calling Φ_r the Poincaré map on a hyperplane transverse to each orbit $\mathscr{C}_i$, we can write a recursion relation

$$x_{n+1} = \Phi_r(x_n)$$

When, for $\rho_{i-1} < r < \rho_i$, we consider the stable orbit $\mathscr{C}_i$, Φ_r has a stable 2^i-point limit cycle. In this way we have a recursion function similar to the one in Ref. 9 since in the ith bifurcation point ρ_i a 2^i-point limit cycle becomes unstable and a stable 2^{i+1}-point limit cycle appears, with the stable orbit $\mathscr{C}_{i+1}$ appearing.

A comparison with Feigenbaum's scale factors α_n can be attempted once something corresponding to $\bar{x}$ is found. Considering our limit cycles at the bifurcation points, we have observed the following. Calling $P_j^{(i)}$, $i = 0, 1, \ldots, 5$, $j = 1, \ldots, 2^i$, the points of the 2^i-point cycle, and $P_{2j-1}^{(i+1)}$ and $P_{2j}^{(i+1)}$ the two points bifurcating from $P_j^{(i)}$ in the 2^{i+1}-point cycle, we let

$$Q_1^{(i)} = P_{2k_0-1}^{(i)}, \qquad Q_2^{(i)} = P_{2k_0}^{(i)}$$

where k_0 is the index for which $d(P_{2k-1}^{(i)}, P_{2k}^{(i)})$ is maximum,[8] $k = 1, \ldots, 2^{i-1}$. Then, for $i \geqslant 2$, we have that $Q_1^{(i)}$ and $Q_2^{(i)}$ correspond either to $Q_1^{(i-1)}$ or to $Q_2^{(i-1)}$. This is equivalent to saying that if we consider the binary tree with the points $P_j^{(i)}$ as nodes of level $i + 1$, a path from $P_1^{(0)}$ to a certain $P_j^{(5)}$ exists, along which the points $P_j^{(i)}$ reproduce themselves with a scale factor that is maximum.

In Ref. 9 the scale factor, by definition $1/\alpha_i$, by which a cluster about a point of a 2^i-cycle reproduces itself is maximum if the point is $\bar{x}$. For this reason it seems relevant to compute the α_i along the path we have specified before, proposing in this way a correspondence between $Q^{(i)}$ and $\bar{x}$. In Table IV we list the values of the α_i for the Poincaré map on the hyperplane $x_3 = 1.0$ for each coordinate and the Euclidean distance.

The same procedure has been followed for the sequence $\mathscr{C}_i^*$, the corre-

[8] d is the usual Euclidean distance in R^5.

Table IV

	x_1	x_2	x_4	x_5	d
α_1	4.20	4.21	5.05	5.09	4.51
α_2	3.43	3.43	3.29	3.30	3.39
α_3	2.90	2.90	2.94	2.94	2.91
α_4	2.53	2.53	2.52	2.52	2.53

Table V

	x_1	x_2	x_4	x_5	d
α_1^*	2.31	2.31	2.38	2.38	2.34
α_2^*	2.65	2.65	2.60	2.60	2.63
α_3^*	2.43	2.43	2.45	2.45	2.44

sponding limit cycle being $m \cdot 2^i$-point, where m is dependent on the plane chosen for the Poincaré map. The scale factors α_i^* for this case are given in Table V, also for the Poincaré map on $x_3 = 1.0$, limited to $x_1 > 0$, with $m = 3$.

For different choices of the hyperplane for the Poincaré map the results are essentially unchanged.

The compatibility of our numerical values for α_i and α_i^* with the asymptotic value of α computed by Feigenbaum seems evident, even if we could compute only a few terms.

Our results make possible the hypothesis that universal metric properties of one-dimensional mappings also hold in dynamical systems with infinite sequences of bifurcations. The fact that complicated n-dimensional phenomena possess characteristics in some sense "one-dimensional" appears significant and very suggestive.[9]

These arguments give more support to our hypothesis of Section 2 on the two sequences being infinite and allow us now to estimate the asymptotic values for the critical Reynolds number ρ_i and ρ_i^*. We obtain for them

$$\rho_\infty = 28.668652, \qquad \rho_\infty^* = 28.720135$$

A complete numerical definition of these values from the model appears impossible, however.

5. ONSET OF TURBULENCE

In Ref. 1 it is shown that for $R_{12} < r < R_{13}$, randomly chosen initial data lead to two attractors (see Fig. 5), on which the motion appears to be completely random, the trajectories looking exactly like the ones found by Lorenz in his model.[(2)] These two "strange attractors" are localized in two

[9] A detailed study generalizing Feigenbaum's results has been carried out by Derrida *et al.*[(10)] Moreover, they have pointed out that also in the Hénon two-dimensional mapping[(5)] the bifurcation rate δ for the sequence of stable periods 2^n is the same as in Ref. 9. An intuitive explanation for this is indicated by the contracting nature of the transformation.

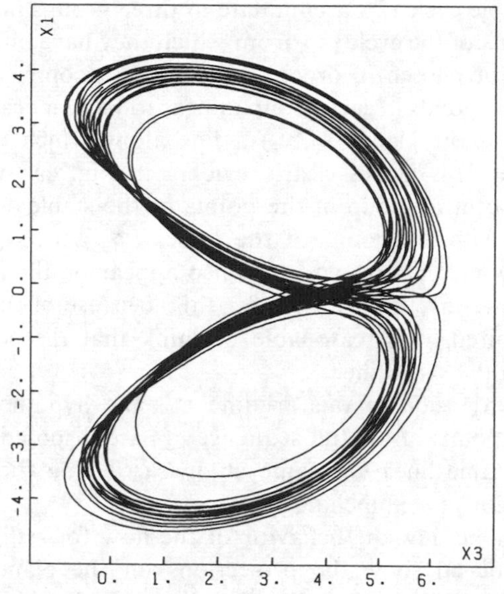

Fig. 5. One of the two "strange attractors" for $r = 33.0$.

symmetric regions, each of them surrounding two fixed points, two sequences of orbits $\mathscr{C}_i$, two sequences $\mathscr{C}_i{}^*$,[10] and two orbits $\bar{\mathscr{C}}_0{}^*$, all unstable in this range of the Reynolds number r.

In the following we give a detailed analysis of the transition from the periodic behavior of the sequences $\mathscr{C}_i{}^*$ to the turbulent behavior on the two attractors.

We consider the motion with random initial data for different values of r, starting from $r = 28.72$, studying the Poincaré map on the hyperplane $x_3 = 1.4$, limited to the region $x_1 \geqslant 0$, $x_4 \geqslant 0$ for simplicity.[11] The orbits of only one of the sequences $\mathscr{C}_i$, the orbits of two of the sequences $\mathscr{C}_i{}^*$, and two orbits $\bar{\mathscr{C}}_0{}^*$, all intersecting the hyperplane $x_3 = 1.4$, are present in the region considered. These intersections are the elements of n-point limit cycles for the Poincaré map. Denoting by c_i a 2^i-point limit cycle related to an orbit $\mathscr{C}_i$, and by $c^*_{m,i}$ an $m \cdot 2^i$-point cycle related to an orbit $\mathscr{C}_i{}^*$, we find the Poincaré map then has one sequence of cycles c_i, one $c^*_{2,i}$, one $c^*_{3,i}$, one cycle $\bar{c}^*_{2,0}$, and one $\bar{c}^*_{3,0}$. The complexity of the situation is evident from Fig. 6a, where for $r = 28.72$ we have represented only the cycle $c^*_{3,4}$, stable for this value of r, and the unstable cycles c_0, c_1, $c^*_{2,0}$, $c^*_{3,0}$, $\bar{c}^*_{2,0}$, $\bar{c}^*_{3,0}$. The

[10] For a better understanding see Ref. 1, especially Figs. 1a, 7a–7d, and 9–11.

[11] Suitable changes in sign, allowed for by the symmetries present in the model, make it possible to study any orbit in this region.

points of the stable cycle $c_{3,4}^*$ accumulate in three groups in a neighborhood of the three points of the cycle $c_{3,0}^*$ from which they have bifurcated. Because of the scale factor chosen in order to represent a complete picture of the Poincaré map, the points in each group appear to be practically indistinguishable, but show in the plane (x_1, x_4) a line along which they duplicate at each bifurcation. This fact is clearly evident in Fig. 6a′, where on an enlarged scale the central group of the points of the stable cycle $c_{3,4}^*$ is represented together with the points of the cycles $c_{3,0}^*$ and $c_{3,1}^*$, now unstable, from which they have bifurcated: they also appear on the line of the points of $c_{3,4}^*$. Even if we give no evidence for this, because of the high computational time required, it is reasonable to think that the points of $c_{3,2}^*$ and $c_{3,3}^*$ also stay on the same line.

A natural extension of this argument is the hypothesis that also for $r > \rho_\infty{}^*$ all the points of the full sequence $c_{3,i}^*$ are disposed in an analogous way along the same line. The same study carried on for the sequence c_i shows an analogous phenomenology.

Let us examine now the behavior of the flow for r slightly larger than 28.72. Figures 6*b*–6d show the projections on the plane (x_1, x_4) of the hyperplane chosen for the Poincaré map for $r = 28.721$, $r = 28.723$, and $r = 28.730$, respectively. In all three figures we see the results of 400 intersections of the solution curve with our codimension-one section. It is observed that the behavior of the flow is still very much analogous to that observed for $r = 28.72$ when the stable orbit $\mathscr{C}_4{}^*$ is present. The numerical data obtained do not allow us at all to state whether the observed motion is periodic, possibly with a very long period, or not. The figures show, however, that with increasing r the intersection points keep disposing themselves only on arcs along the direction identified by the points of the cycle $c_{3,4}^*$, now unstable, but become more and more spread, and they tend to approach the points of the cycle $\bar{c}_{3,0}^*$ (see Figs. 6a–6d). An examination of the values of the coordinates of all these points seems to show more and more randomness with increasing r, confirmed by Fig. 6e, where we see that the behavior of the flow for $r = 28.732$ is definitely changed. In fact it is possible to observe that now also the points of the cycles $c_{2,i}^*$ due to the second sequence $\mathscr{C}_i{}^*$ and of the cycle $\bar{c}_{2,0}^*$ due to the second orbit $\overline{\mathscr{C}}_0{}^*$ contribute to the behavior of the solution curve. Moreover, some intersection points now seem to be arranged rather randomly and not to be connected with any one of the unstable cycles of the Poincaré map. With continued increasing r, we observe a more and more chaotic behavior, due to a gradual involvement of the orbits $\mathscr{C}_i$ also, since the intersection points are now also close to the fixed points of the orbits $\mathscr{C}_0$ and $\mathscr{C}_1$. Figure 6f shows the motion for $r = 28.80$, appearing fully random around the points of the unstable n-cycles present in the region.

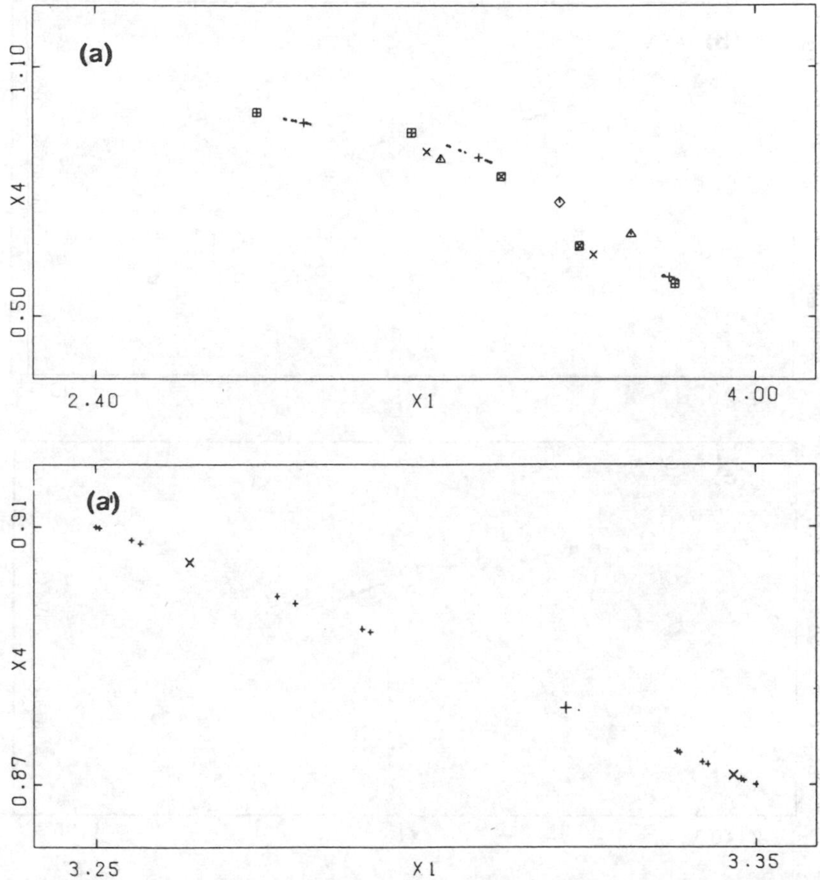

Fig. 6. Projections of the Poincaré map on the plane (x_1, x_4) for r = (a) 28.720; (b) 28.721; (c) 28.723; (d) 28.730; (e) 28.732; (f) 28.800. (+) the three points of the cycle $c_{3,0}^*$; (×) the two points of the cycle $c_{2,0}^*$; ($\boxplus$) the three points of the cycle $\bar{c}_{3,0}^*$; ($\boxtimes$) the two points of the cycle $\bar{c}_{2,0}^*$; ($\diamond$) the fixed point of c_0; ($\triangle$) the two points of the cycle c_1. (a′) Points of the central group [part (a)] of the stable cycle $c_{3,4}^*$ (+) on an enlarged scale, with the points of $c_{3,0}^*$ (+) and $c_{3,1}^*$ (×) from which they have bifurcated.

The phenomenology described above does not allow a rigorous definition of the mechanism of the onset of turbulence. In fact we are unable to evaluate $\rho_\infty{}^*$ exactly, i.e., the critical value of the Reynolds number r for which the sequence $\mathscr{C}_i{}^*$ exhausts itself, and we cannot state definitely what happens for r slightly greater than $\rho_\infty{}^*$. The existence of more infinite sequences of stable periodic orbits in very small ranges of r or with very long periods then cannot be rejected.

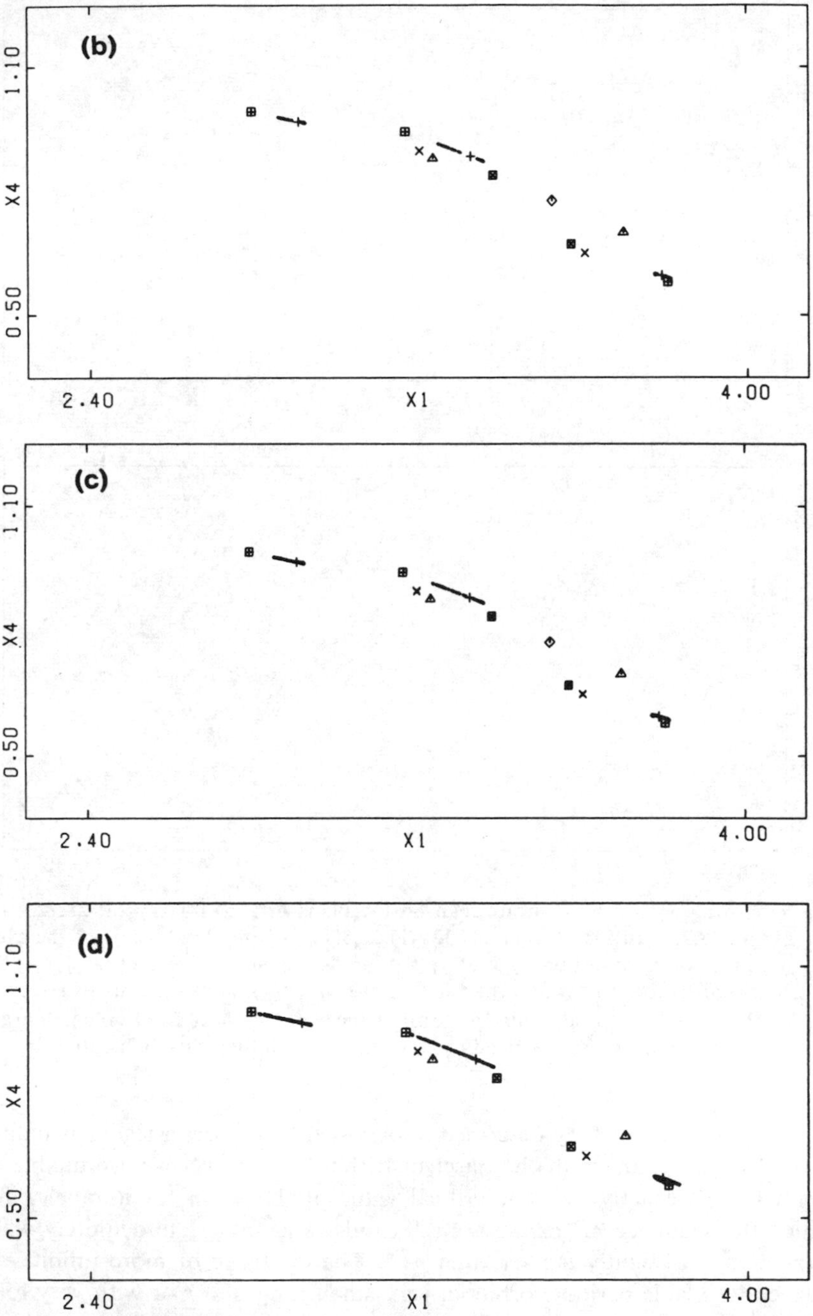

Fig. 6. Continued.

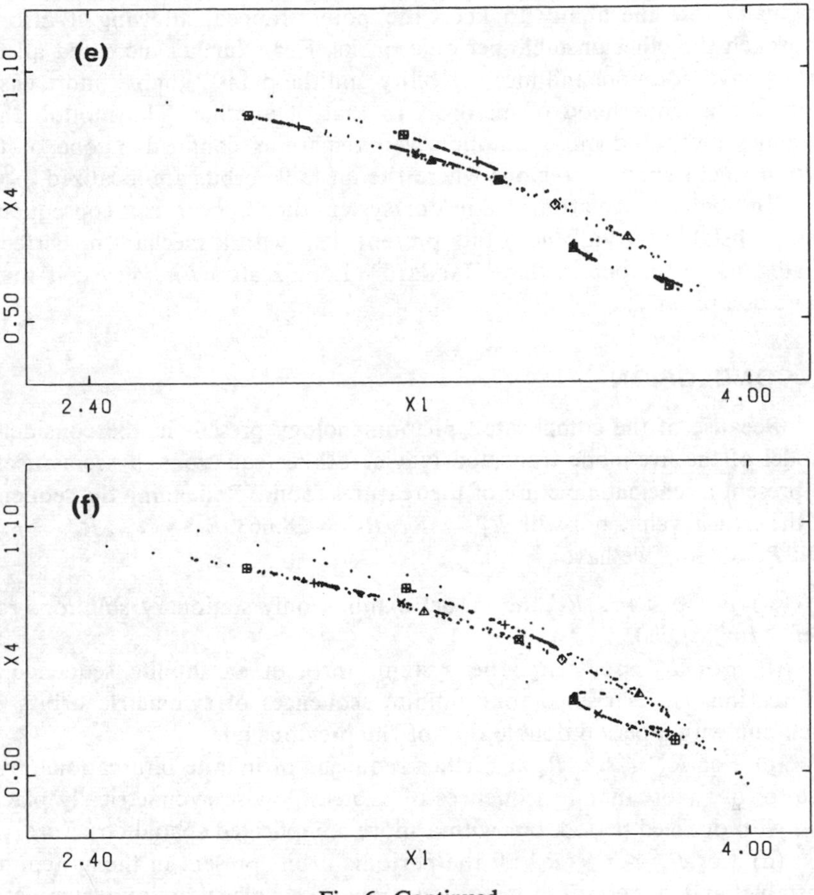

Fig. 6. Continued.

We think, however, that we can interpret the numerical results in the following way. The characteristics of the sequence $\mathscr{C}_i{}^*$ indicate that it tends rapidly to exhaust itself when r approaches a value very likely to be quite close to $\rho_\infty{}^*$, computed in the previous section according to Feigenbaum's theory. For $r > \rho_\infty{}^*$, when all the orbits have become unstable, the system possesses only unstable fixed points and periodic orbits: the seven fixed points (see Ref. 1), the four sequences of orbits $\mathscr{C}_i$, the four sequences $\mathscr{C}_i{}^*$, and the four orbits $\overline{\mathscr{C}}_0{}^*$. Up to $r \cong 28.73$ any random initial value is attracted by the orbits of one of the four sequences $\mathscr{C}_i{}^*$, which are definitely unstable, but being less unstable than the others, succeed in "catching" the point and keep it trapped in their neighborhood, at least for the long time intervals observed. With increasing r, the orbits $\mathscr{C}_i{}^*$ become more unstable and gradually lose the ability to keep the point trapped, allowing it also to approach the other unstable periodic orbits. For r further increased all the orbits have lost more and more stability and the point "jumps" more easily from the neighborhood of an orbit to that of another. The motion then becomes more and more chaotic, even remaining confined in one of the two distinct symmetric regions where the unstable orbits are localized.

The two strange attractors in our system then appear as a consequence of the instability of all the orbits present, i.e., with a mechanism perfectly analogous to the one in the "standard" Lorenz attractor,[3] even if much more complicated.

6. CONCLUSION

Because of the complicated phenomenology present in the considered model of the five-mode truncated Navier–Stokes equations, it seems useful to present a schematic picture of the features found. Redefining the sequence of the critical values of r with $R_1' = R_3$, $R_2' = 28.663$, $R_3' = \rho_\infty$, $R_4' = \rho_\infty{}^*$, and $R_5' = R_{13}$, we have:

(a) For $0 < r \leqslant R_1'$ the model exhibits only stationary solutions (see Ref. 1 for details).

(b) For $R_1' < r < R_3'$ the system, through an infinite sequence of bifurcations, gives rise to four infinite sequences of symmetric orbits $\mathscr{C}_i$, each one with a period double that of the previous one.

(c) For $R_2' \leqslant r < R_4'$ a further sequence of infinite bifurcations gives rise to four more infinite sequences of orbits $\mathscr{C}_i{}^*$, also symmetrically placed and with doubled period, but with a more complicated spatial structure.

(d) For $R_4' \leqslant r < R_5'$ all the periodic orbits present in the system are unstable and an erratic, chaotic motion takes place on two symmetric "strange" attractors, analogous to the Lorenz model ("turbulence").

(e) For $r \geqslant R_5'$ two stable periodic orbits are present.

At this point a detailed knowledge of the phenomenology of the model seems to have been reached. In particular we remark the fact that the turbulent behavior is reached gradually when no stable fixed point or periodic orbit is present and all the unstable ones keep losing stability with increasing r. Also a relevant feature is the fact that the two infinite sequences of bifurcations present seem to possess certain characteristics or universality analogous to the ones found by Feigenbaum in nonlinear transformations of an interval in itself.

We conclude by proposing two basic questions: How does a different choice of the five modes for the truncated Navier–Stokes equations effect the behavior of the model, and how does an increase in the number of modes in the truncation affect the model? Concerning the last question, an ongoing study of a seven-mode truncation obtained by adding two more modes to the five used in this study seems to show a rather different phenomenology, with much more variety and strong features of hysteresis.

ACKNOWLEDGMENTS

We are deeply indebted to G. Gallavotti for his interest in this work and his continuous help and to P. Collet and J. P. Eckmann for informing us of the Feigenbaum conjecture and for suggesting its test. We are also grateful to V. Grecchi and F. Marchetti for many useful conversations.

REFERENCES

1. C. Boldrighini and V. Franceschini, *Comm. Math. Phys.* **64**:159 (1979).
2. E. N. Lorenz, *J. Atmos. Sci.* **20**:130 (1963).
3. J. Marsden, in *Lecture Notes in Mathematics*, No. 615 (1976).
4. D. Ruelle, in *Lecture Notes in Mathematics*, No. 565 (1976).
5. M. Hénon, *Comm. Math. Phys.* **50**:69 (1976).
6. M. Hénon and Y. Pomeau, in *Lecture Notes in Mathematics*, No. 565 (1976).
7. D. Ruelle and F. Takens, *Comm. Math. Phys.* **20**:167 (1977).
8. O. E. Lanford, in *Proceedings of Corso CIME held in Bressanone* (June 1976).
9. M. J. Feigenbaum, Quantitative Universality for a Class of Nonlinear Transformations, Preprint, Los Alamos (1977).
10. B. Derrida, A. Gervois, and Y. Pomeau, Universal Metric Properties of Bifurcations of Endomorphisms, Preprint.
11. J. E. Marsden and M. McCracken, *The Hopf Bifurcation and its Applications* (Applied Mathematical Sciences, No. 19; 1976).
12. J. H. Curry, *Comm. Math. Phys.* **60**:193 (1978).
13. V. Franceschini and C. Tebaldi, in preparation.
14. P. Brunowsky, in *Lecture Notes in Mathematics*, No. 206 (1971).

Journal of Statistical Physics, Vol. 28, No. 4, 1982

Hierarchy of Chaotic Bands

Bai-lin Hao[1,2] and Shu-yu Zhang[1,3]

Received November 5, 1981

Results of a detailed numerical study on the structure of chaotic bands in a forced limit cycle oscillator (the Brusselator) are presented. Embedded in the chaotic bands of primary bifurcation sequence there are many secondary sequences with both direct and inverse segments. Within secondary chaotic bands tertiary sequences of similar structure exist. This has been shown by using subharmonic stroboscopic sampling combined with power spectra analysis.

KEY WORDS: Period-doubling bifurcation; chaos; strange attractor.

1. INTRODUCTION

The occurrence of chaotic behavior in simple dynamical systems has been raising more and more interest. It is hoped to give more insight into the difficult and long-standing problem of the onset of turbulence as well as to diminish the gap between deterministic and stochastic description in physics.

For the time being most of our knowledge of chaotic behavior comes from studies on discrete nonlinear mappings. In particular, direct period-doubling bifurcation sequence, leading to inverse sequence of chaotic bands, has been observed in many one- and two-dimensional iterated maps (for a recent review see Ref. 1 and references therein). Many properties related to these mappings, such as convergence rate of the bifurcation sequences and some characteristics of the power spectra, have been shown to be universal,(1-4) i.e., independent of the detailed structure of the mapping themselves. Moreover, there have been indications that such universal properties persist for higher-dimensional systems(5) and systems described by differential equations as well.

---

[1] Service de Physique Theorique, CEA—Saclay, 91191 Gif-sur-Yvette Cedex, France.
[2] Permanent address: The Institute of Theoretical Physics, Academia Sinica, P.O. Box 2735, Beijing, China.
[3] The Institue of Physics, Academia Sinica, P.O. Box 603, Beijing, China.

In spite of the importance of differential equations in physics, only relatively few cases have been reported. So far two groups of systems have been studied. The first group includes autonomous differential equations with three or more variables. These equations have been obtained either by truncating the Navier–Stokes equations[6,7] or by artificial construction.[8,9] The second group consists of nonlinear oscillators driven by external periodic force. Examples are anharmonic oscillator,[10] parametrically excited pendulum,[11] or the forced Brusselator,[12,13] which is also the subject of our paper. These two groups are related.[14,15] In particular, periodically driven systems can be written as systems of autonomous differential equations by introducing new variables.

We would like to emphasize that the second group mentioned above has the advantage that bifurcations in them may be understood as subharmonic entrainment (frequency-locking) of nonlinear oscillators and the existence of an external frequency as control parameter opens the possibility to use subharmonic stroboscopic sampling[16] to reach much higher resolution (up to 8192th subharmonic in this paper) than any present-day power spectra analysis would give.

2. THE MODEL

We study the following system of ordinary differential equations:

$$\begin{aligned} \dot{X} &= A - (B+1)X + X^2Y \\ \dot{Y} &= BX - X^2Y \end{aligned} \tag{1}$$

adding a period external force $\alpha\cos(\omega t)$ to the first equation. System (1) describes a hypothetical three-molecular chemical reaction with autocatalytic step under far from equilibrium conditions (it is usually called the Brusselator[17]). For parameter values, satisfying inequality

$$B > A^2 + 1 \tag{2}$$

it displays a limit cycle oscillation. With diffusion term added, system (1) shows a variety of spatial and temporal patterns and there exists a wide literature devoted to its study (cf. Ref. 18 and references therein).

By introducing new variables Z, U and fixing the initial conditions $Z(0) = 1$, $U(0) = 0$, the periodically forced Brusselator is equivalent to the following system of autonomous differential equations

$$\begin{aligned} \dot{X} &= A - (B+1)X + X^2Y + \alpha Z \\ \dot{Y} &= BX - X^2Y \\ \dot{Z} &= -\omega U \\ \dot{U} &= \omega Z \end{aligned} \tag{3}$$

Of course, there are many other ways to write this system, but we prefer (3), because it saves computer time owing to the absence of cosine.

In system (3) the nonlinear oscillator (1) is coupled to a linear oscillator by a linear term αZ and parameters α, ω enter the system on the same footing. Systems studied in Refs. 10 and 11 can be treated similarly. This leads us to a physical way of understanding period-doubling bifurcations in such systems. A nonlinear oscillator tends to follow the external frequency ω exactly. When the frequency difference between the limit cycle and the external force increases it adapts to 1/2, 1/4, 1/8, etc. of ω (subharmonic entrainment) and finally falls into a wandering regime.

The forced Brusselator was studied first by Tomita and Kai.[12,13] They discovered a region of chaotic response on the α–ω plane, surrounded by period-doubling bifurcations. Since their work had been completed before the upsurge of papers, triggered by Feigenbaum's discovery of universality and scaling,[2] many questions remain unclear, e.g., whether the convergence rate is governed by the same Feigenbaum constant $\delta = 4.66920\ldots$, does there exist inverse sequence of chaotic bands, what is the systematics of periods 12, 40, or 44, indicated in Ref. 13 as being embedded in the chaotic region, etc. ... Furthermore, no power spectra were given in Ref. 13, which was specially devoted to the chaotic region. In view of recent accumulation of knowledge on discrete mappings it is desirable to have a deeper understanding of systems, described by differential equations. This was our motivation to undertake a more detailed study of system (3).

3. THE METHODS

To explore the very subtle structure of chaotic bands we have to rely heavily on numerical studies. In doing so one should always be cautious not to be deceived in phenomena caused by the numerical algorithm or the computer itself. (There was a recent report[19] that two computers had given qualitatively different results for one and the same mapping.)

The main difficulty with differential equations lies in getting high enough resolution of subharmonic frequencies within reasonable computing time. This is a much more complicated task compared to the case of iterated maps, where bifurcations of rather high orders have been identified with remarkable precision. For each set of parameter values one has to integrate the system and wait for the transients to die out. The closer the bifurcation point, the longer the transient (for this critical slowing-down, see Ref. 20). Then both truncation and round-off errors may come into play.

We shall devote the methodological aspects to a separate publication elsewhere and only sketch here the methods used. Briefly speaking, we have been working in three different domains and the results are expressed in various sections of the parameter space (A, B, α, ω) of system (3).

(a) Subharmonic stroboscopic sampling in the time domain.[16] This is a simple extension of the usual stroboscopic idea.[21] If the system undergoes period-doubling bifurcations with period $T = pT_0$, where $p = m \cdot 2^n$, $n = 0, 1, 2, \ldots$ and $T_0 = 2\pi/\omega$, then sampling at the original period T_0 would give p points on the X, Y plane. For sufficiently large p it would be very difficult to distinguish isolated points from clusters of randomly distributed points, which come from a chaotic band with a certain period, but sampling at the pth subharmonic frequency would give only one point for periodic orbit and an island of points for a chaotic band of period pT. More generally, if the system period T and sampling period T_0 are related by

$$T = \frac{p}{q} T_0 \tag{4}$$

p and q being incommensurable integers, then on the stroboscopic portrait one sees only p points (or islands) for all $q > 1$. This is a common feature of all discrete sampling methods: one can never recognize frequencies higher than the sampling frequency, but in principle can determine any subharmonics provided the total sampling time is long enough and sampled points could be resolved. If p decomposes into product of two integers i and j, then sampling at iT_0 or jT_0 essentially increases the resolution, but misuse of an integer i, which is not a factor of p, will add a spurious multiplier i in the number of periods. To be safe, one should always go from low-order subharmonics to higher ones and compare the results with power spectra analysis whenever possible.

This simple extension appears to be very powerful in reaching high resolution, the only limitation being computer time and precision. Figure 1 shows a clearly resolved period 4096 orbit, obtained by sampling at $256T_0$ and using a double precision (29 decimal digits on a Cyber computer) to avoid accumulation of round-off errors. Such high resolution surpasses any power spectra analysis available on present-day computers.

(b) Power spectra analysis in the frequency domain based on averaging 10 series of 8192 (rarely 16384) points fast Fourier transform. Although using 8192 points power spectra analysis one can hardly go beyond the 64th subharmonic without serious aliasing, power spectra are still very useful in telling chaotic bands from periodic orbits (cf. Fig. 4) and exploring fine structures of periodic orbits embedded in chaotic bands of different periodicity (cf. Fig. 8).

(c) Inspection of the trajectories in the X, Y plane or the projection of trajectories into one of the three-dimensional subspaces of X, Y, Z, U space. Having even lower resolution than the other two methods, it gives an

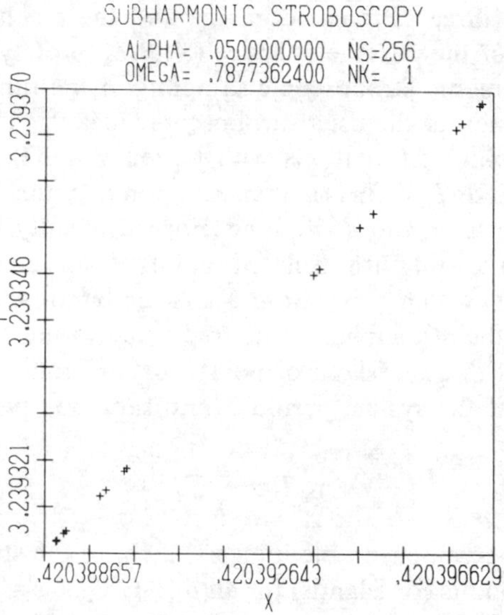

Fig. 1. Subharmonic stroboscopic portrait of a 4096P orbit ($A = 0.4$, $B = 1.2$). 256 points, sampled at $256T_0$.

intuitive feeling on the bifurcation process for low-order subharmonics, say, $p \leqslant 16$.

The results to be reported in the following sections have cost quite a large amount of computer time. To convey them as much as possible to the readers in a concise way, we need in what follows a shorthand notation. A period 32 orbit will be denoted by 32P, P standing for periodic orbit or points in the stroboscopic portrait, and a period 32 chaotic band in the inverse sequence by 32I, I standing for inverse band or islands in the portrait. What we mean by period for chaotic band will be explained in the following section, after Fig. 5.

4. HIERARCHY OF CHAOTIC BANDS

To have more comparable data we started with the same parameter values $A = 0.4$, $B = 1.2$ as in Refs. 12 and 13. Figure 2 shows the chaotic region on the α–ω plane, surrounded by period-doubling bifurcations. We shall not pay attention to the beating regime, separated by broken lines in the lower-right corner of Fig. 2. (We have discussed the nature of beating-entrainment transition elsewhere[22]). Two lines crossing the chaotic region have been examined in detail, i.e., $\alpha = 0.05$ and $\omega = 0.80$. The $\alpha = 0.05$ line was studied also in Ref. 13 and we put together our results in Table I,

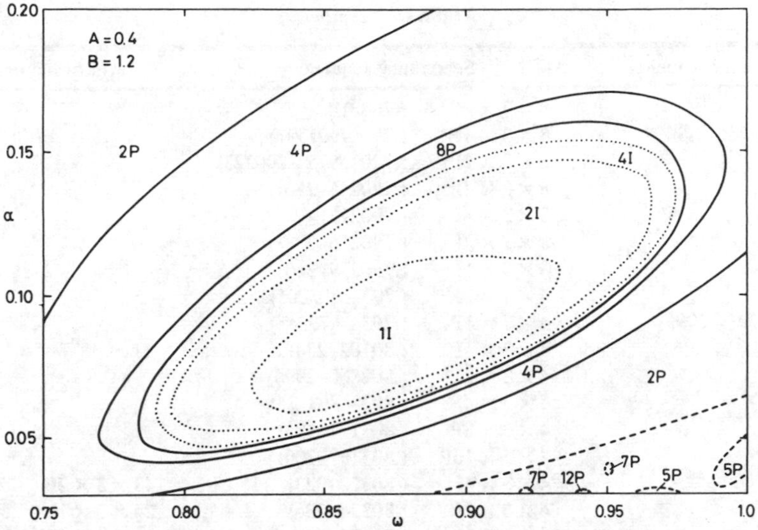

Fig. 2. Location of chaotic region in the α–ω plane: ———, boundary of periodic regime; · · · · · · ·, boundary of chaotic regime; ------, boundary between periodic and beating regimes.

Table I. Hierarchy of Bifurcation Sequences along $\alpha = 0.05$ Line ($A = 0.4$, $B = 1.2$). Values of ω Are Included in Parentheses.

Primary sequence	Secondary sequences		Tertiary sequences
1P–8192P (0.2–.78773625)			
256I (.787737)			
128I (.7877380–428)	$128 \times 3 \times$ 1P	(.7877408–410)	
	$128 \times 3 \times$ 2P	(.7877412)	
	$128 \times 3 \times$ 1I	(.7877414)	
64I (.787743–770)	$64 \times 5 \times$ 1P	(.7877496–498)	
	$64 \times 5 \times$ 1I	(.78775)	
	$64 \times 3 \times$ 1P	(.7877575–580)	
	$64 \times 3 \times$ 1I	(.7877600–605)	
32I (.78778–.78789)	$32 \times 5 \times$ 1P	(.787800–801)	
	$32 \times 3 \times$ 1P	(.787836–842)	
	$32 \times 3 \times$ 2I	(.787843–846)	
	$32 \times 3 \times$ 1I	(.787847–849)	
16I (.787895–.7885)			

[a] The existence of these frequencies has been indicated in Ref. 13.

Table I. (*continued*)

Primary sequence	Secondary sequences		Tertiary sequences	
8I	8 × 7 × 2P	(.7888)		
(.7886–.7909)	8 × 3 × 1P[a]	(.79–.7901597)		
	8 × 3 × 2P[a]	(.79015975–.790223)		
	8 × 3 × 4P[a]	(.790225–250)		
	8 × 3 × 4I	(.790275)		
	8 × 3 × 2I	(.790276–278)		
	8 × 3 × 1I	(.790279–360)		
4I	4 × 11 × 1P[a]	(.793–.793001)		
(.791–.8295)	4 × 7 × 1P	(.7934–.79345)		
	4 × 3 × 1P[a]	(.80102–234)		
	4 × 3 × 2P[a]	(.802360–2996)		
	4 × 3 × 4P[a]	(.803–.803175)		
	4 × 3 × 8P	(.803177–195)		
	4 × 3 × 16P[a]	(.803197–200)		
	4 × 3 × 4I	(.80322–323)	48 × 3 × 2P	(.80321)
	4 × 3 × 2I	(.80324–336)	24 × 3 × 1I	(.80331)
	4 × 3 × 1I	(.80338–440)	24 × 3 × 2I	(.80332)
	4 × 3 × 1I	(.818–.8192)	12 × 5 × 2I	(.81925)
	4 × 3 × 2I	(.81928–934)		
	4 × 3 × 4I	(.81936–938)		
	4 × 3 × 8P[a]	(.8194–.81943)		
	4 × 3 × 4P[a]	(.819433–728)		
	4 × 3 × 2P[a]	(.819731–.8202)		
	4 × 3 × 1P[a]	(.82025–.8214)		
	4 × 3 × 1P	(.8245–.82599)		
8I	8 × 3 × 1I	(.831–.83102)		
(.83–.8325)	8 × 3 × 2I	(.83104)		
	8 × 3 × 4P	(.83106)		
	8 × 3 × 2P	(.8311–.83112)		
	8 × 3 × 1P	(.83114–.83120)		
	8 × 5 × 1I	(.83194–.83196)		
	8 × 5 × 4P	(.83198)		
	?			
	8 × 5 × 1P[a]	(.832–.832005)		
16I				
(.8327–.83304)				
32I				
(.83306–.83314)				
64I				
(.83315–.83317)				
128I				
(.833175)				
256I				
(.83318)				
128P–2P				
(.83319–.950475)				

Table II. A Period-Doubling Bifurcation Sequence in Eqs. (3)

n	Period	Range in ω		ω_n	δ_n
1	1		– 0.39820	0.398205	5.53
2	2	0.39821	– 0.71305	0.7130625	4.24
3	4	0.713075	– 0.769996	0.7699998	4.02
4	8	0.770000	– 0.78337	0.783435	4.46
5	16	0.78350	– 0.786752	0.786776	4.41
6	32	0.78680	– 0.78752	0.787525	5.40
7	64	0.78753	– 0.78769	0.787695	4.04
8	128	0.78770	– 0.787726	0.7877265	4.88
9	256	0.787727	– 0.787734	0.78773425	4.92
10	512	0.7877345	– 0.7877358	0.78773585	
11	1024	0.7877359	– 0.78773615	0.787736175	
12	2048	0.78773620			
13	4096	0.78773624			
14	8192	0.78773625			

where frequencies, whose existence was indicated in Ref. 13, were marked by a superscript *a*. With our high resolution we succeeded in recognizing systematically many inverse bands within the chaotic region and in seeing a lot of secondary bifurcation sequences with both direct and inverse segments.

As regards the primary direct sequence we have located rather high-order bifurcations up to $p = 8192$ (see Table II). To our knowledge, this is the longest period-doubling sequence ever identified for a system of nonlinear differential equations. The estimate of the convergence rate

$$\delta_n = \frac{\omega_n - \omega_{n+1}}{\omega_{n+1} - \omega_{n+2}} \tag{5}$$

given in the last column of Table II, shows that most probably we have here the same Feigenbaum constant δ. Critical slowing-down near every bifurcation point has prevented us from locating the boundaries of each period with high precision and we simply take the middle point between two last seen consecutive periods to be the boundary. Thus we cannot expect a better estimate of δ in this way.

In the third column of Table I there are a few indications on the existence of perodicities of the third level, embedded in chaotic bands of the secondary inverse sequence. This is indeed the case, as it can be seen more clearly from the results along the $\omega = 0.80$ line, summarized in Table III.

This hierarchy structure is visualized in Fig. 3. It shows schematically the bifurcations at fixed external frequency $\omega = 0.80$. When the amplitude

Table III. Hierarchy of Bifurcation Sequences along $\omega = 0.8$ line ($A = 0.4$, $B = 1.2$). Values of α Are Included in Parentheses.

Primary sequence	Secondary sequences		Tertiary sequences	
2P–128P				
(.01–.04696)				
32I	32 × 3 × 1P	(.04700)		
(.04698–.04701)				
16I	16 × 3 × 1P	(.0471000–055)		
(.047025–.047197)	16 × 3 × 2P	(.0471100–114)		
	16 × 3 × 4P	(.0471116–128)		
	16 × 3 × 2I	(.047113)	96 × 3 × 1I	(.0471135)
	16 × 3 × 1I	(.047114–118)		
8I	8 × 3 × 1P	(.047591–637)		
(.047198–.0484)	8 × 3 × 2P	(.0476380–575)		
	8 × 3 × 4P	(.0476580–625)		
	8 × 3 × 8P	(.0476630–640)		
	8 × 3 × 16P	(.0476641–642)		
	8 × 3 × 32P	(.0476643)		
	8 × 3 × 8I	(.0476644–649)		
	8 × 3 × 4I	(.0476651–654)	96 × 3 × 1P	(.0476650)
	8 × 3 × 2I	(.0476656–699)	48 × 5 × 1P	(.0476662)
			48 × 3 × 1P	(.0476673–75)
			48 × 3 × 2P	(.0476676)
			48 × 3 × 1I	(.0476677–80)
	8 × 3 × 1I	(.047670–688)	24 × 3 × 1I	(.047679–680)
4I	4 × 7 × 1P	(.048495)		
(.04845–.0535)	4 × 5 × 1P	(.048945–959)		
	4 × 5 × 2P	(.04896–.04900525)		
	4 × 5 × 4P	(.049010–011)		
	4 × 5 × 8P	(.049012)		
	4 × 5 × 2I	(.049015)		
	4 × 5 × 1I	(.049016–025)	20 × 5 × 1P	(.0490218–20)
2I	2 × 7 × 1P	(.05580–585)		
(.0545–.0725)	2 × 7 × 2P	(.055860–875)		
	2 × 7 × 4P	(.05588)		
	2 × 7 × 8P	(.055885)		
	2 × 7 × 2I	(.05589)		
	2 × 7 × 1I	(.055895–920)		
	2 × 7 × 8P	(.0704)		
	?			
	2 × 7 × 1P	(.0705)		
4I	4 × 3 × 1I	(.0770–772)		
(.0735–.0809)	4 × 3 × 2P	(.0773)		
	4 × 3 × 1P	(.0775–776)		
	4 × 5 × 1P	(.0795)		
8I	8 × 5 × 1P	(.08145)		
(.08092–.0825)	8 × 3 × 1I	(.0817)		
	8 × 3 × 2P	(.08175)		
	8 × 3 × 1P	(.08180–185)		

Table III. (*continued*)

Primary sequence	Secondary sequences		Tertiary sequences
16I	16 × 3 × 1I	(.082671–682)	48 × 7 × 1P (.08268)
(.08255–.08285)	16 × 3 × 8P	(.082683)	
	16 × 3 × 4P	(.082684–685)	
	16 × 3 × 2P	(.0826900–925)	
	16 × 3 × 1P	(.082695–706)	
32I	32 × 5 × 1P	(.082902)	
(.08286–.082914)	32 × 7 × 1P	(.08291)	
64I			
(.082916–.08293)			
128I	128 × 3 × 1P	(.0829310–312)	
(.0829302–327)	128 × 7 × 1P	(.0829318)	
256I	256 × 3 × 1P	(.082933)	
(.0829332–338)			
512P–1P			
(.082934–20.0)			

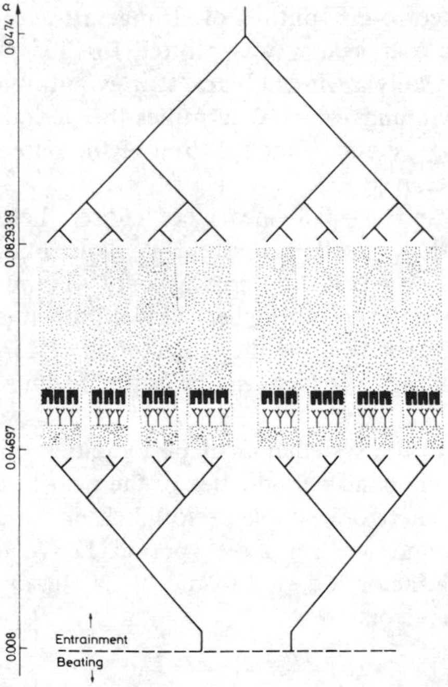

Fig. 3. Hierarchy of period-doubling bifurcation sequences (schematic, not to scale). Only one secondary sequence embedded in a 8I band is shown.

α is very small the periodic force cannot entrain the system and there exist two independent frequencies (beating regime, not shown in Fig. 3). With α increasing the system passes through a sharp beating-entrainment transition (horizontal broken line in Fig. 3) and a period-doubling sequence takes place with $p = 2$. Its inverse sequence merges with that of another period-doubling sequence, which ends with only one period when the external force takes over. Therefore the chaotic behavior appears as a compromise between two trends: at small α the limit cycle tries to show itself up, at large α the external force dominates, and chaotic behavior exists in between as a new regime of nonlinear oscillation. This is the physical understanding of period-doubling bifurcation and associated chaotic behavior, which we mentioned at the end of the Introduction.

Periodic and chaotic orbits can be distinguished best by subharmonic stroboscopic sampling.[16] For not-very-high-order subharmonics the distinctions are clearly expressed in the power spectra and trajectories in the X, Y plane. In Figs. 4 and 5 we compare the spectra and trajectories of 2P, 4P, 8P, 16P, and 32P orbits with that of 2I, 4I, 8I, 16I, and 32I orbits. What is remarkable are the sharp peaks in the broad-band spectra of chaotic orbits. This has been seen before[9] and called phase coherence in Ref. 15. It seems to be related to the splitting of strange attractor, seen in Fig. 5f–j. In all these figures a trajectory was plotted for 305 periods. If we drew many more periods, only a winded black strip would remain in Fig. 5f–j. It is this number of windings which determines the period of a chaotic band. (Notice: in both Fig. 5e and 5j the splitting of the outermost loop into two has not been well resolved.)

The $\alpha = 0.05$ and $\omega = 0.8$ lines were chosen before we located the position of 1I, 2I, 4I, . . . regions in the α–ω plane, marked by dotted lines in Fig. 2. Thus the $\alpha = 0.05$ line crosses only 4I and higher bands, the $\omega = 0.8$ line crosses only 2I and higher bands, both missing the 1I region. By increasing α to 0.08 we drew the trajectories of 1I, 2I, and 4I chaotic orbits, as shown in Fig. 6, which give a better feeling of how a "strange attractor"[4] splits.

Now we turn to the systematics of periodicities, embedded in chaotic bands. In a kI chaotic band periodicities of the next level always appear to have k as a factor. Therefore, simple period 3, 5, or 7 may be found only in 1I band. In a 8I band one never sees period 12, 20, or 28, because the smallest period with factor 3, 5, or 7 would be 24, 40, or 56, respectively. In general, one may have period

$$P = k \cdot m \cdot 2^n \tag{6}$$

[4] We use this term to name the apparently chaotic object seen numerically.

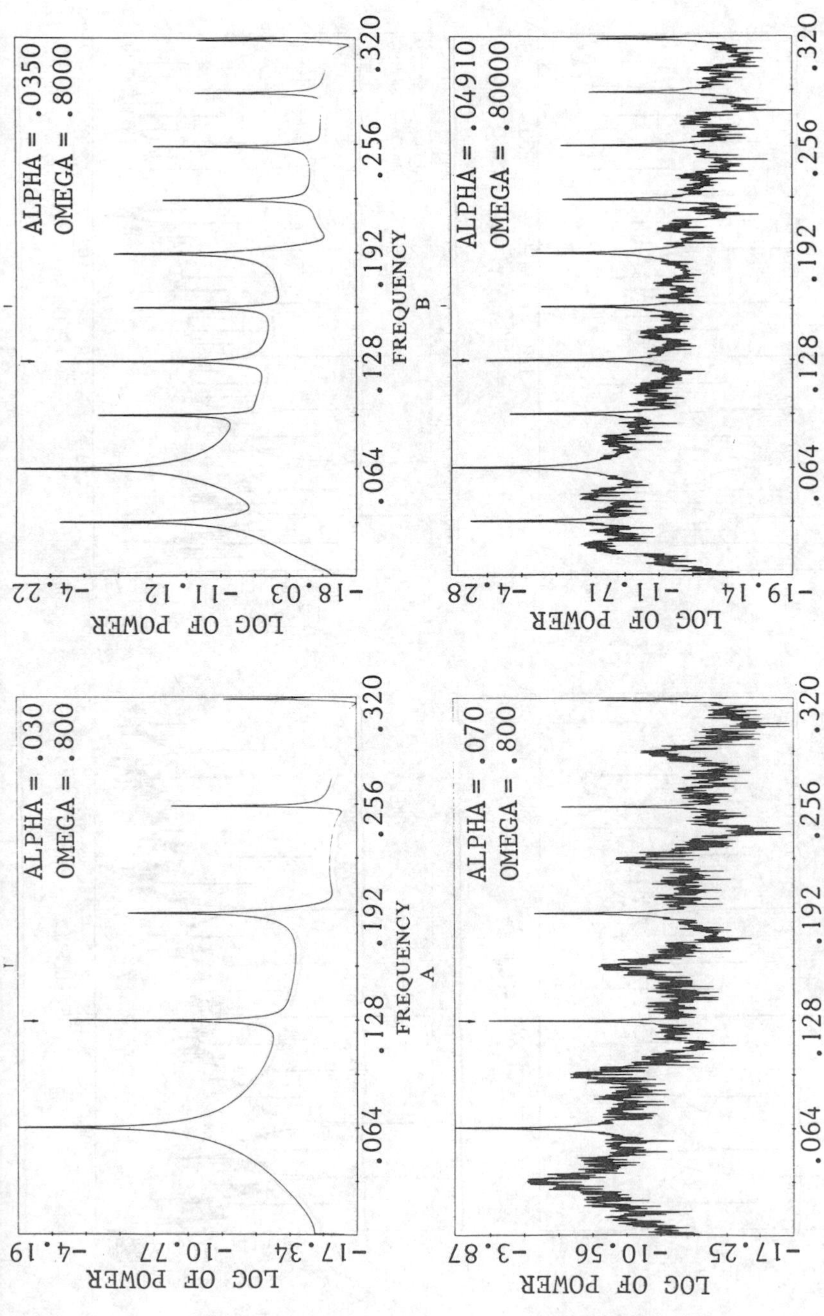

Fig. 4. Power spectra of periodic and chaotic orbits. (a) 2P, (b) 4P, (c) 8P, (d) 16P, (e) 32P, (f) 2I, (g) 4I, (h) 8I, (i) 16I, (j) 32I.

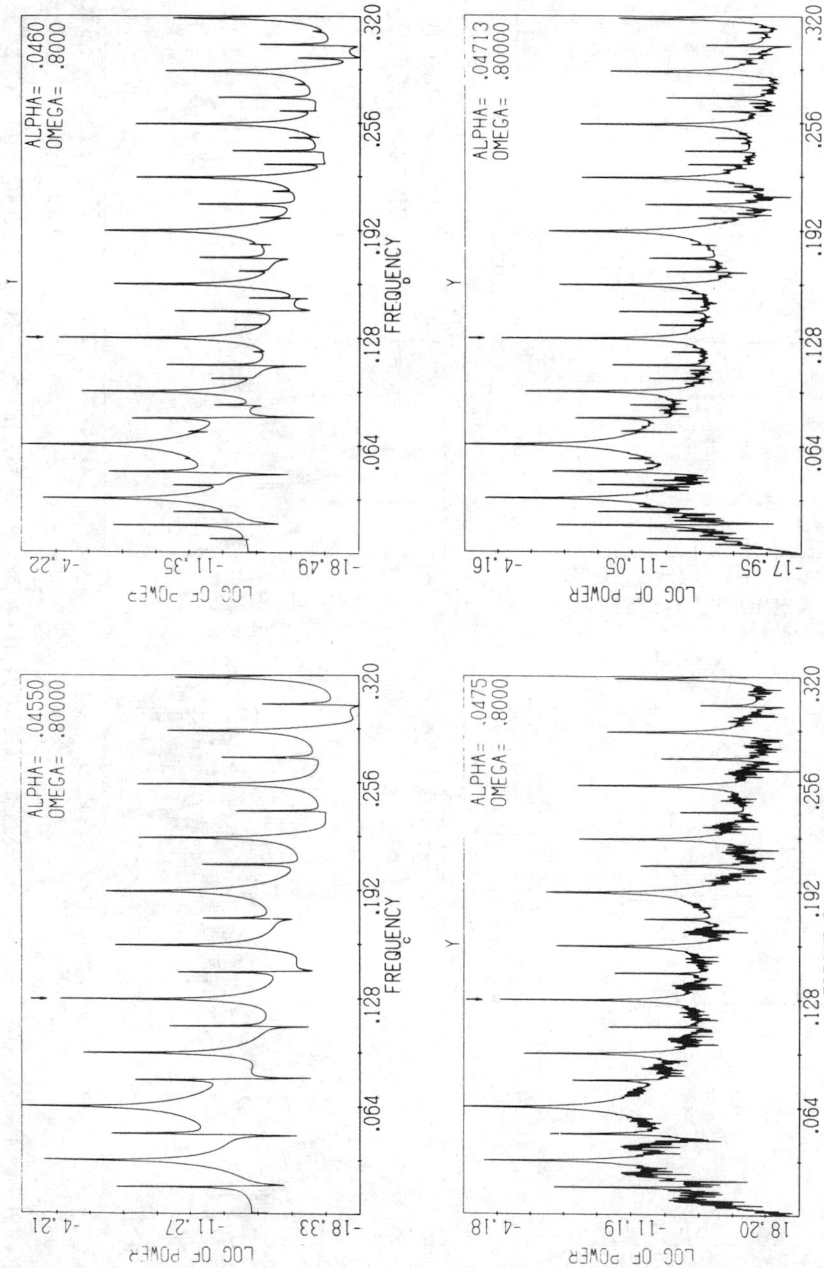
ALPHA= .0460
OMEGA= .8000
LOG OF POWER
FREQUENCY
ALPHA= .04713
OMEGA= .80000
ALPHA= .0455O
OMEGA= .80000
ALPHA= .0475
OMEGA= .8000

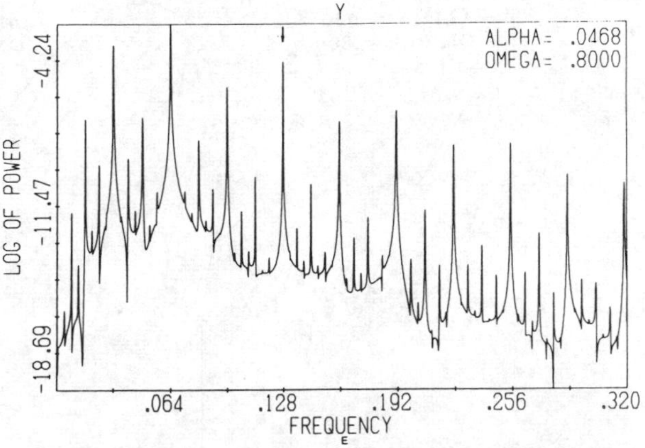

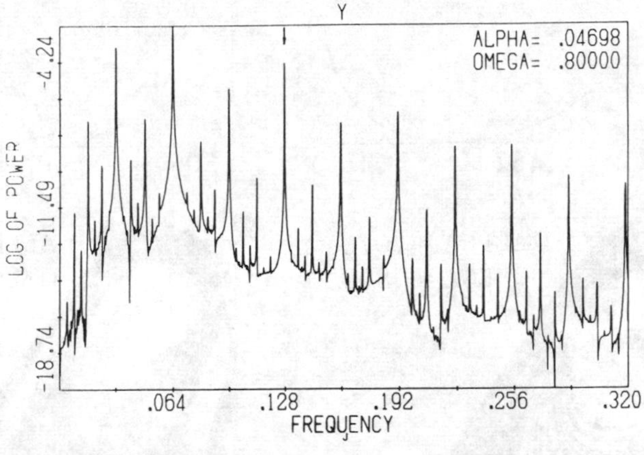

where k is the period of the underlying chaotic band, $m = 3, 5, 7, 11, \ldots$, and $n = 0, 1, 2, \ldots$. This refers to the periodic orbits in the direct sequence as well as to the period of chaotic bands in the inverse sequence. In all cases studied so far, we have not encountered any exception to this rule.

Figure 7 shows schematically the structure of a 8I band which is the one emphasized in Fig. 3 and has been searched in more detail. A direct bifurcation sequence with period $8.3.2^n$, $n = 0$–5, and the associated inverse sequence with period $8.3.2^n$, $n = 0$–3, were located explicitly. Within the 24I and 48I chaotic bands of this secondary sequence, shorter tertiary sequences with 24 or 48 as factor in their period were identified.

This hierarchy structure may be interpreted by splitting and shrinking of the strange attractor at certain parameter values. After primary splitting one can imagine an arm of the strange attractor as a rope made of infinite number of threads. At certain parameter values threads in each arm shrink

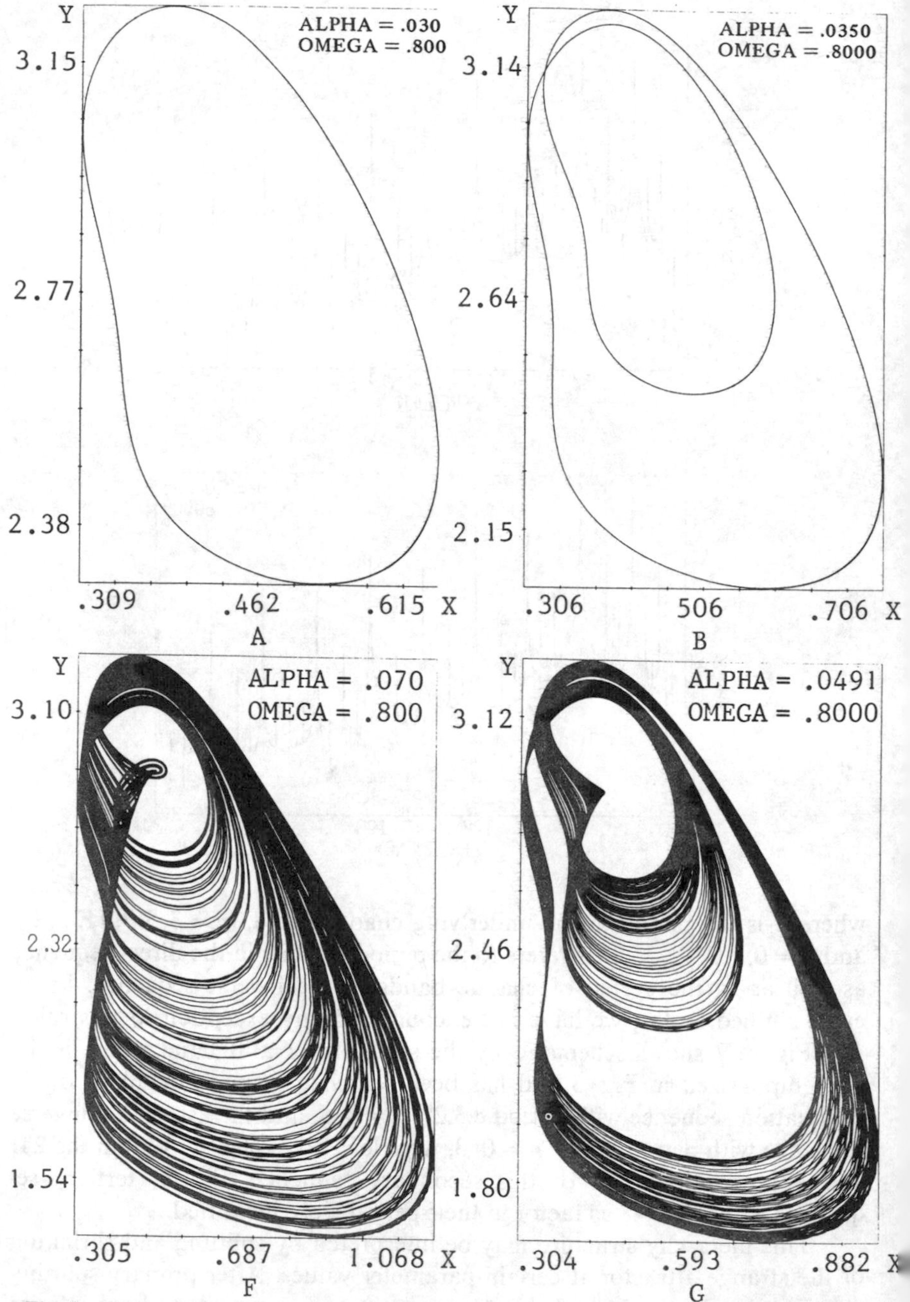

Fig. 5. Trajectories in the *X*, *Y* plane. (a) 2P, (b) 4P, (c) 8P, (d) 16P, (e) 32P, (f) 2I, (g) 4I, (h) 8I, (i) 16I, (j) 32I.

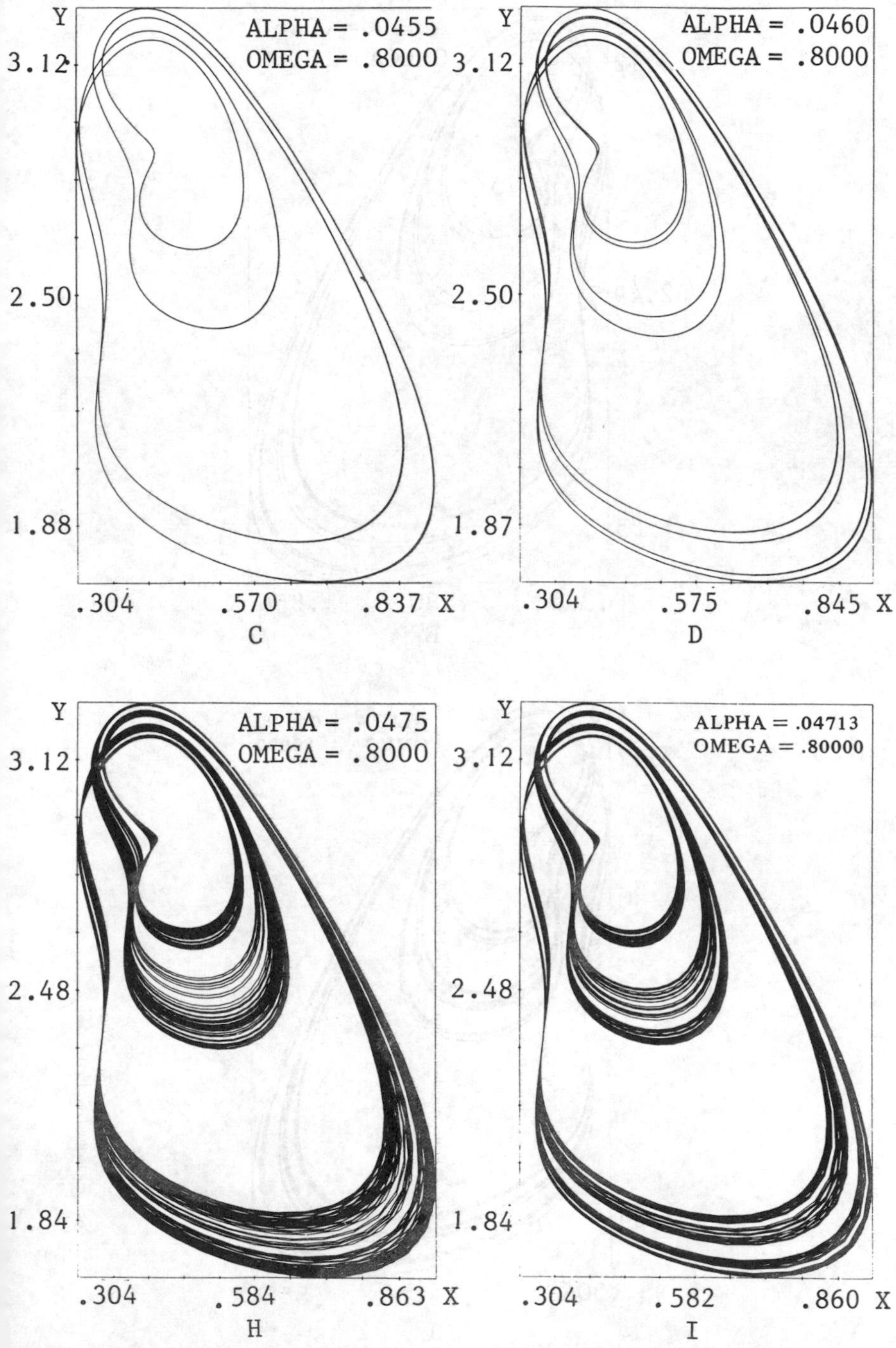
Y
ALPHA = .0455
OMEGA = .8000
3.12
2.50
1.88
.304
.570
.837
X
C
Y
ALPHA = .0460
OMEGA = .8000
3.12
2.50
1.87
.304
.575
.845
X
D
Y
ALPHA = .0475
OMEGA = .8000
3.12
2.48
1.84
.304
.584
.863
X
H
Y
ALPHA = .04713
OMEGA = .80000
3.12
2.48
1.84
.304
.582
.860
X
I

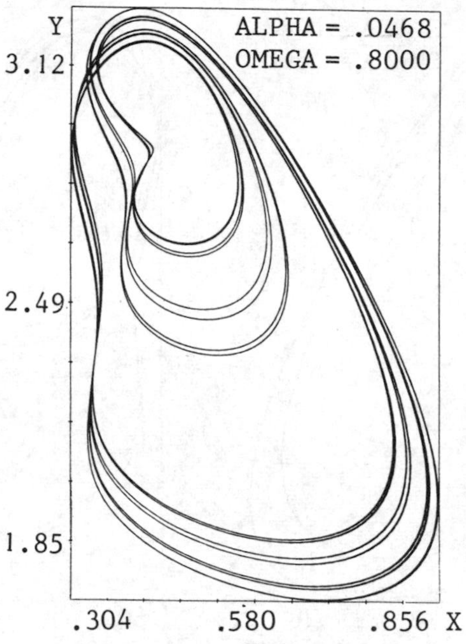
Y
ALPHA = .0468
OMEGA = .8000
3.12
2.49
1.85
.304
.580
.856
X

E

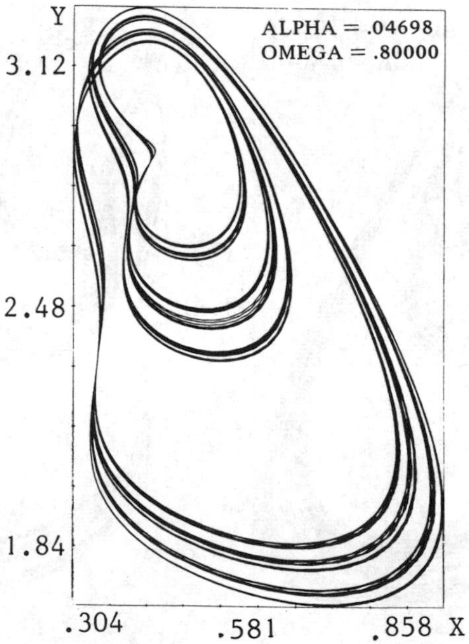
Y
ALPHA = .04698
OMEGA = .80000
3.12
2.48
1.84
.304
.581
.858
X

J

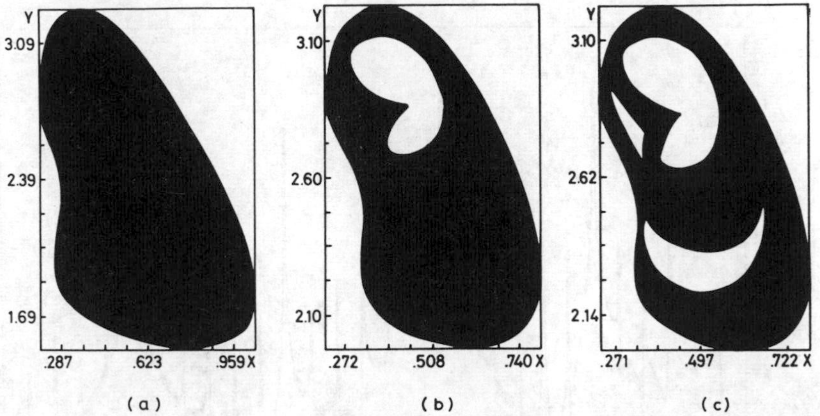

Fig. 6. Strange attractors ($A = 0.4$, $B = 1.2$). (a) 1I ($\alpha = 0.08$, $\omega = 0.86$); (b) 2I ($\alpha = 0.08$, $\omega = 0.91$); (c) 4I ($\alpha = 0.08$, $\omega = 0.915$).

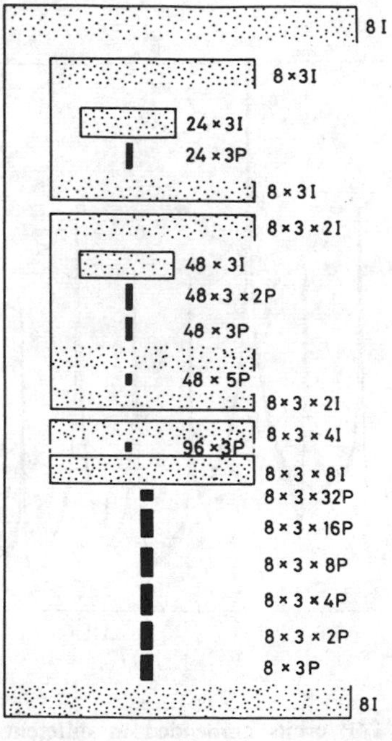

Fig. 7. Secondary and tertiary bifurcation sequences, embedded in a 8I band (schematic).

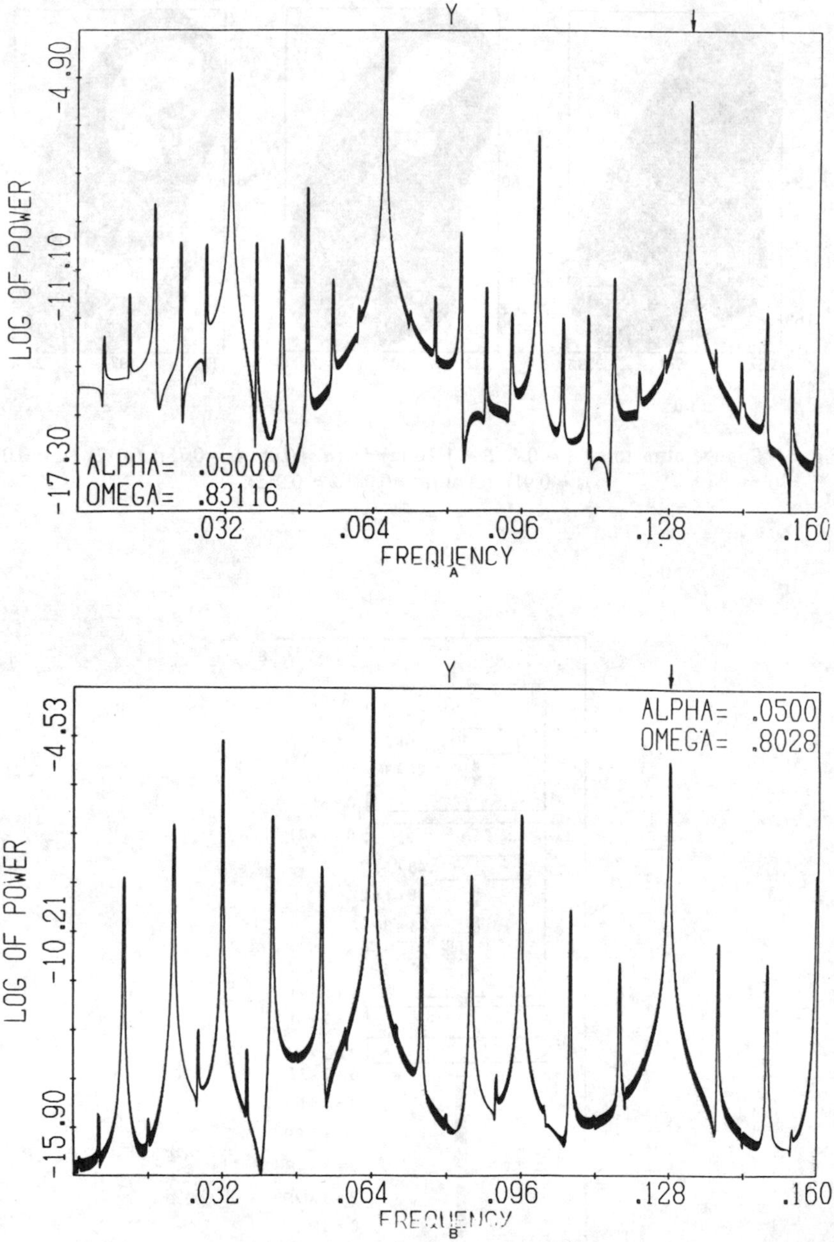

Fig. 8. Power spectra of 24P orbits embedded in different chaotic bands. An arrow indicates the fundamental frequency. (a) $8 \times 3P$ in 8I band; (b) $4 \times 3 \times 2P$ in 4I band.

simultaneously into a finite number of threads, producing thereby a periodic orbit. This process repeats at each level on smaller and smaller scales. It reflects in the power spectra as well. For example, 24P orbits embedded in 8I or 4I chaotic bands can easily be identified for having a 8×3 or $4 \times 3 \times 2$ fine structure (see Fig. 8).

To conclude this section, we make two remarks. First, one can estimate the convergence rate of primary chaotic bands and secondary direct sequences, using data given in Tables I and III. Everywhere we find almost the same range of δ value. Second, being impossible to make an overall search by very small steps, we certainly have missed many periodicities. $m = 3, 5, 7$ in formula (6) correspond to most easily found periods. This explains why there are so many of them in Tables I and III. There is only one case with $m = 11$ and we have nothing to say about $m = 9$.

5. FINITENESS OF CHAOTIC REGION IN PARAMETER SPACE

It is interesting to ask what happens with the chaotic region if we make the original system (1) nearer to or further away from the time-independent

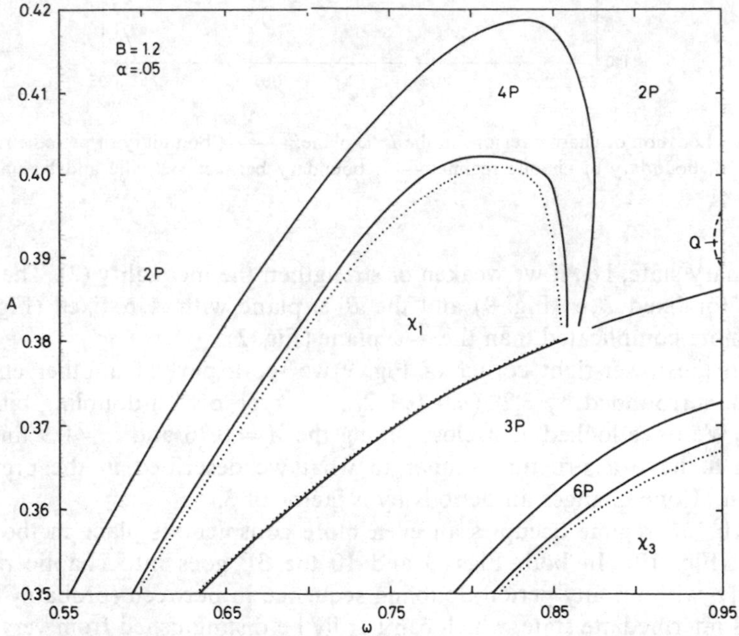

Fig. 9. Location of chaotic regions in the A–ω plane: ———, boundary of periodic regime; · · · · · ·, boundary of chaotic regime; ------, boundary between periodic and beating (Q) regime.

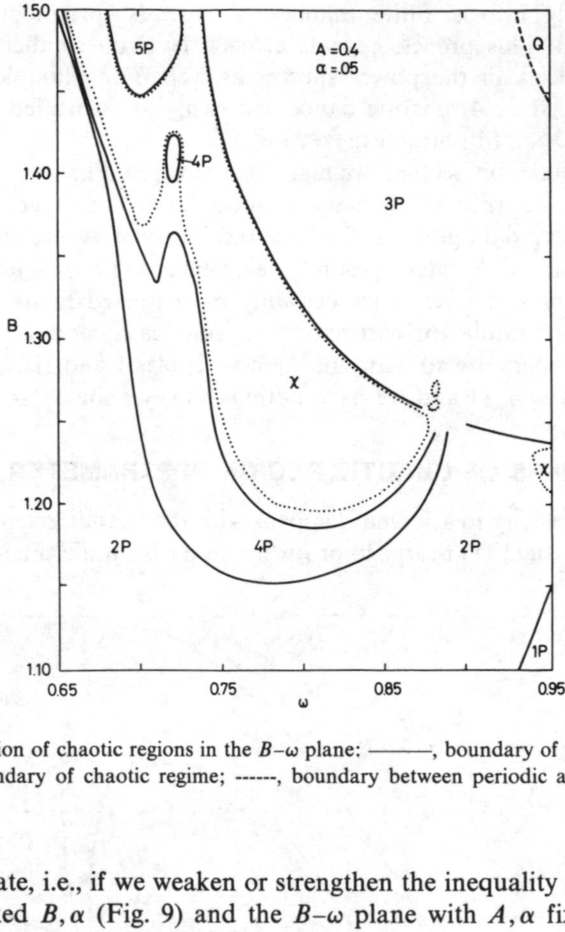

Fig. 10. Location of chaotic regions in the B–ω plane: ———, boundary of periodic regime; · · · · · ·, boundary of chaotic regime; ------, boundary between periodic and beating (Q) regime.

stationary state, i.e., if we weaken or strengthen the inequality (2). The A–ω plane for fixed B, α (Fig. 9) and the B–ω plane with A, α fixed (Fig. 10) look more complicated than the α–ω plane (Fig. 2).

In the lower-right corner of Fig. 9 we see a part of another chaotic region, surrounded by 3.2^n ($n = 0, 1, 2, \ldots$) type period-doubling bifurcations. We have looked at it closer along the $A = 0.36$ and $\omega = 0.9$ lines. It seems to have a structure similar to what we described in the previous section, if one changes all periods by a factor of 3.

The 3P regime occupies an even more conspicuous place in the B–ω plane (Fig. 10). In both Figs. 9 and 10 the 3P goes into chaotic region directly, without any period-doubling sequence in between (probably there is a 3I intermediate state, which can hardly be distinguished from very long transients of 3P). It is unclear to us whether this fact is relevant to the cubic nonlinearity in the particular model or it is a common feature shared by other systems too. It might be a manifestation of some analog to the

Li–Yorke theorem,[23] which was proved onlv for one-dimensional mappings and says that period 3 (then generalized to 5, 7, etc.) always implies chaos. Anyhow, the role of period 3 in the case of differential equations requires further scrutiny. The same comment holds for 5P in Fig. 10.

The structure of a small region near the intersection of 3P, 2^nP, and chaotic regimes on both Figs. 9 and 10 remain unresolved. We leave this question for future study.

A comparison of Figs. 2, 9, and 10 suggests to us a conjecture that the chaotic regime occupies finite region in the parameter space. This seems not to be in disagreement with recent observations on the Lorenz model.[24] If, in addition, chaotic bands and periodicities embedded in them are nested one into another, then so-called "window structure" of periodic orbits, mentioned by many authors, e.g., Ref. 13, would be a simple consequence of this nesting. Our data in Tables I and III support this nested character of secondary bifurcation sequences, e.g., the order of all direct and inverse sequences are reversed at two opposite sides of the parameter range. The same seems to be true for tertiary sequences, but one needs finer data to be confident.

6. DISCUSSION

The forced Brusselator shows very complicated bifurcation and chaotic behavior in certain regions of the control parameter space. Still if only one parameter varies, we see many features common to one-dimensional discrete mappings. Phenomena related to the role of 3P orbits require further study. The hierarchy structure of bifurcation sequences should be observable also in mappings and other nonlinear differential equations. In other words, we expect not only the existence of a few universal numerical characteristics, but also the universality of the essential overall structure of the bifurcation scheme.

Having in mind the results reported in this paper, we can review other systems in retrospect.

In the 5-mode truncated Navier–Stokes equations,[7] most probably what was observed was a primary direct sequence up to 32P and a secondary sequence with $m = 3$, embedded in 1I or 2I chaotic band. A power spectrum analysis would help to reveal its identity.

In the Rössler model the presence of 1P to 8P and 1I to 8I bands has been observed clearly.[9] The authors of Ref. 9 have seen also 16P and 16I bands. Owing to limited precision of the analog computer, no boundaries and δ values were estimated.

In the parametrically excited pendulum,[11] a direct sequence from 1P to 32P was reported and two points belong to 2I and 4I bands observed. It was noticed in Ref. 11 that at the far end of the parameter axis 1P orbit there appears again what would be another indication on finiteness of chaotic region.

In all cases cited only one parameter has been varied. It would be very useful to enlarge the parameter space, as our results have shown, but what is the best (or minimal) parameter space for a given dynamical system to explore the bifurcation and chaotic behavior in all its complexity and variety?

To conclude this paper we would like to emphasize once more our understanding of period-doubling bifurcations and the associated chaos as a new chapter in the theory of nonlinear oscillation, although it certainly throws new lights on fundamental problems of statistical physics.

ACKNOWLEDGMENT

The main body of this research has been done during our stay at the Instituts Internationaux de Physique et de Chimie, fondés par E. Solvay, in Brussels, and the final manuscript written while visiting the Service de Physique Théorique, CEA–Saclay. The first author (H.) thanks these institutions for hospitality and support. He thanks Drs. I. Prigogine, G. Nicolis, L. Reichl, D. K. Kondepudi, and M. Feigenbaum for discussions and comments at different occasions. Dr. J. B. McLaughlin and V. Franceschini sent their reprints to H. Dr. E. Brézin kindly read the manuscript. We express our gratitude to all of them.

REFERENCES

1. P. Collet and J.-P. Eckmann, *Iterated Maps on the Interval as Dynamical Systems* (Birkhäuser, Basel, 1980).
2. M. J. Feigenbaum, *J. Stat. Phys.* **19**:25 (1978); **21**:669 (1979).
3. B. A. Huberman and J. Rudnick, *Phys. Rev. Lett.* **45**:154 (1980).
4. M. J. Feigenbaum, *Commun. Math. Phys.* **77**:65 (1980); B. A. Huberman and A. B. Zisook, *Phys. Rev. Lett.* **46**:626 (1981); P. Collet, J.-P. Eckmann, and L. Thomas, *Commun. Math. Phys.* **81**:261 (1981); M. Nauenberg and J. Rudnick, Universality and the power spectrum at the onset of chaos, UCSC preprint 81/137.
5. P. Collet, J.-P. Eckmann, and H. Koch, *Commun. Math. Phys.* **25**:1 (1981).
6. E. N. Lorenz, *J. Atmos. Sci.* **20**:130 (1963); V. Franceschini, *J. Stat. Phys.* **22**:397 (1980).
7. V. Franceschini and C. Tebaldi, *J. Stat. Phys.* **21**:707 (1979).
8. O. E. Rössler, *Ann. N.Y. Acad. Sci.* **316**:376 (1979).
9. J. Crutchfield, D. Farmer, N. Parcard, R. Shaw, G. Jones, and R. J. Donnelly, *Phys. Lett.* **76A**:1 (1980).
10. B. A. Hubermann and J. P. Crutchfield, *Phys. Rev. Lett.* **43**:1743 (1979).
11. J. B. McLaughlin, *J. Stat. Phys.* **24**:377 (1981).
12. T. Kai and K. Tomita, *Progr. Theor. Phys.* **61**:54 (1979).
13. K. Tomita and T. Kai, *J. Stat. Phys.* **21**:65 (1979).

14. P. Coullet, C. Tresser, and A. Arnéodo, *Phys. Lett.* **72A**:268 (1979).
15. D. Farmer, J. Crutchfield, H. Froehling, N. Parcard, and R. Shaw, *Ann. N.Y. Acad. Sci.* **357**:453 (1980).
16. B.-L. Hao and S.-Y. Zhang, *Phys. Lett.* **87A**:267 (1982).
17. J. J. Tyson, *J. Chem. Phys.* **58**:3919 (1973).
18. G. Nicolis and I. Prigogine, *Self-Organization in Non-Equilibrium Systems* (Wiley, New York, 1977).
19. J. H. Curry, *Commun. Math. Phys.* **68**:129 (1979).
20. B.-L. Hao, *Phys. Lett.* **86A**:267 (1981).
21. N. Minorsky, *Théorie des Oscillations* (Gauthier-Villars, Paris, 1967); K. Tomita and T. Kai, *Phys. Lett.* **66A**:91 (1978).
22. B.-L. Hao, Two kinds of entrainment-beating transitions in a driven limit-cycle oscillator, *J. Theor. Biol.* (to appear).
23. T. Y. Li and J. A. Yorke, *Am. Math. Monthly* **82**:985 (1975).
24. I. Simada and T. Nagashima, *Progr. Theor. Phys.* **59**:1033 (1978); K. A. Robbins, *SIAM J. Appl. Math.* **36**:457 (1979).

Commun. in Theor. Phys. (Beijing, China) *Vol. 2, No.3 (1983) 1075–1080*

U-SEQUENCES IN THE PERIODICALLY FORCED BRUSSELATOR

HAO Bai-lin (郝柏林)
The Institute of Theoretical Physics, Academia Sinica, Beijing, China
WANG Guang-rui (王光瑞) and ZHANG Shu-yu (张淑誉)
The Institute of Physics, Academia Sinica, Beijing, China
Received February 17, 1983

Abstract

We show numerical evidence of exact U-sequences in the periodically forced trimolecular model (the forced Brusselator). Interspersed among period-doubling bifurcation sequences starting with RL^n *type periods, there are chaotic regions bounded on one side by period-doubling bifurcation sequences and on the other side by intermittent transitions. Along certain directions in the parameter space the most clearly seen periods appear in the same order as that in the logistic map, but along other directions the U-sequences may fold and give rise to deviations from the standard patterns. Our results show the coexistence of different "routes to chaos" in one and the same mathematical model and the necessity to enlarge the parameter space in both real and computer experiments on chaotic transitions.*

I. Introduction

A nonlinear system, described by deterministic evolution equations, can undergo sharp transitions into either of two categories of states. The first category consists of states with temporal and/or spatial order and has been studied more or less thoroughly since mid-60's under the name of non-equilibrium phase transition, dissipative structure, synergetics and the like. The second category includes various transitions into more chaotic states which would rather be compared to order without periodicity than disorder. Although known before in many particular cases, the chaotic transitions have been considered a new category of natural phenomena only since the recent discovery of universality and scaling properties associated with them [1,2].

The iterated map on the interval has been the paradigm of universality and scaling properties in chaotic transitions. For unimodal maps universal metric properties such as the Feigenbaum convergence rate δ and scaling factor α are determined by the analytic form of the map near its maximum. Structural properties, e.g. the order of occurrence of stable periods along the parameter axis, the so-called U-sequences [3], depend on there being only one maximum in the map.

While many real systems are modeled by nonlinear differential equations, only metric properties have been tested so far on a few systems of ordinary differential equations (ODE's), for recent reviews see Refs. [4,5]. In particular, we have studied in detail the bifurcation scheme of the forced Brusselator in various sections of the parameter space, discovered the hierarchy structure of the chaotic bands and estimated the Feigenbaum constant δ [6]. To our knowledge there has not been any report on structural universality in computer experiments with ODE's, but U-sequence does have been observed in multicomponent chemical reactions [7]. The reason for the slow-footed advance on ODE's lies in the time-consuming nature of this sort of computer work due to, in particular, slowing-down phénomena near every bifurcation point [8] and in the lacking of numerical means with sufficiently high resolution power. Since we have overcome the latter difficulty by introducing the subharmonic stroboscopic sampling (SSS) method [9], enough data have been accumulated by now to look for systematics in the order of occurrence of various periodic orbits. We report below what is believed to be the first case of U-sequences discovered so far in systems of nonlinear ODE's.

II. The U-sequence

Before going to our model and results it is suitable to put together some useful notions related to U-sequences in one-dimensional unimodal mappings. We skip all mathematical details and send the interested reader to the book of Collet and Eckmann [10] for further references.

A unimodal map with negative Schwarzian derivative (so-called S-unimodal map) can have at most one stable periodic orbit. Let us take that point on the interval which corresponds to the unique maximum of the map as the "central" point and follow the subsequent iterations. We denote a point by the letter R or L according to whether it falls to the Right or Left of the central point. Therefore, each iteration sequence corresponds to a word made of R's and L's. It is possible to introduce an ordering for all admissible words, i.e., to compile a dictionary of them. Then the significant result of Metropolis, Stein and Stein [3] states that there is a correspondence between the position of a word in the dictionary and the order of occurrence of a period on the parameter axis for one-dimensional mappings. Since this sequence of words is universal for a wide class of mappings it was suggested in Ref. [3] to call it U-sequence.

We cite below the Table of all periods equal to or less than 7 from Ref. [3] with a few remarks added. The shorthand notation in the remarks requires some explanation: nP stands for period n, nI stands for chaotic band with period n, PDB is the abbreviation of period-doubling bifurcation sequence, for the meaning of secondary PDB see Ref. [6].

Table 1 U-sequence with period ≤ 7

Word	Period	Remark
R	2	Only 2P, beginning of $2*2^n$ PDB
RLR	4	4P in $2*2^n$ PDB
RLR^3	6	3P embedded in 2I, beginning of secondary $2*3*2^n$ PDB
RLR^4	7	1st 7P
RLR^2	5	1st 5P, embedded in 1I, beginning of secondary $1*5*2^n$ PDB
RLR^2LR	7	2nd 7P
RL	3	Only 3P, beginning of $3*2^n$ PDB
RL^2RL	6	next in $3*2^n$ PDB
RL^2RLR	7	3rd 7P
RL^2R	5	2nd 5P, embedded in 1I, beginning of another $1*5*2^n$ PDB
RL^2R^3	7	4-th 7P
RL^2R^2	6	3rd 6P
RL^2R^2L	7	5-th 7P
RL^2	4	2nd and last 4P, beginning of $4*2^n$ PDB
RL^3RL	7	6-th 7P
RL^3R	6	4-th 6P
RL^3R^2	7	7-th 7P
RL^3	5	3rd and last 5P, beginning of $5*2^n$ PDB
RL^4R	7	8-th 7P
RL^4	6	5-th and last 6P, beginning of $6*2^n$ PDB
RL^5	7	9-th and last 7P, beginning of $7*2^n$ PDB

Table 1 will be compared to our results on the forced Brusselator (Fig.1).

III. Model and Methodology

We present the periodically forced Brusselator as an autonomous system of ODE's:

$$\left.\begin{aligned} \dot{x} &= A-(B+1)x+x^2y+\alpha z, \\ \dot{y} &= Bx-x^2y, \\ \dot{z} &= \omega u, \\ \dot{u} &= -\omega z. \end{aligned}\right\} \qquad (1)$$

This form has the advantage of having direct physical interpretation as two coupled oscillators and of saving computer time due to the absence of any elementary function in it.

We study system (1) by combined use of three methods, i.e., following the trajectories in phase space, stroboscopic and subharmonic stroboscopic sampling (SSS) in the time domain, and power spectra analysis in the frequency domain. For a comparison of these three methods and an exposition of a few subtleties in using SSS we refer to Ref. [9]. It is worthy to emphasize here that the SSS method is very effective in getting high frequency resolution. This explains why we now have a much more detailed knowledge of the bifurcation and chaos structure in parameter space for system (1) than, say, the original autonomous Lorenz system which has been studied during twenty years by many authors.

IV. Results

Our main results are summarized in Fig.1. This is the B=1.2, α=0.05 section of the (A, B, α, ω) parameter space, of which a small part, namely, the upper left corner, appears as Fig.9 in Ref. [6].

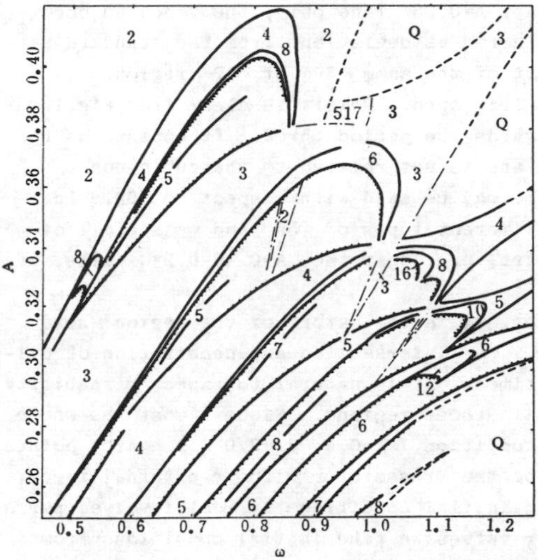

Fig.1 A-ω phase diagram (B=1.2, α=0.05)
- ―――― boundary between periods
- ..... boundary of intermittency
- ----- boundary between periodic and quasiperiodic regimes
- -.-.- boundary of unclear nature
- ııııı PDB sequences

Number indicates period. Q stands for quasiperiodic. All unlabelled regions are chaotic with embedded periodicities.

First of all, if one counts the periods along the slanting line A=0.46-0.2*ω in Fig.1, then all periods of the form RL^n occur successively for n=0, 1,...,6. They are the leading periods in $m*2^n$ type PDB sequences for m=2 to 8. In fact, a more detailed search has been made along this line using Δω=0.001 and all but one periods from 2P to 6P have been encountered in the order they appear in Table 1. The only missing period was the RL^3R type 6P which may easily be overlooked using a Δω=0.001 mesh.

A striking feature of the A-ω phase diagram is the folding and intersection of main PDB sequences. The "window" or "nested" structure of periods mentioned in Ref. [11] or Ref. [6] happened to be the consequence of folding. Since the entire paper Ref. [11] and most of Ref. [6] were devoted to a small segment of A=0.4 horizontal in Fig.1, it is quite understandable now why the systematics of periods could not be recognized as that of the U-sequence. This tells once again about the necessity to enlarge the parameter space in similar research.

Some additional remarks follow on the folding of U-sequences. The original order of words in the U-sequence, as tabulated in Ref. [3], corresponds exactly to the occurrence of different periods in the logistic map

$$x_{n+1} = 1-\mu x_n^2 , \tag{2}$$

as μ varies from 0 to 2. However, it is incorrect to think that this order itself is universal. Collet and Eckmann pointed out in their book [10] that it suffices to substitute μ in Eq.(2) by a non-monotonic function, e.g.

$$x_{n+1} = g(\lambda, x_n) = 1-9/4\lambda(\lambda-\frac{1}{3})^2 x_n^2 , \tag{3}$$

the resulting map Eq.(3) will lead to folding of the U-sequence when λ goes from 0 to 1. Fig.1 gives a clear view of how U-sequence may fold. If one looks for different periods along, say, A=0.33 line only, the repeated occurrence of 3P and 4P would be considered as deviations from the standard pattern, though they are an integral part of the same 3P or 4P region.

In Ref. [6] the role of 3P was left open. Now it is clear from Fig.1 that there is nothing specific as regards the period three. It appears as the only possible 3P in the U-sequence and is not related to the cubic nonlinearity in the system (1). The same may be said with respect to 5P. In Table 1 there are altogether three different types of 5P and we see all of them in the more or less "regular" part, e.g. near the A=0.46-0.2*ω line, of Fig.1.

Another question left open in Ref. [6] is the nature of the region where different regimes intersect. In Fig.1 the intersection and penetration of different periods repeat at least four times. It is natural to expect bistability and dependence on initial conditions in those regions. Since almost the whole Fig.1 was obtained by using initial condition x_o=0.4, y_o=3.0 or nearby points, i.e., very close to the fixed point of the Brusselator without external force, we have tested the effect of changing initial conditions in most involved parts of this diagram. For given parameter values we take initial conditions from a square in the (x,y) plane that is larger than the nearby attractor. For the intersection of, say, 2P and 3P periods, no dependence on initial conditions has been observed. Most probably what has been observed is an interpenetrated distribution of 2P and 3P points, showing a very complicated relation between different basins.

In regions nearer to the quasiperiodic regime we did find dependence on initial conditions. For example, at parameter values A=0.385, ω=0.94, we got 7P for (x_o,y_o)=(10,1), (1,10), but 5P for (9,9). It seems that the bifurcation scheme and chaotic regime of this nonlinear system can be fully exposed only in the even larger space of $(x,y)\otimes(A, B, \alpha, \omega)$.

V. On different routes to chaos

In recent attempts to relate chaotic behaviour to the onset of turbulence many different routes to chaos have been suggested, see, e.g. Ref. [12]. Among these routes the most thoroughly studied are transitions via PDB cascade [1] and the intermittent transition [13], both described by the same renormalization group functional equation with different boundary conditions. Another route to chaos via quasiperiodic motion with two or three incommensurable frequencies has often been observed in hydrodynamical experiments, but less understood theoretically. Some experimentalists tend to name other routes to chaos, e.g. transition via alternating periodic and chaotic regimes [14]. In real experiments it is quite common to see various routes to chaos in the same physical system under slightly different conditions. Therefore, these routes to chaos may happen to be diverse facets of one and the same phenomenon of transition to turbulence.

Our system (1) provides a good mathematical model for observing various routes to chaos along different directions in the parameter space. In Fig.1 regions between any pair of type RL^{n-1} and RL^{n} PDB sequences correspond to chaotic regime with many periodic orbits embedded in them. Every chaotic region is bounded on one side by PDB sequence and on the other by intermittent transitions. Intermittency associated with tangent bifurcations of a not very long period can easily be distinguished from chaotic or transient behaviour by their specific SSS diagrams (for details see a separate paper on intermittent transitions in the forced Brusselator[15]).

If one varies the parameter values along, say, the $A=0.46-0.2*\omega$ line in the $A-\omega$ plane and one does not have sufficiently high resolution, then the impression of an alternating periodic and chaotic regime may occur. We believe this is related to the "alternating route" observed experimentally. In Fig.1 the less "regular" part, where folding and intersection of various periods take place, in most directions goes into quasiperiodic regime. Intuitively this is where to look for transition to chaos via quasiperiodic motion with incommensurable frequencies, but so far we have not been able to make any definite conclusion on this point. A search in the $\alpha-\omega$ plane, corresponding to more favourable (compared with the old $A=0.4$) sections in Fig.1, is being carried on.

Acknowledgement

We thank our colleague CHEN Shi-gang for many discussions.

References

1. M.J. Feigenbaum, J. Stat, Phys., 19 (1978) 25; 21 (1979) 669.
2. P. Coullet, J. Tresser, C.R. Acad. Sci. (Paris), 287A (1978) 577; J. de Phys., C5 (1978) 25.
3. N. Metropolis, M.L. Stein, P.R. Stein, J. Combinatorial Theor., 15 (1973) 25.
4. E. Ott, Rev. Mod. Phys., 53 (1981) 655.
5. HAO Bai-lin, Progress in Physics (China), to appear.
6. B.-L. HAO, S.-Y. ZHANG, Commun. in Theor. Phys., 1 (1982) 111; J. Stat. Phys., 28 (1982)769.
7. R.H. Simoyi, A. Wolf, H.L. Swinney, Phys. Rev. Lett., 49 (1982) 245.
8. B.-L. HAO, Phys. Lett., 86A (1981) 267.
9. HAO Bai-lin, ZHANG Shu-yu, Phys. Lett., 87A (1982) 267; Acta Physica Sinica, 32 (1983)198.
10. P. Collet, J.-P. Eckmann, Iterated Maps on the Interval as Dynamical Systems, Birkhauser, 1980.
11. K. Tomita, T. Kai, J. Stat. Phys., 21 (1979) 65.
12. J.-P. Eckmann, Rev. Mod. Phys., 53 (1981) 643.
13. Y. Pomeau, P. Manneville, Commun. Math. Phys., 74 (1980) 189; J.E. Hirsch, M. Nauenberg, D.J. Scalapino, Phys. Lett., 87A (1982) 391.
14. J.S. Turner, J.C. Roux, W.D. McCormick, H.L. Swinney, Phys. Lett., 85A (1981) 9.
15. WANG Guang-rui, CHEN Shi-gang, HAO Bai-lin, Acta Physica Sinica, to appear.

VOLUME 50, NUMBER 12 PHYSICAL REVIEW LETTERS 21 MARCH 1983

Chaos in a Nonlinear Driven Oscillator with Exact Solution

Diego L. Gonzalez and Oreste Piro
Laboratorio de Física Teórica, Departamento de Física, Universidad Nacional de La Plata, La Plata, Argentina

(Received 6 December 1982)

A nonlinear oscillator externally driven by an impulsive periodic force is investigated. An exact analytical expression is obtained for the stoboscopic or Poincaré map for all values of parameters. The model displays period-doubling sequences and chaotic behavior. The convergence rate of these cascades is in very good agreement with Feingenbaum theory.

PACS numbers: 05.40.+j, 03.40.-t

The existence of period-doubling bifurcation and chaos in nonlinear externally driven oscillators has been studied by many authors. A lot of models involving both additive excitations[1] and parametrical ones[2] were numerically investigated with the restrictions imposed by the numerical integration of the equations of motion. As a consequence the results do not present the precision that is habitual in simulation involving unidimensional and multidimensional discrete mappings.[3] In this Letter we present a model of a forced nonlinear oscillator which allows us to obtain an exact analytical expression for the stroboscopic map[4] (analog to the Poincaré surface of section for autonomous systems). To our knowledge this is the first model that is not piecewise linear in which such exact derivation has been performed. By making use of the mentioned map we are able to investigate the stability of periodic solutions as a function of parameters and to calculate the convergence rate of the cascade of period-doubling bifurcations which leads to chaos in some regions of the parameter space.

Consider the equation

$$\ddot{x}+\dot{x}(4bx^2-2a)+b^2x^5-2abx^3+(\omega_0{}^2+a^2)x=V_E\omega_E\,\delta(\cos\omega_E t)=V_E\textstyle\sum_n\delta(t-n\tau_E) \tag{1}$$

which can be thought of as modeling an electronic oscillator tuned with nonlinear elements and having its frequency synchronized by means of an external impulsive periodic signal. In (1) $\omega_0=2\pi/\tau_0$ is the proper frequency of the system, a and b are constants, and $\omega_E=2\pi/\tau_E$ and V_E are the frequency and amplitude of the driving pulses, respectively. If $V_E=0$, (1) has the general solution[5]

$$x(t)=\cos(\omega_0t+\varphi)\left\{Ae^{-2at}+\frac{b}{2a}\left[\frac{1}{1+a^2/\omega_0{}^2}\right]\left\{1+\frac{a}{\omega_0}\cos(\omega_0t+\varphi)\left(\frac{2a}{\omega_0}\cos(\omega_0t+\varphi)+2\sin(\omega_0t+\varphi)\right)\right.\right.$$
$$\left.\left.-e^{-2at}\left[1+\frac{a}{\omega_0}\cos\varphi\left(\frac{2a}{\omega_0}\cos\varphi+2\sin\varphi\right)\right]\right\}\right\}^{-1/2}, \tag{2}$$

where A and φ are integration constants which are expressed as

$$A=\omega_0{}^2[x_0\omega_0+(\omega_0a-x_0{}^2-\dot{x}_0)^2]^{-1},$$
$$\sin\varphi=A^{1/2}\omega_0{}^{-1}[x_0a-x_0{}^3b-\dot{x}_0],\quad \cos\varphi=x_0\omega_0[x_0\omega_0{}^2+(\omega_0a-x_0{}^2-\dot{x}_0)^2]^{-1/2}, \tag{3}$$

where $x_0=x(0)$ and $\dot{x}_0=\dot{x}(0)$ are the initial conditions. Thus we have for the autonomous system the following two parametrical equations as solution:

$$x(t)=f(x_0,\dot{x}_0,t),\quad \dot{x}(t)=g(x_0,\dot{x}_0,t), \tag{4}$$

where f and g are given in terms of elementary functions by replacing (3) in (2) and its derivative. When $V_E\neq0$ the only effect of the external force is to produce a discontinuity in the first derivative of $x(t)$. The height of this jump is V_E. Therefore the solution of (1) in the driven case is

$$x_F(t)=\sum_{n=0}^{\infty}f(x_F(n\tau_E),\dot{x}(n\tau_E)+V_E,t-n\tau_E)H(t-n\tau_E)H((n+1)\tau_E-t),$$
$$\dot{x}_F(t)=\sum_{n=0}^{\infty}g(x_F(n\tau_E),\dot{x}(n\tau_E)+V_E,t-n\tau_E)H(t-n\tau_E)H((n+1)\tau_E-t), \tag{5}$$

VOLUME 50, NUMBER 12 PHYSICAL REVIEW LETTERS 21 MARCH 1983

where H is the unit step function. From (5) it is straightforward to obtain the stroboscopic map by sampling the trajectory at regular time intervals coincident with τ_E. Then the map can be written

$$x_F((n+1)\tau_E)=f(x_F(n\tau_E),\dot{x}_F(n\tau_E)+V_E,\tau_E), \tag{6}$$
$$\dot{x}_F((n+1)\tau_E)=g(x_F(n\tau_E),\dot{x}_F(n\tau_E)+V_E,\tau_E).$$

With the aid of this map we develop a numerical investigation. The first interesting result is the subharmonic entrainment spectrum shown in Fig. 1. There the zones in the (V_E,τ_E) plane are drawn where the stable output of the oscillator is periodic with period commensurable with τ_E. At this regime of dissipation this spectrum is invariant under the shift $\tau_E \to \tau_E + n\tau_0$. Each zone which touches the τ_E axis at the value τ_j is characterized by the rational number $\tau_j/\tau_0 = p/q$ (p and $q \in N$ are, respectively, primes).[6] q is equal to the periodicity of the response and p is the number of oscillations that the system would perform, if the external force vanished, in a time interval τ_j. The sizes of the various entrainment regions

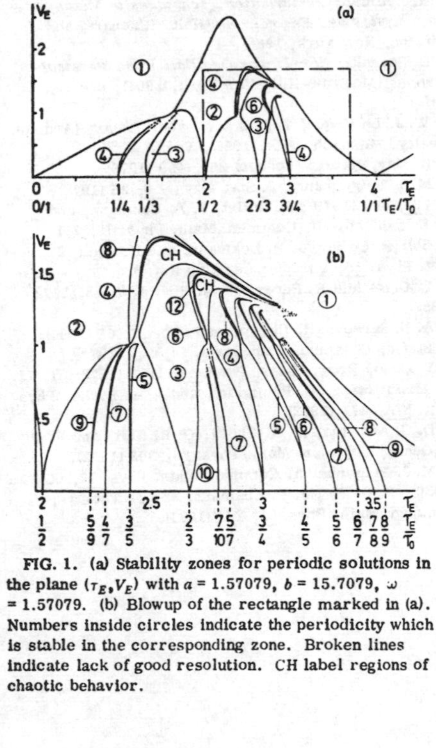

FIG. 1. (a) Stability zones for periodic solutions in the plane (τ_E, V_E) with $a = 1.57079$, $b = 15.7079$, $\omega = 1.57079$. (b) Blowup of the rectangle marked in (a). Numbers inside circles indicate the periodicity which is stable in the corresponding zone. Broken lines indicate lack of good resolution. CH label regions of chaotic behavior.

are ordered in a way related to a concept from number theory, Farey sequences. An n-Farey sequence F_n is the increasing succession of rational numbers whose denominators are less than or equal to n.[7] We call two rational numbers "adjacent" if they are consecutive in F_n for any n. A necessary and sufficient condition for p_1/q_1 and p_2/q_2 to be adjacent is $|p_1q_2 - p_2q_1| = 1$.

A rational number p'/q' belonging to the open interval $(p_1/q_1, p_2/q_2)$ where p_1/q_1 and p_2/q_2 are adjacent will be called "mediant" if there is no other rational in the interval having smaller denominator. It is known that $p'/q' = (p_1+p_2)/(q_1+q_2)$ and is unique. The observation of Fig. 1 and other enlargements not shown here leads us to the following conjecture: The synchronization zone characterized by a mediant number of two adjacent rationals is the greatest of all the zones situated in between those characterized by them. In addition it has, obviously, the least period.

Figure 1(b) shows that the more important entrainment regions have a similar form. They resemble cornucopias, each with the tip attached to the horizontal axis and the other end converging to a point inside of the perfect entrainment region. It is remarkable that all these entrainment regions are of width increasing with V_E when it is less than about 1.0 and above this value each of them decreases in width and folds on itself surrounding a region [marked CH in Fig. 1(b)] which presents a period-doubling route to chaos.[8,9]

Having found such a behavior, we test the Feingenbaum universality. We make use of the fact that, because of the high dissipation, the Jacobian of the stroboscopic map is very small, which ensures that the map neighbors a unidimensional map of the limit cycle on itself. Thus we search for the values of the parameters at which the "critical" point of this "unidimensional" map is periodic.[10] In Table I the values of τ_E (with $V_E = 1.625$) are displayed at which $\partial x_{n+q}/\partial x_n$ evaluated at a stable q-periodic point vanishes, for all $q = 2^r$ with r from 1 to 10. We have named them τ_r. The next two columns list the $\delta_r = |\tau_r - \tau_{r-1}| / |\tau_{r+1} - \tau_r|$ and $\delta_r' = |\delta_r - \delta_{r-1}| / |\delta_{r+1} - \delta_r|$. Since δ' is the convergence rate of δ_r, it allows us to extrapolate the value of the $\lim_{r\to\infty} \delta_r = \delta$. The result is shown in the last column. Notice that it agrees with the exact value to five decimal digits and this accuracy has been achieved without refined numerical techniques.

For most of the points inside the regions CH the response of the oscillator becomes chaotic.

TABLE I. Values obtained through calculations of the convergence rate of the period-doubling sequence at $V_B = 1.625$ and the other values of parameters as in Fig. 1 but with $10.6 < \tau_B \leq 10.8$. Compare with the exact value of the Feigenbaum constant $\delta = 4.669\,2016\ldots$.

r	τ_r	δ_r	δ_r'	δ_{extrap}
1	10.743 617 135 97			
2	10.703 113 442 83	3.310 197		
3	10.690 877 398 97	4.231 186	2.825	4.735 833 1
4	10.687 985 528 98	4.557 198	3.731	4.676 595 5
5	10.687 350 957 63	4.644 584	4.535	4.669 306 1
6	10.687 214 331 59	4.663 855	4.586	4.669 243 2
7	10.687 185 036 92	4.668 057 3	4.658	4.669 179 6
8	10.687 178 761 369	4.668 958 9	4.855	4.669 195 6
9	10.687 177 417 267	4.669 144 6		
10	10.687 177 129 398			

However, there also exist regions contained in CH for which the response is entrained. We have found periodicities other than 2^r. The occurrence of these periodicities ensures, as is known,[11,12] the occurrence of chaos for iterated unidimensional maps.

Finally we stress the compatibility of the geometry of the stability regions shown in Fig. 1 with the idea that period-doubling bifurcations are related to the overlap of synchronization horns in the parameter space. It is in fact remarkable that the beginnings of all period-doubling chains in Fig. 1(b) are situated in the neighborhoods of the points at which prolongations of the boundaries of two adjacent regions of stability would intersect. The conjecture of this connection has been made already in Ref. 13 in a different context (see also Glass and Perez[10]).

As a conclusion we want to stress the adequacy of the model presented here as a numerical laboratory for continuous dynamical system. For example, it is easy to check any universal constant (as has been done with δ) as well as the behavior of the power spectrum[14] and statistical properties of the chaotic regime. Details of this calculation will be given elsewhere.

We wish to thank all members of the Laboratorio de Física Teórica for helpful discussions and encouragement and especially Professor H. Fanchiotti and Professor H. Vucetich for bringing our attention to Eq. (1). This work was supported in part by Consejo Nacional de Investigaciones Científicas y Técnicas, Comisión de Investigaciones Cientificas Provincia de Buenos Aires, and Subsecretaria de Ciencia y Tecnica, Argentina.

[1]B. A. Huberman and J. P. Crutchfield, Phys. Rev. Lett. 43, 1743 (1981); K. Tomita and T. Kai, J. Stat. Phys. 21, 65 (1979); B. Hao and S. Zhang, to be published; W. H. Steeb, W. Erig, and A. Kunic, "Chaotic behavior and limit cycle behavior of anharmonic systems with periodic order and perturbations," to be published; R. Shaw, Z. Naturforsch. 36a, 80 (1981).

[2]J. B. McLaughlin, J. Stat. Phys. 24, 375 (1981); P. Coullet, C. Tresser, and A. Arnéodo, Phys. Lett. 72A, 268 (1979).

[3]B. Derrida, A. Gervois, and Y. Pomeau, J. Phys. A 12, 169 (1979); A. B. Zissok, Phys. Rev. A 24, 1640 (1981).

[4]N. Minorsky, *Non-Linear Oscillations* (Van Nostrand, Princeton, N.J., 1962); T. Kai and K. Tomita, Prog. Theor. Phys. 61, 54 (1979).

[5]R. Bellman, *Perturbation Techniques in Mathematics, Physics and Engineering* (Holt, Rinehart, and Winston, New York, 1966).

[6]C. Hayashi, *Non-Linear Oscillations in Physical Systems* (McGraw-Hill, New York, 1964); see also Ref. 4.

[7]W. J. Le Veque, *Topics in Number Theory* (Addison-Wesley, Massachusetts, 1956), Vol. 1.

[8]R. May, Nature (London) 261, 459 (1979).

[9]M. J. Feigenbaum, J. Stat. Phys. 19, 25 (1978), and 21, 669 (1979); P. Collet, J. P. Eckmann, and O. E. Lanford, III, Commun. Math. Phys. 76, 211 (1980); P. Collet, J. P. Eckmann, and H. Koch, J. Stat. Phys. 25, 1 (1980); see also Ref. 3.

[10]L. Glass and R. Perez, Phys. Rev. Lett. 48, 1772 (1982).

[11]A. N. Sarkovskii, Ukrain. Mat. Zh. 16, 61 (1964); P. Stefan, Commun. Math. Phys. 54, 237 (1977).

[12]Y. Oono, Prog. Theor. Phys. 59, 1028 (1978); T. Li, M. Misiurewicz, G. Pianigiani, and J. A. Yorke, Phys. Lett. 87A, 271 (1982).

[13]D. G. Aronson, M. A. Chory, G. R. Hall, and R. P. McGehee, Commun. Math. Phys. 83, 303 (1982).

[14]M. J. Feigenbaum, Commun. Math. Phys. 77, 65 (1980); P. Collet, J. P. Eckmann, and L. Thomas, Commun. Math. Phys. 81, 261 (1981).

ERRATA

CHAOS IN A NONLINEAR DRIVEN OSCILLATOR WITH EXACT SOLUTION

Diego L. Gonzalez and Oreste Piro (*Phys. Rev. Lett.* 50, 870(1983))

1. There are several misprints in Eq. (3), the correct form is:

$$A = \omega_0^2\,[x_0^2\omega_0^2 + (x_0 a - x_0^3 b - \dot{x}_0)^2]^{-1}$$

$$\sin\varphi = A^{\frac{1}{2}}\,\omega_0^{-1}\,[x_0 a - x_0^3 b - \dot{x}_0]$$

$$\cos\varphi = x_0\,\omega_0[x_0^2\omega_0^2 + (x_0 a - x_0^3 b - \dot{x}_0)^2]^{-\frac{1}{2}} \;.$$

2. A subindex F is missing in the second argument of the functions f and g in Eq. (5). There $\dot{x}$ should be $\dot{x}_F$ instead.

3. The last line above FIG. 1 must be read:

"$q\tau_j$. The sizes..."

NATURE VOL. 303 23 JUNE 1983 ARTICLES 663

Period doubling and chaos in partial differential equations for thermosolutal convection

D. R. Moore* & J. Toomre

Joint Institute for Laboratory Astrophysics and Department of Astro-Geophysics, University of Colorado, Boulder, Colorado 80309, USA

E. Knobloch* & N. O. Weiss

Department of Applied Mathematics and Theoretical Physics, University of Cambridge, Cambridge CB3 9EW, UK

Numerical experiments on two-dimensional thermosolutal convection reveal a transition from periodic oscillations to chaos through a sequence of period-doubling bifurcations. Within the chaotic region there are narrow periodic windows. This is the first example of period-doubling in solutions of partial differential equations. A truncated model indicates that this behaviour is associated with heteroclinic explosions.

A LIGHTLY damped system displaced from stable equilibrium exhibits gradually decaying oscillations. If, in addition, the system is acted on by a small destabilizing force that lags in phase behind the restoring force, the oscillations may grow exponentially with time until nonlinearities become important. While the destabilizing force is sufficiently weak the oscillations will be strictly periodic but, as the driving force is increased, behaviour becomes increasingly nonlinear and there may be a transition to aperiodic, or chaotic, oscillations. One route to chaos that has been extensively studied is by a sequence of bifurcations, at each of which the period doubles[1-3]. Beyond the accumulation point of this sequence the oscillations are typically aperiodic. Period doubling has been found in various experiments, as well as in solutions of difference equations and ordinary differential equations but has not hitherto been demonstrated for partial differential equations.

Thermosolutal convection provides an example of such a system that also has extensive applications (refs 4–6 and refs therein). Consider motion driven by heating from below a layer of fluid (such as water) containing a stabilizing concentration of a solute (such as salt) that diffuses less rapidly than heat. Oscillatory convection can then occur even when the mean density decreases upwards. Oscillations are found in laboratory experiments, though they give way to multilayered convection[7] if the fluid is heated too vigorously. Previous theoretical studies have shown that there can be a transition from periodic to chaotic behaviour as the rate of heating is increased.

Steady and oscillatory convection

In two dimensions, thermosolutal convection in a boussinesq fluid is described by the nondimensionalized equations[8,9]

$$\frac{1}{\sigma}[\partial_t\nabla^2\psi+\partial(\psi,\nabla^2\psi)/\partial(x,z)]=R_T\theta_x-R_SS_x+\nabla^4\psi \quad (1)$$

$$\partial_t\theta+\partial(\psi,\theta)/\partial(x,z)=\psi_x+\nabla^2\theta \quad (2)$$

$$\partial_tS+\partial(\psi,S)/\partial(x,z)=\psi_x+\tau\nabla^2S \quad (3)$$

where σ is the Prandtl number, τ $(0<\tau<1)$ is the ratio of the solutal to the thermal diffusivity, and R_T, R_S are respectively the thermal and solutal Rayleigh numbers. The quantities ψ, θ, S are the streamfunction and the deviations of the temperature and solute concentration from the nonconvecting state ($\psi=\theta=S=0$). We adopt the boundary conditions

$$\psi=\psi_{zz}=\theta=S=0 \quad \text{on } z=0,1 \quad (4)$$

$$\psi=\psi_{xx}=\theta_x=S_x=0 \quad \text{on } x=0,\lambda \quad (5)$$

which correspond to periodic rolls.

Linear theory shows that for $R_S>\tau^2(1+\sigma)R_0/\sigma(1-\tau)$, where $R_0=\pi^4\lambda^{-4}(1+\lambda^2)^3$, there is a Hopf bifurcation from the static state to oscillatory convection at $R_T=R_T^{(o)}$, followed by a simple bifurcation at $R_T=R_T^{(e)}$ that gives rise to a branch of subcritical (unstable) steady solutions[8,10]. This branch eventually turns round and acquires stability, since stable steady solutions with larger amplitudes have been found numerically for $R_T>R_T^{(min)}$ (where $R_T^{(min)}<R_T^{(e)}$) (refs 8–11). The oscillatory branch apparently terminates on the unstable portion of the steady branch where the period of the oscillations becomes infinite[12,13] (see Fig. 3).

For certain values of the parameters, nonlinear oscillatory solutions of equations (1)–(3) show a transition from periodic to aperiodic behaviour before the end of the oscillatory branch[10,11]. To understand the nature of this transition we have studied it in considerable detail, using numerical solutions of the system (1)–(5) obtained by finite difference methods[11,14]. For all these calculations we imposed symmetry about the centre of the convection roll, so that $\psi(x,z)=\psi(\lambda-x,1-z)$, and so on, and set $\sigma=1$, $\tau=10^{-1/2}$, $R_S=10^4$. These values were chosen here, as in earlier work[9,11], to avoid undue computational effort; nonlinear thermohaline convection ($\tau\approx10^{-2}$) produces more disparate thermal and solutal structures, which are less easily resolved[7]. The aspect ratio λ was either $2^{1/2}$ (the value that minimizes both $R_T^{(o)}$ and $R_T^{(e)}$ and was used in refs 9, 11) or 1.5 (for convenience in defining a computational mesh) and the mesh intervals were varied, though there were always at least 188 independent mesh points. The bifurcation pattern that we shall describe was unaffected by halving the mesh intervals or by altering λ (though changing λ from 1.41 to 1.5 may displace the positions of specific bifurcations by up to 1.5%).

Period doubling into chaos

When $\lambda=2^{1/2}$, $R_T^{(o)}=7{,}720$ and $R_T^{(e)}=32{,}280$. Huppert and Moore[10,11] followed the oscillatory branch and discovered a bifurcation at $R_T\approx9{,}200$. Nonlinear oscillations can be described by plotting some global property, such as the spatially averaged kinetic energy, E, or the Nusselt numbers N_S, N_T that measure the rates of transport of solute or heat across the layer, as functions of time. In Fig. 1*a* and *b*, N_S and N_T are shown

* Permanent addresses: Department of Mathematics, Imperial College, London SW7 2BZ, UK (D.R.M.); Department of Physics, University of California, Berkeley, California 94720, USA (E.K.).

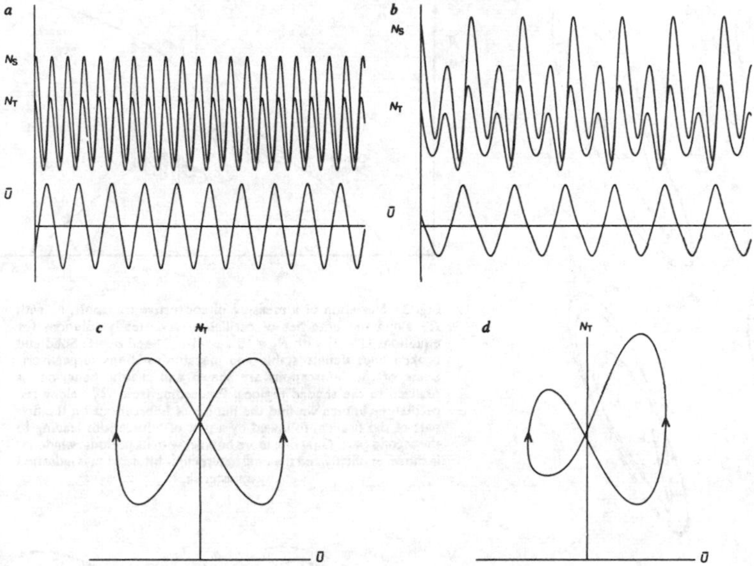

Fig. 1 Symmetrical and asymmetrical oscillations in solutions of the partial differential equations ($\lambda = 2^{1/2}$): plots showing N_S, N_T and $\bar{u}$ as functions of time for *a*, $R_T = 8{,}600$; *b*, $R_T = 9{,}800$, with *c*, *d* the corresponding limit cycles projected onto the $\bar{u}$–N_T plane.

for $R_T = 8{,}600$, 9,800 respectively, together with $\bar{u}$, the mean value of the horizontal velocity at the base of the cell. N_S and N_T are quadratic quantities (like the kinetic energy) and in Fig. 1*a*, where the oscillations are symmetrical, they vary with half the period of $\bar{u}$. In Fig. 1*b* the oscillations are asymmetrical and N_S, N_T and $\bar{u}$ all have the same period. Thus the first bifurcation is from symmetrical to asymmetrical oscillations[12], as shown by the trajectories in the $\bar{u}$–N_T phase plane that are plotted in Fig. 1*c* and *d*.

This is followed by a series of period-doubling bifurcations. Figure 2*a* shows a limit cycle in the N_T–N_S plane for an asymmetrical oscillation (period 1), followed in Fig. 2*b* and *c* by limit cycles with period 2 and period 4. Three successive period-doublings have been identified and the ratio of the increments in R_T, $\delta \approx 5$, is consistent with that found for one-dimensional maps[15–17]. For $10{,}150 \leq R_T \leq 10{,}300$ all solutions were aperiodic but at $R_T = 10{,}325$ there was a solution with period 4; subsequent bifurcations led to solutions with period 2, followed by asymmetrical and, finally, symmetrical oscillations of period 1, as shown in Fig. 2*d–f*. As R_T increases, there is apparently a complete Feigenbaum period-doubling cascade, followed by an inverse sequence of bifurcations at which the period is halved. The overall bifurcation pattern is sketched in Fig. 3.

Transitions in a simpler system

Similar behaviour has also been found[18] in a truncated fifth-order model due originally to Veronis[8]. The model equations[12]

$$\dot{a} = \sigma[-a + r_T b - r_S d] \tag{6}$$

$$\dot{b} = -b + a(1-c) \tag{7}$$

$$\dot{c} = \varpi(-c + ab) \tag{8}$$

$$\dot{d} = -\tau d + a(1-e) \tag{9}$$

$$\dot{e} = \varpi(-\tau e + ad) \tag{10}$$

where $\varpi = 4\lambda^2/(1+\lambda^2)$, $r_S = R_S/R_o$, $r_T = R_T/R_0$, and so on, are an asymptotically exact consequence of the full system (1)–(5) to $0(a^3)$. (Note that when $r_S = 0$ equations (6)–(8) reduce to the Lorenz system, which also exhibits double Feigenbaum cascades when ϖ is sufficiently small[19].) The truncated model is not a valid approximation to the full two-dimensional problem when a is finite. Nevertheless, the overall bifurcation structures of both systems are almost identical[12] and the fifth-order system can be used to gain a qualitative understanding of solutions to the full equations. As the value of $|a|$ (that is, of $r_T - r_T^{(o)}$) is increased, the fifth-order model typically exhibits a bifurcation from symmetry to asymmetry followed by a period-doubling cascade terminating in aperiodic solutions. Figure 4 shows some results for parameters corresponding approximately to those of Fig. 2, the modes *c* and *e* being linearly related to the Nusselt numbers N_T and N_S. As might be expected, the truncated system has less structure in its limit cycles; there is, however, a complete period-doubling cascade, followed by chaotic behaviour beyond $r_T \approx 13.564$ ($R_T \approx 8{,}918$). The bifurcation sequence is relatively compressed and occurs at lower Rayleigh numbers for the truncated model, while the oscillatory branch terminates shortly after the accumulation point. Still, the qualitative resemblance between the transitions to chaos in the two systems is close enough to indicate that we are dealing with the same phenomenon.

Heteroclinic explosions

The model system (6)–(10) has the advantage that the steady branch can be found analytically and the eigenvalues describing its stability can be calculated[12]. Knowledge of these eigenvalues allows us to understand why chaos occurs. Shil'nikov[20] studied a three-dimensional system (that is with three independent variables) containing a homoclinic orbit connecting a saddle-focus to itself. If the eigenvalues at the saddle-focus are s_1, s_2, s_3, where $s_1 > 0$ and s_2, s_3 are complex conjugates with negative real parts satisfying

$$|Re\, s_2| < s_1 \tag{11}$$

Shil'nikov showed that the Poincaré return map for nearby trajectories contains a countable infinity of horseshoes. Each of these horseshoes contains an invariant Cantor set with a countably infinite set of (nonstable) periodic orbits of arbitrarily long periods and an uncountable set of bounded (nonstable)

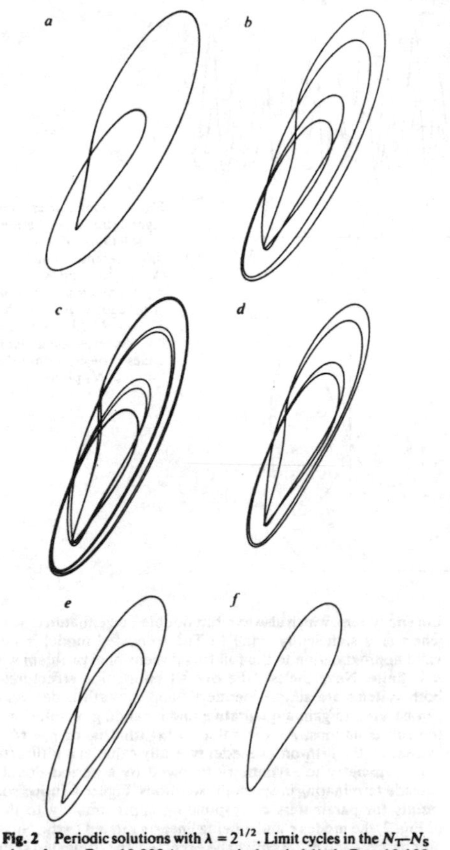

Fig. 2 Periodic solutions with $\lambda = 2^{1/2}$. Limit cycles in the N_T–N_S plane for *a*, $R_T = 10{,}000$ (asymmetrical period 1); *b*, $R_T = 10{,}100$ (period 2); *c*, $R_T = 10{,}120$ (period 4). Further period doublings and an interval of chaos are followed by period halving leading to *d*, $R_T = 10{,}350$ (period 2); *e*, $R_T = 10{,}400$ (asymmetrical) and *f*, $R_T = 10{,}500$ (symmetrical). Orbits are described in an anticlockwise sense.

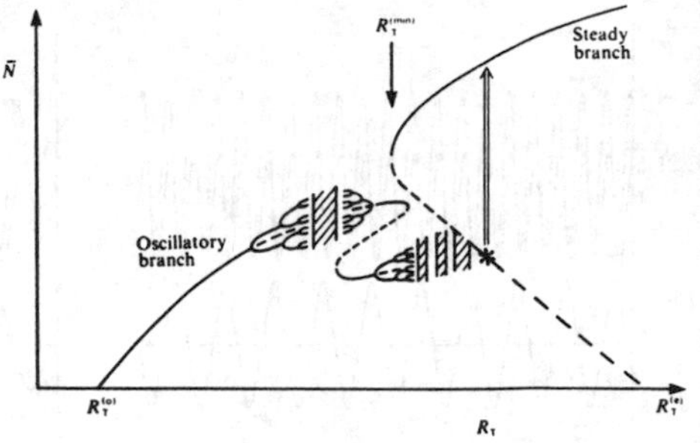

Fig. 3 Variation of a measure of convective transport, $\bar{N}$, with R_T along the branches of oscillatory and steady solutions for equations (1)–(3) with $R_S = 10^4$, $\tau = 10^{-1/2}$ and $\sigma = 1$. Solid and broken lines denote stable and unstable solutions respectively. Some of the bifurcations are shown and chaotic behaviour is realized in the shaded regions. Proceeding from $R_T^{(o)}$ along the oscillatory branch we find the bubble of bifurcations on the first part of the branch, followed by a pair of bifurcations leading to the second part. Only one more bubble, with its periodic windows, is shown explicitly and the final heteroclinic bifurcation is indicated by an asterisk.

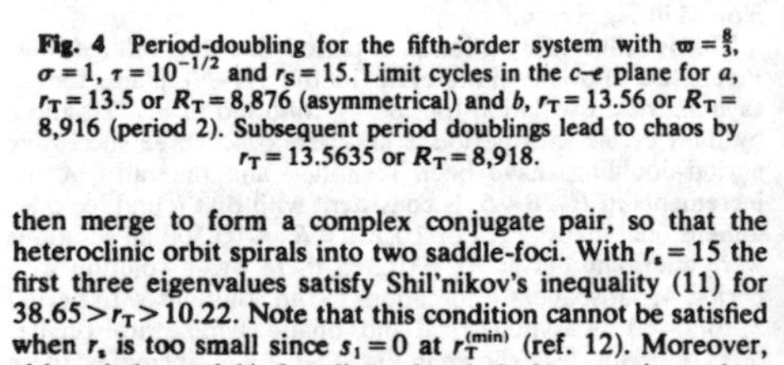

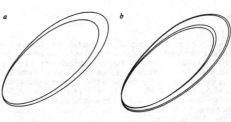

Fig. 4 Period-doubling for the fifth-order system with $\varpi = \frac{8}{3}$, $\sigma = 1$, $\tau = 10^{-1/2}$ and $r_S = 15$. Limit cycles in the *c*–*e* plane for *a*, $r_T = 13.5$ or $R_T = 8{,}876$ (asymmetrical) and *b*, $r_T = 13.56$ or $R_T = 8{,}916$ (period 2). Subsequent period doublings lead to chaos by $r_T = 13.5635$ or $R_T = 8{,}918$.

nonperiodic orbits. Sparrow[19] calls the bifurcation that leads to the creation of so many orbits (with a single horseshoe in the return map) a homoclinic explosion. The Lorenz equations provide good examples of such behaviour[19]. In the situation described by Shil'nikov, multiple bifurcations occur as the stability parameter is varied so as to approach the homoclinic orbit, and in small parameter ranges there may be many attracting orbits. In the presence of symmetry, the same behaviour will also be associated with heteroclinic orbits[21].

The fifth-order system (6)–(10) possesses the reflectional symmetry $(a, b, d) \rightarrow -(a, b, d)$, together with at least one heteroclinic orbit, where the branch of oscillatory solutions meets the branch of (unstable) steady solutions[12]. To investigate behaviour at this point, we compute the five eigenvalues s_i along the branch of subcritical solutions that bifurcates from $r_T^{(e)}$, following ref. 22. For $r_T^{(e)} > r_T > r_T^{(min)}$, s_4 and s_5 are real and negative (that is stable). For small r_S all the eigenvalues are real, with $s_1 > 0$ and $s_5 < s_4 < s_3 < s_2 < 0$, and the oscillatory branch terminates in a heteroclinic orbit connecting the two saddle-points[13]. For larger r_s, the small negative eigenvalue s_2 decreases along the branch until $s_2 = s_3$; the two eigenvalues then merge to form a complex conjugate pair, so that the heteroclinic orbit spirals into two saddle-foci. With $r_s = 15$ the first three eigenvalues satisfy Shil'nikov's inequality (11) for $38.65 > r_T > 10.22$. Note that this condition cannot be satisfied when r_s is too small since $s_1 = 0$ at $r_T^{(min)}$ (ref. 12). Moreover, although the model is five-dimensional, the last two eigenvalues s_4 and s_5 are sufficiently contracting that the system is effectively three-dimensional and the Shil'nikov mechanism is unchanged. This suggests that chaos appears as the result of heteroclinic explosions. Similar, but not identical, behaviour is found in the Lorenz equations with $\varpi = \frac{1}{4}$, $\sigma = 10$ (ref. 19).

The Shil'nikov mechanism has been shown to be present in another system of partial differential equations[23]. Although the full system (1)–(5) is infinite-dimensional, we conjecture that all but the first three eigenvalues are sufficiently large and negative for the dynamics to remain effectively three-dimensional in the neighbourhood of the unstable steady branch. Moreover, the full system contains a heteroclinic orbit[13]; hence we expect that the Shil'nikov mechanism is responsible for heteroclinic explosions in equations (1)–(3) as well.

Periodic windows

As R_T is increased, symmetrical oscillations of the type shown in Fig. 2*f* persist in the partial differential equations until

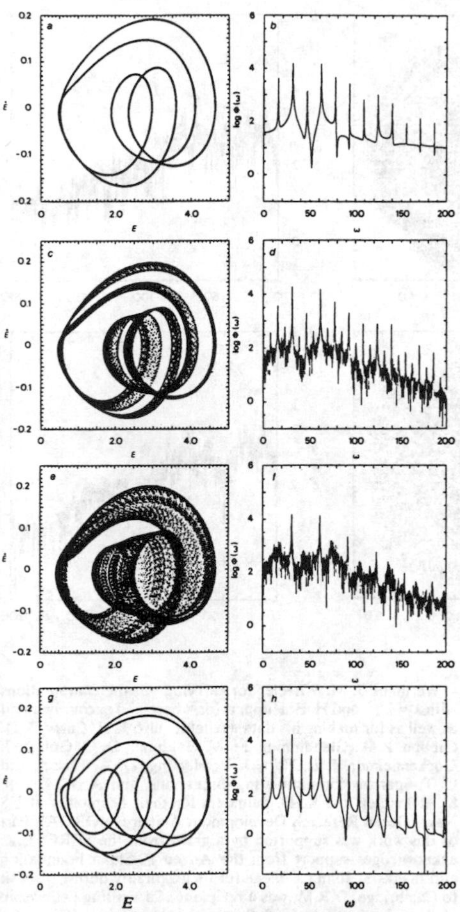

Fig. 5 Solutions on the second oscillatory branch, with $\lambda = 1.5$, for the partial differential equations. Trajectories in the kinetic energy phase plane (kinetic energy, E, plotted against its time derivative, $\dot{E}$) and the semilogarithmic plots of the kinetic energy power spectrum $\phi(\omega)$, as a function of frequency, ω, for *a*, *b*, $R_T = 10{,}450$ (asymmetrical); *c*, *d*, $R_T = 10{,}475$ (semiperiodic); *e*, *f*, $R_T = 10{,}500$ (semiperiodic) and *g*, *h*, $R_T = 10{,}510$ (period 3, symmetrical). Orbits are described in a clockwise sense.

$R_T \approx 10{,}500$, when they lose stability. Solutions are then attracted to a second oscillatory branch which, for $\lambda = 1.5$, appears when $R_T \approx 10{,}300$; the hysteresis suggests that the two branches are connected by a pair of saddle-node bifurcations. We have investigated the behaviour of solutions along the whole oscillatory branch by increasing R_T and using nearby solutions for initial conditions. The relationship between the two branches and the bifurcation patterns that were found are shown schematically in Fig. 3. There may in fact be many branches but we were only able to identify these two. At the beginning of the second branch the oscillations are symmetrical with period 1 but of different form from those on the first branch. Subsequent bifurcations lead to asymmetry and another period-doubling cascade. Figure 5*a* shows a limit cycle for an asymmetrical solution with period 1, projected onto the kinetic energy phase plane. The corresponding power spectrum, $\Phi(\omega)$, where ω is the frequency, is illustrated in Fig. 5*b*: there are sharp peaks at the basic frequency $\omega_0 = 4\pi/P \approx 30$ (where P is the full period of the basic cycle) and at $\frac{1}{2}\omega_0 \approx 15$. Note that the detailed appearance of the power spectra may be sensitive to changes in the interval T for which solutions are obtained, since the ratio T/P is not generally an integer. Beyond the accumulation point of the period-doubling sequence semiperiodic trajectories[24] are found, like that in Fig. 5*c*. The corresponding spectrum is noisy but with peaks clearly visible at frequencies ω_0, $\frac{1}{2}\omega_0$, $\frac{1}{4}\omega_0$, $\frac{1}{8}\omega_0$ and combinations of those values. As R_T increases, so does the noise, and the peaks are gradually submerged: Fig. 5*e* and *f* show chaotic behaviour with only peaks at ω_0 and its harmonics visible in the spectrum.

Within the range of values of R_T that yields aperiodic solutions on the second oscillatory branch, there are various narrow windows with periodic solutions. The most prominent of these exhibit symmetrical limit cycles with period 3. Figure 5*g* shows such a solution for $R_T = 10{,}510$, with the corresponding sharply peaked spectrum in Fig. 5*h*. As R_T is increased, the appearance of this solution is preceded by intermittency[2] and is followed by bifurcations to asymmetry and then to period 6, and so on, until chaos reasserts itself. Eventually there is an inverse cascade of period-halving bifurcations leading to solutions with period 1, followed by another transition back into chaos through a period-doubling sequence as R_T is increased. Although solutions are typically aperiodic, further periodic windows can be found and we have located two more with period 3. Such behaviour appears to be characteristic of the dynamics associated with the Shil'nikov mechanism and detailed investigation of a fifth-order model of magnetoconvection has revealed an intricate sequence of period-doubling windows interspersed with aperiodic bands[22]. The windows can be ordered to provide a natural sequence that ends with one based on a symmetrical limit cycle of period 3. As with thermosolutal convection[18], the actual sequence of bifurcations has a bubble structure, produced by a double-valued passage through the natural sequence and back again as R_T is increased. We have seen that a similar bubble structure is present on the first oscillatory branch, though no subsidiary windows were detected. We have identified at least two complete bubbles on the second branch, with indications that there are more. Possibly there are infinitely many bubbles, culminating in a heteroclinic orbit, as in the Lorenz equations[19]. Note that a second oscillatory branch can be found for the truncated system (6)–(10) as well, though for parameter values that are distinctly different from those considered here.

For $R_T^{(min)} < R_T < R_T^{(e)}$ there are two steady solution branches, of which the upper is stable and the lower unstable. The second oscillatory branch terminates when $R_T \approx 11{,}060$ (for $\lambda = 1.5$). For $R_T \geq 11{,}065$ all trajectories are attracted to the stable fixed point on the upper steady branch. Towards the end of the oscillatory branch solutions become more strikingly chaotic. Figure 6 shows results for $R_T = 10{,}625$ and $R_T = 11{,}000$. These values yield aperiodic solutions but belong to different bubbles: the trajectories fill much more of phase space than those in Fig. 5 and peaks have been eliminated from the spectra. Near the end of the oscillatory branch the basic cycle tends to be interrupted while solutions hover in the neighbourhood of the unstable steady solution; in phase space, trajectories spiral slowly in towards the unstable fixed points before escaping rapidly from their neighbourhoods. This is consistent with the existence of a heteroclinic connection between fixed points with eigenvalues that satisfy Shil'nikov's inequality (11).

Conclusions

It is well known that dissipative systems of difference or ordinary differential equations can exhibit period-doubling bifurcations, leading to chaos, interrupted by periodic windows[1]. We have shown here for the first time that this route to chaos[2] can be found for partial differential equations too. In addition, we have demonstrated how truncated model systems, derived from physical intuition or by asymptotic methods, can be used to interpret the dynamics of more complicated configurations. However,

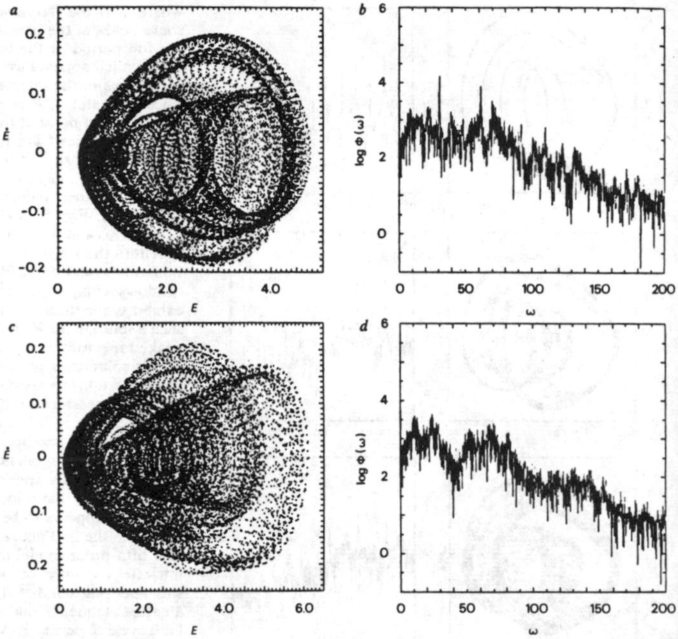

Fig. 6 Chaotic behaviour on the second oscillatory branch. As Fig. 5 but with *a*, *b*, $R_T = 10{,}625$ and *c*, *d*, $R_T = 11{,}000$. Comparison with Fig. 5 will help in interpreting these trajectories.

the robustness of the attractor that we have isolated requires further study, for it is not yet clear how far the behaviour outlined above will persist if the constraints of symmetry and two-dimensionality are relaxed.

Previous studies of partial differential equations have included a proof of the existence of horseshoes in reaction–diffusion equations[23] and for magnetoelasticity[25], while other computations have revealed a quasiperiodic route to chaos[26–30]. The presence of period-doubling sequences, with examples of periodic windows, bubbles and hysteresis, has been established experimentally for circuits[31–33], for lasers[34] and for the chemical Belousov–Zhabotinsky reaction[35,36]. More important, successive period-doubling sequences have been unambiguously observed in experiments on Rayleigh–Bénard convection in liquid helium[37], mercury[38,39] and water[40]. Though its realization may require small aspect ratios or other lateral constraints, the rich behaviour that we have described appears to be a property of real fluid systems.

We thank J. M. Wheeler for carrying out the computations with $\lambda = 2^{1/2}$, and H. E. Huppert for advice and encouragement as well as for making his data available, also J. H. Curry, J. D. Gibbon, P. A. Glendinning, M. V. Goldman, D. O. Gough, J. Guckenheimer, J. E. Hart, J. R. Herring, D. A. Russell and C. T. Sparrow for stimulating discussions, and A. W. Green, S. A. Piacsek and K. D. Saunders for their hospitality at US Naval Ocean Research Development Activity (NORDA). Part of this work was supported by a grant from the SERC; E.K. acknowledges support from the Alfred P. Sloan Foundation and thanks St John's College for its hospitality during his visit to Cambridge. D.R.M. was a recipient of a Visiting Fellowship at the Joint Institute for Laboratory Astrophysics (JILA) for one year, with JILA operated jointly by the University of Colorado and the NBS. The work in Boulder by J.T. and D.R.M. was supported in part by NASA through grants NSG-7511 and NAGW-91 and by the NSF through grant ATM-8020426.

Received 8 February; accepted 12 April 1983.

1. May, R. M. *Nature* **261,** 459–467 (1976).
2. Eckmann, J.-P. *Rev. mod. Phys.* **53,** 643–654 (1981).
3. Lanford, O. E. *A. Rev. Fluid Mech.* **14,** 347–364 (1982).
4. Turner, J. S. *Buoyancy Effects in Fluids* (Cambridge University Press, 1973).
5. Huppert, H. E. *Springer Lect. Notes Phys.* **71,** 239–254 (1977).
6. Huppert, H. E. & Turner, J. S. *J. Fluid Mech.* **106,** 299–329 (1981).
7. Gough, D. O. & Toomre, J. *J. Fluid Mech.* **125,** 75–97 (1982).
8. Veronis, G. *J. mar. Res.* **23,** 1–17 (1965).
9. Veronis, G. *J. Fluid Mech.* **34,** 315–336 (1968).
10. Huppert, H. E. *Nature* **263,** 20–22 (1976).
11. Huppert, H. E. & Moore, D. R. *J. Fluid Mech.* **78,** 821–854 (1976).
12. DaCosta, L. N., Knobloch, E. & Weiss, N. O. *J. Fluid Mech.* **109,** 25–43 (1981).
13. Knobloch, E. & Proctor, M. R. E. *J. Fluid Mech.* **108,** 291–316 (1981).
14. Moore, D. R., Peckover, R. S. & Weiss, N. O. *Comp. Phys. Commun.* **6,** 198–220 (1973).
15. Feigenbaum, M. J. *J. Stat. Phys.* **19,** 25–52 (1978).
16. Coullet, P. H. & Tresser, C. *J. Phys.* **39,** C5:25-28 (1978).
17. Tresser, C. & Coullet, P. H. *C.r. hebd. Séanc., Acad. Sci., Paris* **207A,** 577–580 (1978).
18. Knobloch, E. & Weiss, N. O. *Phys. Lett.* **85A,** 127–130 (1981).
19. Sparrow, C. T. *The Lorenz Equations: Bifurcations, Chaos and Strange Attractors* (Springer, New York, 1982).
20. Shil'nikov, L. P. *Soviet Math. Dokl.* **6,** 163–166 (1965).
21. Coullet, P. H. in *Geophysical Fluid Dynamics* (ed. Mellor, F. K.) 94–108 (Woods Hole Oceanographic Institution Tech. Rep. 81-102, 1981).
22. Knobloch, E. & Weiss, N. O. *Physica D* (in the press).
23. Guckenheimer, J. in *Dynamical Systems and Turbulence* (eds Rand, D. A. & Young, L. S.) 99–142 (Springer, Berlin, 1981).
24. Lorenz, E. N. in *Global Analysis* (eds Grmela, M. & Marsden, J. E.) 53–75 (Springer, New York, 1979); *Ann. N.Y. Acad. Sci.* **357,** 282–291 (1980).
25. Holmes, P. J. & Marsden, J. E. *Arch. Rat. Mech. Anal.* **76,** 135–165 (1981).
26. Moon, H. T., Huerre, P. & Redekopp, L. G. *Phys. Rev. Lett.* **49,** 458–460 (1982).
27. Schubert, G. & Straus, J. M. *J. Fluid Mech.* **121,** 301–313 (1982).
28. McLaughlin, J. B. & Orszag, S. A. *J. Fluid Mech.* **122,** 123–142 (1982).
29. Toomre, J., Gough, D. O. & Spiegel, E. A. *J. Fluid Mech.* **125,** 99–122 (1982).
30. Curry, J. H., Herring, J. R., Loncaric, J. & Orszag, S. A. *J. Fluid Mech.* (submitted).
31. Linsay, P. S. *Phys. Rev. Lett.* **47,** 1349–1352 (1981).
32. Testa, J., Pérez, J. & Jeffries, C. *Phys. Rev. Lett.* **48,** 714–717 (1982).
33. Cascais, J., Dilão, R. & Noronha da Costa, A. *Phys. Lett.* **93A,** 213–216 (1983).
34. Arecchi, F. T., Menucci, R., Puccioni, G. & Tredicce, J. *Phys. Rev. Lett.* **49,** 1217–1220 (1982).
35. Turner, J. S., Roux, J. C., McCormick, W. D. & Swinney, H. L. *Phys. Lett.* **85A,** 9–12 (1981).
36. Simoyi, R. H., Wolf, A. & Swinney, H. L. *Phys. Rev. Lett.* **49,** 245–248 (1982).
37. Libchaber, A. & Maurer, J. *J. Phys.* **41,** C3:51-56 (1980); in *Nonlinear Phenomena and Phase Transition* (ed. Riste, T.) 259–286 (Plenum, New York, 1981).
38. Libchaber, A., Laroche, C. & Fauve, S. *J. Phys.* **43,** L211–216 (1982).
39. Libchaber, A., Fauve, S. & Laroche, C. *Physica D* (submitted).
40. Giglio, M., Musazzi, S. & Perini, U. *Phys. Rev. Lett.* **47,** 243–246 (1981).

6. The Intermittent Transition

Commun. Math. Phys. 74, 189–197 (1980)

Communications in Mathematical Physics

Intermittent Transition to Turbulence in Dissipative Dynamical Systems

Yves Pomeau and Paul Manneville*

Commissariat à l'Énergie Atomique, Division de la Physique. Service de Physique Théorique, F-91190 Gif-sur-Yvette, France

Abstract. We study some simple dissipative dynamical systems exhibiting a transition from a stable periodic behavior to a chaotic one. At that transition, the inverse coherence time grows continuously from zero due to the random occurrence of widely separated bursts in the time record.

Introduction

A number of investigators [1] have observed in convective fluids an intermittent transition to turbulence. In these experiments the external control parameter, say r, is the vertical temperature difference across a Rayleigh-Bénard cell. Below a critical value r_T of this parameter, measurements show well behaved and regular periodic oscillations. As r becomes slightly larger than r_T the fluctuations remain apparently periodic during long time intervals (which we shall call "laminar phases") but this regular behavior seems to be randomly and abruptly disrupted by a "burst" on the time record. This "burst" has a finite duration, it stops and a new laminar phase starts and so on. Close to r_T, the time lag between two bursts is seemingly at random and much larger than – and not correlated to – the period of the underlying oscillations. As r increases more and more beyond r_T it becomes more and more difficult and finally quite impossible to recognize the regular oscillations (see Fig. 1).

This sort of transition to turbulence is also present in simple dissipative dynamical systems [2] such as the Lorenz model [2a]. We present here the results of some numerical experiments on this problem.

When a burst starts at the end of a laminar phase this denotes an instability of the periodic motion due to the fact that the modulus of at least one Floquet multiplier [3] is larger than one. This may occur in three different ways: a real Floquet multiplier crosses the unit circle at $(+1)$ or at (-1) or two complex conjugate multipliers cross simultaneously. To each of these three typical crossings we may associate one type of intermittency that we shall call for convenience type

* DPh. G. PSRM, Cen Saclay, Boîte Postale 2, F-91190 Gif-sur-Yvette, France

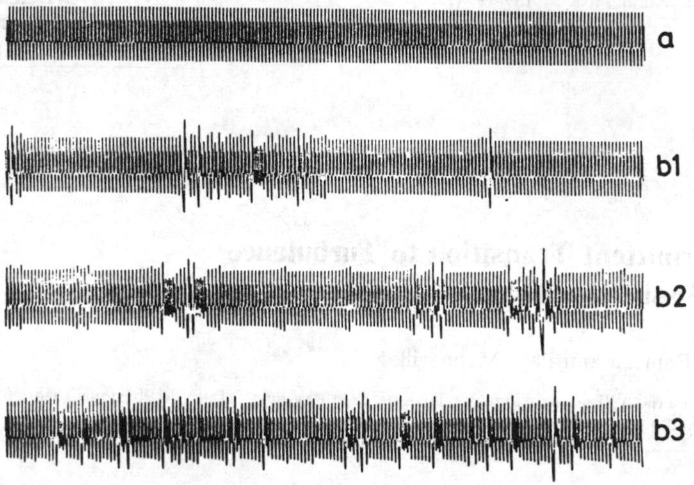

Fig. 1a and b. Time record of one coordinate (z) in the Lorenz model. **a** Stable periodic motion for $r=166$. **b** Above the threshold the oscillations are interrupted by bursts which become more frequent as r is increased

1: crossing at $(+1)$; type 2: complex crossing; and type 3: crossing at (-1) respectively. In all these three cases our numerical studies show that the Lyapunov number grows continuously from zero beyond the onset of turbulence. In what follows we shall present some simple estimates for the "critical behavior" of this Lyapunov number in the vicinity of the turbulence threshold and compare them with the results of numerical experiments.

Type 1. Intermittency in the Lorenz Model

The Lorenz system reads [4]:

$$\frac{dx}{dt}=\sigma(y-z)\,; \qquad \frac{dy}{dt}=-xz+rx-y\,; \qquad \frac{dz}{dt}=xy-bz\,, \tag{1}$$

where σ, b, and r are parameters. We have kept b and σ fixed at their original values ($\sigma=10$, $b=8/3$). Integrating system (1) around $r=166$ one finds for r slightly less than $r_T(\simeq 166.06)$ regular and stable oscillations for a random choice of initial condition (Fig. 1a). For r slightly larger than r_T these oscillations are interrupted by bursts (Fig. 1b). This can be explained quite simply by studying the Poincaré map (restricted here to be 1-dimensional without loss of significance). Let f be the function such that $y_{n+1}=f(y_n, r)$ where y_n is the y-coordinate of the nth crossing of the plane $x=0$. Near $r=r_T$ the curve of equation $y'=f(y,r)$ is nearly tangent to the first bissectrix (Fig. 2). For r slightly below r_T, this curve has two intersections with the bisectrix, they collapse into a single point at $r=r_T$ while for $r>r_T$ the curve is lifted up and no longer crosses the first bisectrix so that a "channel" appears between them (Fig. 3). Hence the successive iterates generated by the map

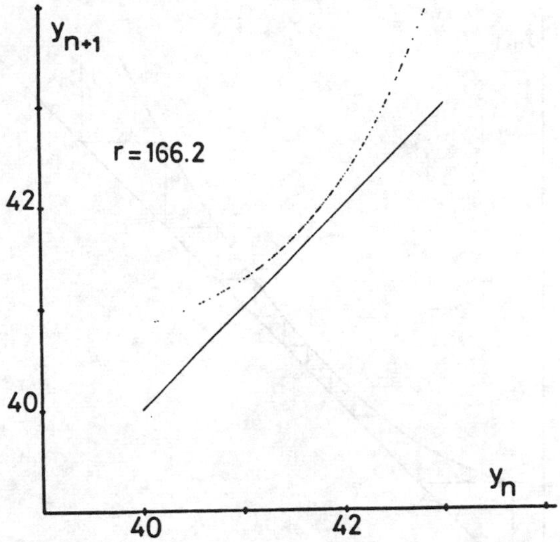

Fig. 2. A part of the Poincaré map along the y-coordinate for $r = 166.2$ slightly beyond the intermittency threshold ($r_T \simeq 166.06$)

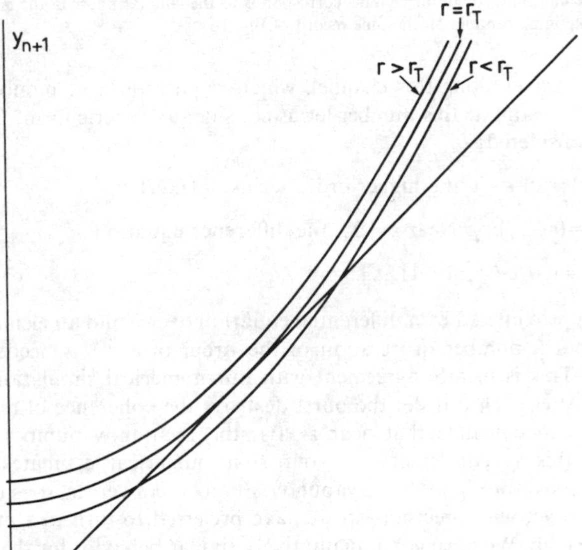

Fig. 3. Idealized picture of the deformation of $y_{n+1}(y_n)$ explaining the transition via intermittency. For $r < r_t$ two fixed points coexist one stable the other unstable. They collapse at $r = r_T$ and then disappear leaving a channel between the curve and the first bisectrix

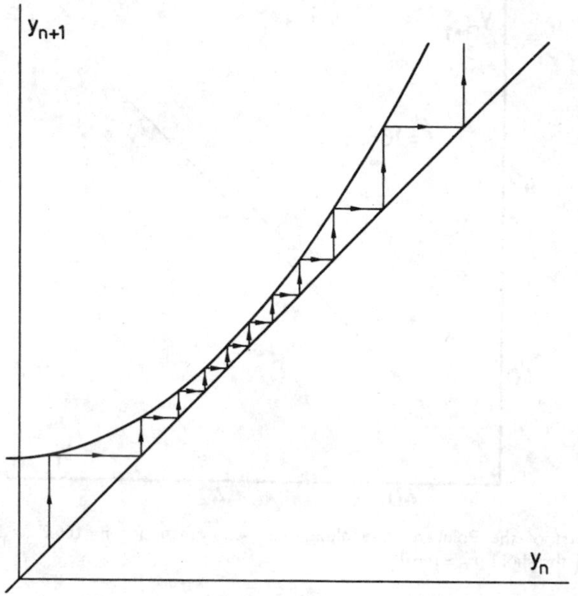

Fig. 4. The motion through the channel corresponds to the laminar phase of the movement. The slow drift is quite imperceptible on the time record of Fig. 1b

$y \to f(y, r)$ travel along this channel, which requires a large number of iterations (Fig. 4). To estimate this number let us consider a "generic form" for $f(y, r)$ in the region considered:

$$f(y, r) = y + \varepsilon + y^2 \text{ (+ higher order terms – H.O.T.)},$$

where $\varepsilon = (r - r_T)/r_T$. Near $\varepsilon = 0_+$ the difference equation

$$y_{n+1} = y_n + \varepsilon + y_n^2 \text{ (+ H.O.T.)}$$

can be approximated by a differential equation over n and an elementary estimate shows that a number of iteration of the order of $\varepsilon^{-1/2}$ is needed to cross the channel. This is in nice agreement with our numerical simulation of system (1) (Fig. 5). After each transfer the burst destroys the coherence of the motion. This leads one to conclude that near $\varepsilon = 0_+$ the Lyapunov number varies as $\varepsilon^{1/2}$. Though this is consistent with our first numerical estimates, close to the intermittency threshold the Lyapunov number converges so slowly that it is difficult to get with precision, so we have preferred to turn to a modelling of the Poincaré map. We have got a qualitatively similar behavior for the following map of $S^1 = [0, 1[$

$$\theta \to 2\theta + r \sin 2\pi\theta + 0.1 \sin 4\pi\theta \pmod 1). \tag{2}$$

As shown in Fig. 6 this applies S^1 twice on itself and it is intermittent at $r_T \simeq -0.24706$. In this model, as well as those we shall consider later, the

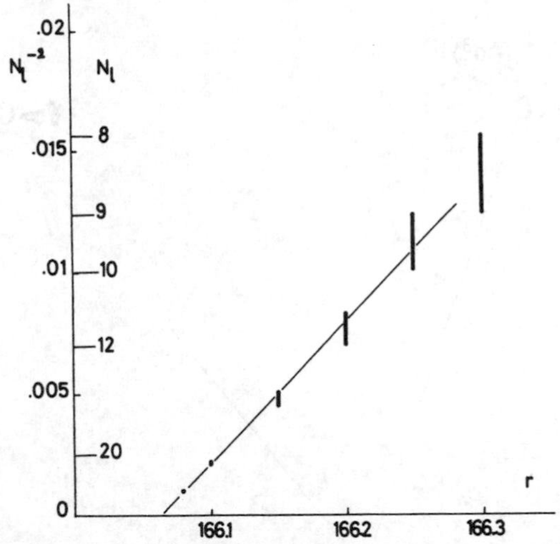

Fig. 5. The square of N_l the largest number of cycles during a laminar period is inversely proportional to the distance from the threshold $r-r_T$. N_l is given within 1 cycle to account for the uncertainty in the definition of the beginning/end of a laminar phase

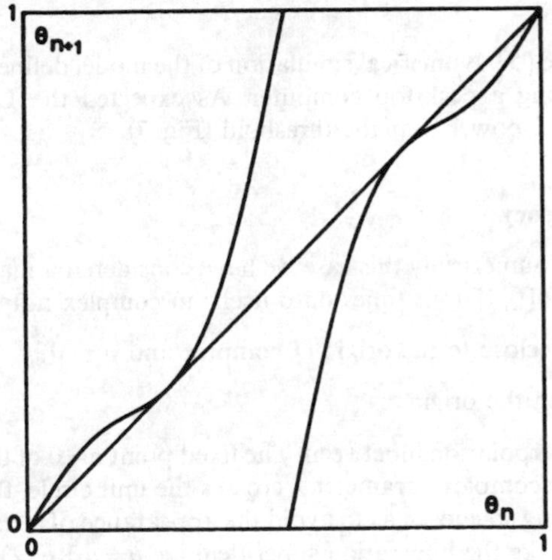

Fig. 6. Model mapping displaying qualitatively the same behavior as the Lorenz model around $r=166$

possibility of starting a laminar phase after a burst comes from the fact that the map is not invertible. In diffeomorphisms the "relaminarization" cannot occur in this way due to the uniqueness of preimages. However dynamical systems for which the "reduced" Poincaré map takes a form similar to (2) [and later to (3) or (5)] can be constructed simply by adding other dimensions along which fluc-

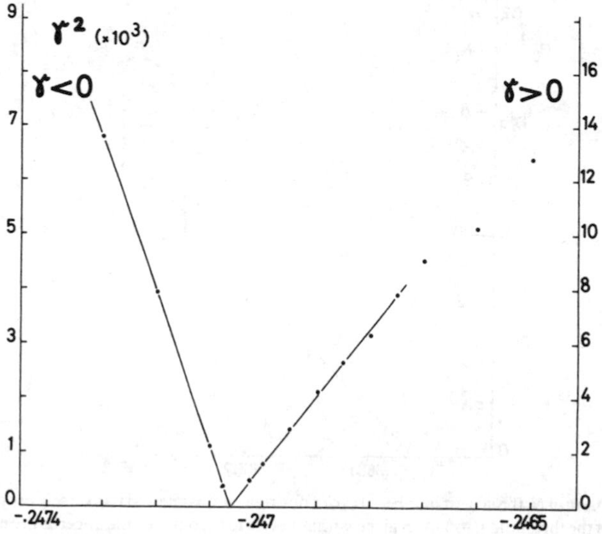

Fig. 7. For the model mapping $r_T \simeq -0.24706$. For $r < r_T$ the Lyapunov number γ is negative and varies as $-\sqrt{r_T - r}$ while for $r > r_T$ it is positive and grows like $\sqrt{r - r_T}$

tuations are stable [5]. Numerical simulation of the model defined by (2) can easily be performed using a desk-top computer. As expected the Lyapunov number grows with the 1/2-power near the threshold (Fig. 7).

Type 2. Intermittency

In order to study numerically this case we have considered a map that applies the torus $T^2 = [0, 1[\times [0, 1[$ four times onto itself: in complex notations $z = x + iy$

$$z' = \lambda z + \mu |z|^2 z \text{ close to the origin } (\lambda \text{ complex and } \mu \text{ real}), \tag{3a}$$

$$z' = 2z \text{ far from the origin} \tag{3b}$$

with a smooth interpolation inbetween. The fixed point $z = 0$ of this map looses its stability when the complex parameter λ crosses the unit circle, the coefficient μ of the cubic term being so chosen as to avoid the appearance of a stable limit cycle or equivalently to make the bifurcation subcritical i.e. $\mu = \mu \ \mathrm{Re}\{\lambda\} > 0$. Iterations of the above map show intermittency when $|\lambda| = 1 + \varepsilon$ and $\varepsilon \to O_+$. Once an iterate falls near $z = 0$, it enters a laminar phase and a large number of further iterations are needed to expell it towards the "bursting region" (where correlations are broken) defined by $|z| > \varrho^*$, ϱ^* being fixed and ε-independent, roughly in the interpolating region. To find how the Lyapunov number grows near the intermittency threshold one may reason as follows: Let $\varrho_j = |z_j|$ be the distance of the jth iterate to the fixed point. The iterates rotate around the fixed point due to the complex nature of λ but we shall neglect the angular variation and only consider the growth of the modulus

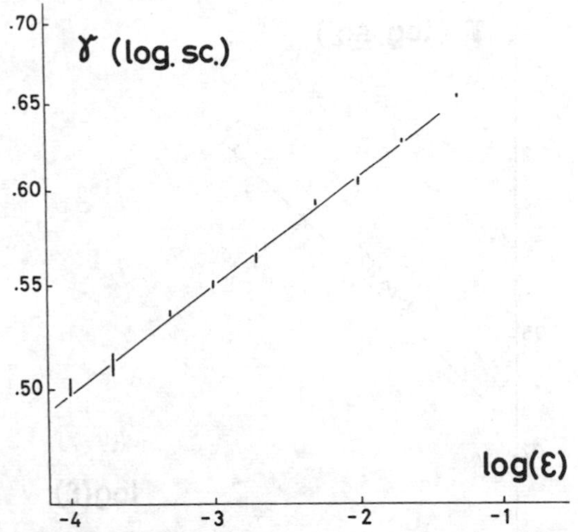

Fig. 8. On the torus T_2 for $\lambda=(1+\varepsilon)\exp i\varphi$, $\varphi=0.05$ rd and $\mu=20$ "mean field" theory predicts $\gamma \sim \ln(1/\varepsilon)$ while the numerical simulation gives $\gamma \sim \varepsilon^{\alpha}$ with $\alpha \sim 0.04$

ϱ. Near $\varrho=0$ it is approximatively given by

$$\varrho_{n+1}=(1+\varepsilon)\varrho_n+\bar{\mu}\varrho_n^3+\text{H.O.T.} \tag{4}$$

Now let us examine a laminar cycle with starting point at $\varrho=\tilde{\varrho} \ll \varrho^*$. If $\tilde{\varrho} \gg \varepsilon^{1/2}$ one easily sees that a number of iteration of order $1/\tilde{\varrho}^2$ are needed to reach ϱ^* and enter a turbulent burst. On the other hand if $\tilde{\varrho} \ll \varepsilon^{1/2}$ the laminar cycle ends after $\varepsilon^{-1}\ln\varrho$ iterations approximately. Assuming then that $\tilde{\varrho}$ is at random with probability $\tilde{\varrho}\,d\tilde{\varrho}$ in the circle of center 0 and radius ϱ^* the estimates given above yield $\ln(1/\varepsilon)$ as an order of magnitude for both the mean duration of a laminar period and the inverse Lyapunov number near $\varepsilon=0_+$. This is in slight disagreement with our computer experiments which seem to indicate rather a power-like growth of the Lyapunov number $\gamma \sim \varepsilon^{\alpha}$ α small and positive (Fig. 8). This descrepancy between the naive theory presented above and computer results may come from the neglect of fluctuations about the mean length of the laminar cycles, which makes the procedure used sound much like a "mean field theory" in the usual jargon of phase transitions (it may also come from the neglect of the rotation of iterates affecting the statistics in an unknown way).

Type 3. Intermittency

The last type of intermittency we shall examine may occur when the Floquet multiplier is real and crosses the unit circle at (-1). Although a differential system has been found which displays this kind of behavior [2b] we shall report here on the simulation of the following mapping of the circle S_1 onto itself:

$$\theta \to 1-2\theta-\frac{1}{2\pi}(1-\varepsilon)\cos\left[2\pi\left(\theta-\frac{1}{12}\right)\right] (\text{mod } 1). \tag{5}$$

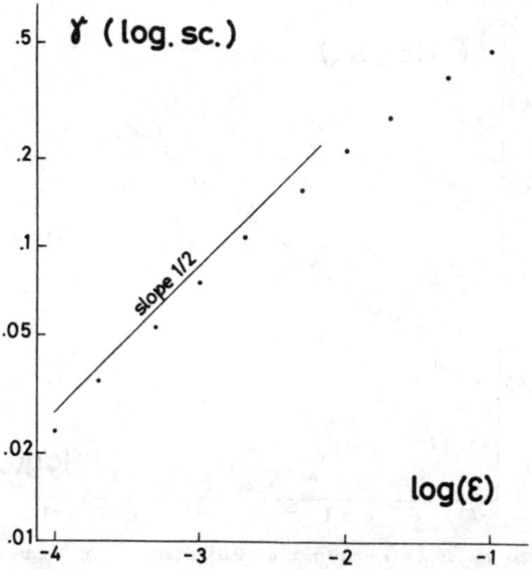

Fig. 9. On the torus T_1 for type 3 intermittency the Lyapunov number γ grows as $\varepsilon^{1/2}$

This map applies S^1 twice onto itself and it reverse the orientation so that the eigenvalue of the map linearized near the fixed point $\theta_F = 1/3$ can easily be made negative. Near the fixed point the map expands as

$$\bar{\theta}_{n+1} = -(1+\varepsilon)\bar{\theta}_n - \frac{(2\pi)^2}{6}\bar{\theta}_n^3 + \text{H.O.T.}\,(\bar{\theta} = \theta - \theta_F). \tag{6}$$

The most general form would be

$$\bar{\theta}_{n+1} = -(1+\varepsilon)\bar{\theta}_n + a\theta_n^2 + b\theta_n^3, \tag{7}$$

a and b being constant. If the r.h.s. of (7) has a positive Schwarzian derivative that is here $b + a^2 < 0$ then the bifurcation at $\varepsilon = 0$ is subcritical and type 3 intermittency can occur. This is precisely the case with (5) since $a = 0$ and $b < 0$. To estimate the mean length of a laminar phase one considers instead of (6) or (7) the equation giving $\bar{\theta}_{n+2}$ in function of $\bar{\theta}_n$. This relation is basically of the same form as Eq. (4) (quadratic terms vanish at $\varepsilon = 0$ and are in inessential for ε small enough). Thus one reasons as for type 2 intermittency with the difference that now the problem is strictly unidimensional so that the probability measure for for the starting point of a laminar cycle is now the usual Lebesgue measure instead of $\tilde{\varrho}\, d\tilde{\varrho}$ previously. An elementary calculation shows that the Lyapunov number should grow like $\varepsilon^{1/2}$ near threshold, this time in agreement with the computer experiment (Fig. 9).

Conclusion

Intermittency is a quite common phenomenon in experimental turbulence. The theory sketched in this paper is more especially related with the case of convection in confined geometries [1] but intermittency is also well known in boundary layers and pipe flows [6] and even in $1/f$ – noise theory [7]. Despite the different meanings of the term "intermittency", the possibility remains that the kind of dynamics described by the models we have studied could afford a qualitative understanding of all these phenomena.

References

1. Maurer, J., Libchaber, A., Bergé, P., Dubois, M.: Personal communications
2. (a) Manneville, P., Pomeau, Y.: Phys. Lett. **75**A, 1 (1979)
   (b) Arneodo, A., Coullet, P., Tresser, C.: Private communication
3. Iooss, G.: Bifurcation of maps and applications. In: North-Holland Math. Studies 36. Amsterdam, New York: North-Holland 1979
4. Lorenz, E.N.: J. Atmos. Science **20**, 130 (1963)
5. Pomeau, Y.: Intrinsic stochasticity in plasmas, Cargèse 1979. (eds. G. Laval, D. Gresillon). Orsay: Editions de Physique 1979
6. Tritton, D.J.: Physical fluids dynamics. New York: Van Nostrand-Reinhold 1977
7. Mandelbrot, B.: Fractals form chance and dimension. San Francisco: Freeman 1977

Communicated by D. Ruelle

Received January 17, 1980

PHYSICAL REVIEW A VOLUME 25, NUMBER 1 JANUARY 198

Theory of intermittency

J. E. Hirsch
Institute for Theoretical Physics, University of California, Santa Barbara, California 93106

B. A. Huberman
Xerox Palo Alto Research Center, Palo Alto, California 94304

D. J. Scalapino
Department of Physics, University of California, Santa Barbara, California 93106

(Received 29 June 1981)

The aperiodic or chaotic behavior for one-dimensional maps just before a tangent bifurcation occurs appears as intermittency in which long laminarlike regions irregularly separated by bursts occur. Proceeding from the picture proposed by Pomeau and Manneville, numerical experiments and analytic calculations are carried out on various models exhibiting this behavior. The behavior in the presence of external noise is analyzed, and the case of a general power dependence of the curve near the tangent bifurcation is studied. Scaling relations for the average length of the laminar regions and deviations from scaling are determined. In addition, the probability distribution of path lengths, the stationary distribution of the maps, the correlation function and power spectrum of the map in the intermittent region, and the Lyapunov exponent are obtained.

I. INTRODUCTION

There has recently been considerable interest in the properties of nonlinear discrete maps.[1-3] Such maps arise in a variety of nonlinear field theories. Perhaps the best known of these is the Poincaré sections for dynamical systems which can be usefully approximated by such maps relating a continuous time process to a discrete process.[4] Spatial order for commensurate-incommensurate phase transitions on a lattice have recently been discussed in terms of discrete maps.[5,6] In addition, the phenomena of electron localization on a lattice with an incommensurate potential has been treated as a discrete mapping problem generated by the Hamiltonian acting on a tight-binding state.[7] In all these cases, there is a physical parameter which enters the map in an essential way. In a hydrodynamic system it could be the Reynolds or Prandtl number, for a phase transition it could be the ratio of near to next-near coupling constants, while in the quantum-mechanics problem it was the ratio of the hopping matrix element to the incommensurate potential. As this parameter is changed, the sequence of iterates generated from the map may alter going from a regular periodic to an irregular aperiodic or chaotic behavior at some critical value of the parameter. The nature of this transition to chaotic behavior forms the focus of much of the recent interest.[8] In particular, the concepts of scaling and universality[9] so successful in the theory of phase transitions play a central role, encouraging the detailed study of simple models.

Here we are interested in the onset of chaotic behavior characterized by the occurrence of regular or laminar phases separated by intermittent bursts. This intermittent transition to turbulenc was discussed by Pomeau and Manneville[10] in cor nection with the Lorenz model. They pointed out that it arises when a tangent bifurcation occurs. A simpler model which exhibits this type of phen omena[11] is the well-known logistic map

$$x_{n+1}=Rx_n(1-x_n) \qquad (1.1$$

with $0 \leq x \leq 1$ and $0 \leq R \leq 4$. The attractor for this map in the region $3 \leq R \leq 4$ is shown in Fig. 1. Past the first period doubling bifurcation cascad at $R=3.57$, various open regions with odd number of fixed points appear which arise from tangent bifurcations. The three-cycle region appears last at $R_c=1+\sqrt{8}$.[12] As R increases further one can see that this three-cycle behavior undergoes the usual cascade bifurcation to chaos. However

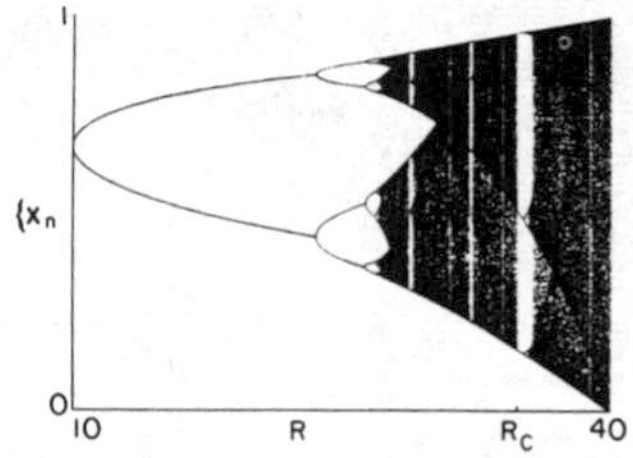

FIG. 1. The attractor vs R for the logistic map $x_{n+1}=Rx_n(1-x_n)$ (from Ref. 16). The intermittent behavior discussed in this paper arises when R decreases below the threshold for an odd-limit cycle. In particular, we focus on the region just below $R_c=1+\sqrt{8}$ which is the threshold for three-cycle orbits.

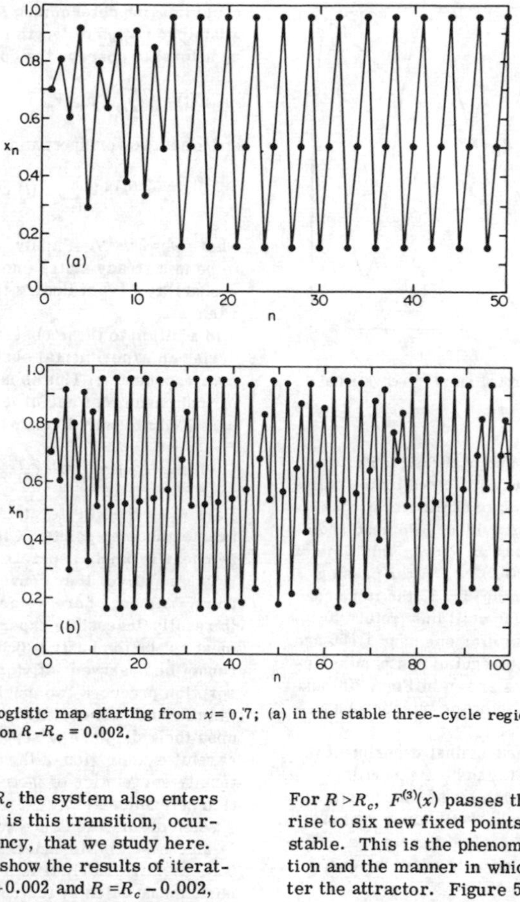

FIG. 2. Iterates of the logistic map starting from $x = 0.7$; (a) in the stable three-cycle region $R_c - R = -0.002$ and
 in the intermittent region $R - R_c = 0.002$.

s R decreases below R_c the system also enters
 chaotic regime and it is this transition, ocur-
ng through intermittency, that we study here.
 Figures 2(a) and 2(b) show the results of iterat-
g Eq. (1.1) for $R = R_c + 0.002$ and $R = R_c - 0.002$,
 spectively. For $R > R_c$, after an initial trans-
nt, the iterates settle down to threefold limit
cle. However, for $R < R_c$, approximate three-
d cycles are interrupted by irregular behavior.
gure 3 shows every third iterate for $R_c - R = 10^{-4}$
d nearly repeating segments of varying length
ppear. These form the "laminar" regions sep-
ated by irregular behavior which appears as
rsts if one magnifies the scale.
The geometric picture of Pomeau and Manne-
le which explains this phenomena can be illus-
ated by constructing the three-iterate map

$$F^{(3)}(x) = F(F(F(x))) \tag{1.2}$$

th $F(x) = Rx(1-x)$. This map is plotted in Fig.
or $R = R_c$. At this critical value the map is just
gent to the line x at $x = (0.160,\ 0.514,\ 0.956)$.

For $R > R_c$, $F^{(3)}(x)$ passes through the line x giving rise to six new fixed points of which three are stable. This is the phenomena of tangent bifurcation and the manner in which odd-limit cycles enter the attractor. Figure 5 shows a blowup of the region near $x_c = 0.514$ for $R = R_c - 5 \times 10^{-3}$. The

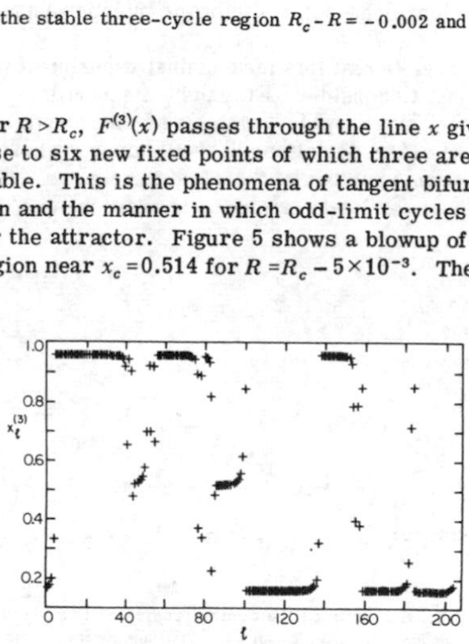

FIG. 3. Third iterate of the logistic map for $R_c - R = 0.0001$ showing regions of laminar behavior by intermittent irregularities.

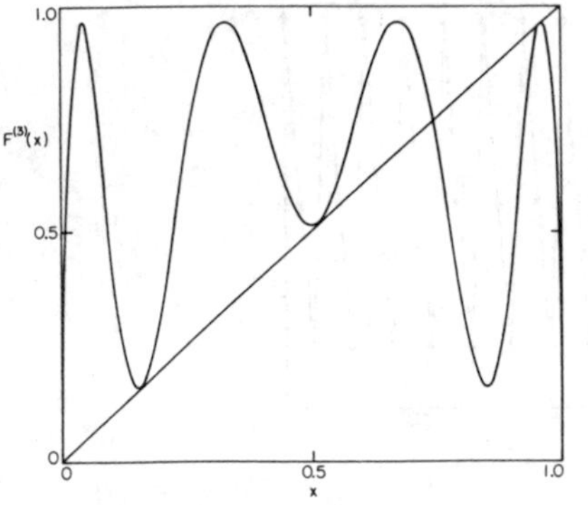

FIG. 4. The threefold iterated map $F^{(3)}(x)$ vs x for $R = R_c$.

points represent a sequence of threefold iterates which map in the usual way indicated by the staircase path. The slow passage of the iterates as x approaches $x_c = 0.514$ is evident. Pomeau and Manneville pointed out that as $R \to R_c$, this time of passage diverges as $(R_c - R)^{-1/2}$. After passing through the region shown in Fig. 5, the iterates move wildly under the map until they return to this neighborhood or a similar one near 1.160 and 0.956. This produces the regular laminarlike regions separated by bursts shown in Figs. 2(b) and 3 which are characteristic of the intermittent region.

In order to test this idea against experiment it is useful to consider what can be measured. Clearly, given a gate G which sets an acceptance $|x - x_c| < G$ on deviations in the laminar region, the average number of threefold iterates or length $\langle l \rangle$ can be determined. In addition, one

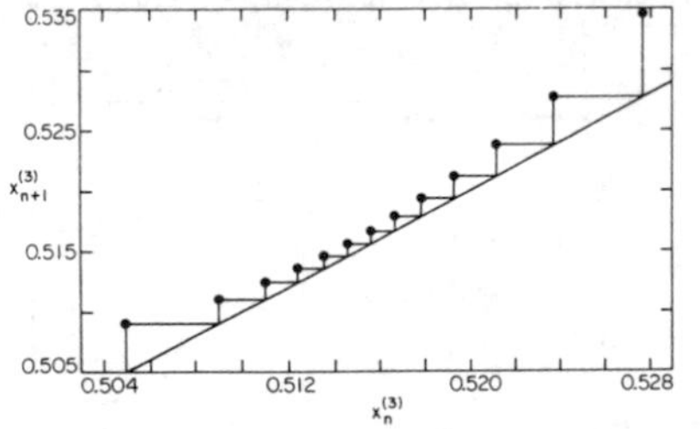

FIG. 5. A blowup of the central region of Fig. 4 showing $F^{(3)}(x)$ for $R_c - R = 0.005$. Successive third iterates of a point show the phenomena of slow passage which gives rise to the laminarlike regions in Figs. 2(b) and 3.

could imagine determining $P(l)$, the probability of a laminar region of length l. Beyond this it would be natural to observe the correlation function

$$C(l) = \frac{1}{N} \sum_{n=1}^{N} \langle x_{n+l}\, x_n \rangle \qquad (1.3$$

and/or its power spectrum

$$S(\omega_k) = C(0) + 2 \sum_{l=1}^{N-1} C(l) \cos(l\omega_k) + C(N) \cos N\omega_k\,, \qquad (1.4)$$

where $\omega_k = 2\pi k/N$. Finally, if the system appeare to be in a steady state, one might try to find the probability $W(x)$ of finding the system with a give value x.

In addition to these observables, one might eve devise an experimental observation of the Lyapunov exponent λ. This is usually introduced as a formal parameter useful in characterizing the sensitivity to initial conditions:

$$\lim_{N\to\infty} e^{N\lambda} = \lim_{N\to\infty} \lim_{\Delta\to 0} \{[x_N(x_0+\Delta) - x_N(x_0)]/\Delta\}\,. \quad (1.5$$

Here x_N is the Nth iterate and $x_0 + \Delta$ and x_0 are two neighboring points. Clearly, if λ is positive, two nearby points separate at an exponential rate if λ is negative, they converge towards a fixed point. Although here we are primarily intereste in readily accessible experimental quantities, on must not be too hasty to judge what can and what cannot be observed. If, for example, light absorption produced two neighboring triplet states which could subsequently recombine depending upon their distance of separation, one might by a careful examination of the tail of the intensity-intensity correlation of their recombination radiation determine λ.

When observations are compared with theory it is also essential to understand the role of noise. This is particularly true in the case of chaotic phenomena which by their nature may appear noisy. There has in fact been considerable progress recently on understanding the effects of stochastic noise on maps near the bifurcation cascade to chaos.[13-16] For example, the Lyapunov exponent has been shown to satisfy a scaling behavior characterized by universal exponents. He we will study the stochastic difference equation

$$x_{n+1} = R x_n (1 - x_n) + g\xi_n \qquad (1.$$

with ξ_n a Gaussian random variable with $\langle \xi_n \rangle = 0$ and $\langle \xi_n \xi'_n \rangle = \delta'_{nn}$. In the presence of noise we find that as $\epsilon = (R_c - R) \to 0$, the ratio of the average length in the presence of noise to that with $g = 0$ scales as[17]

$$\frac{\langle l \rangle_\alpha}{\langle l \rangle_0} = f(\alpha) \qquad (1.$$

with α proportional to $g^2/\epsilon^{3/2}$ for Eq. (1.6) or more generally for $F^{(3)}(x)$ maps which have a quadratic term when expanded about their point of contact x_c (see Fig. 4). If this term vanishes and the lowest-order term beyond the linear $(x-x_c)$ term is $(x-x_c)^z$, then we find that

$$\langle l\rangle_0 \sim \epsilon^{-(z-1)/z} \quad \text{and} \quad \alpha \sim g^2/\epsilon^{(z+1)/z}.$$

The layout of the paper proceeds as follows. In Sec. II, the results of a variety of numerical experiments on the logistic map are presented. With these results in mind, a simple theoretical model is proposed and analyzed in Sec. III. Results for $\langle l\rangle$, $P(l)$, and $W(x)$ are obtained. In addition, the effect of noise on these results is investigated. Exact results as well as the $\epsilon \to 0$ scaling limits are derived. In addition, these theoretical results are compared with results of numerical experiments on a simple map constructed to exhibit the important features of the intermittent transition to chaos. A brief conclusion is given in Sec. IV.

II. NUMERICAL RESULTS FOR THE LOGISTIC MAP

In order to explore the phenomena of intermittency, we carried out a number of numerical experiments using the logistic map. In Fig. 6, observations of the average number $\langle l\rangle$ of threefold iterations which begin and end inside an acceptance gate $G=|x-x_c|=10^{-2}$ are plotted versus ϵ. These data were obtained from runs in which 10^4 laminar regions were observed. A laminar region of length l occurs if an iterate falls inside the gate $|x_n-x_c|<G$ and leaves after l threefold iterates. Deviations of runs starting from different initial values of x were of order 1%. For $\epsilon<10^{-4}$, the measured values of $\langle l\rangle$ clearly follow the expected $\epsilon^{-1/2}$ behavior. The solid and dashed lines are the results of analytic calculations for a model discussed in Sec. III.

For $\epsilon=2.5\times10^{-4}$, the probability distribution $P(l)$ for lengths of a given run was computed and is plotted as the histogram versus l in Fig. 7. The

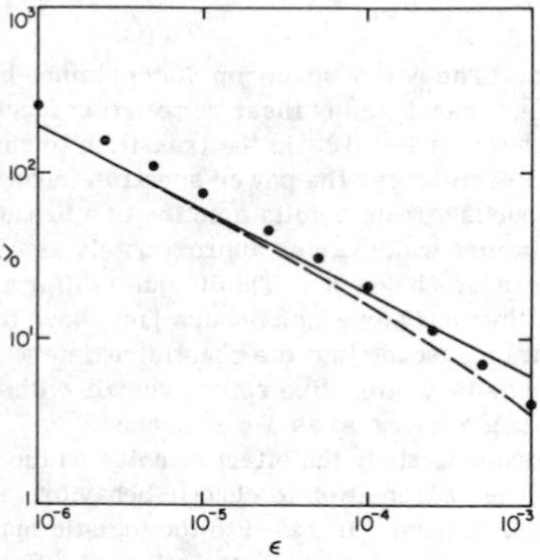

FIG. 6. The dots show the average number $\langle l\rangle$ of threefold iterations measured between bursts for the logistic map vs ϵ for an acceptance gate $|x-x_c|<10^{-2}$. The dashed line corresponds to Eq. (3.9) and the solid line is its asymptotic limit $\epsilon/|x-x_c|\to 0$ given by Eq. (3.10).

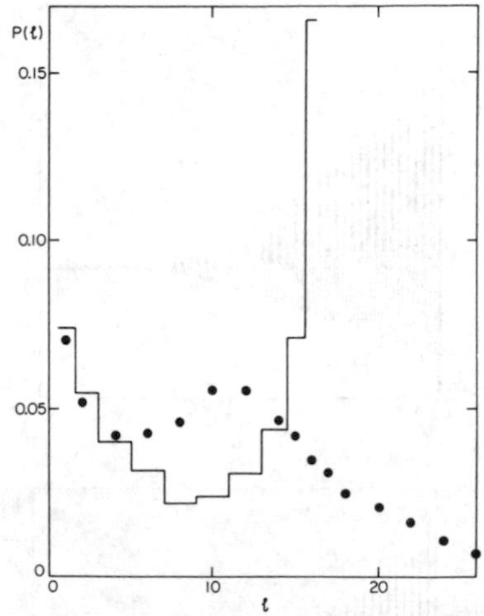

FIG. 7. The histogram shows the measured probability distribution $P(l)$ for having laminar regions $3l$ iterates long for $\epsilon=2.5\times10^{-4}$ in the absence of stochastic noise, $g=0$. The dots show $P(l)$ for the same value of ϵ but with $g=5\times10^{-4}$. In both cases the acceptance gate was $|x-x_c|<10^{-2}$.

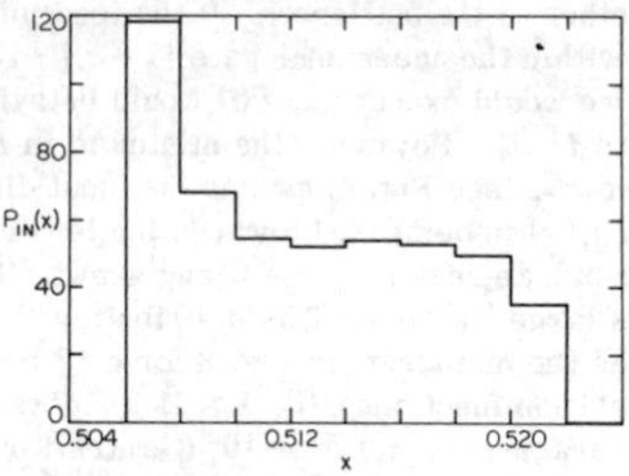

FIG. 8. The probability $P_{\text{in}}(x)$ of a given starting value of x inside the acceptance gate for $\epsilon=2.5\times10^{-4}$. Note that this peaks at the lower end of the gate skewing $P(l)$ shown in Fig. 7 towards larger l values.

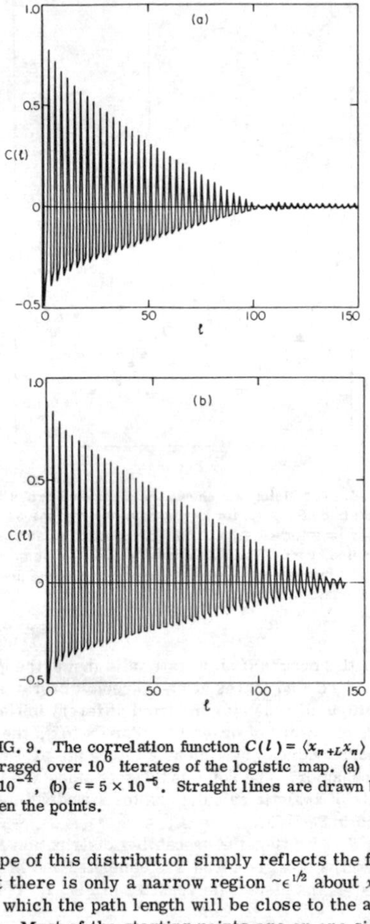

FIG. 9. The correlation function $C(l) = \langle x_{n+L}x_n\rangle$ averaged over 10^6 iterates of the logistic map. (a) $\epsilon = 10^{-4}$, (b) $\epsilon = 5\times 10^{-5}$. Straight lines are drawn between the points.

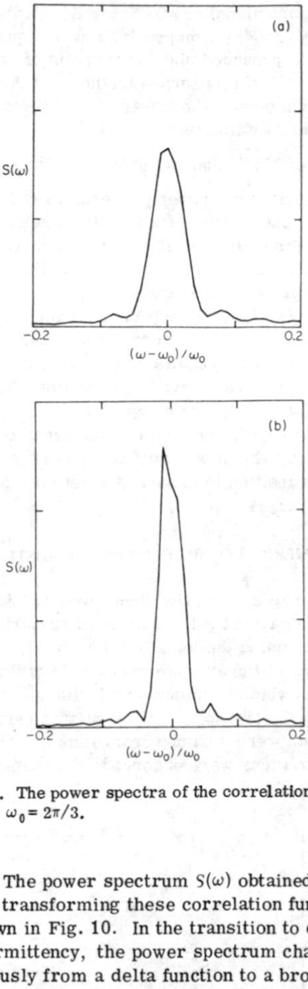

FIG. 10. The power spectra of the correlation functions of Fig. 9, $\omega_0 = 2\pi/3$.

shape of this distribution simply reflects the fact that there is only a narrow region $\sim\epsilon^{1/2}$ about x_c for which the path length will be close to the average. Most of the starting points are on one side or the other of the bottleneck. If the feed into the region within the acceptance gate $|x - x_c| \leq G$ were white, we would expect that $P(l)$ would be symmetric about $l = \langle l\rangle$. However, the minimum in $F^{(3)}(x)$ just below x_c (see Fig. 4) causes the input distribution $P_{in}(x)$ to be peaked towards the lower end of the acceptance region $x = x_c - G$ and skews $P(l)$ towards large l values. This distribution $P_{in}(x)$ is shown as the histogram in Fig. 8 for $\epsilon = 2.5\times 10^{-4}$.

Correlation functions $C(l)$, Eq. (1.3), obtained from averaging $\langle x_{n+l}x_n\rangle$ over 10^6 iterates for two values of ϵ are shown in Fig. 9. These were normalized so that $C(0) = 1$. Clearly as ϵ decreases the correlation length characteristic of the decay increases. It follows an $\epsilon^{-1/2}$ behavior as one would expect. The power spectrum $S(\omega)$ obtained by Fourier transforming these correlation functions are shown in Fig. 10. In the transition to chaos via intermittency, the power spectrum changes continuously from a delta function to a broadened peak whose width varies approximately as $\epsilon^{1/2}$ as R decreases below R_c. This is quite different from the behavior which occurs just above the bifurcation cascade into the chaotic regime where sharp delta-funtion-like spikes remain with broadband noise rising up as a background.[16]

In order to study the effect of noise on the intermittency transition to chaotic behavior, a stochastic term was added to the logistic map, Eq. (1.6). By selecting g, the effect of different noise levels could be studied. Here ξ_n was a pseudo random variable selected in the region $(-\frac{1}{2}, \frac{1}{2})$ so that the variance of the noise term was $g^2/12$. According to the theory discussed in Sec. III, the appropriate scaling variable is $g^2/\epsilon^{3/2}$. In

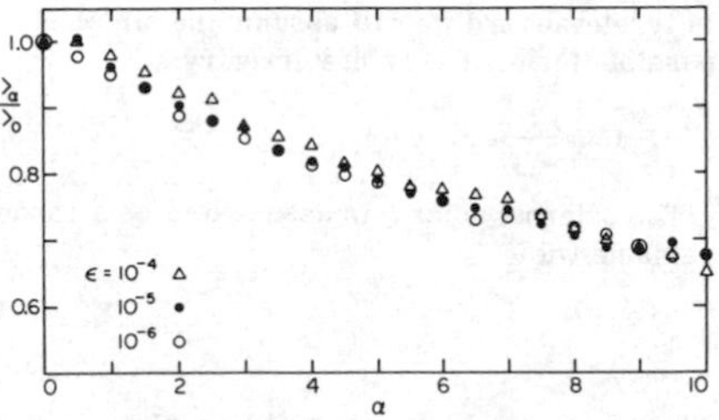

FIG. 11. The average length of the three-iterate laminar regions $\langle l\rangle_{\alpha'}$ for the logistic map in the presence of noise $\alpha' = 10g^2/\epsilon^{3/2}$ normalized to its zero noise value $\langle l\rangle_0$ for three different values of ϵ. Note the approximate scaling of the data when it is plotted in this way.

Fig. 11 the results for the ratio of the average length $\langle l\rangle'_{\alpha}$ in the presence of noise $\alpha' = 10g^2/\epsilon^{3/2}$ to its value $\langle l\rangle_0$ in the absence of noise is plotted as a function of α' for different values of ϵ. Once again 10^4 laminar regions were obtained in a given run leading to rms errors of order 1%. It is clear that the results for $\epsilon = 10^{-4}$ to $\epsilon = 10^{-6}$ are very similar to each other when plotted in this way. There are in fact expected to be small differences proportional to ϵ but we will postpone discussions of deviations from scaling until Sec. III. Just as in the theory of phase transitions, the existence of scaling results in a great economy in dealing with data and, moreover, provides the framework for extracting the intrinsic chaotic behavior from

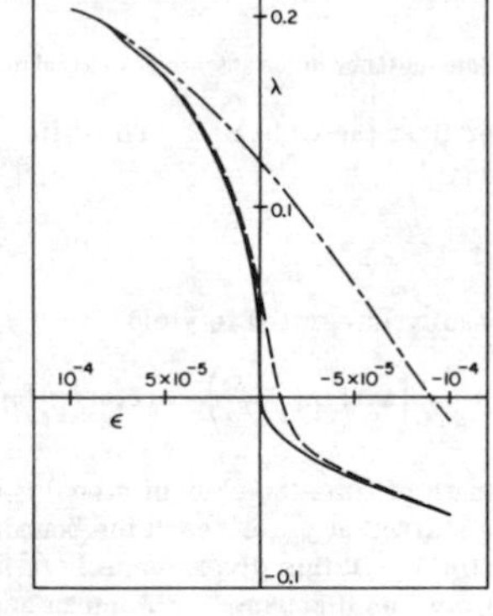

FIG. 12. The Lyapunov exponent λ vs ϵ for the logistic map. The solid curve is without noise $g = 0$ while the dashed and dash-dot curve are with noise $g = 10^{-4}$ and 5×10^{-4}, respectively.

the fluctuations produced by external noise.

The effect of noise on the distribution $P(l)$ of laminar lengths is shown as the dots in Fig. 7. The tail giving values of l exceeding $l_{\max}$ for $g = 0$ results from a passage in which the noise shifts the point back of the bottleneck region after it has already passed through it.

Writing $x_N(x_0) = F^{(N)}(x_0)$ and using the chain rule for differentiation one has for the Lyapunov exponent, Eq. (1.5),

$$\lambda = \lim_{N\to\infty} \frac{1}{N} \sum_{i=1}^{N} \ln[R(1-2x_i)] \,. \tag{2.1}$$

For $\epsilon < 0$, the system settles down to the three-cycle behavior shown in Fig. 2(a) and

$$\lambda = \tfrac{1}{3} \sum_{i=1}^{3} \ln[R(1-2x_i)] \,. \tag{2.2}$$

Here x_1, x_2, and x_3 are the stable three-cycle fixed points for a given $R > R_c$. Results obtained for λ versus ϵ for the logistic map are shown in Fig. 12. The solid line shows λ in the absence of any stochastic noise, $g = 0$. Figure 13 shows $\ln|\lambda|$ versus $\ln|\epsilon|$ for $g = 0$. Here the solid dots correspond to the divergent, intermittent region with $\epsilon > 0$ ($R < R_c$) where $\lambda > 0$ and the open circles correspond to the stable three-cycle region with $\epsilon < 0$ ($R > R_c$) where $\lambda < 0$. For $\epsilon < 0$ it is straightforward to show that

$$\lambda = -\tfrac{2}{3}(a_c b_c)^{1/2}|\epsilon|^{1/2} + O(|\epsilon|) \,, \tag{2.3}$$

where a_c and b_c are given by Eq. (3.2). This is shown as the dashed line in Fig. 13. When $\epsilon > 0$, we know that the number of steps in the bottleneck region varies as $\epsilon^{-1/2}$ so that we expect that $\lambda \sim \epsilon^{1/2} + O(\epsilon)$. The solid line in Fig. 13 varies as $\epsilon^{1/2}$ and

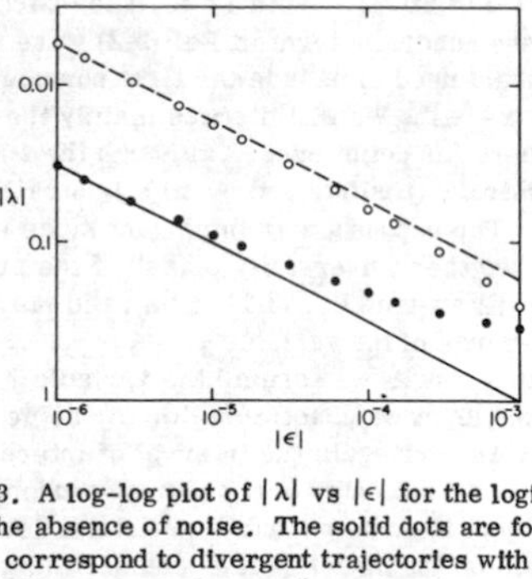

FIG. 13. A log-log plot of $|\lambda|$ vs $|\epsilon|$ for the logistic map in the absence of noise. The solid dots are for $\epsilon > 0$ and correspond to divergent trajectories with $\lambda > 0$ while the open circles are for $\epsilon < 0$ and correspond to stable three-cycle trajectories with $\lambda < 0$.

has been adjusted to fit the data for small ϵ. Note that the "critical" region in which $\epsilon \sim \epsilon^{1/2}$ for $\epsilon>0$ is very narrow due to the contribution from the burst regions. Compare this with the behavior of $\langle l\rangle_0$ shown in Fig. 6 which has the asymptotic form $\epsilon^{1/2}$ for $\epsilon \lesssim 10^{-4}$. The effect of noise is to wash out the $\epsilon=0$ singularity as shown in Fig. 12. In the critical region we find that λ scales with $g^2/\epsilon^{3/2}$ just as $\langle l\rangle_\alpha$ does.

Having explored some of the basic phenomena exhibited by a particular map near the intermittency threshold, we turn in the next section to an analysis of a simple model which contains the essential features of the geometrical picture introduced by Pomeau and Manneville.

III. ANALYTIC FORMULATION AND RESULTS

In this section we discuss a simple analytic formulation of the intermittency problem in the presence of external noise; we consider the recursion relation

$$y'=y+ay^z+\epsilon+g\xi\,. \tag{3.1}$$

Such a recursion relation arises in the logistic map when expanding around the contact point. Near the region of the tangent bifurcation, $F^{(3)}(x,R)$ can be expanded about x_c and R_c,

$$F^{(3)}(x,R)\cong x_c+(x-x_c)+a_c(x-x_c)^2 +b_c(R_c-R)\,. \tag{3.2}$$

Here x_c is one of the three contact points where $R=R_c$. Although the parameters a_c and b_c are in fact different from the middle contact point and the two outer contact points, their product $a=a_cb_c=68.5$ which enters in various physical predictions is identical for all three regions. Setting $y=(x-x_c)/b_c$ and including an external noise gives a recursion relation for the threefold iterate having the form of Eq. (3.1) with $z=2$. If the coefficient a_c of the quadratic term in Eq. (3.2) were to vanish one would need to include the first nonvanishing term $(x-x_c)^z$. We will discuss mainly the case $z=2$ here for definiteness, although the extension to arbitrary (positive and even) z is straightforward. The exponent z is important since it determines the "universality class" of the problem.

We will assume Eq. (3.1) to be valid only over a small range of the variable y, $-y_0<y<y_0$. Outside this range, we assume the variable y to undergo some unknown chaotic motion for some time and then to reenter again the interval of interest. In general, this reentry will occur at random points y_{in} subject to some probability distribution which depends on the model under consideration; a particular example was discussed in Sec. II. For purposes of the theory the detailed form of $P_{in}(y)$ is irrelevant and we will assume the simplest possible form, i.e., "white" reentry:

$$P_{in}(y)=\frac{1}{2y_0}\quad -y_0<y<y_0\,. \tag{3.3}$$

The external noise ξ is assumed to be a random variable with

$$\langle\xi\rangle=0\,, \tag{3.4a}$$

$$\langle\xi^2\rangle=1\,. \tag{3.4b}$$

The quantities of interest in the problem do not depend on the form of the distribution of ξ, so that we will take it to be uniformly distributed for our numerical experiments.

In order to treat the problem analytically, we approximate the recursion relation (3.1) by a differential equation on which the step spacing is dt,

$$\frac{dy}{dt}=ay^z+\epsilon+g\xi(t)\,. \tag{3.5}$$

Here $\xi(t)$ is a Gaussian white noise source, satisfying

$$\langle\xi(t)\xi(t')\rangle=\delta(t-t')\,. \tag{3.6}$$

The passage from (3.1) to (3.5) is justified if the changes in y at each step are small. This will always be the case for small ϵ and y_0. Since we are interested in the limit of small ϵ, and the value of y_0 is at our disposal, this approximation is not essential for our results. This is confirmed by the results of our numerical experiments using Eq. (3.1), which closely agree with the analytic result obtained from (3.5).

We now discuss some of the results one can derive from these recursion relations, first in the absence of external noise.

A. Intermittency in the absence of external noise

Consider first the case $z=2$. The differential equation is

$$\frac{dy}{dt}=ay^2+\epsilon\,. \tag{3.7}$$

This is readily integrated to yield

$$l(y_{in})=\frac{1}{\sqrt{a\epsilon}}\left[\arctan\left(\frac{y_0}{\sqrt{\epsilon/a}}\right)-\arctan\left(\frac{y_{in}}{\sqrt{\epsilon/a}}\right)\right] \tag{3.8}$$

for the length of time (number of steps) it takes a process started at y_{in} to reach the boundary y_0. Note that for $y_{in}<0$ this diverges as $1/\sqrt{\epsilon}$ in the small-ϵ limit, as discussed by Pomeau and Manneville. The average time of passage is obtained by averaging (3.8) over the probability distribution of y_{in}. For the particular example of $P_{in}(y)$ chosen one gets

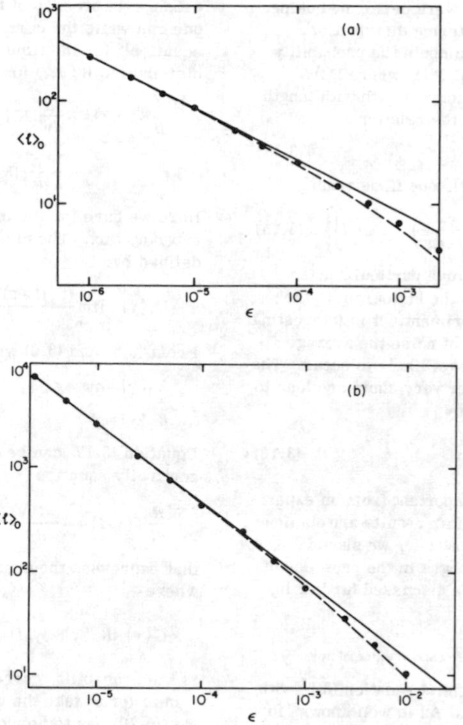

FIG. 14. Average time of passage $\langle l \rangle$ vs ϵ in the absence of external noise for (a) $z=2$ and (b) $z=4$. The dashed lines are the theoretical results from Eq. (3.11), the full lines are limiting forms for $\epsilon/y_0 \to 0$. The dots are results of numerical experiments with 10 000 passes per point. In this and the following figures, the numerical simulations were done using the map (3.1) with $a=34$ and white reinjection, with a gate $y_0 = 0.01$ for $z=2$ and $g_0 = 0.1$ for $z=4$.

$$\langle l \rangle = \frac{1}{\sqrt{a\epsilon}} \arctan\left(\frac{y_0}{\sqrt{\epsilon/a}}\right) \tag{3.9}$$

and, in the limit of $\sqrt{\epsilon/a} \ll y_0$,

$$\langle l \rangle = \frac{\pi}{2} \frac{1}{\sqrt{a\epsilon}}. \tag{3.10}$$

For the case of general z, an identical calculation yields

$$\langle l \rangle = \frac{1}{2} \frac{1}{a^{1/z}} \frac{1}{\epsilon^{1-1/z}} \int_{-y_0/(\epsilon/a)^{1/z}}^{y_0/(\epsilon/a)^{1/z}} \frac{dx}{x^z+1} \tag{3.11}$$

so that the asymptotic behavior for small ϵ is

$$\langle l \rangle \propto 1/\epsilon^{(1-1/z)} \tag{3.12}$$

and in particular, for $z=4$

$$\langle l \rangle = \frac{\pi}{\sqrt{8}} \frac{1}{a^{1/4}} \frac{1}{\epsilon^{3/4}}. \tag{3.13}$$

In Figs. 14(a) and 14(b) we plot the average time of passage for $z=2$ and $z=4$, respectively, in a log-log plot. The full lines are the "scaling limit" results (3.10) and (3.13) while the dashed lines are the "exact" results (3.11). The dots were obtained from numerical experiment on the discrete recursion relation Eq. (3.1) with the "boundary condition" that each time y goes out of the interval, it is reinjected randomly and uniformly between $-y_0$ and y_0. The results were averaged over 10 000 passes. It can be seen that the agreement between theory and "experiment" is excellent. Similarly, in Fig. 6, the solid line is the asymptotic form Eq. (3.10) and the dashed line corresponds to Eq. (3.9). In the case of the full logistic map, while the numerical data for $\langle l \rangle$ vary as $\epsilon^{-1/2}$, they have been shifted to larger values because the input into the acceptance region is not white (see Fig. 8). In the present section, all numerical experiments were performed using Eq. (3.1) with a white reentering distribution.

It is also of interest to compute the probability distribution of path lengths, $P(l)$, where $P(l)dl$ gives the probability to observe a path with length between l and $l+dl$. Using the relation

$$P(l)dl = P_{in}(y_{in})dy_{in} \tag{3.14}$$

together with (3.8) and (3.3), one finds readily

$$P(l) = \frac{\epsilon}{2y_0}\left\{1+\tan^2\left[\arctan\left(\frac{y_0}{\sqrt{\epsilon/a}}\right) - \sqrt{\epsilon a}\, l\right]\right\}. \tag{3.15}$$

This is plotted in Fig. 15 for a particular case, $\epsilon = 10^{-5}$, as a dashed line: the histogram is the result of a numerical experiment. It is interesting to note that in the absence of noise the average value of l is actually the least likely to occur. The most likely paths are either very short or close to the maximum value given by

$$l_{max} = \frac{2}{\sqrt{\epsilon a}}\arctan\left(\frac{y_0}{\sqrt{\epsilon/a}}\right). \tag{3.16}$$

The observation may be important from an experimental point of view. Similar results are obtained for the general z case. However, we should note that this behavior changes in the presence of external noise; this will be discussed further in the following subsection.

B. Intermittency in the presence of external noise

Consider now the differential equation (3.5) with the external noise present. As is well known, for

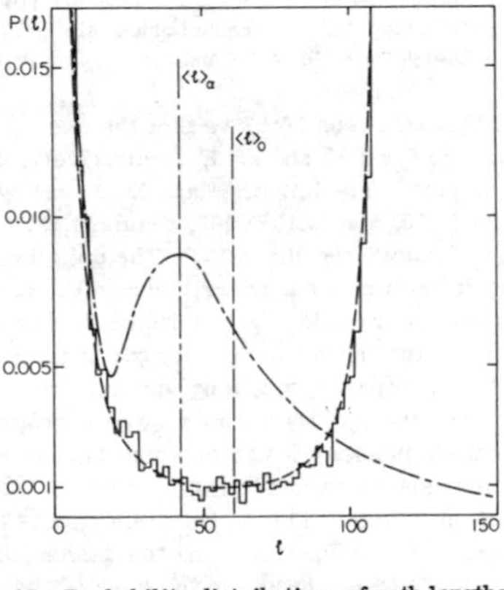

FIG. 15. Probability distributions of path lengths $P(l)$ vs l for $z=2$, $\epsilon=10^{-5}$. Dashed line: theoretical results without noise. Histogram: results of a numerical simulation without noise. Dash-dotted line: numerical simulation in the presence of noise $\alpha = 100$.

a Langevin process of the type described by (3.5) one can write the corresponding Fokker-Planck equation[18] for the time evolution of the probability distribution $W(y,t)$ for y:

$$\frac{d}{dt}W(y,t) = -\frac{d}{dy}[K_1(y)W(y,t)] + \frac{1}{2}\frac{d^2}{dy^2}[K_2(y)W(y,t)]. \tag{3.17}$$

Here we have for the moment neglected the reentering flux. The functions $K_1(y)$ and $K_2(y)$ are defined by

$$K_i(y) = \lim_{\tau\to 0}\frac{\langle[y(t+\tau)-y(t)]^i\rangle}{\tau}. \tag{3.18}$$

From (3.5) and (3.6) we obtain (for the case $z=2$)

$$K_1(y) = ay^2+\epsilon, \tag{3.19a}$$

$$K_2(y) = g^2. \tag{3.19b}$$

Equation (3.17) can be simply interpreted as a continuity equation

$$\frac{d}{dt}W(y,t) = -\frac{dG(y)}{dy} \tag{3.20}$$

that expresses the conservation of probability, where

$$G(y) = K_1(y)W(y,t) - \frac{1}{2}\frac{d}{dy}[K_2(y)W(y,t)] \tag{3.21}$$

is the probability current.

In order to take the reentering flux into account, Eq. (3.20) has to be modified to

$$\frac{d}{dt}W(y,t) = -\frac{dG(y)}{dy} + r. \tag{3.22}$$

Here, r is the rate at which the points reenter the interval per unit length, which is independent of y in our model. Equation (3.22) simply expresses the fact that $W(y,a)$ changes both due to the flow from neighboring points and the flow from outside.

First let us consider stationary solutions to (3.22). We have for our problem

$$(ay^2+\epsilon)W(y) - \tfrac{1}{2}g^2\frac{dW(y)}{dy} = G_0 + ry, \tag{3.23}$$

where G_0 is an integration constant arising from integrating (3.22) once. The appropriate boundary conditions are

$$W(-y_0) = W(y_0) = 0 \tag{3.24}$$

plus a normalization condition to determine r. Note that the current as a function of y is

$$G(y) = G_0 + ry. \tag{3.25}$$

The net number of points that leave the interval $(-y_0, y_0)$ per unit time is

$$N=G(y_0)-G(-y_0)=2ry_0 \tag{3.26}$$

which equals the reentering density rate (r) times the length of the interval ($2y_0$). This expresses the conservation of probability in the more general sense discussed here.

It is convenient to rescale (3.23) as follows. Define

$$\begin{aligned} x &= y/\sqrt{\epsilon}\,, \\ W(x) &= \sqrt{\epsilon}W(y(x)), \\ \alpha &= g^2/\epsilon^{3/2}, \\ g_0 &= G_0/\sqrt{\epsilon}\,. \end{aligned} \tag{3.27}$$

Equation (3.23) then becomes

$$(1+ax^2)W(x)-\tfrac{1}{2}\alpha\frac{dW}{dx}=g_0+rx\,, \tag{3.28}$$

which is readily integrated to yield

$$W(x)=\frac{2}{\alpha}r\int_x^{x_2}dx'(x-x_1+c) \times\exp\{-(2/\alpha)[x'-x+(a/3)(x'^3-x^3)]\}$$

with

$$x_2=-x_1=y_0/\sqrt{\epsilon}\,. \tag{3.30}$$

Here, c is determined by the boundary condition

$$W(x_1)=0 \tag{3.31}$$

while r is determined by the normalization condition

$$\int_{x_1}^{x_2}dx\,W(x)=1\,. \tag{3.32}$$

In Fig. 16 the stationary probability distribution is plotted for a particular case ($\epsilon=10^{-5}$), with and without noise. The agreement between the results of the integral (3.29) and results of numerical experiments is very good. In the absence of noise, the probability distribution is simply

$$W(y)=\frac{\sqrt{\epsilon a}}{\pi y_0}\frac{y+v_0}{ay^2+\epsilon} \tag{3.33}$$

and the most probable value of y is

$$y_m=\frac{\epsilon}{2y_0 a}\,. \tag{3.34}$$

Note that in the presence of noise the most probable value shifts appreciably to negative values of y.

Next, we consider the evaluation of the average length of passage in the presence of noise.[19] Consider a process $y(t)$ starting at a given $y_{\rm in}$ at time $t=0$; the probability density for the process satisfies

$$W(y,t=0)=\delta(y-y_{\rm in})\,. \tag{3.35}$$

We consider now in $W(y,t)$ only processes that have not reached the boundary in the time interval $(0,t)$. This probability density will satisfy the Fokker-Plank equation (3.17) in the interval $(-y_0,y_0)$. The integral

$$W(t)=\int_{-y_0}^{y_0}dy\,W(y,t) \tag{3.36}$$

gives the total probability that the process $y(t)$ did not reach the boundary until time t. Clearly, $W(t)$ satisfies

$$\begin{aligned} &W(0)=1\,, \\ &W(t)\underset{t\to\infty}{\longrightarrow}0\,. \end{aligned} \tag{3.37}$$

The dependence of $W(t)$ on $y_{\rm in}$ is understood. The probability of having a time of passage between t and $t+dt$ is clearly

$$w(t)dt=W(t)-W(t+dt)=-\frac{dW}{dt}dt \tag{3.38}$$

so that the mean first-passage time for a process that started at $y_{\rm in}$ at $t=0$ is

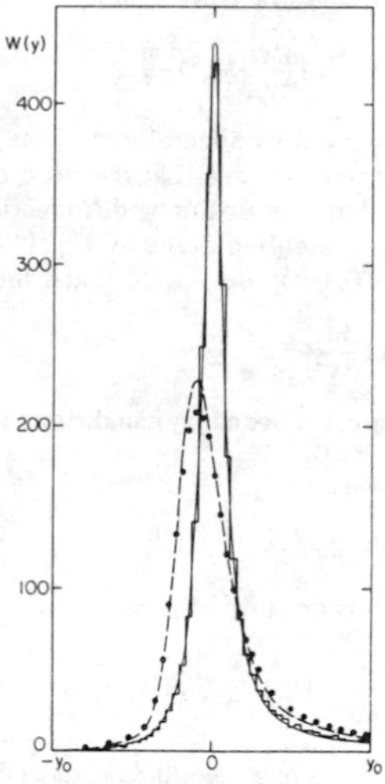

FIG. 16. Stationary probability distribution $W(y)$ for $z=2$, $\epsilon=10^{-5}$. The full and dashed lines are the theoretical results for $\alpha=0$ and $\alpha=100$, respectively. The histogram and the dots are the corresponding results from numerical experiments.

$$M(y_{in}) = \int_0^\infty dt\, w(t)t$$
$$= -\int_0^\infty dt \frac{dW}{dt} t$$
$$= \int_0^\infty dt\, W(t) . \quad (3.39)$$

Finally, we will be interested in the averaged first-passage time, obtained by averaging over y_{in}:

$$\langle l \rangle = \frac{1}{2y_0} \int_{-y_0}^{y_0} dy\, M(y). \quad (3.40)$$

To find an equation satisfied by $M(y)$ it is useful to consider the adjoint equation to (3.17), the Kolmogoroff equation,[19] given by

$$\frac{d}{dt} W(y,t) = K_1(y_{in}) \frac{dW(y,t)}{dy_{in}} + \tfrac{1}{2} K_2(y_{in}) \frac{d^2 W(y,t)}{dy_{in}^2} , \quad (3.41)$$

where $W(y,0)$ satisfies the initial condition (3.35). Integrating (3.41) with respect to y we obtain, using (3.36),

$$\frac{d}{dt} W(t) = K_1(y_{in}) \frac{dW(t)}{dy_{in}} + \tfrac{1}{2} K_2(y_{in}) \frac{d^2 W(t)}{dy_{in}^2} \quad (3.42)$$

and integrating (3.42) in time from $t=0$ to $t=\infty$ and using (3.39) and (3.37) we find

$$-1 = (ay^2 + \epsilon) \frac{dM}{dy} + \tfrac{1}{2} g^2 \frac{d^2 M}{dy^2} , \quad (3.43)$$

where we have also specialized to the problem of interest here. We see that the problem reduces again to solving an ordinary differential equation.

Using the rescaling given by Eq. (3.27), with $m(x) = \sqrt{\epsilon} M(y(x))$, Eq. (3.43) takes the form

$$(ax^2 + \epsilon) \frac{dm}{dx} + \tfrac{1}{2} \alpha \frac{d^2 m}{dx^2} = -1 . \quad (3.44)$$

The appropriate boundary conditions are

$$m(x_1) = m(x_2) = 0 ,$$
$$x_2 = -x_1 = y_0/\sqrt{\epsilon} , \quad (3.45)$$

and the average time of passage is

$$\langle l \rangle_\alpha = \frac{1}{2y_0} \int_{-y_0/\sqrt{\epsilon}}^{y_0/\sqrt{\epsilon}} dx\, m(x,\alpha) . \quad (3.46)$$

For $\alpha = 0$ one has, as discussed in Sec. III A,

$$m(x, \alpha=0) = \frac{1}{\sqrt{a}} (\arctan x_2 - \arctan x) \quad (3.47)$$

leading to the form (3.9) for $\langle l \rangle$. For α not equal to zero, one has the approximate form

$$\langle l \rangle = \frac{1}{\sqrt{\epsilon}} f(\alpha) . \quad (3.48)$$

Equation (3.48) expresses a scaling property of the problem. If we plot $\sqrt{\epsilon}\langle l \rangle$ versus $g^2/\epsilon^{3/2} = \alpha$ for different values of ϵ, the points should fall on a universal curve. An example of this, for the full logistic map, was shown in Fig. 11. In reality, Eq. (3.48) is not entirely correct since there are corrections to scaling in the problem. More precisely,

$$\langle l \rangle_\alpha = \frac{1}{\sqrt{\epsilon}} F(\alpha, \epsilon) , \quad (3.49)$$

$$f(\alpha) = \lim_{\epsilon \to 0} F(\alpha, \epsilon) . \quad (3.50)$$

The limit in Eq. (3.50) is well-behaved. For small α, $f(\alpha) \sim$ const. For large α, we expect $\langle l \rangle$ to be independent of ϵ; this is only possible if

$$f(\alpha) \sim \frac{1}{\alpha^{1/3}} \text{ for } \alpha \to \infty , \quad (3.51)$$

giving

$$\langle l \rangle \sim \frac{1}{g^{2/3}} . \quad (3.52)$$

Note that this is different from what one naively would expect for a random walk

$$y_0 \sim l^{1/2} g \Rightarrow l \sim \frac{y_0}{g^2} . \quad (3.53)$$

The difference is due to the drift term in (3.1), and can be understood as follows: for large y, the drift term dominates while for small y one can assume a pure random walk. To be more precise, the drift term is more important than the noise when $|y| \gtrsim y_c$, while the noise term is more important for $|y| \lesssim y_c$. The crossover value y_c is easily found to be of order $g^{2/3}/a^{1/3}$. The time it takes to go from $-y_0$ to y_c due to drift only is

$$l_1 \sim 1/y_c \sim a^{1/3}/g^{2/3} , \quad (3.54)$$

while the time it takes to go from $-y_c$ to y_c by a pure random walk is

$$l_2 \sim \frac{y_c}{g^2} \sim \frac{1}{a^{1/3} g^{2/3}} . \quad (3.55)$$

Both contributions are of the same order and give the behavior (3.52). Clearly, this will be valid in the regime

$$\epsilon^{3/2} \ll g^2 \ll y_0^3 a . \quad (3.56)$$

For still larger values of g, there is a crossover to a pure random-walk regime satisfying (3.53). We will not consider that regime in what follows.

The solution of Eq. (3.44) yields

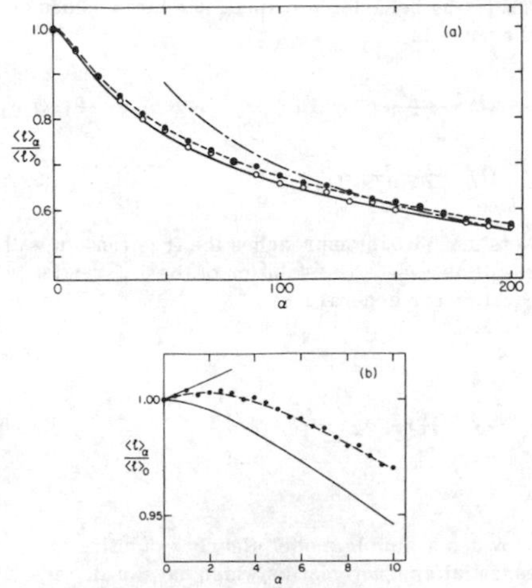

FIG. 17. Average length of passage $\langle l \rangle_\alpha$ normalized to zero noise value $\langle l \rangle_0$ as a function of $\alpha = g^2/\epsilon^{3/2}$, for $z = 2$. The full line is the scaling limit result Eq. (3.60). (a) The dashed line is the theoretical result for $\epsilon = 10^{-5}$, the dots and open circles are results of numerical experiments for $\epsilon = 10^{-5}$ and 10^{-6}, respectively (10 000 passes), rms deviation $\lesssim 0.01$. The dash-dotted line is the asymptotic behavior $\alpha^{-1/3}$. (b) Detailed behavior for small α. Dashed line: theory, $\epsilon = 5\times 10^{-5}$. Dots: numerical simulation, 200 000 passes, rms deviation $\lesssim 0.002$. Straight line: slope at $\alpha = 0$, from Eq. (3.61).

$$m(x) = \frac{2}{\alpha}\int_x^{x_2} d\bar{x} \int_{x_1}^{\bar{x}} dx' e^{-(2/\alpha)[\bar{x} - x' + (\alpha/3)(\bar{x}^3 - x'^3)]} - f_0 \int_x^{x_2} d\bar{x}\, e^{-(2/\alpha)[\bar{x} + (\alpha/3)\bar{x}^3]} . \quad (3.57)$$

f_0 is determined by the condition $m(x_1) = 0$; from (3.57), one finds that f_0 is exponentially small so that we will neglect it in what follows. The scaling function is given by

$$f(\alpha) = \frac{1}{2y_0} \lim_{\epsilon \to 0} \sqrt{\epsilon} \int_{-y_0/\sqrt{\epsilon}}^{y_0/\sqrt{\epsilon}} dx\, m(x) = \tfrac{1}{2} m(-\infty) \quad (3.58)$$

or

$$f(\alpha) = \frac{1}{\alpha} \int_x^{\infty} d\bar{x} \int_{-\infty}^{x} dx' e^{-(2/\alpha)[x - x' + (\alpha/3)(x^3 - x'^3)]} . \quad (3.59)$$

One can evaluate one of the integrals in (3.59) analytically by changing variables to get

$$f(\alpha) = \left(\frac{\pi}{2a}\right)^{1/2} \int_0^{\infty} \frac{du}{u^{1/2}} e^{-2[u + (a\alpha^2/12)u^3]} . \quad (3.60)$$

This is plotted in Fig. 17 as a full line and the dash-dot curve shows the large α behavior $1/\alpha^{1/3}$. We show also in Fig. 17 the results of numerical experiments for $\epsilon = 10^{-5}$ and $\epsilon = 10^{-6}$. The statistical error in those data (obtained from averaging over 10 000 passages) was $\lesssim 0.01$. As we can see, scaling is approximately satisfied. The dashed line is the theoretical result obtained using the full solution Eq. (3.57) for a finite ϵ, $\epsilon = 10^{-5}$. The corresponding numerical results fit this curve extremely well. It can be seen that small deviations from the scaling limit Eq. (3.60) are evident even for this small value of ϵ.

It is interesting to expand the result (3.57) to lowest order around $\alpha = 0$. One finds

$$\frac{F(\alpha, \epsilon)}{F(0, \epsilon)} = 1 + \frac{1}{8} \frac{\sqrt{\epsilon}\,\alpha}{y_0} + O(\alpha^2) + O(\epsilon) . \quad (3.61)$$

Thus, we see that corrections to scaling in fact give an *increase* in $F(\alpha, \epsilon)$ with respect to its $\alpha = 0$ value for small α. This implies that a small amount of noise actually *increases* the average time of passage, contrary to what one might have expected. This effect is shown in Fig. 17(b) for $\epsilon = 5\times 10^{-5}$. The straight line is the slope at $\alpha = 0$ given by (3.61), the dashed line is obtained from the integral (3.57), and the points are a numerical experiment with 200 000 passes each. The statistical error here is $\lesssim 0.002$. Again, the agreement between theory and experiment is excellent, and one can clearly notice the initial rise of the curve for small α. On this scale of resolution, the difference between the finite ϵ results and the limiting scaling function [given by the full line in Fig. 17(b)] is substantial. Note that in the $\epsilon = 0$ limit, the scaling function $f(\alpha)$ is a monotonically *decreasing* function of α.

We have also computed numerically the probability distribution of path lengths in the presence of external noise. An example is shown as the dash-dotted line in Fig. 15. Note that for the value of $\alpha = 100$ used in this calculation, the average path length is shorter than in the absence of noise. Nevertheless, the probability distribution has a long tail, and it is now possible to observe arbitrarily long path lengths. Note also that in the presence of noise the average path length is close to a peak in the distribution, so that experimentally one will observe predominantly either very short passes or lengths close to the average path length. Thus, although the increase in $P(l)$ for small l is not removed by the presence of noise, the increase for l near $l_{max}(\alpha = 0)$ disappears.

Finally, we discuss briefly the extension of these results to arbitrary values of z in Eq. (3.1). It is easy to show that the appropriate variable in this case is

$$\alpha = g^2/\epsilon^{(1+1/z)} \tag{3.62}$$

and the scaling form for the average time of passage is

$$\langle l \rangle = \frac{1}{\epsilon^{(1-1/z)}} f(\alpha) \tag{3.63}$$

with

$$f(\alpha) \xrightarrow[\alpha \to 0]{} \text{const}, \tag{3.64a}$$

$$f(\alpha) \xrightarrow[\alpha \to \infty]{} \frac{1}{\alpha^{(z-1)/(z+1)}} . \tag{3.64b}$$

Thus, the behavior for small and large values of the noise is

$$\langle l \rangle \sim \frac{1}{\epsilon^{(1-1/z)}}, \quad \alpha \ll 1 \tag{3.65a}$$

$$\langle l \rangle \sim \frac{1}{g^{2(z-1)/(z+1)}}, \quad \alpha \gg 1 . \tag{3.65b}$$

Note that (3.65b) approaches the free random walk result as $z = \infty$. The solution of the differential equation for general z is

$$m(x) = \frac{2}{\alpha} \int_x^{x_2} dy \int_{x_1}^{x} dy' \exp(-(2/\alpha)\{y - y' + [a/(z+1)](y^{z+1} - y'^{z+1})\}), \quad x_2 = -x_1 = y_0/\epsilon^{1/z} \tag{3.66}$$

and the average time of passage is

$$\langle l \rangle = \frac{1}{2y_0} \frac{1}{\epsilon^{1-2/z}} \int_{x_1}^{x_2} dx\, m(x) . \tag{3.67}$$

In Fig. 18 we show an example of results for $z = 4$. Here, times of passage are much longer so that it is more difficult to get good statistics. The statistical error in the numerical results is ~ 0.03, with runs of 1000 passes each. As before, the dashed-dot curve in Fig. 16 shows the asymptotic behavior of the scaling function, in this case $1/\alpha^{3/5}$. Again we obtain excellent agreement with the theoretical prediction Eq. (3.67).

IV. CONCLUSION

The approach to aperiodic or chaotic behavior via intermittency which has been discussed here provides a simple model clearly exhibiting the essential characteristics which have made the analysis of one-dimensional maps interesting. First there are a set of clearly defined features for the purely deterministic map which signal the onset of chaos. The average length, as well as the maximum length, of steps associated with a fixed acceptance gate vary as $(R_c - R)^{(1/z-1)}$ as $R_c - R \to 0$. Here z is the parameter characterizing the universality class of the map. Furthermore, in the presence of noise, we have shown that this average length of the laminar regions scales so that

$$\lim_{\epsilon \to 0} \epsilon^{(1-1/z)} \langle l \rangle_\alpha = f(\alpha) \tag{4.1}$$

with $\alpha = g^2/\epsilon^{(1+1/z)}$, $f(0) = \text{const}$, and $f(\alpha) \sim \alpha^{-(z-1)/(z+1)}$ for $\alpha \gg 1$. Here, as opposed to the bifurcation cascade the "critical indices" are rational numbers. One should not however think of this as a mean-field behavior. In fact for the case of the logistic map where $z = 2$, the scaling relations are analogous to those of Pott's model near its multicritical point.[20]

We have also studied how corrections to scaling enter and in particular how they give rise to an initial increase in $\langle l \rangle_\alpha$ for small α. It is important to have an idea of the size of these effects since any real system is bound to have some external noise acting on it.

Besides the characteristic length and the correlation function and its power spectrum, we have studied the behavior of the distribution of lengths $P(l)$ and the stationary probability distribution $W(x)$. In the absence of external noise, the deep valley in $P(l)$ near $\langle l \rangle$ provides a possible flag for

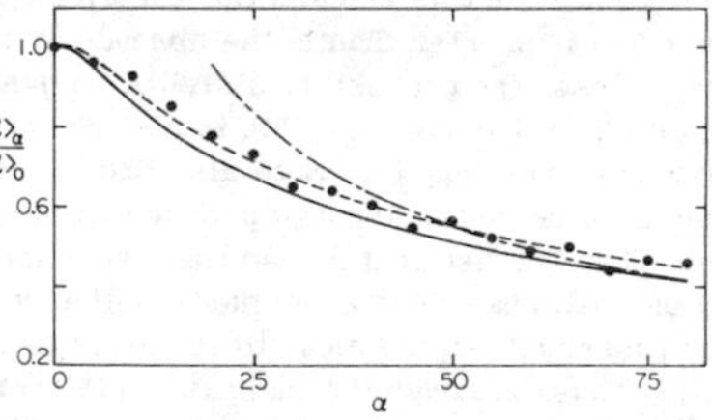

FIG. 18. For $z = 4$ the average length of passage $\langle l \rangle_\alpha$ normalized to the zero noise value $\langle l \rangle_0$ is plotted as a function of $\alpha = g^2/\epsilon^{5/4}$. Here $y_0 = 0.1$. Dashed line: theory, $\epsilon = 10^{-5}$. Dots: numerical simulation, $\epsilon = 10^{-5}$, 1000 passes, rms deviation $\lesssim 0.03$. Full line: scaling limit. Dash-dotted line: asymptotic behavior $\alpha^{-3/5}$.

this type of intermittency. However, in the presence of external noise, it is clear from Figs. 7 and 15 that real care must be exercised in using $P(l)$ to prove that observed intermittency[21] is or is not of the type discussed here.

Finally, just as in the case of phase transitions, we have seen that one must be careful to be inside the critical region in order to compare data with the various leading exponents and scaling relations derived here. In the case of $\langle l \rangle_0$, this critical region was set by requiring that $\sqrt{\epsilon/a}$ be sufficiently small compared to the gate y_0 that $(\pi/2)\arctan(y_0/\sqrt{\epsilon/a}) \cong 1$. However the Lyapunov exponent exhibited a substantially narrower critical region for $\epsilon > 0$.

Clearly, in spite of the numerical tests reported here which provide strong confirmation for the analytic results, the real test of these ideas will be experimental. To the authors present knowledge, the intermittent behavior so far reported does not in fact fit the type of results described here.[22] Hopefully, physical systems falling within this universality class will be found so that these ideas can be tested further.

ACKNOWLEDGMENTS

One of the authors (D.J.S.) would particularly like to thank M. Nauenberg for point out to him the similarity between the ideas of Y. Pomeau and P. Manneville and the behavior of the two-dimensional Potts model near its multicritical point which stimulated his interest in this problem. We would also like to acknowledge helpful discussions with H. Abarbanel, G. Ahlers, M. Feigenbaum, L. Kadanoff, and A. Libchaber. We are also grateful to J.-P. Eckman, L. Thomas, and P. Wittwer for sending us a copy of their paper prior to publication. J. Hirsch would like to acknowledge support from the NSF under Grant No. PHY77-27084, and D. J. Scalapino would like to acknowledge the support of the ONR under Contract No. N00014-79-C-0707. One of us (D.J.S.) would also like to thank Xerox PARC for their generous hospitality and the use of their computing facilities where the numerical work on the logistic map was carried out. He would also like to thank A. Zisook for his help.

---

[1] R. May, Nature (London) 261, 459 (1976).
[2] M. Feigenbaum, J. Stat. Phys. 19, 25 (1978); 21, 669 (1979).
[3] P. Collet and J. P. Eckmann, *Iterated Maps on the Interval as Dynamical Systems* (Birkhäuser, Basel, 1980).
[4] E. N. Lorenz, J. Atmos. Sci. 20, 130 (1963).
[5] P. Bak, Phys. Rev. Lett. 46, 791 (1981).
[6] E. Fradkin and B. A. Huberman (unpublished).
[7] S. Aubry, *Annals of the Israel Physical Society*, edited by C. G. Kuper (Adam Hilger, Bristol, 1979), Vol. 3, p. 133.
[8] M. Feigenbaum, Los Alamos Sci. Mag. 1, 4 (1980).
[9] B. A. Huberman and J. Rudnick, Phys. Rev. Lett. 45, 154 (1980).
[10] P. Manneville and Y. Pomeau, Phys. Lett. 75A, 1 (1979); Commun. Math. Phys. 74, 189 (1980).
[11] G. Mayer-Kress and H. Haken (unpublished).
[12] T. Li and J. A. Yorke, Am. Math. Monthly 82, 985 (1972).
[13] J. P. Crutchfield and B. A. Huberman, Phys. Lett. 77A, 407 (1980).
[14] J. P. Crutchfield, M. Nauenberg, and J. Rudnick, Phys. Rev. Lett. 46, 933 (1981).
[15] B. Shraiman, C. E. Wayne, and P. C. Martin, Phys. Rev. Lett. 46, 935 (1981).
[16] B. A. Huberman and A. Zisook, Phys. Rev. Lett. 46, 626 (1981); J. P. Crutchfield, J. D. Farmer, and B. A. Huberman (unpublished).
[17] After this work was completed we received some unpublished work from J. P. Eckmann, L. Thomas, and P. Wittwer in which some of the results reported here were also obtained.
[18] R. Graham, in *Quantum Statistics in Optics and Solid-State Physics*, Vol. 66 of *Springer Tracts in Modern Physics* (Springer, Berlin, 1973), Vol. I.
[19] R. L. Stratonovich, *Topics in the Theory of Random Noise* (Gordon and Breach, New York, 1963), Vol. I.
[20] J. Cardy, M. Nauenberg, and D. J. Scalapino, Phys. Rev. B 22, 2560 (1980).
[21] Intermittency in certain parameter regions of the Raleigh-Benard experiments of A. Libchaber and J. Mauer evidently does not exhibit the characteristics of the model discussed here (A. Libchaber, private communication).
[22] G. Ahlers (unpublished).

Volume 87A, number 8 PHYSICS LETTERS 1 February 1982

INTERMITTENCY IN THE PRESENCE OF NOISE: A RENORMALIZATION GROUP FORMULATION

J.E. HIRSCH and M. NAUENBERG [1]
Institute for Theoretical Physics, University of California, Santa Barbara, CA 93106, USA

and

D.J. SCALAPINO
Institute for Theoretical Physics and Department of Physics, University of California, Santa Barbara, CA 93106, USA

Received 14 November 1981

A renormalization group (RG) formulation of the transition to chaotic behavior via intermittency in one-dimensional maps is presented. The known scaling behavior of the length of the laminar regions in the presence of external noise is obtained from the leading relevant eigenvalues of the RG transformation. In addition, the complete spectrum of eigenvalues and corresponding eigenfunctions is found.

The renormalization group equations describing a phase transition are regular functions of the parameters such as temperature and external field which determine the state of the system. As these parameters are continuously varied through a phase transition, the singular behavior of the system arises from an infinite iteration of the regular renormalization group equations. Recent work suggests that a similar point of view provides a useful framework for understanding the onset of irregular or chaotic behavior of dynamical systems as a parameter is continuously varied. In particular, iterates of the one-dimensional logistic map

$$x_{n+1} = Rx_n(1 - x_n) , \tag{1}$$

can change from a regular to an irregular pattern as R is varied. The logistic map exhibits two types of such transitions. One of these involves an infinite cascade of period-doubling or pitchfork bifurcations, while the other arises from a saddle or tangent bifurcation leading to intermittency. Feigenbaum [1] and others [2] have developed a renormalization approach to describe the scaling and universal properties of the transition to chaos through period-doubling for the class of one-dimensional maps $x_{n+1} = f(x_n)$, with

$$f(x) = 1 - a|x|^z , \tag{2}$$

where $z = 2$ for the logistic map. A scaling theory describing the effect of external noise on the period-doubling cascade has also been developed [3,4].

Here we are interested in the second type of transition exhibited by the logistic map. Following the initial ideas of Pomeau and Manneville [5], the onset of chaotic behavior characterized by the occurrence of regular or "laminar" sequences of x_n values separated by intermittent bursts has also been shown to scale [6,7]. For the class of saddle point maps

$$f(x) = x + a|x|^z + \epsilon , \tag{3}$$

with $z > 1$, the length of the laminar regions l varies for small ϵ as $\epsilon^{-(1-1/z)}$. In the presence of a stochastic noise source of amplitude g, l satisfies the scaling equation

$$l(\epsilon, g) = \epsilon^{-(1-1/z)} f(g/\epsilon^{(z+1)/2z}) . \tag{4}$$

These relations were established by considering a Langevin equation describing the map near the saddle

[1] Permanent address: Natural Science, University of California, Santa Cruz, CA 95060, USA.

point, and using Fokker–Planck techniques to determine the time of passage in the presence of noise.

Here we develop a renormalization approach for saddle point maps which puts the known scaling results for intermittency in the same framework as Feigenbaum's treatment of the period-doubling cascade. We consider the class of maps given by eq. (3) in the presence of external noise,

$$x' = f(x) + g\xi \,, \tag{5}$$

where ξ is a random variable of unit standard deviation. The idea of the renormalization approach is to evaluate the map $x \to x''$ associated with two consecutive iterations of eq. (5) and by rescaling cast it back into the original form. This requires new parameters, ϵ', g' which in the limit $\epsilon, g \to 0$ satisfy the relation

$$\epsilon' = \lambda_\epsilon \epsilon, \quad g' = \lambda_g g \,, \tag{6}$$

where λ_ϵ and λ_g are the largest relevant eigenvalues of the linearized renormalization group transformation. Then the length $l(\epsilon, g)$ satisfies the homogeneity relation

$$l(\epsilon, g) = 2l(\epsilon', g') \,, \tag{7}$$

which leads in the usual way to the scaling relation

$$l(\epsilon, g) = \epsilon^{-\nu} f(g/\epsilon^{\mu}) \,, \tag{8}$$

with exponents

$$\nu = \log 2/\log \lambda_\epsilon, \quad \mu = \log \lambda_g/\log \lambda_\epsilon \,. \tag{9}$$

The functional recursion relation we use to define our renormalization procedure is the same as in Feigenbaum's case:

$$T\{f(x)\} = \alpha f(f(x/\alpha)) \,, \tag{10}$$

where α is a rescaling factor, but with boundary conditions

$$f(0) = 0 \,, \quad f'(0) = 1, \tag{11}$$

appropriate to a saddle point bifurcation at $x = 0$. It can be readily verified that

$$f^*(x) = x/(1 - ax) \,, \tag{12}$$

is a fixed point of the transformation (10) with $\alpha = 2$ and a an arbitrary constant. For small x this solution corresponds to eq. (3) for $\epsilon = 0$ and $z = 2$. For $z \neq 2$, we can find the fixed point of (10) by series expansion, and to third non-vanishing order obtain

$$f^*(x) = x + a|x|^z + \tfrac{1}{2} z a^2 |x|^{2z-1} + \dots \,, \tag{13}$$

with the scale factor $\alpha = 2^{1/(z-1)}$.

The next step is to consider the effect of small perturbations around the fixed point. We write

$$f(x) = f^*(x) + \epsilon h(x) \,, \tag{14}$$

where ϵ is a small parameter, and determine the eigenfunction from the usual condition of form invariance after rescaling:

$$f^{*\prime}(f^*(x)) h_n(x) + h_n(f^*(x)) = (\lambda_n/\alpha) h_n(\alpha x) \,, \tag{15}$$

where λ_n is the nth eigenvalue. For the case $z = 2$ we find $\lambda_n = 4/2^n$ and obtain the eigenfunctions h_n by series expansions. The relevant eigenfunction with eigenvalue $\lambda_\epsilon = 4$ is, to second order in x,

$$h_\epsilon(x) = 1 + ax + \tfrac{4}{3} a^2 x^2 + \dots \,. \tag{16}$$

The other relevant eigenfunction, with eigenvalue $\lambda_1 = 2$, does not correspond to the physical situation of interest here and will not be considered [‡1]. The marginal eigenfunction, with $\lambda_2 = 1$, is associated with the arbitrary constant a in the map eq. (12) and can be found in closed form:

$$h_2(x) = x^2/(1 - ax)^2 \,. \tag{17}$$

Finally, the irrelevant eigenfunctions with eigenvalue $\lambda_n, n > 2$, have the leading behavior $h_n(x) = x^n + \dots$.

In the general case $z > 1$, the form of the eigenvalues is $\lambda_n = 2^{(z-n)/(z-1)}$, and the leading behavior of the eigenfunctions is x^n. The largest relevant eigenvalue is $\lambda_\epsilon = 2^{z/(z-1)}$. The case $n = 1$ is again not of interest here. For $1 < n < z$ the perturbation is still relevant and gives a crossover to a behavior described by the map eq. (3) with z replaced by n. The marginal eigenvalue, for $n = z$, is again associated with the arbitrary constant a, and the eigenfunctions with $n > z$ are irrelevant.

We consider now the effect of adding a small amount of external noise to the fixed point function:

$$f(x) = f^*(x) + g(x)\xi \,. \tag{18}$$

Under iteration, this leads to the eigenvalue equation [4]

‡1 The eigenfunction corresponding to $\lambda = 2$ has the leading behavior $h_1(x) = x$ which changes the character of eq. (3), eliminating the intermittent behavior.

Volume 87A, number 8 PHYSICS LETTERS 1 February 1982

$$f^{*\prime}(f^*(x))g^2(x) + g^2(f^*(x)) = (\lambda_g^2/\alpha^2)g^2(\alpha x)\,. \quad (19)$$

For the leading eigenvalue λ_g one obtains the exact result

$$\lambda_g = 2^{(z+1)/2(z-1)}\,, \quad (20)$$

and the corresponding eigenfunction is

$$g(x) \propto 1 + \tfrac{1}{2} z a |x|^z/x + \dots\,. \quad (21)$$

Using the above results for λ_ϵ and λ_g we obtain for the exponents defined in eq. (9),

$$\nu = (z-1)/z\,, \quad \mu = (z+1)/2z\,, \quad (22)$$

and from eq. (8) the scaling behavior eq. (4) follows.

In conclusion, we have shown that the scaling behavior for the average length of the laminar regions in the transition to chaos via intermittency can be easily derived from a renormalization group formulation of the problem. We have obtained exact results for the complete spectrum of eigenvalues and the leading terms of the corresponding eigenfunctions of the renormalization group transformation.

Two of us (J.E.H. and M.N.) would like to acknowledge support by the NSF under PHY77-27084 and PHY78-22253. D.J.S. would like to acknowledge the support of the ONR under N0014-79-C-0707.

References

[1] M. Feigenbaum, J. Stat. Phys. 19 (1978) 25; 21 (1979) 669.
[2] P. Collet and J.-P. Eckmann, in: Iterated maps on the interval as dynamical systems (Birkhäuser, Boston, 1980).
[3] J.P. Crutchfield, M. Nauenberg and J. Rudnick, Phys. Rev. Lett. 46 (1981) 933.
[4] B. Shraiman, C.E. Wayne and P.C. Martin, Phys. Rev. Lett. 46 (1981) 935.
[5] P. Manneville and Y. Pomeau, Phys. Lett. 75A (1979) 1; Commun. Math. Phys. 74 (1980) 189.
[6] J.-P. Eckmann, L. Thomas and P. Wittwer, to be published.
[7] J.E. Hirsch, B.A. Huberman and D.J. Scalapino, to be published.

VOLUME 48, NUMBER 24 PHYSICAL REVIEW LETTERS 14 JUNE 1982

Exact Solutions to the Feigenbaum Renormalization-Group Equations for Intermittency

Bambi Hu
Department of Physics, University of Houston, Houston, Texas 77004

and

Joseph Rudnick[a]
Department of Physics, University of California, Davis, California 95616

(Received 19 March 1982)

Exact solutions to the Feigenbaum renormalization-group recursion relation, and the associated eigenvalue equations describing deterministic as well as stochastic perturbations, are found for the case of intermittency. These solutions are generated by a reformulation of the one-dimensional iterated map that exploits its topological equivalence to a translation. Direct resummation of series expansions gives the same results.

PACS numbers: 05.40.+j, 02.50.+s

The study of bifurcation and the transition to chaos has attracted intense interest recently, and considerable progress has been made. The three most commonly discussed scenarios,[1] associated, respectively, with the works of Feigenbaum,[2] Manneville and Pomeau,[3] and Ruelle and Takens,[4] are based on three different types of bifurcations: the pitchfork, tangent, and Hopf bifurcations. The much discussed period-doubling route to chaos is based on the pitchfork bifurcation.

The tangent bifurcation, on the other hand, offers a different route to chaos via intermittency. In this scenario, intermittency is a precursor to periodic behavior. It consists of long-lived episodes of nearly periodic behavior, the duration of which becomes arbitrarily long as the transition, via a tangent bifurcation [Fig. 2(a)], is approached.

Recently, Hirsch, Huberman, and Scalapino,[5] following the initial ideas of Manneville and Pomeau, proposed a detailed theory of intermittency. Scaling relations for the length of laminarity in the presence of noise were established[5,6] by considering a Langevin equation describing the map near the saddle point, and using the Fokker-Planck techniques to determine the time of passage. Very remarkably, Hirsch, Nauenberg, and Scalapino[7] later found that the same results can be simply explained by using the same functional renormalization-group equations first proposed by Feigenbaum in his study of period doubling—with a mere change of boundary conditions appropriate to the tangency condition. Thus the renormalization group provides a unified and elegant approach to both period doubling and intermittency.

The renormalization-group approach as formulated by Feigenbaum postulates the existence of a universal map, obtained by repeated compositions and rescalings of the original map, at the onset of chaos. The rescaling factor needed to generate the universal map yields one universal exponent. Eigenvalues describing the rate at which perturbations of this map grow provide the others.

To find the spectrum of eigenvalues and corresponding eigenfunctions, Hirsch, Nauenberg, and Scalapino used series-expansion techniques. In the simplest $z=2$ case they were able to sum the series and obtain a closed-form solution to the universal function. However, the universal function for arbitrary z and all eigenfunctions were only computed to the first few orders.

We have found that it is possible to obtain not only all the exponents for intermittency, but also closed-form results for the universal functions and all eigenfunctions corresponding to deterministic as well as stochastic perturbations for arbitrary z. This was achieved by a simple transformation that recasts the map near a tangent bifurcation into a simple translational map $x_{i+1} = x_i + b$, with b a constant. Direct resummation of series expansions corroborates our results. Whether this technique will prove to be of general utility remains to be seen, but the remarkable simplification it leads to in the renormalization-group study of intermittency induces us to believe that it may well prove a useful tool in the study of other dynamical transitions.

The tangent bifurcation as it occurs in iterated one-dimensional maps is illustrated in Fig. 1. Here the map $f(x) = rx(1-x)$ and its third iterate $f^{(3)}(x) = f(f(f(x)))$ are shown at $r = r_3 = 1+\sqrt{8}$. For

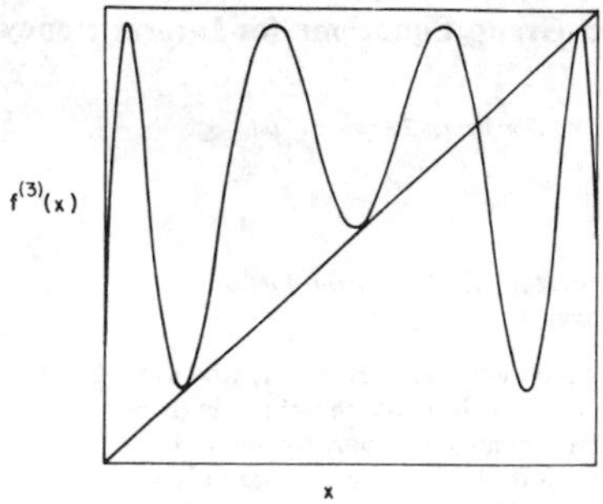

FIG. 1. The third-iterated map $f^{(3)}(x)$ at r_3.

$r \lesssim r_3$, $f^{(3)}(x)$ has two unstable fixed points, at $x = 0$ and $x = (r-1)/r$. These are the unstable fixed points of $f(x)$. As r passes through r_3 [see Fig. 2(a)], $f^{(3)}(x)$ acquires six new fixed points, three stable and three unstable. The three stable fixed points are the three elements of a stable period-three limit cycle of $f(x)$. Even though the map has no stable period-three cycle when $r \lesssim r_3$, it is evident [Fig. 2(b)] that under repeated iterations of $f^{(3)}(x)$, x_i's spend several iterations in the immediate vicinity of the points of closest approach to the 45° line. This behavior corresponds to orbits under $f(x)$ that look nearly periodic for a sizable number of iterations, i.e., they almost repeat themselves every third iteration, but eventually slip out of this pattern, and, shortly thereafter, establish another pattern of near periodicity. This sequence of long-lived episodes is the phenomenon of intermittency.

To study the transition to periodicity of order n we consider the nth iterated map in the immediate vicinity of one of the n points at which it achieves tangency to the 45° line at the transition. Shifting the origin of coordinates to that point, we have for the map at tangency

$$f^n(x) = x(1 + ux^{z-1}) + O(x^z), \tag{1}$$

where u is the coefficient of expansion. The exponent z determines the "universality classes." Most commonly z will be equal to 2. Here we keep it general with the understanding that

$$x^{z-1} \equiv |x|^{z-1}\operatorname{sgn}(x). \tag{2}$$

The universal map $f^*(x)$ has a power-series expansion in x whose two lowest-order terms match the right-hand side of Eq. (1). The map furthermore satisfies

$$f^*(f^*(x)) = \alpha^{-1} f^*(\alpha x), \tag{3}$$

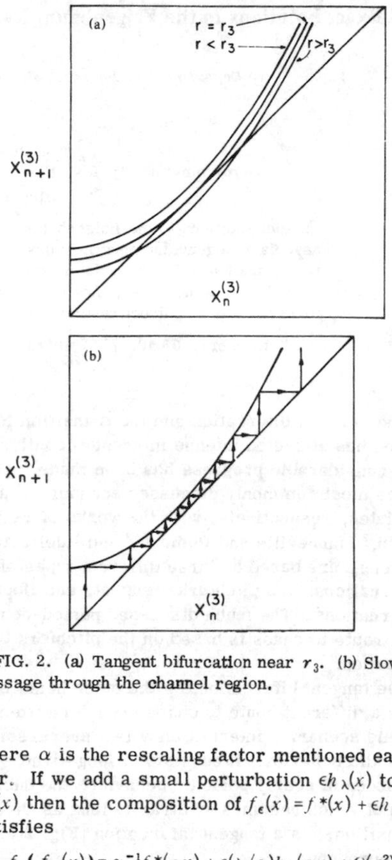

FIG. 2. (a) Tangent bifurcation near r_3. (b) Slow passage through the channel region.

where α is the rescaling factor mentioned earlier. If we add a small perturbation $\epsilon h_\lambda(x)$ to $f^*(x)$ then the composition of $f_\epsilon(x) = f^*(x) + \epsilon h_\lambda(x)$ satisfies

$$f_\epsilon(f_\epsilon(x)) = \alpha^{-1} f^*(\alpha x) + \epsilon(\lambda/\alpha) h_\lambda(\alpha x) + O(\epsilon^2) \tag{4}$$

when the eigenfunction $h_\lambda(x)$ satisfies

$$f^{*\prime}(f^*(x)) h_\lambda(x) + h_\lambda(f^*(x)) = (\lambda/\alpha) h_\lambda(\alpha x). \tag{5}$$

There will actually prove to be a *spectrum* of eigenvalues λ and corresponding eigenfunctions $h_\lambda(x)$. Stochastic exponents[8,9] are associated with the rate of growth of stochastic perturbations of the form $\xi g_{\lambda_g}(x)$, with ξ a random variable controlled by a probability distribution of unit width. Here the eigenfunctions satisfy

$$f^{*\prime 2}(f^*(x)) g_{\lambda_g}{}^2(x) + g_{\lambda_g}{}^2(f^*(x)) = (\lambda_g/\alpha)^2 g_{\lambda_g}{}^2(\alpha x). \tag{6}$$

Consider now the following recursion relation:

$$G(x') = G(x) - a \tag{7}$$

with $G(x)$ a function to be determined shortly. Iterating this recursion relation we have

$$G(x'') = G(x') - a = G(x) - 2a. \qquad (8)$$

We can generate the universal map for intermittency by choosing a $G(x)$ for which a rescaling of x yields the original recursion relation Eq. (7) from the iterated recursion relation Eq. (8). Such a function satisfies

$$G(x) = 2G(\alpha x). \qquad (9)$$

The form we need is $G(x) = x^{-(z-1)}$, with

$$\alpha = 2^{1/(z-1)}. \qquad (10)$$

Note that here the quantity z is arbitrary. Thus, a function satisfying Eq. (3) is obtained by recasting the recursion relation Eq. (7) into explicit form. Using $G(x) = x^{-(z-1)}$ we obtain

$$x' = f^*(x) = (x^{-(z-1)} - a)^{-1/(z-1)}. \qquad (11)$$

Replacing a by $(z-1)u$ in Eq. (11) we have

$$f^*(x) = x[1-(z-1)ux^{z-1}]^{-1/(z-1)}. \qquad (12)$$

It can be verified explicitly that $f^*(x)$ in Eq. (12) satisfies Eq. (3) and reproduces the correct power-series expansion. Furthermore the scale factor α given by Eq. (10) is correct.

We now consider the effect of a perturbation to $G(x)$ in Eq. (7). Our new implicit recursion relation is

$$x'^{-(z-1)} + \epsilon H(x') = x^{-(z-1)} + \epsilon H(x) - u(z-1). \qquad (13)$$

If $H(x) = x^{-p}$, then iterating the recursion relation Eq. (13) and rescaling by α as given by Eq. (10), we obtain our original recursion relation except that the coefficient ϵ has been increased by the factor λ, where

$$\lambda = 2^{p-z+1/(z-1)}. \qquad (14)$$

The associated eigenfunction is obtained by recasting Eq. (13) into an explicit recursion relation. Solving for x' in terms of x to order ϵ we obtain

$$x' = x[1-u(z-1)x^{z-1}]^{-1/(z-1)} - \frac{\epsilon}{z-1}[x^{-(z-1)} - u(z-1)]^{-z/(z-1)}\{x^{-p} - [x^{-(z-1)} - u(z-1)]^{p/(z-1)}\} + O(\epsilon^2)$$

$$\equiv f^*(x) - \frac{\epsilon u p}{z-1} h_\lambda(x) + O(\epsilon^2). \qquad (15)$$

The eigenfunction $h_\lambda(x)$ has been normalized so that its lowest-order term in x is x^{2z-1-p}. If we want an eigenfunction corresponding to a shift from tangency, that lowest-order term must be 1, and so we must choose $p = 2z-1$, which means than λ in Eq. (14) is equal to $2^{z/(z-1)}$. This matches with the relevant eigenvalue of Hirsch, Nauenberg, and Scalapino.[7]

The stochastic eigenfunctions are variants of the nonstochastic ones. They are

$$g_{\lambda_g}{}^2(x) = (1/uq)[x^{-(z-1)} - u(z-1)]^{-2z/(z-1)}\{x^{-q} - [x^{-(z-1)} - u(z-1)]^{q/(z-1)}\}$$

$$= (x/uq)^{2z-q}\{[1-u(z-1)x^{z-1}]^{-2z/(z-1)} - [1-u(z-1)x^{z-1}]^{-(2z-q)/(z-1)}\}. \qquad (16)$$

with

$$\lambda_g = 2^{[q-2(z-1)]/2(z-1)}. \qquad (17)$$

The lowest-order term in $g_{\lambda_g}{}^2$ is x^{3z-1-q}. If we want that term to be a constant we must choose $q = 3z-1$, in which case λ_g in Eq. (17) is equal to $2^{(z+1)/2(z-1)}$. All these results can also be obtained directly by resumming the series expansions.

The fact that a reformulation of the recursion relation leads to an immediate and complete solution of the renormalization-group equations for the iterated map near a tangent bifurcation is highly intriguing. Whether or not this kind of reformulation proves useful in the study of other transitions in dynamical systems remains to be seen. It certainly deserves to be considered as a viable approach.

This complete set of exact solutions provides a rare laboratory where ideas and theories can be experimented with and tested. The underlying mathematical structure, physical implications, and experimental consequences are still to be ruminated. However, since the method employed here depends crucially on the fact that the map for intermittency is topologically equivalent to a translation, most likely it will not prove to be fruitful for the study of period doubling.

One of us (B.H.) would like to thank Professor C. N. Yang for his interest in this work and encouragement. This work was supported in part by the National Science Foundation.

[(a)]Permanent address: Department of Physics, University of California, Santa Cruz, Cal. 95064.

[1]J.-P. Eckmann, Rev. Mod. Phys. 53, 643 (1981).

[2]M. J. Feigenbaum, J. Stat. Phys. 19, 25 (1978), and 21, 669 (1979).

[3]P. Manneville and Y. Pomeau, Phys. Lett. 75A, 1 (1979), and Commun. Math. Phys. 74, 189 (1980).

[4]D. Ruelle and F. Takens, Commun. Math. Phys. 20, 167 (1971).

[5]J. E. Hirsch, B. A. Huberman, and D. J. Scalapino, Phys. Rev. A 25, 519 (1982).

[6]J.-P. Eckmann, L. Thomas, and P. Wittwer, J. Phys. A 14, 3153 (1981).

[7]J. E. Hirsch, M. Nauenberg, and D. J. Scalapino, Phys. Lett. 87A, 391 (1982).

[8]J. P. Crutchfield, M. Nauenberg, and J. Rudnick, Phys. Rev. Lett. 46, 933 (1981).

[9]B. Shraiman, C. E. Wayne, and P. C. Martin, Phys. Rev. Lett. 46, 935 (1981).

VOLUME 49, NUMBER 2 PHYSICAL REVIEW LETTERS 12 JULY 1982

Universal Transition from Quasiperiodicity to Chaos in Dissipative Systems

David Rand[(a)]
Institute for Theoretical Physics, University of California, Santa Barbara, California 93106

and

Stellan Ostlund, James Sethna, and Eric D. Siggia
Laboratory of Atomic and Solid State Physics, Cornell University, Ithaca, New York 14853, and Institute for Theoretical Physics, University of California, Santa Barbara, California 93106

(Received 5 April 1982)

An exact renormalization-group transformation is developed which describes how the transition to chaos may occur in a universal manner if the frequency ratio in the quasiperiodic regime is held fixed. The principal low-frequency peaks in an experimental spectrum are universally determined at the transition. Our approach is a natural extension of Kolmogorov-Arnold-Moser theory to strong coupling.

PACS numbers: 47.20.+m, 02.30.+q

Our understanding of the onset of turbulence, within the context of low-order dynamical systems, has been extended from a qualitative description of various temporal regimes to a quantitative and universal set of predictions in the case of successive period doublings.[1,2] Although period doubling is not even the most prevalent route to chaos in low-aspect-ratio experiments, Feigenbaum's analysis has aroused great interest because the theory predicts that the Navier-Stokes equations are rigorously modeled by a one-dimensional map at the transition. His arguments are analogous to those used to describe scaling in critical phenomena. In this paper we show by renormalization-group methods how the transition from quasiperiodicity (flow with two incommensurate frequencies) to chaos can be made to proceed in a quantitatively universal manner.[3] This universality has not yet been seen experimentally, even though quasiperiodicity is a common percursor to turbulence, because it is associated with a critical point that can only be probed by varying two parameters in a consistent way.

On the mathematical side, our study suggests a means of realizing the strong-coupling limit of the small-divisor perturbation theory of Kolmogorov, Arnold, and Moser, who examined weakly nonlinear Hamiltonian and dissipative systems.[4] They realized that it is essential to work at a fixed frequency ratio or winding number which will become the second relevant variable at our fixed point. Conventional assumptions about the topology of the flows in function space generated by our renormalization group, we believe, imply the existence of a piecewise analytic variable change (a conjugacy), back to unperturbed quasiperiodic motion.[5] It is more interesting to study dissipative systems rather than Hamiltonian ones in this context because a fixed point found in one dimension is likely to carry over, rigorously, to an arbitrary number of dimensions just as one found for period doubling.[2]

Our calculations resemble Feigenbaum's treatment of period doubling in that we construct a transformation on a space of functions and find a nontrivial fixed point with the requisite linearized eigenvalues.[2] The alternative procedure, suggested by the Kolmogorov-Arnold-Moser proof, of successive variable changes is useful to keep in mind.[6]

The following map of an annulus illustrates succinctly the transition we propose to study.

Let

$$r_{i+1}-1=\lambda(r_i-1)-(a/2\pi)\sin(2\pi\varphi_i),$$
$$\varphi_{i+1}=\varphi_i+\omega+r_{i+1}-1, \tag{1}$$

where (r,φ) are polar coordinates and $0\le\lambda<1$. When $\lambda=1$ we obtain the "standard" area-preserving map of Chirikov.[4] Otherwise, since (1) contracts areas at a rate λ, there can be at most one invariant circle, $r=r(\varphi)$. Orbits on the invariant circle may have either a rational winding number $\rho(\omega,a)=p/q$ corresponding to mode locking in the differential system or an irrational ρ corresponding to a flow with two incommensurate frequencies. [In general $\rho=\lim_{i\to\infty}(\varphi_i-\varphi_0)/i$.] While a is analogous to the Reynolds number, the second relevant parameter ω should be thought of as a "bare" winding number that is adjusted to keep $\rho(\omega,a)$ fixed as a increases.

Since the nth iterate of (1) contracts areas at a uniform rate λ^n, this suggests that one can work with $\lambda=0$. This defines the one-dimensional diffeomorphism,

$$\varphi_{i+1}=f(\varphi_i)=\varphi_i+\omega-(a/2\pi)\sin(2\pi\varphi_i). \qquad (2)$$

The renormalization group may then be used to justify the neglect of the radial contraction. The strong-coupling fixed point occurs at $a=1$ where f has an inflection point. For $a>1$, (2) is noninvertible and shows all the complexity of a one-dimensional map, while for $a<1$ and almost all irrational winding numbers, the orbits are analytically conjugate to a simple rotation: $\varphi'=\varphi+\rho(\omega,a)$.[7]

We will consistently denote an irrational value of $\rho(\omega,a)$ by σ and write its continued-fraction representation as $1/[n_1+1/(n_2+\ldots)]$. The lth rational approximate to σ, obtained by setting $n_i=0$ for $i>l$, is p_l/q_l.

The universal features of the quasiperiodic-to-turbulent transition are restricted to low frequencies or long times so that the renormalization group is essentially functional composition. Our construction, however, will preserve both the character of f as a homeomorphism of the circle and its rotation number which will have important consequences when we consider spectra.

Both to define and to implement our renormalization group requires consideration of a larger class of functions than just analytic homeomorphisms of the circle. Specifically, let S_n be the class of pairs (ξ,η) of analytic homeomorphisms of the real line subject to the conditions (a) $\xi(\eta(0))=\eta(\xi(0))$, $(\xi\eta)'(0)=(\eta\xi)'(0)$, (b) $\xi(0)=\eta(0)+1$, (c) $0<\xi(0)<1$, (d) if $\xi'(0)=0$ or $\eta'(0)=0$ for $x\in[\eta(0),\xi(0)]$ then $\xi'(0)=\eta'(0)=\xi''(0)=\eta''(0)=0$ but $\xi'''(0)\neq 0$ and $(\xi\eta)'''(0)=(\eta\xi)'''(0)$, (e) $\xi^n(\eta(0))>0$, $\xi^{n-1}(\eta(0))<0$ ($\xi\eta\ldots$, etc., denotes composition).

Define a mapping T_n on S_n by

$$T_n(\xi,\eta)=(\alpha\xi^{n-1}\eta\alpha^{-1},\alpha\xi^{n-1}\eta\xi\alpha^{-1}), \qquad (3)$$

where $\alpha=1/[\xi^{n-1}\eta(0)-\xi^n\eta(0)]$ and obeys $\alpha<-1$. The image under T_n of a member of S_n satisfies conditions (a)–(d), and in a neighborhood of the fixed point, (e) in addition.

Define a homeomorphism of the circle $f=f_{\xi,\eta}$ with $(\xi,\eta)\in S_n$ by identifying f with ξ on $[\eta(0),0]$ and with η on $(0,\xi(0)]$ and denote the set of all such homeomorphisms by $\tilde{S}_n$. All analytic circle homeomorphisms [e.g., Eq. (2)] belong to $\tilde{S}_n$. The mapping T_n then induces a mapping $\tilde{T}_n$ on $\tilde{S}_n$ for which it may be proven that

$$\rho(\tilde{T}_n(f))=1/\rho(f)-n. \qquad (4)$$

$\tilde{T}_{n_1}$ then simply removes the first term in the continued fraction of $\sigma=\rho(f)$. Thus for any periodic continued fraction, $n_{i+s}=n_i$, $s\geq 1$, one could string together $T=T_{n_s}\circ\cdots\circ T_{n_2}\circ T_{n_1}$ and sensibly search for a fixed point of T. We will henceforth consider only $\sigma=\sigma_G=(5^{1/2}-1)/2$ because then $n_i\equiv 1$ and $T=T_1$. The corresponding rational approximates are $p_l/q_l=F_{l-1}/F_l$ where F_l are Fibonacci numbers and satisfy $F_{l+1}=F_l+F_{l-1}$, $F_0=0$, $F_1=1$.

In this case, the fixed-point equations, $(\xi^*,\eta^*)=T(\xi^*,\eta^*)$, simplify considerably; in particular $\alpha=\xi^*(0)/[\xi^*(0)-1]$. There is then a trivial fixed point corresponding to a pure rotation with $\xi^*(x)=x+\sigma_G$, $\alpha=-\sigma_G-1=-1/\sigma_G$, which has one unstable direction with eigenvalue $\delta=-\alpha^2$. The physical meaning of δ is clear from (4) since if $\rho(f)=F_{l-1}/F_l$ then $\rho(\tilde{T}f)=F_{l-2}/F_{l-1}$. In analogy with period doubling, δ measures the accumulation rate of the ω_l in (2) for which f has a F_{l-1}/F_l cycle. However, here ω_∞ corresponds to quasiperiodicity and not to incipient chaos.

At the nontrivial fixed point we can consistently assume that ξ^* and η^* are analytic functions of x^3. By retaining terms up to x^{36} and using a numerical technique suggested by Feigenbaum we find $\alpha=-1.288\,575$. Note that even if f is analytic, $\tilde{T}^kf$ is nonanalytic but continuous at the origin and end points of the unit interval. However, there is good numerical evidence that the conjugacy that relates $\tilde{T}^kf$ to any $\tilde{T}^lf$ is smooth for any f on the critical surface, i.e., $\lim_{l\to\infty}T^lf=f^*$. This fact, together with the existence of a fixed point f^*, establishes that spectra are universal in the sense discussed below.

Within the space of eigenfunctions, expandable in x^3 and consistent with conditions (a)–(e), we find one unstable eigenvalue $\delta=-2.833\,61$ with the same meaning as before. When arbitrary powers of x are allowed, we are able to show analytically from (a)–(e) that there is only one additional relevant mode with eigenvalue $\gamma=\alpha^2=\sigma_G^{-\nu}$, which corresponds to the addition of a small linear term to the fixed point. (The relation $\gamma=\alpha^2$ is true for all winding numbers for which a fixed point exists.) Depending on its sign, the map either iterates into the chaotic regime where it initially has two critical points and is noninvertible, or remains invertible and iterates toward the trivial fixed point. In the space of all $f\in\tilde{S}$ with $\rho(f)=\sigma$ the nontrivial fixed point is a saddle and the trivial one a sink under $\tilde{T}$. If one assumes globally that there are no other fixed points then we conjecture that Kolmogorov-

Arnold-Moser–like results follow for all maps attracted to the trivial fixed point.

Another way of displaying the scaling behavior of (1) or (2) is to examine how points map back close to the origin after q_l iterations of f [n.b.: $p_l/q_l \to \sigma = \rho(f)$]. We therefore define

$$\zeta_l(x) = \alpha^l [f^{q_l}(\alpha^{-l}x) - p_l]. \tag{5}$$

For a suitable choice of α, $\lim_{l\to\infty}\zeta_l = \zeta$ exists as an analytic function on the line and to within a scale change is identical with ξ^*. The recursion formula for q_l generates nonlinear equations for ζ.[8]

We believe that a transformation like ours which preserves the circle homeomorphism structure and keeps track of the winding number is preferable to a straightforward iteration of f for the following reasons: (1) We can identify the universal numbers δ, γ, etc., as eigenvalues of T and make clear their physical meaning. (2) The minimal fixed-point functional equation is obvious, which, especially for $\sigma \neq \sigma_G$, is not clear otherwise. (3) There exist scaling equations for the conjugacy associated with $f^* = f_{\xi^*,\eta^*}$ which fix the experimental spectra. (4) It allows for mathematically rigorous proofs; e.g., Jonker and Rand have proven that the fixed-point and hyperbolic structure of T exist when ξ and η are analytic functions of $x|x|^\epsilon$, $0 \le \epsilon \ll 1$.

Our renormalization group can readily be extended to higher dimensions and we believe that the fixed-point structure will remain unchanged.[9] Specifically, for annular maps [e.g., Eq. (1)], let $\Lambda(r,\varphi) = (\alpha^3 r, \alpha\varphi)$ and for $\sigma = \sigma_G$ define

$$T^{(2)}(E,F) = (\Lambda F \Lambda^{-1}, \Lambda F E \Lambda^{-1}),$$

where E and F are analytic homeomorphisms of the plane subject to conditions analogous to (a)–(e). A nontrivial fixed point of $T^{(2)}$ is $E^*(r,\varphi) = (0, \xi^*([r+\varphi^3]^{1/3}))$, $F^*(r,\varphi) = (0, \eta^*([r+\varphi^3]^{1/3}))$. Our *Ansatz* becomes plausible if one considers the family of curves along which the flow contracts exponentially onto the invariant curve, Γ. Below the critical point they intersect Γ transversally, but at the critical point after suitable iterations and rescalings they are tangent and converge to the family $r = -\varphi^3 + \text{const}$.

We will assume that our one-dimensional fixed point applies in higher dimensions and by invoking universality, analyze only the conjugacy h^* that reduces f^* to a pure rotation, i.e., $h^{*-1}f^* h^*(\theta) = \theta + \sigma_G$. The quantity of greatest physical interest is the periodic part of h^*, $\chi = h^* - \theta$, which is continuous but nondifferentiable. Its Fourier transform, $\bar\chi(n)$, has prominent peaks at all the low-order Fibonacci series and scales as n^{-1}.[8]

If we use the natural range for h^* set by $\bar T$, $1/(\alpha-1) \le \varphi \le \alpha/(\alpha-1)$, and allow $-\sigma_G^2 \le \theta \le \sigma_G$, then h^* is continuous on the interior of its domain and satisfies

$$h^*(\theta) = \begin{cases} \alpha^{-1}h^*(-\theta/\sigma_G), & -\sigma_G^2 \le \theta \le \sigma_G^3, \quad (6a) \\ f^{*-1}(h^*(\theta+\sigma_G-1)), & \sigma_G^3 \le \theta < \sigma_G. \quad (6b) \end{cases}$$

Equation (6a) is a consequence of how $\bar T$ acts on the invariant measure of any f with $\rho(f) = \sigma_G$ and may be generalized to other winding numbers. Once h^* is known on (σ_G^3, σ_G), (6a) recursively determines the rest of h^* on a sequence of intervals that converge onto the origin. The second equation smoothly relates h^* on $(-\sigma_G^4, \sigma_G^3)$ back to (σ_G^3, σ_G) so that (6a) and (6b) together should determine h^*.

To proceed further analytically it is convenient to use a piecewise linear approximation to f^* which utilizes a geometrically convergent set of vertices to approximate x^3 near the origin. The equation $Tf = f$ is satisfied everywhere. The corresponding value of α obeys $\alpha^6 + \alpha^5 - 1 = 0$, $\alpha = -1.285\,3$. The principal amplitudes in $\bar\chi$ differ by no more than 10% from their exact values. Over the interval required by (6b), $f^{-1}(y) = (\alpha^2 + \alpha)y + 1/(1-\alpha)$. The relation $\bar\chi(n) \sim n^{-1}$ for $n = m_1F_i + m_2F_{i-1}$, $i \gg 1$ follows rigorously.

The same scaling relation for $\bar\chi(n)$ at the exact fixed point is proven from the lemma that any sufficiently smooth periodic function of h^* has a Fourier series bounded by η^{-2} for large η. The same lemma is used in the demonstration that spectra are universal. We conclude by suggesting how a laboratory experiment may be done to test the theory developed here.

In the quasiperiodic regime, an experimental spectrum is a series of peaks at all integer combinations of two incommensurate frequencies ω_1, ω_2. Any other choice of reference frequencies is related to (ω_1, ω_2) by an integer-valued matrix with determinant ± 1. Precisely the same condition is necessary and sufficient for the tails of the continued fractions of two irrational winding numbers σ and σ' to agree. Our renormalization group implies that any choice of (ω_1, ω_2) for a given experimental spectrum is associated with the same fixed point.

The easiest way to control the frequency ratio $\sigma = \omega_1/\omega_2$ in an experiment, as the Rayleigh number varies, is to introduce ω_2 by means of an external force. The optimal choice of σ experimentally is σ_G since it is the least susceptible to mode locking and one can expect to see the largest number of self-similar bands for a given level of noise.

There is a one-to-one relation between the low frequencies in the spectrum of a time series and $\tilde{\chi}$ which completely determines the former to within an overall scale. Specifically, the complex amplitude at $\omega = m_1\omega_1 - m_2\omega_2$, $0 \leqslant \omega < \omega_2$, is proportional to $\tilde{\chi}(m_1)$. The principal peaks in $\tilde{\chi}$ correspond to $(m_1, m_2) = (F_l, F_{l-1})$ $(\sigma = \sigma_G)$, and their power scales as σ_G^{-2l}. The condition that a given m be asymptotic is that it may be represented in the form $m_1F_l + m_2F_{l-1}$ with $l \gg 1$. An experimental determination of δ from the accumulation rate of periodic orbits will be difficult since to be useful it must distinguish between the trivial value of $\sigma_G^{-2} \sim 2.61803$ and 2.83361.

The authors benefitted from a number of conversations with P. C. Hohenberg and J. Guckenheimer in the early stages of this work. We are particularly indebted to S. Shenker, L. Kadanoff, and M. Feigenbaum for communicating their results prior to publication and freely sharing their ideas and insights. Our research was supported by the National Science Foundation under Grants No. PHY77-27084, ATM80-05796, DMR80-20429, DMR79-24008, and the Sloan Foundation.

[(a)]Permanent address: Mathematics Institute, University of Warwick, Coventry CV4 7AL, England.

[1]A. Libchaber and J. Maurer, in Proceedings of the Conference on Nonlinear Phenomena at Phase Transitions and Instabilities, Geilo, Norway, 1981, edited by T. Riste (to be published); J. P. Gollub and S. V. Benson, J. Fluid Mech. 100, 449 (1978).

[2]P. Collet and J. P. Eckmann, *Iterated Maps on the Interval as Dynamical Systems* (Birkhauser, Boston, 1980); M. Feigenbaum, J. Stat. Phys. 19, 25 (1978), and 21, 669 (1979).

[3]Many of the results to follow were obtained in parallel by M. Feigenbaum, L. P. Kadanoff, and Scott J. Shenker, private communication.

[4]J. Moser, *Stable and Random Motions in Dynamical Systems* (Princeton Univ. Press, Princeton, N.J., 1973); N. N. Bogoliubov and J. A. Mitropolski, *Methods of Accelerated Convergence in Nonlinear Mechanics* (Springer, New York, 1976); B. V. Chirikov, Phys. Rep. 52, 264 (1979).

[5]For a related approach to Hamiltonian problems, see L. P. Kadanoff, Phys. Rev. Lett. 47, 1641 (1981); Scott J. Shenker and L. P. Kadanoff, J. Stat. Phys. 27, 631 (1982).

[6]D. F. Escande and F. Doveil, J. Stat. Phys. 26, 257 (1981).

[7]M. R. Herman, in *Geometry and Topology*, edited by J. Palis, Lecture Notes in Mathematics Vol. 597 (Springer-Verlag, Berlin, 1977), pp. 271–293.

[8]This scheme as well as the spectra were examined numerically by Scott J. Shenker, to be published, and independently by ourselves.

[9]Strong evidence for this conjecture has been provided by Scott J. Shenker, who found that the critical indices of (1) and (2) agree; see Ref. 3.

Physica 5D (1982) 370–386
North-Holland Publishing Company

QUASIPERIODICITY IN DISSIPATIVE SYSTEMS: A RENORMALIZATION GROUP ANALYSIS

Mitchell J. FEIGENBAUM
Center for Nonlinear Studies, Los Alamos National Laboratory, Los Alamos, New Mexico 87545, USA

Leo P. KADANOFF*
The Enrico Fermi and James Franck Institutes of The University of Chicago, 5630–5640 South Ellis Avenue, Chicago, Illinois 60637, USA

Scott J. SHENKER*
The James Franck Institute of The University of Chicago, Chicago, Illinois 60637, USA

Received 29 March 1982
Revised 1 June 1982

Dynamical systems with quasiperiodic behavior, i.e., two incommensurate frequencies, may be studied via discrete maps which show smooth continuous invariant curves with irrational winding number. In this paper these curves are followed using renormalization group techniques which are applied to a one-dimensional system (circle) and also to an area-contracting map of an annulus. Two fixed points are found representing different types of universal behavior: a trivial fixed point for smooth motion and a nontrivial fixed point. The latter represents the incipient breakup of a quasiperiodic motion with frequency ratio the golden mean into a more chaotic flow. Fixed point functions are determined numerically and via an ϵ-expansion and eigenvalues are calculated

1. Introduction

In recent years there has been much interest in the transition to chaotic behavior in dynamical systems. Experimental results, on fluid systems for example, suggest the existence of several distinct "routes to chaos" [1]. Significant progress has been made on the period-doubling [2–3] and intermittency [4–5] routes by focusing on low-dimensional attractors and studying their dynamics through the use of Poincaré return maps. Renormalization group analysis of these maps has provided important insight into the scaling behavior at the onset of chaos.

It is our purpose here to apply this paradigm to another "route to chaos," that of quasiperiodic behavior (two incommensurate frequencies) followed by broadband noise. While this general scenario is commonly observed experimentally, no comparison can yet be made between experiment [6] and the theory presented here.

Consider a dissipative dynamical system depending on some parameter μ having an attracting stationary solution for $\mu = 0$. As μ is varied, one commonly observed sequence of bifurcations is a Hopf bifurcation to a periodic solution followed by a secondary Hopf bifurcation to a quasiperiodic solution. Immediately after this second bifurcation, the system's dynamiçs are described by a diffeomorphic (differentiable with differentiable inverse) return map T_μ of the plane [7]

$$\begin{bmatrix}\theta_{i+1}\\ r_{i+1}\end{bmatrix} = T_\mu\begin{bmatrix}\theta_i\\ r_i\end{bmatrix}. \tag{1.1}$$

Identifying θ and $\theta + 1$, the periodicity in θ is

* Supported in part by the Central Facilities of the Materials Research Laboratory Program at the University of Chicago under Grant No. NSF-MRL 79-24007 and Grant No. NSF-DMR 80-20609.

expressed by introducing the map S,

$$S\begin{bmatrix}\theta\\ r\end{bmatrix}=\begin{bmatrix}\theta-1\\ r\end{bmatrix}, \tag{1.2}$$

and demanding that S and T_μ commute. Close to the second bifurcation there will be an attracting invariant curve in the plane topologically equivalent to a circle. As long as this invariant circle exists, the dynamics of the associated attractor must be smooth; i.e., there can be no broadband noise. However, the pioneering numerical studies of Curry and Yorke [8] on diffeomorphisms of the plane suggest that as μ is increased further the invariant circle breaks up and chaotic behavior appears soon thereafter. Thus, the emergence of chaos after quasiperiodic motion is linked to the destruction of invariant circles.

Each attracting invariant circle is characterized by a winding number W,

$$W\equiv\lim_{i\to\infty}\frac{(\theta_i-\theta_0)}{i}, \tag{1.3}$$

where $[{\theta_0 \atop r_0}]$ can be any point in the basin of attraction. The winding number represents the ratio of the two frequencies in the quasiperiodic regime. When the frequencies are commensurate then W is rational, say $W=P/Q$, with P and Q relatively prime integers. In the typical case considered here, the attractor will be a stable Q-cycle; i.e., a set of points $[{\theta_i \atop r_i}]$ such that

$$T\begin{bmatrix}\theta_i\\ r_i\end{bmatrix}=\begin{bmatrix}\theta_{i+1}\\ r_{i+1}\end{bmatrix} \tag{1.4}$$

and

$$\theta_Q=\theta_0+P,\qquad r_Q=r_0. \tag{1.5}$$

One could also view these as Q distinct attracting fixed points of the map T^QS^P. Associated with this stable Q-cycle will be an unstable Q-cycle whose unstable manifold flows smoothly into the stable cycle, forming the invariant circle. Aronson et al. [9] have recently given a beautifully detailed description of the disappearance of invariant circles with rational winding numbers.

When the frequencies are incommensurate W is irrational and the attractor is the entire invariant curve. We ask the question how does an invariant circle with irrational winding number break up? This is the dissipative analogue of the subject of an earlier series of papers exploring the disappearance of KAM curves in the area-preserving "standard map" [10–14]. As in the previous work, we choose to study a particular irrational winding number, the reciprocal of the golden mean,

$$\bar{W}=\frac{\sqrt{5}-1}{2}, \tag{1.6}$$

because of its extremely simple continued fraction expansion [15]

$$\begin{aligned}\bar{W}&=\langle 1111\ldots.\rangle\\ &=\cfrac{1}{1+\cfrac{1}{1+\cfrac{1}{1+\cdots}}}\end{aligned} \tag{1.7}$$

Using a method introduced by J. Greene [10] we study this irrational winding number by considering a series of rational winding numbers $W_n=P_n/Q_n$ that coverate to $\bar{W}$

$$\lim_{n\to\infty}W_n=\bar{W}. \tag{1.8}$$

We assume, following Greene, that the properties of the stable Q_n cycles with winding number W_n will, in the limit of large n, accurately represent the properties of the invariant curve with winding number $\bar{W}$. It turns out that the optimal choice for the W_n are successive truncations to the continued fraction expansion which, in the case of $\bar{W}$, yields

$$W_n = F_n/F_{n+1}, \tag{1.9}$$

where F_n is the nth Fibonacci number. The Fibonacci numbers satisfy the recursion relation

$$F_{n+1} = F_n + F_{n-1} \tag{1.10}$$

and have initial conditions

$$F_0 = 0, \qquad F_1 = 1. \tag{1.11}$$

Thus, our study of invariant curves reduces to the study of the fixed points of the sequence of planar maps

$$T^{(n)} = T^{F_{n+1}} S^{F_n}, \tag{1.12}$$

which obey the recursion relations

$$T^{(n+1)} = T^{(n)} \circ T^{(n-1)}, \tag{1.13a}$$

$$T^{(n+1)} = T^{(n-1)} \circ T^{(n)}. \tag{1.13b}$$

The particular map we study is

$$T\begin{bmatrix}\theta \\ r\end{bmatrix} = \begin{bmatrix}\theta + \Omega + br - (K/2\pi)\sin 2\pi\theta \\ br - (K/2\pi)\sin 2\pi\theta\end{bmatrix}, \tag{1.14}$$

which has constant Jacobian b. For the area-preserving case, $b = 1$, this is merely the standard map and in the singular limit $b = 0$ it reduces to the one-dimensional map on a circle,

$$f(\theta) = \theta + \Omega - \frac{K}{2\pi}\sin 2\pi\theta. \tag{1.15}$$

The circle map is a diffeomorphism for $|K| < 1$ and acquires a cubic inflection point at $|K| = 1$. Intuitively, one might expect the longtime behavior of the planar map with $0 < b < 1$ will, for many purposes, be modelled by the $b = 0$ limit. Following that reasoning, one of us has recently investigated the analogous quasi-periodic phenomena in the circle map [16]. Two interesting classes of scaling behavior were found, one for $|K| < 1$ and the other for $|K| = 1$. Denoting by Ω_n the value of Ω such that there is a cycle with winding number W_n passing through $\theta = 0$ (this will, of course, depend on K) and letting Ω_∞ denote the accumulation point of that sequence, it was found that for both cases

$$\Omega_n - \Omega_\infty \approx \delta^{-n}. \tag{1.16}$$

Furthermore, the distance from $\theta = 0$ to the nearest point on the cycle, which is given by $f^{F_n}(0) - F_{n-1}$, also converged geometrically;

$$f^{F_n}(0) - F_{n-1} \approx \alpha^{-n} \tag{1.17}$$

with $\Omega = \Omega_n$. When $|K| < 1$, α and δ took on their "trivial" values

$$\alpha = -\bar{W}^{-1} = -1.6180339\ldots, \quad \delta = -\bar{W}^{-2} = -2.6180339\ldots, \tag{1.18}$$

whereas for $|K| = 1$

$$\alpha = -1.28857 \pm 0.00002, \quad \delta = -2.83360 \pm 0.00003. \tag{1.19}$$

Similar to eq. (1.12) define the map $f^{(n)}$ by

$$f^{(n)}(\theta) = f^{F_{n+1}}(\theta) - F_n \tag{1.20}$$

and note that the $f^{(n)}$ obey two recursion relations identical to (1.13)

$$f^{(n+1)} = f^{(n)} \circ f^{(n-1)}, \tag{1.21a}$$

$$f^{(n+1)} = f^{(n-1)} \circ f^{(n)}. \tag{1.21b}$$

At $\Omega = \Omega_\infty$ and n large, it was found that for $|\theta| \ll 1$

$$f^{(n)}(\theta) \approx \alpha^{-n}\bar{f}(\alpha^n\theta). \tag{1.22}$$

For $|K| < 1$, $\bar{f}$ is merely the linear map

$$\bar{f}(x) = -1 + x. \tag{1.23}$$

For $K = 1$, however, $\bar{f}$ is a universal and nontrivial function. Near the origin it is apparently an analytic function of θ^3.

Combining eq. (1.21) with (1.22) we find that the function $\bar{f}$ must satisfy two equations,

$$\bar{f}(x) = \alpha\bar{f}(\alpha\bar{f}(\alpha^{-2}x)), \tag{1.24a}$$

$$\bar{f}(x) = \alpha^2\bar{f}(\alpha^{-1}\bar{f}(\alpha^{-1}x)). \tag{1.24b}$$

Ref. 16 provided strong numerical evidence for the existence of the function $\bar{f}$ in eq. (1.22) but did not investigate directly the equations (1.24). Our aim here is to analyze these equations using a renormalization group framework akin to that used to study period doubling [2].

In section 2 we formulate this renormalization group framework, setting up eqs. (1.24) as fixed point equations. In sections 3 and 4 we find and analyze solutions to (1.24) corresponding to $|K| < 1$ and $K = 1$ respectively. We return to the original subject of this paper in section 5 where we study the incipient break-up of invariant circles in the map (1.14). We find that for the values of K and Ω where the invariant circle with winding number $\bar{W}$ breaks up, the maps $T^{(n)}$ of eq. (1.13) exhibit a scaling behavior

$$T^{(n)} \approx L^{-n}\bar{T}L^{n}, \tag{1.25}$$

where

$$L\begin{bmatrix} x \\ y \end{bmatrix} = \begin{bmatrix} \alpha x \\ \alpha^3 y \end{bmatrix} \tag{1.26}$$

and $\bar{T}$ is the function

$$\bar{T}\begin{bmatrix} x \\ y \end{bmatrix} = \begin{bmatrix} \bar{f}((x^3+y)^{1/3}) \\ 0 \end{bmatrix}. \tag{1.27}$$

Beyond this fixed point, we believe that chaotic motion can occur. Thus, we argue that one way to drive a transition from quasiperiodic to chaotic behavior is to have the motion of the circle become singular in the sense of the nontrivial fixed point of equations (1.24).

Many of the results of the present paper were obtained in parallel by D. Rand, S. Ostlund, J. Sethna, and E. Siggia, with whom we have had frequent and fruitful exchanges [17]. Tsuda [18] has discussed piecewise continuous solutions of (1.24).

2. Formulation of renormalization group

Choosing a value for α, we define a sequence of functions f_n,

$$f_n(x) = \alpha^n f^{(n)}(\alpha^{-n}x). \tag{2.1}$$

These functions obey the recursion relations

$$f_{n+1}(x) = \alpha f_n(\alpha f_{n-1}(\alpha^{-2}x)), \tag{2.2a}$$

$$f_{n+1}(x) = \alpha^2 f_{n-1}(\alpha^{-1}f_n(\alpha^{-1}x)), \tag{2.2b}$$

which are similar in spirit to the period-doubling recursion relation

$$g_{n+1}(x) = \alpha g_n(g_n(\alpha^{-1}x)) \tag{2.3}$$

found in ref. 2. There are two crucial differences between the two recursion schemes, both related to the second-order nature of the recursion relation (1.10) of the Fibonacci numbers. Here, one must specify both f_n and f_{n-1} in order to produce f_{n+1}. If F is the space of functions we are considering, then our recursion relations (2.2) are really transformations on $F \times F$ to F. To properly formulate a renormalization group transformation we must enlarge this to a transformation taking $F \times F$ back to itself. One way to do this is to consider transformations that map $[{f_n \atop f_{n-1}}]$ onto $[{f_{n+1} \atop f_n}]$, such as

$$R_a\begin{bmatrix} u(x) \\ v(x) \end{bmatrix} = \begin{bmatrix} \alpha u(\alpha v(\alpha^{-2}x)) \\ u(x) \end{bmatrix}, \tag{2.4a}$$

$$R_b\begin{bmatrix} u(x) \\ v(x) \end{bmatrix} = \begin{bmatrix} \alpha^2 v(\alpha^{-1}u(\alpha^{-1}x)) \\ u(x) \end{bmatrix}. \tag{2.4b}$$

Eqs. (2.4) will clearly have the same fixed point structure as (2.2).

The second difference is that there are two distinct and inequivalent recursion relations (2.2), each sufficient to define f_{n+1} in terms of f_n and f_{n-1}. However, if we choose as our initial conditions that

$$f_0(\alpha^{-1}f_1(\alpha x)) = \alpha^{-1}f_1(\alpha f_0(x)) \tag{2.5}$$

then the two recursion relations become equivalent.

Direct iteration of the function f (eq. (1.15)), which automatically ensures (2.5), indicates that for certain values of α and Ω this series of functions reaches a limiting fixed point function $\bar{f}(x)$, which then clearly satisfies eqs. (1.24) [16]. We would like to know how initially small deviations about this fixed point $\bar{f}$ increase or decrease upon iterating (2.2). Set

$$f_n(x) = \bar{f}(x) + \epsilon h_n(x). \tag{2.5}$$

Then, eqs. (2.2) demand that to first order in ϵ,

$$h_{n+1}(x) = \alpha h_n(\alpha\bar{f}(\alpha^{-2}x)) + \alpha^2\bar{f}'(\alpha\bar{f}(\alpha^{-2}x))h_{n-1}(\alpha^{-2}x), \tag{2.6a}$$

$$h_{n+1}(x) = \alpha^2 h_{n-1}(\alpha^{-1}\bar{f}(\alpha^{-1}x)) + \alpha\bar{f}'(\alpha^{-1}\bar{f}(\alpha^{-1}x))h_n(\alpha^{-1}x). \tag{2.6b}$$

There are two kinds of solutions to these equations. Those in the commuting subspace which are at once solutions to both of these equations, and those in the noncommuting subspace which are solutions to only one of the two equations. We must emphasize this seemingly obvious distinction because we can, in practice, only analyze one of the two equations at a time, and must use the other equation as a guide to which solutions to keep.

We can also investigate deviations about the fixed point using the renormalization group transformation R_μ (μ = a or b). Let DR_μ denote the linearization of R_μ about the fixed point $\begin{bmatrix}\bar{f}\\ \bar{f}\end{bmatrix}$. Then, rewriting (2.6)

$$DR_{\mathrm{a}}\begin{bmatrix}u(x)\\ v(x)\end{bmatrix} = \begin{bmatrix}\alpha u(\alpha\bar{f}(\alpha^{-2}x)) + \alpha^2\bar{f}'(\alpha\bar{f}(\alpha^{-2}x))v(\alpha^{-2}x)\\ u(x)\end{bmatrix}, \tag{2.7a}$$

$$DR_{\mathrm{b}}\begin{bmatrix}u(x)\\ v(x)\end{bmatrix} = \begin{bmatrix}\alpha^2 v(\alpha^{-1}\bar{f}(\alpha^{-1}x)) + \alpha\bar{f}'(\alpha^{-1}\bar{f}(\alpha^{-1}x))u(\alpha^{-1}x)\\ u(x)\end{bmatrix}. \tag{2.7b}$$

Note that

$$DR_\mu\begin{bmatrix}h_n\\ h_{n-1}\end{bmatrix} = \begin{bmatrix}h_{n+1}\\ h_n\end{bmatrix}. \tag{2.8}$$

Thus, the renormalization group transformation R_μ accurately embodies all important facets of the recursion relations (2.2), having exactly the same fixed point structure and linearization equations.

Finding an eigenfunction of eqs. (2.7) is equivalent to searching for an initial deviation $h_0(x)$ such that $h_n(x) = \lambda^n h_0(x)$. Borrowing a technique from ref. 2, we find a family of such solutions generated by coordinate transformations. Consider the sequence of functions $f_n(x)$ defined by

$$f_n(x) = H_n \circ \bar{f} \circ H_n^{-1}(x), \tag{2.9}$$

where $H_n(x)$ is close to the identity

$$H_n(x) = x + \epsilon\sigma_n(x). \tag{2.10}$$

We demand that

$$H_{n+1}(x) = \alpha H_n(\alpha^{-1}x), \tag{2.11}$$

or, equivalently,

$$\sigma_{n+1}(x) = \alpha\sigma_n(\alpha^{-1}x), \tag{2.12}$$

which then implies that f_n satisfy the recursion

relations (2.2). Expanding to first order in ϵ, the functions then become

$$f_n(x) = \bar{f}(x) + \epsilon h_n(x), \tag{2.13}$$

with

$$h_n(x) = \sigma_n(\bar{f}(x)) - \bar{f}'(x)\sigma_n(x). \tag{2.14}$$

It is then clear that any sequence of functions $\sigma_n(x)$ satisfying eq. (2.12) and such that

$$\sigma_{n+1}(x) = \lambda\sigma_n(x) \tag{2.15}$$

produces an eigenfunction with eigenvalue λ. Eqs. (2.15) and (2.12) taken together imply that

$$\alpha\sigma_0(\alpha^{-1}x) = \lambda\sigma_0(x) \tag{2.16}$$

The smooth solutions to eq. (2.16) are simple monomials

$$\sigma_0(x) = x^p, \quad \lambda = \alpha^{1-p}, \tag{2.17}$$

with $p \geq 0$.

The only relevant eigenvalue, $\lambda = \alpha$, corresponds to a translation while the marginal eigenvalue, $\lambda = 1$, corresponds to a magnification. There will, of course, be other eigenfunctions besides those generated by coordinate changes; those must be found by directly solving eqs. (2.6).

3. Trivial solutions and perturbations

3.1. *Linear solution*

One obvious solution to eqs. (1.24), which corresponds to $|K| < 1$, is

$$\bar{f}(x) = -1 + x, \tag{3.1}$$

with α given by $-\bar{W}^{-1}$. For notational clarity, this value of α will be denoted by β. The spectrum of DR_μ around this fixed point certainly includes those eigenfunctions generated by coordinate changes, producing eigenvalues

$$\lambda = \beta^{1-p} \tag{3.2}$$

with $p \geq 1$. The $p = 0$ eigenvalue is eliminated because the corresponding eigenfunction, given by eq. (2.14), is identically zero.

There are other eigenfunctions. Let $h_n(x) = \lambda_m^n \Psi_m^\mu(x)$, μ taking on the value a or b depending on whether $h_n(x)$ solves eqs. (2.6a) or (2.6b). The resulting equations are

$$\Psi_m^a(x) = (\beta\lambda_m^{-1})\Psi_m^a(-\beta + \beta^{-1}x) + (\beta\lambda_m^{-1})^2\Psi_m^a(\beta^{-2}x), \tag{3.3a}$$

$$\Psi_m^b(x) = (\beta\lambda_m^{-1})^2\Psi_m^b(-\beta^{-1} + \beta^{-2}x) + (\beta\lambda_m^{-1})\Psi_m^b(\beta^{-1}x). \tag{3.3b}$$

If we look only in the space of polynomials, and let $\Psi_m^\mu(x)$ have as its highest power x^m, then both eqs. (3.3) yield the equation

$$1 = \beta^{1-m}\lambda_m^{-1} + (\beta^{1-m}\lambda_m^{-1})^2. \tag{3.4}$$

Eq. (3.4) has two solutions,

$$\lambda_m = \beta^{-m}, \tag{3.5}$$

which corresponds to the eigenfunctions generated by coordinate changes already considered, and

$$\lambda_m = -\beta^{2-m} \tag{3.6}$$

which is a new solution. For $m = 0$, the new solution is

$$h_n(x) = (-\beta^2)^n, \tag{3.7}$$

which is certainly relevant and in the commuting space (satisfies both eqs. (3.3)). This perturbation corresponds to a change in winding number (change in Ω in eq. (1.15)) and is responsible for the geometric convergence rate of $\delta = -\beta^2$ when $|K| < 1$.

One can solve for the $\Psi_m^{\mu}(x)$ with $m \geq 1$ but these solutions depend on μ so they are in the noncommuting space and will be neglected.

3.2. *More general solutions*

In this section we shall look at functions $f(x)$ which have the following properties:

(a) the x derivative of these functions is always nonnegative;

(b) as $x \to 0$, they are analytic functions of x^{ν_0}, i.e., as $x \to 0$

$$f(x) = a + bx^{\nu_0} + \cdots, \tag{3.8}$$

where x^{ν} means $x|x|^{\nu-1}$,

(c) the unique zero of $f(x)$ is at $x = 1$, and is a point of singularity of the form $(x-1)^{\nu_1}$, i.e., as $x \to 1$

$$f(x) = c(x-1)^{\nu_1} + \cdots. \tag{3.9}$$

Therefore, the $|K| < 1$ case considered previously corresponds to $(\nu_0, \nu_1) = (1, 1)$ and the $K = 1$ case to be considered in section 4 corresponds to $(\nu_0, \nu_1) = (3, 1)$. In general we shall write the fixed point solution with these characteristic singularities as $\bar{f}_{\nu_0, \nu_1}(x)$ and the corresponding α-values as $\alpha(\nu_0, \nu_1)$. These solutions are related, as can be seen by considering the function

$$g(x) = H_\lambda^{-1} \circ \bar{f}_{\nu_0, \nu_1} \circ H_\lambda(x), \tag{3.10}$$

where

$$H_\lambda(x) = x^\lambda. \tag{3.11}$$

Clearly, $g(x)$ has singularity signature $(\lambda\nu_0, \nu_1/\lambda)$ and is a fixed point solution of eq. (1.24) with $\alpha = (\alpha_{\nu_0, \nu_1})^{1/\lambda}$. Therefore if $\bar{f}_{\nu_0, \nu_1}(x)$ is uniquely defined by eqs. (1.24) and the boundary conditions (3.8) and (3.9) then

$$\begin{aligned} \bar{f}_{\lambda\nu_0, \nu_1/\lambda}(x) &= [\bar{f}_{\nu_0, \nu_1}(x^\lambda)]^{1/\lambda}, \\ \alpha_{\lambda\nu_0, \nu_1/\lambda} &= [\alpha_{\nu_0, \nu_1}]^{1/\lambda}, \\ \alpha_{\lambda\nu_0, \nu_1/\lambda} &= [a_{\nu_0, \nu_1}]^{1/\lambda}, \end{aligned} \tag{3.12}$$

where a_{ν_0, ν_1} is merely $\bar{f}_{\nu_0, \nu_1}(0)$.

Let us define a standard situation in which $\nu_0 = \nu_1 = \nu$ and call the quantities defined in this standard situation by the names $\bar{f}(\nu, x) = \bar{f}_{\nu, \nu}(x)$, $\alpha(\nu) = \alpha_{\nu, \nu}$ and $a(\nu) = a_{\nu, \nu}$. Then, according to eq. (3.12)

$$\begin{aligned} \bar{f}_{\nu_0, \nu_1}(x) &= [\bar{f}(\sqrt{\nu_0\nu_1}, x^{\sqrt{\nu_0/\nu_1}})]^{\sqrt{\nu_1/\nu_0}}, \\ \alpha_{\nu_0, \nu_1} &= [\alpha(\sqrt{\nu_0\nu_1})]^{\sqrt{\nu_1/\nu_0}}, \\ a_{\nu_0, \nu_1} &= [a(\sqrt{\nu_0\nu_1})]^{\sqrt{\nu_1/\nu_0}}. \end{aligned} \tag{3.13}$$

Furthermore, consideration of the inverse function

$$\bar{f}(\nu^{-1}, x) = \bar{f}^{-1}(\nu, a(\nu)x)/a(\nu) \tag{3.14}$$

yields the relations

$$\alpha(\nu) = \alpha(\nu^{-1}), \qquad a(\nu) = [a(\nu^{-1})]^{-1}. \tag{3.15}$$

These relations enable one to discuss solutions for ν close to 1. Write $\nu = e^\epsilon = 1 + \epsilon + \cdots$ and

$$\bar{f}(\nu, x) = \bar{f}(1, x) + \epsilon\bar{h}(x) + \cdots. \tag{3.16}$$

From eqs. (3.15) it follows that for small ϵ

$$\alpha(\nu) = \beta + \mathcal{O}(\epsilon^2), \tag{3.17}$$

$$a(\nu) = -e^{\epsilon a_1} + \mathcal{O}(\epsilon^3), \tag{3.18}$$

while eq. (3.14) implies that in order ϵ

$$\bar{h}(x) + \bar{h}(1-x) + a_1 = 0. \tag{3.19}$$

Since α has no first order corrections, eqs. (2.6) imply that $\bar{h}(x)$ obeys the two equations

$$\bar{h}(x) = \beta\bar{h}(-\beta + \beta^{-1}x) + \beta^2\bar{h}(\beta^{-2}x), \tag{3.20a}$$

$$\bar{h}(x)=\beta^2\bar{h}(-\beta^{-1}+\beta^{-2}x)+\beta\bar{h}(\beta^{-1}x). \qquad (3.20b)$$

According to eqs. (3.8) and (3.9) $\bar{h}(x)$ has logarithmic singularities at $x=0$ and $x=1$;

$$\bar{h}(x)\simeq -a_1+x\ln|x|+\mathcal{O}(x), \quad \text{for } x\to 0, \qquad (3.21a)$$

$$\bar{h}(x)\simeq -(1-x)\ln|1-x|+\mathcal{O}(1-x), \quad \text{for } x\to 1. \qquad (3.21b)$$

Notice that if $\bar{h}(x)$ obeys eq. (3.20b) and the symmetry condition (3.19) then the identity $1=\beta+\beta^2$ enables one to derive the fact that $\bar{h}(x)$ will automatically satisfy eq. (3.20a). Also notice that the boundary conditions (3.21) are consistent with eq. (3.19). Hence, in order to get the first order perturbation theory correct we need only solve eq. (3.20b) using boundary conditions (3.21). In the appendix we construct the solution $\bar{h}(x)$, demonstrating that to first order in ϵ the eqs. (1.24) indeed have a solution.

4. K = 1 solution

4.1. *Calculation of fixed point*

We now want to calculate the fixed point solution of eqs. (1.24) which correspond to the $K=1$ scaling found in ref. 16. First, notice that when $K=1$, all functions f_n in eq. (2.1) including the limiting fixed point $\bar{f}(x)$ have no linear or quadratic terms in their Taylor series about $x=0$. It is straightforward to show that any analytic function $\bar{f}(x)$ satisfying either of eqs. (1.24) and having $f'(0)=0$ and $f''(0)=0$ must have a Taylor series in x^3 about $x=0$. Accordingly, we will limit our search to the space of smooth functions of x^3.

Conceivably, both eqs. (1.24) must be used to determine the fixed point we seek. This turns out not to be the case with eq. (1.24a) being the only necessary one. Basically, a functional equation is an infinite dimensional equation, and the initial conditions that determine a unique solution can be the specification of the solution on some interval. Such functional equations produce a global solution given its restriction to this interval. A "serious" functional equation demands self-consistency on this determining interval and so determines a unique solution. Accordingly, we must find a self-determining equation and interval. This is most easily done by considering the function

$$g(x)=\alpha f(x), \qquad (4.1)$$

which is monotone decreasing, since $\bar{f}$ is monotone increasing, and possesses a unique fixed point. Applying eqs. (1.24) we find two new fixed point equations,

$$g(x)=\alpha g(g(x/\alpha^2)), \qquad (4.2a)$$

$$g(x)=\alpha^2 g(\alpha^{-2}g(x/\alpha)). \qquad (4.2b)$$

Since these equations are all invariant under magnification, we may also specify the constraint

$$g(0)=1, \qquad (4.3)$$

which then implies, by (4.2a),

$$g(1)=\alpha^{-1}. \qquad (4.4)$$

Thus, the fixed point lies in the interval (0, 1). Substituting $x=1$ in eq. (4.2a) yields the result

$$g^2(\alpha^{-2})=\alpha^{-2}, \qquad (4.5)$$

so that α^{-2} is either the fixed point of g or a point of period two. The fact that $g\circ g(x)$ is monotonic eliminates the possibility of a two-cycle so α^{-2} must indeed be the fixed point of $g(x)$.

Consider eq. (4.2a) for $x\in[0,1]$. $g(x/\alpha^2)$ maps $[0,1]$ onto $[\alpha^{-2},1]$ so that nowhere in the equation is $g(x)$ needed for x outside of $[0,1]$. Thus, eq. (4.2a) has a self-determining interval $[0,1]$. We expect then that eq. (4.2a) restricted to $[0,1]$ has an isolated monotonic solution in terms of real analytic functions of x^3. That is, we expect that (4.2a) alone determines the fixed point solution $g(x)$ on [0, 1]. With this restriction obtained, eq. (4.2b) then determines $g(x)$ on $[-\alpha g^{-1}(0), 0]$ through the evaluation of right-hand side terms on the obtained solution to (4.2a). Repeated use of eqs. (4.2) finally allows a global determination of g. A solution for g, then, reduces to a solution of eq. (4.2a) on [0, 1], which we now numerically obtain.

As a functional equation (4.2) is of course infinite dimensional. We seek a solution of the form

$$g(x) = 1 + \sum_{n=1}^{\infty} c_n x^{3n}. \tag{4.6}$$

Employing (4.4), we require (4.6) to satisfy

$$F[g](x) \equiv g(1)g(x) - g(g(x[g(1)]^2)) = 0 \tag{4.7}$$

for $x \in [0, 1]$. Loosely speaking, the vanishing of F at an infinite number of points x determines the infinite number of coefficients c_n. We obtain an Nth order approximation by writing

$$g_N(x) = 1 + \sum_{n=1}^{N} c_n^{(N)} x^{3n} \tag{4.8}$$

and demanding that (4.7) be satisfied at N points in [0, 1]. The accuracy of this approximation is then determined by

$$\epsilon_N = \sup_{x\in[0,1]} |F[g_N](x)|, \tag{4.9}$$

where ϵ_N depends upon the choice of the N points x_i at which F vanishes. The determination of the period-doubling fixed point [2] which utilized exactly the same method, was rather insensitive to the choice of points. In the present case, however, a careful choice is mandatory. Had we taken

$$x_i = i/N, \quad i = 1, \ldots, N, \tag{4.10}$$

then for $N > 3$ the numerical satisfaction of $F(x_i) = 0$ is difficult to obtain, and ϵ_N fails to decrease below 10^{-3} with increasing N. However, the choice

$$x_i = (i/N)^{1/3}, \quad i = 1, \ldots, N, \tag{4.11}$$

results in swift convergence of a Newton's method solution of $F(x_i) = 0$ for all N, with ϵ_N decreasing rapidly with $N (\epsilon_N \sim 10^{-8}$ with $N = 12$ on a machine with 16 significant figures). The spectrum of DF at this solution possesses a unique eigenvalue outside of the unit circle (at 8.675) and all others well inside the unit circle. We regard this as numerical evidence that (4.2a) indeed possesses an isolated fixed point of the desired character (see fig. 1). From eq. (4.4) we can determine α from our solution,

$$\alpha = -1.288575 \pm 0.000001, \tag{4.12}$$

which is in evident agreement with the iteration data.

Using an essentially identical procedure, we can find the fixed points having singularity structure $(\nu, 1)$ by considering polynomials in x^ν and points $x_i = (i/N)^{1/\nu}$. These fixed points, while not physically relevant, will prove helpful in identifying the nature of eigenfunctions in the next section.

4.2. *Perturbations about the fixed points*

When studying perturbations about the fixed point, it is important to determine the function

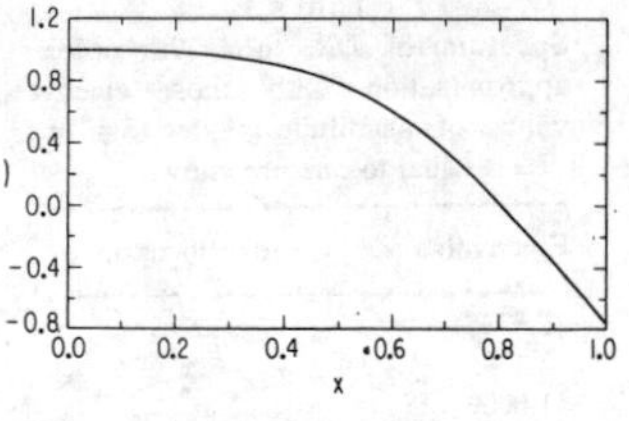

Fig. 1. The fixed point function $g(x)$ calculated in the $N = 12$ approximation over the interval [0, 1].

space of allowed perturbations. Since the fixed point solution was obtained in the space of functions of x^3, we will first consider perturbations in that space. Furthermore, we will concentrate only on the spectrum of the linearized operator DR_a (eq. (2.7a)), as the fixed point was obtained for the operator R_a (eq. (2.4a)).

To solve, numerically, the infinite dimensional eigenvalue equation

$$DR_a\begin{bmatrix} u(x) \\ v(x) \end{bmatrix} = \lambda \begin{bmatrix} u(x) \\ v(x) \end{bmatrix} \tag{4.13}$$

(with DR_a evaluated at the fixed point), we consider N-dimensional approximates in the spirit of our previous considerations about such approximations to eq. (1.24a) (see also ref. 2). Namely, we replace (4.13) by the same equation but restricted to $\{x_i\}$ $i = 1, \dots, N$ with u and v polynomials of x^3 of degree N. In determining the spectrum, we can use any N points in [0, 1], although the most natural choice is the same set used to compute the fixed point g_N. Table 1 lists the numerical results that follow from this method, with $N = 12$. Only those eigenvalues of magnitude greater than or equal to one are shown, and we now seek to identify each entry.

From the analytic considerations of section 2, we know that certain coordinate changes generate perturbations which are eigenfunctions of both DR_a and DR_b. These eigenfunctions have eigenvalues α^{1-p} with $p \geq 0$. However, if we consider only those eigenfunctions which are polynomials of x^3, then we find only the spectral elements α^{-3p} with $p \geq 0$. Thus, the entry α^0 in table 1 results from coordinate changes due to magnifications. This marginal eigenvalue is not seen in the numerical work on cycles (ref. 16), as the periodic nature of $f(x)$ in eq. (1.15) automatically rules out magnifications. The other entries can be identified if we consider the fixed points $\bar{f}_{\nu,1}$ and their spectra. When $\nu = 1$, we have the trivial solution. There, in addition to the marginal magnification, we had an eigenvalue of $-\beta^2$ corresponding to change in winding number, and eigenvalues of $-\beta$ and $-\beta^0$ corresponding to noncommuting eigenfunctions. As ν changes from 1 to 3, these three eigenvalues are connected continuously to the eigenvalues δ, $-\alpha^3$, and $-\alpha^0$ in table I. We can then conclude that the eigenfunction corresponding to δ is responsible for changes in winding number. This explains the geometric convergence rate of the Ω_n's found in ref. 16. The other two eigenfunctions, corresponding to $-\alpha^3$ and $-\alpha^0$, are not in the commuting space and can be ignored.

TABLE I

Spectrum of DR_a in a 12th order approximation. Only those eigenvalues of magnitude greater than or equal to one are shown

Eigenvalue	Identification
−2.83361	δ
2.1395	$-\alpha^3$
1.0000	α^0
−1.0000	$-\alpha^0$

We have found a unique relevant eigenvalue δ when only polynomials in x^3 are considered. We now want to consider the case

$$1 - K = \epsilon > 0, \tag{4.14}$$

in which case $f_0(x)$ has a perturbation of the form ϵx. Since ϵx destroys the leading cubic behavior, such a perturbation is singular in that f_n is no longer in the space of functions of x^3. It is easy to see that the coordinate transformation

$$H(x) = (x^3 + \epsilon x)^{1/3} \tag{4.15}$$

provides the required perturbation. Expanding (4.15)

$$H(x) = x + \frac{\epsilon}{3x} + \cdots, \tag{4.16}$$

we have $p = -1$ in (2.17), so that the relevant

eigenvalue for this fixed-point destroying perturbation is α^2 again in agreement with the cycle iteration data (i.e., this result would imply $\nu = 2x$ in the notation of ref. 16, which is true to numerical accuracy).

Of course, once we consider these more general perturbations we generate the entire spectrum α^{1-p} with $p \geq 0$. The only relevant perturbation, $p = 0$, corresponds to a translation. This eigenvalue is not observed in the cycle data [16] because all the cycles begin at $\theta = 0$.

5. Invariant circles in the planar map

We now return to our original problem, that of invariant circles of the map T (eq. (1.14)) having winding number $\bar{W}$. In particular, we are interested in how the structure of this curve depends on K. As described in the introduction, this invariant curve is seen as the limiting case of the set of fixed points of the maps $T^{(n)}$ defined in eq. (1.13). Consider b, $0 < b < 1$, to be given. For each value of K there are a series of Ω values, Ω_n, and initial points $\left[\begin{smallmatrix}\theta_0(n)\\ r_0(n)\end{smallmatrix}\right]$ such that

$$T^{(n)}\begin{bmatrix}\theta_0(n)\\ r_0(n)\end{bmatrix} = \begin{bmatrix}\theta_0(n)\\ r_0(n)\end{bmatrix} \tag{5.1}$$

(we suppress the K dependence of all these quantities). Eq. (5.1) imposes only two conditions on the three variables $r_0(n)$, $\theta_0(n)$, Ω_n, leaving one variable, say $\theta_0(n)$, undetermined. This ambiguity can be removed by defining a quantity $D_n = \mathrm{Tr}\, M_n$, where M_n is the tangent matrix of $T^{(n)}$ evaluated at $\left[\begin{smallmatrix}\theta_0(n)\\ r_0(n)\end{smallmatrix}\right]$, and then choosing the value of $\theta_0(n)$ by demanding that D_n be a minimum. Note that this criterion is equivalent to demanding that cycles start at $\theta = 0$ in the circle map with $K = 1$, as was done in ref. 16.

Letting $n \to \infty$, we empirically find three cases

$$\text{I: } D_n \to 1, \quad 0 \leq K < K_c,$$

$$\text{II: } D_n \to 0, \quad K = K_c, \tag{5.2}$$

$$\text{III: } D_n \to -\infty, \quad K > K_c,$$

where K_c depends on b. J. Greene, in his work on the standard map, found three analogous situations and concluded that a smooth continuous invariant curve with winding number $\bar{W}$ existed for $0 \leq K < K_c$, this curve was no longer smooth at $K = K_c$, and was no longer continuous for $K > K_c$ (see Mather [19]). Our numerical results suggest that these conclusions apply to the dissipative case as well. With $b = 0.5$ and using eq. (5.2) to define K_c, it appeared that for $K < K_c$ the fixed points of $T^{(n)}$ were converging onto a smooth invariant curve as $n \to \infty$. This can be seen in fig. 2a where the fixed points of $T^{(13)}$ are shown for $K = 0.3$ ($Q_{13} = 377$, $P_{13} = 233$). The density of fixed points gives a discrete approximation to the invariant distribution curve, and for $K < K_c$ this appears to be a smoothly varying function. At $K = K_c = 0.978837778 \pm 0.000000002$, the invariant curve appears to have an infinite number of cubic kinks (the phenomena of kinks in invariant

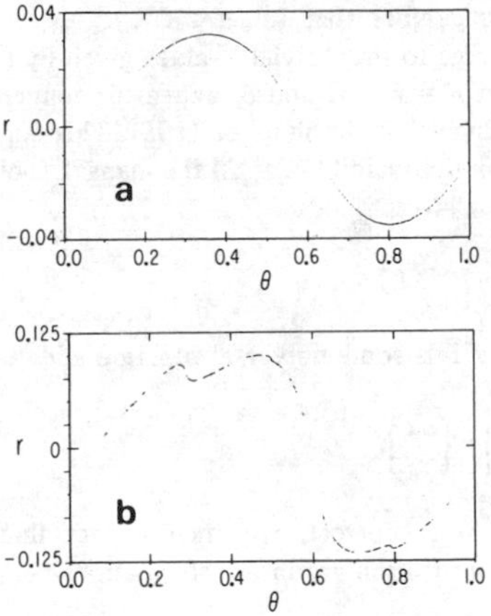

Fig. 2. The fixed points of the map $T^{(13)}$ with $b = 0.5$ for (a) $K = 0.3$; and (b) $K = K_c = 0.978837778$.

curves was first pointed out by Curry and Yorke [8]). Fig. 2b shows the fixed points of $T^{(13)}$ for $K = K_c$, and it is clear that in addition to the kinks the invariant distribution as seen from the density of points on this curve has developed singularities which appear to include both zeroes and infinities.

When $K > K_c$, the fixed points become unstable and period-double and there are numerous tangent bifurcations producing extraneous fixed-point pairs. This makes the numerical determination of Ω_n, $\theta_0(n)$, and $r_0(n)$ both impractical and ambiguous. However, we feel that one can safely assume that no continuous invariant curve with winding number $\bar{W}$ exists for $K > K_c$.

Similar to refs. 16 and 11, the convergence ratios α_n and δ_n can be defined,

$$\delta = \frac{\Omega_{n-1} - \Omega_{n-2}}{\Omega_n - \Omega_{n-1}}, \tag{5.3}$$

$$\alpha_n = \frac{\theta_{F_{n-1}}(n-1) - \theta_0(n-1) - F_{n-2}}{\theta_{F_n}(n) - \theta_0(n) - F_{n-1}}. \tag{5.4}$$

In Table II we list the values of D_n, δ_n, and α_n for $b = 0.5$ and two values of K: $K = 0.3$ and $K = K_c$. Note that for $0 \leq K < K_c$, α_n and δ_n converge to the "trivial" values given by (1.18). When $K = K_c$, α_n and δ_n appear to converge to the "nontrivial" values of (1.19). This suggests that in the vicinity of $\left[\begin{smallmatrix}\theta_0(n)\\ r_0(n)\end{smallmatrix}\right]$ the maps $T^{(n)}$ obey a scaling law

$$T^{(n)} \approx L^{-n}\bar{T}L^n, \tag{5.5}$$

where $\bar{T}$ is some universal function and

$$L\begin{bmatrix}x\\ y\end{bmatrix} = \begin{bmatrix}\alpha x\\ \gamma y\end{bmatrix}. \tag{5.6}$$

If (5.5) is correct, we then expect that the renormalization group transformations

$$R_a^{(2)}\begin{bmatrix}u\binom{x}{y}\\ v\binom{x}{y}\end{bmatrix} = \begin{bmatrix}L\circ u\circ L\circ v\circ L^{-2}\binom{x}{y}\\ u\binom{x}{y}\end{bmatrix}, \tag{5.7a}$$

TABLE II

Data for maps $T^{(n)}$

Q_n	P_n	D_n	α_n	δ_n
(a) $b = 0.5$, $K = 0.3$				
1	0	1.2000000	0.00000	0.00000
1	1	1.2000000	0.00000	0.00000
2	1	1.1600000	−2.00000	−2.00000
3	2	1.0950413	−0.41094	−3.20115
5	3	1.0286961	−1.62866	−2.50618
8	5	1.0038341	−1.54713	−2.66325
13	8	1.0001219	−1.61415	−2.60091
21	13	1.0000005	−1.59975	−2.62457
34	21	1.0000000	−1.61779	−2.61554
55	34	1.0000000	−1.61470	−2.61899
89	55	1.0000000	−1.61845	−2.61767
144	89	1.0000000	−1.61723	−2.61817
233	144	1.0000000	−1.61828	−2.61798
377	233	1.0000000	−1.61780	−2.61805
810	377	1.0000000	−1.61814	−2.61803
987	610	1.0000000	−1.61794	−2.61804
1597	987	1.0000000	−1.61807	−2.61803
2584	1597	1.0000000	−1.61801	−2.61803
4181	2584	1.0000000	−1.61805	−2.61803
(b) $b = 0.5$, $K = K_c = 0.978837778$				
1	0	0.5211622	0.00000	0.00000
1	1	0.5211622	0.00000	0.00000
2	1	0.2918766	−2.00000	−2.00000
3	1	0.0727570	−1.27893	−3.21162
5	3	0.0432264	−1.47022	−2.62633
8	5	−0.0229573	−1.29011	−2.84237
13	8	0.0063467	−1.35497	−2.79914
21	13	−0.0064649	−1.28119	−2.83307
34	21	0.0031614	−1.31881	−2.83061
55	34	−0.0019829	−1.28074	−2.83406
89	55	0.0011349	−1.30364	−2.83393
144	89	−0.0006707	−1.28220	−2.83380
233	144	0.0003922	−1.29659	−2.83382
377	233	−0.0002294	−1.28413	−2.83365
610	377	0.0001355	−1.29306	−2.83368
987	610	−0.0000775	−1.28569	−2.83361
1597	987	0.0000486	−1.29117	−2.83363
2584	1597	−0.0000232	−1.28677	−2.83361
4181	2584	0.0000224	−1.29010	−2.83361

$$R_b^{(2)}\begin{bmatrix}u\binom{x}{y}\\ v\binom{x}{y}\end{bmatrix} = \begin{bmatrix}L^2\circ v\circ L^{-1}\circ u\circ L^{-1}\binom{x}{y}\\ u\binom{x}{y}\end{bmatrix}, \tag{5.7b}$$

will have a fixed point, namely $\begin{bmatrix}\bar{T}\binom{x}{y}\\ \bar{T}\binom{x}{y}\end{bmatrix}$.

It is easy to construct fixed points to $R_\mu^{(2)}$

given the results for the circle map. Corresponding to the $0 \leq K < K_c$ case, we have

$$\bar{T}\begin{bmatrix} x \\ y \end{bmatrix} = \begin{bmatrix} -1+x \\ 0 \end{bmatrix}, \tag{5.8}$$

with $\alpha = \beta$ and γ arbitrary. The $K = K_c$ solution is represented by the fixed point

$$\bar{T}\begin{bmatrix} x \\ y \end{bmatrix} = \begin{bmatrix} \bar{f}((x^3+y)^{1/3}) \\ 0 \end{bmatrix}, \tag{5.9}$$

with α and $\bar{f}$ being the $K = 1$ circle map quantities and $\gamma = \alpha^3$. This fixed point is similar to the generalization of the one-dimensional period-doubling fixed point to planar maps [20]. The fixed point (5.9) will have two important relevant eigenvalues; δ, corresponding to variations in Ω and α^2 corresponding to decreasing K. There will of course be other relevant eigenvalues, generated by coordinate transformations, but which can be eliminated by appropriate normalizations. But perhaps the major physics of deviations from this incipient breakup will be caught by the two relevant directions of the one-dimensional map.

6. Conclusions

We have presented a theory that predicts certain scaling laws at the onset of chaotic behavior in a quasiperiodic dynamical system. While this theory applies only to the irrational frequency ratio $\bar{W}$, it can be generalized in a straightforward way to any irrational winding number with a repeating continued fraction expansion. The crucial question is can this scenario be observed experimentally? The theory requires that the winding number be held constant as the nonlinear coupling is increased. At present most experiments have only one adjustable parameter, so the winding number cannot be controlled independently from the nonlinear coupling, preventing any comparison with the present theory. It is our hope that in the future, experiments with two adjustable parameters will display this "route to chaos."

Acknowledgements

We are particularly indebted to D. Rand, S. Ostlund, J. Sethna, and E. Siggia for communicating their results prior to publication and freely sharing their ideas and insights. We have also had helpful discussions with A. Zisook, M. Widom, J. Greene, and R. MacKay. One of us (SJS) would like to acknowledge his NSF and McCormick Fellowships.

Appendix A

Exact first order solution

To avoid the difficulties produced by the inequivalence of eqs. (2.2a) and (2.2b), we choose initial conditions for these recursion relations which will guarantee the necessary commutation relations. As described by eq. (2.5), choose

$$\begin{aligned} f_0(x) &= -1 + x + \epsilon\beta P(x/\beta), \\ f_1(x) &= -1 + x + \epsilon\beta^2 P(x/\beta^2), \end{aligned} \tag{A.1}$$

where $P(x+1) = P(x)$. Equivalently, take

$$\begin{aligned} h_0(x) &= \beta P(x/\beta), \\ h_1(x) &= \beta^2 P(x/\beta^2). \end{aligned} \tag{A.2}$$

To get a solution with $\nu_0 = \nu_1 = 1 + \epsilon + \cdots$ choose $P(z)$ to have a behavior

$$\begin{aligned} P(z) &= c + z \ln z + \cdots, \qquad z \to 0, \\ P(z) &= (z - \beta^{-1}) \ln|z - \beta^{-1}|, \quad z \to \beta^{-1}. \end{aligned} \tag{A.3}$$

One choice which will certainly do this is to

take

$$P(x/\beta) = Q\left(\frac{x}{\beta}\right) + Q\left(\frac{x-1}{\beta}\right) - Q(\beta^{-1}), \tag{A.4}$$

with

$$Q(z) = \frac{\sin 2\pi z}{2\pi} \ln\left|\frac{\sin \pi z}{\pi}\right|. \tag{A.5}$$

Given any periodic function $P(z)$ the solution to eqs. (2.6) is direct and relatively simple. Take as an Ansatz the proposed solution

$$h_n(x) = \beta^{n+1} \sum_{y \in S_n} P\left(\frac{x-y}{\beta^{n+1}}\right). \tag{A.6}$$

For $n = 0$ and $n = 1$ we have the set S_n containing but one element, 0,

$$S_0 = S_1 = \{0\}. \tag{A.7}$$

Now substitute eq. (A.6) into eq. (2.6b) to find

$$\beta^{n+2} \sum_{y \in S_{n+1}} P\left(\frac{x-y}{\beta^{n+2}}\right) = \beta^{n+2}\left[\sum_{y \in S_n} P\left(\frac{x-\beta y}{\beta^{n+2}}\right) + \sum_{y \in S_{n-1}} P\left(\frac{x-\beta^2 y-\beta}{\beta^{n+2}}\right)\right]. \tag{A.8}$$

Eq. (A.6) is then a solution if the set S_{n+1} is properly chosen. Here S_{n+1} contains two types of elements. If $y \in S_n$, then $(\beta y) \in S_{n+1}$. If $y \in S_{n-1}$, then $(\beta^2 y + \beta) \in S_{n+1}$. To represent this result in a convenient notation, say that the sets A and B are related by

$$B = uA + v, \tag{A.9}$$

if for each y in A there is one and only one corresponding y' in B, with $y' = uy + v$. In this notation, then, the set S_n is defined recursively via

$$S_{n+1} = (\beta S_n) \cup (\beta^2 S_{n-1} + \beta) \tag{A.10}$$

and the initial condition (A.7).

The properties of the set S_n are crucial to our analysis since if $y \in S_n$ then $h_n(x)$ has singularities at $x = y$ and at $x = y + \beta^n$. The singularities will, of course, determine the solution.

Notice that each S_n includes all the elements of S_{n-1}. For this reason we can define S_n as the union of S_{n-1} and the new elements of S_n by writing

$$S_n = S_{n-1} \cup T_n. \tag{A.11}$$

The elements in T_n increase in magnitude as n increases. Hence lower n tends to give singularities for smaller x-values.

Note also that the second term on the right-hand side of eq. (A.4) may be interpreted as a $y = 1$ term. If we started from $S'_0 = S'_1 = \{1\}$ and generated successive S'_n by using recursion relation (A.10) then the new sets S'_n thereby generated would be identical to the sets S_n, except for one element. S_n always includes 0; S'_n always includes a corresponding 1, instead of the 0. Therefore, except for an x-independent term, which we shall deal with later on, the solution to the recursion problem is

$$h_n(x) = \beta^{n+1}\left[\sum_{y \in S_n} Q\left(\frac{x-y}{\beta^{n+1}}\right) + \sum_{y \in S'_n} Q\left(\frac{x-y}{\beta^{n+1}}\right)\right]. \tag{A.12}$$

The set S_n is relatively easy to characterize. From the definitions (A.10) and (A.11) one can show that $y \in T_n$ if and only if y can be written as

$$y = \sum_{k=1}^{n-1} m_k \beta^k, \tag{A.13}$$

with each m_k chosen to be either zero or one and to satisfy

$$m_{n-1} = 1, \quad m_k m_{k-1} = 0, \tag{A.14}$$

so that no two successive values of m_k are unity. The elements of S_n are distributed between the extreme values $-\beta^n+1$ and $-\beta^{n-1}+\beta$ for n odd and the extreme values $-\beta^n+\beta$ and $-\beta^{n-1}+1$ for n even. For each n the number of elements in S_n is F_n, where F_n is the Fibonacci number defined by

$$F_n=\frac{(-\beta)^{n+1}-\beta^{-(n+1)}}{-(\beta+\beta^{-1})}. \tag{A.15}$$

When the set S_n is arranged in order of size the spacing between successive elements of S_n is either $-\beta$ or β^2. Hence roughly speaking, S_n contains elements spaced uniformly between $-\beta^{n-1}$ and β^n with an average number of elements per unit length being

$$\rho=|\beta+\beta^{-1}|^{-1}. \tag{A.16}$$

In this rough representation

$$\sum_{y\in S_n}H(y)=(-1)^{n+1}\int_{-\beta^{n-1}}^{-\beta^n}\mathrm{d}y\rho H(y). \tag{A.17}$$

Eq. (A.17) is useful for estimating the rate of convergence of sums over S_n.

It is possible to prove one more result about S_n and S_n' which is crucial to the further analysis, namely

$$S_n\equiv 1-S_n' \mod \beta^{n+1}. \tag{A.18}$$

That is to say that for each element y of S_n there is a corresponding element y' of S_n' with $y=1-y'$ (mod β^{n+1}). Consequently, eq. (A.12) reduces to

$$h_n(x)=\beta^{n+1}\sum_{y\in S_n}\left[Q\left(\frac{x-y}{\beta^{n+1}}\right)+Q\left(\frac{x+y-1}{\beta^{n+1}}\right)\right]. \tag{A.19}$$

The most useful way of writing this result is to take a set

$$U_n=S_n\cup(1-S_n) \tag{A.20}$$

and to write

$$h_n(x)=\beta^{n+1}\sum_{y\in U_n}{}'\ Q\left(\frac{X-Y}{\beta^{n+1}}\right). \tag{A.21}$$

Here U_n is characterized by a recursion relation identical to eq. (A.10),

$$U_{n+1}=(\beta U_n)\cup(\beta^2U_{n-1}+\beta),\qquad U_0=U_1=\{0,1\}. \tag{A.22}$$

Each of the elements (save 0) in S_∞ appear exactly twice in U_∞. The prime in eq. (A.21) is a reminder to calculate the sum in order of increasing size of $|y|$. This reminder is necessary to ensure that conditionally convergent sums like

$$\sum_{y\in U_\infty}{}'\ \frac{1}{y-\frac{1}{2}}=0 \tag{A.23}$$

converge and converge to zero.

One could analyze the properties of the sum (A.21) quite directly. However, we are only interested in the singularities in $h_n(x)$ generated by the singularities in $Q(z)$, which are for small z like $z\ln|z|$. Hence as an ansatz, let us replace the sum in (A.21) by the form which arises if we replace $Q(z)$ by its small z form, namely

$$h_n(x)=\sum_{y\in U_n}\{(x-y)\ln|x-y|+\cdots\}. \tag{A.24}$$

The ... includes analytic terms in x which are added to ensure that the sum converges in the small y region. Clearly, the sum does not converge as it stands. In fact, it is hard to analyze directly. Consequently, we instead analyze the behavior of the x-derivative to eq. (A.24). We try the ansatz

$$\dot{h}(x)=\sum_{y\in U_\infty}{}'(\ln|x-y|-\ln|x_0-y|)+A. \tag{A.25}$$

Because of the results (A.17) and (A.24) we are

sure that the definition (A.25) gives a convergent expression for a quantity which we hope to identify as the x-derivative of $\bar{h}(x)$. There are two undetermined parameters in eq. (A.25), A and x_0.

We now set out to determine these parameters. By differentiating eq. (3.20b) we find

$$\dot{\bar{h}}(x) = \dot{\bar{h}}\left(\frac{x}{\beta}\right) + \dot{\bar{h}}\left(\frac{x-\beta}{\beta^2}\right). \tag{A.26}$$

Let $x \to 0$, then the only term which contributes to $\bar{h}(x) - \bar{h}(x/\beta)$ is the $y = 0$ term in eq. (A.25) which gives this difference as $\ln|x| - \ln|x/\beta| = \ln|\beta|$. Hence

$$\dot{\bar{h}}(-\beta^{-1}) = \ln|\beta|. \tag{A.27}$$

For this reason, we can pick the parameters in eq. (A.25) to be

$$x_0 = -\beta^{-1}, \quad A = \ln|\beta|. \tag{A.28}$$

Now substitute the result (A.25) into eq. (A.26) to obtain

$$\begin{aligned} -A + \sum_{y\in U_\infty}{}' (\ln|x-y| - \ln|x_0 - y|) &= \sum_{y\in U_\infty}{}' (\ln|x - \beta y| - \ln|\beta x_0 - \beta y|) \\ &+ \sum_{y\in U_\infty}{}' (\ln|x - \beta^2 y - \beta| \\ &- \ln|\beta^2 x_0 + \beta - \beta^2 y - \beta|). \end{aligned} \tag{A.29}$$

Use the definiton (A.22) of U_∞ to rewrite the last sum as

$$\sum_{y\in U_\infty}{}' H(\beta^2 y + \beta) = \sum_{y\in U_\infty}{}' H(y) - \sum_{y\in U_\infty}{}' H(\beta y). \tag{A.30}$$

After a rearrangement, we find

$$\begin{aligned} A = \sum_{y\in U_\infty} (&\ln|\beta^2 x_0 - y + \beta| - \ln|x_0 - y| \\ &- \ln|\beta^2 x_0 + \beta - \beta y| + \ln|\beta x_0 - \beta y|). \end{aligned} \tag{A.31}$$

As x_0 approaches $-\beta^{-1}$ the only term left in the sum is the $y = 0$ term which leaves

$$\begin{aligned} A &= \ln|\beta^2 x_0 + \beta| - \ln|x_0| - \ln|\beta^2 x_0 + \beta| - \ln|\beta x_0| \\ &= \ln|\beta|. \end{aligned} \tag{A.32}$$

Hence, eq. (A.25) and (A.28) together provide a solution

$$\dot{\bar{h}}(x) = \sum_{y\in U_\infty}{}' (\ln|x-y| - \ln|\beta^{-1} + y|) + \ln|\beta|. \tag{A.33}$$

Eq. (A.33) may then be integrated to yield our formula for the first correction to the trivial fixed point

$$\begin{aligned} \bar{h}(x) = (x-1)\ln|\beta| + \sum_{y\in U_\infty}{}' (&(x-y)\ln|x-y| \\ &- (1-y)\ln|1-y| - (x-1)\ln|\beta^{-1} + y|). \end{aligned} \tag{A.34}$$

Here we used $\bar{h}(1) = 0$ to set a constant of integration.

To verify this solution we need examine two points: Does $\bar{h}(x)$ equally well satisfy eqs. (3.20a) and (3.20b)? Does it satisfy either equation? The second question could have a negative answer if the integration of eq. (A.26) to yield eq. (3.20b) led to a nonzero constant of integration. To check this point, we need only check to see that eq. (3.20b) is satisfied for one value of x. Hence, set $x = 1$, and notice that the condition is $\bar{h}(\beta^{-1}) = 0$. A calculation of the sum in eq. (A.34) with 5186 terms and an estimate of the remainder gives $\bar{h}(\beta^{-1}) = 2 \times 10^{-7}$, which is consistent with zero. Thus, eq. (3.20b) is indeed satisfied. Next, eq. (3.20a) will also be satisfied if eq. (3.19) holds. This is the statement that $\bar{h}(x) + \bar{h}(1-x)$ is x independent. However, since the sum in eq. (A.34) remains invariant under the replacement of y by $1-y$, we find

$$\bar{h}(x)+\bar{h}(1-x)=-\ln|\beta| - \sum_{y\in U_\infty}{}'\,(y\ln|y|+(1-y)\ln|1-y| - \ln|\beta^{-1}+y|), \tag{A.35}$$

which then verifies eq. (3.19).

Note added in proof

M. Nauenberg (private communication) has constructed an elegant analysis which indicates when the solutions to equations (1.24a) and (1.24b) are identical.

References

[1] For a recent review see J.-P. Eckmann, Rev. Mod. Phys. **53** (1981) 643.
[2] M.J. Feigenbaum, J. Stat. Phys. 19 (1978) 25; 21 (1979) 669.
[3] M.J. Feigenbaum, Comm. Math. Phys. 77 (1980) 65.
[4] P. Manneville and Y. Pomeau, Physica D 1 (1980) 219.
[5] Y. Pomeau and P. Manneville, Comm. Math. Phys. 77 (1980) 189.
[6] See, for example, H.L. Swinney and J.P. Gollub, Physics Today 31 (1978) 41.
[7] D. Ruelle and F. Takens, Commun. Math. Phys. 20 (1971) 167.
[8] J.H. Curry and J.A. Yorke, Lecture Notes in Mathematics, vol. 668 (Springer, Berlin, 1978).
[9] D.G. Aronson, M.A. Chory, G.R. Hall and R.P. McGehee, Comm. Math. Phys. 83 (1982) 303.
[10] J.M. Greene, J. Math. Phys. 9 (1968) 760; 20 (1979) 1183.
[11] Scott J. Shenker and L.P. Kadanoff, J. Stat. Phys. 27 (1982) 631.
[12] L.P. Kadanoff, Phys. Rev. Lett. 47 (1981) 1641.
[13] L.P. Kadanoff, Proceedings of the 9th Midwestern Solid State Theory Seminar (Argonne), in press.
[14] R. MacKay, Princeton Preprint (1982).
[15] I. Niven, Irrational Numbers (Mathematical Ass. Amer., Wisconsin, 1982).
[16] Scott J. Shenker, Physica D (1982), to be published.
[17] D. Rand, S. Ostlund, J. Sethna and E. Siggia, ITP Preprint 82-32, April 1982.
[18] I. Tsuda, Prog. Theor. Phys. 66 (1981).
[19] J.N. Mather, Topology, to appear.
[20] P. Collet, J.-P. Eckmann and H. Koch, J. Stat. Phys. 25 (1981) 1.

PHYSICAL REVIEW A VOLUME 14, NUMBER 6 DECEMBER 1976

Kolmogorov entropy and numerical experiments

Giancarlo Benettin
Istituto di Fisica dell'Università, and Gruppo Nazionale di Struttura della Materia del Consiglio Nazionale delle Ricerche, Padova, Italy

Luigi Galgani
Istituto di Matematica and Istituto di Fisica dell'Università, Milano, Italy

Jean-Marie Strelcyn
Département de Máthematiques Centre Scientifique et Polytechnique, Université Paris-Nord, Paris, France
(Received 8 June 1976)

Numerical investigations of dynamical systems allow one to give estimates of the rate of divergence of nearby trajectories, by means of a quantity which is usually assumed to be related to the Kolmogorov (or metric) entropy. In this paper it is shown first, on the basis of mathematical results of Oseledec and Piesin, how such a relation can be made precise. Then, as an example, a numerical study of the Kolmogorov entropy for the Hénon-Heiles model is reported.

I. INTRODUCTION

In recent years many attempts have been made in order to investigate the so-called stochasticity properties of dynamical systems, in particular Hamiltonian systems, by numerical computations. However, stochasticity is generally defined and tested in a rather qualitative way, and the connection between the empirical parameters introduced to describe it and rigorous theoretical concepts is far from being clear.

One of the most powerful empirical tools has always been the study of the divergence of nearby trajectories in phase space. Such a method allows one to define a quantitative parameter (the "entropylike quantity"), which is supposed to be strictly related to the Kolmogorov (or metric) entropy for associated flow.[1-4]

The aim of the present paper is to analyze this entropylike quantity, deriving its precise connection with the metric entropy, and to explain certain properties observed in the numerical computations. This connection turns out to be particularly simple for the case of Hamiltonian systems with two degrees of freedom. As an example, for one of them, the well-known Hénon-Heiles model,[5-7] we compute the entropylike quantity and test its properties; moreover, we are able to draw a tentative curve for the entropy itself as a function of energy.

In Sec. II we collect first the necessary mathematical tools, i.e., the results of Oseledec[8] and the fundamental results of Piesin.[9,10] [We are very grateful to Dr. A. B. Katok (Moscow) for the communication of the latter results.] We then recall the definition of the entropylike quantity and explain its empirically observed properties. The numerical example for the Hénon-Heiles model is treated in Sec. III.

This paper has been written, as far as possible, in a self-contained way; however, a certain familiarity with ergodic theory and in particular with entropy is necessary (see, for example, Refs. 11 and 12). The elementary notions on differentiable manifolds used here can be found, for example, in Ref. 13.

II. THEORETICAL ANALYSIS OF THE NUMERICAL COMPUTATIONS

A. Mathematical preliminaries: Lyapunov characteristic numbers and entropy

Let us give first the main definitions and fix the notation.

Let M be a differentiable, n-dimensional, compact, connected Riemannian manifold of class C^2. If $x \in M$, the tangent space to M at x and the norm induced in it by the Riemannian metric on M will be denoted by E_x and $\| \cdots \|$, respectively. Let X be a vector field of class C^2 defined on M and $\{T^t\}$ the flow induced by X, i.e., for any t let $T^t x = x(t)$, where $\{x(t)\}$ is an integral curve of the vector field X such that $x(0) = x$. The tangent mapping of E_x onto $E_{T^t x}$ induced by the diffeomorphism T^t will be denoted by dT^t_x. It will also be assumed that the flow $\{T^t\}$ preserves a normalized measure μ which is equivalent to the Lebesgue measure on M and whose density in local coordinates is of class C^2, i.e., that the flow $\{T^t\}$ admits an integral invariant of order n and class C^2.

The following theorems A and B, which partially summarize theorems 2 and 4 of Ref. 8, are the basis for all further considerations of the present paper.

Theorem A. There exists a measurable set $M_1 \subset M$, $\mu(M_1)=1$, such that for every $x \in M_1$ the following properties hold:

(a) For every vector $e \in E_x$, $e \neq 0$, the limit

$$\lim_{t\to\infty}(1/t)\ln\|dT_x^t(e)\| = \lambda(x,e)$$

exists and is finite. M being compact, such a limit is independent of the Riemannian metric chosen on M.

(b) There exists a basis $(e_1,\ldots,e_n)$ of E_x such that

$$\sum_{i=1}^{n}\lambda(x,e_i)=\inf_{\Pi}\sum_{i=1}^{n}\lambda(x,\tilde{e}_i),$$

where

$$\Pi=\{(\tilde{e}_1,\ldots,\tilde{e}_n):\ (\tilde{e}_1,\ldots,\tilde{e}_n)\text{ is a basis of }E_x\}.$$

As e varies in E_x, $\lambda(x,e)$ takes only the values $\{\lambda(x,e_i)\}_{1\leq i\leq n}$.

The number $\lambda(x,e)$ is called the Lyapunov characteristic number of the vector $e \in E_x$, and the numbers $\lambda(x,e_i)$, which depend only on the flow $\{T^t\}$ and the point x, are called the Lyapunov characteristic numbers of the flow $\{T^t\}$ at x. We will use the notation $\lambda(x,e_i)=\lambda_i(x)$, $1\leq i\leq n$, and suppose $\lambda_1(x)\leq\lambda_2(x)\leq\cdots\leq\lambda_n(x)$. Then the functions $\lambda_1,\ldots,\lambda_n$ are defined on M and can be shown to be measurable with respect to the Lebesgue measure on M.

We will also denote $\lambda_n(x)$ by $\lambda_{\max}(x)$. This is the quantity of main interest for the present paper, since it will be shown below to correspond to the entropylike quantity.

Given $x\in M_1$, the numbers $\{\lambda_i(x)\}_{1\leq i\leq n}$ are not all necessarily distinct. Denote by $\{\nu_j(x)\}_{1\leq j\leq s(x)}$ the distinct values taken by the numbers $\{\lambda_i(x)\}_{1\leq i\leq n}$, and by $k_j(x)$ the multiplicity of $\nu_j(x)$. Also, let $\nu_i<\nu_j$ if $i<j$.

Theorem B. For every $x\in M_1$ there exist linear subspaces $H_1,\ldots,H_s$ of E_x, with $s=s(x)$, such that (a) $E_x=H_1\oplus\cdots\oplus H_s$ ($\oplus$ denotes, as usual, direct sum); (b) $\dim H_j=k_j(x)$, $1\leq j\leq s$; (c) if $e\neq 0$, $e\in H_j$, then

$$\lim_{t\to\pm\infty}(1/|t|)\ln\|dT_x^t(e)\|=\pm\nu_j(x),\quad 1\leq j\leq s;$$

(d) if $e\neq 0$, $e\in H_1\oplus\cdots\oplus H_j$ but $e\notin H_1\oplus\cdots\oplus H_{j-1}$, then $\lambda(x,e)=\nu_j(x)$, $1\leq j\leq s$.

From this theorem it immediately follows that if one chooses in E_x a vector e "at random" then one may expect to find $\lambda(x,e)=\nu_s(x)\equiv\lambda_{\max}(x)$. Indeed, the vectors e, such that $\lambda(x,e)<\nu_s(x)$, constitute a subspace of E_x which has positive codimension and thus vanishing Lebesgue measure.

We recall now the fundamental theorem of Piesin connecting Lyapunov characteristic numbers with metric entropy:

Theorem C. Under the conditions given above, one has

$$h_\mu\{T^t\}=\int_M\Big[\sum_{\lambda_i(x)>0}\lambda_i(x)\Big]\,d\mu(x)$$

where $h_\mu(\{T^t\})$ is the entropy of the flow $\{T^t\}$ with respect to the invariant measure μ.

The following remark is an immediate consequence of the very definition of the Lyapunov characteristic numbers: Suppose the vector field X does not have singular points, i.e., for every $x\in M$ it is $X(x)\neq 0$. Then for any $x\in M_1$ one has $\lambda(x,X(x))=0$. As a consequence, at least one of the Lyapunov characteristic numbers of the flow $\{T^t\}$ vanishes for every $x\in M_1$, so that one has $\lambda_{\max}(x)\geq 0$ for $x\in M_1$.

From the above remark and Piesin's theorem one thus gets the inequality

$$\int_M\lambda_{\max}(x)\,d\mu(x)\leq h_\mu(\{T^t\})\leq(n-1)\int_M\lambda_{\max}(x)\,d\mu(x). \tag{1}$$

Let us now consider the particular case of a Hamiltonian flow of N degrees of freedom. We assume we are given a function $H(q,p)$ of class C^2 defined on an open subset U of R^{2N}, with $N\geq 1$ and $q=(q_1,\ldots,q_N)$, $p=(p_1,\ldots,p_N)$, such that $\operatorname{grad}H(q,p)\neq 0$ for $(q,p)\in U$. The Hamiltonian flow $\{T^t\}$ on U is thus defined by the Hamilton equations of motion with Hamiltonian H.

The energy surfaces $\Omega_E=\{(q,p)\in U{:}H(q,p)=E\}$ are not necessarily compact. We will suppose that there exists an interval of energies E such that Ω_E contains a $(2N-1)$-dimensional submanifold Γ_E which is compact, connected, and $\{T^t\}$ invariant.

Denote by $\{T_E^t\}$ the restriction of the flow $\{T^t\}$ to Γ_E. It is well known that the Lebesgue measure on R^{2N} induces on Γ_E an invariant measure μ_L (strictly positive and of class C^2 in local coordinates) which can be supposed to be normalized by $\mu_L(\Gamma_E)=1$. Under such conditions inequality (1) becomes

$$\int_{\Gamma_E}\lambda_{\max}(q,p)\,d\mu_L(q,p)\leq h_{\mu_L}(\{T_E^t\})\leq 2(N-1)\int_{\Gamma_E}\lambda_{\max}(q,p)\,d\mu_L(q,p). \tag{2}$$

Such an inequality will be the basis for the discussion of the connections between the entropylike quantity and entropy. A more stringent inequality, which holds under rather particular assumptions and will be of interest in Sec. III B, is deduced in Sec. II D.

In closing this subsection, we may remark that the results of Oseledec and Piesin recalled here for a flow $\{T^t\}$ are easily formulated also in the case of a diffeomorphism T.

B. Entropylike quantity

Here, we give the description of the quantities $k_n(\tau, x, d)$ and $k(\tau, x, d)$ which were defined in Ref. 4 and also in the cited works by Chirikov and coworkers.[1-3]

We make reference here to the case of a Hamiltonian flow $\{T_E^t\}$ on Γ_E, but the cases of a flow or a diffeomorphism on a smooth manifold would be treated in a similar way.

Given $\tau > 0$, consider the diffeomorphism T^τ of Γ_E onto itself. Fix a point $x \in \Gamma_E$ and another point $y \in \Gamma_E$ very close to x, not on the same trajectory. Denote by d the segment relaying x to y and by $|d|$ its length.

Let $x_1 = T^\tau x$ and $|d_1| = \|T^\tau x - T^\tau y\|$, where $\|\cdots\|$ is the Euclidean norm. Denote by y_1 the unique point of the half-line issuing from x_1 and containing $T^\tau y$ such that $\|y_1 - x_1\| = |d|$. Then one can iterate such a procedure and define $x_2 = T^\tau x_1 = T^{2\tau} x$, $|d_2| = \|T^\tau x_1 - T^\tau y_1\|$; by y_2 we will denote the unique point of the half-line issuing from $x_2 = T^\tau x_1$ and containing $T^\tau y_1$, such that $\|y_2 - x_2\| = |d|$, and so on (see Fig. 1).

One thus gets a sequence of positive numbers $\{|d_i|\}$, $i = 1, 2, \ldots$, and one can so define the quantity

$$k_n(\tau, x, d) = \frac{1}{n\tau} \sum_{i=1}^{n} \ln \frac{|d_i|}{|d|} .$$

From the numerical computations described in Ref. 4 it appears that for the model there considered, if $|d|$ is not too big, one finds (i) the limit $\lim_{n\to\infty} k_n(\tau, x, d) = k(\tau, x, d)$ seems to exist; (ii) $k(\tau, x, d)$ is independent of τ; and (iii) $k(\tau, x, d)$ is independent of d.

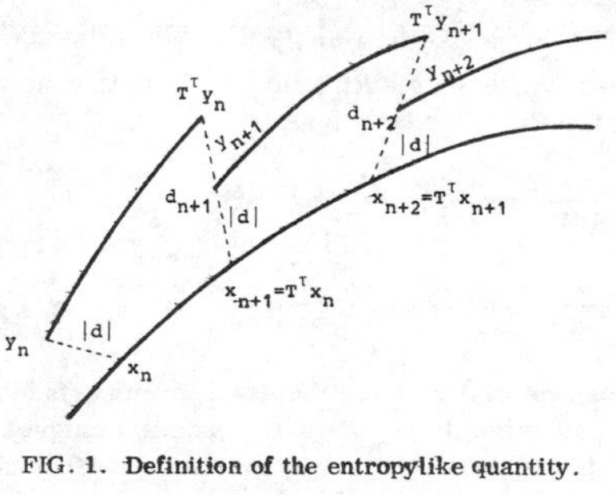

FIG. 1. Definition of the entropylike quantity.

In general, all of the numerical computations on a large class of Hamiltonian systems[1-4,7,14] show that for those systems, given an energy E, Γ_E decomposes, roughly speaking, into two regions which are $\{T_E^t\}$ invariant. One of those regions, which is called the ordered region (or sometimes the stable region), is characterized by the property that the trajectories of the flow $\{T_E^t\}$ have a behavior which seems similar to quasiperiodic motions. The other region, which is called the stochastic one, is instead characterized by a very irregular behavior of the trajectories of the flow, which seems similar to the behavior of the Anosov flows (see Ref. 1, and particularly Refs. 7 and 14). Such characterizations are not rigorous, but have nevertheless an undeniable heuristic value. It appears moreover that (iv) $k(\tau, x, d) = 0$ if x is taken in the ordered region of Γ_E; and (v) $k(\tau, x, d)$ is independent of the choice of x if x is taken in the stochastic region of Γ_E. In such a case, $k(\tau, x, d)$ is always positive. Property (v) allows one to speak simply of the quantity $k = k(E)$ instead of $k(\tau, x, d)$, with x belonging to the stochastic region of Γ_E. The number $k(E)$ was considered as an entropylike quantity. In Sec. II C such a statement will be given a precise meaning.

C. Identification of the entropylike quantity

We are now able to explain (i)–(iii) of Sec. II B and give some heuristic remarks about (iv) and (v). This will be based on the identification of $k(\tau, x, d)$ with the Lyapunov characteristic number $\lambda(x, e)$, where $e = y - x$.

(a) It is clear that, τ being fixed and $|d|$ sufficiently small, one has $|d_1|/|d| \cong \|dT_x^\tau(e)\|/\|e\|$, and moreover

$$\frac{|d_2|}{|d|} \cong \frac{1}{\|e\|} \left\| dT^\tau_{T^\tau_x}\left(\frac{\|e\|}{\|dT_x^\tau(e)\|} dT_x^\tau(e)\right)\right\| = \frac{\|dT_x^{2\tau}(e)\|}{\|dT_x^\tau(e)\|},$$

where the property $dT_x^{t+s} = dT^t_{T^s_x} dT_x^s$, following from $T^{t+s} = T^t T^s$, has been used. In general one has

$$\frac{|d_i|}{|d|} = \frac{\|dT_x^{i\tau}(e)\|}{\|dT_x^{(i-1)\tau}(e)\|}, \quad i = 1, 2, 3, \ldots .$$

As a consequence,

$$k_n(\tau, x, d) = \frac{1}{n\tau} \sum_{i=1}^{n} \ln \frac{|di|}{|d|}$$

$$\cong \frac{1}{n\tau} \sum_{i=1}^{n} \ln \frac{\|dT_x^{i\tau}(e)\|}{\|dT_x^{(i-1)\tau}(e)\|}$$

$$= \frac{1}{n\tau} \ln \frac{\|dT_x^{n\tau}(e)\|}{\|(e)\|},$$

and this allows, by note (a) of Theorem A, to identify $k(\tau, x, d) = \lim_{n\to\infty} k_n(\tau, x, d)$ with the Lyapunov characteristic number $\lambda(x, e)$. This identification is made with an error which tends to zero with $|d|$.

(b) Property (ii) of Sec. II B is then an immediate consequence of the very definition of the Lyapunov characteristic numbers.

(c) In (a) above $k(\tau, x, d)$ has been identified with $\lambda(x, e)$. Then, by the remark following Theorem B, one sees that if one chooses d at random, one may expect to find $k(\tau, x, d) = \lambda_{\max}(x)$.

(d) While properties (i)–(iii) of Sec. II B are consequences of the Oseledec theorem on the Lyapunov characteristic numbers and of the very definition of $k(\tau, x, d)$, properties (iv) and (v) instead have to be considered as empirical, and are far from being well understood. The following considerations are then, by necessity, of a heuristic character:

For the ordered region, the vanishing of the Lyapunov numbers should be related to the fact that its numerically computed trajectories are apparently of quasiperiodic type; this much we can say about (iv).

Property (v) supports instead the idea that the stochastic region either is ergodic or contains an ergodic component the measure of which is close to the measure of the stochastic region itself. In this connection we recall that an example of a flow, preserving Lebesgue measure, has been found[15] which is not ergodic, while all of its Lyapunov characteristic numbers but one are nonvanishing almost everywhere.

D. Connection with entropy

For the connection between entropy and the entropylike quantity, the only rigorous relation that can be given is essentially inequality (2), which follows from Piesin's theorem (Theorem C here) and the assumption that the vector field considered does not have singular points.

However, use may be made of the heuristic picture described in Secs. II B and II C concerning those particular models for which stochastic and ordered regions appear to exist. Indeed one may then assume that $\lambda_{\max}(x)$ vanishes in the ordered region of Γ_E and is equal to a positive constant, $k(E)$, in the stochastic region S_E of Γ_E, so that inequality (2) gives

$$\mu_L(S_E)k(E) \le h_{\mu_L}(\{T_E^t\}) \le 2(N-1)\mu_L(S_E)k(E)\,. \quad (3)$$

A more stringent relation, to be used in Sec. III B, 7, can, however, be given for the Hamiltonian systems such that $H(q,p) = H(q,-p)$, a condition which is satisfied in all models to which reference has been made in this paper. Indeed, defining $\psi : \Gamma_E \to \Gamma_E$ by $\psi(q,p) = (q, -p)$, if one assumes that the stochastic region S_E is ψ invariant and that the functions $\lambda_{N+1}, \dots, \lambda_{2N-1}$ are μ_L constant almost everywhere on S_E, then one easily deduces, on S_E, the relations

$$\lambda_{2N-k} = -\lambda_k\,, \quad k = 1, \dots, N \quad (4)$$

(in particular $\lambda_N = 0$). This is seen as follows: Let $x = (q,p)$; from $H(\psi(x)) = H(x)$, as is well known, one gets the time-reversal property $T_E^{-t} = \psi T_E^t \psi^{-1}$ of the flow. Then from part (c) of Theorem B one gets

$$\lim_{t\to\infty} \frac{1}{|t|} \ln \|dT_x^t(e)\| = -\lim_{t\to-\infty} \frac{1}{|t|} \ln \|dT_x^t(e)\|$$

for all $0 \ne e \in H_j$, $1 \le j \le s$, so that one immediately deduces $\lambda(x,e) = -\lambda(\psi(x), d\psi(e))$; the stated property then follows.

As a consequence, inequality (3) can then be strengthened and one gets

$$\mu_L(S_E)k(E) \le h_{\mu_L}(\{T_E^t\}) \le (N-1)\mu_L(S_E)k(E)\,. \quad (5)$$

This is of particular interest in the case $N = 2$, because it then gives the approximate equality

$$h_{\mu_L}(\{T_E^t\}) \cong \mu_L(S_E)k(E)\,. \quad (6)$$

This relation will allow us to produce, for the Hénon-Heiles model, an approximate curve for entropy as a function of energy, on the basis of numerical estimates for $k(E)$ and $\mu_L(S_E)$.

For this model it was possible to check numerically the two additional hypotheses introduced to obtain relation (4).

III. NUMERICAL EXAMPLE: THE HÉNON-HEILES MODEL

A. Hénon-Heiles model

Here, we briefly recall the definition and describe the main properties of the Hénon-Heiles model.[5] This is a dynamical system of two degrees of freedom characterized by the Hamiltonian

$$H(q_1, q_2, p_1, p_2) = \tfrac{1}{2}(p_1^2 + p_2^2 + q_1^2 + q_2^2) + q_1^2 q_2 - \tfrac{1}{3} q_2^3\,,$$

where $q_1, q_2, p_1, p_2 \in R$; a flow on R^4 is thus defined by the Hamilton equations

$$\frac{dq_1}{dt} = p_1\,, \quad \frac{dq_2}{dt} = p_2\,,$$

$$\frac{dp_1}{dt} = -q_1 - 2q_1q_2\,, \quad \frac{dp_2}{dt} = -q_2 + q_2^2 - q_1^2\,.$$

It can be easily shown that the energy surfaces Ω_E are such that for $0 < E < \frac{1}{6}$ they admit a unique nonvoid invariant compact three-dimensional mani-

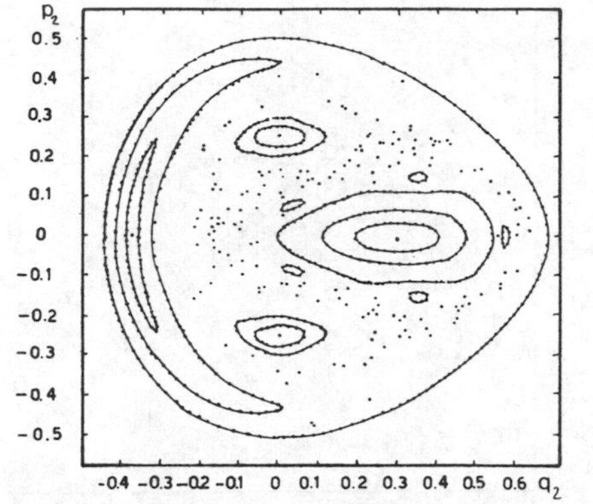

FIG. 2. Surface of section for the Hénon-Heiles model, at the intermediate energy $E = 0.125$, where ordered and stochastic regions have comparable measure (from Ref. 5).

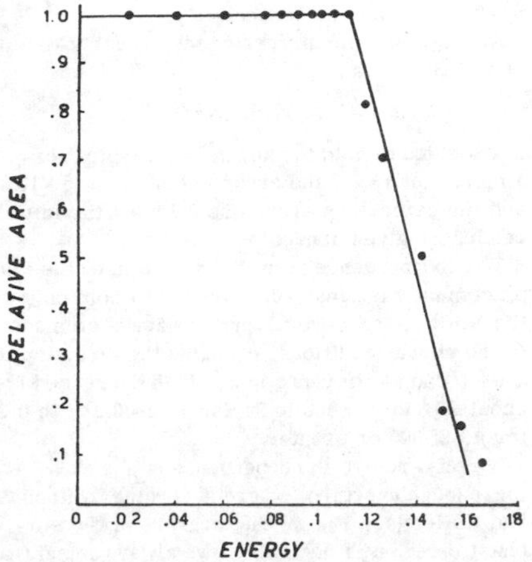

FIG. 3. Graph of the function $1-\tilde{\mu}(E)$ (from Ref. 5).

fold Γ_E and that the vector field X_H there defined does not have singular points.

This was the first model for which numerical computations indicated the existence of ordered and stochastic regions on Γ_E, according to the general qualitative description mentioned in Sec. II B.

This picture is clearly illustrated by Fig. 2, which is reproduced from Ref. 5. Such a figure gives a graphical presentation for the numerical integration of the equations of motion, by a standard device which reduces the study of a three-dimensional flow on Γ_E to the study of a two-dimensional plane mapping, as it will be now recalled.

Considering in Γ_E the two-dimensional surface given by $q_1 = 0$, one plots the successive points at which a particular solution intersects this surface with $p_1 > 0$. If one eliminates p_1 with the help of $H(q,p) = E$ and sets $q_1 = 0$, one can use q_2 and p_2 as coordinates on this two-dimensional surface; since $p_1^2 \geq 0$, they are restricted to belong to the region

$$\tilde{\Gamma}_E = \{(q_2, p_2): \tfrac{1}{2}(p_2^2 + q_2^2) - \tfrac{1}{3}q_2^3 \leq E\}.$$

As is seen from Fig. 2, which refers to $E = 0.125$, it appears that for some initial conditions these successive points are organized on various closed curves in the plane q_2, p_2. Other initial conditions (in the stochastic region) give instead successive points which are scattered around without any apparent order; actually, all such points in the figure refer to a single particular solution.

Let $\tilde{\mu}_L$ denote the (two-dimensional) Lebesgue measure on $\tilde{\Gamma}_E$ normalized by $\tilde{\mu}_L(\tilde{\Gamma}_E) = 1$. Let $\tilde{\mu}(E) = \tilde{\mu}_L(\tilde{S}_E)$, where $\tilde{S}_E$ is the stochastic region in $\tilde{\Gamma}_E$. Hénon and Heiles gave a numerical estimate for $\tilde{\mu}(E)$ as a function of E and found for $1 - \tilde{\mu}(E)$ the graph of Fig. 3. In this connection one may recall that as has been recently proven[16] one has rigorously $\tilde{\mu}(E) < 1$ for $0 < E < \frac{1}{6}$.

B. Results of numerical computations

We come now to the description of our numerical results. The computations were performed on a CDC 7600 computer, with a precision of 14 digits. The integration algorithm was the so-called central-point method,[17] correct up to the third order of the time step; the time step was typically 0.004.

The aim was to compute the quantity $k_n(\tau, x, d)$ described in Sec. II B. Given a value of E, the initial point x was chosen in $\tilde{\Gamma}_E$, i.e., by arbitrarily fixing q_2 and p_2, taking $q_1 = 0$, and determining p_1 by the condition $H(q,p) = E$. The displaced point y was chosen at a distance $|d|$ from x, with typically $|d| = 3\times10^{-4}$. Typically it was $\tau = 0.2$ and n up to 10^5.

The properties (i)–(v) of k_n discussed in Sec. II B and II C were checked with good accuracy.

For the independence from τ [property (ii) of Sec. II B], we remark that for any n the property

$$k_{jn}(\tau/j, x, d) \cong k_n(\tau, x, d), \quad j = 2, 3, \ldots,$$

should be expected to hold, from the considerations of Sec. II C. For example, this property was satisfied with an error of about 5×10^{-5} for $j = 2$ and 5×10^{-4} for $j = 10$, for any n up to 10^5, in a computation with x_0 in the stochastic region and $E = 0.125$. These values correspond to percentage errors of about 0.1% and 1%, respectively.

Analogously, the property [see property (iii) of Sec. II B]

$$k_n(\tau, x, \alpha d) \cong k_n(\tau, x, d), \quad \alpha \neq 0,$$

is expected to hold for any n. As a typical example, for $\alpha = 2.2$ the error was of about 5×10^{-5}, and the percentage error was 0.1% for the initial conditions given above.

The independence from the direction of the displacement d is, instead, expected to hold only in the limit $n\to\infty$. Several checks have been made: In the above conditions, changing the direction of d we found a difference in k_n which decreased from about 20% for $n = 250$ to 2% for $n = 2500$ and to 0.2% for $n = 25\,000$ or greater.

We pass now to the dependence on n and x. At low enough energies, where according to Hénon and Heiles (see Fig. 3) the measure of the stochastic region is negligible, we always found that k_n decreased with n (at least for large enough n), as illustrated in Fig. 4, which refers to various initial conditions at $E = 0.08$. With large enough n all curves seem to approach a straight line in the log-log scale, which corresponds to a behavior of the type $k_n \cong \alpha n^{-\beta}$ $(\alpha, \beta > 0)$. This very regular behavior strongly supports the conjecture that $\lim_{n\to\infty} k_n = 0$. In particular it seems to be a rather general rule that in this model one has $\beta \cong 1$, as can be seen by comparison with the dotted line in Fig. 4.

This general behavior for k_n does not change by increasing the energy, provided one chooses appropriately the initial point x in such a way that it is in the ordered region. This is illustrated in Fig. 5.

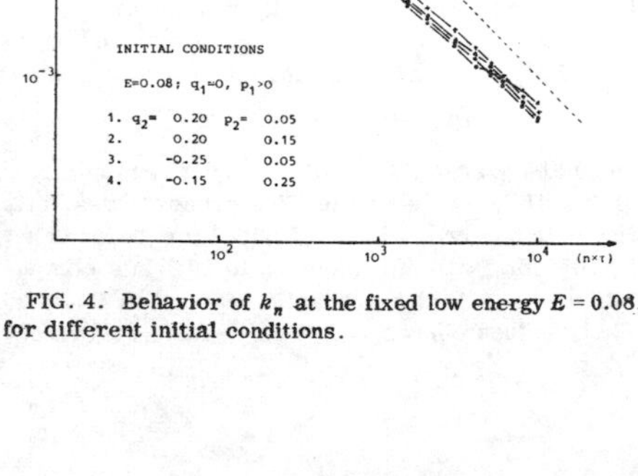

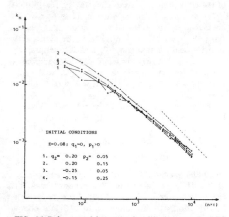

FIG. 4. Behavior of k_n at the fixed low energy $E = 0.08$. for different initial conditions.

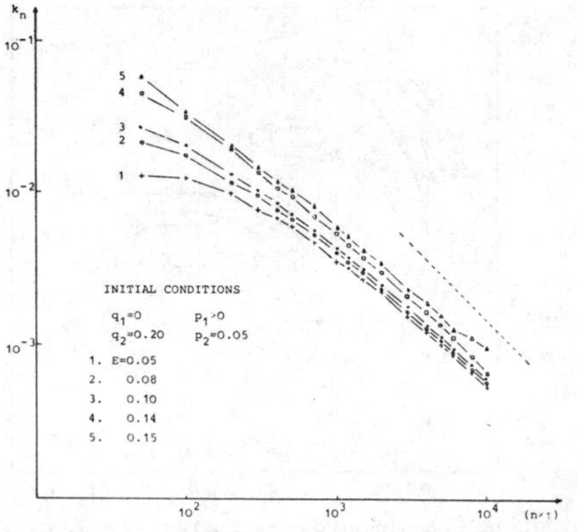

FIG. 5. Behavior of k_n at different energies, when the initial point is taken in the ordered region.

Let us now examine the case of a typical energy, $E = 0.125$, where the ordered and the stochastic regions have comparable measures (see Fig. 3). In Fig. 6 six curves are reported for $E = 0.125$. Curve 1, which has the same general feature of Figs. 4 and 5, corresponds to an initial point x lying in the large ordered region around the q_2 axis with $q_2 > 0$. Curve 2 refers to an initial point in one of the small islands surrounding such a region, while curve 3 corresponds to an initial point in one of the two symmetric ordered regions near the p_2 axis. For these curves, too, the limit seems to be zero, being approached, however, with a less regular behavior. Such a feature ap-

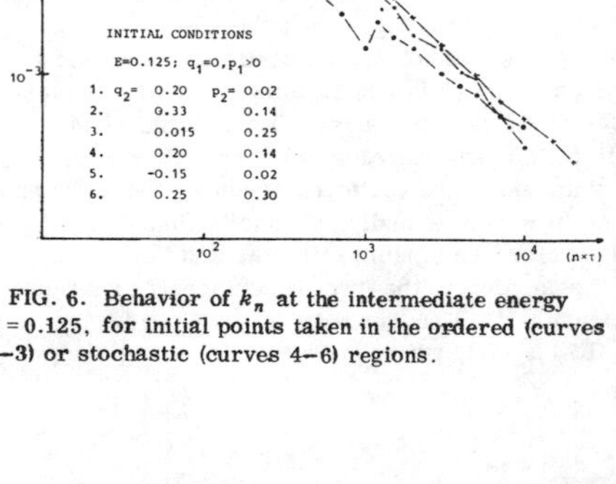

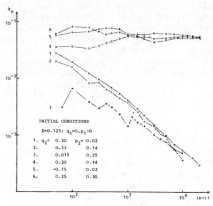

FIG. 6. Behavior of k_n at the intermediate energy $E = 0.125$, for initial points taken in the ordered (curves 1–3) or stochastic (curves 4–6) regions.

pears to be characteristic of these components of the ordered region, and is difficult to explain it rigorously at the moment. We hope to return to this problem in future. Curves 4–6 refer to initial points x in the stochastic region and rather clearly appear to approach a limit which is independent of x. This property has been checked for many other initial points at the same energy.

In general, for any energy $E>0$ we always found for the quantity $k=\lim_{n\to\infty} k_n$ either zero or a positive value which depends only on E. Such a positive value will be denoted, as in Sec. II B, by $k(E)$. The values found for $k(E)$ are reported in Fig. 7 by asterisks; the vanishing values found in the ordered region have also been marked (dots). All of the positive values between $E=0.105$ and 0.1666 are rather well defined with the same accuracy as that of Fig. 6. In particular, it may be remarked that two of them (i.e., those for $E=0.105$ and 0.110) refer to energies for which, according to Hénon and Heiles, the measure of the stochastic region is negligible (see Fig. 3). For the two points at $E=0.975$ and 0.1, they are defined with a much smaller accuracy, and a very careful search was necessary in order to find initial conditions with nonvanishing k at such energies. One can also note that apart from these two points the positive values are rather well fitted by an exponential, $k(E)=3.4e^{22E}$, as shown in Fig. 7 (continuous line).

We come now to an estimate for the metric entropy $h(E)=h_{\mu_L}(\{T_E^t\})$ as a function of energy E, on the basis of relation (6). The entropylike quantity $k(E)$ being known from the numerical computations described above, one needs only a knowledge of $\mu_L(S_E)$ as a function of E. For want of a direct estimate, we assume it to be given approximately by the function $\tilde{\mu}(E)=\tilde{\mu}_L(\tilde{S}_E)$ defined in Sec. IIIA and estimated numerically by Hénon and Heiles (Fig. 3). One thus has, approximately,

$$h(E)\cong\tilde{\mu}(E)k(E)\,. \tag{7}$$

For example, if as a rough interpolation one takes for $\tilde{\mu}(E)$ the function given by $\tilde{\mu}(E)=0$ for $E<0.11$, $\tilde{\mu}(E)=1-17.6(E-0.11)$ for $0.11\le E\le\frac{1}{6}$, and for $k(E)$ the exponential $k(E)=3.4e^{22E}$, one gets the function $h(E)=0$ for $0<E<0.11$, $h(E)=60e^{22E}(E-0.11)$ for $0.11\le E\le\frac{1}{6}$, represented by a dotted line in Fig. 7. One may note that very probably one should have $h(E)>0$ for $0<E<\frac{1}{6}$, although this is still unproven, to our knowledge.

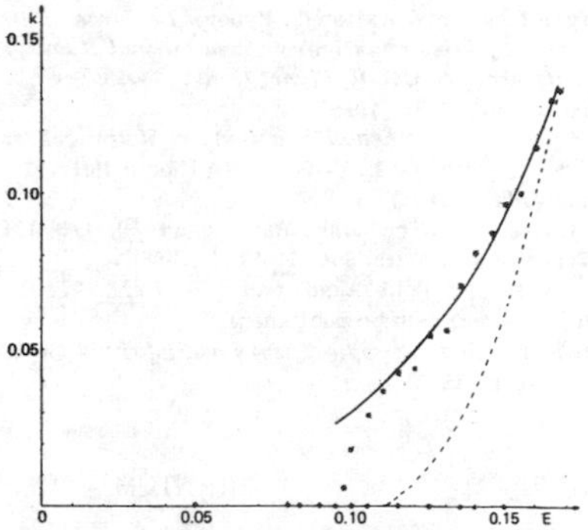

FIG. 7. Nonvanishing values found for k_n at different energies in the stochastic region (asterisks), vanishing values found in the ordered region (dots), exponential curve interpolating the nonvanishing values (continuous line), and tentative curve for entropy as a function of energy (dotted line).

IV. CONCLUDING REMARKS

We conclude by mentioning some open problems, strictly related to those considered in the present paper: (i) How can one calculate numerically the Lyapunov characteristic numbers other than the maximal one? (ii) Does the stochastic region in the Hénon-Heiles model contain more than one ergodic components, as suggested to us by Dr. M. Hénon? (iii) Does one have for the function $\tilde{\mu}(E)$ in the Hénon-Heiles model $\tilde{\mu}(E)>0$ for $0<E<\frac{1}{6}$, from both a numerical and a theoretical point of view? This last question is related to the already mentioned problem of whether one has $h(E)>0$ for $0<E<\frac{1}{6}$.

Note added in proof. Theorems *A* and *B* have also been proved by V. M. Millionščikov; see Mat. Sbornik 78, 179 (1969) [Math. USSR Sbornik 7, 171 (1969)].

---

[1]B. V. Chirikov: *Researches concerning the theory of nonlinear resonance and stochasticity.* CERN Transcript 71-40 Geneva, 1971 (unpublished).

[2]B. V. Chirikov and F. M. Izrailev, in Colloque International CNRS sur les Transformations Ponctuelles et leurs Applications, 10–14 Sept. (Toulouse, France, 1973) (unpublished).

[3]B. V. Chirikov, F. M. Izrailev, and V. A. Tayurski, Comput. Physics Commun. 5, 11 (1973).

[4]M. Casartelli, E. Diana, L. Galgani, and A. Scotti, Phys. Rev. A 13, 1921 (1976).

[5]M. Hénon and C. Heiles, Astron. J. 69, 73 (1964).

[6]Jurgen Moser and Walter T. Kyner, *Lectures on Hamiltonian Systems and Rigorous and Formal Stability of Orbits about an Oblate Planet* (Amer. Math. Soc., Providence, R. I., 1968).

[7]J. Ford, in *Fundamental Problems in Statistical Mechanics*, edited by E. D. G. Cohen (North-Holland, Amsterdam, 1975), Vol. 3.

[8]V. I. Oseledec, Tr. Mosk. Mat. Obsch. 19, 179 (1968) [Trans. Mosc. Math. Soc. 19, 197 (1968)].

[9]Ya. B. Piesin, Dokl. Akad. Nauk. SSSR 226, 774 (1976).

[10]Ya. B. Piesin, (to be published).

[11]P. Billingsley. *Ergodic Theory and Information* (Wiley, New York, 1965).

[12]P. Walters. *Ergodic Theory—Introductory Lectures: Lecture Notes in Mathematics No. 458* (Springer, Berlin, 1975).

[13]S. Sternberg. *Lectures on Differential Geometry* (Prentice Hall, Englewood Cliffs, 1964).

[14]J. Ford, *Stochastic Behavior in Nonlinear Oscillator Systems: Lectures in Statistical Physics No. 28* (Springer. Berlin, 1974).

[15]Ya. B. Piesin, Funkt. Anal. Jego Prilog 8, 81 (1974) [English translation in Funct. Anal. Applic. 8, 263 (1974)].

[16]M. Braun. J. Diff. Eq. 13. 300 (1973).

[17]See, for example, L. Verlet, Phys. Rev. 159, 98 (1967).

ERRATUM

In section III B, in the expressions for $k(E)$ and $h(E)$, a factor 10^{-3} is missing.

Progress of Theoretical Physics, Vol. 61, No. 6, June 1979

A Numerical Approach to Ergodic Problem of Dissipative Dynamical Systems

Ippei SHIMADA[†] and Tomomasa NAGASHIMA*

Department of Physics, Hokkaido University, Sapporo 060
**Department of Precision Engineering, Faculty of Engineering Hokkaido University, Sapporo 060*

(Received November 13, 1978)

Based on the Lyapunov characteristic exponents, the ergodic property of dissipative dynamical systems with a few degrees of freedom is studied numerically by employing, as an example, the Lorenz system. The Lorenz system shows the spectra of $(+,0,-)$ type concerning the 1-dimensional Lyapunov exponents, and the exponents take the same values for orbits starting from almost of all initial points on the attractor.

This result suggests that the ergodic property for general dynamical systems not necessarily belonging to the category of the axiom-A may also be characterized in the framework of the spectra of the Lyapunov characteristic exponents.

§ 1. Introduction

Recently, chaotic motions that arise due to non-linearities of dissipative dynamical systems have received a great concern in physical and non-physical fields.[1] However, in general dynamical systems which do not satisfy the axiom-A, little progress has been made to analyse those chaotic motions by theoretically well-established methods.[2)~4)]

One of the purposes of this paper is to present numerical methods, by which wide-spread chaotic motions in dissipative dynamical systems would be characterized in a systematic manner. Our basic idea for this aim is to utilize the complete set of 1-dimensional Lyapunov exponents, which characterize the asymptotic orbital instability of dynamical systems.[5)~9)] The second purpose is to show that the concept of Lyapunov exponents presents a practical tool to discuss problems of bifurcation of those chaotic solutions.

For these purposes, it becomes an important problem to estimate the Lyapunov exponents by some numerical methods, because it may not be expected, in general, that the equations for orbits exhibiting chaotic motions have globally single-valued analytic solutions. In case of measure preservig diffeomorphisms, Benettin et al.[11),*)] have recently pointed out almost the same method as developed in this paper.

[†] Present address: Department of Physics, College of Science and Technology, Nihon University, Tokyo.

*) The reader should not confuse two reference 7) and 11) by Benettin et al..

In § 2, brief discussion on the existence and properties of the k-dimensional Lyapunov exponent is presented. The relations of the characteristic exponents to the invariant measure and also to the measure-theoretic entropy (Kolmogorov entropy[10]) are commented. In the Appendix, a general scheme for estimating numerically the k-dimensional Lyapunov exponent is briefly developed, and discussion on the relation between our method and another approximate method[7] for estimating the Lyapunov exponent is also given.

In § 3, the famous turbulent model, i.e., the Lorenz model,[12] is studied in the light of our method, and the result of the complete set of the 1-dimensional Lyapunov exponents for this system is explicitly given. As an application of our approach based on the Lyapunov exponents, § 4 is devoted to problems of bifurcation of chaotic solutions in the Lorenz system.

§ 2. A partial summary of Lyapunov characteristic exponents for irregular motions

It is considered that dynamical systems exhibiting chaotic motions without any contact with external disturbance may possess some unstable properties of orbits. From this point of view, it is worth while noting that there is a fundamental method for investigating the time-dependent behavior of small deviations from an orbit. The method is called Lyapunov's method which uses the first variational equation of orbits.*)

Now, let us consider the system of which time evolution is described by a set of differential equations in N-dimensional Euclidian space,

$$\dot{\boldsymbol{x}} = \boldsymbol{F}(\boldsymbol{x}). \tag{1}$$

The solution of Eq. (1) under the initial condition $\boldsymbol{x}(0) = \boldsymbol{x}_0$ is written as

$$\boldsymbol{x}(t) = T^t \boldsymbol{x}_0 , \tag{2}$$

where T^t is the map which describes time-t evolution of all phase points.

On the other hand, the time evolution equation for the first variation of the orbit obeys the following set of non-autonomous linear differential equations:

$$\delta\dot{\boldsymbol{x}} = \frac{\partial \boldsymbol{F}}{\partial \boldsymbol{x}}(T^t \boldsymbol{x}_0)\,\delta \boldsymbol{x} . \tag{3}$$

The solution of Eq. (3) can be written as

$$\delta \boldsymbol{x}(t) = U^t_{\boldsymbol{x}_0} \delta \boldsymbol{x}_0 , \tag{4}$$

where $U^t_{\boldsymbol{x}_0}$ is the fundamental matrix[5] of Eq. (3), and $\delta\boldsymbol{x}_0$ is an initial deviation at $t=0$. The fundamental matrix in Eq. (4) satisfies the following chain rule:

*) As a general introduction to this chapter, one may refer to the book cited in Ref. 5).

$$U^{t+s}_{x_0}=U^{t}_{T^s x_0}\circ U^{s}_{x_0}\,. \tag{5}$$

It is apparent that the asymptotic behavior of a small deviation is described by the asymptotic behavior of the fundamental matrix for $t\to\infty$. Now, the asymptotic behavior of this matrix for $t\to\infty$ can be characterized by the following exponents:[5),6),11)]

$$\lambda(e^k, x_0)=\lim_{t\to\infty}\frac{1}{t}\log\frac{\|U^t_{x_0}e_1\wedge U^t_{x_0}e_2\wedge\cdots\wedge U^t_{x_0}e_k\|}{\|e_1\wedge e_2\wedge\cdots\wedge e_k\|} \tag{6}$$

for $k=1, 2, \cdots, N$. The symbols in (6) have the following meanings: e^k is a k-dimensional subspace in the tangent space E_{x_0} at x_0, $\{e_i\}$ $(i=1, 2, \cdots, k)$ are a set of bases of e^k, $\wedge$ is an exterior product and $\|\circ\|$ is a norm with respect to some Riemannian metric. The exponent defined by (6) represents an expanding rate of volume of the k-dimensional parallelepiped in the tangent space along the orbit which starts at x_0, and is called the k-dimensional Lyapunov exponent. It is clear from this definition that the exponent does not depend on a choise of a set of bases nor norms, but depends only on the k-dimensional subspace e^k.[13)]

It may be useful to summarize the properties of the Lyapunov exponents, which will be utilized in the subsequent discussion.

1) 1-dimensional exponent $\lambda(e^1, x)$ may take, at most, N distinct values, and we will use the notations $\{\lambda_i\}_{1\leq i\leq N}$ and suppose $\lambda_1\geq\lambda_2\geq\cdots\geq\lambda_N$.

2) k-dimensional exponent $\lambda(e^k, x)$ may take, at most, ${}_NC_k$ distinct values, and each value is connected with a sum of k distinct 1-dimensional exponents. For instance, in the case $N=3$, the k-dimensional exponents $\lambda(e^k, x)$ $(k=1, 2, 3)$ may take the following values respectively:

$$\lambda(e^1, x)=\text{one of the values in } \{\lambda_1, \lambda_2, \lambda_3\},$$

$$\lambda(e^2, x)=\text{one of the values in } \{(\lambda_1+\lambda_2), (\lambda_1+\lambda_3), (\lambda_2+\lambda_3)\},$$

$$\lambda(e^3, x)=(\lambda_1+\lambda_2+\lambda_3).$$

3) If a set bases $\{e_i\}$ $(i=1, 2, \cdots, N)$ is chosen at random in tangent space, then the k-dimensional exponents $\lambda(e^k, x)$ for $k=1, 2, \cdots, N$ converge respectively, with probability 1, to the maximal values among sets of values which are allowed to possess ${}_NC_k$ distinct values. (This proposition was proved by Benettin et al. in case of diffeomorphisms.[11)])

It must be mentioned here that the above discussion on the Lyapunov exponents becomes meaningful only if the existence of the limit of the quantity defined on the r.h.s. of (6) would be guaranteed.

The proof of the existence of such limits has been made by Oseledec. Here we describe his theorem in the form suited to our discussion.

The multiplicative ergodic theorem of Oseledec:[6)] If there is a T^t-invariant measure μ and $\|\partial F/\partial x\|\in L^1(\mu)$, then the k-dimensional Lyapunov exponents

$\lambda(e^k, x_0)$ $(k=1, 2, \cdots, N)$ exist for μ-almost all x_0. In this theorem, notations have the same meaning as the one used in the preceding discussion.

Hereafter, we would like to remark some relations between the Lyapunov exponents and the measure-theoretic entropy (K-entropy) of dynamical systems. It has been known that the existence of the Lyapunov exponents is directly related to the K-entropy. Of the relations obtained so far, the weakest relation may be the following:

$$H(\mu) - \int \sum_{\lambda_i>0} \lambda_i(x)\, d\mu \leq 0\,, \tag{7}$$

where $H(\mu)$ is the K-entropy of the dynamical system with invariant measure μ.

This inequality has been proved by Ruelle.[14),15)] In cases of Hamiltonian systems and axiom-A dynamical systems, the equality in (7) does hold.[17)] If the equality in (7) would be assured, then the phase average of the sum of positive Lyapunov exponents becomes the K-entropy itself. Furthermore, it is expected that the category of dynamical systems which do satisfy the equality in (7) would be much more extensive than the dynamical systems mentioned.

§ 3. 1-, 2- and 3-dimensional Lyapunov exponent of the Lorenz system

In this section, based on the numerical method developed in the Appendix, we present the explicit result of a complete set of the 1-dimensional Lyapunov exponents for the famous turbulent model due to Lorenz.

The Lorenz model is described by the following set of differential equations:

$$\frac{d}{dt}\begin{pmatrix} X \\ Y \\ Z \end{pmatrix} = \begin{pmatrix} -\sigma X & +\sigma Y & \\ (\gamma - Z)X & -Y & \\ XY & & -bZ \end{pmatrix} = \boldsymbol{F}(\boldsymbol{x}), \tag{8}$$

where (σ, b, γ) are the parameters. The Lorenz model is a dissipative system, and therefore it does not have a priori invariant measure in contrast to Hamiltonian systems. It is believed in this system that the high-dimensional attractor with very complicated geometrical structure like Cantor set comes forth beyond a certain value of the parameter γ (σ and b are suitably chosen), and orbits on the attractor are non-periodic.

It should be mentioned that the high-dimensional attractor of the Lorenz system does not belong to a well-established mathematical category like the axiom-A strange attractor[17)] because on the edge of the attractor, there is a fixed point (0, 0, 0), and therefore the uniform hyperbolic structure of the attractor is not materialized.

In our calculations, parameters are set as $\sigma=16.0$, $b=4.0$ and $\gamma=40.0$. These values of parameters are not equal to the original Lorenz's values, i.e., $\sigma=10.0$, $b=8/3$ and $\gamma=28.0$, but the geometrical structure of orbits on the attractor is

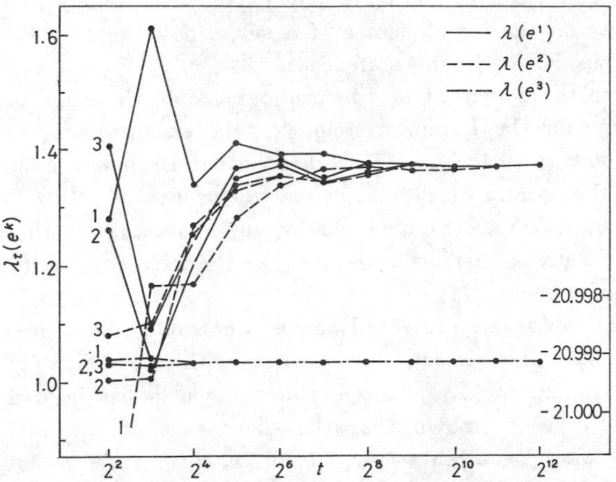

Fig. 1. Temporal convergence of the 1-, 2- and 3-dimensional Lyapunov exponent for the Lorenz system ($\sigma=16.0$, $b=4.0$ and $\gamma=40.0$). Exponents are calculated, based on the formula (A·2) given in the Appendix, under the following initial conditions ($\boldsymbol{x}_0, \boldsymbol{e}_1, \boldsymbol{e}_2, \boldsymbol{e}_3$):
1. ((10.0, 0.0, 30.0), (1.0, 0.0, 0.0), (0.0, 1.0, 0.0), (0.0, 0.0, 1.0))
2. ((10.0, 10.0, 30.0), (1.0, 0.0, 0.0), (0.0, 1.0, 0.0), (0.0, 0.0, 1.0))
3. ((10.0, 10.0, 30.0), (0.0, 1.0, 1.0), (1.0, 0.0, 1.0), (1.0, 1.0, 0.0)).

qualitatively the same as the original one within the region $50 \gtrsim \gamma \gtrsim \gamma_T \equiv \sigma(\sigma+b+3)/(\sigma-b-1)$. Our numerical integration scheme is the usual Runge-Kutta-Gill method in double precision, and typical time difference is taken as 0.01. By integrating the orbit equations and the first variational equations, according to the method investigated in the Appendix (the renormalization time was chosen as $\tau=1.0$), the 1-, 2- and 3-dimensional Lyapunov exponent have been calculated. They converged respectively to certain definite values. It should be noted here that three exponents $\lambda(e^1, \boldsymbol{x}_0)$, $\lambda(e^2, \boldsymbol{x}_0)$ and $\lambda(e^3, \boldsymbol{x}_0)$ for an orbit starting at $\boldsymbol{x}_0$ did not depend on the initial choice of the subspace e^k $(k=1, 2, 3)$. The numerical results are shown in Fig. 1. Following the statement for the Lyapunov exponents described in § 2·3), the estimated values of the exponents $\lambda(e^k, \boldsymbol{x}_0)$ $(k=1, 2, 3)$ select, respectively, the maximal values undoubtedly. Therefore, we can expect that the following relations hold:

$$\lambda(e^1, \boldsymbol{x}_0) = \lambda_1,\ \lambda(e^2, \boldsymbol{x}_0) = \lambda_1 + \lambda_2 \text{ and } \lambda(e^3, \boldsymbol{x}_0) = \lambda_1 + \lambda_2 + \lambda_3, \tag{9}$$

where λ_i $(i=1, 2, 3)$ are the 1-dimensional Lyapunov exponents and are assumed to be $\lambda_i \gtrsim \lambda_j$ for $j>i$. Therefore we can estimate the 1-dimensional Lyapunov exponents as

$$\lambda_1 = \lambda(e^1, \boldsymbol{x}_0),\ \lambda_2 = \lambda(e^2, \boldsymbol{x}_0) - \lambda(e^1, \boldsymbol{x}_0) \text{ and } \lambda_3 = \lambda(e^3, \boldsymbol{x}_0) - \lambda(e^2, \boldsymbol{x}_0). \tag{10}$$

Figure 1 indicates that the Lyapunov exponents $\lambda(e^k, \boldsymbol{x}_0)$ $(k=1, 2, 3)$ do not depend not only on the initial choice of a set of bases in the subspace e^k, but also on the initial position $\boldsymbol{x}_0$ in the state space.

Contrary to the statement of the last paragraph, there are other trivial Lyapunov exponents for the Lorenz system, i.e., the exponents for orbits which tend to the fixed point at (0, 0, 0). The 1-dimensional Lyapunov exponents for these orbits become the spectra of the linearized vector field at (0, 0, 0) and take different values from $\lambda_i (i=1, 2, 3)$ obtained by our numerical experiment. Therefore, it should be considered that orbits tending to the fixed point at (0, 0, 0) would be negligible in some sense.

Relating to the result presented above, we would like to make the following two remarks. The first remark is concerned with the ability of our numerical scheme for estimating the Lyapunov exponents, which has been developed in the Appendix. It has been known that the divergence of the vector field of the Lorenz system takes a constant value, i.e., $\mathrm{div}\, \boldsymbol{F}(\boldsymbol{x}) = -(\sigma+b+1) = -21.0$. This property of the Lorenz system can be utilized to test the ability of our numerical scheme.

That is, as the 3-dimensional exponent $\lambda(e^3, \boldsymbol{x}_0)$ represents the rate of expansion of a volume element, it is directly comparable to the divergence of the vector field. The explicit result representing the temporal convergence of the 3-dimensional exponent, together with that of the 1- and 2-dimensional characteristic exponents, is shown in Table I. From the result of the 3-dimensional exponents $\lambda(e^3, \boldsymbol{x}_0)$, it might be concluded that we can expect, in principle, the same order

Table I. An explicit data representing temporal convergence of the 1-, 2- and 3-dimensional Lyapunov exponent for the Lorenz system ($\sigma=16.0$, $b=4.0$ and $\gamma=40.0$). In this table, it should be mentioned that the 3-dimensional exponent $\lambda(e^3)$ has converged to -21.0 within the accuracy of 0.01%.

t	$\lambda(e^1)$	$\lambda(e^2)$	$\lambda(e^3)$
2.	1.475172	−0.219607	−20.99911
4.	1.281321	0.543842	−20.99913
8.	1.610772	1.166704	−20.99910
16.	1.339753	1.171347	−20.99914
32.	1.411494	1.283690	−20.99913
64.	1.391993	1.340821	−20.99914
128.	1.392513	1.364981	−20.99914
256.	1.378710	1.371627	−20.99914
512.	1.371741	1.365655	−20.99914
1024.	1.370685	1.367358	−20.99914
2048.	1.373692	1.371871	−20.99914
4096.	1.374207	1.373337	−20.99914

Table II. The complete set of the 1-dimensional Lyapunou exponent for the Lorenz system ($\sigma=16.0$, $b=4.0$ and $\gamma=40.0$). The value in ( ) represents the maximal Lyapunov exponent obtained by Benettin's procedure.[7] In the latter procedure, we have set $\|\alpha\|=0.005$ and $\tau=0.11$.

λ_1	1.37	(1.36)
λ_2	0.00	
λ_3	−22.37	

accuracy in the calculations of $\lambda(e^2, \boldsymbol{x}_0)$ and $\lambda(e^1, \boldsymbol{x}_0)$ as that for $\lambda(e^3, \boldsymbol{x}_0)$, if we take a sufficiently long time to estimate these exponents.

According to the relation (10), we can estimate the complete set of the 1-dimensional Lyapunov exponents $\lambda_i (i=1, 2, 3)$ for the Lorenz system. The result is presented in Table II, together with the result for the maximal 1-dimensional exponents estimated by the method which uses the orbit equations (A·4) given in the Appendix.[7] As is understood from the discussion in the Appendix, two values for the maximal 1-dimensional exponent estimated by the present method and that uses the orbit equations showed a good coincidence, when conditions mentioned in the Appendix were satisfied. This makes our second remark.

§ 4. The attractor of another type and the Lyapunov exponents in the Lorenz model

It can easily be imagined that the spectral type of the Lyapunov exponents and the type of attractor are closely related with each other. Non-periodic motions of the Lorenz system possess a positive exponent, and the spectra of the 1-dimensional exponents $\lambda_i (i=1, 2, 3)$ show the hyperbolicity of $(+, 0, -)$ type. It is clear that stable periodic orbit is characterized by the Lyapunov exponents of $(0, -, -)$ type. In this sense, the spectral type of the 1-dimensional Lyapunov exponents is a very useful tool for investigating the appearance of attractors of a new type. From this point of view, we reported in a previous paper that the Lorenz system (σ and b fixed) ends up with a stable periodic attracted for large γ.[18]

Hereafter, we would like to add some new features that were not reported in the previous paper. In the intermediate region of γ, i.e., $50 \lesssim \gamma \lesssim 330$, the solutions in the Lorenz system show a very complicated behavior of bifurcation, when the parameter γ is changed ($\sigma=16.0$ and $b=4.0$).[19] The situations are illustrated in Fig. 2.

In order to discuss or predict the global bifurcation scheme of attractors, one needs the precise knowledge of the Poincaré mapping for this system. In the range of the parameter γ discussed here, there might happen to occur a violation of the tansversality of orbits in the sense that if we make the Poincaré mapping on a certain surface of the section, i.e., $Z=\gamma-1$, the orbit does not cross the surface transeversely. Therefore we cannot state, at the present stage, the global structure of bifurcation scheme for the Lorenz system in a precise manner.

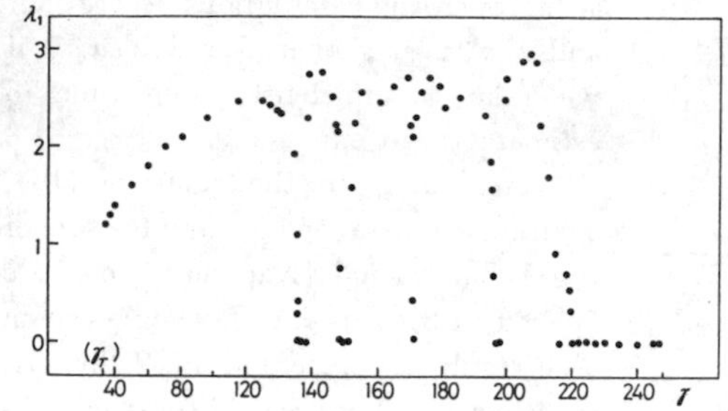

Fig. 2. γ-dependence of the maximal exponent λ_1 for the Lorenz system ($\sigma=16.0$ and $b=4.0$). In this figure, it means that if λ_1 would converge to zero at a certain value of γ, the corresponding dyamical system has a stable limit cycle.

However, for periodic attractors, there are the following clearly distinct bifurcations in some restricted ranges of the parameter γ. The first type might be called the symmetry breaking type and is illustrated in Fig. 3(a). After the original stable limit cycle becomes unstable, a pair of stable limit cycles come forth in this bifurcation. Under the map $(X, Y, Z) \overset{S}{\mapsto} (-X, -Y, Z)$, one of the limit cycles is mapped into the other limit cycle. The orbit is attracted to one of these limit cycles depending on initial conditions. Therefore the symmetry of the Lorenz equation for the transformation S breaks after the bifurcation of this type. The second is the usual Brunovsky bifurcation[20] and is also illustrated in Fig. 3(b). After the Brunovsky bifurcation of this type, the period of the limit cycles becomes twice the original period ω. The symmetry does not change in this bifurcation.

If we decrease the parameter γ from a certain value, where the system possesses an attracting periodic orbit with some periodicity and the symmetry under the transformation S, there occurs at first a symmetry breaking bifurcation. After the bifurcation of this type, there occurs a series of the Brunovsky bifurcations, through which periods of the limit cycles become longer and longer in such manner as $\omega \cdot 2^n \, (n=0, 1, \cdots)$.

Chaotic solutions are considered to appear beyond the limiting value of the parameter γ, at which the interval of the parameter consisting of a stable limit cycle with period n seems to vanish. Analogous bifurcation phenomena mentioned above have been also observed in such dynamical systems as Rössler's model[21] and a certain chemical reaction model.[22] This series of bifurcations leading to chaos may be considered to be an example of the generalized catastrophe introduced by Thom.[23]

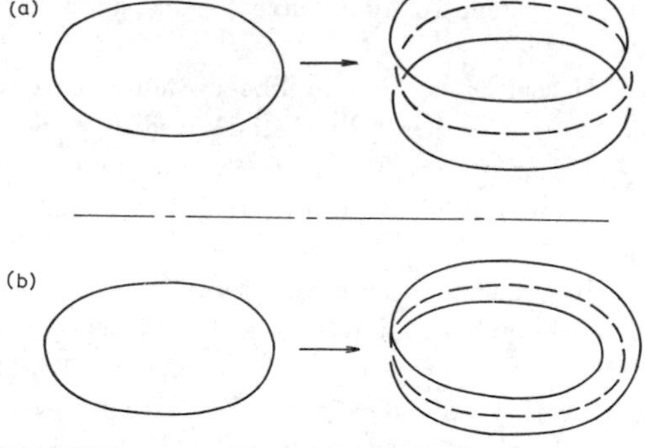

Fig. 3. Bifurcations of periodic orbits in the Lorenz system;
a) a symmetry breaking bifurcation,
b) a Brunovsky bifurcation.

It should be noted here that it includes a very delicate problem in analyzing the series of these bifurcations quantitatively. Namely, as the spectral type of the Lyapunov exponents $\lambda_i (i=1, 2, 3)$ for stable limit cycles and that of chaotic motions must be different from each other, it is clear that there is a critical point at which the spectral type changes between $(+, 0, -)$ and $(0, -, -)$. At this critical point, one of the exponents changes its sign and degenerates to zero. When this degeneracy of the exponents occurs, the dynamical system becomes structurally unstable. So, near this critical point, we must take scrupulous care of analyzing problems by any numerical method.

§ 5. Conclusions and discussion

It is shown numerically that the complete set of the 1-dimensional Lyapunov exponents $(\lambda_1, \lambda_2, \lambda_3)$ for the Lorenz attractor should exist, and the exponents take the same values for almost all orbits stating at the neighbourhood of the Lorenz attractor. The above result supports strongly that there should exist an invariant measure on the attractor, and the Lorenz system should be ergodic on the attractor with respect to this invariant measure. Furthermore, from the fact that the spectra of the 1-dimensional exponents $(\lambda_1, \lambda_2, \lambda_3)$ is of $(+, 0, -)$ type, the motion on the attractor should take the positive-definite K-entropy and therefore possesses the property of mixing.

It should be mentioned here that there is another trivial set of the 1-dimensional exponents, which implies a violation of the uniform hyperbolicity of the non-wandering set for the Lorenz system. This contradictory fact might be considered as follows: Although the Lorenz system does not satisfy the axiom-A in the strict sense,[2] but the trouble is not so severe in a measure theoretic meaning, i.e., orbits, which belong to the trivial Lyapunov exponents, would occupy a space of measure zero in the non-wandering set.

The result for the Lorenz system described here has already pointed out in the previous paper.[8] However, in the previous paper, we have employed the method which uses the orbit equation (A·4) and is more appropriate to visualizing the existence of the exponential orbital separation in state space.[24)~26)] By solving the tangent equation, it becomes more apparent in this paper that the existence of positive Lyapunov exponent is related, on a mathematical basis, to the theory of ergodicity of dynamical systems.

As stated in § 4, the method of the Lyapunov exponents employed in this paper is influential not only in characterizing irregular motions in a quantitative manner, but also in applying to problems of bifurcation of attractors.

In closing this paper, we would like to propose that chaotic motions in 4 and /or higher dimensional dissipative dynamical systems should be classified according to the spectral type of the complete set of the Lyapunov exponents $\lambda_i (i=1, 2, \cdots)$. In 4-dimensional case, it is easily considered that there should exist two kinds of chaotic motions clearly distinguishable from each other, because there are two types of the Lyapunov spectra such as $(+,+,0,-)$ and $(+,0,-,-)$.

Acknowledgements

This work is partly financially supported by the Scientific Research Fund of the Ministry of Education, Science and Culture.

Appendix A

——Method for Numerical Estimation of the k-Dimensional Lyapunov Exponent——

There is a technical problem, when we try to evaluate Lyapunov exponents directly, on the basis of the definition (6), by integrating 1-st variational equations. Namely, there occurs a overflow trouble in computer calculations, because 1-st variational equations have, at least, an exponentially diversing solution.

In order to keep calculations from this trouble, we exchange the base, after each time integration, in the following manner:

$$\begin{aligned}
\boldsymbol{e}_1^{j+1} &= U^{\tau}_{x_0}\boldsymbol{e}_1^{j}/\|U^{\tau}_{x_0}\boldsymbol{e}_1^{j}\|,\\
\boldsymbol{e}_2^{j+1} &= U^{\tau}_{x_0}\boldsymbol{e}_2^{j}-(\boldsymbol{e}_1^{j+1}\cdot U^{\tau}_{x_0}\boldsymbol{e}_2^{j})\cdot\boldsymbol{e}_1^{j+1}/\|U^{\tau}_{x_0}\boldsymbol{e}_2^{j}-(\boldsymbol{e}_1^{j+1}\cdot U^{\tau}_{x_0}\boldsymbol{e}_1^{j+1})\cdot\boldsymbol{e}_1^{j+1}\|,\\
\boldsymbol{e}_3^{j+1} &= U^{\tau}_{x_0}\boldsymbol{e}_3^{j}-(\boldsymbol{e}_1^{j+1}\cdot U^{\tau}_{x_0}\boldsymbol{e}_3^{j})\cdot\boldsymbol{e}_1^{j+1}-(\boldsymbol{e}_2^{j+1}\cdot U^{\tau}_{x_0}\boldsymbol{e}_3^{j})\cdot\boldsymbol{e}_2^{j+1}/\|U^{\tau}_{x_0}\boldsymbol{e}_3^{j}\\
\vdots\qquad &\qquad -(\boldsymbol{e}_1^{j+1}\cdot U^{\tau}_{x_0}\boldsymbol{e}_3^{j})\cdot\boldsymbol{e}_1^{j+1}-(\boldsymbol{e}_2^{j+1}\cdot U^{\tau}_{x_0}\boldsymbol{e}_3^{j})\cdot\boldsymbol{e}_2^{j+1}\|,\\
\boldsymbol{e}_k^{j+1} &= U^{\tau}_{x_0}\boldsymbol{e}_k^{j}-(\boldsymbol{e}_1^{j+1}\cdot U^{\tau}_{x_0}\boldsymbol{e}_k^{j})\boldsymbol{e}_1^{j+1}\cdots-(\boldsymbol{e}_{k-1}^{j+1}\cdot U^{\tau}_{x_0}\boldsymbol{e}_k^{j})\cdot\boldsymbol{e}_{k-1}^{j+1}/\|U^{\tau}_{x_0}\boldsymbol{e}_k^{j}\\
&\qquad -(\boldsymbol{e}_1^{j+1}\cdot U^{\tau}_{x_0}\boldsymbol{e}_k^{j})\cdot\boldsymbol{e}_1^{j+1}\cdots-(\boldsymbol{e}_{k-1}^{j+1}\cdot U^{\tau}_{x_0}\boldsymbol{e}_k^{j})\cdot\boldsymbol{e}_{k-1}^{j+1}\|.
\end{aligned}\tag{A·1}$$

Using the chain rule (5) and exchanging the base $\{U^{\tau}_{x_j}\boldsymbol{e}_i^{j}\}_i$ into $\{\boldsymbol{e}_i^{j+1}\}_i$ $(j=0,1,\cdots,(n-1)$ and $i=1,2,\cdots,k)$, we obtain the following equation:

$$\lim_{n\to\infty}\frac{1}{n\tau}\log\frac{\|\bigwedge_i U^{n\tau}_{x_0}\boldsymbol{e}_i^{0}\|}{\|\bigwedge_i \boldsymbol{e}_i^{0}\|}
=\lim_{n\to\infty}\frac{1}{n\tau}\sum_{j=0}^{n-1}\log\frac{\|\bigwedge_i U^{\tau}_{x_j}\boldsymbol{e}_i^{j}\|}{\|\bigwedge_i \boldsymbol{e}_i^{j}\|}\,. \tag{A·2}$$

The procedure of exchange of bases leading to Eq. (A·2) is justified by using the following property of exterior product; if $\{\boldsymbol{e}_i\}$ and $\{\boldsymbol{f}_i\}$ generate the same k-dimensional subspace, then the relation

$$\frac{\|\bigwedge_i U\boldsymbol{e}_i\|}{\|\bigwedge_i \boldsymbol{e}_i\|}=\frac{\|\bigwedge_i U\boldsymbol{f}_i\|}{\|\bigwedge_i \boldsymbol{f}_i\|}$$

holds.

Now, it may be useful to note the relation between our present method and the approximate method developed in Ref. 7) by Benettin et al. Benettin's procedure in Ref. 7) will converge to our method under the restrictions $\tau \ll 1$ and $\|e_i^j\| \ll 1$. Under these restrictions, an approximate relation

$$U^{\tau}_{x_j} e_i^j \cong T^{\tau}(x_j + e_i^j) - T^{\tau}_{x_j} \tag{A·3}$$

holds. We can, therefore, obtain approximately the Lyapunov exponents as follows:

$$\lambda(e^k, x_0) \cong \lim_{n\to\infty} \frac{1}{n\tau} \sum_{j=0}^{n-1} \log \frac{\|\bigwedge_i (T^{\tau}(x_j + e_i^j) - T^{\tau} x_j)\|}{\|\bigwedge_i e_i^j\|}, \tag{A·4}$$

where T is the map which has appeared in Eq. (2).

The special case ($k=1$) of the expression (A·4) has been utilized first by Benettin et al.,[7] who gave discussion for estimating the K-entropy of Hamiltonian systems. The authors applied their method, in the previous paper,[8] to a dissipative dynamical system, and pointed out the possibility that turbulent phenomena in dissipative dynamical systems would be discussed in a generalized framework of the ergodic theory of classical dynamical systems.

References

1) See, for instance, *Synergetics*, edited by H. Haken (Springer, Berlin, 1977).
2) S. Smale, Bull. Am. Math. Sod. **73** (1967), 747.
3) D. Ruelle, Am. J. Math. **98** (1976), 619.
4) R. Bowen and D. Ruelle, Inventions Math. **29** (1975), 181.
5) V. V. Nemytskii and V. V. Stepanov, *Qualitative Theory of Differential Equations* (Princeton Univ. Press, Princeton, 1960).
6) V. I. Oseledec, Trans. Moscow Math. Soc. **19** (1968), 197.
7) G. Benettin, L. Galgani and J. M. Strelcyn, Phys. Rev. **A14** (1976), 2338.
8) T. Nagashima and I. Shimada, Prog. Theor. Phys. **58** (1977), 1318.
9) S. D. Feit, Comm. Math. Phys. **61** (1978), 249.
10) V. I. Arnold and A. Avez, *Ergodic Problems of Classical Mechanics* (Benjamin, New York, 1968).
11) G. Benettin, L. Galgani, A. Giorgilli and J. M. Strelcyn, C. R. Acad. Sci. Paris **286** (1978), A-431.
See also, G. Contopoulos, L. Galgani and A. Giorgilli, Phys. Rev. **A** (to appear).
12) E. N. Lorenz, J. Atmos. Sci. **20** (1963), 130.
13) For details of the mathematical terminologies used in this paragraph, see, for instance, Y. Matsushima, *Introduction to Manifold* (in Japanese) (Shokabo, Tokyo, 1965).
14) D. Ruelle, preprint.
15) D. Ruelle, *Proceedings of the International Conference on Bifurcation Theory and Its Applications in Scientific Disciplines* (New York, 1977).
16) Ja. B. Pesin, Soviet Math.-Doklady **17** (1976), 196.
17) J. Guckenheimer, G. Oster and A. Ipaktchi, J. Math. Biol. **4** (1977), 101.
18) I. Shimada and T. Nagashima, Prog. Theor. Phys. **59** (1978), 1033.
19) T. Shimizu and N. Morioka, Phys. Letters **66A** (1978), 182.
20) P. Brunovsky, *Symposium on Differential Equations and Dynamical Systems, Warwick, 1968 & 69,*

21) T. Nagashima, Prog. Theor. Phys. Suppl. No. 64 (1978), 368.
22) K. Tomita and T. Kai, Phys. Letters **66A** (1978), 91.
23) R. Thom, *Structual Stability and Morphogenesis* (W. A. Benjamin, New York, 1974), chap. 6.
24) H. Fujisaka and T. Yamada, Phys. Letters **66A** (1978), 450.
25) K. Nakamura, Prog. Theor. Phys. **59** (1978), 74.
26) H. Yahata, Prog. Theor. Phys. **61** (1979), 791.

VOLUME 45, NUMBER 14 PHYSICAL REVIEW LETTERS 6 OCTOBER 1980

Dimension of Strange Attractors

David A. Russell
School of Electrical Engineering, Cornell University, Ithaca, New York 14853

and

James D. Hanson and Edward Ott
Laboratory for Plasma and Fusion Energy Studies, University of Maryland, College Park, Maryland 20742

(Received 4 August 1980)

A relationship between the Lyapunov numbers of a map with a strange attractor and the dimension of the strange attractor has recently been conjectured. Here, the conjecture is numerically tested with use of several different maps, one of which results from a system of ordinary differential equations occurring in plasma physics. For the cases tested, the conjecture is verified to within the obtained accuracy.

PACS numbers: 47.25.-c, 02.90.+p, 52.35.Ra

An *attractor* is a subspace of some ordinary N-dimensional space to which the solution of an N-dimensional dynamical system of equations asymptotes for large time. Two cases of dynamical systems will be considered here: maps (discrete time variable j),

$$\vec{x}_{j+1} = \vec{f}(\vec{x}_j), \tag{1}$$

where j is an integer, and autonomous ordinary differential equations (continuous time),

$$d\vec{X}(t)/dt = \vec{F}(\vec{X}), \tag{2}$$

where $\vec{F}$, $\vec{f}$, $\vec{X}$, and $\vec{x}$ are N-dimensional vectors. Equation (1) generates a sequence $\vec{x}_1, \vec{x}_2, \ldots$ if an initial $\vec{x}_1$ is given, while Eq. (2) generates an orbit $\vec{X}(t)$, if $\vec{X}(0)$ is given. A *strange* attractor may, for most purposes, be thought of as an attractor with dimension $d<N$, where d is noninteger. The relevant definition of dimension is that due to Hausdorff[1]

$$d = \lim_{\epsilon\to 0}[\ln n(\epsilon)][\ln(\epsilon^{-1})]^{-1}, \tag{3}$$

where $n(\epsilon)$ is the number of N-dimensional cubes of side ϵ needed to cover the attracting subset. Alternatively, $n(\epsilon) \cong K\epsilon^{-d}$ for small ϵ, where K is a constant.

Strange attractors have received special attention in recent years because of the possibility that they occur in a wide variety of physical situations. An interesting attribute of strange attractors in that they lead to chaotic or turbulent orbits. For example, the onset of turbulence in fluids in currently thought to coincide with the appearance of a strange attractor.[2]

A possible reason for interest in the Hausdorf dimension of a strange attractor is that it says something about the amount of information necessary to specify the attracting set to within an accuracy ϵ. More concretely, if one wanted to give a coarse-grained distribution function $\hat{f}_\epsilon(\vec{X})$ for the approximate calculation of a time average over the turbulent evolution of a given function g of $\vec{X}(t)$, then one would write

$$\langle g(\vec{X})\rangle \cong \int \hat{f}_\epsilon(\vec{X}) g(\vec{X})\, d^N X, \tag{4}$$

where $\langle g(\vec{X})\rangle$ is the time average of $g(\vec{X}(t))$, ϵ is the coarse-graining scale, and (4) results from the ergodic hypothesis if $g(\vec{X})$ is assumed to be slowly varying in the scale ϵ. One way of constructing $\hat{f}_\epsilon$ is to divide the original N-dimensional space into cubes of side ϵ and then specify the fraction of time that the orbit on the strange attractor spends in each cube. Only $n(\epsilon)$ cubes will have nonzero values of $\hat{f}_\epsilon$. Thus, the information necessary to specify $\hat{f}_\epsilon$ is the coordinates of the $n(\epsilon)$ cubes in which $\hat{f}_\epsilon \neq 0$ and the value of $\hat{f}_\epsilon$ in each of these cubes. Hence, in principle, the information necessary to specify $\hat{f}_\epsilon$ is contained in $n(\epsilon)(N+1)$ numbers. Thus, the dimension says something about the amount of information necessary to characterize the attractor.

Recently, a relationship between the dimension of a strange attractor of an N-dimensional map [Eq. (1)] and the *Lyapunov numbers* of the map has been conjectured.[3] Let $\lambda_1, \lambda_2, \ldots, \lambda_N$ be the Lyapunov numbers of the map ordered so that $\lambda_1 > \lambda_2 > \lambda_3 > \cdots > \lambda_N$. Then Kaplan and Yorke[3] conjecture that (the result of Mori and Fujisaka[3] is different for $N>2$)

$$d = j + [\ln(\lambda_1\lambda_2\cdots\lambda_j)][\ln \lambda_{j+1}^{-1}]^{-1}, \tag{5}$$

where j is the largest number for which $\lambda_1\lambda_2\cdots\lambda_j$

>1. The Lyapunov numbers λ_i are defined to be

$$\lambda_i = \lim_{q\to\infty} \{\text{magnitude of the eigenvalues of } \underline{J}(\vec{x}_q)\cdots\underline{J}(\vec{x}_2)\underline{J}(\vec{x}_1)\}^{1/q}, \tag{6}$$

where $\vec{x}_1, \vec{x}_2, \ldots, \vec{x}_q$ is an orbit generated by (1), and $\underline{J}(\vec{x})$ is the Jacobian matrix of (1), $J_{ij}(\vec{x}) = \partial f_i(\vec{x})/\partial x_j$. For the special case $N=2$ with $\lambda_1 > 1 > \lambda_1\lambda_2$, Eq. (5) becomes

$$d = 1 + (\ln\lambda_1)/(\ln\lambda_2^{-1}). \tag{7}$$

As a way to intuitively motivate Eq. (7), we have constructed a simple special map for which (7) is satisfied exactly,

$$x_{n+1} = \lambda_2 x_n + y_n - \lambda_1^{-1} y_{n+1},$$

$$y_{n+1} = \lambda_1 y_n \pmod 1, \tag{8}$$

where $\lambda_1 > 1 > \lambda_1\lambda_2 > 0$ is assumed and we take λ_1 to be an integer. This map may be viewed as resulting from two operations, illustrated in Fig. 1 for $\lambda_1 = 3$, which map the unit square, $0 \le x \le 1$, $0 \le y \le 1$, into itself. Application of (8) M times will map the unit square into λ_1^M vertical bands each of width along x of λ_2^M. Furthermore, it can be shown that these λ_1^M bands are contained within the $\lambda_1^{(M-1)}$ bands that result from $M-1$ applications of (8) to the unit square. Clearly the dimension of the attractor along y is 1. The dimension along x can be obtained from Eq. (3) by noting that the necessary number of coverings of length $\epsilon = \lambda_2^p$ is λ_1^p (p is an integer). Equation (7) then follows.

In order to apply Eq. (5) to the case of ordinary differential equations, Eq. (2), we introduce the *Lyapunov exponents*, $h_1, h_2, \ldots, h_N$, where N denotes the dimension of the system (2). Viewing the ordinary differential equations as generating a map advancing $\vec{X}$ forward by some fixed arbitrary increment in time, τ, we can identify $\lambda_i = \exp(h_i\tau)$, and insert the λ_i in (5). The result is independent of τ. For example, for $N=3$, we have (see also Mori[3])

$$d = 2 - h_1/h_3, \tag{9}$$

where we have assumed $h_1 > 0 > h_3$ and $h_1 + h_3 < 0$ and made use of the fact that[4] $h_2 = 0$.

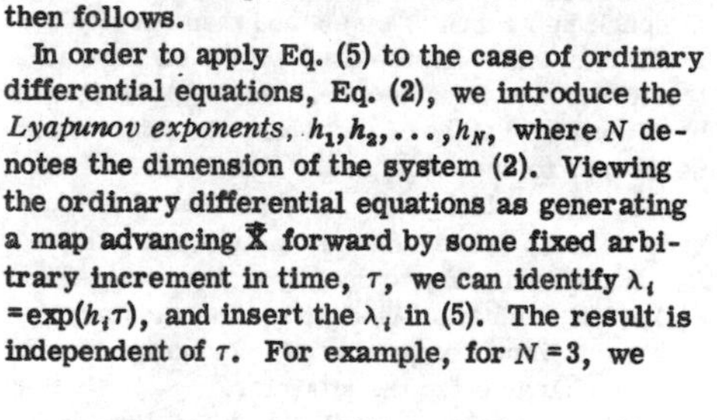

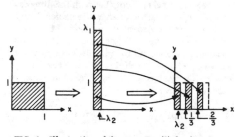

FIG. 1. Illustration of the map Eq. (8) for $\lambda_1 = 3$.

In what follows, we will first describe some numerical experiments designed to test Eq. (7). The technique will then be applied to a test of Eq. (9) with use of a system of ordinary differential equations.[5]

The Hausdorf dimension for a strange attractor of a two-dimensional map is calculated using a computer program based on Eq. (3). To calculate $n(\epsilon)$ the space is divided into boxes of side ϵ. An initial vector is chosen, and the map is iterated a sufficient number of times (i.e., much greater than $\ln\epsilon/\ln\lambda_2$) that the subsequently generated points can be considered to be on the attractor. A list is made of those boxes containing at least one point on the attractor. Each newly generated point on the attractor is checked to see if its box is on the list. If not, it is added to the list. After many iterations, the number of boxes on the list approaches $n(\epsilon)$. For small ϵ, we expect that $n(\epsilon) \sim K\epsilon^{-d}$. Thus, defining $d_\epsilon \equiv \{\ln[n(\epsilon)]\}\{\ln\epsilon^{-1}\}^{-1}$, we see that $d_\epsilon - d \cong (\ln K)/(\ln\epsilon^{-1})$. It is difficult to make $d_\epsilon - d$ small by making ϵ small since the dependence is logarithmic. Note, however, that for small ϵ a plot of d_ϵ vs $[\ln\epsilon^{-1}]^{-1}$ will be approximately linear. This is, in fact, observed numerically. Our "measured" values of d are determined by least-squares fitting a straight line to d_ϵ vs $[\ln\epsilon^{-1}]^{-1}$ for several small values of ϵ, and then extrapolating the result to $\epsilon \to 0$. The accuracy of the result is estimated from the standard deviation of the points from the fitted line. The above-described dimension measuring program has been tested, with good results by use of several sets of known dimension [an area, a line, a Cantor set, and Eqs. (8), among others].

Tests of Eq. (7) with use of three different two-dimensional maps were performed. The three maps are one originally studied by Henon[6] ($x_{j+1} = y_j + 1 - ax_j^2$, $y_{j+1} = bx_j$), one introduced by Kaplan and Yorke[3] [$x_{j+1} = 2x_j \pmod 1$, $y_{j+1} = \alpha y_j + \cos 4\pi x_j$], and one studied by Zaslavskii[7] as a model of the effect of dissipation on a Hamiltonian system $\{x_{j+1} = [x_j + \nu(1+\mu y_j) + \epsilon\nu\mu\cos 2\pi x_j] \pmod 1$, $y_{j+1} = \exp(-\Gamma)(y_j + \epsilon\cos 2\pi x_j)$, where μ

VOLUME 45, NUMBER 14 PHYSICAL REVIEW LETTERS 6 OCTOBER 1980

TABLE I. Summary of test data.

System tested	d from Lyapunov numbers	d from program based on Eq. (3)
Henon map, $a = 1.2$, $b = 0.3$	1.200 ± 0.001	1.202 ± 0.003
Henon map, $a = 1.4$, $b = 0.3$	1.264 ± 0.002	1.261 ± 0.003
Kaplan and Yorke map, $\alpha = 0.2$	1.430 676 6	1.4316 ± 0.0016
Zaslavskii map, $\Gamma = 3.0$, $\epsilon = 0.3$, $\nu = 10^2 \times 4/3$	1.387 ± 0.001	1.380 ± 0.007
Ordinary differential equations of Ref. 5	2.317 ± 0.001	2.318 ± 0.002

$\equiv [1 - \exp(-\Gamma)] \Gamma^{-1}\}$. For all of these maps $\det \underline{J}(\vec{x})$ is a constant independent of $\vec{x}$ [$-b$ for the Henon map, 2α for the Kaplan-Yorke map, and $\exp(-\Gamma)$ for the Zaslavskii map]. Thus, all of these maps lead to uniform contraction of areas on each iteration, and $\lambda_1\lambda_2 = |\det \underline{J}(\vec{x})|$. Furthermore, for the map of Kaplan and Yorke the Lyapunov numbers may be calculated analytically, $\lambda_1 = 2$ and $\lambda_2 = \alpha$. To find the Lyapunov numbers for the Henon and Zaslavskii maps we utilize Eq. (6) to calculate the largest eigenvalue, λ_1, and then find λ_2 from $\lambda_1\lambda_2 = |\det \underline{J}(\vec{x})|$. [$\lambda_2$ is usually inaccurately determined from (6) unless a very large number of decimal places is retained in the calculation.]

The first four rows of Table I summarize results from tests with use of the above-mentioned two-dimensional maps. The second column is the value of d predicted from Eq. (7). The third column gives the value of d calculated using our dimension measuring program. The fifth row of the table gives results of a similar test of Eq. (9) with use of a system of three ordinary differential equations that describes the saturation of a linearly unstable plasma wave via cubicly nonlinear coupling to linearly damped waves.[5,8] For the system studied, the divergence of $\vec{F}(\vec{X})$ [cf. Eq. (2)] is a negative constant independent of $\vec{X}$, $\partial F_1/\partial X_1 + \partial F_2/\partial X_2 + \partial F_3/\partial X_3 = -k$. Thus, by the divergence theorem, phase-space volumes evolved according to the given system of equations will shrink exponentially in time, and $h_1 + h_3 = -k$. Thus, Eq. (9) yields $d = 2 + h_1(h_1 + k)^{-1}$. h_1 can be determined by numerically computing the average exponential divergence of two infinitesimally close-by points. Thus, we obtain a prediction of d. To compute d based on Eq. (3), we associate a Poincaré map with the differential equation system.[5] This gives a picture of the intersection of the strange attractor with a surface (the "surface of section"). The dimension-measuring program is applied to this intersection. The dimension of the strange attractor is then the dimension of the intersection plus one. The last row of Table I compares results from this procedure with $d = 2 + h_1(h_1 + k)^{-1}$.

It is evident from Table I that the predicted and measured values of d agree to within the accuracy obtained for all cases considered. We note that calculation of the dimension from the Lyapunov numbers is computationally much less costly than using a routine based upon Eq. (3). As an example, for the second row of Table I, the calculation of the measured value of d required about 5 min on the Cray computer and about 4×10^5 words of memory. The calculation of λ_1, however, required about 0.3 min, and a relatively insignificant amount of memory.

Based on further development of the theory,[9] it is now clear that (7) and (9) cannot be true in general. In particular, they should not be expected to hold for cases where the eigenvalues of the Jacobian depend on $\vec{x}$ (i.e., all the maps we have tested except for that of Kaplan and Yorke). The results in Table I, however, indicate that, even if (7) and (9) do not hold exactly, they must still yield a surprisingly good approximation to the dimension for typical cases where the contraction rate is constant. The reason for this close agreement is currently under investigation.

This work was supported by the U. S. Department of Energy. We wish to thank Professor J. A.

Yorke for several extremely useful discussions.

[1]B. B. Mandelbrot, *Fractals: Form, Chance, and Dimension* (Freeman, San Francisco, 1977).

[2]Y. Treve, in *Topics in Nonlinear Dynamics*, edited by S. Jorna, AIP Conference Proceedings No. 46 (American Institute of Physics, New York, 1978), p. 147.

[3]J. L. Kaplan and J. A. Yorke, in *Functional Differential Equations and Approximation of Fixed Points*, edited by Heinz-Otto Peitten and Heinz-Otto Walther, Lecture Notes in Mathematics, Vol. 730 (Springer-Verlag, New York, 1979), p. 228; P. Frederickson, J. L. Kaplan, and J. A. Yorke, to be published; H. Mori, Prog. Theor. Phys. 63, 1004 (1980); H. Mori and H. Fujisaka, to be published.

[4]In order to see that $h_2 = 0$, note that the separation between two neighboring points on the same orbit is proportional to $|\vec{F}(\vec{X})|^{-1}$.

[5]D. A. Russell and E. Ott, to be published; M. I. Rabinovich and A. L. Fabrikant, Zh. Eksp. Teor. Fiz. 77, 617 (1979) [Sov. Phys. JETP 50, 311 (1979)].

[6]M. Henon, Commun. Math. Phys. 50, 69 (1976).

[7]G. M. Zaslavskii, Phys. Lett. 69A, 145 (1978); G. M. Zaslavskii and Kh.-R. Ya. Rachko, Zh. Eksp. Teor. Fiz. 76, 2052 (1979) [Sov. Phys. JETP 49, 1039 (1979)].

[8]A similar instability saturation process based upon quadratic rather than cubic nonlinearities also results in a strange attractor [J.-M. Wersinger, J. M. Finn, and E. Ott, Phys. Rev. Lett. 44, 453 (1980), and Phys. Fluids 23, 1142 (1980)].

[9]J. A. Yorke, private communication.

Physica 3D (1981) 605–617
North-Holland Publishing Company

ON DETERMINING THE DIMENSION OF CHAOTIC FLOWS

Harold FROEHLING, J.P. CRUTCHFIELD, Doyne FARMER, N.H. PACKARD and Rob SHAW
Physics Department, University of California, Santa Cruz, CA 95064, USA

Received 20 August 1980

We describe a method for determining the approximate fractal dimension of an attractor. Our technique fits linear subspaces of appropriate dimension to sets of points on the attractor. The deviation between points on the attractor and this local linear subspace is analyzed through standard multilinear regression techniques. We show how the local dimension of attractors underlying physical phenomena can be measured even when only a single time-varying quantity is available for analysis. These methods are applied to several dissipative dynamical systems.

1. Introduction

Recent progress in dynamical systems theory has strengthened the connection between "strange" or "chaotic" attractors and aperiodic behavior found in nature [1, 2, 3, 4]. Furthermore, Couette flow experiments performed near the transition from laminar to turbulent behavior suggest that fluid motion in the weakly turbulent regime can be understood in terms of low-dimensional chaotic attractors [5, 6]. Thus it is possible that weakly turbulent fluid flow, which in principle must be considered as an infinite-dimensional system, can be modeled by a system with relatively few phase-space dimensions. Many natural phenomena, in contrast to this, exhibit aperiodic behavior that can only be explained by a model with a very large number of dimensions. A model of thermal noise in a resistor, for instance, must account for the motions of the individual electrons within the resistor; the number of electrons involved is so large that any signal derived from the system appears "noisy". As a first step in modeling systems exhibiting aperiodic behavior, then, we must distinguish between those having an underlying low-dimensional chaotic attractor, and those requiring a large number of phase-space dimensions for a dynamical description.

Current techniques of data analysis, applied to aperiodic physical phenomena, cannot resolve these two fundamentally different sources of behavior. Power spectral analysis, for example, characterizes aperiodic behavior by the presence of broadband noise in the power spectrum, but broadband noise can be produced by systems requiring either a small or large number of phase-space dimensions. Thus the power spectrum fails to make this distinction. Low- and high-dimensional aperiodicity must instead be distinguished by a direct measurement of the number of coordinates needed to specify the state of the physical system under observation. Applied to fluid turbulence, this measurement might indicate that a chaotic attractor underlies weakly turbulent fluid flow. If a chaotic attractor exists for such flows, then the dimension would provide an experimental classification of turbulent flows.

We will describe a technique that measures the dimension of attractors in dissipative systems. More precisely, the technique measures the *approximate fractal dimension* by examining small regions of an attractor and determining whether or not the points of the attractor in each small region lie in or close to linear subspaces of dimension less than that of the phase space used to represent the system's states. From this perspective, the limit cycle attractor of the Van der Pol oscillator (a two-dimensional dynamical system) appears one dimensional, since lines (one-dimensional linear sub-spaces) are a suitable approximation to local regions of the attracting limit cycle. The simplest chaotic dynamical systems in three dimensions (such as those of Lorenz [1] and Rössler [7]) have attractors that appear locally two-dimensional. The method we introduce relies on standard multilinear regression techniques that measure the goodness of fit of linear subspaces to local regions of the attractor.

The application of this method requires an experimentalist to first develop a phase space representation of the dynamics under observation. Many experiments provide no clue at all as to which measured quantities might correspond to useful phase space coordinates. We will discuss these problems, and give examples for which unambiguous results can be obtained using phase space coordinates reconstructed from a single time series [8, 9].

2. Geometry of strange attractors

Since we are seeking a description of experimentally observable chaotic dynamics, we now review some geometrical and topological features of simple chaotic systems.

The phase space of a dissipative dynamical system can be divided into regions in which motion is unbounded and regions in which the motion is attracted into compact subsets. These compact subsets are called *attractors*, the set of all phase space points which asymptotically tend to an attractor is called its *basin of attraction*.

Certain asymptotic properties of a dynamical system's attractor are characterized by the attractor's spectrum of Lyapunov characteristic exponents (LCE's). There are as many characteristic exponents as there are dimensions in the phase space of the dynamical system. The LCE's measure the average rate of exponential convergence of trajectories onto the attractor when negative, and the average rate of exponential divergence of nearby trajectories within the attractor when positive. The magnitude of an attractor's positive exponents is a measure of its "degree of chaos" [10, 11].

The spectrum of LCE's yields a useful classification of attractors. For example, in three dimensions a dynamical system with all negative exponents is a fixed point: the LCE spectrum is denoted by (−−−). A limit cycle attractor has an LCE spectrum of (0−−). A two-torus attractor has an LCE spectrum of (00−). Chaotic attractors in three dimensions have an LCE spectrum of (+0−). In this case the positive exponent indicates exponential spreading within the attractor in the direction transverse to the flow and the negative exponent indicates exponential contraction onto the attractor. Under the action of such a flow, phase space volumes evolve into sheets, as illustrated in fig. 1. For attractors in three dimensions then, the spectrum of characteristic exponents gives a rough measure of dimension. As a first step toward a dimensional classification of attractors, we can identify the dimension of an attractor with the number of non-negative characteristic exponents (we will make this notion more precise later): the (−−−) fixed point is zero-dimensional, the (0−−) limit cycle is one-dimensional, and the (00−) two-torus is two dimensional.

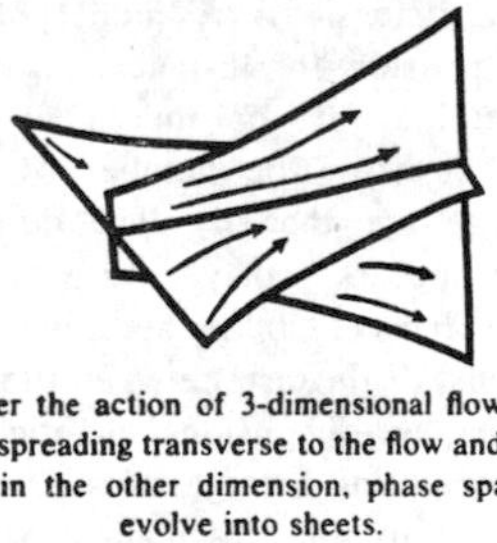

Fig. 1. Under the action of 3-dimensional flows with local exponential spreading transverse to the flow and exponential contraction in the other dimension, phase space volumes evolve into sheets.

The (+0−) chaotic attractor is also two dimensional, but its structure is actually more complicated than simple sheets: exponential divergence of nearby trajectories within a compact object requires the "folding" of sheets. A simple example of this process is illustrated in fig. 2. Trajectories diverge exponentially within a sheet; then the sheet folds and connects back to itself, forming an attractor which bears a striking resemblance to the attractor (shown in fig. 3) in a system due to Rössler [7],

$$\dot{x} = -y - z,$$
$$\dot{y} = x + ay, \tag{1}$$
$$\dot{z} = b + xz - cz,$$

with parameter values of $a = 0.2$, $b = 0.2$, and $c = 5.7$. The attractor is not simply a sheet with a single fold, but a sheet folded and refolded infinitely by the flow. A line segment which cuts

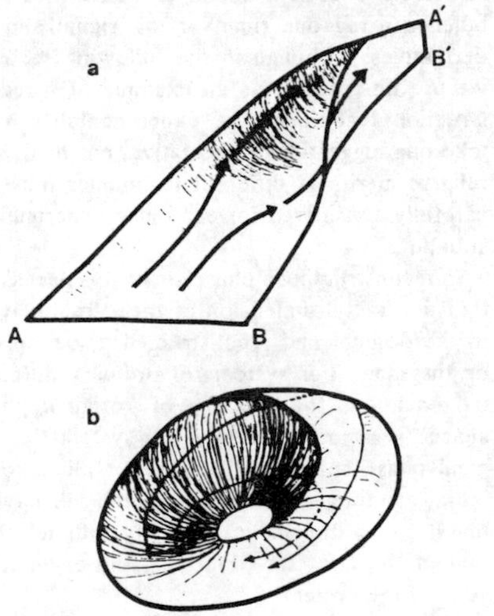

Fig. 2. (a) Exponential divergence of nearby trajectories within a compact object requires folding of sheets: connecting points A and A' together and B and B' together results in the object of (b).

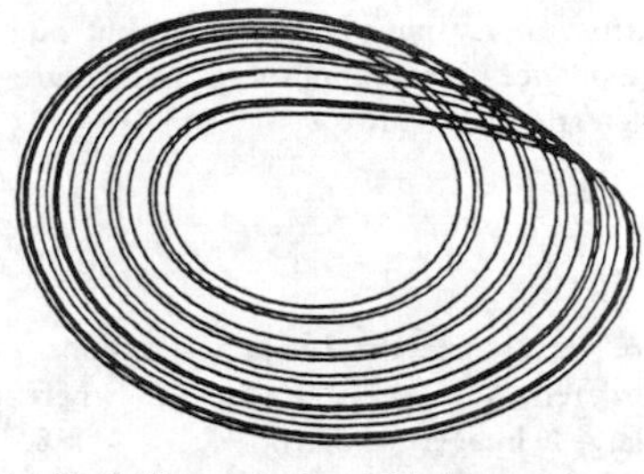

Fig. 3. X–y projection of Rössler attractor.

the attractor transverse to these sheets will intersect the attractor in a Cantor set. The attractor has topological dimension two, but a fractal dimension [12] greater than two.

To clarify the notion of fractal dimension, imagine a box which contains a small region of an attractor. If this box is subdivided into smaller boxes, some fraction of the smaller boxes will contain pieces of the attractor, while the rest won't. For example, if the small region of the attractor is a simple plane, and if a 3-dimensional box which contains it is divided by 10 in each dimension (for a total of 1000 smaller boxes), then roughly 100 of these smaller boxes will contain pieces of the plane. The number of piece-containing boxes will scale as L^d, where L is the factor by which each dimension of the box is divided. This construction defines the fractal dimension [13]:

$$d = \lim_{\epsilon \to 0} \frac{\log N(\epsilon)}{\log(1/\epsilon)}, \tag{2}$$

where $N(\epsilon)$ is the number of boxes, whose sides have length ϵ, necessary to cover the attractor. For a plane d is 2 by this construction, the same as the topological dimension. The simplest chaotic attractors (those lying in a 3-dimensional phase space) have fractal dimension between 2 and 3; d then measures how "closely packed" the sheets of an attractor are.

Mori [14] conjectured one relationship between an attractor's fractal dimension and its spectrum of Lyapunov exponents, but numerical evidence [15] supports a conjecture by Kaplan and Yorke [16],

$$d = j + \frac{\sum_{i=1}^{j} \lambda_i}{-\lambda_{j+1}}, \tag{3}$$

where we assume the Lyapunov exponents to be ordered, $\lambda_1 > \lambda_2 > \ldots . > \lambda_N$, and where j is the largest integer so that $\lambda_1 + \cdots + \lambda_j > 0$. The Rössler attractor, with parameter values $a = 0.2$, $b = 0.2$, $c = 5.7$, has an LCE spectrum of (0.075, 0, −5.372) [17]. According to eq. (3) then, this attractor has a fractal dimension of 2.014.

In discussing global properties, Williams [19] and Shaw [20] "collapse" the fractal structure to simple folded sheets by taking a neighborhood of the attractor and identifying all points within the neighborhood that become arbitrarily close under the action of the flow. They call the resulting object the attractor's "branched manifold". Any experiment, either physical or numerical, through its finite resolution automatically makes this identification: at any degree of resolution an observer sees only a branched manifold.

3. Phase space reconstruction

Certain experimental systems can be modeled by systems of ordinary differential equations. Stirred chemical reactions [9] are an example: the concentrations of various compounds and intermediates serve as coordinates in phase space. In these cases one can make an unambiguous identification between the available experimental quantities and the phase space coordinates of our dynamical systems approach. With these coordinates, determination of the attractor's dimension should be straightforward.

An experimental system modeled by partial differential equations presents greater difficulties. To obtain a complete dynamical description of a fluid its velocity at every point must, in principle, be known. But to determine the dimension of a low-dimensional attractor only a few independent quantities are needed: as few as the fractal dimension of the attractor rounded to the next high integer. Even if a low-dimensional attractor underlies fluid flow, the experimentally accessible quantities which contain the information necessary to reconstruct a phase space picture of the dynamics are not given a priori. The notion of obtaining a picture of a system's dynamics by viewing attractors projected onto a space of experimentally accessible quantities was first introduced by R. Abraham [21], who called these projections "macrons".

The problem of reconstructing a finite-dimensional phase space picture that gives a faithful representation of fluid motion (after transients have died away) is not yet settled. Packard et al. [8] mention as possibilities time derivatives, time delays [22], and even spatially separated sampling. J.C. Roux et al. [9] reconstructed a chaotic attractor from chemical turbulence using one time-varying signal and its derivatives. Although in the following sections we use time delays as an example of a reconstruction technique, the experimentalist must take our suggestions as tentative, not final. The relative merits of different techniques must be carefully considered for each new experimental situation.

In reconstructing a phase space it is necessary that the fractal dimension of the attractor lying in the original and reconstructed phase spaces be the same. For systems of ordinary differential equations the meaning of "original phase space" is clear. For a physical system the original phase space is the space of all possible initial conditions, which in general will have a much larger dimension (possibly infinite) than that of the attractor describing the asymptotic state of the system.

A simple and adequate technique for reconstructing an N-dimensional phase space, from a single time series with time delay Δt between samples, equates the reconstructed phase space point with N successive points of the time series separated by a delay $n\Delta t$. The next phase space point is found by advancing each previous point by a time Δt. Unfortunate choices for N, n, or Δt may grossly distort the phase space portrait of the original attractor: although the dimension may not change, it may for practical purposes become nearly impossible to measure.

Fig. 4 shows the Rössler attractor reconstructed from a single time series using this method. There is an obvious similarity between this attractor and the original attractor shown in fig. 3. In section 5 we show that the two attractors have the same approximate fractal dimension. As an example of an unfortunate choice of variables, the reconstructed Rössler attractor would be unrecognizable if the time delay Δt were simply rationally related to the average period of this system.

This technique reconstructs an object in an N-dimensional phase space from a single time series. But in experimental situations N need not equal the dimension of the phase space of the phenomenon studied. In the next section we will show that, as long as N is chosen greater than the dimension of the attractor in the original phase space, the attractor's approximate fractal dimension can be determined unambiguously.

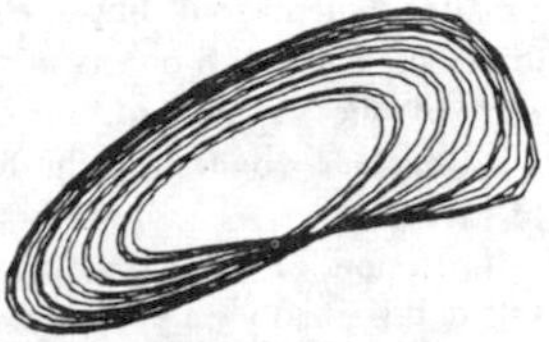

Fig. 4. X–y projections of Rössler attractor reconstructed from a time series: $\Delta t = 0.2$, $N = 3$, $n = 1$. This reconstructed attractor fits in a three dimensional box roughly 20 units on a side.

4. Local determination of the approximate fractal dimension

The concept of dimension evolved slowly from the turn of the century when Poincare set out its basic definitions [23]. Practical application of these notions, however, flourished only with the advent of modern computers, which allowed for the development of techniques for dimension measurement. Questions about the dimension associated with a given data set were first addressed by Shepard in 1962 [24] to measure the number of significant parameters in psychological experiments. Since then, techniques have been refined to measure the "intrinsic" dimension of data [25] in a wide range of fields including signal analysis [26] and pattern recognition [27]. We shall now extend some ideas related to this earlier line of inquiry to the context of dynamical behavior.

Packard et al. [8] discuss several techniques that measure the approximate fractal dimension of attractors. Takens [28] has also recently proposed a technique for measuring the fractal dimension as well as the topological entropy of an attractor. Unfortunately, these methods appear to be sensitive to instrumental noise, and they require an unduly large number of data points. In what follows we present a practical algorithm which is substantially less sensitive to noise.

After collecting N-dimensional phase space points that presumably lie on an attractor with approximate fractal dimension N-1 or less, we require for the following discussion that the phase space be partitioned so that each portion of the imbedded attractor is approximately flat. This partition is accomplished by sorting the phase space data into N-dimensional "boxes" according to their coordinate value in each dimension. If the phase space points in each box lie close to a piece of a manifold or branched manifold we can describe these points as lying approximately in a linear space of dimension less than N. Multilinear regression is the natural analysis to apply to these points: on N-dimensional phase space point it finds the best hyperplane (dimension N-1 or less) that fits the data. By using smaller dimensional phase spaces we can fit smaller-dimensional linear subspaces. The goodness of fit in each box is measured by χ^2, the sum of the squares of the deviations from this hyperplane divided by the number of degrees of freedom.

If the dimension of the hyperplane is too small the fit will typically be poor and χ^2 will be large. When the dimension is increased so that the hyperplane gives a good fit to the data, χ^2 will drop sharply. The dimension of this hyperplane provides the closest integer approximation to the fractal dimension, which we call the *approximate fractal dimension.*

We display the results of this Local Linear Regression (LLR) in a histogram of the logarithm of χ^2, showing the number of occurrences of ranges of χ^2 for regions of the attractor where the analysis is applicable. In the following histograms five columns represent a factor of ten difference in χ^2; every tick mark on the vertical axis represents ten occurrences of a particular range of χ^2.

Figure 5 shows the χ^2 histograms for Rössler attractor. The greatest number of occurrences of χ^2 for planes occurs lower by a factor of one million than the peak in occurrences for points and lines, indicating that the Rössler attractor looks locally planar. That some local regions have poor plane fits is due to the branched manifold structure mentioned above.

There is another quantity arising in LLR which also characterizes the dimension of an attractor. If too many phase space coordinates are chosen the system of equations determining the hyperplane is overdetermined: the resulting correlation matrix will be singular. (With noise in the system the correlation matrix will be approximately singular.) This fact can be used to determine the number of independent quantities needed to specify a physical system, although we will not develop this approach here.

χ^2 of multilinear regression measures the deviation of the "dependent variable" from the best fit hyperplane, not the perpendicular deviation of the points from a hyperplane. Thus χ^2 would be measured as overly large if one or more slopes were appreciable, and the χ^2 histogram would not accurately reflect the actual deviations from a hyperplane. As a practical remedy we perform multilinear regression twice for each local region. The first calculation of best fit hyperplane is used to rotate the points in the local region so that all slopes are zero; then for the second multilinear regression the deviation of the dependent variable from a hyperplane is approximately the true perpendicular deviation.

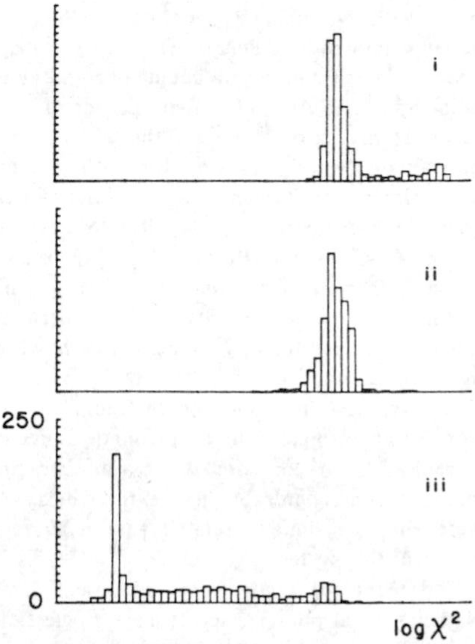

Fig. 5. Chi-squared histograms for Rössler attractor using natural coordinates, $\Delta t = 0.2$, box size = 0.5, noise $\sigma = 0.01$ (= 0.1%) added, 10,000 3-D points: (i)–(iii) best fit points, lines, and planes. On this and all succeeding histograms, every five columns on the horizontal axis represent a factor of 10 difference in χ^2; every tick mark on the vertical axis represents 10 occurrences of a particular range of χ^2.

5. Choosing a delay time

We have given evidence that the system of phase space coordinates reconstructed from a time series can be a suitable replacement for the original (possibly unknown) system. The time interval between successive samples in a time series is as yet undetermined. For the state space construction from a time series to be of practical use, we must either specify a way to choose an optimum time interval, or show that the results are relatively insensitive to the time interval chosen. Systems with chaotic attractors offer an upper limit to this time interval, corresponding roughly to the average time between "folding" of adjacent sheets of the attractor. If trajectories from different sheets approach one another exponentially, then in some finite time they will for all practical purposes become identified; a bit of information is lost, corresponding to which sheets the trajectories were on. Thus if we choose too long a time interval we can no longer make a one-to-one correspondence with points of the time series and the original attractor. For attractors that are not chaotic (solutions lying on an n-torus, for instance) there is no upper limit on the time interval. In our numerical studies we typically chose the delay time to be approximately 10% of the folding time.

A qualitative lower limit also exists. If the time interval is chosen too small, then the N successive points are all approximately equal; the attractor appears "stretched out" along the $x = y = z = \ldots$ direction. This problem with short time intervals could be alleviated by taking appropriate differences between coordinates and dividing by the time interval (analogous to taking derivatives), but this procedure introduces noise into the phase space picture. Choosing $n\Delta t$ of our reconstruction technique sufficiently large obviates the need for this noise-producing procedure.

Our experience indicates that any time interval between these extremes is a suitable one. Of course, if the system is periodic or quasiperiodic and one chooses a time interval simply rationally related to a period, the results will be misleading.

As a check on the relative insensitivity of results with time interval (within extremes), we have performed LLR on the Rössler attractor reconstructed from a time series by the method in section 3. Fig. 6 shows the histograms for the Rössler attractor for time increments of 0.1, and delays of 5, 10, and 20 time increments between coordinates. The results show that for each time interval, a local 2-dimensional surface (plane) fits the data much better than lines or points.

Fig. 6 also illustrates the effect of choosing too many coordinates to describe a system. A four-dimensional phase space was reconstructed from the time series using delays, although only three independent quantities are needed to describe points on the attractor. The correlation matrices needed to calculate χ^2 are approximately singular when trying to fit 3-D linear subspaces to 4-D points; still, the χ^2 histograms for these indicate fits as good as those for planes. Normally one would check the values of the correlation matrix to avoid any possible problems.

Notice that the time interval we recommend is of the same order as typical sampling times used in power spectral analysis. In fact, the same time series used for power spectral analysis may be used for LLR with absolutely no change.

6. Noise and curvature

The technique of LLR naturally accommodates deviation from a hyperplane. But two sources (excluding branched manifold structure) contribute to this deviation: (1) observational noise in the data and (2) curvature of the attractor.

It is possible to distinguish these two sources of deviation. If the box size is decreased the deviation from a hyperplane will decrease, since the deviation due to curvature is decreased. As the box size decreases, eventually observational noise becomes the only source of deviation.

The ability to measure the dimension of an

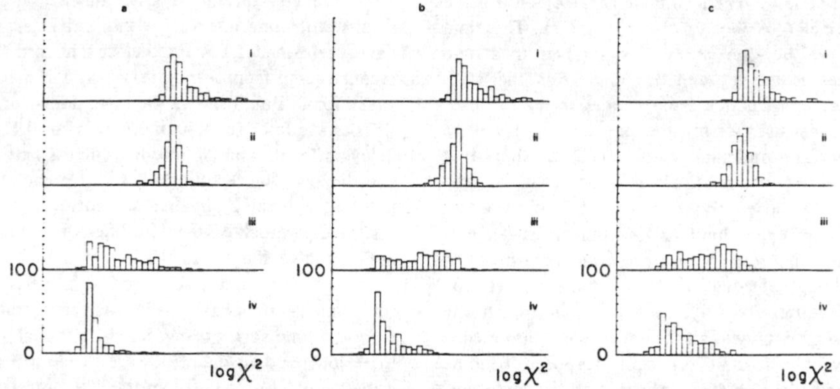

Fig. 6. Chi-squared histograms for Rössler attractor reconstructed from a time series, $\Delta t = 0.2$, box size = 1.0, 10,000 4-D points, noise $\sigma = 0.01$ (= 0.1%) added: (a) delay = 5 time intervals: (i)–(iv) best fit subspaces of 0, 1, 2, and 3 dimensions; (b) delay = 10 time intervals: (i)–(iv) best fit subspaces of 0, 1, 2, and 3 dimensions; (c) delay = 20 time intervals: (i)–(iv) best fit subspaces of 0, 1, 2 and 3 dimensions.

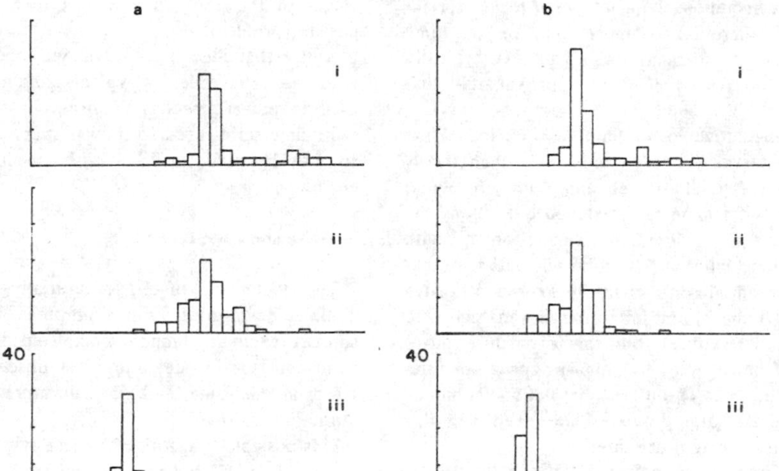

Fig. 7. Chi-squared histograms of Rössler attractor reconstructed from a time series, $\Delta t = 0.2$, time delay = 5 box size 2.0, 5,000 points: (a) noise $\sigma = 0.1$ (= 1%) added: (i)–(iii) best fit points, lines, and planes; (b) noise $\sigma = 0.2$ (= 2%) added: (i)–(iii) best fit points, lines, and planes.

attractor in the presence of noise depends on both the magnitude of the noise compared to the size of the attractor, and the box size. To reliably detect a hyperplane the size of the box must be larger than the noise. But if the noise is too large we must choose a box whose size is of the same order of magnitude as (or larger than) the curvature of the attractor and any attempt to fit a hyperplane must result in a large χ^2; the hyperplane fit fails.

By this method the dimension of chaotic attractors can be reliably measured even when observation noise is appreciable, if the geometry of these attractors is simple. Fig. 7 shows χ^2 histograms for the Rössler attractor, with zero-mean Gaussian random noise of standard deviations 0.1 and 0.2 added to the signals. These correspond roughly to 1% and 2% noise relative to the size of the attractor. In both cases planes fit the data significantly better than points or lines. For noise levels much greater than this a determination of the dimension becomes difficult.

7. The effect of fractal structure

The Rössler attractor shown in fig. 3 has strong contraction of phase space volumes effectively truncating its fractal structure to a branched manifold after a small fraction of a revolution around the attractor. To examine the effect of fractal structure on determining the dimensionality of chaotic attractors we have studied the scaling properties of the Hénon map [29] which has weak convergence of adjacent folds of the attractor.

$$x_{i+1} = y_i + 1 - ax_i$$
$$y_{i+1} = bx_i. \tag{4}$$

This 2-dimensional system clearly reveals many leaves of the fractal structure. Fig. 8 shows the Hénon map and χ^2 histograms for the Hénon map. The histograms are qualitatively the same for the different box sizes, showing that even for substantial fractal structure we can confidently determine the approximate fractal dimension of an attractor.

8. Direct calculation of the fractal dimension

The sorting necessary to use LLR offers the information necessary to estimate the fractal dimension using the scaling construction of eq. (2). For 20,000 2-dimensional points we measure

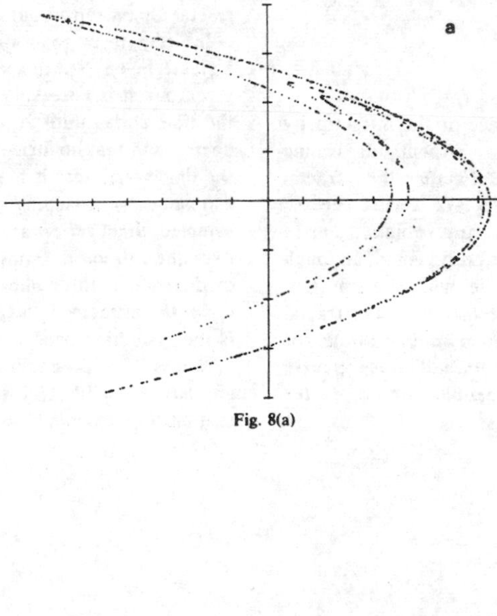

Fig. 8(a)

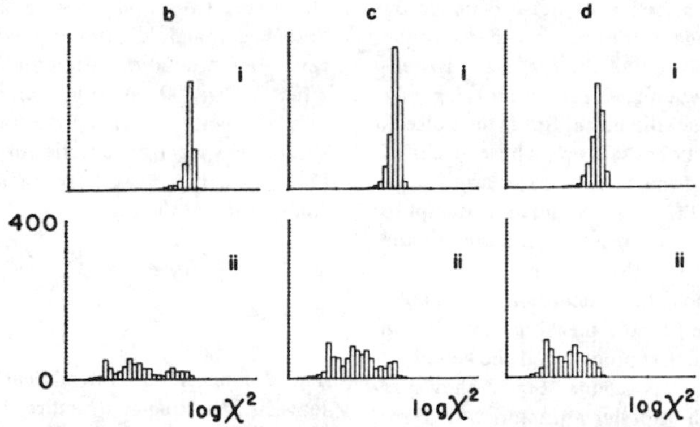

Fig. 8. (a) Hénon map, $a = 1.4$, $b = 0.3$, 1000 2-D points, noise $\sigma = 0.0001$ (= .01%) added (b) box size = 0.005: (i)–(ii) best fit points and lines (c) box size = 0.01: (i)–(ii) best fit points and lines; (d) box size = 0.02: (i)–(ii) best fit points and lines

a fractal dimension

$$d = 1.19,$$

while eq. (3) gives a fractal dimension of

$$d = 1.26,$$

and Simó [30] finds

$$d = 1.2365.$$

Since we define the surface of the attractor by the points lying on it, we expect our scaling construction to underestimate the fractal dimension because as we take smaller boxes, some will not contain any points, simply because they have not yet been visited, although pieces of the attractor lie within them. Unfortunately, this simple estimate of the fractal dimension becomes even more inaccurate for higher-dimensional attractors, and its use appears impractical for even sheet-like structures for this reason.

9. Data acquisition requirements

In principle, the LLR technique will work for attractors of any dimension. Unfortunately, there are practical limitations that prevent this from being feasible for attractors of dimension greater than about five. These limitations are also present, and in fact they are more severe, for the straightforward determination of the fractal dimension as outlined in section 8.

In order to fit an m-dimensional hyperplane, at least $m + 1$ points are required. Thus for a given box it is necessary to collect points from the time series until at least $m + 1$ points have been found that lie inside the box. The length of the time series that is needed to do this may be estimated by assuming that the points of the sampled time series are uniformly distributed over the attractor. Using D-dimensional boxes of diameter ϵ, the number of boxes required to cover the attractor is $N(R) \approx R^D$, where $R = 1/\epsilon$ is the resolution used in constructing the cover of boxes. The probability of finding a point in any box is roughly $1/N$, so the number of points that must be examined to make it likely to find a single point in any given box is $\approx N$. For a successful determination of the dimension of the best fit hyperplane for the points in a box, at least $D+1$ points are needed. The number of data points n that must be taken is therefore

$$n \approx (D+1)N \approx (D+1)R^D .$$

For reasonable values of R, this number becomes very large when D exceeds 5. For example, if $R=20$ and $D=5$, $n \approx 2\times 10^7$. For typical rates of data acquisition, it is often impractical to gather or process this many data points.

For LLR the situation is improved slightly because it is not necessary to gather sufficient points in every box, but rather only to gather points in enough boxes to gain a representative sample of boxes covering the attractor. Since the points on an attractor do not typically have uniform distribution, the more probable boxes will fill out much earlier than others, reducing the number of data points that must be examined.

A nonuniform distribution of points works *against* a direct measurement of the fractal dimension based on eq. (2), such as that made for the Hénon map in section 8. To measure the fractal dimension directly from the definition it is necessary to count the number of boxes required to cover the attractor at several levels of resolution. For a uniform distribution, this typically takes considerably more than $n = R^D$ points; if the distribution is uneven, this problem becomes worse. In addition, for a direct determination, sufficient computer memory must be allocated to cover the attractor. For many applications this proves to be a more severe constraint than does the problem of gathering and processing sufficient data.

In summary, we have shown that this technique may be feasibly applied to physical systems that are described by low dimensional attractors. Thus, this may be useful for studying fluid flows such as those found near the transition to turbulence. For fully developed turbulence this technique, or any technique that requires sufficient data points to cover the attractor with moderate resolution, will certainly be impractical.

10. Other systems

Fig. 9 shows the results of LLR for x–y–z coordinates of the Lorentz system [1]

$$\dot{x} = -10x + 10y,$$
$$\dot{y} = -xz + 28x - y, \qquad (5)$$
$$\dot{z} = xy - (8/3)z.$$

We interpret the results as indicating that local 2-D surfaces are almost everywhere good fits; not all plane fits are good, indicating the branched manifold structure of the attractor.

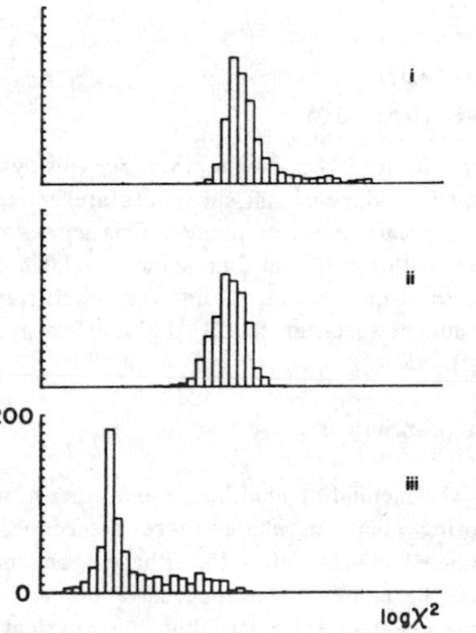

Fig. 9. Chi-squared histograms for Lorentz system, $\Delta t = 0.2$, time delay = 5, box size = 1.0, noise $\sigma = 0.01$ (= 0.1%) added, 10,000 3-D points: (i)–(iii) best fit points, lines, and planes.

It is difficult for LLR to distinguish between LCE signatures of (+0−), a branched manifold attractor, and (00−), a torus. However, LLR analysis, in conjunction with power spectral analysis, does distinguish these two very different types of attractor. A power spectrum of a 2-torus has sharp peaks corresponding to two incommensurate frequencies (along with harmonics and sums and differences of frequencies) but no broadband noise, while a chaotic system is characterized by broadband noise in its power spectrum. Thus we see that power spectral analysis and LLR give complementary information.

For 4-D systems LLR apparently cannot distinguish–even with power spectral analysis–between a (++0−) system and a (+00−) system: both will exhibit branched manifold structure and have broadband noise in their power spectra. An example of a (++0−) system is the Hyperchaos system of Rössler [31]:

$$\begin{aligned} \dot{x} &= -y - z, \\ \dot{y} &= x + 0.25y + w, \\ \dot{z} &= 3 + xz, \\ \dot{w} &= -0.5z + 0.05w. \end{aligned} \tag{6}$$

Fig. 10 shows the χ^2 histograms for this system. Locally 3-dimensional subspaces are better fits than points, lines, or planes. This agrees quite well with the fractal dimension $d = 3.006$ computed with eq. (3) from the characteristic exponent spectrum (0.121, 0.021, 0.005, −23.7) [17].

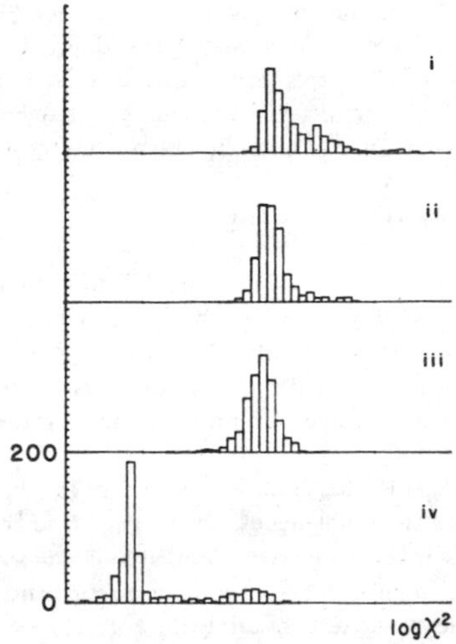

Fig. 10. Chi-squared histograms for hyperchaos system reconstructed from a time series, $\Delta t = 0.2$, time delay = 5, box size = 2.0, noise $\sigma = 0.01$ (= 0.1%) added, 20,000 4-D points: (i)–(iv) best fit local subspaces of 0, 1, 2, and 3 dimensions.

11. Conclusions

The method of multilinear analysis on neighboring points in phase space determines the dimension of an attractor. Phase space points may be chosen from successive points in the time series of some variable in a physical system when no a priori choice is clear. The dimension of simple attractors we have studied (as measured by a histogram of the logarithm of χ^2) is independent of this choice of phase space points; it is also relatively independent of the choice of time interval between successive points in a time series, and of the definition of propinquity of phase space points (size of box chosen).

The information obtained in performing local linear regression may also give quantitative information concerning the fractal nature of strange attractors. Furthermore, we hope to extend the method of LLR to calculate directly the Lyapunov characteristic exponents–either from known phase space coordinates or from a time series–in order to provide a classification of dynamical behavior found in physical systems. In light of these results LLR analysis may prove useful in fluid, chemical, and solid state turbulence experiments and in the understanding of other chaotic phenomena for which power spectral analysis proves inadequate to characterize a system's behavior.

Acknowledgements

DF would like to thank the Hertz Foundation for their support; JC NP, and RS gratefully acknowledge NSF support.

References

[1] E.N. Lorentz and J. Atmos. Sci. **20** (1963) 130.

[2] D. Ruelle and F. Takens, Comm. Math. Phys. **50** (1976) 69.

[3] B.A. Huberman and J.P. Crutchfield, Phys. Rev. Lett. **43** (1979) 1743.

[4] B.A. Huberman, J.P. Crutchfield and N.H. Packard, Appl. Phys. Lett. **37** (1980) 750.

[5] J. Gollub and H. Swinney, Phys. Rev. Lett. **35** (1975) 927.

[6] R.W. Walden and R.J. Donnelly, Phys. Rev. Lett. **42** (1979) 301.

[7] O.E. Rössler, Phys. Lett. **57A** (1976) 196.

[8] N.H. Packard, J.P. Crutchfield, J.D. Farmer and R.S. Shaw Phys. Rev. Lett. **45** (1980) 712.

[9] J.C. Roux, A. Rossi, S. Bachelart and C. Vidal, Phys. Lett. **77A** (1980) 391.

[10] I. Shimada and T. Nagashima, Prog. Theor. Phys. **61** (1979) 1605.

[11] G. Bennetin, L. Galgani and J.M. Strelcyn, Phys. Rev. **A14** (1976) 2338.

[12] B.B. Mandelbrot, Fractals: Form, Chance, and Dimension W.H. Freeman, San Francisco, (1977).

[13] Eq. (2) defines a set's *capacity*, which we will take to be the same as the Haussdorf–Besicovitch dimension for the attractors we study.

[14] H. Mori, Prog. Theor. Phys. **63** (1980) 1044.

[15] J.D. Farmer, N.H. Packard, Chaotic Attractors in Infinite-Dimensional Systems I: Differential Delay Equations, UCSC preprint, submitted to Physica D (1980).

[16] J. Kaplan and J. Yorke, Chaotic Behavior of Multi-dimensional Difference Equations, Springer Lecture Notes in Mathematics **730** (1979) 204.

[17] J.P. Crutchfield, Senior Thesis, University of California at Santa Cruz (1979).

[18] J.D. Farmer, J.P. Crutchfield, H. Froehling, N.H. Packard and R.S. Shaw, Annals, N.Y. Acad. Sci. **357** (1980) 453.

[19] R.F. Williams, Berkeley Turbulence Seminar, 1976–77, P. Benard, and T. Ratiu, eds., Springer Lecture Notes in Mathematics **615** (1977).

[20] R. Shaw, Z. Naturforsch. 36a (1981) 80.

[21] R. Abraham, in Evolution and Consciousness E. Jantsch and C.H. Waddington, eds., (Addison Wesley, Reading, Mass., 1976).

[22] D. Ruelle, private communication.

[23] H. Poincaré, The Foundations of Science (Science, Garrison, N.Y. 1913).

[24] R.N. Shepard, Psychometrika **27** (1962) 125; **27** (1962) 219.

[25] K. Fukunaga and D.R. Olsen, IEEE Trans. Comp. **C-20** (1971) 176.

[26] G.V. Trunk, IEEE Trans. Comp. **C-25** (1976) 165.

[27] L.J. White and A.A. Ksienski, Pattern Recognition **6** (1974) 35.

[28] F. Takens, Detecting Strange Attractors in Turbulence, preprint (1980).

[29] M. Hénon, Comm. Math. Phys. **50** (1976) 69.

[30] C. Simó, On the Hénon–Pomeau Attractor, preprint (1979).

[31] O.E. Rössler Phys. Lett. **71A** (1979) 155.

VOLUME 50, NUMBER 5 PHYSICAL REVIEW LETTERS 31 JANUARY 1983

Characterization of Strange Attractors

Peter Grassberger[a] and Itamar Procaccia
Chemical Physics Department, Weizmann Institute of Science, Rehovot 76100, Israel
(Received 7 September 1982)

A new measure of strange attractors is introduced which offers a practical algorithm to determine their character from the time series of a single observable. The relation of this new measure to fractal dimension and information-theoretic entropy is discussed.

PACS numbers: 47.25.-c, 52.35.Ra

Dissipative dynamical systems which exhibit chaotic behavior often have an attractor in phase space which is *strange*.[1-3] Strange attractors are typically characterized by fractal dimensionality[4] D which is smaller than the number of degrees of freedom F, $D<F$. So far, this fractal (or Hausdorff) dimension has been the most commonly used measure of the "strangeness" of attractors.[5-10] Several attempts to compute this number directly from box-counting algorithms, which stem from the definition of this dimensionality, have been presented.[7-10] It turns out that it is very difficult[10] to compute D whenever $D>2$. Most importantly, the use of a single time series of any observable to extract this measure of the attractor has been found to be impractical for dynamical systems which possess attractors whose $D>2$.[10] An important question is then how to analyze experimental signals. In this Letter we suggest a different measure for the strangeness of attractors, a measure which can be easily obtained from any time series without resorting to Poincaré maps,[3] and which is closely related to the fractal dimension. We shall attempt to argue that in fact this measure is more relevant in many cases than D itself.

The measure is obtained by considering correlations between points of a long-time series on the attractor. Denote the N points of such a long-time series by $\{\vec{X}_i\}_{i=1}^{N} \equiv \{\vec{X}(t+i\tau)\}_{i=1}^{N}$, where τ is an arbitrary but fixed time increment. The definition of the correlation *integral* is

$$C(r) \equiv \lim_{N\to\infty} \frac{1}{N^2} \sum_{i,j=1}^{N} \theta(r - |\vec{X}_i - \vec{X}_j|)$$

$$\equiv \int_0^r d^d r' \, c(\vec{r}\,'), \qquad (1)$$

where $\theta(x)$ is the Heaviside function and $c(\vec{r})$ is the standard correlation function. The main point of this paper is that $C(r)$ behaves as a power of r for small r:

$$C(r) \propto r^{\nu}. \qquad (2)$$

Moreover, the exponent ν is closely related to D as well as to a properly defined entropy which is discussed below. Before continuing the analysis we show in Fig. 1 two examples of the behavior (2).

Shown are the logarithms of the correlation integrals for the Hénon map[11] [Fig. 1(a)] and the Lorenz model[1] [Fig. 1(b)] as a function of $\log r$. Equally convincing power laws were obtained for the Kaplan-Yorke map,[5] the Rabinovich-Fabrikant[12] equations, and the logistic map[13] at the onset of chaos (see Table I). For the Zaslavskii map[15] no clear power law is obtained even for longer runs.

One sees from Table I that ν is in all cases very close to D (the errors quoted are "educated guesses"), but is never greater than D (with the exception of Zaslavskii's map where no good power law is seen). Below, we shall argue that

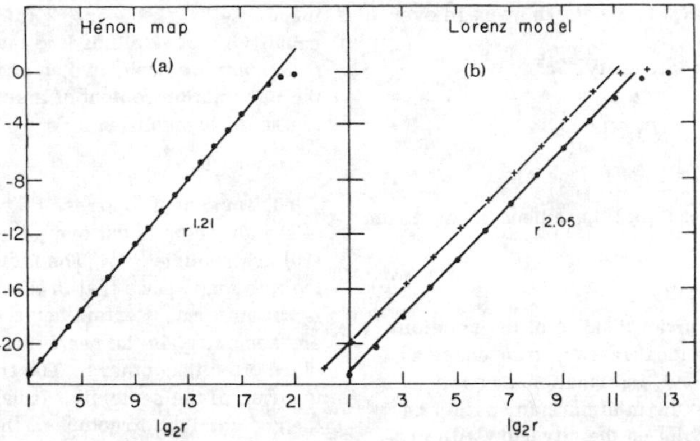

FIG. 1. Correlation integrals for (a) Hénon map and (b) Lorenz model on doubly logarithmic scales. In (b) the upper line was computed from a single variable time series. In both panels the scale of r is arbitrary.

the value $\nu = 1.21$ obtained for the Hénon map is wrong as a result of systematic errors. An improved method yields indeed $\nu = 1.25 \pm 0.02$, such that $\nu = D$ within the errors. We now discuss the relation between various measures of the strangeness of attractors.

Consider a coverage of the attractor by hypercubes of edge length l. If the attractor is a fractal, then the number $M(l)$ of cubes that contain a piece of the attractor is[4]

$$M(l) \sim l^{-D}. \tag{3}$$

Denote now by μ_i ($i = 1, 2 \ldots$) the number of points from the set $\{\vec{X}_i\}_{i=1}^{N}$ which are in the ith nonempty cube. Up to a factor of $O(1)$ (i.e., the number of nearest-neighbor cubes) we can write

$$C(l) \sim \frac{1}{N^2} \sum_{i=1}^{M(l)} \mu_i^{\,2} = \frac{M(l)}{N^2} \langle \mu^2 \rangle , \tag{4}$$

TABLE I. Maps used in evidence, with values of corresponding parameters.

	ν	No. of iterations, time increment τ	D	σ
Hénon map,	1.21 ± 0.01[d]	15 000	1.26[g]	...
$a = 1.4$, $b = 0.3$	1.25 ± 0.02[e]	20 000		
Kaplan-Yorke map,				
$\alpha = 0.2$	1.42 ± 0.02	15 000	1.431[g]	...
Logistic equation,	0.500 ± 0.005			
$b = 3.569\,945\,6\ldots$	$0.4926 < \nu < 0.5024$[f]	25 000	0.538[h]	0.517 097 6[i]
Lorenz equation[a]	2.05 ± 0.01	15 000; $\tau = 0.25$	2.06 ± 0.01[g]	...
Rabinovich-				
Fabrikant equation[b]	2.18 ± 0.01	15 000; $\tau = 0.25$	...	...
Zaslavskii map[c]	(~ 1.5)	25 000	1.39[g]	...

[a] Parameters as in Refs. 10 and 6.
[b] Parameters as in Sec. 3 of Ref. 12.
[c] Parameters as in Ref. 7.
[d] From Eqs. (1) and (2).
[e] From single-variable time series, with $f = 3$.
[f] Exact analytic bounds (Ref. 14).
[g] Ref. 7.
[h] Ref. 9.
[i] Ref. 14.

where angular brackets denote an average over all occupied cells.

By the Schwartz inequality,

$$C(l) \geq \frac{M(l)}{N^2}\langle\mu\rangle^2 = \frac{1}{N^2 M(l)}\left[\sum_{i=1}^{M(l)}\mu_i\right]^2 = 1/M(l) \sim l^D, \qquad (5)$$

where $\sum\mu_i = N$ and Eq. (3) have been used. From this it follows that

$$\nu \leq D. \qquad (6)$$

To gain further understanding of the relation between these two measures of strangeness we introduce a third one, information-theoretic entropy.[16] This is the minimal information needed to pin down a point on the attractor with precision l:

$$S(l) = -\sum_{i=1}^{M(l)} p_i \ln p_i, \qquad (7)$$

where p_i is the probability for a point to fall in the ith cube (for $N\to\infty$, $p_i = \mu_i/N$). For a *uniform* coverage [i.e., $p_i = 1/M(l)$] the entropy is $S^0(l) = \ln M(l) = \text{const} - D\ln l$. In the general case $S(l) < S^0(l)$. If we adopt the *Ansatz*

$$S(l) = S_0 - \sigma \ln l \qquad (8)$$

(with σ called "information dimension" by Farmer, and "dimension" by Renyi[16]), we are led to the inequality

$$\sigma \leq D. \qquad (9)$$

Finally we want to prove that $\nu \leq \sigma$, thus estimating σ from above and below. A sketch of the proof is as follows. Consider two nested coverings with cubes of lengths l and $2l$, respectively. Evidently, $M(l) = 2^D M(2l)$. Define p_i as the probability for a point to fall in cube i of the fine covering and P_j the probability for it to fall in the jth cube of the coarser covering, $P_j = \sum_{i\in j} p_i$. Define now ω_i by $p_i = \omega_i P_j$. Evidently $\sum_{i\in j}\omega_i = 1$.

According to Eq. (4) the correlation integral $C(l)$, up to a factor of $O(1)$, is

$$C(l) \sim \sum_{i=1}^{M(l)} p_i^2 = \sum_{j=1}^{M(2l)} P_j^2 \sum_{i\in j}\omega_i^2.$$

If we assume now that ω_i is independent of j, we can write $C(l)/C(2l) = \langle\omega^2\rangle/\langle\omega\rangle$. On the other hand we consider

$$S(2l) - S(l) = \sum_j P_j \sum_{i=j}\omega_i \ln\omega_i = \langle\omega\ln\omega\rangle/\langle\omega\rangle.$$

Defining the quantity $W = \omega/\langle\omega\rangle$ we can employ the inequality[17,18] $\langle W^2\rangle \geq \exp[\langle W\ln W\rangle]$ to prove the inequality $\nu \leq \sigma$. We thus find that, combining ν and D *together*, we can have an excellent estimate of the information content of a strange attractor via the set of inequalities

$$\nu \leq \sigma \leq D. \qquad (10)$$

It is important to stress that when the covering of the attractor is uniform, the equalities in Eq. (10) are realized.[18,19] The fact that $\nu \neq D$ in the logistic map shows that in this case the coverage is not uniform. Certain neighborhoods have higher "seniority" in the sense that they are visited more often than others. The fractal dimension is ignorant of the seniority. It has to do only with the geometrical structure of the attractor. Regions of the attractor which are rarely visited contribute to D with equal weight as regions of high visiting rate. The correlation integral (and the entropy) are, however, sensitive to this effect. In this sense ν may be a more relevant measure of the attractor than D because it is sensitive to the *dynamical* process of coverage of the attractor. The difference between ν and D gives a measure of the importance of different seniority of different neighborhoods.

Although the data of Table I were obtained using ~20 000 points, convergence in all cases expect Zaslavskii's map was already apparent with a few thousand points. (This should be compared with 200 000 points needed to obtain convergence of the box-counting algorithm used to compute D in the case of the Hénon map, and the lack of convergence with the same number of points in the case of the Lorenz model.)

A variant of our method, inspired by Refs. 20 and 21, consists of measuring instead of $\vec{X}_i$ only one component, say X_i. A new f-dimensional phase space is then constructed by using vectors

$$\vec{\xi}_i = (X_i, X_{i+\tau}, X_{i+2\tau}, \ldots, X_{i+f\tau}),$$

which are then inserted in Eq. (1) instead of the $\vec{X}_i$. An example of the results obtained from this procedure with $f = 3$ is shown in Fig. 1(b) for the Lorenz model. The + marks were obtained by following the x variable only. The power law is still satisfactory. The value of ν calculated from the one-variable data is 2.06 ± 0.02. From an experimental point of view this procedure is much preferable. In fact, it allows a consistent determination of ν in a high-dimensional dynamical system, as will be shown in a forthcoming publication.[14] Also, it allows elimination of systematic errors due to corrections to scaling, by choosing the "embedding dimension" f larger than strictly necessary. Choosing $f=3$ for the Hénon map we get, e.g., the value $\nu = 1.25 \pm 0.02$ quoted above. More details will be presented in Ref. 14.

To summarize, we have introduced the exponent ν of the power-law dependence of the correlation integral as a new measure for the strangeness of attractors.[22] The value of this exponent can be obtained from a time series of one or more variables. The computation is relatively easy and converges rapidly. The relation of the exponent ν to the fractal dimension and information entropy have been discussed. If the attractor is visited by the trajectory uniformly, all these measures coalesce. Otherwise, we attempted to argue that ν might be more dynamically relevant than D. It is our hope that this new characteristic exponent would be actually measured in experimental systems whose dynamics is governed by strange attractors.

This paper has been supported in part by the Israel Commission for Basic Research and by the Minerva Foundation. We thank Dr. H. G. E. Hentschel and Professor R. M. Mazo for some useful discussions.

[a]Permanent address: Physics Department, University of Wuppertal, Wuppertal, West Germany.

[1]E. N. Lorenz, J. Atmos. Sci. 20, 130 (1963).

[2]D. Ruelle and F. Takens, Commun. Math. Phys. 20, 167 (1971).

[3]E. Ott, Rev. Mod. Phys. 53, 655 (1981).

[4]B. B. Mandelbrot, *Fractals—Form, Chance and Dimension* (Freeman, San Francisco, 1977).

[5]J. C. Kaplan and J. A. Yorke, in *Functional Differential Equations and Approximations of Fixed Points*, edited by H.-O. Peitgen and H.-O. Walther, Lecture Notes in Mathematics Vol. 730 (Springer, Berlin, 1979).

[6]H. Mori, Prog. Theor. Phys. 63, 1044 (1980).

[7]D. A. Russel, J. D. Hanson, and E. Ott, Phys. Rev. Lett. 45, 1175 (1980).

[8]H. Froehling, J. P. Crutchfield, D. Farmer, N. H. Packard, and R. Shaw, Physica (Utrecht) 3D, 605 (1981).

[9]P. Grassberger, J. Stat. Phys. 26, 173 (1981).

[10]H. S. Greenside, A. Wolf, J. Swift, and T. Pignataro, Phys. Rev. A 25, 3453 (1982).

[11]M. Hénon, Commun. Math. Phys. 50, 69 (1976).

[12]M. I. Rabinovich and A. L. Fabrikant, Zh. Eksp. Teor. Fiz. 77, 617 (1979) [Sov. Phys. JETP 50, 311 (1979)].

[13]R. M. May, Nature 261, 459 (1976).

[14]P. Grassberger and I. Procaccia, to be published.

[15]G. M. Zaslavskii, Phys. Lett. 69A, 145 (1978).

[16]J. Balatoni and A. Renyi, Publ. Math. Inst. Hung. Acad. Sci. 1, 9 (1956), and in *Selected Papers of A. Renyi, Vol. 1* (Academiai Budapest, Budapest, 1976); J. D. Farmer, Physica (Utrecht) 4D, 366 (1982).

[17]W. Feller, *An Introduction to Probability Theory and its Applications* (Wiley, New York, 1971), Vol. 2, 2nd ed.

[18]B. B. Mandelbrot, in *Turbulence and Navier-Stokes Equations*, edited by R. Temam, Lecture Notes in Mathematics, Vol. 565 (Springer, Berlin, 1975).

[19]T. A. Witten, Jr., and L. M. Sander, Phys. Rev. Lett. 47, 1400 (1981).

[20]N. H. Packard, J. P. Crutchfield, J. D. Farmer, and R. S. Shaw, Phys. Rev. Lett. 45, 712 (1980).

[21]F. Takens, in *Proceedings of the Symposium on Dynamical Systems and Turbulence, University of Warwick, 1979–1980*, edited by D. A. Rand and L. S. Young (Springer, Berlin, 1981).

[22]After this paper had been submitted for publication we learned about recent reports by L. S. Young and J. D. Farmer, F. Ott, and J. A. Yorke in which smaller measures are discussed. We thank Dr. Farmer for bringing these papers to our attention.

Volume 77A, number 6 PHYSICS LETTERS 23 June 1980

FLUCTUATIONS AND THE ONSET OF CHAOS

J.P. CRUTCHFIELD [1] and B.A. HUBERMAN
Xerox Palo Alto Research Center, Palo Alto, CA 94304, USA

Received 3 April 1980

We consider the role of fluctuations on the onset and characteristics of chaotic behavior associated with period doubling subharmonic bifurcations. By studying the problem of forced dissipative motion of an anharmonic oscillator we show that the effect of noise is to produce a bifurcation gap in the set of available states. We discuss the possible experimental observation of this gap in manv svstems which display turbulent behavior.

It has been recently shown that the deterministic motion of a particle in a one-dimensional anharmonic potential, in the presence of damping and a periodic driving force, can become chaotic [1]. This behavior, which appears after an infinite sequence of subharmonic bifurcations as the driving frequency is lowered, is characterized by the existence of a strange attractor in phase space and broad band noise in the power spectral density. Furthermore, it was predicted that under suitable conditions such turbulent behavior may be found in strongly anharmonic solids [2]. Since condensed matter is characterized by many-body interactions, one may ask about the effects that random fluctuating forces have on both the nature of the chaotic regime and the sequence of states that lead to it. This problem is also of relevance to the behavior of stressed fluids, where it has been suggested that strange attractors play an essential role in the onset of the turbulent regime [3]. Although there are experimental results supporting this conjecture [4–6], other investigations have emphasized the possible role of thermodynamic fluctuations directly determining the chaotic behavior [7].

With these questions in mind, we study the role of fluctuations on the onset and characteristics of chaotic behavior associated with period doubling subharmonic bifurcations. We do so by solving the problem of forced dissipative motion in an anharmonic potential with the aid of an analog computer and a white-noise generator. As we show, although the structure of the strange attractor is very stable even under the influence of large fluctuating forces, their effect on the set of available states is to produce a symmetric gap in the deterministic bifurcation sequence. The magnitude of this bifurcation gap is shown to increase with noise level. By keeping the driving frequency fixed we are also able to determine that increasing the random fluctuations induces further bifurcations, thereby lowering the threshold value for the onset of chaos. Finally, the universality of these results is tested by observing the effect of random errors on a one-dimensional map, and suggestions are made concerning the possible role of temperature in experiments that study the onset of turbulence.

Consider a particle of mass m, moving in a one-dimensional potential $V = a\eta^2/2 - b\eta^4/4$, with η the displacement from equilibrium and a and b positive constants. If the particle is acted upon by a periodic force of frequency ω_d and amplitude F, and a fluctuating force $f(t)$, with its coupling to all other degrees of freedom represented by a damping coefficient γ, its equation of motion in dimensionless units reads

$$\frac{d^2\psi}{dt^2} + \alpha\frac{d\psi}{dt} + \psi - 4\psi^3 = \Gamma\cos\left(\frac{\omega_d}{\omega_0}\right)t + f(t) \qquad (1)$$

with $\psi = \eta/2\eta_0$, the particle displacement normalized to the distance between maxima in the potential (η_0

[1] Permanent address: Physics Department, University of California, Santa Cruz, CA 95064, USA.

Volume 77A, number 6 PHYSICS LETTERS 23 June 1980

$= (a/b)^{1/2}$), $\alpha = \gamma/(ma)^{1/2}$, $\Gamma = Fb^{1/2}/2a^{3/2}$, $\omega_0 = (a/m)^{1/2}$ and $f(t)$ a random fluctuating force such that

$$\langle f(t) \rangle = 0 \quad (2a)$$

and

$$\langle f(0) f(t) \rangle = 2A\delta(t) \quad (2b)$$

with A a constant proportional to the noise temperature of the system.

The range of solutions of eq. (1), in the case where $f(t) = 0$ (the deterministic limit) has been investigated earlier [1]. For values of Γ and ω_d such that the particle can go over the potential maxima, as the driving frequency is lowered, a set of bifurcations takes place in which orbits in phase space acquire periods of 2^n times the driving period, T_d. At a threshold frequency ω_{th}, a chaotic regime sets in, characterized by a strange attractor with "periodic" bands. Within this chaotic regime, as the frequency is decreased even further, another set of bifurcations takes place whereby 2^m bands of the attractor successively merge in a mirror sequence of the 2^n periodic sequence that one finds for $\omega \to \omega_{th}^+$. The final chaotic state corresponds to a single band strange attractor, beyond which there occurs an irreversible jump into a periodic regime of lower amplitude.

In order to study the effects of random fluctuations on the solutions we have just described, we solved eq. (1) using an analog computer in conjunction with a white-noise generator having a constant power spectral density over a dynamical range two orders of magnitude larger than that of the computer. Time series and power spectral densities were then obtained for different values of Γ, A and ω_d. While we found that the folding structure of the strange attractor is very stable under the effect of random forces, the bifurcation sequence that is obtained in the presence of noise differs from the one encountered in the deterministic limit.

Our results can be best summarized in the phase diagram of fig. 1, where we plot the observed set of bifurcations (or limiting set) as a function of the noise level, N, normalized to the rms amplitude of the driving term, Γ. The vertical axis denotes the possible states of the system, labeled by their periodicity $P = 2^n$, which is defined as the observed period normalized to the driving force, T_d. As can be seen, with increasing noise level a symmetric bifurcation gap appears, depleting states both in the chaotic and periodic phases. This set of inaccessible states is characterized by the fact that the longest periodicity which is observed before a strange attractor appears is a decreasing function of N, with the maximum number of bands which appear in the strange attractor behaving in exactly the same fashion. This gap extends over a large range of noise levels (up to $N = 1.5$), beyond which the motion either becomes unstable (i.e., $|\psi| \to \infty$; lower dashed line) or an amplitude jump takes place from the chaotic regime to a limit cycle of period 1 (upper dashed line).

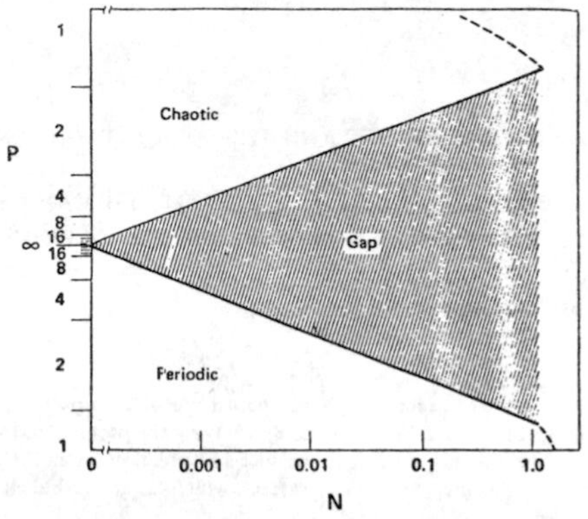

Fig. 1. The set of available states of a forced dissipative anharmonic oscillator as a function of the noise level. The vertical axis denotes the periodicity of a given state with $P = T/T_d$. The noise level is given by $N = A/\Gamma_{rms}$. The shaded area corresponds to inaccessible states.

We can illustrate this behavior by looking at the power spectral densities, $S(\omega)$ at fixed values of the driving frequency while increasing N. Fig. 2 shows such a sequence for $\omega_d/\omega_0 = 0.6339$, $\Gamma = 0.1175$, and $\alpha = 0.4$. Fig. 2(a) corresponds to $S(\omega)$ near the deterministic limit which, for the parameter values used, displays a limit cycle of period four. As N is increased, a transition takes place into a chaotic regime characterized by broad band noise with subharmonic content of periodicity $P = 4$ (fig. 2(b)) [‡1]. As the noise is increased even further, a new bifurcation occurs from which a new

‡1 We should mention that the Poincare map corresponding to this state clearly shows a four-band strange attractor with a single fold.

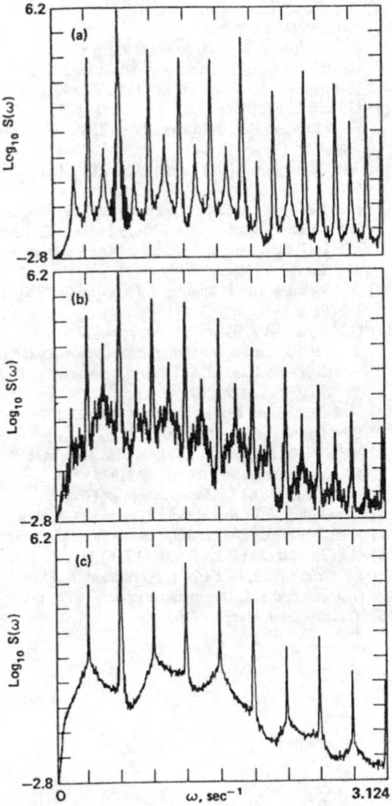

Fig. 2. Power spectral densities at increasing values of the effective noise temperature, for $\Gamma = 0.1175$, $\alpha = 0.4$, and $\omega_d = 0.6339\ \omega_0$. Fig. 2(a): $N = 10^{-4}$. Fig. 2(b): $N = 0.005$. Fig. 2(c): $N = 0.357$.

chaotic state with $P = 2$ emerges. Physically, this sequence reflects the fact that a larger effective noise temperature (and hence a larger fluctuating force) makes the particle gain enough energy so as to sample increasing nonlinearities of the potential, with a resulting motion which in the absence of noise could only occur for longer driving periods [‡2].

A different set of states appears if the noise level is kept fixed while changing the driving frequency. In this case the observed states of the system correspond to vertical transitions in the phase diagram of fig. 1, with the threshold value of the driving period, T_{th}^n, at which one can no longer observe periodicities $P \geqslant 2^n$, behaving like

$$T_{th}^n = T_{th}^{\infty}(1 - N_n^{\gamma}) \qquad (3)$$

for $0 \leqslant N_n \leqslant 1$, with N_n the corresponding noise level, T_{th}^{∞} the value of the driving period for which the deterministic equation undergoes a transition into the chaotic state, and γ a constant which we determined to be $\gamma = \sim 1$ for $P \geqslant 2$ [‡3].

In order to test the universality of the bifurcation gap we have just described, we have also studied the bifurcation structure of the one-dimensional map described by

$$x_{L+1} = \lambda x_L(1 - x_L) + n_L(0, \sigma^2) \qquad (4)$$

where $0 \leqslant x_L \leqslant 1$, $0 \leqslant \lambda \leqslant 4$, and n_L is a gaussian random number of zero mean and standard deviation σ. For $n_L = 0$, eq. (4) displays a set of 2^n periodic states universal to all single hump maps [8–9], with a chaotic regime characterized by 2^m bands that merge pairwise with increasing λ [10]. For $n_L \neq 0$ and a given value of σ, the effect of random errors on the stability of the limiting set is to produce a bifurcation gap analogous to the one shown in fig. 1.

The above results are of relevance to experimental studies of turbulence in condensed matter, for they show that temperature plays an important role in the observed behavior of systems belonging to this same universality class. In particular, Belyaev et al. [11], Libchaber and Maurer [12] and Gollub et al. [13] have reported that under certain conditions the transition to turbulence is preceded and followed by different finite sets of 2^n subharmonic bifurcations. It would therefore be interesting to see if temperature changes or external sources of noise in the fluids can either reduce or increase the set of observed frequencies, thus providing for a test of these ideas. In the case of solids such as superionic conductors, the expo-

‡2 In the regime of subharmonic bifurcation the dependence of response amplitude on driving frequency is almost linear.

‡3 Using the scaling relation $(T_{th} - T_n)/(T_{th} - T_{n+1}) = \delta$ [1] this implies that the threshold noise level scales like $N_n/N_{n+1} = \delta$, with $\delta = 4.669201609$ ... .

Volume 77A, number 6 PHYSICS LETTERS 23 June 1980

nential dependence on temperature of their large diffusion coefficients might provide for an easily tunable system with which to study the existence of bifurcation gaps. Last, but not least, these studies can serve as useful calibrations on the relative noise temperature of digital and analog simulations.

In concluding we would like to emphasize the wide applicability of the effects that we have reported. Beyond the experimental studies of turbulence, there exist other systems which belong to the same universality class as the anharmonic oscillator and one-dimensional maps. These systems range from the ordinary differential equations studied by Lorenz [10], Robbins [14], and Rossler [15] to partial differential equations describing chemical instabilities [16]. Since period doubling subharmonic bifurcation is a universal feature of all these models, our results provide a quantitative measure of the effect of noise on their non-linear solutions.

The authors wish to thank D. Farmer, N. Packard, and R. Shaw for helpful discussions and the use of their simulation system.

References

[1] B.A. Huberman and J.P. Crutchfield, Phys. Rev. Lett. 43 (1979) 1743.
[2] See also, C. Herring and B.A. Huberman, Appl. Phys. Lett. 36 (1980) 976.
[3] See, D. Ruelle, in Lecture notes in physics, eds. G. Dell'Antonio, S. Doplicher and G. Jona-Lasinio (Springer-Verlag, New York, 1978), Vol. 80, p. 341.
[4] G. Ahlers, Phys. Rev. Lett. 33 (1975) 1185; G. Ahlers and R.P. Behringer, Prog. Theor. Phys. (Japan) Suppl. 64 (1978) 186.
[5] J.P. Gollub and H.L. Swinney, Phys. Rev. Lett. 35 (1975) 927; P.R. Fenstermacher, H.L. Swinney, S.V. Benson and J.P. Gollub, in: Bifurcation theory in scientific disciplines, eds. D.G. Gorel and D.E. Rossler (New York Academy of Sciences, 1978).
[6] A. Libchaber and J. Maurer, J. Physique Lett. 39 (1978) L-369.
[7] G. Ahlers and R.W. Walden, preprint (1980).
[8] T. Li and J. Yorke, in: Dynamical systems, an International Symposium, ed. L. Cesari (Academic Press, New York, 1972), Vol. 2, 203.
[9] M. Feigenbaum, J. Stat. Phys. 19 (1978) 25.
[10] E.N. Lorenz, preprint (1980).
[11] Yu.N. Belyaev, A.A. Monakhov, S.A. Scherbakov and I.M. Yavorshaya, JETP Lett. 29 (1979) 295.
[12] A. Libchaber and J. Maurer, preprint (1979).
[13] J.P. Gollub, S.V. Benson and J. Steinman, preprint (1980).
[14] K.A. Robbins, SIAM J. Appl. Math. 36 (1979) 451.
[15] O.E. Rossler, Phys. Lett. 57A (1976) 397; J.P. Crutchfield, D. Farmer, N. Packard, R. Shaw, G. Jones and R.J. Donnelly, to appear in Phys. Lett.
[16] Y. Kuramoto, preprint (1980).

VOLUME 46, NUMBER 14 PHYSICAL REVIEW LETTERS 6 APRIL 1981

Scaling for External Noise at the Onset of Chaos

J. Crutchfield, M. Nauenberg, and J. Rudnick
Physics Department, University of California, Santa Cruz, California 95064
(Received 8 December 1980)

The effect of external noise on the transition to chaos for maps of the interval which exhibit period-doubling bifurcations are considered. It is shown that the Liapunov characteristic exponent satisfies scaling in the vicinity of the transition. The critical exponent for noise is calculated with the use of Feigenbaum's renormalization group approach, and the scaling function for the Liapunov characteristic exponent is obtained numerically by iterating a map with additive noise.

PACS numbers: 64.60.Fr, 02.90.+p, 47.25.Mr

The notion that the transition to turbulence in fluids has universality properties similar to those of critical phenomena has been suggested by Feigenbaum[1] on the basis of the scaling behavior of mathematical models near the onset of chaos.[2] A further impetus for an analogy between the transition to chaos and critical point phase transitions was given[3] by the observation that as a control parameter r in these models increases past a critical value r_c into the chaotic regime the measure-theoretic entropy—the Liapunov characteristic exponent $\bar{\lambda}$—has an envelope curve of the form $(r-r_c)^\tau$. The universal exponent τ is given by $\tau=\ln 2/\ln\delta=0.449\,806\,9\ldots$, where δ is the maximum eigenvalue associated with perturbations about the invariant map[1] of the interval. The transition to chaos in these models is heralded by a cascade of period-doubling bifurcations,[2] which is also of interest to an understanding of the onset of turbulence in physical systems.[4]

Motivated by the interpretation of experiments in fluids[5] and solids and by some recent numerical calculations,[6,7] we have considered theoretically the effect of added external noise on the transition to chaos in maps of the interval. The main result to be reported here is that the noise amplitude behaves as a *scaling variable* and that the dependence of the Liapunov characteristic exponent $\bar{\lambda}$ on the noise amplitude σ and $\bar{r}=(r-r_c)/r_c$ is of the scaling form

$$\bar{\lambda}(r,\sigma)=\sigma^\theta L(\bar{r}/\sigma^\gamma) \tag{1}$$

with $L(y)$ a universal function, and θ and γ universal exponents. In the limit of vanishing noise $\sigma\to 0$ we have $\bar{\lambda}\propto\bar{r}^\tau$ which implies that as $y\to\infty$, $L(y)\propto y^\tau$, and leads to the exponent relation $\theta=\gamma\tau$.

The idea that the noise plays a role parallel to that of the ordering field in a ferromagnetic transition was conjectured previously in Ref. 7. The noise exponent θ is a new critical exponent which we evaluate from an extension of Feigenbaum's scaling theory. Our result agrees with the recently observed value[7] of θ to within the limits of accuracy of the measurement. We also report on the measured form of the scaling function $L(y)$.

We start out by specifying the form of the one-dimensional map with additive noise. It is defined by the stochastic recursion relation

$$x_{K+1}=f(x_K;r)+\xi_K\sigma \tag{2}$$

with $f(x;r)$ a continuous function of x in a finite interval having a parabolic maximum, and r a parameter that controls the shape of the function.[2] A common example is the function $rx(1-x)$ with $0\leq r\leq 4$, and $0\leq x\leq 1$. The quantity ξ_K is a random variable controlled by an even distribution of unit width, and σ is a variable that controls the width (or amplitude) of the noise. Note that when $\sigma=0$ the map is perfectly deterministic.

We consider successive iterations of the stochastic map, Eq. (2) with r at the critical value r_c, following techniques introduced by Feigenbaum. Setting the origin of coordinates to the x for which the function $f(x;r)$ is a maximum and rescaling this maximum to 1, the 2^nth iterate of $f(x;r_c)$ converges to $(-\alpha)^{-n}g(\alpha^n x)$, where $g(x)$ is a universal map satisfying the equation

$$g(g(x))=-\alpha^{-1}g(\alpha x) \tag{3}$$

with $\alpha=-1/g(1)$. Adding a small amount of noise $\xi\sigma$, we assume that the corresponding 2^nth iterate of the map converges to $(-\alpha)^{-n}[g(\alpha^n x)+\xi\sigma\kappa^n D(\alpha^n x)]$ with $D(x)$ a universal x-dependent noise amplitude function and κ a constant. When σ is small enough, we have

$$\begin{aligned}g(g(x)+\xi\sigma D(x))+\xi'\sigma D(g(x)+\xi\sigma D(x)) &\approx g(g(x))+\xi\sigma g'(g(x))D(x)+\xi'\sigma D(g(x))+O(\sigma^2)\\ &=g(g(x))+\xi''\sigma\{[g'(g(x))D(x)]^2+[D(g(x))]^2\}^{1/2}.\end{aligned} \tag{4}$$

In going to the last line we used the fact that ξ and ξ' are independent random variables, and that ξ'' is also a random variable. This and our above assumption implies that $D(x)$ must satisfy the eigenvalue equation

$$KD(\alpha x)=\alpha\{[g'(g(x))D(x)]^2+[D(g(x))]^2\}^{1/2}. \quad (5)$$

We have solved Eq. (5) for the eigenvalue κ and the corresponding eigenfunction $D(x)$ using the known results[1] for α and $g(x)$. Carrying out a calculation involving a polynomial interpolation for $D(x)$ we have found $\kappa=6.61903\ldots$.

In the immediate vicinity above the transition to chaos the invariant probability distribution associated with the stochastic map will consist of 2^n bands, where n is an integer that grows in the case of the deterministic map by unit steps to infinity as the transition is approached.[8,9] In the case of the stochastic map, n grows to a finite value—and then decreases by unit steps as one passes to the other side of the transition. This modification of the deterministic bifurcation sequence is called a bifurcation gap.[6]

We now extend to the present case the previous discussion in Ref. 2 of the scaling behavior of the Liapunov characteristic exponent $\bar{\lambda}$, given by

$$\bar{\lambda}=\lim_{N\to\infty}\frac{1}{N}\sum_{k=1}^{N}\ln|f'(x_k;r)|, \quad (6)$$

or alternatively

$$\bar{\lambda}=\int p(x)\ln|f'(x;r)|\,dx, \quad (7)$$

where $p(x)$ is the invariant probability distribution associated with the map. Applying the above-mentioned considerations we obtain[10]

$$\bar{\lambda}=2^{-n}L(\delta^n\bar{r},\kappa^n\sigma). \quad (8)$$

Now, we assume that there will be 2^n bands in the chaotic regime when $\kappa^n\sigma$ is of order unity so that $n=-\ln\sigma/\ln\kappa$. Substituting this result into Eq. (9) we obtain Eq. (1) for $\bar{\lambda}$ with the two exponents θ and γ given in terms of Feigenbaum's eigenvalue δ and the new eigenvalue κ by $\theta=\ln 2/\ln\kappa=0.366754\ldots$ and $\gamma=\ln\delta/\ln\kappa=0.815359\ldots$. The appearance of a bifurcation gap implies that $L(y)$ vanish at some $y=y_0$ which in turn implies that the maximum number n of bifurcations is determined by the relation $\bar{r}_{n\,\max}=y_0\sigma^\gamma$. This behavior has been observed numerically.[6]

Measurements of the behavior of $\bar{\lambda}$ as a function of σ at $\bar{r}=0$ have already been made by numerically calculating $\bar{\lambda}$ according to Eq. (6) with varying amounts of noise.[7] The measured value for θ is 0.37 ± 0.01. This agrees with our theoretical value for θ to within the experimental error.

To verify the existence of the scaling function $L(y)$ of Eq. (1) we used our values of θ and γ to plot $\bar{\lambda}\sigma^{-\theta}$, with $\bar{\lambda}$ the result of numerical calculations of Eq. (6), as a function of $\bar{r}\sigma^{-\gamma}$. The results are shown in Figs. 1 and 2 for three different noise levels: $\sigma=10^{-6}$, 10^{-8}, and 10^{-10}. The results for those three different noise levels all fall on a universal curve in the chaotic regime,

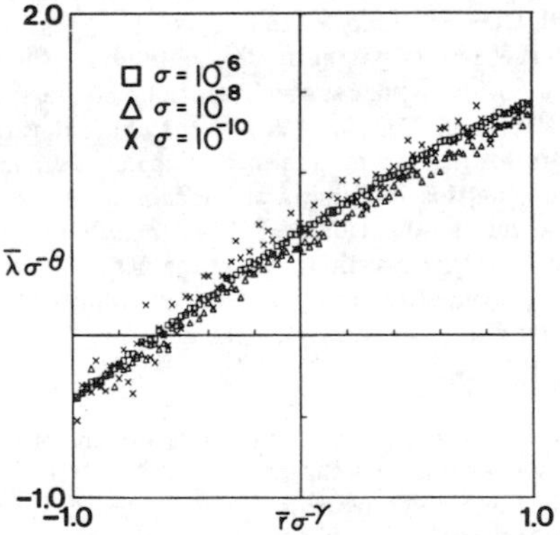

FIG. 1. Numerical determination of the scaling function $L(y)$, Eq. (1). The quantity $\bar{\lambda}\sigma^{-\theta}$ is plotted against 100 values of $y=\bar{r}\sigma^{-\gamma}$ at each of three noise levels: $\sigma=10^{-6}$, 10^{-8}, and 10^{-10}. $\bar{\lambda}$ was calculated with use of Eq. (6), with $N=10^6$ and with ξ_κ a uniformly distributed random number of standard deviation σ.

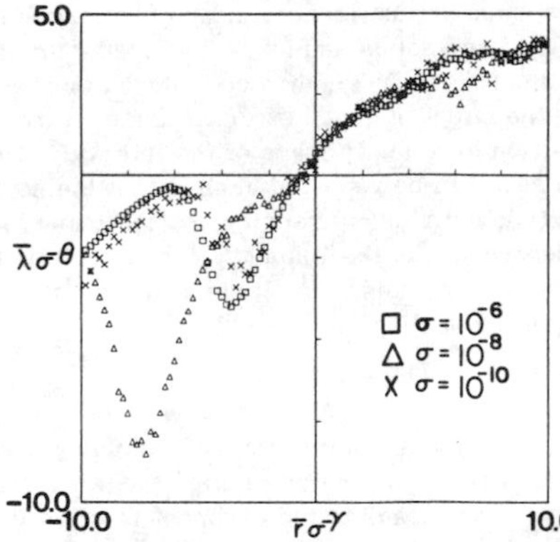

FIG. 2. $\bar{\lambda}\sigma^{-\theta}$ is plotted again, but over a wider range of $y=\bar{r}\sigma^{-\gamma}$ to illustrate the scaling regime. See text for discussion of various features. The details are the same as in Fig. 1, expect that $\bar{\lambda}$ was calculated with $N=10^5$ in Eq. (6).

and in its immediate vicinity, Fig. 1, and fit the asymptotic behavior $L(y) \sim y^{\tau}$ for large y. The results do *not* coincide in the periodic regime, Fig. 2, but they could have been made to agree if we had chosen noise amplitudes differing by factors of κ, instead of factors of 100. This more restricted scaling follows from considerations of the type enunciated above.

These results appear to us to be both exciting and highly provocative. A theoretical picture of the transition to turbulence is just beginning to emerge; the analogy to critical phenomena should lead to new and important insights into the nature and characteristics of this transition.

The authors have benefited from conversations with B. A. Huberman and wish to thank him for the use of computing facilities at Xerox Palo Alto Research Center. One of us (J.C.) would also like to acknowledge useful discussions with N. Packard and the receipt of a University of California Regents Fellowship. This work is supported by the National Science Foundation.

[1]M. J. Feigenbaum, J. Statist. Phys. 19, 25 (1978).

[2]For a recent monograph on this subject, see P. Collet and J. P. Eckmann, *Iterated Maps of the Interval as Dynamical Systems* (Birkhäuser, Boston, 1980).

[3]B. A. Huberman and J. Rudnick, Phys. Rev. Lett. 45, 154 (1980). The exponent τ appears as t in this reference. We have replaced the latin by a greek letter for consistency with other critical exponents.

[4]A. Libchaber and J. Maurer, J. Phys. (Paris), Colloq. 41, C3-51 (1980); M. J. Feigenbaum, Phys. Lett. 74A, 375 (1979); J. P. Gollub, S. V. Benson, and J. Steinman, Ann. N.Y. Acad. Sci. (to be published); B. A. Huberman and J. P. Crutchfield, Phys. Rev. Lett. 43, 1743 (1979).

[5]G. Ahlers, private communication.

[6]J. P. Crutchfield and B. A. Huberman, Phys. Lett. 77A, 407 (1980).

[7]J. P. Crutchfield, J. D. Farmer, and B. A. Huberman, to be published.

[8]J. P. Crutchfield, J. D. Farmer, N. Packard, R. Shaw, G. Jones, and R. J. Donnelly, Phys. Lett. 76A, 1 (1980).

[9]E. N. Lorentz, Ann. N.Y. Acad. Sci. (to be published).

[10]The details of the derivation of the result (8), which involves a careful consideration of the structure of the bands, will be presented in a future paper.

VOLUME 46, NUMBER 14 PHYSICAL REVIEW LETTERS 6 APRIL 1981

Scaling Theory for Noisy Period-Doubling Transitions to Chaos

Boris Shraiman, C. Eugene Wayne, and Paul C. Martin

Department of Physics and Division of Applied Sciences, Harvard University, Cambridge, Massachusetts 02138

(Received 19 December 1980)

The effect of noise on systems which undergo period-doubling transitions to chaos is studied. With the aid of nonequilibrium field-theoretic techniques, a correlation-function expression for the Lyapunov parameter (which describes the sensitivity of the system to initial conditions) is derived and shown to satisfy a *scaling theory*. Since these transitions have previously been shown to exhibit *universal behavior*, this theory predicts *universal effects* for the noise. These predictions are in good agreement with numerical experiments.

PACS numbers: 64.60.Fr, 02.90.+p, 47.25.Mr

During the past few years, the onset of chaotic behavior, after a sequence of period-doubling transitions, has been extensively studied. Feigenbaum[1] has observed that these transition sequences exhibit "universal" features akin to those of phase transitions; Collet and Eckmann[2] have noted that these universal features are shared by differential equations and multidimensional maps in which chaos is preceded by a sequence of period doublings; and Libchaber and Maurer[3] have observed this phenomenon in a convective cell with small aspect ratio. Recently, Huberman and Rudnick[4] have related one of the pretransitional parameters identified by Feigenbaum with the growth of disorder (i.e., the Lyapunov parameter) in the chaotic regime, and Huberman and Crutchfield[5] have examined numerically the effect of external noise on the onset of chaos. Nevertheless, many connections between period-doubling chaotic transitional phenomena and the critical phenomena at second-order phase transitions remain unclear.

The purpose of this Letter is the following: (1) to present a scaling theory (in which "noise" and "stress" play the role of external field and temperature) for systems that become chaotic via the period-doubling mechanism, and (2) to compare the dependence on noise and stress of the Lyapunov exponent predicted by this theory with numerical experiments performed by Huberman and collaborators.[6] Our results, found by field-theoretic methods designed for nonequilibrium systems, clarify some connections between phase transitions and the onset of chaos. These methods identify the Lyapunov exponent with the "long-time" limit of the nonequilibrium response function introduced by Martin, Siggia, and Rose.[7]

The Lyapunov exponent, λ, describes how solutions that were initially close to one another evolve after a long time (or many steps). Its sign and magnitude provide a measure of the "sensitivity of the system to initial conditions"; a large negative value implies great insensitivity to initial differences, a vanishing value implies that initial differences neither grow nor decay, and a large positive value implies rapid separation and great sensitivity.

Our principal quantitative result is that, at corresponding stress points between successive period doubling transitions, as a function of the noise amplitude, σ, of the noise and the magnitude of the difference between the stress, r, and the stress, r_∞, at which the onset to chaos occurs without noise, the Lyapunov parameter satisfies

$$\lambda(r_\infty - r; \sigma) = (r_\infty - r)^t \, \Phi((r_\infty - r)^{-t} \sigma^u); \tag{1}$$

with

$$t = (\ln 2)/\ln\delta = 0.4498\ldots \tag{2a}$$

and

$$u = (\ln 2)/\ln\beta \approx 0.34\ldots. \tag{2b}$$

The quantity, $\delta = 4.669\ldots$, is Feigenbaum's universal scaling parameter for functions $f(x)$ with quadratic maxima and β is a scaling parameter, associated with the noise, whose value we have calculated in a second-order approximation to be $\beta \approx 7.7\ldots$.

Let us consider the one-dimensional difference equation

$$x_{m+1} = f_r(x_m) + \xi_m. \tag{3}$$

The dynamical variable x_m ranges over the interval $[-1, 1]$, the function $f_r(x)$ has its maximum value, $f_r(0) = 1$, at $x = 0$, and ξ_m is a Gaussian random variable with $\langle \xi_m \rangle = 0$ and $\langle \xi_m \xi_{m'} \rangle = \sigma^2 \delta_{mm'}$. We analyze this stochastic difference equation, using a discrete version of the path-integral formulation[8] developed for stochastic nonlinear Langevin equations. The average of the functional $F[\{x\}]$ over sequences $\{x\}$ which obey Eq. (3) is given by

$$\langle F[\{x\}] \rangle = Z^{-1} \langle \pi_m \int dy_m \, \delta(y_m - x_m) F[\{y\}] \rangle = Z^{-1} \int [\mathfrak{D}y][\mathfrak{D}s] F[\{y\}] \langle \exp\{ i \sum_m s_m [y_{m+1} - f_r(y_m) - \xi_m] \} \rangle$$
$$= Z^{-1} \int [\mathfrak{D}x][\mathfrak{D}s] F[\{y\}] \exp \Omega_r(x, s, \sigma)$$

with

$$\Omega_r(x, s, \sigma) \equiv \sum_m \{ i s_m [x_{m+1} - f_r(x_m)] - \tfrac{1}{2}\sigma^2 s_m^2 \}. \tag{4}$$

Specifically, the correlation function is given by

$$\langle x_m x_{m'} \rangle = Z^{-1} \int [\mathfrak{D}x][\mathfrak{D}s] x_m x_{m'} \exp \Omega_r(x, s, \sigma). \tag{5}$$

We also introduce the Martin-Siggia-Rose response function, $R(r, \sigma; m - m') \equiv i \langle x_m s_{m'} \rangle$,

$$R(r, \sigma; m - m') = Z^{-1} \int [\mathfrak{D}x][\mathfrak{D}s] \, i x_m s_{m'} \exp \Omega_r(x, s, \sigma) \tag{6}$$

which depends on both x and s, the variable "conjugate" to x, to define the Lyapunov parameter in the presence of noise. Let $X_N(y)$ be the expectation value of x_N in the ensemble with the initial condition $x_0 = y$. We then have for large N

$$\exp(\lambda N) \equiv \lim_{N \to \infty, \, \epsilon \to 0} [X_N(y + \epsilon) - X_N(y)]/\epsilon = Z^{-1} \epsilon^{-1} \int [\mathfrak{D}x][\mathfrak{D}s] x_N \exp \Omega_r(x, s, \sigma) \{ \exp[i \epsilon s_0 f_r'(x_0)] - 1 \}$$
$$\approx i f_r'(y) \langle x_N s_0 \rangle = f_r'(y) R(r, \sigma; N). \tag{7}$$

As observed originally by Feigenbaum, it is sometimes preferable to study the iterated equation which, for weak noise, takes the form

$$x_{m+1} = f_r(x_m; n) + \xi_m' g_r(x_m; n)$$

in terms of the 2^n-th iterate of f_r

$$f_r(x;n) = \underbrace{f_r \circ f_r \circ f_r \circ \cdots \circ f_r(x)}_{2^n \text{ terms}}. \tag{8}$$

We call this iterative process "coarse-graining." If $f_r(x) = f_r(x;0)$ has a period-2 limit cycle, then $f_r(x;1)$ has a pair of isolated fixed points; similarly, a four-cycle becomes a pair of two-cycles, etc. As a result of coarse graining, the function $g_r(x;n)$ appears, which, along with $f_r(x;n)$, is assumed to approach a fixed point through the coarse graining and rescaling transformations. Approximate expressions for the scaling variables are obtained by explicitly integrating over every other x_m variable in the functional integral.

With these ideas in mind, let us introduce $\langle\,\rangle_n$ to describe expectations computed at the nth coarse-graining level, i.e.,

$$R(r,\sigma;N;n) \equiv i\langle x_N s_0\rangle_n = Z^{-1}\int[\mathfrak{D}x][\mathfrak{D}s]\,ix_N s_0 \exp\Omega_r(x,s,\sigma;n)$$

with

$$\Omega_r(x,s,\sigma;n) \equiv \sum_m \{is_m[x_{m+1} - f_r(x_m;n)] - \tfrac{1}{2}\sigma^2 s_m^2 g_r^2(x;n)\}. \tag{9}$$

We can also express the Lyapunov parameter in terms of the coarse-grained response function,

$$\exp(\lambda N) = f_r'(y;n) R(r,\sigma;N;n). \tag{10}$$

Note that the quantity $f_r'(y;n)$ is the "stability parameter" for a 2^n-cycle with given r.

To find the desired recursion relations for $f_r(x)$ and $g_r(x)$ we first integrate over every even s_m in the "partition function"

$$Z \equiv N\int[\mathfrak{D}x][\mathfrak{D}s]\exp\Omega_r(x,s,\sigma;n) = N\int[\mathfrak{D}x][\mathfrak{D}s]_{\text{odd}}\exp\bar{\Omega}_r(x,s,\sigma;n)$$

with

$$\bar{\Omega}_r(x,s,\sigma;n) \equiv \sum_{\text{odd } m}\{is_m[x_{m+1} - f_r(x_m;n)] - \tfrac{1}{2}\sigma^2 s_m^2 g_r^2(x_m;n)\} - \tfrac{1}{2}\sigma^{-2}\sum_{\text{even } m}\{g_r^{-2}(x_m;n)[x_{m+1} - f_r(x_m;n)]^2\}. \tag{11}$$

We next calculate the x_m integrals for odd values of m in the saddle-point approximation, obtaining

$$Z = N\int[\mathfrak{D}x]_{\text{even}}[\mathfrak{D}s]_{\text{odd}}\exp\Omega_r'(x,s,\sigma;n)$$

with

$$\Omega_r'(x,s,\sigma;n) \equiv \sum_{\text{odd } m}\{is_m[x_{m+1} - f_r(f_r(x_{m-1};n);n)] - \tfrac{1}{2}\sigma^2 s_m^2[g_r^2(f_r(x_{m-1};n);n) + f_r'^2(f_r(x_{m-1};n);n)g_r^2(x_{m-1};n)]. \tag{12}$$

Equation (12) leads to the coarse-graining recursion relations,

$$f_r(x;n+1) = f_r(f_r(x;n);n),\quad g_r^2(x;n+1) = f_r'^2(f_r(x;n);n)g_r^2(x;n) + g_r^2(f_r(x;n);n). \tag{13}$$

Rescaling Eq. (13) by

$$x_m' = -\alpha^{-1}x_{2m} \text{ and } s_m' = -\alpha s_{2m+1},$$

using Feigenbaum's result,

$$f_{r_{n+1}}(x;n+1) = -\alpha^{-1}f_{r_n}(-\alpha x;n),$$

and assuming the existence of a fixed point for

$$g_{r_{n+1}}(x;n+1) = \beta\alpha^{-1}g_{r_n}(\alpha x;n)$$

(where β is a multiplicative renormalization constant for the noise amplitude, and r_n and r_{n+1} are, respectively, points with corresponding stability for the 2^n- and 2^{n+1}-cycle), we find

$$\Omega_{r_{n+1}}(x, s, \sigma; n+1) = \Omega_{r_n}(x', s', \beta\sigma; n). \tag{14}$$

This leads to a scaling form for the response function

$$R(r_n, \sigma; N; n) = Z^{-1}\int[\mathfrak{D}x][\mathfrak{D}s]\, i x_N s_0 \exp\Omega_{r_n}(x, s, \sigma; n)$$
$$= Z^{-1}\int[\mathfrak{D}x'][\mathfrak{D}s']\, i x_{N/2}' s_0' \exp\Omega_{r_{n-1}}(x', s', \beta\sigma; n-1) = R(r_{n-1}, \beta\sigma; \tfrac{1}{2}N; n-1). \tag{15}$$

The factor $\frac{1}{2}$ multiplies N since coarse graining doubles the length of the unit iteration step.

We have searched for the approximate fixed points of Eq. (13) by using linear and quadratic approximations for $g_r(x)$. The linear approximation leads to an expression for $\beta^2 = \alpha^2 + \delta^2$, entirely in terms of Feigenbaum's universal constants. In the quadratic approximation we obtain the value[9] $\beta \approx 7.7$. For large enough n, Eqs. (10) and (15) imply that $\lambda(r_n; \sigma) = 2\lambda(r_{n+1}; \beta^{-1}\sigma)$ which, upon iteration, yields

$$\lambda(r_n; \sigma) = 2^m \lambda(r_{n+m}; \beta^{-m}\sigma). \tag{16}$$

Let us first examine the behavior of λ as a function of σ at r_∞. From Eq. (16), we see that

$$\lambda(r_\infty; \sigma) = 2^m \lambda(r_\infty; \beta^{-m}\sigma);$$

whence, assuming that $\lambda(r_\infty; \sigma)$ is proportional to σ^u, we find

$$u = (\ln 2)/\ln\beta \approx 0.34\ldots. \tag{2b$'$}$$

Since $r_k - r_\infty$ is proportional to δ^{-k}, we can rewrite Eq. (16) as

$$\lambda(r_\infty - r_n; \sigma) = 2^m \lambda(\delta^{-m}(r_\infty - r_n); \beta^{-m}\sigma). \tag{17}$$

Introducing $t = (\ln 2)/\ln\delta = 0.4498\ldots$, and fixing the first argument on the right-hand side of Eq. (17) at some small constant value, we are led to

$$\lambda(r_\infty - r_n; \sigma) = (r_\infty - r_n)^t \Phi((r_\infty - r_n)^{-t}\sigma^u), \tag{1$'$}$$

and the zero-noise scaling relation,

$$\lambda(r_\infty - r_n; 0) \sim (r_\infty - r_n)^t. \tag{18}$$

Equation (18) gives the same scaling exponent for the Lyapunov parameter below threshold that Huberman and Rudnick previously obtained for this parameter beyond threshold in the chaotic regime. Our relation, which complements theirs, holds only when the Lyapunov exponent is calculated for corresponding values of the stability parameter, i.e., it describes the curve connecting the points in each 2^n-cycle which have equal stability parameters. With this understanding, we see that the same power law describes the Lyapunov exponent above and below threshold.

From Eq. (1), we can calculate how, as the external noise is varied, the point r_c, at which the Lyapunov parameter changes sign, is shifted. Since $(r_\infty - r_c)^{-t}\sigma^u$ must be constant for $\sigma \neq 0$, we see that γ, defined by $(r_\infty - r_c) \sim \sigma^\gamma$, satisfies

$$\gamma = u/t \approx 0.75\ldots. \tag{19}$$

In careful numerical experiments, Huberman and collaborators[6] have found

$$u = 0.37 \pm 0.02 \text{ and } \gamma = 0.82 \pm 0.02$$

which agree satisfactorily with our simple approximate values, 0.34 and 0.75.

This work is supported in part by the National Science Foundation under Grant No. DMR-77-10210. One of us (C. E. W.) is a National Science Foundation Predoctoral Fellow.

[1]M. Feigenbaum, J. Statist. Phys. 19, 25 (1978), and 21, 669 (1979).

[2]P. Collet and J.-P. Eckmann, *Iterated Maps on the Interval as Dynamical Systems* (Birkhäuser, Boston, 1980), and references therein.

[3]A. Libchaber and J. Maurer, J. Phys. (Paris), Colloq. 41, C3-51 (1980).

[4]B. A. Huberman and J. Rudnick, Phys. Rev. Lett. 45, 154 (1980).

[5]J. P. Crutchfield and B. A. Huberman, Phys. Lett. 77A, 407 (1980).

[6]We are grateful to Dr. B. A. Huberman (private communication) for carrying out more detailed and accurate calculations akin to those in Ref. 4, for informing us of the results prior to publication and permitting us to quote them, and for several helpful comments.

[7]P. C. Martin, E. D. Siggia, and H. A. Rose, Phys. Rev. A 8, 423 (1973).

[8]R. Graham, *Quantum Statistics in Optics and Solid-State Physics*, Springer Tracts in Modern Physics Vol. 66 (Springer-Verlag, New York, 1973), p. 1; C. DeDominicis, J. Phys. C 1, 247 (1976); R. Phythian, J. Phys. A 9, 269 (1976), and J. Phys. A 10, 777 (1977); H. K. Janssen, Z. Phys. B 23, 377 (1977); B. Jouvet and R. Phythian, Phys. Rev. A 19, 1350 (1979).

[9]An additional interesting property of β has been brought to our attention: β is approximately equal to

μ, the ratio of the intensity of successive subharmonics in the power spectrum studied by M. Feigenbaum [Phys. Lett. 74A, 375 (1979)], who found $\mu \simeq 4\alpha[2(1+\alpha^{-2})]^{1/2}$. The current best values of β and μ are, respectively, 6.618 and 6.557. That β and the "noise-free" parameter μ are related can be made plausible by requiring the ratio of the intensities of the noise-induced power spectra for chaotic transitions at r_n and r_{n+1} to coincide with the ratio of the spectral peaks at corresponding values of the control parameter, and performing a field-theoretic calculation which identifies the former ratio with the ratio of noise levels causing the transition.

J. Fluid Mech. (1980), *vol.* 100, *part* 3, *pp.* 449–470 449
Printed in Great Britain

Many routes to turbulent convection

By J. P. GOLLUB AND S. V. BENSON†
Physics Department, Haverford College, Haverford, Pa 19041, U.S.A.

(Received 25 September 1979)

Using automated laser-Doppler methods we have identified four distinct sequences of instabilities leading to turbulent convection at low Prandtl number (2·5–5·0), in fluid layers of small horizontal extent. Contour maps of the structure of the time-averaged velocity field, in conjunction with high-resolution power spectral analysis, demonstrate that several mean flows are stable over a wide range in the Rayleigh number R, and that the sequence of time-dependent instabilities depends on the mean flow. A number of routes to non-periodic motion have been identified by varying the geometrical aspect ratio, Prandtl number, and mean flow. Quasi-periodic motion at two frequencies leads to phase locking or entrainment, as identified by a step in a graph of the ratio of the two frequencies. The onset of non-periodicity in this case is associated with the loss of entrainment as R is increased. Another route to turbulence involves successive subharmonic (or period doubling) bifurcations of a periodic flow. A third route contains a well-defined regime with three generally incommensurate frequencies and no broadband noise. The spectral analysis used to demonstrate the presence of three frequencies has a precision of about one part in 10^4 to 10^5. Finally, we observe a process of intermittent non-periodicity first identified by Libchaber & Maurer at lower Prandtl number. In this case the fluid alternates between quasi-periodic and non-periodic states over a finite range in R. Several of these processes are also manifested by rather simple mathematical models, but the complicated dependence on geometrical parameters, Prandtl number, and mean flow structure has not been explained.

1. Introduction

The behaviour of a thin fluid layer heated from below has been of great importance in the development and gradually increasing sophistication of nonlinear hydrodynamics. The Rayleigh–Bénard instability is well understood, and convecting solutions of the nonlinear hydrodynamic equations have been found for layers of infinite horizontal extent. The linear stability of this flow with respect to various disturbances has been examined, and various secondary instabilities predicted and observed (Busse 1980). Some of these secondary instabilities change the spatial structure of the velocity field without the addition of time dependence, while others result in an oscillatory flow. However, more complex time-dependent flows and the onset of turbulence have remained out of reach theoretically.

Oscillatory instabilities have been observed experimentally, but secondary instabilities leading to quasi-periodic flows with several distinct frequencies have not been

† Present address: Physics Department, Stanford University, Stanford, Ca 94305.

identified. Furthermore, such flows cannot be easily distinguished from turbulent flows by qualitative observations. Many elementary questions have not yet been answered. How many instabilities occur as the temperature gradient across a fluid layer is increased? What is the nature of the time-dependent flows that occur? Is there a well-defined temperature gradient at which non-periodic flow begins? How does the sequence of instabilities depend on the parameters of the system?

The experiments reported in this paper utilize significant innovations in order to address these questions. A non-perturbative laser-Doppler probe with computer control and data acquisition allows the space and time structure of the velocity field to be examined in great detail. The statistical properties of the flow are determined by power spectral analysis capable of detecting secondary flows far too weak to observe visually. Precision control of the thermal environment allows external perturbations to be largely eliminated, and changes in the dynamical behaviour resulting from small parameter increments to be detected. Using these methods, we are able to distinguish clearly between periodic, quasi-periodic, and non-periodic flows. By varying the geometry and Prandtl number, we have observed instabilities leading to quasi-periodic flows with two and three independent frequencies; phase-locking phenomena involving these various frequencies; subharmonic or period-doubling bifurcation of periodic convective flows; and intermittent non-periodicity. At least four qualitatively different sequences of instabilities can lead to turbulence in this system.

The paper is organized in the following fashion. In § 2 we review previous theoretical and experimental work on the transition to turbulent convection. Section 3 contains a description of the fluid system, the laser-Doppler technique, signal processing electronics, and the methods used to compute spectra and contour maps. Our results for the various routes to turbulent flow are presented in § 4, and discussed in the light of other experiments and theoretical models in § 5.

2. Background

The flow of a fluid confined to a horizontal layer and heated from below is defined by a small number of parameters. We denote the separation between highly conducting parallel plates by d, and the temperature difference between them by ΔT. The state of the fluid depends on the boundary conditions and two dimensionless parameters, the Rayleigh number $R = g\alpha d^3 \Delta T/\kappa\nu$, and the Prandtl number $P = \nu/\kappa$, where ν is the kinematic viscosity of the fluid, α is the thermal expansion coefficient, κ is the thermal diffusivity, and g is the gravitational acceleration. This paper is concerned with the range $2 \leqslant P \leqslant 5$. It is often convenient to use the relative Rayleigh number $R/R_c = \Delta T/\Delta T_c$, where ΔT_c is the temperature difference at which convection begins in an infinite layer. For a fluid layer of finite horizontal extent, a third important parameter is the aspect ratio Γ, defined for rectangular boundaries as the ratio of the largest horizontal dimension to the layer thickness.†

2.1. *Time-independent convection*

The form of the time-independent convection that precedes the onset of time dependence is discussed by Busse (1978, 1980), Koschmieder (1974), Normand, Pomeau & Velarde (1977) and Dubois & Bergé (1978). Of particular relevance is the fact that, in

† One can also define a *horizontal* aspect ratio as the ratio of the two horizontal dimensions.

a rectangular parallelepiped of small aspect ratio, both calculations (Davis 1967) and experiments (Stork & Müller 1972) show that convection near its onset usually has the form of rolls oriented parallel to the short side of the cell. The onset is delayed somewhat beyond the value R_c which characterizes an infinite fluid layer. Furthermore, the flows are not generally unique. Several different flows have been observed experimentally to be stable for a given geometry, Rayleigh number and Prandtl number. For example, the number of rolls may vary (i.e. the dominant wave-number may have several different values). Any given flow pattern is stable over a wide range in R if the aspect ratio is small. Because of this non-uniqueness, which is incompletely understood theoretically, it is important to supplement local measurements of time-dependent phenomena with a method capable of yielding the overall spatial structure of the flow. In practice, we found that the multiplicity of stable states was substantially greater than calculations and previous experiments suggest.

2.2. *The onset of time dependence*

A stability theory for the onset of time dependence was presented by Clever & Busse (1974) and Busse & Clever (1978), who superimposed infinitesimal disturbances on the parallel convection rolls for a layer of infinite extent. Whereas the onset of convection is independent of P, the onset of time dependence is strongly dependent on P. For large P, secondary instabilities involving complex three-dimensional stationary structures precede the onset of time dependence. However, for sufficiently low P, Clever & Busse predicted that the onset of time dependence would be a transverse oscillation of the basic convective rolls. This instability was observed by Willis & Deardorff (1970), Krishnamurti (1970, 1973), and Ahlers & Behringer (1979). Even when a transition to three-dimensional convection precedes the onset of time dependence, the oscillatory instability seems to be qualitatively similar to the predictions for low P (Busse & Whitehead 1974).

2.3. *Transition to turbulent convection*

The possible existence of further time-dependent instabilities has not been investigated by stability analysis because of mathematical complexity. However, there have been several numerical studies based on highly truncated normal mode expansions. The first of these is the well-known study by Lorenz (1963) of a model originally due to Saltzman (1962). The model assumes a two-dimensional flow and contains only three time-dependent amplitudes from the expansions of the velocity and temperature fields. Numerical integration of the equations shows non-periodic behaviour for certain parameter values even though there are only three degrees of freedom.

Curry (1978) has recently generalized Lorenz' model by retaining additional variables for a total of 14 time-dependent amplitudes. This model is still not realistic because of its two-dimensionality and lack of lateral boundaries. Nevertheless, it manifests a much greater variety of phenomena: a bifurcation to a limit cycle or periodic oscillation, a subharmonic bifurcation in which the dominant oscillation frequency is halved, an instability leading to quasi-periodic motion with two distinct frequencies, and non-periodic motion.

McLaughlin & Martin (1975) have extended the idea of Lorenz in a way that dispenses with the unrealistic requirement of two-dimensionality. This model of 39 coupled variables shows both periodic and chaotic states for reasonable values of the parameters of the problem. The calculations were not extensive enough to provide evidence for quasi-periodic flows.

Lipps (1976) simulated stationary and time-dependent convection at $P = 0{\cdot}7$ using the full hydrodynamic equations in the Boussinesq approximation (Busse 1978). Periodic boundary conditions at lateral boundaries were assumed, along with perfectly conducting vertical boundaries. Examples of both periodic and apparently non-periodic convection were found, but the numerical solutions did not extend to large enough times to permit reliable discrimination between periodic and non-periodic flows to be made.

Overall, the theoretical picture of the transition to turbulent convection is quite muddy. It is not known what occurs beyond the onset of oscillations, and how the phenomena depend on P and the aspect ratio.

Recently Ahlers (1974) and Ahlers & Behringer (1978, 1979) have used precision thermal methods at cryogenic temperatures to study the heat transport in convecting liquid helium with cylindrical lateral boundaries. They observed a striking dependence of the qualitative behaviour on the aspect ratio (the ratio of radius to depth). For small aspect ratio, distinct periodic and quasi-periodic regimes precede the onset of non-periodic motion. However, for large aspect ratio non-periodic motion begins very near R with no intervening periodic regimes. Some aspect ratio dependence has also been found by Libchaber & Maurer (1978) using a local temperature probe.

The present study employs somewhat higher Prandtl numbers (2·5 and 5·0), rectangular cells of relatively small aspect ratio and a laser-Doppler probe for both local measurements and flow mapping. Some preliminary results of our work were reported in Gollub *et al.* (1977), Fenstermacher, Swinney, Benson & Gollub (1979) and Gollub & Benson (1978, 1979).

3. Experimental techniques

3.1. *Convection cell*

In order to permit a range of Prandtl numbers to be investigated in a single system, we selected water as the working fluid. The strong temperature dependence of the viscosity of water allows P to be varied between 9 and 2 by varying the mean working temperature between 10 and 90 °C. High Rayleigh numbers can be reached with a relatively thin cell, a fact that permits the entire system to be small and thermally stable. However, the Oberbeck–Boussinesq approximation (Busse 1978) was not particularly well satisfied. The viscosity (or locally defined Prandtl number) varied by $\pm 10\,\%$ over the fluid layer under the most extreme conditions we employed; more typically the variation was $\pm 2\,\%$.

The cell was designed with an emphasis on temperature stability and horizontal temperature uniformity, because external perturbations can obscure the onset of intrinsic dynamical noise in the fluid, and non-uniformities can blur sharp transitions. The system design is shown in figure 1. Two copper plates of horizontal dimensions 9·3 by 6·7 cm and thickness 2·54 cm (upper) and 1·27 cm (lower), are separated by a Plexiglas spacer of variable interior size containing the fluid. The space around the cell is evacuated in order to eliminate convective heat loss from the sides of the cell, and the surrounding vacuum can is kept at the mean working temperature to minimize radiative heat loss. The temperature gradient at the lateral fluid boundary cannot be

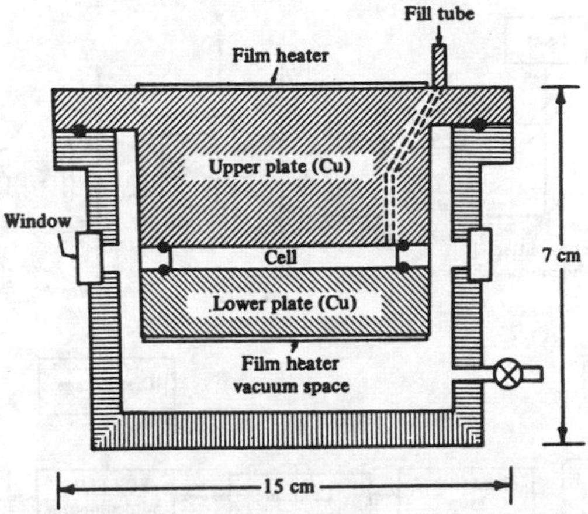

FIGURE 1. Schematic diagram of convection cell showing vacuum space and optical access.

assumed to be uniform, although good thermal contact between the copper and Plexiglas was maintained by silicone grease.

Results for two aspect ratios are reported in this paper. The first cell has interior dimensions 16·42 by 27·72 by 7·90 mm high (aspect ratio $\Gamma = 3{\cdot}51$) and the second is 14·66 by 28·85 by 11·94 mm high ($\Gamma = 2{\cdot}42$). Some preliminary observations on a larger rectangular cell ($\Gamma = 10{\cdot}0$) and a circular cell (radius to height ratio of 3·14) were previously reported (Gollub *et al.* 1977; Fenstermacher, Swinney, Benson & Gollub 1979).

The temperatures of the two copper plates are controlled by a.c bridges using sensitive thermistors and lock-in amplifiers. The d.c. output of each lock-in drives a resistive film heater via an operational power supply. There are three feedback loops altogether: two a.c. bridges for the copper plates, and a commercial d.c. bridge controller for the vacuum can and heat shield. Temperatures were measured using four additional matched and calibrated thermistors embedded in the copper near the fluid.

The long term (4 h) thermal stability of this system is within about 1–2 mK, corresponding to effective fluctuations in ΔT of about 0·05 %. (At $R/R_c = 25$, ΔT was typically several k.) It is difficult to do much better than this in a system that cannot be completely isolated owing to the necessary heat flux through it. The readings of the different thermistors on each plate indicate a horizontal temperature uniformity of better than 0·5 % of ΔT.

3.2. *Laser-Doppler velocimetry (LDV)*

A dual beam forward scatter LDV system is used for both flow mapping and local velocity measurements in real time. The optical arrangement is shown in figure 2 (*a*). A vertically polarized 15 mW laser beam at 6328 Å is first split into two beams of equal intensity, and then each part is frequency shifted by scattering from an acoustic wave in a liquid. Each beam is shifted by about 40 MHz, but the frequency difference between the two shifts is accurately maintained at 2000·0 Hz by electronic feedback techniques. The two shifted beams are then focused by a lens of focal length 12 cm to a point inside the convecting fluid, which is doped with polystyrene latex spheres of

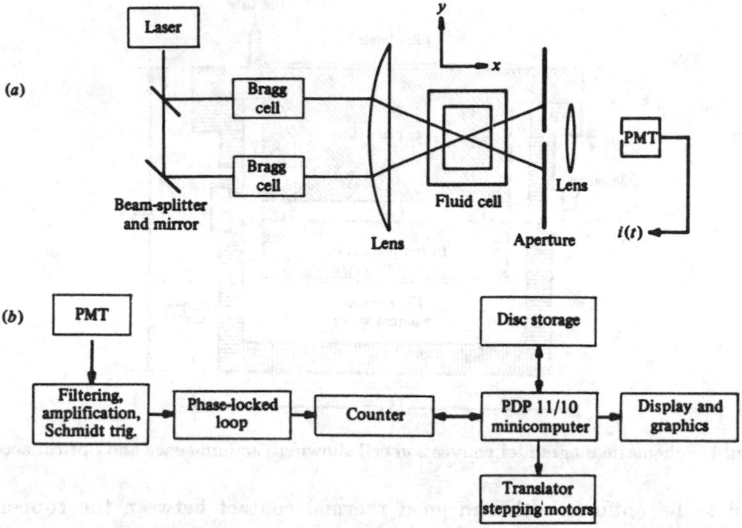

FIGURE 2. Schematic diagrams of (*a*) laser-Doppler optics, and (*b*) signal-processing electronics and data-acquisition system. The photocurrent $i(t)$ in (*a*) oscillates at a frequency proportional to the fluid velocity component V_y.

diameter 0·369 μm. The scattering volume is the ellipsoidal region of intersection of the two beams, and has major (x) and minor (y) axes of about 500 μm and 60 μm respectively. Since the beam profiles are actually gaussian, the boundaries of the scattering volume are not precisely defined. Scattered light from a small solid angle of about 5×10^{-5} sr is collected by a lens and focused on a 200 μm pinhole in front of a photomultiplier.

The optical system is sensitive to the velocity component $V_y(\mathbf{r},t)$ in the y direction in figure 2. Light scattered from one beam by particles moving in this direction is advanced in phase, while light scattered from the other beam is retarded in phase. Since the photocurrent is proportional to the square of the optical field, it contains a component oscillating at the difference frequency of the light scattered by the two beams:

$$\nu = 2000\,\text{Hz} + V_y\, n\lambda_0^{-1}\sin(\tfrac{1}{2}\theta),$$

where λ_0 is the vacuum wavelength, n is the index of refraction of the water, and $\theta = 13{\cdot}6°$ is the angle between the two beams. The acoustic frequency shifters permit velocities of stationary or slow particles to be measured easily. The velocity component measured is horizontal and generally perpendicular to the axes of the convective rolls. The optical system is non-perturbative since very little of the light is absorbed by the fluid.

3.3. *Data acquisition*

The photomultiplier signal is amplified and bandpass filtered to remove most of the high-frequency shot noise and the low-frequency amplitude fluctuations. A Schmidt trigger then precedes the input to a phase-locked loop, which locks the frequency of an external voltage-controlled oscillator (VCO) to the Doppler frequency, with some reduction in the bandwidth of the frequency fluctuations and consequent noise reduction. Finally, the VCO frequency is determined by counting cycles over a time interval of 0·5–2 s with a commercial counter controlled by a Digital Equipment Corporation PDP 11/10 minicomputer.

The minicomputer is used for experimental control, data acquisition, and analysis in two distinct modes. The fluid system is mounted on computer-controlled stepping-motor-driven translation stages with a resolution of about 0·005 mm for motion in either the x or y direction. Contour maps of the velocity field are obtained by measuring V_y at each point in a rectangular grid of up to 450 points in a horizontal plane. Software is used both to perform the sampling and to plot contour maps of constant Doppler shift using linear interpolation between points. About 60 min are required to acquire the data to construct a contour map if the velocity field is time independent. If the field is oscillating in a periodic state, the velocity is averaged over one cycle at each grid point in order to yield a map of the mean velocity. If the fluid motion is non-periodic, mean velocity contour maps are less precise unless very long averaging times are used to eliminate fluctuations due to low-frequency noise in the fluid.

The second mode of data acquisition involves measuring the time-dependent velocity at a single location. Sampling times Δt are long enough (0·5–2 s) to obtain accurate measurements of the Doppler frequency, which is in the range 1500–2500 Hz, but much shorter than the characteristic frequencies of the fluid motion. The accuracy of the timing routines was found to be better than 10^{-5} s, and fluctuations in timing are less than this. Typical runs consist of $N = 4096$ sequential data points. Longer runs require an inordinate amount of time without yielding much more information.

Velocity power spectra are computed from the squared modulus of the fast Fourier transform of the sampled velocity. We use a Cooley–Tukey FFT algorithm, and apply a GEO window (Otnes & Enochson 1972) to suppress side lobes caused by the finite duration $T = N\Delta t$ of the record. Spectral estimates are obtained at frequency intervals of $1/T = 1/N\Delta t$ up to the Nyquist frequency $f_N = (2\Delta t)^{-1}$. The methods used to minimize aliasing are discussed by Fenstermacher, Swinney & Gollub (1979).

The resolution of the computed spectra is approximately equal to the interval $1/T$ between spectral estimates. Often, the spectra show peaks with approximately this resolution. If one assumes that they are actually discrete spectral lines, their frequencies can be computed to an accuracy much better than $1/T$. A power-weighted average of the frequencies of the two largest spectral estimates within a peak can be used to extract this information, and is accurate to about 5 % of $1/T$. The deviations are a systematic function of the frequency of the peak and the frequencies at which the spectral estimates are computed, a function which we obtained from artificial data of known frequency. By employing this correction to the power-weighted average, we are able to obtain discrete frequencies to an accuracy of about 1 % of $1/T$ ($10^{-5}f_N$) if there is little noise. We tested our procedure for computing spectral peaks on artificial data with and without background noise. The standard deviation σ_f of the frequency determination was found to depend on the ratio A_n/A_p of the integrated noise (up to f_N) to the area under the peak, as shown in figure 3. The precision of frequency determination is degraded when the peak height is not far above the background noise level.

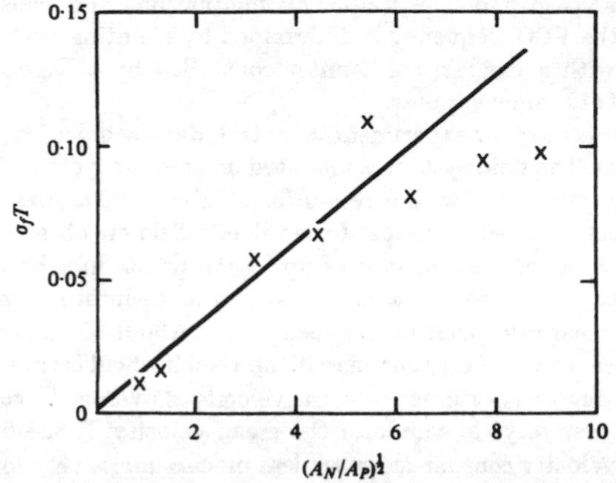

FIGURE 3. Standard error σ_f (in units of the spacing T^{-1} between spectral estimates) for frequency determination, as a function of the ratio A_N/A_P of the integrated area of the background noise to that of the peak. The crosses were obtained from empirical tests on artificial data having a single known frequency and broadband noise. The line is the estimated standard error actually used in analysing data. As an example, a peak 2 orders of magnitude above the noise typically yields $A_N/A_P = 10$ so $\sigma_f = 0{\cdot}05/T$ in this case.

4. Experimental results

4.1. *Mean flow patterns*

Although the main purpose of this work was to study the time-dependent regimes, the non-uniqueness of spatial states (§ 2.1) made it necessary to monitor the mean flow using the contour mapping technique of § 3.3. Several distinct mean flows were found to be stable in each of the convection cells, and examples are shown in figure 4. Contours of constant Doppler shift represent the velocity component parallel to the y axis (figure 2a), in a horizontal plane above the centre of the cell. The geometry is such that the rolls generally align parallel to x, which is the short cell dimension. Positive and negative shifts correspond to motion in the positive and negative y directions, respectively. In the flow of figure 4 (a) fluid rises (out of the page) in the centre of the cell, moves toward the edges and descends. The line of zero shift (dashed) marks the vertical boundary between convective rolls. The contour line labelled 400 Hz corresponds to a velocity component of 0·082 cm s^{-1}. (The lines do not extend to the edges of the cell because optical distortions reduce the accuracy of measurement there.) These contour maps are reproducible after a day with very little change.

In both of the cells that were studied in detail, the most stable mean flow (figure 4a) can be qualitatively summarized as two symmetrical rolls oriented with axes parallel to the short side of the cell. For $\Gamma = 3{\cdot}5$ it is stable over the entire range of Rayleigh number studied ($2 < R/R_c < 100$). For $\Gamma = 2{\cdot}4$ this flow is stable up to $R/R_c = 85$ when $P = 5$, and stable up to $R/R_c = 110$ when $P = 2{\cdot}5$. The velocity field is approximately two-dimensional away from the ends of the rolls. The variation of V_y with y (figure 5a) is a distorted sinusoid with harmonics that become relatively more prominent as R is increased. However, even at $R/R_c = 17{\cdot}7$, only the first four terms of the Fourier expansion for $V_y(y)$ are significant, as shown in figure 5 (b). This observation

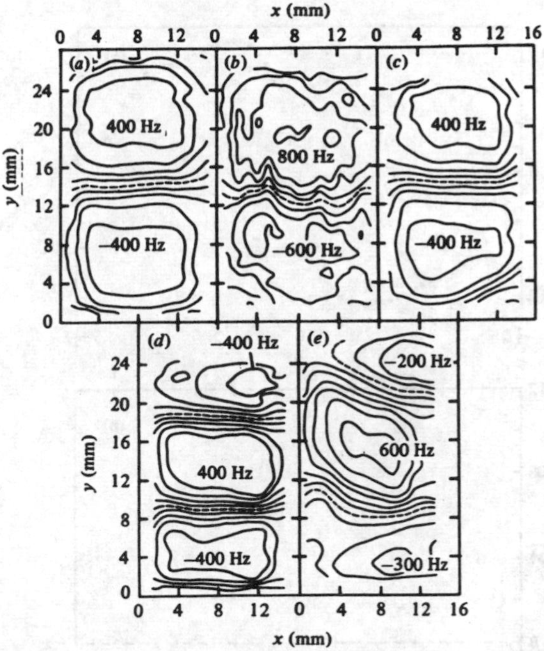

FIGURE 4. Contour maps of constant Doppler shift for the mean velocity component $\bar{V}_y$, measured in a horizontal plane slightly above the centre of the convection cell (see text). (*a*) Two symmetrical rolls ($P = 2{\cdot}5$, $\Gamma = 3{\cdot}5$, $R/R_c = 17{\cdot}1$); (*b*) two rolls persist in the mean flow, although the fluid is strongly non-periodic ($P = 5{\cdot}0$, $\Gamma = 3{\cdot}5$, $R/R_c = 65{\cdot}4$); (*c*) a distorted but stable two-roll flow ($P = 2{\cdot}5$, $\Gamma = 3{\cdot}5$, $R/R_c = 18{\cdot}1$); (*d*) three rolls ($P = 5{\cdot}0$, $\Gamma = 3{\cdot}5$, $R/R_c = 30{\cdot}9$); (*e*) a complex mean flow for a periodic state in which the rolls are also inclined to the vertical ($P = 5{\cdot}0$, $\Gamma = 2{\cdot}4$, $R/R_c = 84{\cdot}7$).

indicates that a truncated normal-mode expansion would probably not require an inordinate number of terms to represent even the full three-dimensional velocity field.† Even at $R/R_c = 84{\cdot}7$, where the flow is chaotic, the contour map of figure 4(*b*) shows that the two-roll flow persists in a time-averaged sense. Some of the irregularities in this figure may be artifacts of the finite averaging time (8 s) at each grid point. Further detailed study of these maps is intended.

Other mean flows that are also stable over a wide range in R can be obtained from different initial conditions. For example, cooling the upper plate rather than heating the lower one to attain the desired Rayleigh number yields a mean flow almost identical to figure 4(*a*) but with signs reversed, indicating that the fluid descends at the centre of the cell and rises at the edges.

Somewhat surprising is the observation of stable mean flows differing in relatively small but still significant ways from the basic two-roll flow. By stability, we mean that a flow persists without noticeable change for a time long compared to the horizontal

† The functional form of the velocity field in time-independent convection has recently been studied by Dubois & Bergé (1978). The amplitudes of the first few spatial Fourier components were found to be in good agreement with theoretical predictions for $R/R_c < 10$.

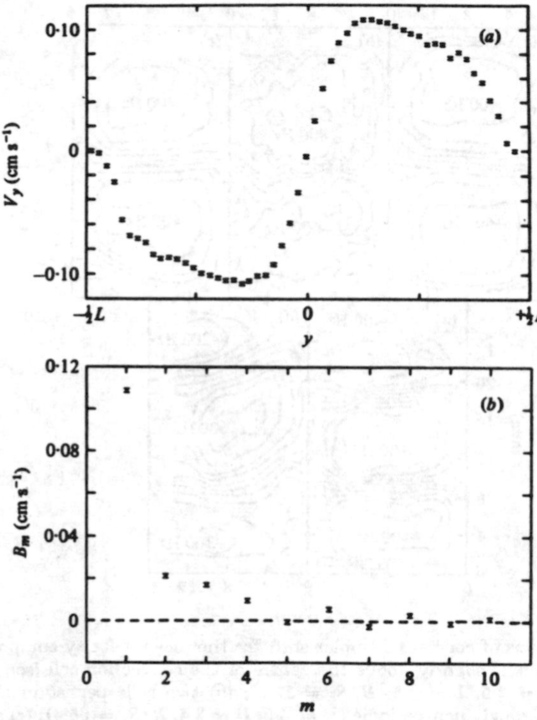

FIGURE 5. (*a*) Fluid velocity component V_y as a function of y for the flow of figure 4(*a*). (*b*) Fourier coefficients B_m of the curve in (*a*). Only the first four terms are significant. $R/R_c = 17{\cdot}7$.

thermal diffusion time, which is about 10^4 s for our cells. In this sense, the flow of figure 4(*c*), in which one roll appears distorted, is stable at least over the range $18 < R/R_c < 40$ (at $\Gamma = 3{\cdot}5$). This flow evolves into a sequence of time-dependent states substantially different from those associated with figure 4(*a*).

A mean flow with three rolls was observed in the larger cell ($\Gamma = 3{\cdot}5$), and its contour map appears in figure 4(*d*). It loses stability with respect to the two-roll flow when R/R_c exceeds a threshold somewhat above 50 at $P = 5$.

In the cell with smaller aspect ratio ($\Gamma = 2{\cdot}4$) yet another distinct mean flow was observed in which the two rolls become highly inclined to the vertical, and an eddy forms in a corner. This behaviour is shown in figure 4(*e*), and occurs for $R/R_c > 85$ at $P = 5$. It can also appear with signs reversed.

Other mean flows can be obtained in these cells in addition to the ones we have discussed, but they are not stable over a large range in R. Because of the multiplicity of spatial states, we adopted the following procedure for studying time-dependent phenomena. After selecting the aspect ratio and Prandtl number, one of the stable mean flows was chosen by manipulation of the boundary temperatures, and verified by taking a contour map. Then the time dependence was studied at a single location in the cell by spectral analysis. The Rayleigh number was increased and then decreased in small steps, with occasional verification that the mean flow structure had not changed. Several weeks of data taking were required to complete such a sequence.

4.2. *Transition to turbulence*

We have observed several qualitatively different types of time-dependent flow. It will be useful in describing them to introduce a simple set of abbreviations. Periodic time dependence (P) evolves from a stationary flow (S) and is recognized by a velocity power spectrum containing a single instrumentally sharp peak and perhaps harmonics, but no broadband noise, except for instrumental noise. It is possible for a periodic motion to undergo a subharmonic bifurcation to another periodic state (P_2) or (P_4) in which the fundamental frequency is half or one quarter of the original frequency. Quasi-periodic states (QP_2) and (QP_3) are recognized by spectra containing respectively two or three incommensurate frequencies, along with sums and differences of these frequencies, but no broadband noise. The two frequencies in a state QP_2 can lock to a rational ratio. The resulting flow, denoted (L), is actually periodic. If the velocity spectrum contains any broadband noise, the motion is non-periodic (N) even if relatively sharp spectral peaks are also present. We regard non-periodic motion as being weakly turbulent. However, some workers reserve the latter term for flows that are known to vary randomly in space as well as time (Monin 1979).

The sequence of instabilities leading to non-periodic flow depends on the aspect ratio, Prandtl number, and mean flow. The major sequences we have observed are summarized by the bar graphs in figure 6. Each graph is labelled by a Roman numeral specifying a route to non-periodicity described in the following sections, and by a letter denoting one of the mean flows of figure 4.

4.3. *Route I: quasi-periodicity and phase locking*

Non-periodicity preceded by quasi-periodicity and phase locking is observed under several different conditions, as shown in figure 6. The basic path is specified by the symbolic sequence $P \rightarrow QP_2 \rightarrow L \rightarrow N$. We describe the observations in detail for the case $\Gamma = 3{\cdot}5$ and $P = 5{\cdot}0$. The mean flow is that of figure 4(*a*) throughout the time-dependent regimes studied.

A sequence of velocity records and corresponding power spectra for increasing R is shown in figure 7. The duration of each velocity record is 6144 s, only a small portion of which is shown. Power spectra are plotted on a logarithmic vertical scale. The instrumental background noise is that of figure 7(*a*) and corresponds to a root mean square noise in the Doppler shifts of only 2 Hz. The spectral peaks are instrumentally sharp even when the duration of the run is doubled, and the strongest peak is four orders of magnitude above the instrumental noise. The integrated noise to peak ratio A_n/A_p defined in § 3.3 is 0·1, and figure 3 then implies that the frequencies of the peaks can be determined to an accuracy of $0{\cdot}01T^{-1} = 10^{-6}$ Hz. However, peaks in the plotted spectra are broadened because four adjacent spectral estimates have been averaged in order to reduce the background fluctuations. The observed sequence of instabilities is as follows:

1. A bifurcation leads to a transfer of stability from time-independent flow to periodic flow at $R/R_c = 27{\cdot}2$. This instability (like the others described in this section) has a hysteresis of about $\pm 0{\cdot}15R_c$ depending on the direction in which R is changed. The frequency labelled f_2 (for historical reasons) in figure 7(*a*) appears at all points in

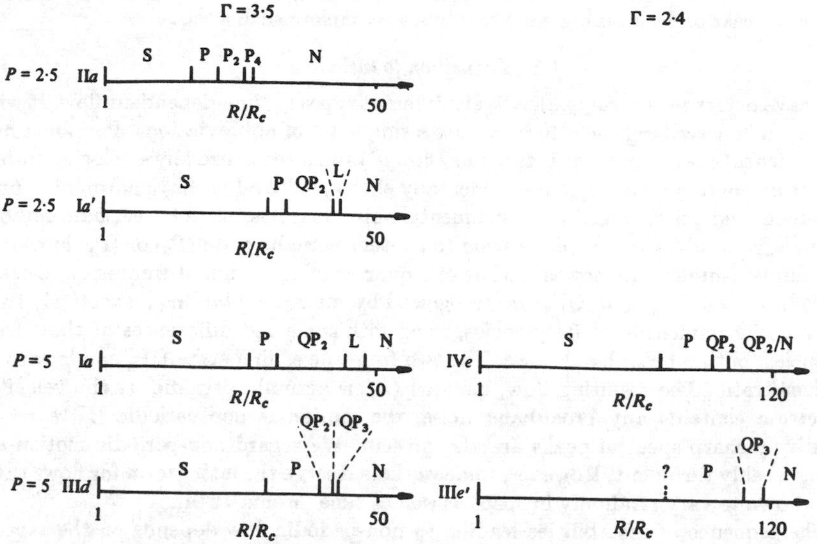

FIGURE 6. Bar graphs of the various instabilities leading to non-periodicity as a function of the aspect ratio Γ, the Prandtl number P, and the mean flow pattern. Roman numerals denote the routes to non-periodicity we have identified, and are keyed to the subheadings of § 4. The lower-case letters refer to the mean flows of figure 4, except that a prime indicates sign reversal.

the cell, along with harmonics at integral multiples of the fundamental. The amplitudes of the fundamental and harmonics vary with position in the cell, but the frequencies are invariant.

2. At $R/R_c = 32$ a second frequency appears in the spectrum (throughout the cell), as shown in figure 7 (*b*). All of the peaks have been identified, to within the estimated standard error σ_f, as linear combinations of two basic frequencies of the form

$$f = m_1 f_1 + m_2 f_2,$$

where m_1 and m_2 are integers, $f_1 = 0.03082$ Hz and $f_2 = 0.08348$ Hz. The presence of high-order mixing components in the spectrum indicates that these time-dependent processes are strongly nonlinear. The ratio f_2/f_1 decreases smoothly with increasing Rayleigh number, suggesting that the two frequencies are at least sometimes incommensurate and hence that the spectrum corresponds to a quasi-periodic motion.†

3. Phase locking of the two frequencies occurs when R/R_c exceeds 44·4. This can be seen as a step in the ratio f_2/f_1, as shown in figure 8. The ratio decreases smoothly until the step is reached, where $f_2/f_1 = 2.333 \pm 0.001 = \frac{7}{3}$. It is interesting to note that the locking ratio is not exactly reproducible from experiment to experiment. The cause of this is unknown. A spectrum corresponding to locking at the ratio $\frac{9}{4}$ is shown in figure 7 (*c*). Because of the locking, the spectrum corresponds to periodic motion at a

† There is some ambiguity in the choice of the second frequency f_1 because it is not the strongest line at all locations in the cell. One could choose the peak at 0·11430 Hz instead as f_1, but this change would not have any important effect.

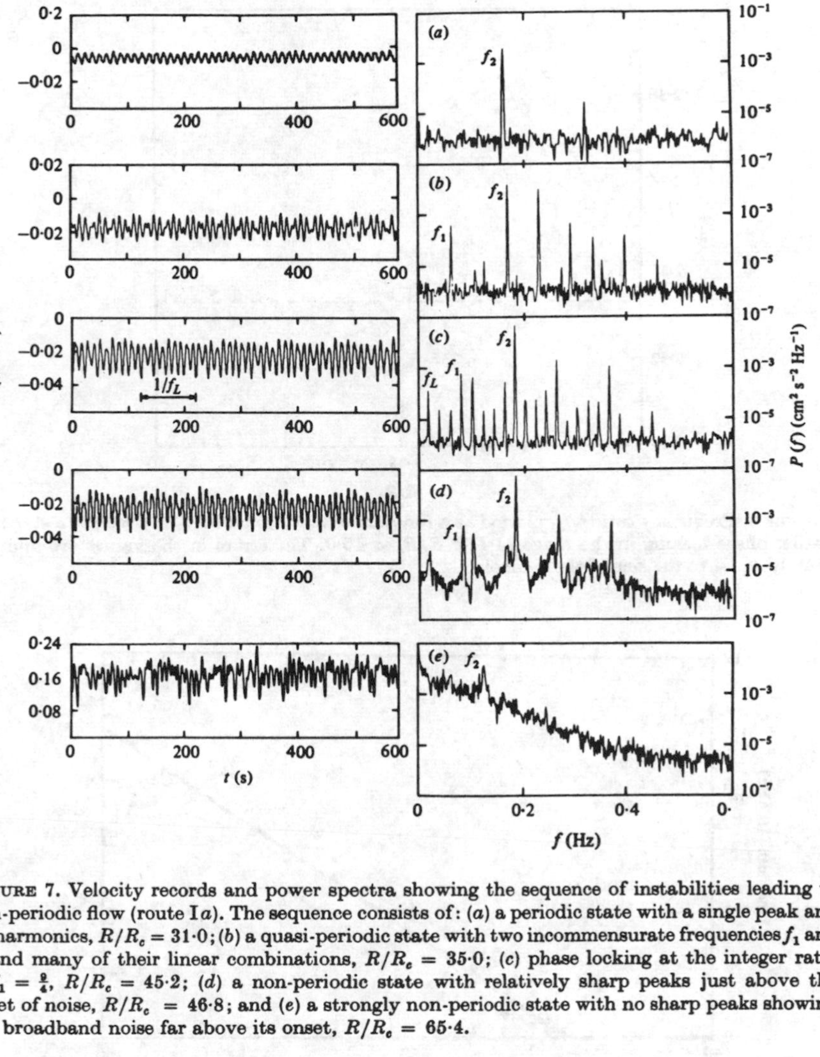

FIGURE 7. Velocity records and power spectra showing the sequence of instabilities leading to non-periodic flow (route Ia). The sequence consists of: (a) a periodic state with a single peak and its harmonics, $R/R_c = 31{\cdot}0$; (b) a quasi-periodic state with two incommensurate frequencies f_1 and f_2 and many of their linear combinations, $R/R_c = 35{\cdot}0$; (c) phase locking at the integer ratio $f_2/f_1 = \frac{9}{4}$, $R/R_c = 45{\cdot}2$; (d) a non-periodic state with relatively sharp peaks just above the onset of noise, $R/R_c = 46{\cdot}8$; and (e) a strongly non-periodic state with no sharp peaks showing the broadband noise far above its onset, $R/R_c = 65{\cdot}4$.

frequency $f_L = \frac{1}{9}f_2 = \frac{1}{4}f_1$. Strong mixing induced by the locking creates strong peaks at all multiples of f_L. No locking has been observed at lower R, suggesting that the strength of the nonlinear interaction between the two oscillations is weaker there.

4. The spectra develop broadband noise and the peaks begin to broaden at $R/R_c = 46{\cdot}0$, indicating the onset of non-periodic motion. At the same Rayleigh number, to within experimental error, the ratio f_2/f_1 starts to decrease again. Thus, the onset of non-periodicity coincides with the loss of entrainment of the two oscillations. The ratio of the two frequencies is not measurable with high accuracy beyond $R/R_c = 46{\cdot}5$,

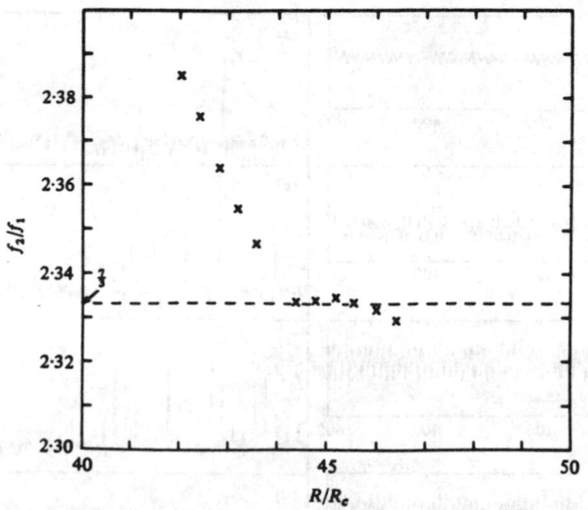

FIGURE 8. Frequency ratio f_2/f_1 plotted as a function of R for route I*a*. The ratio has a step indicating phase locking in the range $44{\cdot}4 < R/R_c < 46{\cdot}0$. The errors in the values are approximately equal to the size of the symbols.

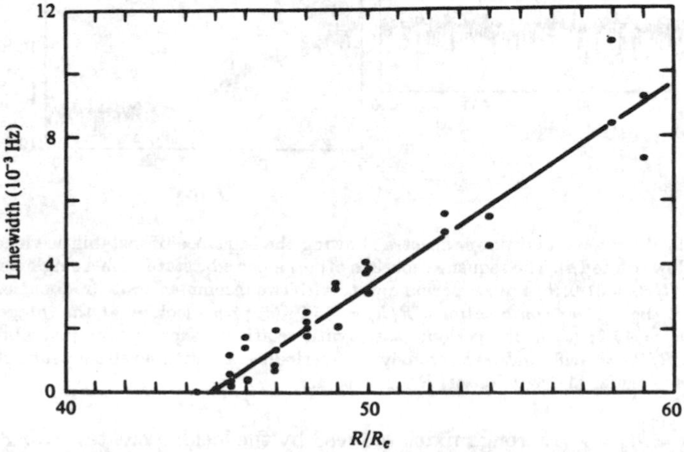

FIGURE 9. Linewidth of the spectral peaks beyond the onset of noise, with the instrumental linewidth subtracted. The variation is linear in R/R_c above the threshold.

so the data in figure 8 end at this point. The spectrum in figure 7 (*d*) shows the broadband noise near its onset, where relatively sharp (but not instrumentally sharp) peaks still exist, but the high-order mixing peaks of the phase-locked regime have disappeared. As R is increased, the peaks gradually broaden and the gaps between them become shallower, until nearly featureless spectra are produced, as in figure 7 (*e*).

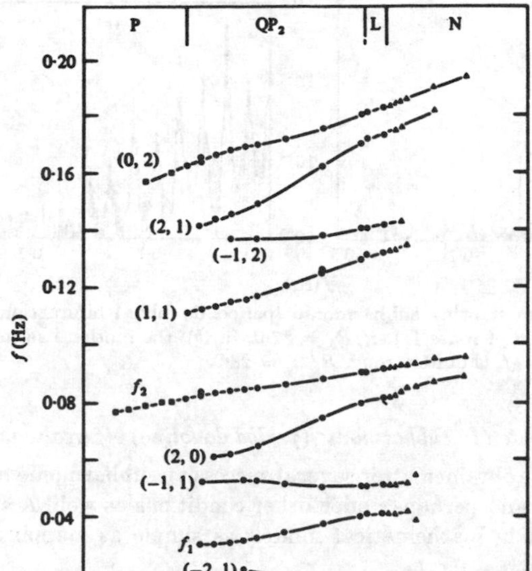

FIGURE 10. Frequencies of all statistically significant spectral peaks as a function of R/R_c for route I*a*. The numbers in parentheses identify sums and differences of f_1 and f_2. Squares, circles and triangles are used in the periodic, quasi-periodic or locked, and non-periodic dynamical regimes, respectively.

In this spectrum, the power falls as $f^{-4\cdot3\pm0\cdot5}$, in agreement with the observations of Ahlers & Behringer (1979) at larger aspect ratio.

In order to describe the growth of noise quantitatively, we determined the linewidth of spectral peaks as a function of R/R_c above the onset of noise. As figure 9 demonstrates, the linewidth is approximately linear in $(R-R_t)/R_c$, where the extrapolated turbulent threshold R_t/R_c is about 45, slightly below the value 46·0 where the non-periodicity becomes obvious from examination of a single spectrum. We conclude from figure 9 that the non-periodic motion has a well-defined onset (answering a question posed in the introduction) and that the noise grows smoothly with Rayleigh number above this onset.

The frequencies f_1 and f_2 increase with R, as shown in figure 10. All of the statistically significant peaks are plotted and labelled with the integers (m_1, m_2) specifying the linear combinations; for example, $(-2, 1)$ denotes the peak at $(-2f_1+f_2)$. The symbols in the figure are differentiated to indicate the distinct dynamical regimes P, QP_2 or L, and N. The frequencies in the non-periodic regime are only known to about 1 % because of the broadening of spectral features.

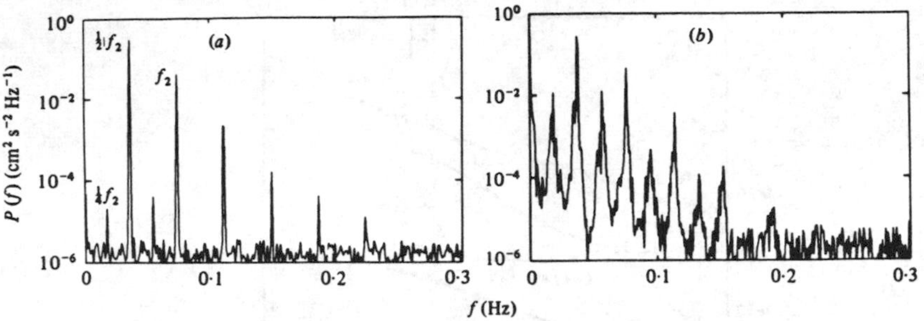

FIGURE 11. (*a*) Spectrum showing subharmonic (period doubling) bifurcation to frequencies $\frac{1}{2}f_2$ and $\frac{1}{4}f_2$ (see discussion of route II), $R/R_c = 27{\cdot}0$. In (*b*), the motion has just become non-periodic and the peak at $\frac{1}{4}f_2$ is quite strong, $R/R_c = 28{\cdot}0$.

4.4. *Route II: subharmonic (period doubling) bifurcations*

Non-periodicity can be obtained after several successive subharmonic bifurcations at $\Gamma = 3{\cdot}5$ and $P = 2{\cdot}5$, and perhaps under other conditions as well. A similar process had been seen earlier in mathematical models as simple as mappings of the unit interval of the form $x_{n+1} = f(x_n)$.

We observe the following sequence of instabilities leading to turbulence via subharmonic bifurcations. The periodic state has an onset at $R/R_c = 17$, substantially lower that that at $P = 5{\cdot}0$ described in §4.3. The amplitudes of the peak (again denoted by f_2) and its harmonics grow with increasing R, until a subharmonic bifurcation at $R/R_c = 21{\cdot}5$ doubles the period of the oscillation. This instability is indicated by a peak at $\frac{1}{2}f_2$ (and harmonics) which grows rapidly with R and eventually dominates f_2. A second subharmonic bifurcation occurs at $R/R_c = 26{\cdot}5$, and produces peaks at $\frac{1}{4}f_2$ and harmonics in the spectrum, as shown in figure 11. The peak at $\frac{1}{4}f_2$ is again very weak at first, but grows quickly and continues to grow even after the onset of non-periodicity at $R/R_c \simeq 28$. There may also be a small region near $R/R_c = 28$ where the motion is quasi-periodic (QP_2). The spectrum shows many closely spaced peaks resulting from a slow modulation in the time domain, but the presence of noise cannot be ruled out. Hence the existence of a quasi-periodic regime is uncertain. At $R/R_c = 29$, the spectra are clearly broadband, and they are nearly featureless at $R/R_c = 40$.

Each of the transitions described in this section has hysteresis of about unity in R/R_c and average values are quoted above.

4.5. *Route III: three frequencies*

Quasi-periodic states characterized by three distinct frequencies were found in several cases (see figure 6). We summarize here the evidence obtained at $\Gamma = 3{\cdot}5$ and $P = 5$. The mean flow in each case is characterized by three rolls, but we do not know if this is significant. The flow becomes periodic at $R/R_c \simeq 30$ with a frequency (0·110 Hz) that is higher than in the previous sections because of the higher wavenumber. Instabilities leading to the second and third frequencies occur at $R/R_c = 39{\cdot}5$ and 41·5, and a typical spectrum containing all three is shown in figure 12. Broadband noise begins to grow at $R/R_c = 43$ and the peaks also become broadened.

In order to verify that three basic frequencies are both necessary and sufficient to

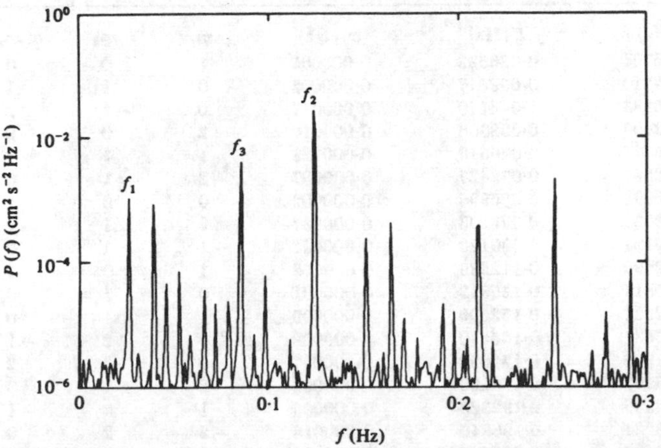

FIGURE 12. Spectrum showing presence of three incommensurate frequencies (route IIId), $R/R_c = 42{\cdot}3$. All peaks are linear combinations of the three frequencies f_1, f_2, and f_3.

describe figure 12, we used the following procedure. First, the three highest peaks were selected and their frequencies f_1, f_2, and f_3 were determined by the method of §3.3. Integers m_1, m_2, and m_3 (less than 20) were chosen for each spectral line by matching its frequency to the equation $F = m_1 f_1 + m_2 f_2 + m_3 f_3$ as closely as possible. Now holding the integers fixed, a least squares fit to all of the lines was performed, with the f_i as adjustable parameters. This slightly modified the f_i from the original estimates, and provided an estimate of the goodness of fit by the chi-squared criterion. We found χ^2 to be 34·5 with 20 degrees of freedom, which is acceptable in view of the limited accuracy with which we know σ_f, the standard error of the measured frequencies. The resulting fit is given in table 1. For each spectral line, the measured frequency f, its estimated standard error, the fitted frequency F, and the three integers m_i are listed.

In order to demonstrate that three frequencies are not only sufficient but also necessary, we followed this fitting procedure with various combinations of two basic frequencies. No value of chi squared less than 1700 could be obtained when the integers were restricted to values less than 20. This clearly rules out the possibility of a satisfactory two-frequency fit to the peaks in figure 12. We also eliminated the possibility of aliasing in the spectrum, and verified that no combination of integer multiples of the f_i was equal to zero, for integers less than 20. These facts, combined with the observation that the three ratios f_2/f_1, f_2/f_3, and f_3/f_1 vary smoothly with R, provide strong evidence that the f_i are in general incommensurate.

A QP_3 regime was also found for $\Gamma = 2{\cdot}4$ and $P = 5$. The mean flow was similar to figure 4 (e) but with signs reversed. In this case we did not attempt to find the onset of time dependence, but the flow is periodic above $R/R_c = 97$. There may be a narrow QP_2 regime near $R/R_c = 109$, and three frequencies are definitely present in the range $110 < R/R_c < 117$. Above this point, the noise grows rapidly and the spectra at $R/R_c = 120$ are almost featureless. Three frequencies were never observed for the mean flow of figure 4 (a) (two rolls).

f (Hz)	F (Hz)	σ_f (Hz)	m_1	m_2	m_3
0·026704	0·026693	0·000005	1	0	0
0·039810	0·039817	0·000002	0	1	−1
0·046183	0·046179	0·000017	0	−1	2
0·053409	0·053386	0·000016	2	0	0
0·066547	0·066510	0·000028	1	1	−1
0·072381	0·072427	0·000037	−2	1	0
0·085991	0·085996	0·000002	0	0	1
0·093232	0·093203	0·000037	2	1	−1
0·099121	0·099120	0·000001	−1	1	0
0·112690	0·112689	0·000072	1	0	1
0·125813	0·125813	0·000001	0	1	0
0·152521	0·152506	0·000009	1	1	0
0·165637	0·165630	0·000008	0	2	−1
0·171985	0·171992	0·000047	0	0	2
0·179197	0·179199	0·000006	2	1	0
0·192336	0·192323	0·000026	1	2	−1
0·198224	0·198240	0·000018	−2	2	0
0·205467	0·205447	0·000074	0	3	−2
0·211805	0·211809	0·000007	0	1	1
0·219048	0·219016	0·000060	2	2	−1
0·251626	0·251626	0·000003	0	2	0
0·278288	0·278319	0·000025	1	2	0
0·291445	0·291443	0·000016	0	3	−1

TABLE 1. Least squares fit of the function $F = m_1 f_1 + m_2 f_2 + m_3 f_3$ to the frequencies f of the peaks in figure 12. The integers m_i and the estimated error σ_f in f are also listed.

4.6. *Route IV: intermittent noise*

A. Libchaber & J. Maurer (1979, private communication) recently observed in liquid helium a type of intermittent noise that is qualitatively different from the non-periodic flows described in earlier sections. We have found that it also occurs at $P = 5{\cdot}0$, in a cell ($\Gamma = 2{\cdot}4$) of the same proportions as theirs. The stable mean flow above $R/R_c = 85$ is that of figure 4 (*e*). It becomes quasi-periodic (QP_2) at $R/R_c = 95$ (see the bar graph labelled IV*d* in figure 5), and the second frequency is visible as an extremely slow but regular modulation in the time record of figure 13 (*a*). The spectrum (not shown) has many closely spaced peaks that overlap somewhat. The low frequency increases slowly with increasing R until intermittent noise appears at $R/R_c = 102$. The dynamical behaviour appears to switch at irregular intervals between a non-periodic state and the quasi-periodic state. Examples of this dynamical discontinuity are clearly visible in figure 13 (*b*). As R is increased, the formerly quasi-periodic regions of the time record become irregular and eventually the entire record appears non-periodic, with a featureless spectrum.

5. Discussion

The various sequences of instabilities are all repeatable, but we have discerned no simple rules for predicting which sequence will occur for a given aspect ratio, Prandtl number, and mean flow. However, several generalizations are supported by the evidence. First, quasi-periodicity with more than two frequencies seems to be associated only with rather complex mean flow patterns. Secondly, phase locking occurs at

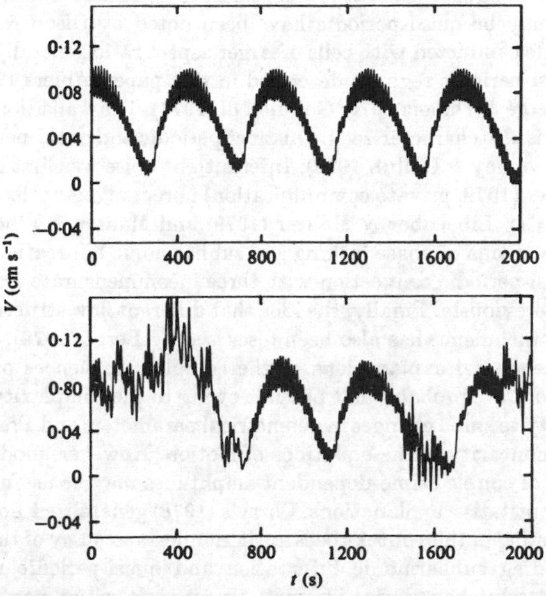

FIGURE 13. Velocity records showing the presence of intermittent noise (route IV). In (*a*), $R/R_c = 100{\cdot}4$, the fluid is quasi-periodic with frequencies $f_1 = 0{\cdot}0023\,3$ and $f_2 = 0{\cdot}0756\,3$ Hz. In (*b*), at $R/R_c = 102{\cdot}8$, intervals of quasi-periodic motion are interspersed between very noisy intervals.

relatively small integer ratios between two characteristic frequencies, and only near the onset of non-periodicity. Thirdly, intermittent noise occurs only for flows that attain rather high Rayleigh number while remaining periodic. Fourthly, quasi-periodicity or subharmonic bifurcation generally precedes non-periodic motion in these experiments.

Not all of our observations fit neatly into the sequences described in §4. For example, the mean flow of figure 4 (*a*) undergoes the sequence $\mathrm{S} \rightarrow \mathrm{P} \rightarrow \mathrm{QP}_2 \rightarrow \mathrm{S}$ at $\Gamma = 2{\cdot}4$. This observation of a reversion to stationary flow at $R/R_c = 60$ is in striking contrast to the general pattern of increasing complexity of time-dependent motion as the Rayleigh number is increased.

The information contained in figure 6 may be incomplete. For example, it is possible that phase locking occurs more frequently than is noted in the bar graphs, but was missed because of the necessary coarseness of the Rayleigh number scans. Furthermore, some instabilities that do not produce dramatic changes in the spectrum may have entirely escaped detection. We can reliably conclude, however, that some phenomena are pervasive in the transition to turbulent convection at small aspect ratio: quasi-periodicity with as many as three independent frequencies; phase locking; and subharmonic bifurcation.

Because of the large number of parameters needed to characterize the fluid, the geometry, and the boundary conditions, it is difficult to make quantitative comparisons between the observations of different investigators. However, quasi-periodicity has also been found in circular cells at low aspect ratio by Ahlers & Behringer (1979). Oscillations that may be quasi-periodic have been noted by Olson & Rosenberger (1979). Experiments conducted with cells of larger aspect ratio generally do not show the noise-free quasi-periodic regimes described in this paper (Ahlers 1974; Ahlers & Behringer 1978; Bergé & Dubois 1976; Gollub *et al.* 1977). The transition to turbulent Taylor vortex flow is also characterized by distinct periodic and quasi-periodic regimes (Fenstermacher, Swinney & Gollub 1979). Intermittent noise was first discovered by Libchaber & Maurer (1979, private communication) in rectangular cells of convecting liquid helium ($P = 1$). Libchaber & Maurer (1979) and Maurer & Libchaber (1979) also noted the phenomena of phase locking and subharmonic bifurcation described in §§4.3 and 4.4. Quasi-periodic convection with three incommensurate frequencies has not been reported previously. Finally, the idea that different flow structures can show different routes to turbulence has also been discussed by Bergé (1979).

Quantitative theoretical explanations of the complete sequences of instabilities observed in this work are probably not possible owing to the complexity of the mean flows, the sensitivity to small changes in geometrical parameters and Prandtl number, and the strong nonlinearity of the equations of motion. However, models consisting of a small number of coupled time-dependent amplitudes may be useful even if they do not provide quantitative explanations. Curry's (1978) generalized Lorenz model is particularly interesting in this context because it manifests so many of the phenomena we observe, including subharmonic bifurcation and quasi-periodic motion. It is possible that the lateral boundaries in small aspect ratio experiments restrict the number of modes enough to justify the use of strongly truncated models.

One phenomenon that can probably be successfully explained with simple models is the onset of non-periodicity via phase locking. The entrainment of one oscillator by another is well known in many contexts. One example is the forced Van der Pol oscillator discussed extensively by Flaherty & Hoppensteadt (1978). A system of two coupled tunnel diode relaxation oscillators was also found to exhibit phase locking (Gollub, Brunner & Danly 1978). This simple electronic system has only four dynamical variables, yet it also manifests non-periodic motion. Both the phase locking and the non-periodic motion have been successfully modelled numerically (Gollub, Romer & Socolar 1980). It seems quite likely that the onset of non-periodic convection can also be explained by modelling it as a system of interacting nonlinear oscillators.

The observation of quasi-periodic convection with three generally incommensurate frequencies may have interesting theoretical implications. This motion can be described by a trajectory covering the surface of a three-dimensional torus T^3 in the phase space spanned by the time-dependent Fourier amplitudes of the velocity field (see Fenstermacher, Swinney & Gollub 1979). An interesting theorem about such tori has been proven recently by Newhouse, Ruelle & Takens (1978). It states (paraphrased) that, in every suitably differentiable neighbourhood of a vector field on the torus T^m, there is a vector field having a non-periodic attractor if $m \geqslant 3$. If this theorem is applicable to the Navier–Stokes equations, it suggests that infinitesimal perturbations of a quasi-periodic motion with three incommensurate frequencies may result in non-periodic motion. Such quasi-periodic motion would then probably not be observable over a significant range in the relevant parameters. The fact that we do observe this behaviour over a finite interval in R means that the physical relevance of this theorem is as yet obscure. However, the basic prediction of Ruelle & Takens

(1971) that non-periodic motion should occur after a small number of time-dependent instabilities is generally consistent with our observations.

Finally, quantitative predictions for the amplitudes of spectral peaks have been obtained recently (Feigenbaum 1980) for the case of a cascade of subharmonic bifurcations. The theory contains universal scaling parameters and may be applicable to a variety of hydrodynamic problems. A more detailed description of the subharmonic route to turbulence, including maps of the spatial structure of the oscillations and a comparison with the theoretical predictions, appears elsewhere (Gollub, Benson & Steinman 1980).

The diversity of processes involved in the transition to turbulent convection is very great. However, the prevalence of relatively simple phenomena (period doubling bifurcations, quasi-periodic motion, and phase locking) does offer some hope for the possibility of achieving qualitative understanding of the basic processes.

It is a pleasure to acknowledge the contributions of H. Swinney to the planning and execution of this work. We also appreciate extremely helpful discussions with G. Ahlers, P. Hohenberg, D. Joseph, and A. Libchaber. J. Steinman assisted us with recent experiments. This work was supported by the National Science Foundation and the Research Corporation.

REFERENCES

Ahlers, G. 1974 Low-temperature studies of the Rayleigh–Bénard instability and turbulence. *Phys. Rev. Lett.* **33**, 1185.

Ahlers, G. & Behringer, R. 1978 Evolution of turbulence from the Rayleigh–Bénard instability. *Phys. Rev. Lett.* **40**, 712–716.

Ahlers, G. & Behringer, R. P. 1979 The Rayleigh–Bénard instability and the evolution of turbulence. *Prog. Theor. Phys. Suppl.* **64**, 186–201.

Bergé, P. 1979 Experiments on hydrodynamic instabilities and the transition to turbulence. In *Dynamical Critical Phenomena and Related Topics* (ed. H. P. Enz). Springer.

Bergé, P. & Dubois, M. 1976 Time-dependent velocity in Rayleigh–Bénard convection: a transition to turbulence. *Opt. Comm.* **19**, 129–133.

Busse, F. H. 1978 Non-linear properties of thermal convection. *Rep. Prog. Phys.* **41**, 1929–1967.

Busse, F. H. 1980 Transition to turbulence in Rayleigh–Bénard convection. In *Hydrodynamic Instabilities and the Transition to Turbulence* (ed. H. L. Swinney & J. P. Gollub). Springer.

Busse, F. H. & Clever, R. M. 1979. *J. Fluid Mech.* **91**, 319–335.

Busse, F. H. & Whitehead, J. A. 1974 Oscillatory and collective instabilities in large Prandtl number convection. *J. Fluid Mech.* **66**, 67–79.

Clever, R. M. & Busse, F. H. 1974 Transition to time-dependent convection. *J. Fluid Mech.* **65**, 625–645.

Curry, J. H. 1978 A generalized Lorenz system. *Comm. Math. Phys.* **60**, 193–204.

Davis, S. H. 1967 Convection in a box: linear theory. *J. Fluid Mech.* **30**, 467–478.

Dubois, M. & Bergé, P. 1978 Experimental study of the velocity field in Rayleigh–Bénard convection. *J. Fluid Mech.* **85**, 641–653.

Feigenbaum, M. J. 1980 The onset spectrum of turbulence. *Phys. Lett.* **A74**, 375–378.

Fenstermacher, P. R., Swinney, H. L., Benson, S. A. & Gollub, J. P. 1979 Bifurcations to periodic, quasiperiodic, and chaotic regimes in rotating and convecting fluids. *Ann. N.Y. Acad. Sci.* **316**, 652–666.

Fenstermacher, P. R., Swinney, H. L., & Gollub, J. P. 1979 Dynamical instabilities and the transition to chaotic Taylor vortex flow. *J. Fluid Mech.* **94**, 103–128.

Flaherty, J. E. & Hoppensteadt, F. C. 1978 Frequency entrainment of a forced Van der Pol oscillator. *Stud. Appl. Math.* **58**, 5–15.

Gollub, J. P. & Benson, S. V. 1978 Chaotic response to periodic perturbation of a convecting fluid. *Phys. Rev. Lett.* **41**, 948–951.

Gollub, J. P. & Benson, S. V. 1979 Phase locking in the oscillations leading to turbulence. In *Pattern Formation* (ed. H. Haken), pp. 74–80. Springer.

Gollub, J. P., Benson, S. V. & Steinman, J. 1980 A subharmonic route to turbulent convection. *Ann. N.Y. Acad. Sci.* (to appear).

Gollub, J. P., Brunner, T. O., & Danly, B. G. 1978 Periodicity and chaos in coupled nonlinear oscillators. *Science* **200**, 48–50.

Gollub, J. P., Hulbert, S. L., Dolny, G. M. & Swinney, H. L. 1977 Laser Doppler study of the onset of turbulent convection at low Prandtl number. In *Photon Correlation Spectroscopy and Velocimetry* (ed. H. Z. Cummins & E. R. Pike), pp. 425–439. Plenum.

Gollub, J. P., Romer, E. J. & Socolar, J. E. 1980 Trajectory divergence for coupled relaxation oscillators: measurements and models. *J. Stat. Phys.* **23**, 321–333.

Koschmieder, E. L. 1974 Bénard convection. *Adv. Chem. Phys.* **26**, 177–212.

Krishnamurti, R. 1970 On the transition to turbulent convection. Part 2. The transition to time-dependent flow. *J. Fluid Mech.* **42**, 309–320.

Krishnamurti, R. 1973 Some further studies on the transition to turbulent convection. *J. Fluid Mech.* **60**, 285–303.

Libchaber, A. & Maurer, J. 1978 Local probe in a Rayleigh–Bénard experiment in liquid helium. *J. Phys. Lett.* **39**, L369–L372.

Libchaber, A. & Maurer, J. 1979 Une expérience de Rayleigh–Bénard de géométrie réduite: multiplication, accrochage et démultiplication de fréquences. *J. Phys.* (to appear).

Lipps, F. B. 1976 Numerical simulation of three-dimensional Bénard convection in air. *J. Fluid Mech.* **75**, 113–148.

Lorenz, E. N. 1963 Deterministic nonperiodic flow. *J. Atmos. Sci.* **20**, 130–141.

Maurer, J. & Libchaber, A. 1979 Rayleigh–Bénard experiment in liquid helium: frequency locking and the onset of turbulence. *J. Phys. Lett.* **40**, L419–L423.

McLaughlin, J. B. & Martin, P. C. 1975 Transition to turbulence in a statically stressed fluid system. *Phys. Rev.* A **12**, 186–203.

Monin, A. S. 1979 On the nature of turbulence. *Sov. Phys. Usp.* **21**, 429–442.

Newhouse, S., Ruelle, D. & Takens, F. 1978 Occurrence of strange axiom *A* attractors near quasi-periodic flows on T^m, $m \geqslant 3$. *Comm. Math. Phys.* **64**, 35–40.

Normand, C. Y., Pomeau, Y. & Velarde, M. G. 1977 Convective instability: a physicist's approach. *Rev. Mod. Phys.* **49**, 581–624.

Olson, J. M. & Rosenberger, F. 1979 Convective instabilities in a closed vertical cylinder heated from below. Part 1. Mono-component gases. *J. Fluid Mech.* **92**, 609–629.

Otnes, R. K. & Enochson, L. 1972 *Digital Time Series Analysis*, p. 286. Wiley.

Ruelle, D. & Takens, F. 1971 On the nature of turbulence. *Comm. Math. Phys.* **20**, 167–192.

Saltzman, B. 1962 Finite amplitude free convection as an initial value problem. *J. Atmos. Sci.* **19**, 329–341.

Stork, K. & Müller, U. 1972 Convection in boxes: experiments. *J. Fluid Mech.* **54**, 599–611.

Willis, G. E. & Deardorff, J. W. 1970 The oscillatory motions of Rayleigh convection. *J. Fluid Mech.* **44**, 661–672.

Physica 6D (1983) 385–392
North-Holland Publishing Company

ON THE OBSERVATION OF AN UNCOMPLETED CASCADE IN A RAYLEIGH–BÉNARD EXPERIMENT

A. ARNEODO, P. COULLET and C. TRESSER
Laboratoire de Physique, Université de Nice, 06034 Nice Cedex, France

and

A. LIBCHABER, J. MAURER and D. d'HUMIÈRES
Groupe de Physique des Solides de l'Ecole Normale Supérieure, 24 rue Lhomond, 75231 Paris Cedex 05, France

Received 1 January 1982
Revised 12 October 1982

In Rayleigh–Bénard experiments in liquid helium and liquid mercury, the scenario leading to chaos via a period-doubling bifurcation is sometimes interrupted by an odd-period bifurcation. This scenario can be understood in the framework of two-dimensional invertible mappings. Such a behavior is tested on a simulated experiment of a parametric pumped pendulum.

1. Introduction

Recent Bénard–Rayleigh experiments in helium [1], water [2] and mercury [3] have shown the relevance of a period-doubling cascade as a route to chaos. But the helium and mercury experiments gave also indications [1, 4] that a somewhat different scenario was possible, where the period–doubling cascade is interrupted by an odd-period multiplication. This puzzling result is not consistent with the simplest one-dimensional dynamical system [5], including even fluctuations and non-adiabaticity in the variation of the control parameter. In such models, the mapping of the interval as a dynamical system, there is a strict hierarchy in the order of appearance of bifurcations as a function of the control parameter μ. As μ increases a cascade of period-doubling bifurcations evolves to its limit μ_∞. Beyond this value a chaotic state sets in, within which laminar windows appear with periods of various lengths. In some experiments the cascade of period doubling on one hand, and period multiplication by odd integers on the other, seem to be competing for the same values of μ.

The aim of this paper is to show that, in the framework of two-dimensional invertible mappings, this experimental scenario becomes a reasonable one. Two-dimensional dynamical systems, such as the Hénon mapping [6], can present, for given values of the parameters, several stable dynamical states, so that taking into account a non-adiabatic variation of the control parameter, we can reproduce scenarii comparable to the experimental findings.

Let us note that in no way do we intend to say that the couple of Navier–Stokes and heat equations of the Rayleigh–Bénard experiment are reducible to such a two-dimensional map. We are only showing the analogies in the scenarii.

As an added confirmation of those ideas, we have also computer simulated the behavior of a dissipative parametrically pumped pendulum as a function of the amplitude of the pump signal. There too we find two competing scenarii, one with period-doubling bifurcations up to a chaotic state, and a second one where, for an abrupt change of the pump amplitude, another period multiplication sets in.

The remainder of the paper is organized as follows. In the first part the essentials of the experimental observations in fluids are presented. In a second part, after a brief introduction to the Hénon mapping, we show that a behavior analogous to the experimental one can be derived from this two-dimensional map. The third part will consider within the same framework, a simulation of the parametric pendulum.

2. Period three in a Rayleigh–Bénard experiment

This part pertains to previous publications by two of the authors (A.L. and J.M.) on a Rayleigh–Bénard experiment in helium [1, 4] and mercury [3], where we observed the period-doubling cascade. We noted there that the cascade is sometimes interrupted by a bifurcation to a period multiplication by integers others than two. Various integers have been observed, occurring at different stages of the evolution of the cascade. For example in mercury [3] we observed that the period 2 and period 3 could compete via intermittency. In earlier experiments in helium the following bifurcations occurred:

$$f_1 \to (f_1, f_2) \to \left(\text{lock in } f_1, \frac{f_1}{2}\right) \to \frac{f_1}{6}.$$

In this paper we come back in detail to this scenario where the cascade is interrupted by a frequency division by three.

The experimental set-up is well documented in our previous publications and we refer to them. The cell size is a box of height 1.25 mm and lateral dimensions 3 mm and 1.5 mm. In the convective regime, two rolls are present with an apparent wave-number $\alpha = 2.62$. The Prandtl number of helium in this experiment is 0.5. The evolution of the fluid temperature field, as one increases the Rayleigh number, is identical to the case of the period-doubling cascade for the first few bifurcations. We end up with two locked oscillators f and $f/2$. If we then increase the control parameter a division by three sets in, followed by a division close to 12. This scenario is shown in fig. 1 as a function of the Rayleigh number. We have also reproduced in fig. 1 the scenario of the period-doubling cascade for comparison. In fig. 2, we present various Fourier spectrum data of the local temperature field, for the characteristic Rayleigh number corresponding to: one oscillator, two unlocked oscillators, locking, period three, a new period close to 12, and chaos. The relevant results of the data are as follows, when we compare them with the period-doubling case:

— In both cases the route to a chaotic state goes through period multiplication.

— The period three appears sooner in Rayleigh number value than the period two of the cascading case.

— After the period three bifurcation, a new bifurcation sets in which looks like a period 12 but is not at the exact frequency value and with a rather broad linewidth. The chaotic state is so close to this last bifurcation that we cannot follow its evolution.

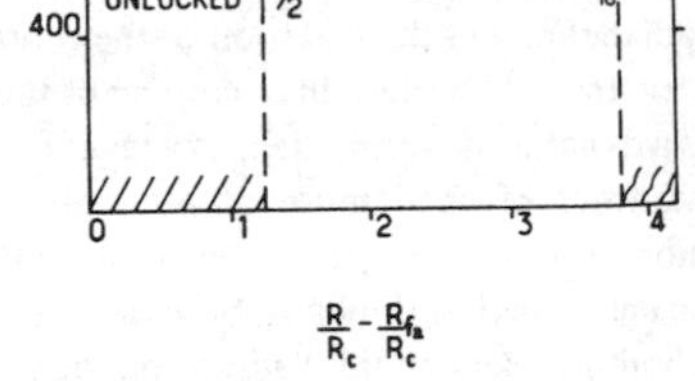

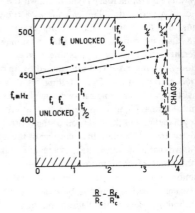

Fig. 1. Comparison between the scenario described in this article (triangles) and the period-doubling cascade observed previously [1] (circles). On the X-axis, we plot R/R_c starting from the onset value of the second oscillator, thus $\Delta R/R_c$. On the Y-axis, we plot the frequency value of the first oscillator f_1.

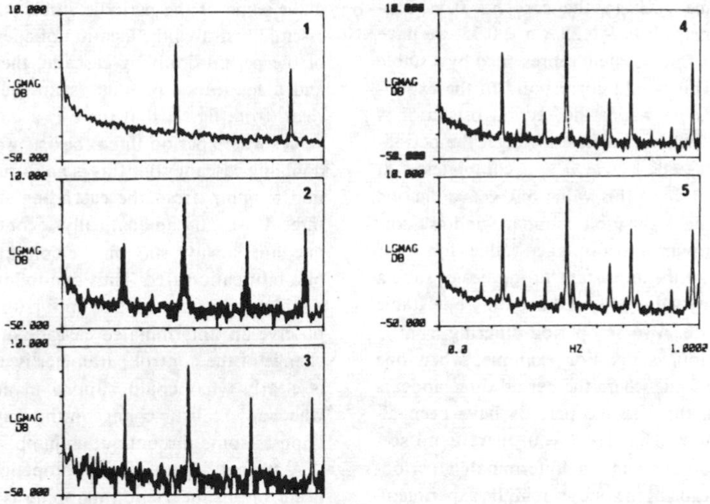

Fig. 2. Power spectrum of the experimental scenario in the Rayleigh–Bénard experiment described in part I: 1) oscillator f_1 and harmonic $2f_1$; 2) f_1 and f_2 unlocked; 3) locking f_1 and $f_1/2$; 4) division by three of the second oscillator f_1, $f_1/2$, $f_1/6$; 5) just before the chaotic state onset of $f_1/24$; (Log Mag DB means $20\log(A_2/A_1)$ where A is the amplitude of the temperature signal).

3. Competing scenarii in the Hénon mapping

In this section, we are going to show, using general arguments and explicit numerical calculations, that an interrupted cascade of period-doubling bifurcations occur in invertible maps of the plane if we allow non-adiabatic variations of the control parameter. For simplicity, we shall focus on the Hénon map [6]. The Hénon mapping models the Poincaré section of the Lorenz system for a high value of the Rayleigh number [7]. Its expression is

$$X_{n+1} = 1 - aX_n^2 + Y_n,$$

$$Y_{n+1} = bX_n. \tag{1}$$

This involves two parameters. The parameter a can be roughly thought as a constraint parameter, like the Rayleigh number in the convection experiment. The parameter b represents the area contraction rate at each iteration. Negative b values are more physical, because they correspond to orientation preserving maps. $b = -1$ represents the conservative case [8], and $|b| < 1$ a dissipative one.

For $b = 0$ the mapping reduces to the non-invertible one-dimensional map:

$$X_{n+1} = 1 - aX_n^2. \tag{2}$$

At first sight the parameter b could be related to the Prandtl number of the fluid in the convection experiment. But the Prandtl number is associated to a global dissipation for the convection equations, whereas b acts as a local dissipation for the dynamical behavior of the few modes involved in the convection experiment. Thus the connection between b and the Prandtl number is difficult, even qualitatively, to formulate.

We must repeat here again that our intention is not to model the previously described experiment using the Hénon map, but only to indicate that a similar type of scenario exists.

Let us come back to the case $b = 0$, a one-dimensional map. For $-0.25 < a < 0.75$, we have a laminar dynamical system represented by a stable fixed point. This would correspond in the experiment to the state where the second oscillator is locked ($f_1/2$). When a increases, we get the period-doubling cascade which accumulates at $a = 1.401\ldots$ Beyond this value one gets a chaotic dynamical state with stable laminar windows corresponding to arbitrary integer value for their periods. One of the important properties of such a map is: for any value of a we have only one stable dynamical state. Also any period differing from 2^n is in the chaotic state. For example, when one reaches the a value where the period three appears ($a = 1.75$) all the possible periods have been already generated. Thus for $b = 0$, there is no scenario corresponding to an unterminated period-doubling cascade as the one seen in the experiment. Even with a non-adiabatic change of the control parameter, an enormous change in values of a is necessary to go from the fixed point to the period three, (from $a = 0.75$ to $a = 1.75$).

The situation is quite different for $b \neq 0$. In this case, there exists several competing dynamical states with distinct basins of attraction [9], for a given value of a and b. Fig. 3 shows in the a–b plane some of the periodic stable states. We have essentially drawn the location of a few bifurcations of the period-doubling cascade, the period three, and a few other periods as an illustration. It is clear, from fig. 3, that there is a large domain of values where period three coexists with the period-doubling cascade. But this is not a sufficient condition to jump from the cascading state to period three. If we vary adiabatically a control parameter one must always stay on the continuous branches of a bifurcation tree. Thus one follows the period-doubling cascade to its end [10]. In order to observe an unterminated cascade a non-adiabatic change of the control parameter is necessary. This is clearly what could happen in an experiment, where any realistic change in the Rayleigh number imposes some discontinuous jump.

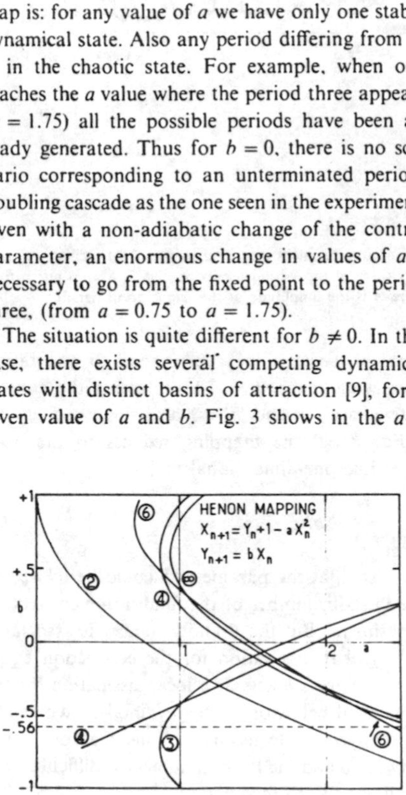

Fig. 3. Hénon mapping. Traces of the period-doubling cascade 2, 4, ∞. Traces of some periodic attractors 3, 4, 6. For $b = -0.56$, one sees (dotted line) that a period 6 is situated between period 2 and 4 of the cascade. This is relevant to the parametric pendulum case.

A typical aspect of the competing phenomena occurring in the Hénon mapping is illustrated on fig. 4. For several values of a and b, we have computed the basins of attraction of period 1 and period 3 cycles. In the conservative case ($b = -1$) period 3 appears as a strong resonance, and occurs necessarily before the bifurcation to period 2 [11]. This is why near the conservative case the basin of attraction of period 3 may become more important that that of period 1. Period multiplication by three should thus be more likely in weakly dissipative systems.

We will now show, performing an explicit computer simulated experiment on the parametric pumped pendulum, that one can reproduce a complete period-doubling cascade, or an interrupted one, depending on how careful we are in varying the control parameter.

4. The parametric pumped pendulum

From a different point of view, the cascade of period-doubling bifurcations can be viewed as a cascade [12] of parametric instabilities [13]. Thus the parametrically driven pendulum comes as a natural computer simulated experiment to test our previous argumentations.

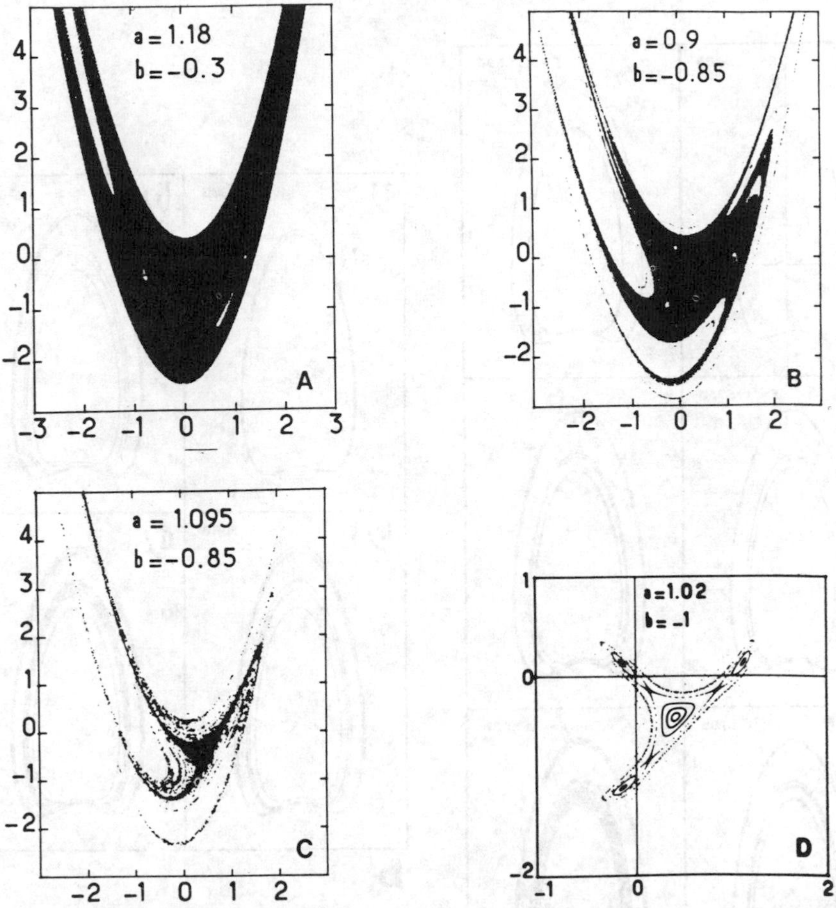

Fig. 4. Evolution of the respective basins of attractions of period 1 and 3 (black region period 1, dots period 3). The small white dots indicate the attractive points. Figure D refers to the conservative case ($b = -1$) where the basins of attraction disappear. We present here, successive iterates of various (X, Y) initial conditions. The ghosts of the basins of attraction appear in the form of the well-known islands. The value of a is close to the resonance one where period 1 loses its stability.

The equation describing the pendulum is

$$\frac{d^2x}{dt^2} + \nu\frac{dx}{dt} + \omega_0^2(1 + F\cos\Omega t)\sin x = 0, \qquad (3)$$

where ν is the viscous coefficient, ω_0 the natural frequency of the pendulum, Ω and F respectively the frequency and amplitude of the external forcing source. Let us choose for convenience $\Omega = 2$. The stability of the rest solution $x = \dot{x} = 0$ is determined by the linearized equation

$$\frac{d^2x}{dt^2} + \nu\frac{dx}{dt} + \omega_0^2(1 + F\cos 2t)x = 0. \qquad (4)$$

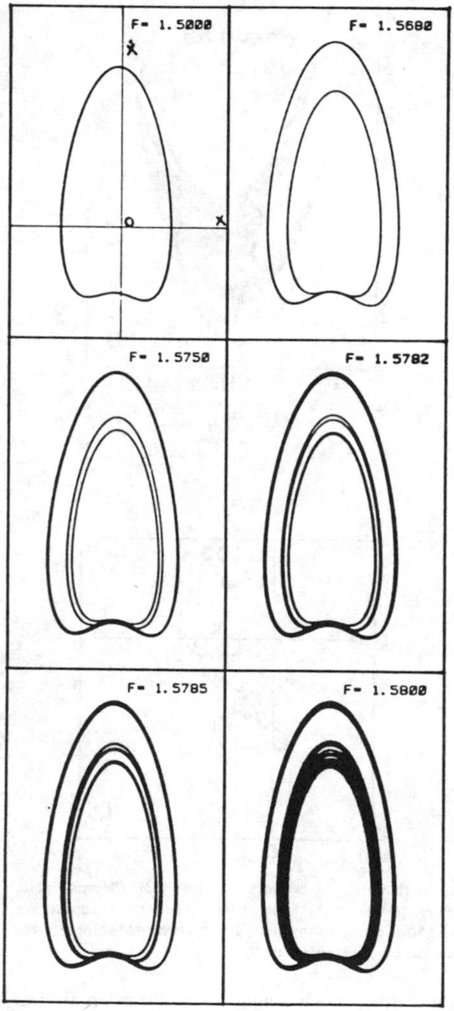

a

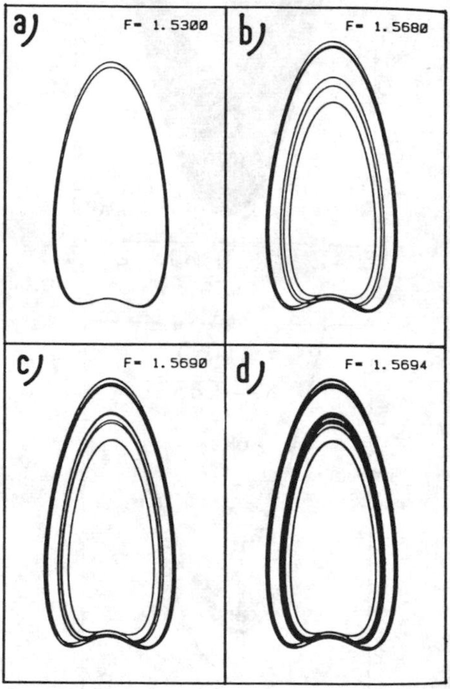

b

Fig. 5 Orbits in phase space (X position, Y velocity) for the parameter **pendulum** for the values of the parameters $\nu = 0.18\mathrm{s}^{-1}$ $\Omega = \omega_0 = 2\mathrm{s}^{-1}$. a) The period doubling cascade; b) The second scenario for an abrupt change of F, from $F = 1.5$ to $F = 1.568$ where period 6 appears. From then on the change is adiabatic and shows in c) the period doubling cascade starting from period 6, and in d) a chaotic case.

which is a damped Mathieu equation. By a change of variable

$$x = e^{-\nu t/2} y \tag{5}$$

one gets the Mathieu equation

$$\frac{d^2 y}{dt^2} + \omega_0^2\left(1 - \frac{\nu^2}{4\omega_0^2} + F\cos 2t\right)y = 0. \tag{6}$$

The instability regions in the F, ω plane (where $\omega = |\omega_0^2 - \nu^2/4|^{1/2}$) are labelled by an integer n which correspond, in the absence of forcing, to the resonant values of ω, $\omega = n$. The effect of the viscosity is to shrink the instability region and to impose a finite driving force at the resonance frequency [14] for the instability threshold.

Going back to the non-linear equation (3) the first effect of the non-linearities is to saturate the instabilities.

Let us start now a description of the simulated experiment. We drive the pendulum at its own natural frequency [15] ($\Omega = \omega_0 = 2$). We take as a value of the viscosity $\nu = 0.18\ s^{-1}$. To compensate viscous damping, we need an amplitude of the driving force $F \approx 0.7$. Above this value, the pendulum oscillation is locked to the pump frequency. It corresponds to what we called period one in the preceding chapter, and is analogous to the locking state of the second oscillator in the fluid experiment. We shall now show that the routes to chaos are very sensitive to how adiabatically the constraint is applied.

In the first simulation, we increase the constraint

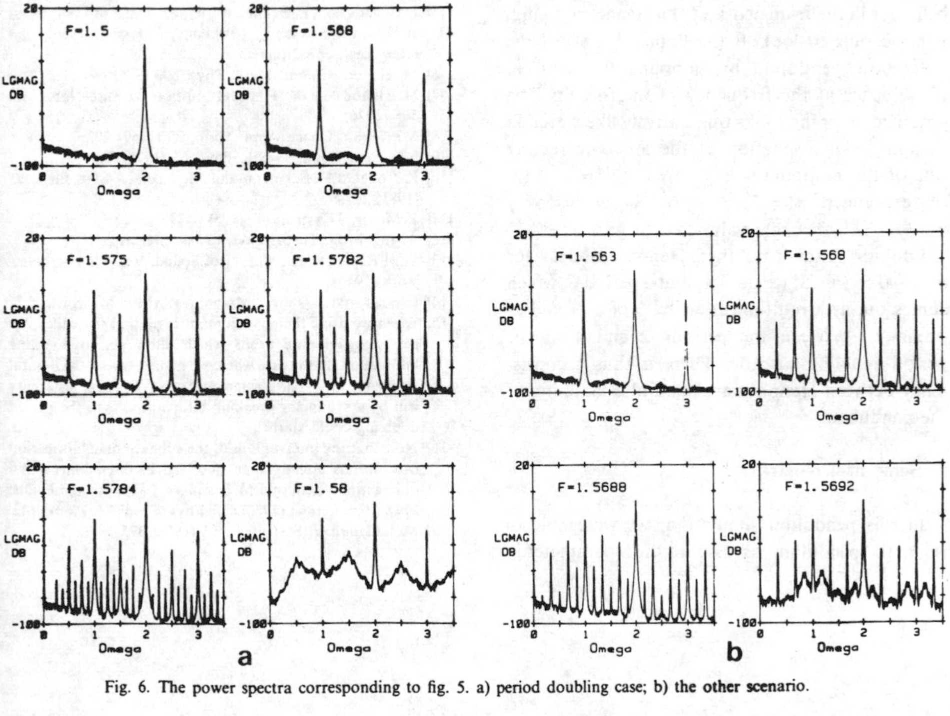

Fig. 6. The power spectra corresponding to fig. 5. a) period doubling case; b) the other scenario.

at a slow rate ($\Delta F = 10^{-4}$). We observe a sequence of period-doubling bifurcations up to $\Omega/16$ and finally a chaotic behavior. This scenario is illustrated in fig. 5a for the orbit in the phase space and in fig. 6a for the corresponding power spectra.

If we perform a second experiment without taking much care to the adiabaticity in turning on the constraint, we observe various possible bifurcations [16]. We show one scenario in figs. 5b and 6b where one observes a division by 2 followed by a division by 3 ($\Omega/6$) if F is increased at a larger rate ($\Delta F = 10^{-2}$). If from there we change again slowly the constraint, ($\Delta F = 10^{-4}$) we see a period-doubling cascade starting from $\Omega/6(\Omega/12, \Omega/24\ldots)$ and leading to chaos. The chaotic state is reached for a value of the forcing term which corresponds, to a situation between period 2 and 4, for the first simulation.

Let us try now to see if one can analyse this behavior in the framework of the Hénon mapping. It is possible to look at the Poincaré map of the parametric pendulum by sampling the orbit in phase space at the frequency of the forcing. The obtained mapping looks qualitatively like a Hénon mapping. An estimation of the area contracting rate of this mapping is given by $e^{-\nu 2\pi/\Omega} = e^{-\nu\pi}$ (in the experiment $\Omega = 2$). For the chosen viscosity $\nu = 0.18\ s^{-1}$, this rate value is 0.56.

Looking now at the Hénon model for $b = -0.56$ (fig. 3) we do find that a period 6 (which comes out as a multiplication by 3 of a period 2) occurs in between the periods 2 and 4 of the period-doubling cascade. There is thus a consistency between Hénon mapping and a mapping of the pendulum.

5. Some final remarks

In this pendulum simulation, we were able to relate the pendulum viscosity to the b parameter in the Hénon model. Unfortunately, we do not see how one could do the same for the fluid experiment.

In this simulation, we could reach a chaotic state in the regime where the pendulum oscillates before the threshold for the rotational behavior of the pendulum [17].

References

[1] A. Libchaber and J. Maurer, J. de Physique, Coll. C3 41 (1980) 51.
[2] M. Giglio, S. Muzatti and U. Perini, Phys. Rev. Lett. 47 (1981) 243.
[3] A. Libchaber and S. Fauve, Proceedings 9th Midwestern Solid State Theory Seminar (Argonne), in press.
[4] A. Libchaber and J. Maurer, Proc. of the Geilo School, "Non-linear phenomena at phase transitions and instabilities", T. Riste, ed. (Plenum, New York, 1981).
[5] J. Collet and J.P. Eckmann, Iterated maps of the interval as dynamical systems, (Birkhauser, Boston, 1980); and references quoted therein.
[6] M. Hénon, Comm. Math. Phys. 50 (1976) 69.
[7] M. Hénon and Y. Pomeau, Springer Lecture Notes vol. 565 (1976).
[8] M. Hénon, Quart. Appl. Math. 27 (1969) 292.
[9] S.D. Feit, Comm. Math. Phys. 61 (1978) 249.
[10] P. Collet, J.P. Eckmann and M. Koch, J. Stat. Phys. 25 (1981) 1.
[11] J. Moser, J. Astrom 63 (1958) 439.
[12] A. Ito, Prog. Theor. Phys, to be published.
[13] Lord Rayleigh, The Theory of Sound, Vol. I (Dover, New York, 1945).
[14] Landau and Lifshitz, "Mecanique" (Mir, Moscow).
[15] When we drive the pendulum at its natural frequency, we get two asymetric orbits which allow period doubling bifurcation. In the case where we pump at twice the natural frequency we obtain a symmetric orbit. The next instability will be a symmetry breaking bifurcation, then the period doubling could start.
[16] Note that the interruption of the subharmonic bifurcation cascade has already been observed in three-dimensional differential system by J.M. Sessinger, J.M. Finn and E. Ott, Phys. Rev. Lett 44 (1980) 453; Phys. Fluids 23 (1980) 1142.
[17] McLaughlin, J. Stat. Phys. 24 (1981) 375.

Physica 8D (1983) 257–266
North-Holland Publishing Company

OBSERVATION OF A STRANGE ATTRACTOR

J.-C. ROUX,* Reuben H. SIMOYI and Harry L. SWINNEY

Department of Physics, The University of Texas, Austin, Texas 78712, USA

Received 14 December 1982

Phase space portraits have been constructed and analyzed for noisy (nonperiodic) data obtained in an experiment on a nonequilibrium homogeneous chemical reaction. The phase space trajectories define a limit set that is an "attractor" – following a perturbation, the trajectory quickly returns to the attracting set. This attracting set is shown to be "strange" – nearby trajectories separate exponentially on the average. Moreover, the Poincaré sections exhibit the stretching and folding that is characteristic of strange attractors.

1. Introduction

Recent observations of nonperiodic behavior in nonlinear chemical, hydrodynamic, and other systems have been described as being characterized by strange attractors [1, 2]. While the data have been shown to exhibit some features of deterministic chaos, the evidence supporting the use of the term "strange attractor" has been incomplete. In this paper we will examine carefully a set of nonperiodic data and describe the construction and analyses of phase portraits that strongly support the strange attractor appellation for these data. Terms such as phase portrait and strange attractor [3–5] will be defined in a simple intuitive manner; readers familiar with these concepts may simply examine the figures to see how the concepts have been implemented with laboratory data.

The data analyzed here were obtained in our experiments on the Belousov–Zhabotinskii reaction in a well-stirred flow reactor [2]. In this reaction, which involves more than 30 chemical constituents, an acidic bromate solution oxidizes malonic acid in the presence of a metal ion catalyst. The concentration of one of the chemicals, the bromide ion, was measured with a specific ion probe, digitized, and recorded as a function of time in a computer in files of 32768 points.

Periodic and nonperiodic states can be distinguished by their power spectra, as fig. 1 illustrates. The spectrum for a periodic state consists of a sharp fundamental frequency component and its harmonics, while the spectrum for a nonperiodic state contains broadened spectral lines or, as in fig. 1, broadband noise. The analysis presented in the following sections indicates that the broadband noise in the spectrum for the nonperiodic state in fig. 1 is primarily deterministic rather than stochastic (e.g., environmental or thermal) in origin.

We will describe phase portraits in section 2, Poincaré sections and maps in section 3, attractors in section 4, the determination of the largest Lyapunov exponent in section 5, and stretching and folding of the attractor in section 6.

* Permanent address: Centre de Recherche Paul Pascal, Université de Bordeaux-I, Domaine Universitaire, 33405 Talence Cedex, France.

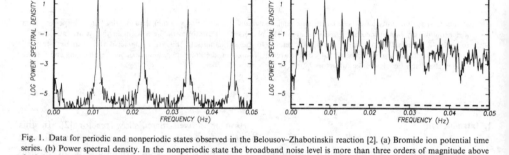

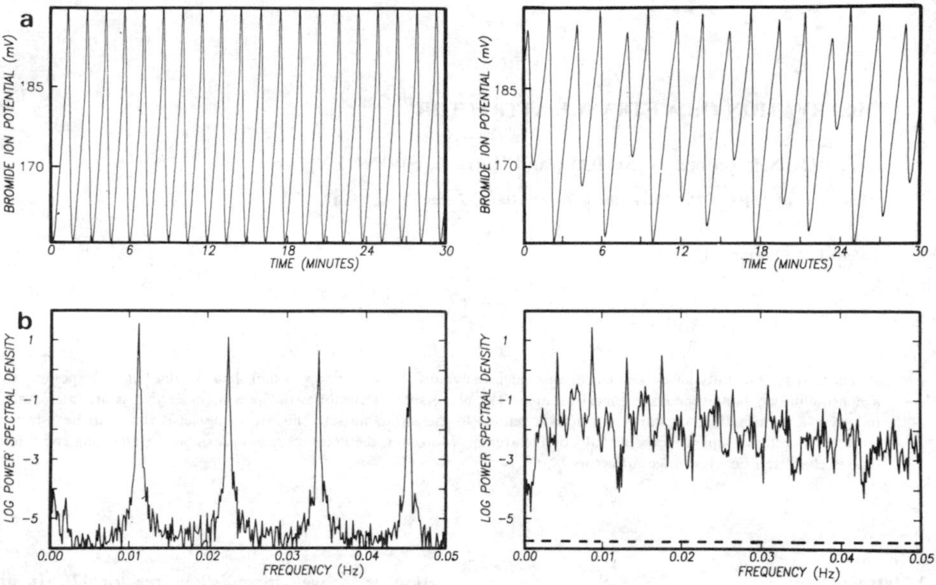

Fig. 1. Data for periodic and nonperiodic states observed in the Belousov–Zhabotinskii reaction [2]. (a) Bromide ion potential time series. (b) Power spectral density. In the nonperiodic state the broadband noise level is more than three orders of magnitude above the instrumental noise level, which is indicated by the horizontal dashed line.

2. Construction of phase portraits

Around the turn of the century Poincaré and others recognized that much could be learned about dynamical behavior from an analysis of system trajectories in a multi-dimensional phase space in which a single point characterizes the entire system at an instant of time. The set of phase space trajectories for all possible initial conditions (for a given set of control parameter values) forms a *phase portrait* of the system.

The N-dimensional phase portrait describing the well-stirred (that is, homogeneous) Belousov–Zhabotinskii system could be constructed from measurements of the time dependence of the concentration of all $N(=30+)$ chemical species in the reaction. Fortunately, such a difficult task is unnecessary – a multi-dimensional phase portrait can be constructed from measurements of a *single* variable by a procedure proposed by Ruelle [6] and Packard et al. [7]. The idea, which is justified by embedding theorems [8, 9], is as follows: For almost every observable $B(t)$ and time delay T an m-dimensional portrait constructed from the vectors $\{B(t_k), B(t_k+T), \ldots, B(t_k+(m-1)T\}$, where $t_k = k\Delta t$, $k = 1, 2, \ldots, \infty$, will have the same properties (for example, the same spectrum of Lyapunov exponents) as one constructed from measurements of N independent variables, if $m \geqslant 2N+1$. Strictly speaking, the phase portrait obtained by this procedure gives an embedding of the original manifold [8, 9].

The choice of the time delay T is almost but not completely arbitrary. An obvious example of the necessity of the phrase "for almost any T" would be a time delay equal to the period of a periodic

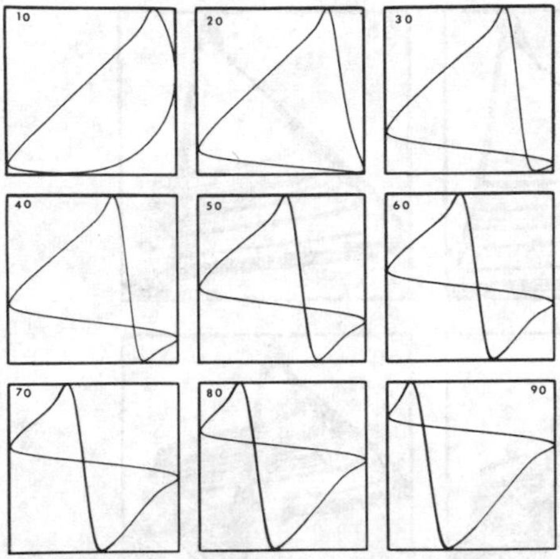

Fig. 2. Two-dimensional phase portraits, $B(t_k + T)$ vs $B(t_k)$, constructed from post-transient data for the periodic state of fig. 1. The different portraits correspond to different time delays T. The values of T, expressed as multiples of the time between successive measurements (0.88 s), are indicated on each portrait; there are 136 measurements per orbit.

state – the portrait would then be a straight line at 45° rather than a closed loop. The effect of varying T is illustrated with data for a periodic state in fig. 2. For small T the trajectories would approach the line $y = x$, while for large T trajectories in a two-dimensional projection cross (see fig. 2). The crossing of trajectories illustrates the necessity for the inequality $m \geqslant 2N + 1$; however, in most cases, including fig. 2, an unambiguous phase portrait can be obtained with many fewer dimensions than the inequality requires. In practice, phase portraits can be constructed with m increased by one at a time until additional structure fails to appear when an extra dimension is added. For the data in fig. 2 the character of the attractor is clear in 2D (two-dimensional) portraits.

The effect of varying T is illustrated for chaotic data in fig. 3. The 3D character of the phase portrait can be seen in fig. 4, which shows 2D projections of 3D portraits rotated about an axis; this 3D character is especially clear when the attractor is viewed as it is continuously rotated about an axis [10]. Thus even for these chaotic data the character of the phase portrait is clear in projections of far fewer dimensions than the 61+ required to satisfy the embedding theorems; this low dimensionality of the phase portrait is revealed even more clearly in the Poincaré section.)

3. Poincaré section and map

Rather than analyze phase portraits directly it is easier to analyze the lower-dimensional Poincaré section which is formed by the intersection of "positively directed" orbits of an m-dimensional phase portrait with an $(m - 1)$-dimensional hypersurface. A 2D Poincaré section constructed for a 3D phase portrait of a nonperiodic state is shown on the left-hand side of fig. 5. (The right-hand side of the figure shows the effect of perturbations, as discussed in the following section.)

The points on the Poincaré section [fig. 5(b)] lie to a good approximation along a parameterizable curve, nor on a higher dimensional curve. (However, the actual dimension of the Poincaré section must be at least slightly greater than unity because of the fractal nature of attractor [11, 12].) Therefore, the coordinate values at successive intersections provide a sequence $\{X_n\}$ which defines a 1D map, $X_{n+1} = f(X_n)$, as shown in fig. 5(c). The data appear to fall on a single-valued curve, indicating that the system is deterministic – for any X_n, the map *determines* X_{n+1}.

4. An attractor: response to perturbations

The right-hand side of fig. 5 illustrates that the post-transient set described by the phase space trajectories is really an *attractor*: the trajectories rapidly return to this limit set after finite perturbations. The *basin of attraction* of an attractor is the set of all initial conditions for which the trajectories asymptotically approach the attractor. For the state illustrated in fig. 5 the trajectories

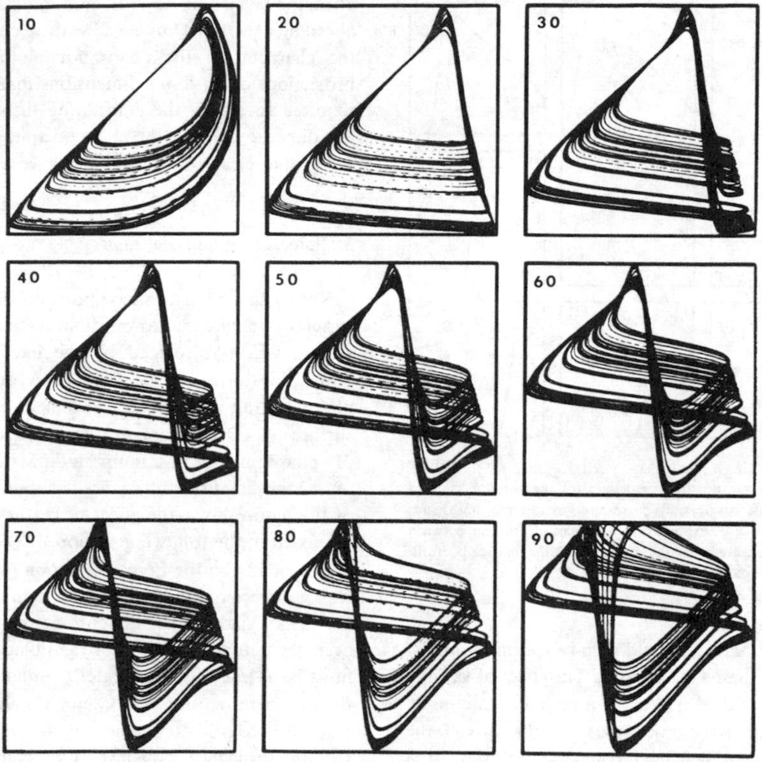

Fig. 3. Two-dimensional phase portraits, $B(t_k + T)$ vs $B(t_k)$, constructed from post-transient data for the nonperiodic state of fig. 1. The different portraits correspond to different time delays T. The values of T, expressed as multiples of the time between successive measurements (0.88 s), are indicated on each portrait; there are 130–140 measurements per orbit.

have been found to return to the attractor for all perturbations we have used, so this attractor could be globally attracting. Perturbations have included injections of bubbles into one of the feed lines, injections of bromide ions into the reactor, and turning off the stirrer or one of the chemical feed lines for a few seconds. For other values of the control parameters multiple stable states have been observed [13], each with its own basin of attraction – the trajectories will then still return to the original attractor for perturbations that are not too large, but a sufficiently large perturbation can send the system trajectory from one basin of attraction into another basin.

5. A Lyapunov exponent

A quantitative measure of nonperiodic (chaotic) behavior is provided by the value of the largest Lyapunov exponent, which characterizes the average rate of separation of nearby trajectories [14]. This exponent is positive for a chaotic state, while for a periodic state it is zero. An attractor with a

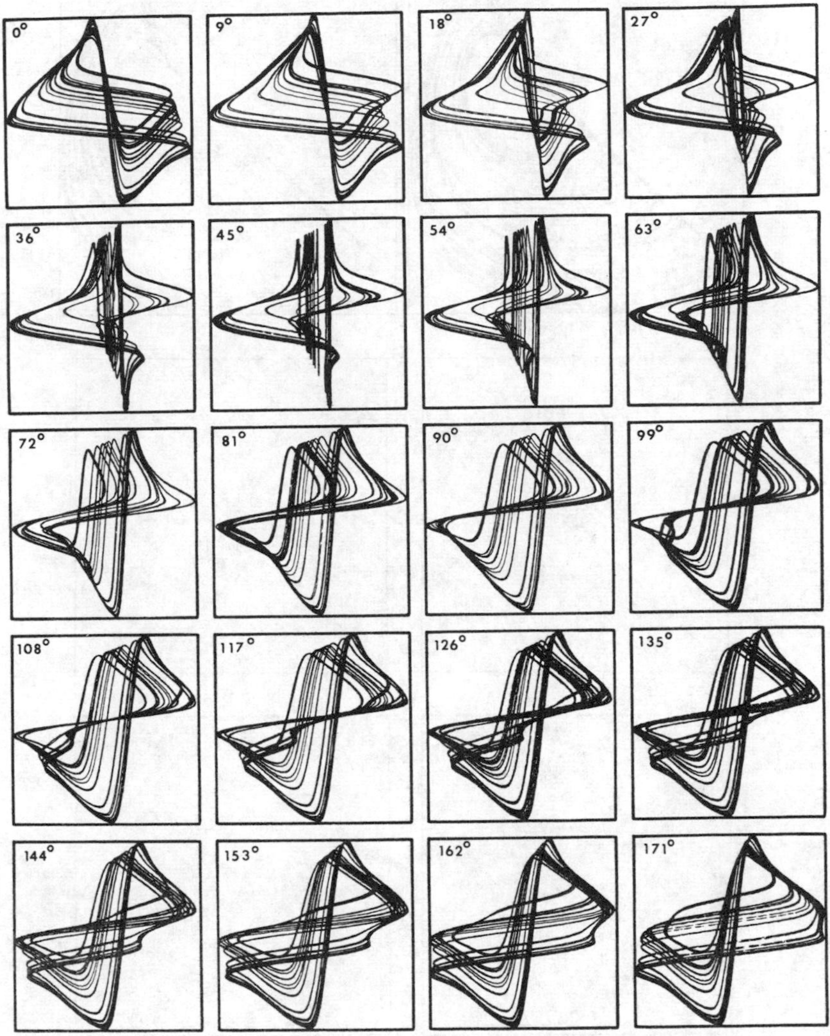

Fig. 4. The 3D character of the phase portrait for the nonperiodic state is illustrated by these 2D projections (of the 3D phase portrait) rotated about the ordinate in angular steps of 9°. The 3D portrait is given by $[B(t_k), B(t_k+T), B(t_k+2T)]$ with $T = 52.8$ s (about one-half the average time per orbit).

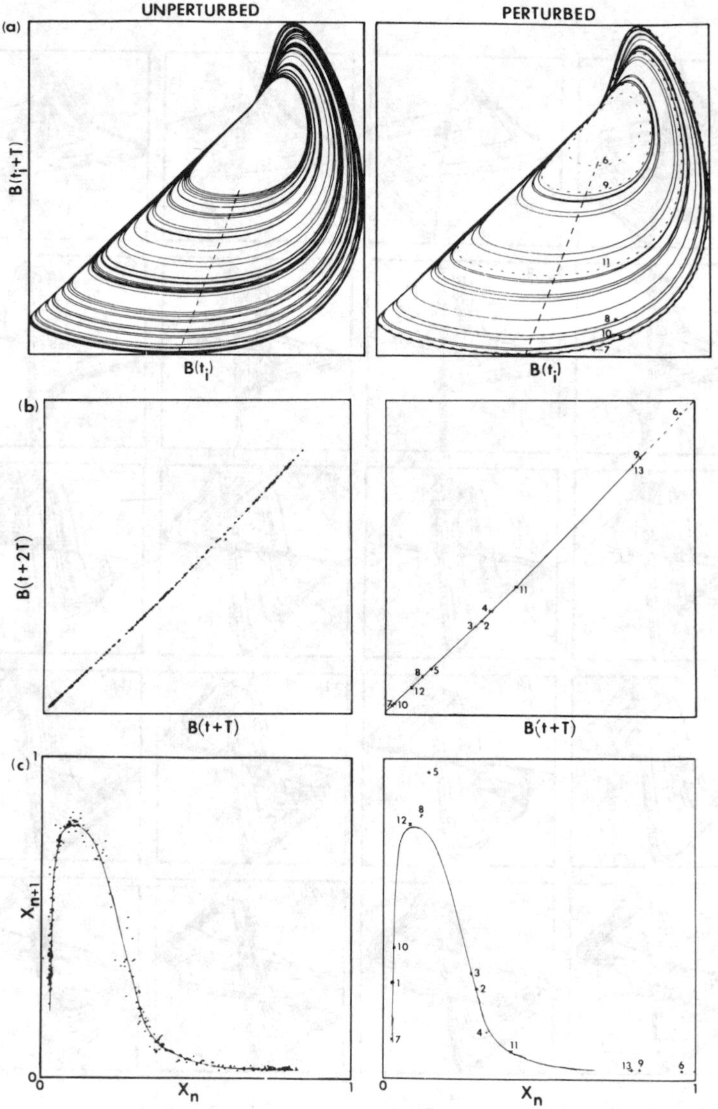

Fig. 5. A chaotic state in the absence of an external perturbation (left-hand side) and with a perturbation (right-hand side): (a) A 2D projection of a 3D phase portrait. (b) A Poincaré section constructed by the intersection of positively directed trajectories with the plane (normal to the paper) passing through the dashed line in (a). (c) a 1D map constructed by plotting as ordered pairs (X_{n+1} X_n) the successive values of the ordinate of trajectories when they cross the dashed line in (a).

positive Lyapunov exponent exhibits *sensitive dependence on initial conditions* – trajectories starting from two nearby points will evolve quite differently in time [15]. Thus all information about the initial conditions is rapidly lost, since any uncertainty in the initial values will be magnified until it becomes as large as the attractor.

The largest Lyapunov exponent for the data discussed here can be computed from the 1D map: the Lyapunov exponent for a set of data $\{X_i\}$, $i = 1, \ldots, n$, described by a 1D map is given by [14]

$$\lambda = \frac{1}{n} \sum_{i=1}^{n} \ln|f'(X_i)| , \quad (1)$$

where $f'(X_i)$ is the derivative of the map at X_i. We have calculated $f'(X_i)$ at each point by fitting the data with cubic splines [16]. Fig. 6(a) shows a spline fit to a set of data $\{X_i\}$ and fig. 6(b) shows $\ln|f'(X)|$ computed from the fit. The resultant Lyapunov exponent value for these data is 0.3 ± 0.1.

We have determined λ for several different sets of data and the results have been found to be alarmingly sensitive to the scatter in the data and the number and placement of the knots in the spline fit. A drift or scatter of several percent in the data defining a map with 100 to 200 points yielded in some cases values of λ ranging from positive to negative, depending on the fitting procedure. The sensitivity to noise and to the data fitting procedure, to be discussed in detail elsewhere [17], has not been adequately emphasized in past reports of the determination of λ [2, 18, 19]. On the other hand, for the data in fig. 6 the value of λ was reasonably robust under changes in the number and positions of the knots in the spline fit and under changes in the number of data points in the fit. From this analysis it is concluded that $\lambda = 0.3 \pm 0.1$ for these data.

The experimentally determined value of λ is positive, yet well below the maximum value for a 1D map, $\lambda_{max} = \ln 2 = 0.69$ [4]. Comparison of an experimental value of λ with theory is difficult since the dependence of λ on bifurcation parameter is extremely complex (see figs. 3 and 4 of [20]). However, the presence of a small amount of external noise, invariably present in any experiment, tends to smooth this function to a simple power law behavior (see fig. 10 in [20]). We have examined the effect of the external noise on our data by

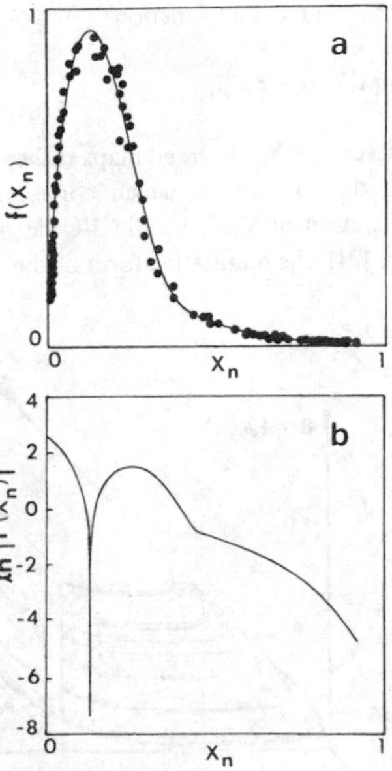

Fig. 6. (a) The data for a 1D map, $f(X)$, fit with cubic splines with four equally spaced knots. (b) The function needed for computing the Lyapunov exponent [eq. (1)], $\ln|f'(X)|$, computed from the spline fit.

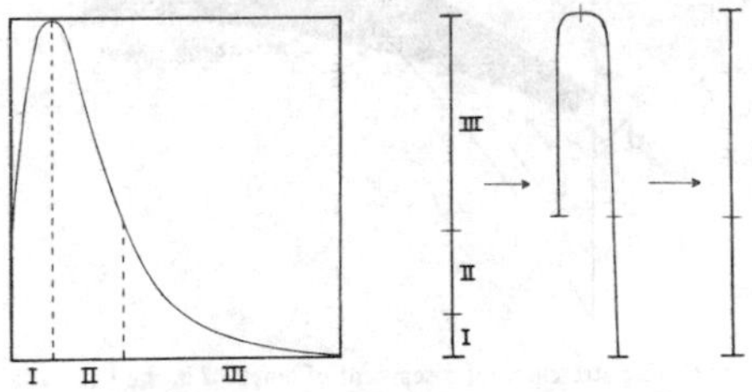

Fig. 7. A schematic representation of the stretching and folding of the X-axis described by the 1D map.

fitting the data to the function

$$X_{n+1} = aX_n \exp(-bX_n), \tag{2}$$

which describes the observed maps rather well [17]. For the data in fig. 6, which correspond to a chaotic state near the periodic 3-cycle of the U-sequence [21], the qualitative form of the scatter in the data and the observed value of λ are both reproduced by adding the same amount of multiplicative noise to eq. (2). Finally, the value 0.3 ± 0.1 may also be compared with the topological entropy of a 3-cycle, $\ln[(1+\sqrt{5})/2] = 0.48$ [22], which is an upper bound for the Lyapunov exponent.

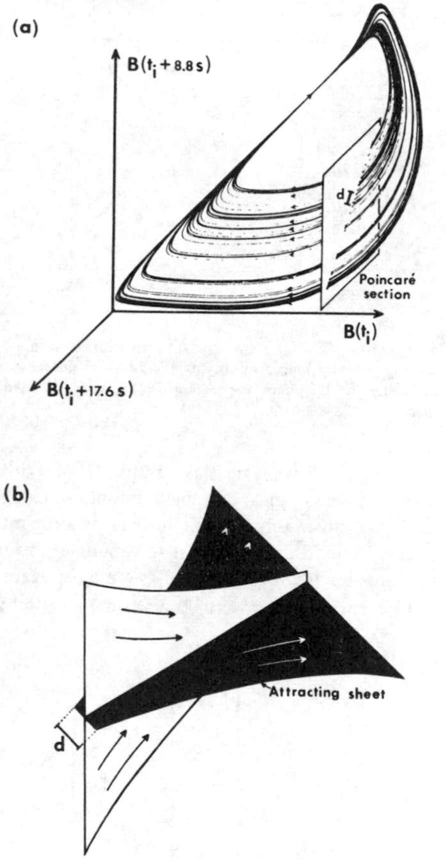

Fig. 8. The stretching of a segment of length d in the Poincaré section in (a) is illustrated schematically in (b). Trajectories in the sheet-like attractor separate exponentially, while trajectories off of the attractor converge exponentially to it.

6. A strange attractor (stretching and folding)

A *strange attractor* is an attractor with at least one positive Lyapunov exponent. Therefore, figs. 3–5 show strange attractors.

The exponential divergence of nearby trajectories implies that there must be a *stretching* of the attractor as trajectories evolve, but since the attractor lies within a bounded region of phase space, the attractor must also exhibit *folding*. Indeed, stretching and folding is a hallmark of strange attractors.

The stretching and folding can be seen in the 1D map, as fig. 7 illustrates. The stretching of the attracting sheet itself is illustrated schematically in

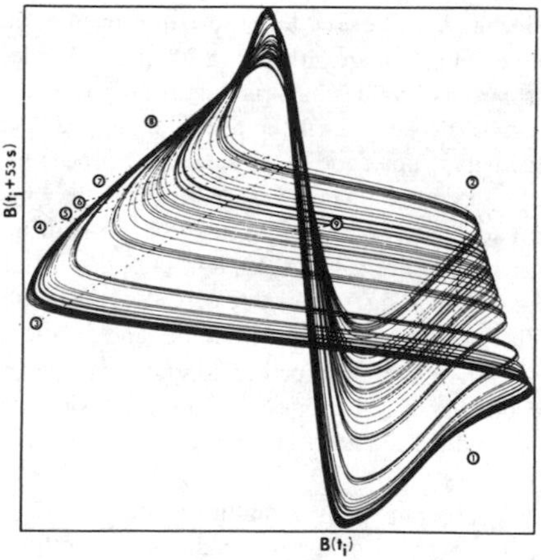

Fig. 9. A 2D projection of a 3D phase portrait, indicating the locations of the Poincaré sections shown in fig. 10. The Poincaré sections are planes normal to the paper, passing through the dashed lines.

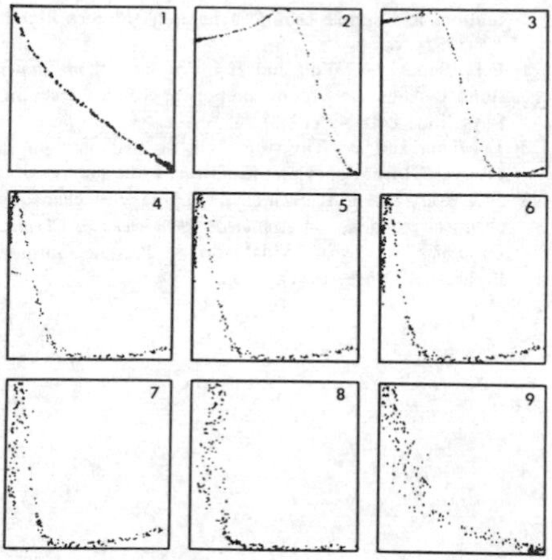

Fig. 10. Successive Poincaré sections illustrating the stretching and folding that is characteristic of strange attractors. For convenience the sections are drawn equal in size; the actual relative sizes (height × width) are: (1) 0.3 × 0.2, (2) 1.1 × 0.9, (3) 1 × 1, (4) 0.5 × 0.9, (5) 0.5 × 0.9, (6) 0.5 × 0.8, (7) 0.4 × 0.5, (8) 0.4 × 0.2, (9) 0.4 × 0.08.

fig. 8 [23]. Figs 9 and 10 demonstrate that the stretching and folding can be directly observed by analyzing the evolution of Poincaré sections at successive positions along the attractor.

In summary, we have shown that a nonperiodic state observed in an experiment on the Belousov–Zhabotinskii reaction is described by a phase portrait that has the properties of a strange attractor.

Acknowledgements

We are very grateful to Alan Wolf for many helpful discussions and for the computation of the Lyapunov exponent. We also acknowledge helpful discussions with D. Farmer, J. Guckenheimer, W.D. McCormick, P. Richetti and J. Swift. This research is supported by National Science Foundation Grants CHE79-23627 and INT81-15672, Robert A. Welch Foundation Grant F-805, and a C.N.R.S. Grant.

References

[1] J.-C. Roux, A. Rossi, S. Bachelart and C. Vidal, "Representation of a strange attractor from an experimental study of turbulence," Phys. Lett. 77A (1980) 391; P.R. Fenstermacher, H.L. Swinney and J.P. Gollub, "Dynamic instabilities and the transition to chaotic Taylor vortex flow", J. Fluid Mech. 94 (1979) 103.

[2] J.S. Turner, J.-C. Roux, W.D. McCormick and H.L. Swinney, "Alternating periodic and chaotic regimes in a chemical reaction – experiment and theory," Phys. Lett. 85A (1981) 9; J.-C. Roux, J.S. Turner, W.D. McCormick and H.L. Swinney, "Experimental observations of complex dynamics in a chemical reaction," in *Nonlinear Problems: Present and Future*, ed. by A.R. Bishop, D.K. Campbell and B. Nicolaenko (North-Holland, Amsterdam, 1982), p. 409.

[3] For a discussion of these terms see, for example, J.P. Eckmann, "Roads to turbulence in dissipative dynamical systems" Rev. Mod. Phys. 53 (1981) 643; see also [4] and [5].

[4] R. Shaw, "Strange attractors, chaotic behavior, and information flow," Z. Naturforsch. 36A (1981) 80.

[5] D. Ruelle, "Strange attractors," The Mathematical Intelligencer 2 (1980) 126.

[6] D. Ruelle, private communication.

[7] N.H. Packard, J.P. Crutchfield, J.D. Farmer and R.S. Shaw, "Geometry from a time series," Phys. Rev. Lett. 45 (1980) 712.

[8] H. Whitney, "Differentiable manifolds," Ann. Math. 37 (1936) 645.

[9] F. Takens, "Detecting strange attractors in turbulence," *Lecture Notes in Mathematics*, ed. by D.A. Rand and L.-S. Young (Springer, Berlin–Heidelberg–New York, 1981) p. 366.

[10] J.C. Roux, H.L. Swinney, and R.S. Shaw, "Strange attractor in a chemical system," a 16 mm movie.

[11] B. Mandelbrot, *Fractals, Form, Chance, and Dimension* (Freeman, San Francisco, 1977).

[12] The different definitions of dimension are discussed and compared by D. Farmer, E. Ott and J.A. Yorke, "The dimension of chaotic attractors," in *Order in Chaos*, Physica 7D (1983) 153.

[13] P. DeKepper and J. Boissonade, "Theoretical and experimental analysis of phase diagrams and related dynamical properties in the Belousov–Zhabotinskii system," J. Chem. Phys. 75 (1981) 189.

[14] G. Benettin, L. Galgani, A. Giorgilli and J.-M. Strelcyn, "Lyapunov characteristic exponents for smooth dynamical systems and for Hamiltonian systems; a method for computing all of them. Part 1: Theory," Meccanica 15 (1980) 9.

[15] D. Ruelle, "Sensitive dependence on initial conditions and turbulent behavior of dynamical systems," Ann. N.Y. Acad. Sci. 317 (1979) 408.

[16] IMSL Cubic splines fit, Routine DCSEVU (IMSL, Sixth Floor, GNB Building, 7500 Bellaire Boulevard, Houston, Texas 77036 USA).

[17] A. Wolf and J. Swift, "Progress in computing Lyapunov exponents from experimental data," in *Statistical Physics and Chaos in Fusion Plasmas*, ed. by C.W. Horton and L. Reichl (Wiley, New York, 1984).

[18] J.L. Hudson and J.C. Mankin, "Chaos in the Belousov–Zhabotinskii Reaction," J. Chem. Phys. 74 (1981) 6171.

[19] H. Nagashima, "Experiment on chaotic response of forced Belousov–Zhabotinskii reaction," J. Phys. Soc. Japan. 51 (1982) 21.

[20] J.P. Crutchfield, J.D. Farmer and B.A. Huberman, "Fluctuations and simple chaotic dynamics," Physics Reports 92 (1982) No. 2.

[21] R.H. Simoyi, A. Wolf and H.L. Swinney, "One dimensional dynamics in a multi-component chemical reaction," Phys. Rev. Lett. 49 (1982) 245.

[22] J. Milner and W. Thurston, "On iterated maps of the interval," Inventiones mathematicae, to be published.

[23] J.C. Roux and H.L. Swinney, "Topology of chaos in a chemical reaction," in *Nonlinear Phenomena in Chemical Dynamics*, ed. by C. Vidal and A. Pacault (Springer, Berlin–New York, 1981), p. 38.

VOLUME 47, NUMBER 16 PHYSICAL REVIEW LETTERS 19 OCTOBER 1981

Subharmonic Sequences in the Faraday Experiment: Departures from Period Doubling

R. Keolian, L. A. Turkevich, S. J. Putterman, and I. Rudnick
Department of Physics, University of California, Los Angeles, California 90024

and

J. A. Rudnick
Department of Physics, University of California, Santa Cruz, California 95064
(Received 6 August 1981)

Subharmonic sequences in the shallow-water surface waves generated in a resonator with one-dimensional properties are shown to exhibit departures from period doubling for long-period subharmonics.

PACS numbers: 47.35.+i, 47.20.+m, 47.25.-i

Recent interest in driven nonlinear systems has focused on the transition from the preturbulent to the "weakly turbulent" regimes. It is thought that such transitions in deterministic systems with limited degrees of freedom might provide insight into turbulence in fluids.[1]

While it has generally been accepted[2] that nonlinearities can lead to higher harmonics (as well as sum and difference frequencies), it has recently been reemphasized that driven nonlinear systems can also generate *sub*harmonics. Feigenbaum[3] has shown that a universal period-doubling sequence arises quite generally in nonlinear one-dimensional maps, appearing as a precursor to chaotic behavior. Recent computer experiments on the driven anharmonic oscillator[4] and Rössler attractor[5] exhibit this universal period-doubling sequence. The early work of Pedersen[6] on strongly driven loudspeakers at frequency f clearly evinces the beginning of a period-doubling sequence ($f/2$, $f/4$).[7] Longer sequences have also been observed in dc-driven Rayleigh-Bénard convective instabilities.[8-10]

Rayleigh[11] already noted that two classes of driven systems possess subharmonic response: driven nonlinear oscillators[12] (described by a Duffing equation, where the drive appears as an external time-varying force) and parametric excitation[13] (described by a Mathieu equation, where the drive appears as a time variation of one of the parameters). The first observation, 150 years ago, of subharmonic parametric excitation is due to Faraday.[14] He studied shallow-water waves when the containing vessel was driven vertically, and observed an $f/2$ response. Rayleigh[11] analyzed and experimentally confirmed these results.[15]

An alternate route to turbulence[16] is followed in some Rayleigh-Bénard[17,18] and couette-flow[19] experiments with the appearance of incommensurate freqencies. The convective instabilities inherently invove more than one degree of freedom, which could account for the presence of these frequencies. In order to isolate the effects of period multiplication, it is important to inhibit this other mechanism. Our version of the Faraday experiment therefore utilizes a resonator with response along only one of its three dimensions. We find a rich spectrum of subharmonic instabilities, which, however, does not generally follow universal period-doubling sequences for long-period subharmonics.[20]

Our resonator is a narrow Plexiglas annulus (Fig. 1) which is, to good approximation, a one-dimensional wave guide closing on itself. The anharmonicity of the modes caused by the curvature is negligible. The annulus has the advantage of allowing arbitrary spatial phase for the various harmonics. However, this makes determina-

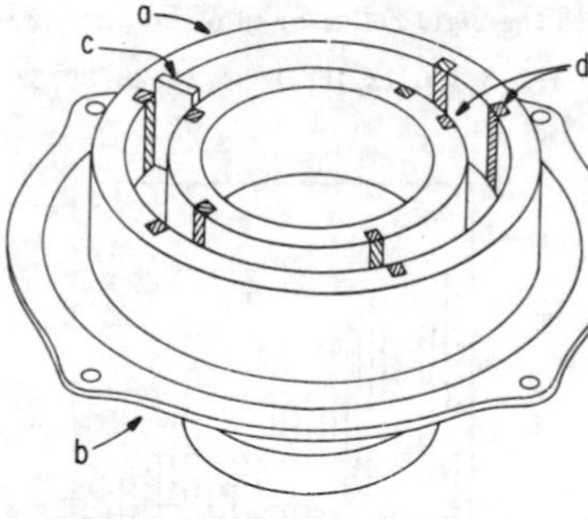

FIG. 1. Sketch of the annular plexiglas resonator, a, sitting on top of a loudspeaker, b. The resonator can be used with or without the rigid reflector, c. The wave height of the salt water in the annulus is determined by measuring the conductance between the copper blocks, d. Annulus average radius is 4.8 cm and the cross section is 0.8×2.5 cm.

tion of the amplitudes more difficult since it is unknown where the antinodes are relative to the transducers (copper-block pairs). For this reason five transducers were installed at various angles. We can also place a rigid reflector at a transducer, causing an antinode in the height oscillation to occur there. Salt and a wetting agent were added to the water. The conductance between the copper blocks gives a measure of the height oscillation.

When we oscillate the resonator with an acceleration $\sim g/3$, large-amplitude standing waves form, with peaks as high as 20 mm and troughs as low as 1 mm. By taking the power spectrum of the height we find that period multiplication occurs but not necessarily by a factor of 2. In approximately 100 cases we find that the allowed subharmonic series are given by pf/m, where p and m are integers and (with two exceptions to be discussed later) m is even. This is unlike the Rayleigh-Bénard period-doubling sequences,[8-10] where m is restricted to powers of 2.

A typical preturbulent series, $m = 14$, is shown in Fig. 2. The presence of $f/14$ implies that the water does not repeat its motion until fourteen cycles of the drive have passed. There is no ambiguity in identifying the subharmonic since *all* of its harmonics are present to beyond $2f$, many at least 40 dB above the base line, and there are no unaccounted-for peaks.

The values of m that we have observed so far at various drive amplitudes and frequencies without the rigid reflector are

$$m = 1, 2, 4, 12, 14, 16, 18, 20, 22, 24, 28, 35$$

and with the rigid reflector in place they are

$$m = 1, 2, 3, 4, 6, 12, 16, 18, 24, 26, 28, 30, 32, 34.$$

We are optimistic that with additional effort the list will grow. Note that all the prime numbers from 1 to 17 are present as factors of m in this list.

If we fix the drive frequency and carefully increase the drive amplitude we see sequences of period multiplications which always (except for $f/3$) start with at least one factor of 2 but later develop larger factors. We have seen, for example, the sequence $m = 1, 2, 4, 24$. The sequence is reversible, with some hysteresis, if the drive is lowered. The jump from $m = 4$ to $m = 24$ is by a factor of 6 without intermediate factors of 2 and 3. The sequence can also drop and pick up factors; once, on the same run, we observed a jump from $m = 24$ to $m = 28$.

When we pick the drive frequency to be near the mth, low-amplitude mode of the resonator we tend to see the mth subharmonic series at large amplitude. The pf/m member of the series then occurs approximately at the pth resonance of the annulus. The ability to make every frequency in a subharmonic series approximately coincide with a mode regardless of m value is a great strength of our resonator. By way of contrast, in a preliminary experiment using a Petri dish we saw only $m = 2$ and a weak $m = 4$ before turbulence set in. In a cylinder there are no low-lying harmonic overtones.

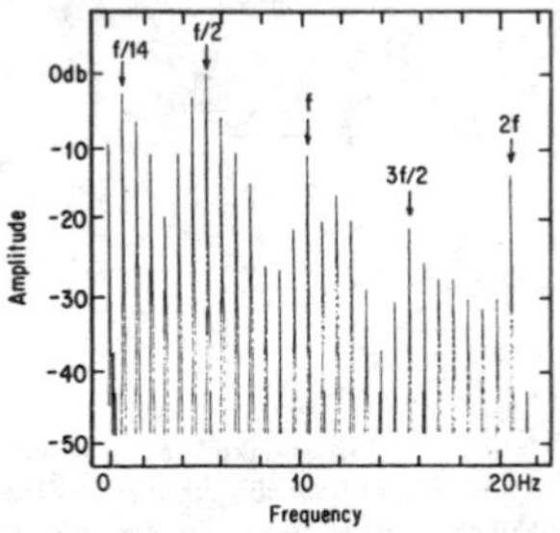

FIG. 2. Power spectrum when the resonator is oscillated at f and response is seen at $f/14$ and its harmonics.

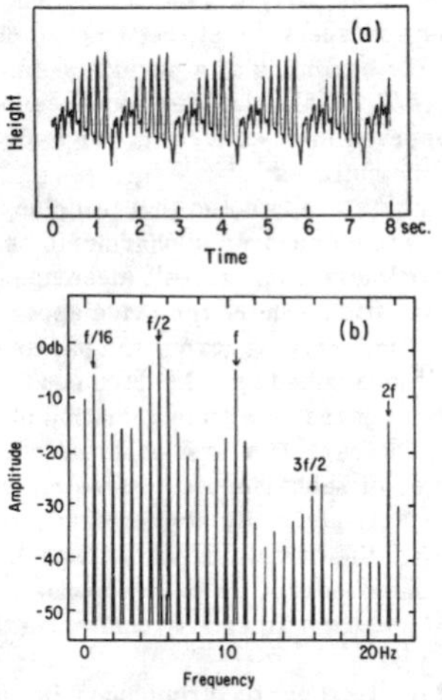

FIG. 3. (a) Height vs time of the $f/16$ periodic (preturbulent) state. (b) Power spectrum of (a).

It is interesting to compare the response in frequency and time of the periodic (preturbulent) and near-periodic (weakly turbulent) states. Figure 3(a) is a wave height versus time plot in the periodic state. The amplitude builds up and suddenly dies in a characteristic "cockscomb" pattern in which even minute details repeat. The state is robust; we have seen it persist for hours, and it is stable to small drive-amplitude changes. The periodicity is reflected in the power spectrum [Fig. 3(b)] by the absence of a visible noise floor. The spectrum also shows that this is an $m = 16$ subharmonic, consistent with the fact that there are eight cycles of $f/2$ per cockscomb. If we raise the drive amplitude by 28% we see an example of the near-periodic state (Fig. 4). Cockscombs still tend to form but end randomly, and the spectrum shows broadband noise along with periodicity at $m = 32$. In the list of observed m values, 2 and 4 were always observed in the periodic state, but the remaining m-value observations were about equally divided between the periodic and quasiperiodic states.

Earlier we noted two exceptions to the even-m series. We have seen a $pf/3$ and a $pf/35$ series in the near-periodic modes. They are the only odd series we have observed although it must be said that no concerted effort in this direction was made. The latter series is unmistakably $pf/35$ rather than $pf/34$ or $pf/36$ since unlike all other spectra, $f/2$ is not only not dominant, it is in the noise and unobservable. It is instinctive to reject odd-m modes as a possibility. However, this must not be dismissed given the observed departure from period doubling. Sum and difference frequencies are the hallmark of nonlinear processes, and the improbable presence of simultaneous excitation of two of these sequences can give such odd modes. For example, $f/10 - f/14 = f/35$.

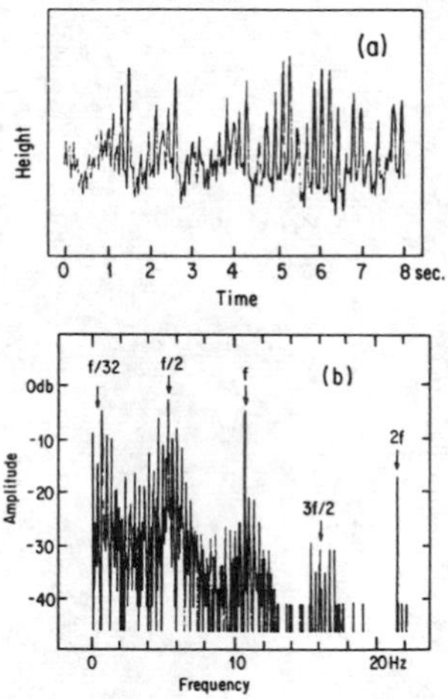

FIG. 4. (a) Height vs time of the $f/32$ quasiperiodic (weakly turbulent) state when the drive amplitude was increased 28% over that used in Fig. 3. (b) Power spectrum of (a). The 0-dB level coincides in Figs. 3 and 4.

We hope in the near future to use superfluid helium as the fluid in this experiment.

This research was supported in part by the U. S. Office of Naval Research and by the National Science Foundation.

[1]E. N. Lorenz, J. Atmos. Sci. 20, 130 (1963).

[2]L. D. Landau and E. M. Lifshitz, *Theory of Elasticity* (Pergamon, Oxford, 1970), Sec. 26.

[3]M. Feigenbaum, J. Stat. Phys. 19, 25 (1978), and 21, 669 (1979), and Phys. Lett. 74A, 375 (1979). That period doubling, universal or not, ought to exist in iterated maps was shown by N. Metropolis, M. L. Stein, and P. R. Stein, J. Combinatorial Theory 15, 25 (1973).

[4]B. A. Huberman and J. P. Crutchfield, Phys. Rev. Lett. 43, 1743 (1979).

[5]J. Crutchfield, D. Farmer, N. Packard, R. Shaw, G. Jones, and R. J. Donnelly, Phys. Lett. 76A, 1 (1980).

[6]P. O. Pedersen, J. Acoust. Soc. Am. 6, 227 (1935), and 7, 64 (1935).

[7]Subharmonics have also been seen extensively in acoustically cavitating fluids. See E. A. Neppiras, J. Acoust. Soc. Am. 46, 587 (1968).

[8]J. P. Gollub, S. V. Benson, and J. Steinman, Ann. N. Y. Acad. Sci. 357, 22 (1980).

[9]A. Libchaber and J. Maurer, J. Phys. (Paris), Colloq. 41, C3-51 (1980).

[10]M. Giglio, S. Musazzi, and V. Perini, Phys. Rev. Lett. 47, 243 (1981).

[11]Lord Rayleigh, Philos. Mag. 15, 229 (1883), and 16, 50 (1883), and 24, 145 (1887), and *The Theory of Sound* (Dover, New York, 1945), Sec. 68b.

[12]K. O. Friedrichs and J. J. Stoker, Q. Appl. Math. 1, 97 (1942); N. Krylov and N. Bogoliubov, *Introduction to Nonlinear Mechanics* (Princeton Univ. Press, Princeton, 1943); and N. N. Bogoliubov and Y. A. Mitropolsky, *Asymptotic Methods in the Theory of Nonlinear Oscillations* (Gordon and Breach, New York, 1961).

[13]N. Minorsky, J. Franklin Inst. 240, 25 (1945).

[14]M. Faraday, Philos. Trans. Roy. Soc. London 299, Sec. 103 (1831).

[15]For subsequent work on this problem, see T. B.

Benjamin and F. Ursell, Proc. Roy. Soc. London, Ser. A 225, 505 (1954).
[16]L. D. Landau and E. M. Lifshitz, *Fluid Mechanics* (Pergamon, Oxford, 1959), Sec. 27; J. Ruelle and F. Takens, Commun. Math. Phys. 20, 67 (1971).
[17]J. P. Gollub and S. V. Benson, Phys. Rev. Lett. 41, 625 (1978).
[18]J. Maurer and A. Libchaber, J. Phys. (Paris), Lett. 41, L515 (1980).
[19]M. Gorman and H. L. Swinney, Phys. Rev. Lett. 43, 1871 (1979).
[20]A. B. Pippard, *The Physics of Vibration* (Cambridge Univ. Press, Cambridge, 1978), Vol. I, Chap. 9, should be consulted by the reader seeking discussion of the physics of subharmonics and stability in specific systems.

VOLUME 47, NUMBER 20 PHYSICAL REVIEW LETTERS 16 NOVEMBER 1981

Subharmonic Route to Chaos Observed in Acoustics

Werner Lauterborn and Eckehart Cramer[(a)]
Third Physical Institute, University of Göttingen, D-3400 Göttingen, Federal Republic of Germany
(Received 14 September 1981)

A subharmonic route to chaos including period-doubling bifurcations up to $f/8$ has been observed in experiments on acoustical turbulence (acoustic cavitation noise). The system also shows signs of reverse bifurcation with increasing control parameter (acoustic driving pressure amplitude). In view of the large variety of phenomena observed and yet to be expected the system investigated may well serve as a further experimental paradigm of nonlinear dynamical systems besides Rayleigh-Bénard and circular couette flow.

PACS numbers: 47.55.Bx, 47.25.Mr, 43.25.+g

There is increasing evidence that period-doubling bifurcations[1] and strange attractors[2] are common phenomena for a large class of nonlinear dynamical systems. Most of this evidence stems from relatively simple mathematical models like the three-variable differential systems of Lorenz[3] and Rössler[4] and one-dimensional iterated maps on the unit interval[5] which show links to dynamical systems via the Poincaré return map. The discovery of universal properties in period-doubling bifurcations of iterated maps by Feigenbaum[1] could be confirmed for the Lorenz model[6] and a five-variable model of the Navier-Stokes equation,[7] and has stimulated the search for additional universal features of nonlinear dynamical systems.[8-11]

Compared with the large body of theoretical work, experiments are rather sparse. Up to now there are only two physical systems where the onset of chaos is studied systematically and which show some analogy to the behavior of the above mathematical models. These are the Rayleigh-Bénard experiment on the flow in a flat convective layer of liquid heated from below[12-14] and experiments on circular couette flow (flow between two cylinders, the inner one rotating).[15,16] Experiments in other fields are just emerging, such as in optics for optically bistable cavities[17] or proposed similar experiments with noisy Josephson junctions,[18] charge-density waves in anisotropic solids and superionic conductors,[19] or pinned dislocation lines.[20]

This paper presents experiments in acoustics which in view of the results to be reported and in analogy to the newly coined terms of "optical turbulence"[17] and "solid-state turbulence"[19,20] may be called experiments on "acoustical turbulence." The experiment consists in irradiating a liquid (water) with sound of high intensity (control parameter is the sound pressure amplitude) and looking for the sound output of the liquid, called acoustic cavitation noise. The physical situation is a somehow fundamental one: The transport of acoustical energy through a liquid is considered. It bears much resemblance with the Rayleigh-Bénard problem where the transport of heat through a liquid is investigated.

To irradiate the liquid a piezoelectric cylinder of 76-mm length, 76-mm inner diameter, and 5-mm wall thickness is used. When driven at its main resonance at 23.56 kHz a high-intensity acoustic field is generated in the interior, and cavitation is easily achieved. The noise is picked up by a broadband microphone[21] and digitized at rates up to 2 MHz after suitable low-pass filtering (to avoid aliasing in the subsequent Fourier analysis) and strong filtering of the driving frequency (to be able to store the noise with just an 8-bit storage). Sound pressure power spectra are calculated via the fast-Fourier-transform algorithm from usually 4K samples out of the 128K storage available. More details of the experimental setup are given elsewhere.[22]

Power spectra of acoustic cavitation noise usually consist of instrumentally sharp lines on a noise background. The lines are related to the driving frequency f_0 and lie at $(n/m)f_0$ $(n, m = 1, 2, 3, \ldots)$. Of special interest are the lines at $m \geq 2$, $n < m$, i.e., in the subharmonic region $f < f_0$ of the spectrum. In early experiments the occurrence of lines at $f_0/2$, $f_0/3$, and $f_0/4$ has already been found,[23,24] but no convincing explanation could be given. The explanation that bubbles in water driven at twice their natural resonance are responsible for the $f_0/2$ line[25] had to be abandoned since bubbles of the necessary size could not be found and are unlikely to be present in the experimental situation.[26] Instead, after an extensive numerical investigation of bubble oscillations, it has been argued that special ultraharmonic resonances of bubbles smaller than resonant size (especially the $\frac{3}{2}$ resonance where two oscillations of the driving sound field match three oscillations of the natural oscillation frequency of the bubble) are responsible for the subharmonic line at $f_0/2$ in the power spectrum,[26] but a direct verification of this hypothesis has not yet been possible.

The present experiments were undertaken to add to our understanding of the subharmonic line problem. They differ from previous experiments in that they are fully computer controlled to realize almost any desired control parameter history.

Figure 1 gives just one example of a pressure power spectrum obtained at a driving voltage of 15 V. The history in this case was to linearly increase the driving voltage to 15 V and then to stay there for some time. These precautions are necessary to arrive at the third period-doubling bifurcation with lines at $nf_0/8$. But even when staying at a constant voltage large fluctuations are observed, and usually only spectra with subharmonics as low as $f_0/4$ can be observed. Figure 1 suggests that our nonlinear acoustical system may follow the period-doubling route to chaos and may belong to the universal class of Feigenbaum systems. Unfortunately this could not yet be proved as the next bifurcation with lines at $nf_0/16$ could not be reached. There may be principal difficulties in achieving this aim because our system can be expected to be a noisy one and this has been shown to limit the bifurcation sequence.[8]

As a result of the total computerization of the experiment we are able to do complex measurements on acoustical turbulence and to gather enormous amounts of data. The question therefore arises of how to present the results. One way that we found very appealing is in the form of grey-scale pictures analogous to "visible speech" where the power spectrum is plotted versus time with the amplitude encoded as grey scale. We have adopted this kind of presentation for our studies of how the system reaches chaos and have plotted the pressure power spectrum as a function of the voltage at the driving transducer. Figure 2 gives an example for the case where the voltage is increased linearly from 0 to 60 V in about 250 ms. During this time 128K samples of the pressure in the liquid are taken at a rate of 500 kHz. The total experiment thus lasts just a quarter of a second. From these data about 1000 overlapping short-time spectra are calculated with 4K data each and a shift of 128 samples from one spectrum to the next. In Fig. 2 three successive spectra are combined to give a total of 370 spectra. The grey level is encoded with the aid of a 3×3 matrix so that a binary plotter can be used.

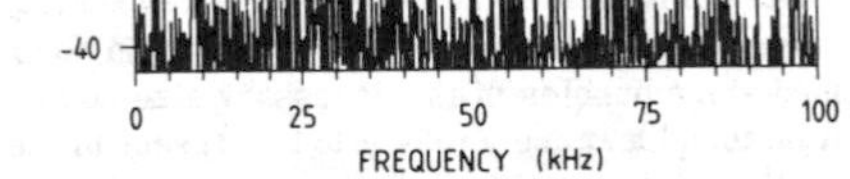

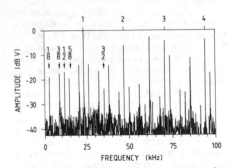

FIG. 1. Example of a pressure power spectrum of acoustic cavitation noise with subharmonic lines as low as $f_0/8$ (f_0 = 22.56 kHz), i.e., three period-doubling bifurcations have taken place. The driving frequency is strongly filtered.

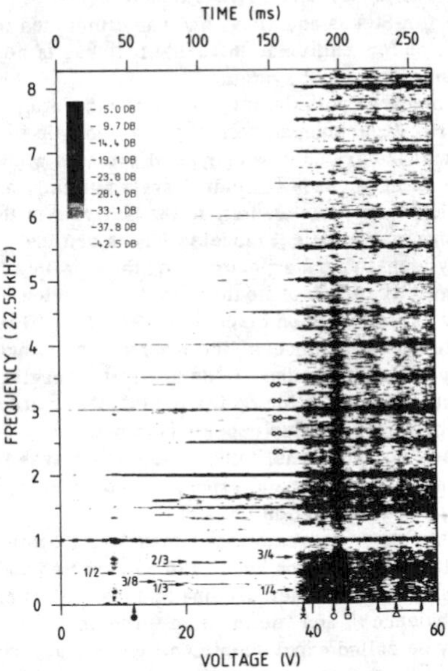

FIG. 2. Sequence of power spectra displayed as a grey-scale picture. The voltage at the driving piezoelectric cylinder is increased linearly from 0 to 60 V in 262 ms.

Many interesting features are immediately visible in Fig. 2:

(1) The first period-doubling bifurcation sets in at about 12 V (see closed circle in Fig. 2).

(2) Further subharmonic lines appear in a process that does not look like period doubling. These lines seem to detach from the $f_0/2$ (and $3f_0/2$) line with stops at $3f_0/8$ ($5f_0/8$) and $f_0/3$ ($2f_0/3$) until suddenly and with large amplitude the $f_0/4$ line and its odd harmonics set in (see 1/2, 3/8, 1/3, 2/3, 1/4, 3/4, and cross in Fig. 2).

(3) The $f_0/4$ bifurcation occurs together with a marked increase in broadband noise and a broadening of the otherwise sharp line spectrum (from cross to open circle in Fig. 2; the white gap around f_0 is due to the band reject filter needed to suppress the otherwise dominating driving frequency).

(4) At about 43 V a further increase in the broadband noise level is observed with the line spectrum still detectable and additional lines at certain odd harmonics of $f_0/8$ (see from open circle to open square and ∞ in Fig. 2). This state of the system ranges from 43 to 46 V only and is the most chaotic state encountered (chaos defined in terms of broadband noise at high level in the power spectrum).

(5) In a process looking like reverse bifurcation,[27] the system returns to a line spectrum with lines only at $f_0/2$ and its harmonics and with much less broadband noise (open square in Fig. 2).

(6) Satellite lines appear around $f_0/2$ and $3f_0/2$ (and also some other lines) which show some periodicity with the control parameter (open triangle in Fig. 2).

A general observation from other experiments is that the state of total chaos seems to be unstable since it cannot be sustained by the driving sound field for a longer period of time. Instead, oscillations are observed between the background noise and the line strength. Also single lines may visibly oscillate. This is best seen in films produced from sequences of up to 2000 power spectra plotted on a graphic display (cathode-ray tube) and filmed with a 16-mm film camera.

The picture of Fig. 2 is a rather condensed form of looking at the properties of a dynamical system. It would be very interesting to see other dynamical systems like those of Lorenz[3] and Rössler[4] as well as the experimental ones of Rayleigh-Bénard and circular couette flow displayed in this way.

Some effort has been spent in modeling the experiments theoretically. As a first step single spherical bubbles have been taken and subjected to a sinusoidal driving pressure of increasing amplitude. The mathematical model is a highly nonlinear ordinary differential equation of second order for the radius of the bubble as a function of time and includes surface tension, viscosity, and compressibility of the liquid (water).[28] Radius-time curves have been calculated numerically for different bubble sizes, sound pressure amplitudes, and frequencies to get response curves for this nonlinear system.[28] To simulate the present experiments bubble wall oscillation power spectra have been calculated and plotted in the manner of Fig. 2 as a function of the driving pressure. Qualitatively similar behavior is observed but strongly depends on the bubble size. Of most importance seems to be the observation that bifurcations are obtained, mostly from f_0 to $f_0/2$ to $f_0/3$ (or also $f_0/4$) to (quickly) $f_0/4$ to chaos and then back to f_0 or $f_0/2$ (directly). Thus both successive bifurcations as well as reverse bifurcations are observed in this simple mathematical model, like those encountered in the experiments.

More sophisticated models must take into account that many bubbles are present in the liquid which all couple via their sound radiation. A fuller discussion of these questions as well as of the physical processes responsible for the observed phenomena is given in Ref. 22.

We arrive at the conclusion that in our acoustic system there is a subharmonic route to turbulence, but one which does not solely proceed via successive period-doubling bifurcations although this route is strongly involved. We therefore propose to make a distinction between a subharmonic and a period-doubling route to chaos. Moreover, our system shows signs of reverse bifurcations which may be worthwhile to study in greater detail.

We are greatly indebted to W. Steinhoff for building the 128K-byte buffer storage out of 1K, 1-bit chips for our experiments, and G. Heinrich for building the microphone. This work was supported by the Fraunhofer-Gesellschaft, Munich.

---

[a]Present address: Physikalisch Technische Bundesanstalt-IB, Abbestr. 2-12, D-1000 Berlin 10, West Germany.

[1]M. J. Feigenbaum, J. Stat. Phys. 19, 25 (1978).

[2]D. Ruelle and F. Takens, Commun. Math. Phys. 20,

VOLUME 47, NUMBER 20 PHYSICAL REVIEW LETTERS 16 NOVEMBER 1981

167 (1971).
[3]E. N. Lorenz, J. Atmos. Sci. 20, 130 (1963).
[4]O. E. Rössler, Phys. Lett. 57A, 397 (1976).
[5]P. Collet and J. P. Eckmann, *Iterated Maps on the Interval as Dynamical Systems* (Birkhäuser, Basel, 1980).
[6]V. Franceschini, J. Stat. Phys. 22, 397 (1980).
[7]V. Franceschini and C. Tebaldi, J. Stat. Phys. 21, 707 (1979).
[8]J. P. Crutchfield and B. A. Huberman, Phys. Lett. 77A, 407 (1980).
[9]B. A. Huberman and A. B. Zisook, Phys. Rev. Lett. 46, 626 (1981).
[10]J. P. Crutchfield, M. Nauenburg, and J. Rudnick, Phys. Rev. Lett. 46, 933 (1981).
[11]B. Shraiman, C. E. Wayne, and P. C. Martin, Phys. Rev. Lett. 46, 935 (1981).
[12]A. Libchaber and J. Maurer, J. Phys. (Paris), Colloq. 41, C3-51 (1980).
[13]J. P. Gollub and S. V. Benson, J. Fluid Mech. 100, 449 (1980).
[14]M. Giglio, S. Musazzi, and U. Perini, Phys. Rev. Lett. 47, 243 (1981).
[15]Yu. N. Belyaev, A. A. Monakhov, S. A. Shcherbakov, and I. M. Yavorskaya, Pis'ma Zh. Eksp. Teor. Fiz. 29, 329 (1979) [JETP Lett. 29, 295 (1979)].
[16]P. R. Fenstermacher, H. L. Swinney, and J. P. Gollub, J. Fluid Mech. 94, 103 (1979).
[17]H. M. Gibbs, F. A. Hopf, D. L. Kaplan, and R. L. Shoemaker, Phys. Rev. Lett. 46, 474 (1981).
[18]B. A. Huberman, J. P. Crutchfield, and N. H. Packard, Appl. Phys. Lett. 37, 750 (1980).
[19]B. A. Huberman and J. P. Crutchfield, Phys. Rev. Lett. 43, 1743 (1979).
[20]C. Herring and B. A. Huberman, Appl. Phys. Lett. 36, 975 (1980).
[21]Our own construction, built in our laboratory by G. Heinrich. It is made up of a thin disk of PZT-4 material bonded on a brass rod, with cutoff frequency of about 1 MHz.
[22]W. Lauterborn and E. Cramer, to be published.
[23]R. Esche, Acustica, Akust. Beihe. 2, 208 (1952).
[24]L. Bohn, Acustica 7, 201 (1957).
[25]W. Güth, Acustica 6, 532 (1956).
[26]W. Lauterborn, J. Acoust. Soc. Am. 59, 283 (1976).
[27]J. Crutchfield, D. Farmer, N. Packard, R. Shaw, G. Jones, and R. J. Donelly, Phys. Lett. 76A, 1 (1980).
[28]E. Cramer, in *Cavitation and Inhomogenities in Underwater Acoustics*, edited by W. Lauterborn (Springer-Verlag, New York, 1980).

VOLUME 48, NUMBER 11 PHYSICAL REVIEW LETTERS 15 MARCH 1982

Evidence for Universal Chaotic Behavior of a Driven Nonlinear Oscillator

James Testa, José Pérez, and Carson Jeffries
Materials and Molecular Research Division, Lawrence Berkeley Laboratory, and Department of Physics, University of California, Berkeley, California 94720
(Received 8 January 1982)

A bifurcation diagram for a driven nonlinear semiconductor oscillator is measured directly, showing successive subharmonic bifurcations to $f/32$, onset of chaos, noise band merging, and extensive noise-free windows. The overall diagram closely resembles that computed for the logistic model. Measured values of universal numbers are reported, including effects of added noise.

PACS numbers: 05.40.+j, 05.20.Dd, 47.25.-c

Our purpose is to report detailed measurements on a driven nonlinear semiconducting oscillator and to make quantitative comparisons with the predictions of a simple model of period-doubling bifurcation as a route to chaos,[1-3] which stems from earlier work in topology.[4] There is surprising agreement, lending support to the belief and the hope that some nonlinear systems can be approximately understood by a universal model, as has been suggested by some experiments.[5,6] This upsurge of interest in nonlinear behavior has been triggered by the remarkable result that deterministic computer iterations of such a simple nonlinear recursion relation as the logistic equation

$$x_{n+1} = \lambda x_n(1 - x_n) \tag{1}$$

yield exceedingly complex pseudorandom or chaotic behavior.[2,3] The results are best summarized by a bifurcation diagram[7-9]: a scatter plot of the iterated value $\{x_n\}$ versus the control parameter λ, which shows that as λ is increased $\{x_n\}$ displays a series of pitchfork bifurcations at λ_n, with period doubling by 2^n, $n = 1, 2, \ldots$. These converge geometrically, as $\lambda_c - \lambda_n \propto \delta^{-n}$, to the onset of chaos at λ_c, where $\{x_n\}$ becomes aperiodic; in the chaotic regime, $\lambda > \lambda_c$, noise bands merge and there exist narrow periodic windows in a specific order and pattern.[4] This model is quantified by universal numbers as $n \to \infty$: $\delta = 4.669\ldots$, and the pitchfork scaling parameter $\alpha = 2.502\ldots$, first computed by Feigenbaum. Other universal numbers characterize the spectral power density[10,11] and effects of noise.[8,12]

Our experimental system is a series *LRC* circuit driven by a controlled oscillator, described by $L\ddot{q} + R\dot{q} + V_c = V_d(t) = V_0 \sin(2\pi f t)$, where V_c is the voltage across a Si varactor diode (type 1N953 supplied by TRW Company), which is the nonlinear element. Under reverse voltage, $V_c = q/C$,

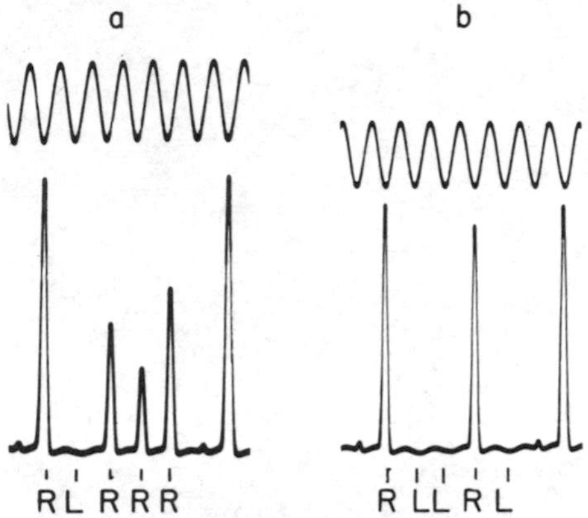

FIG. 1. (a) The varactor voltage $V_c(t)$ and the driving oscillator voltage $V_d(t)$ (upper) for period 6 window at 2.073 V; the pattern is RLRRR, and describes the sequence of visitation of the oscillator to its states according to whether it is to the right or left of zero, following the notation of Ref. 4. (b) Period 6 window at 3.338 V, with different pattern RLLRL.

where $C \simeq C_0/[1+V_c/0.6]^{0.5}$, $C_0 \simeq 300$ pF; under forward voltage the varactor behaves like a normal conducting diode. The coil inductance $L = 10$ mH, the resistance $R = 28\ \Omega$. At low values of V_0, the system behaves like a high-Q resonant circuit at $f_{res} = 93$ kHz; as V_0 is increased, the resonant frequency shifts upward and the Q is lowered. It is not our intention to solve the intractable nonlinear differential equations for this system[13] but rather to do extensive and novel measurements

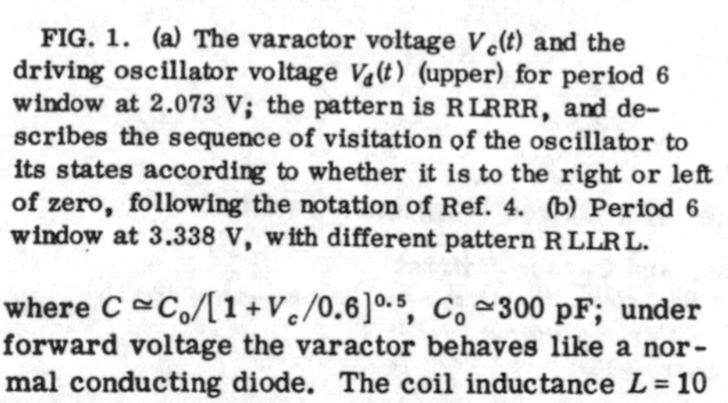

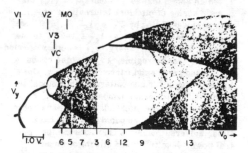

FIG. 2. Bifurcation diagram V_y vs V_0 at $f = 96.85$ kHz, showing thresholds V_1, V_2, and V_3 for periods 2, 4, and 8; threshold for chaos V_c; band merging M_0; and windows of periods 6, 5, 7, 3, 6, 12, 9, and 13. The veiled lines are peaks in the spectral density in the chaotic regime.

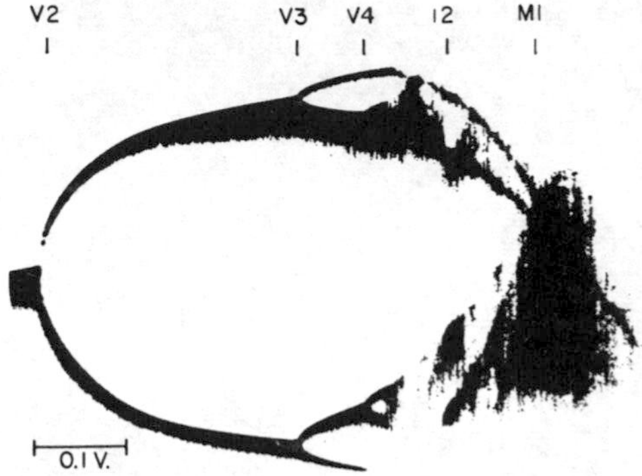

FIG. 3. Expansion of a region of Fig. 2, showing bifurcation thresholds V_2, V_3, and V_4; window of period 12; and band merging M_1.

designed to compare its behavior as fully as possible with the simple logistic model. We fix f near f_{res}, vary the driving voltage V_0, and measure the varactor voltage $V_c(t)$. We assume a correspondence between V_0 and λ and between V_c and x of Eq. (1).

A real-time display, e.g., Fig. 1, of $V_c(t)$ and $V_0(t)$ on a dual-beam oscilloscope, with V_0 as a parameter, clearly revealed threshold values V_{0n} for bifurcation; the bifurcation subharmonics $f/2^n$ up to $f/16$; and the pattern of visitation of the oscillator to its stable points. The data shown at two different windows in the chaotic regime, both for period-6 orbits, show different patterns, as expected.[4] During the diode con-

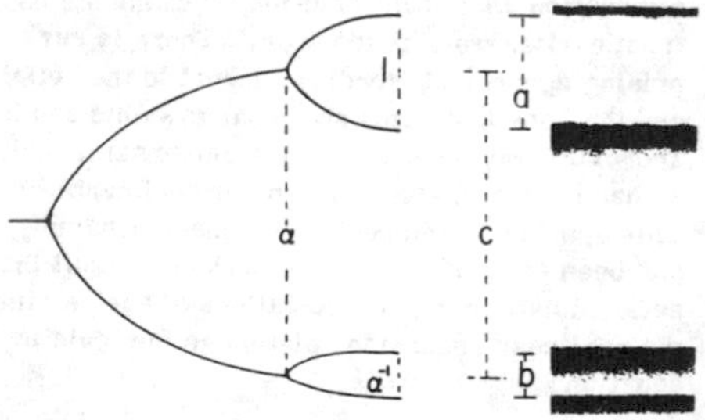

FIG. 4. (a) Schematic of universal metric scaling of pitchfork bifurcation, determined by α (Ref. 2). (b) Data for period 16 between V_4 and V_5, which yield the values $\alpha = a/b = 2.35$ and $\alpha = c/a = 2.61$.

TABLE I. Measured thresholds at 99 kHz.

Period	Threshold V_0 rms volts	Comments
2	0.639	
4	1.567	Threshold for periodic bifurcation
8	1.785	
16	1.836	
32	1.853	
Chaos	1.856	Onset of noise
12	1.901	Window
24	1.902	
6	2.073	Window
12	2.074	
5	2.353	Window
10	2.363	
7	2.693	Window
14	2.696	
3	3.081	Wide Window
6	3.338	
12	3.711	
24	3.821	
9	4.145	Window
18	4.154	

ducting half cycle, V_c is compressed toward the zero line; in the reverse half cycle, V_c has a set of discrete values, which correspond to the upper half of the bifurcation diagram.

To analyze V_c, a window comparator was constructed which selected components between V_y and $V_y + \Delta V$, $\Delta V \simeq 10$ mV. A vertical scan of V_y simultaneously with a slower horizontal scan of V_0 on an oscilloscope yielded Figs. 2 and 3, the first measured bifurcation diagram for a physical system showing subharmonic sequences. It has a striking resemblance to the computed diagram,[7,8] including bifurcation thresholds, onset of chaos, band merging, noise-free windows, and the subtle veiled structure, corresponding to regions of high probability.[8] The diagram allows a direct measurement of the number α; from the expanded region, Fig. 4, the ratio of the pitchfork splittings is directly measured in a series of ten similar measurements:

$$\alpha = 2.41 \pm 0.1 . \tag{2}$$

The diagram shows at least five noise-free windows, which bifurcate within the window: From Fig. 2 and Table I, at $V_0 = 3.081$ V, a noise-free window of period 3 appears, which bifurcates to periods 6, 12, and 24 before onset of chaos again.

The power spectral density of $V_c(t)$ was measured with a spectrum analyzer with 40 dB dynamic range, which showed the expected subharmon-

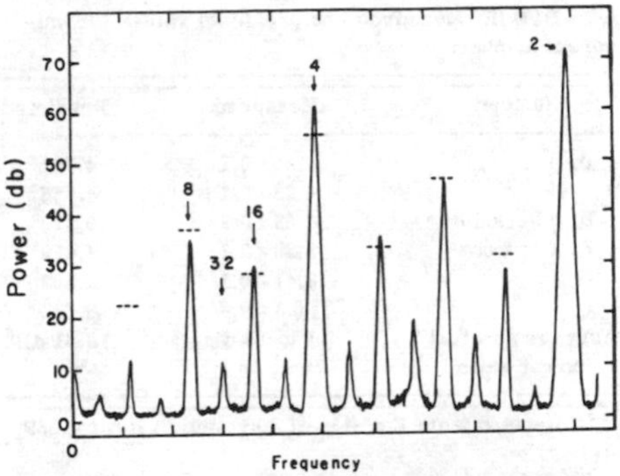

FIG. 5. Power spectral density (dB) vs frequency for $f = 98$ kHz, dynamic range 70 dB, showing subharmonics to $f/32$. The components agree with prediction (dashed bars, Ref. 14) within 2 dB rms deviation, except for the peak at $f/16$.

ics $\frac{1}{2}$, $\frac{1}{4}$, $\frac{3}{4}$; $\frac{1}{8}$, $\frac{3}{8}$, $\frac{5}{8}$, $\frac{7}{8}$, etc., rather symmetrically displayed about $f/2$. The data shown in Fig. 5 were obtained with a more sensitive spectrum analyzer with 85 dB of dynamic range, sensitivity of 300 nV, and range $f = 0$ to 50 kHz $\geq f/2$, thus allowing observation of spectral components 95 dB below V_0 at f. Figure 5 shows periodic subharmonics to $f/32$ at V_0 just below the threshold for chaos V_{0c}; the predicted values of the individual spectral components are shown.[14] It is predicted[10] that the average heights of the peaks for a period is $10 \log 20.963 = 13.21$ dB below the previous period; the data are consistent with this, although the region between $f/2$ and f is not available for exact averaging. Spectral analysis showed other noise-free windows (60 dB above noise) at periods 12, 6, 5, 7, and 9, at thresholds listed in Table I; all show bifurcations within the window. The entire V_0 sequence of Table I, identified by period and pattern, is consistent with the universal U sequence of Metropolis, Stein, and Stein[4] (who limit computation to period ≤ 11). From the first four threshold voltages V_{0n} we calculate the convergence rate

$$\delta_1 = \frac{V_{02} - V_{01}}{V_{03} - V_{02}} = 4.257 \pm 0.1 ;$$
$$\delta_2 = \frac{V_{03} - V_{02}}{V_{04} - V_{03}} = 4.275 \pm 0.1 . \tag{3}$$

We observed the effect on the system of adding a random noise voltage $V_n(t)$ to $V_d(t)$. The bi-

TABLE II. Measured and predicted values for universal numbers.

Number		Measured	Predicted
δ_1	Eq. (3)	4.26 ± 0.1	4.751[a]
δ_2	Eq. (3)	4.28 ± 0.1	4.656[a]
δ_1	Period 3 window	0.69 ± 0.1	0.979[a]
δ_2	Period 3 window	3.38 ± 0.1	4.429[a]
α		2.41 ± 0.1	2.502[b]
ϵ		6.3 ± 0.3	6.55[c]
Average spectral power ratio		11 to 15 dB	13.61 dB[d]

[a]Computed from Eq. (1); cf. asymptotic limit 4.669, Ref. 2.
[b]Ref. 2.
[c]Ref. 12.
[d]Ref. 10.

furcation diagram and the power spectra were observed as $|V_n|$ was increased: periods 16, 8, 4, and 2 were successively obliterated at $V_n = 10$, 62, 400, and 2500 mV_{rms}, respectively, yielding an average value

$$\kappa = 6.3 \qquad (4)$$

for the noise voltage factor required to reduce by one the number of observable bifurcations.

To summarize, Table II compares our measured values with predicted values for some universal numbers. There is overall reasonable quantitative agreement between the data and the logistic model. The likely cause for some discrepancy in δ is that the data cannot be taken in the asymptotic limit $n \to \infty$. These are first direct measurements for α and κ. The strong similarity between the predicted and the observed bifurcation diagram gives further support to the utility of simple models as a key to chaotic behavior of nonlinear systems. The measurement of a bifurcation diagram is a powerful method for assessing the degree to which a particular physical system will follow this route, or other routes[14]; it is not yet known how to predict this in advance.

We thank J. Rudnick, M. Nauenberg, J. P. Crutchfield, M. P. Klein, and H. A. Shugart for helpful conversations. This work was supported by the Director, Office of Energy Research, Office of Basic Energy Sciences, Materials Sciences Division of the U. S. Department of Energy under Contract No. W-7405-ENG-48.

[1]R. M. May, Nature (London) 261, 459 (1976).
[2]M. J. Feigenbaum, J. Stat. Phys. 19, 25 (1978).
[3]P. Collet and J.-P. Eckmann, *Iterated Maps on the Interval as Dynamical Systems* (Birkhauser, Boston, 1980).
[4]N. Metropolis, M. L. Stein, and P. R. Stein, J. Comb. Theory, Ser. A 15, 25 (1973).
[5]A. Libchaber and J. Maurer, J. Phys. (Paris), Colloq. 41, C3-51 (1980); M. Giglio, S. Musazzi, and U. Perini, Phys. Lett. 47, 243 (1981).
[6]P. S. Linsay, Phys. Rev. Lett. 47, 1349 (1981), first reported period doubling in a varactor oscillator, similar to the system studied here; however, our experimental methods differ.
[7]Collet and Eckmann, Ref. 3, pp. 26, 38, and 44.
[8]J. P. Crutchfield, J. D. Farmer, and B. A. Huberman, to be published.
[9]S. Grossman and S. Thomas, Z. Naturforsch. 32A, 1353 (1977).
[10]M. Nauenberg and J. Rudnick, Phys. Rev. B 24, 493 (1981).
[11]B. A. Huberman and A. B. Zisook, Phys. Rev. Lett. 46, 626 (1981).
[12]J. Crutchfield, M. Nauenberg, and J. Rudnick, Phys. Rev. Lett. 46, 933 (1981).
[13]However, B. A. Huberman and J. P. Crutchfield, Phys. Rev. Lett. 43, 1743 (1979), have computed solutions for an anharmonic oscillator with a restoring force $\propto x - 4x^3$.
[14]J.-P. Eckmann, Rev. Mod. Phys. 53, 643 (1981).

VOLUME 49, NUMBER 14 PHYSICAL REVIEW LETTERS 4 OCTOBER 1982

Comment on a Driven Nonlinear Oscillator

In their recent Letter Testa, Pérez, and Jeffries[1] gave experimental evidence for universal chaotic behavior by using a driven nonlinear oscillator. The oscillator was composed of a generator driving a series RLC circuit where the capacitor, a varactor diode, is the nonlinear element. The diode conducts normally in the forward direction and its capacitance decreases with increasing reverse voltage. Their results were confirmed here with a different diode, an ordinary rectifier (1N1221) with a similar capacitance-voltage relation.

The purpose of this Comment is to point out that there is another property of the diode which, at least in our case, is responsible for the observed behavior. A junction diode continues to conduct for a time after the current is reversed. The reverse recovery time, the time for the minority carriers to recombine, increases with increasing forward current,[2] and is a significant fraction of the driving cycle in our case. That this is the major cause of the effects can easily be proved by using the parallel combination of a fast switching diode and the varactor diode biased such that it never conducts. This combination virtually eliminates any reverse-recovery-time effect while maintaining the conduction and nonlinear capacitance properties. No period doubling is observed.

Furthermore, very similar results are obtained when using a pure RLC circuit with the capacitor shunted by a simple transistor circuit simulating only the conducting and recovery properties of the diode. The circuit shown in Fig. 1 shows period doubling, chaotic behavior,

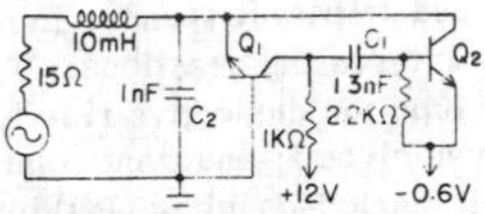

FIG. 1. A circuit that demonstrates the period-doubling route to chaotic behavior. The signal generator has a 50-Ω impedance and the transistors are 2N3393 or equivalent. None of the values of the components are critical.

and stable orbits of the chaotic regime. Transistor Q_1 acts as the diode to ground and provides a current to charge C_1 while Q_1 is conducting. After the conduction portion of the cycle C_1 discharges through Q_2 causing it to conduct for an additional portion of the cycle. The larger the current through Q_1, the longer Q_2 remains conducting. The nonlinear capacities associated with the transistor junctions are orders of magnitude smaller than the varactor diode used in the above experiments, and negligible compared with the 1-nF series capacitor. Therefore, the major cause of the period doubling and chaotic behavior is the reverse-recovery-time effect of the varactor diode rather than its nonlinear capacitance.

E. R. Hunt
Department of Physics, Ohio University
Athens, Ohio 45701

Received 14 May 1982
PACS numbers: 05.40.+j, 05.20.Dd, 47.25.-c

[1]J. Testa, J. Pérez, and C. Jeffries, Phys. Rev. Lett. 48, 714 (1982).
[2]J. Millman and C. Halkias, *Electronic Devices and Circuits* (McGraw-Hill, New York, 1967).

VOLUME 49, NUMBER 14 PHYSICAL REVIEW LETTERS 4 OCTOBER 1982

Testa, Pérez, and Jeffries Respond: Hunt's Comment[1] raises interesting questions: What properties of a p-n junction diode give rise to bifurcation in a driven circuit; and what conditions lead to the specific period-doubling (PD) bifurcation sequence previously reported.[2] Although Hunt implies that we ascribe the nonlinearity of the diode solely to its junction capacitance, this is stated nowhere in our paper. In fact, the exact nonlinearity of the varactor was not our concern. Diodes are characterized by nonlinear conduction (the well-known I-V characteristic curve), by a reverse recovery time τ, and by two types of capacitance[3]: (1) the junction capacitance C_j, dominant under reverse bias, and (2) the charge storage capacitance C_s, dominant under forward bias. Ordinary rectifier diodes usually have $C_s > C_j$ and $\tau \sim 1$ μs. Varactor diodes usually have $C_j > C_s$ and $\tau \sim 0.1$ to 1 μs. Fast signal diodes have $\tau \sim 1$ ns. Experiments done here on many diodes in a series circuit driven near resonance at period $T \approx 10^{-5}$ s gave the following results: (1) Fast signal diodes do not bifurcate. (2) Two diodes in series back to back do not bifurcate. (3) Diodes always reversed biased do not bifurcate. (4) Diodes with $\tau \sim (0.1$ to $0.6)T$ readily bifurcate, but not always in a PD sequence. (5) If the total circuit resistance is made sufficiently large, a PD route is usually observed, as in the case of Ref. 2. From these experiments we conclude that a diode must switch to show bifurcation and that the reverse recovery time must be a reasonable fraction of the driving period. To this extent we do agree with Hunt's Comment.[1] However, since the requisite reverse-recovery-switching characteristic is related to both the nonlinear conduction and the nonlinear charge-storage capacitance which are not separable in a real diode (in contrast to analog simulations), we do not conclude with Hunt that the nonlinear conduction is more "important" than the nonlinear capacitance. From measurements of the resonant frequency dependence on the driving voltage, we find that C_j is the dominant frequency-determining nonlinearity for driven RLC circuits using varactor diodes, and that C_s is dominant for rectifier diodes. However, the nonlinearity giving rise to bifurcation may be different.

We note that this Comment and Response do not in any way invalidate our observations and conclusions[2,4] that the nonlinear oscillator used (diode 1N953) displays a universal period-doubling bifurcation sequence to chaos.[5] That this is so is quite interesting in view of the complexity of the nonlinearity and the higher-order differential equations of the system. Apparently because of the dissipation, it behaves as if it were approximated by a one-dimensional quadratic map.

J. Testa
J. Pérez
C. Jeffries
Materials and Molecular Research Division
Lawrence Berkeley Laboratory, and
Department of Physics, University of California
Berkeley, California 94720

Received 28 June 1982
PACS numbers. 05.40.+j, 05.20.Dd, 47.25.-c

[1]E. R. Hunt, preceding Comment [Phys. Rev. Lett. 49, 1054 (1982)].

[2]J. Testa, J. Pérez, and C. Jeffries, Phys. Rev. Lett. 48, 714 (1982).

[3]B. G. Streetman, *Solid State Electronic Devices* (Prentice Hall, Englewood Cliffs, 1972).

[4]J. Pérez and C. Jeffries, Phys. Rev. B 26, 3460 (1982).

[5]M. J. Feigenbaum, J. Stat. Phys. 19, 25 (1978).

VOLUME 46, NUMBER 7 PHYSICAL REVIEW LETTERS 16 FEBRUARY 1981

Observation of Chaos in Optical Bistability

H. M. Gibbs, F. A. Hopf, D. L. Kaplan, and R. L. Shoemaker
Optical Sciences Center, University of Arizona, Tucson, Arizona 85721
(Received 9 October 1980)

Optical turbulence and periodic oscillations are easily seen with a hybrid optically bistable device with a delay in the feedback. The behavior of these instabilities is in good agreement with the recent work of Ikeda, Daido, and Akimoto, who predicted them for both ring-cavity and delay-line hybrid devices.

PACS numbers: 42.65.-k

In the last few years there has been a substantial theoretical and experimental effort on optically bistable systems (i.e., devices which exhibit two distinct states of optical transmission). Recently, Ikeda, Daido, and Akimoto[1,2] have pointed out that intrinsic bistable devices with use of a ring cavity are described by difference equations and that the stability analysis of steady states in that case is different from the usual criteria applied to differential equations. This causes instabilities to arise in steady-state solutions that have previously been described as stable.[3] Furthermore, the time dynamics in such circumstances can be chaotic. In this Letter we describe experiments on an optically bistable hybrid device which shows instabilities with periodic and chaotic dynamics that are in good agreement with the theory of Ikeda, Daido, and Akimoto.

There are several aspects of this problem that make it interesting. Intrinsic optically bistable devices can be extremely fast optical switches[4]; periodic dynamics could convert them into short-pulse generators. Optical bistability has also received substantial attention as a model problem in nonequilibrium statistical mechanics, and the appearance of chaos in this context is interesting. By comparison with turbulent flows,[5] the problem of optical turbulence (i.e., chaos) is both experimentally and theoretically very simple. Although instabilities of the type described here are not widely known in physics, they are common in ecological modeling[6] since population dynamics is realistically described by difference equations. We therefore begin by explaining how stability analysis works and how it relates to the differential-equation analysis used in most theories of optical bistability.

Let us suppose we have a difference equation $\vec{x}_{n+1}=\vec{f}(\vec{x}_n)$ where $\vec{x}_n$ is a vector, $\vec{f}$ is a nonlinear vector function, and n labels the steps. The steady states, defined by $\vec{x}_{n+1}=\vec{x}_n$, are denoted $\vec{x}^k$. The stability analysis proceeds by defining $\vec{x}_n=\vec{x}^k+\vec{\epsilon}_n$ where $\vec{\epsilon}_n$ is small. One then linearizes the problem by a Taylor's expansion about the steady state and finds the eigenvalues Λ^k of the resulting matrix. The kth state is stable if and only if $|\Lambda^k|<1$ for all eigenvalues. Let us next associate a differential equation with the difference equation. We expand x_{n+1} as $\vec{x}_{n+1}\simeq\vec{x}_n+t_R\dot{\vec{x}}_n$, where t_R is the time-step between $\vec{x}_n$ and $\vec{x}_{n+1}$. Then we drop the meaningless subscript n to obtain $t_R\dot{\vec{x}}=\vec{f}(\vec{x})-\vec{x}$, a differential equation whose steady states are the same as those of the difference equation. The eigenvalues λ^k that are obtained from the stability analysis of this differential equation are related to those of the difference equation by $\lambda^k=(\Lambda^k-1)/t_R$, and the state $\vec{x}^k$ is stable if and only if $\mathrm{Re}\,\lambda^k<0$. One can see that if the differential equation is unstable the difference equation is also unstable, but the *difference equation can be unstable even though the differential equation is stable.* It is in this latter regime that one finds periodic and chaotic time dynamics. We will call this the regime of "Ikeda instability," to differentiate it from other types of instabilities[7,8] that occur in optical bistability. The instability described by McCall[7] is a relaxation oscillation having a period related to the response times of the medium rather than t_R. The self-pulsing instability[8] of absorptive optical bistability can be distinguished experimentally since it has a period of t_R, while periodic pulses of the Ikeda type have period[9] nt_R where $n\geq 2$.

Since the ring cavity, the problem Ikeda first analyzed, is a standard model problem in optical bistability and is the motivation for introducing a delay in our hybrid, we discuss the ring first. In the limit of purely dispersive bistability, by using a cubic nonlinearity inside a ring cavity, the dynamics are given by the difference-differential equations[2,10]

$$\tau\dot{\varphi}=-\varphi+\mathrm{sgn}(n_2)|E(t-t_R)|^2, \qquad (1)$$

$$E(t)=A+BE(t-t_R)\exp[i(\varphi-\varphi_0)]. \qquad (2)$$

Here $E(t)$ is the normalized field in the cavity, φ is the phase lag of the field across the nonlinear medium, n_2 is the nonlinear coefficient, φ_0 is the small-signal phase lag, τ is the response time of the medium, t_R is the delay time, $A = T^{1/2}E_0$, and $B = 1 - T$ where T is the mirror transmission and E_0 is the incident field. In the limit $t_R \gg \tau$, these equations reduce to a continuous-time difference equation[1] which has an Ikeda instability over virtually all of what would normally be referred to as upper stable branches.[1,2] In the limit $t_R \ll \tau$, the difference equation features are largely eliminated, and normal bistable behavior is predicted.[2]

In our experiment, we follow the suggestion in Ref. 2 and modify a hybrid[11] optically bistable device, by introducing an electrical delay line with delay t_R in the feedback [see Fig. 1(a)]. In our hybrid[12] an optical beam passes through a PLZT (Pb-based lanthanum-doped zirconate titanate) piezoelectric crystal sandwiched between crossed polarizers. The transmitted light is monitored by a photodiode detector whose output signal is delayed and then fed back to the electrodes of the PLZT, causing changes in the refractive index and the transmission through the device. This delay-line hybrid device satisfies[2]

$$\tau\dot{\varphi} = -\varphi + A^2\{1 + 2B\cos[\varphi(t - t_R) - \varphi_0]\}, \qquad (3)$$

where the light transmission of the PLZT as a function of voltage V is approximately $(1 - 2B \times \cos\varphi)/2$, $\varphi = \pi V/V_H$, V_H is the half-wave voltage, A^2 is proportional to the input intensity, $\varphi_0 = -\pi V_B/V_H$, V_B is the bias voltage, and τ is the composite detector-feedback-PLZT response time. In our experiment, B has a value close to the 0.5 for an ideal modulator. Figure 3 of Ikeda, Daido, and Akimoto[2] shows periodic oscillations and chaos obtained by solving Eq. (3) numerically with $B = 0.3$. The output of our device is qualitatively indistinguishable from their solutions (and our own). Equation (1) for a ring cavity reduces to Eq. (3) in the limit $B \ll 1$. Since we see Ikeda instability in a delay-line hybrid, it almost surely exists for a ring cavity, and could have important implications.

Since optical or electronic delay lines with $t_R \gg 1$ ms are impractical, we introduce a delay by inserting a microprocessor in the feedback loop. The output from the photodetector is digitized every 160 μs. Next, since the electro-optic device gives a phase shift that goes as V^2 and we need a response proportional to V for the device to be described by Eq. (3), we take the square

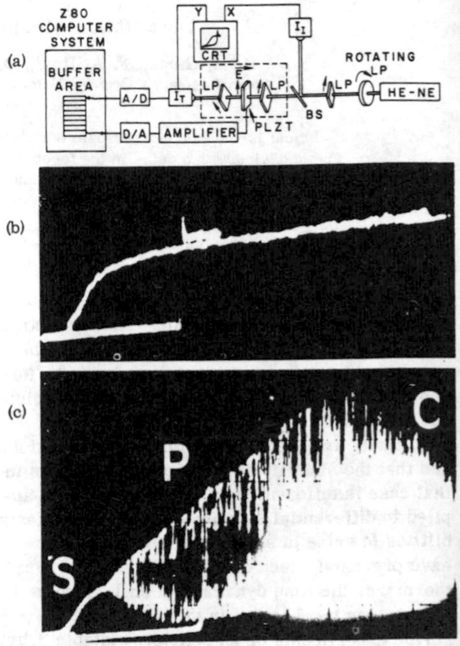

FIG. 1. (a) Block diagram of experimental apparatus. (b) Plot of output intensity (vertical axis) vs input intensity (horizontal axis) as the input intensity is cycled slowly from zero to some maximum and back over a time of 30 s. Here $t_R = 160\ \mu s \ll \tau = 1$ ms. (c) Same as (b) except $t_R = 40$ ms $\gg \tau = 1$ ms. The labels, *S*, *P*, and *C* indicate the stable, periodic, and chaotic domains of the upper branch.

root of the input voltage and store it in a first-in, first-out buffer (queue). We can then adjust the delay by leaving the sampling rate constant and changing the length of the buffer. Whenever a new sample is digitized and stored in the buffer, the oldest sample is retrieved from the memory and fed back to the PLZT through a digital-to-analog converter. The response time τ was typically set at 1 ms. The output characteristics were unchanged by reasonable variations in τ and the digitizing error size, verifying that the use of a microprocessor to introduce the delay does not affect the dynamics.

In Fig. 1(b) we show the *usual* hysteresis curve for the case $t_R = 160\ \mu s \ll \tau$ obtained by slowly cycling the incident laser intensity from zero to a maximum and back to zero. In Fig. 1(c) the only change was to make $t_R = 40\text{ms} \gg \tau$. While the low er branch of the curve is the same as in Fig. 1(b) the upper branch is very different. One can distinguish three qualitatively different regimes in the upper branch, which are labeled S, P, and C in the figure. When the incident power is low, the upper branch is stable (S) and is identical to the upper branch in Fig. 1(b). For higher powers, the upper branch is unstable, and appears as a "wash" rather than as a well-defined line. This comes from the variation of the output intensity with time. Within this unstable regime, two characteristic patterns can be seen. When the wash is continuous (C), as is the case in the right portion of the branch, one has a chaotic output that appears to wander randomly in time between an upper and lower bound as in Fig. 2(c). In the central portion of the upper branch (P) the wash has a dark area in the center, because the output is periodic with period[9] $2t_R$ and close to a square wave [Fig. 2(b)]. The curves in Fig. 2 are in close agreement with Fig. 3 of Ref. 3 which gives a theoretical result for the hybrid with use of Eq. (3).

We observe only periods of $2t_R$ and $4t_R$, and never encounter periods of t_R, which rules out self-pulsing.[8] By varying t_R and τ, we eliminate any relationship between τ and the observed period of $2t_R$, which rules out pulses of the McCall[7] variety. The periodic part of the upper branch is periodic throughout its domain, but the chaotic portion is observed to have very small domains within it which are periodic. In these domains one observes traces with fundamental periods of either $2t_R$ or $4t_R$ with roughly equal probability. At some settings we are able to see a stable domain at the high-intensity end of the upper branch. In the unstable regime that borders this stable domain, the intensity again varies periodically with period $2t_R$, but the modulation depth is much smaller than in the periodic domain shown in Fig. 1. These results are in excellent qualitative agreement with the computations in Refs. 2 and 3 and agree also with our own calculations.

In summary, we report the first experimental observation of Ikeda optical instabilities, resulting in periodic and chaotic outputs. Ikeda instability may be of value in constructing new optical devices. If this instability is not desired, it may be avoided by using round-trip times shorter than the response time of the nonlinear medium. This was most likely the case for previous experiments on short semiconductor etalons.[4] Intrinsic devices have been studied experimentally[13] in the regime $\tau \ll t_R$, but no instability was reported. Two points are worth noting, however. First, the detector response was several t_R in Ref. 13, so that the instability could not have been seen because of time averaging. Secondly, a Fabry-Perot cavity was used rather than a ring, and it is an open question as to how this instability affects that case. Finally, we note that several recent theoretical studies, such as the proposed phase-switching of dispersive systems,[14] have postulated conditions under which the Ikeda instability exists. These proposals will have to be reanalyzed.

We acknowledge useful conversations with K. Ikeda, D. McLaughlin, P. Fife, and H. Riskin. We thank M. Sargent for helping perform the numerical calculations. The support of the U. S. Air Force Office of Scientific Research and the U. S. Army Research Office is gratefully acknowledged.

(a)
(b)
(c)

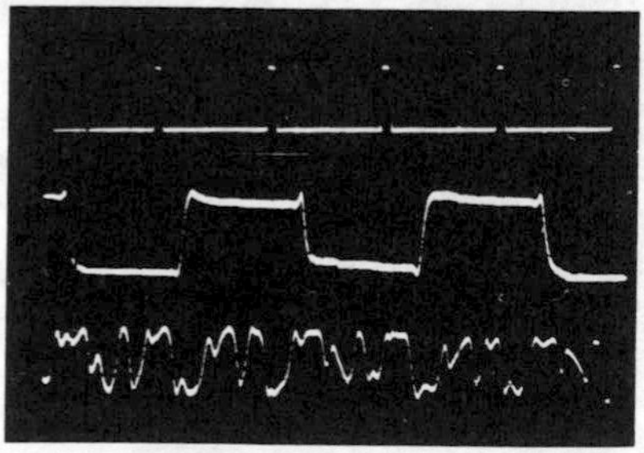

FIG. 2. (a) Time calibration; one pulse every delay time $t_R = 40$ ms. (b) Intensity vs time in the periodic domain. (c) Intensity vs time in the chaotic domain.

[1]K. Ikeda, Opt. Commun. 30, 257 (1979).

[2]K. Ikeda, H. Daido, and O. Akimoto, Phys. Rev. Lett. 45, 709 (1980).

[3]H. Seidel, U. S. Patent No. 3 610 731 (5 October 1971). A. Szoke, V. Daneu, J. Goldhar, and N. A. Kurnit, Appl. Phys. Lett. 15, 376 (1969); S. L. McCall, Phys. Rev. A 9, 1515 (1974); M. Spencer and W. E. Lamb, Jr., Phys. Rev. A 5, 864 (1972); H. M. Gibbs, S. L. McCall, and T. N. C. Venkatesan, Phys. Rev. Lett. 19, 1135 (1976).

[4]H. M. Gibbs, T. N. C. Venkatesan, S. L. McCall, A. Passner, A. C. Gossard, and W. Wiegmann, Appl. Phys. Lett. 35, 451 (1979); D. A. B. Miller, S. D. Smith, and A. Johnston, Appl. Phys. Lett. 35, 658 (1979).

[5]E. N. Lorenz, J. Atmos. Sci. 20, 130 (1963); O. E. Rossler, in *Synergetics, A Workshop*, edited by H. Haken (Springer-Verlag, Berlin, 1977); *Bifurcation Theory and Applications in Scientific Disciplines*, edited by O. Gurel and O. E. Rossler (New York Academy of Sciences, New York, 1979). For the appearance of chaos in optical contexts, see T. Yamada and R. Graham, Phys. Rev. Lett. 45, 1322 (1980).

[6]R. May, Nature (London) 261, 459 (1976), and references cited therein.

[7]S. L. McCall, Appl. Phys. Lett. 32, 284 (1978).

[8]R. Bonifacio, M. Gronchi, and L. A. Lugiato, Opt. Commun. 30, 129 (1979); R. Bonifacio, L. A. Lugiato, and M. Gronchi, in *Proceedings of the Fourth International Conference on Laser Spectroscopy IV, Rottach-Egern, Germany, June, 1979*, edited by H. Walther and K. W. Rother (Springer-Verlag, Heidelberg, 1979). This type of instability was first discussed in other contexts by H. Risken and K. Nummedal, J. Appl. Phys. 49, 4662 (1968), and R. Graham and H. Haken, Z. Phys. 213, 420 (1968).

[9]The period is always slightly larger than nt_R by an amount related to the medium decay time. This effect is analogous to a similar effect in a passively mode-locked laser. We have verified that our observed periods converge on nt_R as $\tau \to 0$, and we prefer to simplify the discussion by using the limiting value in the text.

[10]We follow, for the most part, the notation of Ikeda, Daido, and Akimoto, Ref. 2 for easy comparison. The equations can be derived from the Maxwell-Bloch equations in the limit that T_2 is very short, the system is very far off resonance (i.e., no absorption) and one can expand the nonlinearity to third order. Then $\tau = T_1$.

[11]A. A. Kastal'skii, Fiz. Tekh. Poluprovodn. 7, 935 (1973) [Sov. Phys. Semicond. 7, 635 (1973)]; P. W. Smith and E. H. Turner, Appl. Phys. Lett. 30, 280 (1977); S. L. McCall, Appl. Phys. Lett. 32, 284 (1978); E. Garmire, J. H. Marburger, and S. D. Allen, Appl. Phys. Lett. 32, 320 (1978).

[12]W. P. Greene, H. M. Gibbs, A. Passner, S. L. McCall, and T. N. C. Venkatesan, Opt. News 6, 16 (1980). This is essentially the device of Garmire, Marburger, and Allen (Ref. 11) with PLZT replacing their $LiNbO_3$ crystal.

[13]T. Bischofberger and Y. R. Shen, Phys. Rev. A 19, 1169 (1979).

[14]F. A. Hopf and P. Meystre, Opt. Commun. 33, 225 (1980).

Reprint Series
18 December 1981, Volume 214, pp. 1350–1353

SCIENCE

Phase Locking, Period-Doubling Bifurcations, and Irregular Dynamics in Periodically Stimulated Cardiac Cells

Abstract. *The spontaneous rhythmic activity of aggregates of embryonic chick heart cells was perturbed by the injection of single current pulses and periodic trains of current pulses. The regular and irregular dynamics produced by periodic stimulation were predicted theoretically from a mathematical analysis of the response to single pulses. Period-doubling bifurcations, in which the period of a regular oscillation doubles, were predicted theoretically and observed experimentally.*

The phase of neural and cardiac oscillators can be reset by a single brief depolarizing or hyperpolarizing stimulus (*1–3*). Experimental determination of the dependence of the phase shift on the phase of the autonomous cycle at which the stimulus was delivered allows computation of a mathematical function called the Poincaré map (*4*). Analysis of the Poincaré map is carried out to predict the response to periodic stimulation (*1*, *2*, *4*). This work provides experimental confirmation of a recent theoretical prediction (*4*) that period-doubling bifurcations and irregular dynamics (*5*) should be observable in periodically stimulated oscillators.

The preparation has been described in detail (*6*). Briefly, apical portions of heart ventricles of 7-day-old embryonic chicks were dissociated into their component cells in 0.05 percent trypsin. The cells were transferred to a flask containing tissue culture medium (818A with a potassium concentration of 1.3 m*M*), which was placed on a gyratory shaker. Spheroidal aggregates (100 to 200 μm in diameter) of electrically coupled cells that beat spontaneously with a period between 0.4 and 1.3 seconds form after 48 to 72 hours of gyration. Experiments were performed on aggregates in the same culture medium at 35°C under a gas mixture of 5 percent CO_2, 10 percent O_2, and 85 percent N_2. Intracellular electrical recordings were made with glass microelectrodes filled with 3*M* KCl (resistance, 20 to 60 megohms). Current pulses were delivered through the same electrode and measured with a virtual ground circuit. Impalements were maintained for 2 to 5 hours. This report presents results for two aggregates out of ten studied.

Consider the response of an aggregate to a single current pulse delivered δ msec after the upstroke of the action potential (Fig. 1A). The length of the cycle immediately preceding the perturbation is called τ, and the phase ϕ of the cycle at which the stimulus was delivered is $\phi = \delta/\tau$, $0 \leq \phi < 1$. Control cycles with the phase labeled are shown in Fig. 1B. The cycle time of the perturbed cycle (the time from the upstroke immediately preceding the stimulus to the next upstroke) is called *T*. A stimulus was delivered after every ten beats, with δ increased by 10 msec each time. In Fig. 1C the normalized perturbed cycle length T/τ is plotted for two different preparations. In a single preparation, an increase in stimulus intensity produces a transition from a continuous function such as that shown in Fig. 1C(i) to an apparently discontinuous function such as that shown in Fig. 1C(ii) (*7*).

Current pulses were periodically injected into the aggregate with period t_s ($100 \leq t_s \leq 700$ msec). For most t_s values in this range, phase-locked patterns result. A pattern is called an $N{:}M$ phase-locked pattern if it is periodic in time and if for every N stimuli there are M action potentials, with action potentials occurring at M different times in the stimulus cycle. At first t_s was varied in 50-msec steps to sketch the boundaries of the major (2:1, 1:1, and 2:3) phase-locked regions. Then the intermediate regions were sampled by varying t_s in 10-msec steps. We first describe our experimental observations and then offer an interpretation based on an analysis of the Poincaré map.

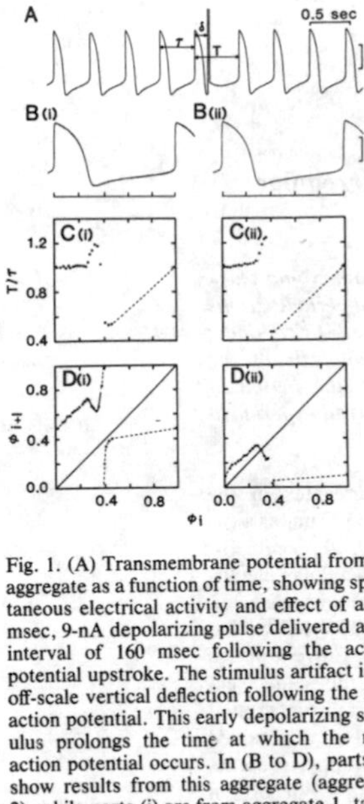

Fig. 1. (A) Transmembrane potential from an aggregate as a function of time, showing spontaneous electrical activity and effect of a 20-msec, 9-nA depolarizing pulse delivered at an interval of 160 msec following the action potential upstroke. The stimulus artifact is an off-scale vertical deflection following the fifth action potential. This early depolarizing stimulus prolongs the time at which the next action potential occurs. In (B to D), parts (ii) show results from this aggregate (aggregate 2), while parts (i) are from aggregate 1, taken from a different culture. (B) Membrane voltage as a function of phase ϕ, $0 \leq \phi < 1$. (C) Phase-resetting data, showing the normalized length T/τ of the perturbed cycle as a function of ϕ. (i) Pulse duration 40 msec, pulse amplitude 5 nA; (ii) pulse duration 20 msec, pulse amplitude 9 nA. For approximately $0.4 < \phi < 1.0$ the action potential upstroke occurs during the stimulus artifact and hence the perturbed cycle length cannot be exactly determined. The dashed line represents a linear interpolation that approximates the data. During collection of these data, the average control interbeat intervals ($\pm$ 1 standard deviation) were (i) $\bar{\tau} = 515 \pm 5.7$ msec and (ii) $\bar{\tau} = 434 \pm 5.5$ msec. (D) Poincaré maps computed from Eq. 1 and the data in Fig. 1C; (i) $t_s = 250$ msec, (ii) $t_s = 480$ msec. The dashed line represents a linear interpolation used in iterating the Poincaré map; the solid line through the data points is a quartic fit for $0.22 < \phi_i < 0.37$.

Characteristic zones of regular and irregular dynamics were seen in both aggregates. Transitions occur at approximately the same values of $t_s/\bar{\tau}$, where $\bar{\tau}$ is the average control interbeat interval. For $0.55 < t_s/\bar{\tau} < 1.05$, 1:1 phase locking [Fig. 2A(ii)] is found. When t_s is decreased to $0.40 < t_s/\bar{\tau} < 0.55$ (zone α), dynamics analogous to the clinically observed Wenckebach phenomenon (*8*) are present. This phenomenon is characterized by a gradual prolongation of the time between a stimulus and the subsequent action potential until an action potential is skipped. This can occur in an irregular fashion (Fig. 2B) as well as in $m + 1{:}m$ phase-locked patterns. As t_s is decreased below $0.4\,\bar{\tau}$, 2:1 phase locking is observed [Fig. 2A(i)].

For $1.05 < t_s/\bar{\tau} < 1.15$ (zone β), the ratio of stimulus frequency to action potential frequency is 1, but the stimulus no longer occurs at one fixed phase of the aggregate cycle as it does in 1:1 locking. Instead, the stimulus falls at two or more phases of the cycle. The dynamics in this narrow zone are highly variable and phase-locked patterns, when they exist, are typically not maintained for long stretches of time. For example, Fig. 2C(i) shows a transition from a 1:1 to a 2:2 phase-locked pattern which spontaneously occurred during stimulation with a fixed frequency. Similarly, brief stretches of 4:4 phase locking [Fig. 2C(ii)] and irregular dynamics [Fig. 2C(iii)] can both be observed at $t_s = 490$ msec. Such transitions may be due to slow drifts in the intrinsic frequency of the aggregate during stimulation.

For $1.15 < t_s/\bar{\tau} < 1.35$ (zone γ) there are irregular patterns with extra or interpolated beats (Fig. 2D). Further increase in t_s leads to a 2:3 phase-locked pattern [Fig. 2A(iii)].

As the stimulus strength increases, the widths of zones α and γ decrease. However, zone β is widest at intermediate stimulus strength. The two examples in this report were selected because there is a relatively broad β zone.

The experimentally derived curves shown in Fig. 1C can be used to predict the effects of periodic stimulation (*1, 2, 4*). If ϕ_i is the phase of the oscillator immediately before the ith stimulus, then

$$\phi_{i+1} = 1 - f(\phi_i) + \phi_i + \frac{t_s}{\bar{\tau}} \pmod 1 \qquad (1)$$

where $f(\phi)$ gives the normalized perturbed interbeat interval (T/τ) as a function of ϕ. The relation

$$\phi_{i+1} = g(\phi_i, t_s) \qquad (2)$$

defined by Eq. 1 is called the Poincaré map (Fig. 1D). For $t_s = 0$, the Poincaré map corresponds to the "new phase–old phase" phase-resetting curve, also called the phase transition curve (*1, 3, 4*). Starting from some initial phase ϕ_0, Eq. 1 can be iterated to compute the sequence of phases $\phi_0, \phi_1, \phi_2, \ldots$ If $\phi_N = \phi_0$ and $\phi_j \neq \phi_0$ for $0 < j < N$, then the Poincaré map has a cycle $\phi_0, \phi_1, \ldots, \phi_{N-1}, \phi_N = \phi_0$ of period N. If

$$\prod_{i=0}^{N-1} \left| \left. \frac{\partial g}{\partial \phi} \right|_{\phi_i} \right| < 1 \qquad (3)$$

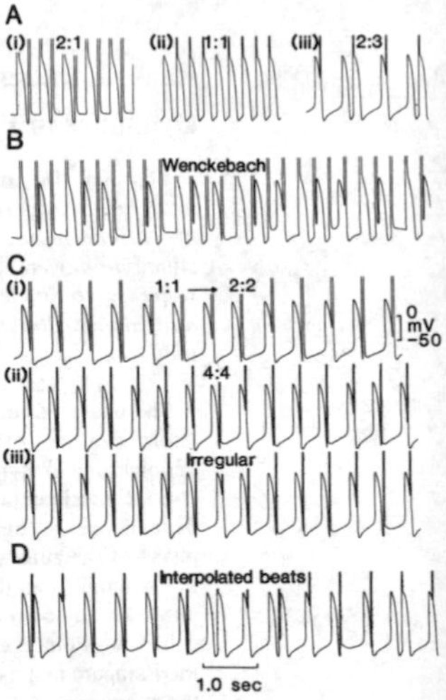

Fig. 2. Representative transmembrane recordings from both aggregates showing the effects of periodic stimulation with the same pulse durations and amplitudes as in Fig. 1C. (A) Stable phase-locked patterns: (i) 2:1 (aggregate 1, $t_s = 210$ msec); (ii) 1:1 (aggregate 2, $t_s = 240$ msec); (iii) 2:3 (aggregate 2, $t_s = 600$ msec). (B) Dynamics in zone α: irregular dynamics displaying the Wenckebach phenomenon (aggregate 1, $t_s = 280$ msec). (C) Dynamics in zone β: (i) 1:1 phase locking spontaneously changing to 2:2 phase locking (aggregate 1, $t_s = 550$ msec). During 2:2 phase locking there are two distinct phases of the cycle at which the stimuli fall. (ii) 4:4 phase locking (aggregate 2, $t_s = 490$ msec). There are four distinct phases of the cycle at which the stimuli fall. (iii) Irregular dynamics with one action potential in each stimulus cycle (aggregate 2, $t_s = 490$ msec). There is a narrow range of phases in which the stimuli fall. (D) Dynamics in zone γ: irregular dynamics displaying extra interpolated beats (aggregate 2, $t_s = 560$ msec).

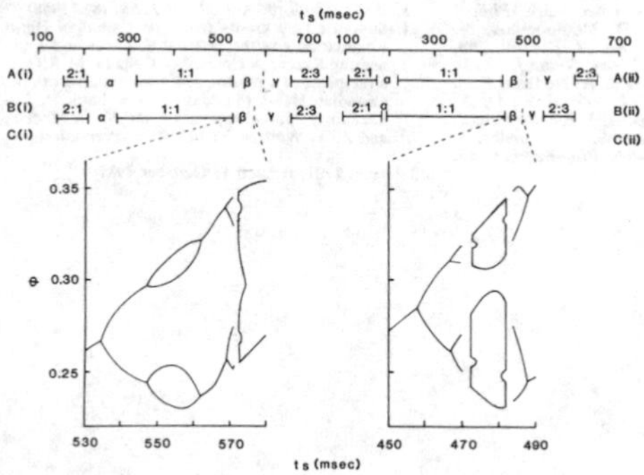

Fig. 3. Experimentally determined and theoretically computed dynamics. Parts (i) refer to aggregate 1, parts (ii) to aggregate 2. (A) Experimentally determined dynamics: there are three major phase-locking regions (2:1, 1:1, 2:3) and three zones of complicated dynamics labeled α, β, and γ (see text). (B) Theoretically predicted dynamics; note agreement with (A). (C) Theoretically predicted dynamics in zone β: curves give phase or phases in the cycle at which the stimuli fall during 1:1, 2:2, and 4:4 locking; stippled regions show the range of phases in which the stimulus falls during irregular dynamics.

then the period N cycle is stable, and the corresponding N:M phase-locked pattern is stable with a period of Nt_s (*4*). A stable pattern is maintained despite small variations in either the oscillator itself or the stimulus parameters. The value of M is given by

$$M = \sum_{i=1}^{N} \left[1 - f(\phi_i) + \frac{t_s}{\bar{\tau}} \right] \quad (4)$$

To iterate the Poincaré map, the function $f(\phi_i)$ in Eq. 1 was approximated by linear interpolation between the data points of Fig. 1C. There is close agreement (*9*) between the experimentally observed ranges of t_s that give simple phase-locked patterns (Fig. 3A) and those theoretically predicted from iteration of the Poincaré map (Fig. 3B). The Poincaré map also predicts the existence of irregular dynamics (dynamics that are not phase locked) in zones α, β, and γ. There are complex changes in the phase-locking patterns as t_s is changed within these zones. We explicitly compute these changes in zone β and briefly discuss the dynamics in the other two zones.

Numerical analysis of the Poincaré maps for t_s in zone β shows that for any ϕ_i in the interval (0.22, 0.37) all iterates of ϕ_i remain in this invariant interval. Moreover, this interval attracts iterates of ϕ_i for all ϕ_0 outside this interval. To determine the theoretically predicted dynamics in zone β, the Poincaré map in the invariant interval was fit to a quartic polynomial by a least-squares method. Numerical iteration of the Poincaré map in zone β shows a sequence of period-doubling bifurcations and irregular dynamics (Fig. 3C) as t_s is increased (*5*). In the zone of irregular dynamics, the phases of the stimuli are restricted to the shaded regions of Fig. 3C (*10*). The irregular dynamics result from a deterministic iterative process, with no added stochastic terms.

We do not experimentally observe all the bifurcations theoretically computed in Fig. 3C (*11*). However, we propose that the transition shown in Fig. 2C(i) from 1:1 phase locking (period of repeating pattern = 550 msec) to 2:2 phase locking (period of repeating pattern = 1100 msec) corresponds to the period-doubling bifurcation theoretically computed at $t_s \approx 535$ msec [Fig. 3C(i)]. The 4:4 phase-locked pattern of Fig. 2C(ii) and the irregular dynamics of Fig. 2C(iii) provide further evidence for the correspondence between the experimental observations and the theoretical computations [Fig. 3C(ii)].

In zones α and γ complex behavior is also observed experimentally and predicted theoretically. For example, in zone α, the dynamics experimentally seen at t_s = 280 msec (Fig. 2B) are very similar to those predicted at t_s = 250 msec from the Poincaré map in Fig. 1D(i). The extra beats characteristic of zone γ (Fig. 2D) are also predicted.

This work has implications for the understanding of normal and pathologic behavior in cardiac tissue. The experimental work supports previous studies showing that periodically forced oscillators display phase-locked dynamics that are similar to clinically observed cardiac dysrhythmias (*2, 4, 12*). Moreover, the work suggests novel explanations for the genesis of 2:2 rhythms (*13*) and irregular dysrhythmias (*4*).

We have observed behavior that we propose arises as a consequence of period-doubling bifurcations in these experiments. Thus, exotic dynamic behavior that was previously seen in mathematical studies and in experiments in the physical sciences (*5*) may in general be present when biological oscillators are periodically perturbed.

MICHAEL R. GUEVARA
LEON GLASS, ALVIN SHRIER
Department of Physiology, McGill University, Montreal, Canada H3G 1Y6

References and Notes

1. D. H. Perkel, J. H. Schulman, T. H. Bullock, G. P. Moore, J. P. Segundo, *Science* **145**, 61 (1964); T. Pavlidis, *Biological Oscillators: Their Mathematical Analysis* (Academic Press, New York, 1973); H. M. Pinsker, *J. Neurophysiol.* **40**, 527 (1977); *ibid.*, p. 544.
2. G. K. Moe, J. Jalife, W. J. Mueller, B. Moe, *Circulation* **56**, 968 (1977); J. Jalife and G. K. Moe, *Am. J. Cardiol.* **43**, 761 (1979); S. Scott, thesis, State University of New York, Buffalo (1979).
3. A. T. Winfree, *Science* **197**, 761 (1977); *The Geometry of Biological Time* (Springer-Verlag, New York, 1980); J. Jalife and C. Antzelevitch, *Science* **206**, 695 (1979); T. Sano, T. Sawanobori, H. Adaniya, *Am. J. Physiol.* **235**, H379 (1978).
4. M. R. Guevara and L. Glass, *J. Math. Biol.*, in press.
5. A period-doubling bifurcation is the doubling of the period of an oscillation due to parametric changes. Period-doubling bifurcations and "chaotic" dynamics are observed in simple mathematical models and in experiments in the physical sciences: T. Y. Li and J. A. Yorke, *Am. Math. Mon.* **82**, 985 (1975); R. M. May, *Nature (London)* **261**, 459 (1976); M. C. Mackey and L. Glass, *Science* **197**, 287 (1977); J. P. Gollub, T. O. Brunner, B. G. Danly, *ibid.* **200**, 48 (1978); K. Tomita and T. Kai, *Prog. Theor. Phys. Suppl. 64* (1978), p. 280; R. H. G. Helleman, Ed., *Nonlinear Dynamics* (New York Academy of Sciences, New York, 1980); Editorial, *Phys. Today* **34**, 17 (March 1981).
6. R. L. DeHaan and L. J. DeFelice, *Theor. Chem.* **4**, 181 (1978); J. R. Clay and R. L. DeHaan, *Biophys. J.* **28**, 377 (1979); J. R. Clay and A. Shrier, *J. Physiol. (London)* **312**, 471 (1981).
7. There are many questions concerning the continuity properties of the graphs in Fig. 1, C and D. It is difficult to demonstrate experimentally that the points in Fig. 1C(ii) actually display a discontinuity. However, in this and many other preparations, repeated stimulation at the apparent discontinuity did not show evidence for intermediate values of T/τ. On theoretical grounds, discontinuities in the plots in Fig. 1C are expected [M. Kawato, *J. Math. Biol.* **12**, 13 (1981)]. However, the plots in Fig. 1D would not be expected to show discontinuities if one constructed the Poincaré map using the eventual phase [M. Kawato, cited above; E. N. Best, *Biophys. J.* **27**, 87 (1979)].
8. W. J. Mandel, Ed., *Cardiac Arrhythmias: Their Mechanisms, Diagnosis and Management* (Lippincott, Philadelphia, 1980).
9. There are many potential reasons for discrepancy between the predictions and experimental results. (i) The use of the Poincaré map is an approximation since it maps the state of a many-dimensional dynamical system to a single variable, the phase. (ii) The stimulation can lead to secondary electrophysiological changes in the aggregate such as changes in intrinsic frequency. (iii) There is beat-to-beat fluctuation in the autonomous cycle length in the absence of stimulation (*6*), and this "noise" tends to decrease the widths of the major stable phase-locked zones (compare Fig. 3A with Fig. 3B).
10. In the irregular region there are strong limitations on the phase at which the stimulus occurs. Compare Fig. 3C in this report with figure 3 in E. N. Lorenz, *Ann. N.Y. Acad. Sci.* **357**, 282 (1980).
11. "Noise" tends to destroy stable phase-locked patterns that either are complex or exist over very small ranges of t_s [L. Glass, C. Graves, G. A. Petrillo, M. C. Mackey, *J. Theor. Biol.* **86**, 455 (1980); J. P. Crutchfield and B. A. Huberman, *Phys. Lett.* **A77**, 407 (1980); R. Guttman, L. Feldman, E. Jakobsson, *J. Membr. Biol.* **56**, 9 (1980)]. Thus, there are fundamental limitations to the experimental observability of such patterns.

12. B. van der Pol and J. van der Mark, *Philos. Mag.* **6**, 763 (1928); S. D. Moulopoulos, N. Kardaras, D. A. Sideris, *Am. J. Physiol.* **208**, 154 (1965); J. V. O. Reid, *Am. Heart J.* **78**, 58 (1969); F. A. Roberge and R. A. Nadeau, *Can. J. Physiol. Pharmacol.* **42**, 695 (1969); D. A. Sideris and S. D. Moulopous, *J. Electrocardiol.* **10**, 51 (1977); C. R. Katholi, F. Urthaler, J. Macy, T. N. James, *Comp. Biomed. Res.* **10**, 529 (1977).

13. R. Langendorff, *Am. Heart J.* **55**, 181 (1958).

14. Supported by grants from the Canadian Heart Foundation and the Natural Sciences and Engineering Research Council of Canada. M.R.G. is a recipient of a predoctoral traineeship from the Canadian Heart Foundation. We thank D. Colizza for technical assistance and M. C. Mackey and A. T. Winfree for helpful conversations.

3 August 1981; revised 13 October 1981

Part Three

BIBLIOGRAPHY

I. Books and Conference Proceedings

B1 Arnold, V.I., and Avez, A., (1968). *Ergodic Problems of Classical Mechanics*, Addison-Wesley, Reading, Mass.

B2 Moser, J., (1973). *Stable and Random Motions in Dynamical Systems*, Princeton University Press.

B3 Marsden, E.J., and McCracken, M., (1976). *The Hopf Bifurcation and its Applications*, Applied Math. Sci. **19**, Springer-Verlag.

B4 Casati, G., and Ford, J., eds., (1977). *Stochastic Behavior in Classical and Quantum Hamiltonian Systems*, Lect. Notes in Phys. **93**, Springer-Verlag.

B5 Abraham, R., and Marsden, J.E., (1978). *Foundations of Mechanics*, 2nd ed., Addison-Wesley, Reading, Mass.

B6 Arnold, V.I., (1978). *Mathematical Methods of Classical Mechanics*, Springer-Verlag.

B7 Jorna, S., ed., (1978). *Topics in Nonlinear Dynamics*, AIP Conf. Proc. **46**.

B8 Markley, N.G., Martin, J.C., and Perrizo, W., eds., (1978). *The Structure of Attractors in Dynamical Systems*, Lect. Notes in Math. **668**, Springer-Verlag.

B9 Mori, H., ed., (1978). *Nonlinear Nonequilibrium Statistical Mechanics*, Proceedings of the 1978 Oji Seminar, Suppl. Prog. Theor. Phys., **64**.

B10 Gurel, O., and Rossler, O.E., eds., (1979). *Bifurcation Theory and its Applications in Scientific Disciplines*, Ann. N.Y. Acad. Sci., **316**.

B11 Iooss, G., (1979). *Bifurcations of Maps and Applications*, North-Holland, Amsterdam.

B12 Month, M., and Herra, J.C., eds., (1979). *Nonlinear Dynamics and the Beam-Beam Interaction*, AIP Conf. Proc., **57**.

B13 Bardos, C., and Besis, D., eds., (1980). *Bifurcation Phenomena in Mathematical Physics and Related Topics*, Riedel.

B14 Collet, P., and Eckmann, J.-P., (1980). *Iterated Maps on the Interval as Dynamical Systems*, Birkhauser, Boston.

B15 Helleman, R.G.H., ed., (1980). *Nonlinear Dynamics*, Ann. N.Y. Acad. Sci., **357**.

B16 Holmes, P., ed., (1980). *New Approaches to Nonlinear Problems in Dynamics*, SIAM Publications, Philadelphia.

B17 Iooss, G., and Joseph, D.D., (1980). *Elementary Stability and Bifurcation Theory, Undergraduate Texts in Math.*, Springer-Verlag.

B18 Laval, G., and Gresillon, D., eds., (1980). *Intrinsic Stochasticity in Plasmas*, Les Editions de Physique Courtboeuf, Orsay.

B19 Nitecki, Z., and Robinson, C., eds., (1980). *Global Theory of Dynamical Systems*, Lect. Notes in Math., **819**, Springer-Verlag.

B20 Osterwalder, K., ed., (1980). *Mathematical Problems in Mathematical Physics*, Lect. Notes in Phys., **116**, Springer-Verlag.

B21 Haken, H., ed., (1981). *Chaos and Order in Nature*, Springer Series in Synergetics, **11**, Springer-Verlag.

B22 Guckenheimer, J., Moser, J., and Newhause, S.E., (1981). *Dynamical Systems*, Prog. in Math., **8**, Birkhauser, Boston.

B23 Rand, D.A., and Young, L.S., ed., (1981). *Dynamical Systems and Turbulence*, Lect. Notes in Math., **898**, Springer-Verlag.

B24 Riste, T., ed., (1981). *Nonlinear Phenomena at Phase Transitions and Instabilities*, Plenum.

B25 Swinney, H.L., and Gollub, J.P., eds., (1981). *Hydrodynamical Instabilities and the Transition to Turbulence*, Springer-Verlag.

B26 Vidal, C., and Pacault, A., eds., (1981). *Nonlinear Phenomena in Chemical Dynamics*, Springer-Verlag.

B27 Barrenblatt, G.I., Iooss, G., and Joseph, D.D., eds., (1982). *Nonlinear Dynamics and Turbulence*, Pittman.

B28 Bishop, A.R., Campbell, D.K., and Nicolaenko, B., eds., (1982). *Nonlinear Problems: Present and Future*, North-Holland.

B29 Haken, H., ed., (1982). *Evolution of Order and Chaos in Physics, Chemistry, and Biology*, Springer Series in Synergetics, **17**, Springer-Verlag.

B30 Kalia, R.K., and Vashishta, P., eds., (1982). *Melting, Localization, and Chaos, Elsevier Sci. Publ. Co.*

B31 Lichtenberg, A.J., and Lieberman, M.A., (1982). *Regular and Stochastic Motion*, Appl. Math. Sci., **38**, Springer-Verlag.

B32 Sparrow, C., (1982). *The Lorenz Equations, Bifurcations, Chaos, and Strange Attractors*, Appl. Math. Sci., **41**, Springer-Verlag.

B33 *Stochastic Processes in Quantum Theory and Statistical Physics*, Proc. Int. Symp., Marseille, June 1981, Springer-Verlag, 1982.

B34 *Nonequilibrium Statistical Physics Problems in Fusion Plasma — Stochasticity and Chaos*, US-Japan Joint Inst. for Fusion Theory Workshop, Nagoya, 1981, Inst. Plasma Phys., 1982.

B35 Benedeck, G., Bliz, H., and Zeyher, R., eds., (1983). *Statics and Dynamics of Nonlinear Systems*, Springer-Verlag.

B36 Guckenheimer, J., and Holmes, P., (1983). *Nonlinear Oscillations, Dynamical Systems and Bifurcations of Vector Fields*, Appl. Math. Sci., **42**, Springer-Verlag.

B37 Horton, J.C.W., Reichl, L.E., and Szebehely, V.G., eds., (1983). *Long-Time Prediction in Dynamics*, Wiley.

B38 *Dynamical Systems and Chaos*, Proc. Sitges Conf. on Stat. Mech., 1982, Springer-Verlag, 1983.

B39 *Order in Chaos*, Proc. Int. Conf. at Los Alamos, 1982, Physica **7D** (1983), and Separate book, North-Holland.

B40 Iooss, G., Helleman, R.H.G., and Stora, R., (1983). *Chaotic Behaviour of Deterministic Systems*, Les Houches 1981, North-Holland.

II. Papers (including reviews)

Abarbanel, H. D. J., and Latham, P. E., (1982). Finite resolution approximation to the asymptotic distribution for dynamical systems, *Phys. Lett.* **89A**, 55.

Abe, S., and Mukamel, S., (1983). Anharmonic molecular spectra-self-consistent mode coupling, nonlinear maps, and quantum chaos, *J. Chem. Phys.* **70**, 5457.

Abraham, E., Firth, W. J., and Carr, J., (1982). Self-oscillation and chaos in nonlinear Fabry Perot resonators with finite response time, *Phys. Lett.* **91A**, 47.

Abraham, N. B., (1983). A new focus on laser instabilities and chaos, *Laser Focus* **19**, 73.

Abraham, N. B., Chyba, T., Coleman, M., Gioggia, R. S., Halas, N. J., Hoffer, L. M., Liu, S.-N., Maeda, M., and Wessen, J. C., (1983). "Experimental evidence for self-pulsing and chaos in CW-excited lasers," in *Laser Physics, Proc. 3rd New Zealand Symposium*, p. 107, Springer Verlag.

Ackerhalt, J. R., Galbraith, H. W., and Milonni, P. W., (1983). Chaos in multiple-photon excitation of molecules, *Phys. Rev. Lett.* **51**, 1259.

Aizawa, Y., (1982). Global aspects of the dissipative dynamical system, I. Statistical identification and fractal properties of the Lorenz chaos, *Prog. Theor. Phys.* **68**, 64.

Aizawa, Y., and Uezu, T., (1982a). Topological aspects in chaos and in 2**k period-doubling cascade, *Prog. Theor. Phys.* **67**, 982.

Aizawa, Y., and Uezu, T., (1982b). Global aspects of dissipative dynamical systems. II. Periodic and chaotic responses in the forced Lorenz system, *Prog. Theor. Phys.* **68**, 1864.

Aizawa, Y., and Murakami, C., (1983). Generalization of Baker's transformation — chaos and stochastic process on a Smale's horseshoe, *Prog. Theor. Phys.* **69**, 1416.

Aizawa, Y., (1983). Symbolic dynamics approach to intermittent chaos, *Prog. Theor. Phys.* **70**, 1249.

Alekseev, V. M., (1968–1969). Quasirandom dynamical systems, I, II and III, *Math. USSR Shornik* **5**, 73; **6**, 505; **7**, 1.

Alekseev, V. M., and Yakobson, M. V., (1981). Symbolic dynamics and hyperbolic dynamic systems, *Phys. Reports* **75**, 287.

Ali, M. K., and Somorjai, R. L., (1982). Reappearance of ordered motion in nonintegrable Hamiltonian systems — the strong coupling case, *Prog. Theor. Phys.* **68**, 1854.

Allwright, D., (1978). Hypergraphic function and bifurcation in recurrence relations, *SIAM J. Appl. Math.* **34**, 687.

Ananthakrishnan, G., and Valsakumar, M. C., (1983). Chaotic flow in a model for repeated yielding,*Phys. Lett.* **95A**, 69.

Angelo, P. M., and Riela, G., (1981). A six-mode truncation of the Navier-Stokes equations on a two-dimensional torus: a numerical study, *Nuovo Cimento* **64B**, 207.

Ankiewicz, A., (1983). Regular and irregular motion: new mechanical results and fibre optics, *J. Phys.* **A16**, 3657.

Anishchenko, V. S., Astakov, V. V., and Letchford, T. E., (1983). Experimental study of the structure of a strange attractor in a model generator with a nonlinear inertial term, *Sov. Phys. Tech. Phys.* **28**, 91; Russian original, *JTF*, **53**, 152.

Aoki, K., and Yamamoto, K., (1983). Firing wave instability of the current filaments in a semiconductor, *Phys. Lett.* **98A**, 72.

Aoki, K., Ikezawa, O., and Yamamoto, K., (1983). Firing wave instability and chaos of the current filament in a semiconductor, *Phys. Lett.* **98A**, 217.

Arecchi, F. T., Meucci, R., Puccioni, P., and Tredicce, J., (1982). Experimental evidence of subharmonic bifurcations, multistability and turbulence in a Q-switched gas laser, *Phys. Rev. Lett.* **49**, 1217.

Arneodo, A., Coullet, P., and Tresser, C., (1979). A renormalization group with periodic behavior, *Phys. Lett.* **70A**, 74.

Arneodo, A., Coullet, P., and Tresser, C., (1980). Occurrence of strange attractors in three-dimensional Volterra equation, *Phys. Lett.* **79A**, 59.

Arneodo, A., Coullet, P., and Tresser, C., (1981a). A possible new mechanism for the onset of turbulence, *Phys. Lett.* **81A**, 197.

Arneodo, A., Coullet, P., and Tresser, C., (1981b). Possible new strange attractors with spiral structure, *Commun. Math. Phys.* **79**, 573.

Arneodo, A., Coullet, P., and Tresser, C., (1982a). Oscillators with chaotic behavior: an illustration of a theorem by Shil'nikov, *J. Stat. Phys.* **27**, 171.

Arneodo, A., Coullet, P., Tresser, A., Libchaber, A., Maurer, J., and d'Humieres, (1983). About the observation of the uncompleted cascade in Rayleigh-Benard experiment, *Physica* **6D**, 385.

Arneodo, A., Coullet, P., and Spiegel, E. A., (1982). Chaos in a finite macroscopic system, *Phys. Lett.* **92A**, 369.

Arneodo, A., Coullet, P., and Spiegel, E. A., (1983). Cascade of period-doublings of tori, *Phys. Lett.* **94A**, 1.

Arnold, V. I., (1962). The classical theory of perturbations and the problem of stability of planetary systems, *Sov. Math. Dokl.* **3**, 1008.

Arnold, V. I., (1963a). Proof of a theorem of A. N. Kolmogorov on the invariance of periodic motions under small perturbations of the Hamiltonian, *Russ. Math. Surv.* **18**, 9; Russian original, *UMN* **18**:5,13.

Arnold, V. I., (1963b). Small denominators and problems of stability of motion in classical and celestial mechanics, *Russ. Math. Surv.* **18**, 85; Russian original, *UMN* **18**:6,91.

Arnold, V. I., (1964). Instability of dynamical systems with several degrees of freedom, *Sov. Math. Dokl.* **5**, 581; Russian original: *Doklady* **156**, 9.

Arnold, V. I., (1965). Small denominators. I: Mappings of the circumference onto itself, *AMS Transl. Ser.* **2**, 46, 213.

Arnold, V. I., (1972). Lectures on bifurcations in versal families: *Russ. Math. Surv.* **27**, 54.

Arnold, V. I., (1977). Loss of stability of self-oscillations close to resonance, *Funct. Anal.* **11**, 85.

Aronson, D. G., Chory, M. A., Hall, G. R., and McGehee, R. P., (1980). "A discrete dynamical system with subtly wild behavior," in *New Approaches to Nonlinear Problems in Dynamics*, ed. P. Holmes p. 339. S.I.A.M. Publications, Philadelphia.

Aronson, D. G., Chory, M. A., Hall, G. R., and McGehee, R. P., (1982). Bifurcations from an invariant circle for two-parameter families of maps of the plane: A computer-assisted study, *Commun. Math. Phys.* **83**, 303.

Aubry, S., (1978). The new concept of transition by breaking of analyticity in a crystallographic model, in *Solitons and Condensed Matter Physics*, ed. A. R. Bishop and T. Schneider, *Solid State Sciences* **8**, p. 64, Springer-Verlag, N. Y.

Aubry, S., (1979). "Some nonlinear physics in crystallographic structures," in *Stochastic Behavior in Classical and Quantum Hamiltonian Systems*, ed. G. Casati and J. Ford, *Lect. Notes in Phys.* **93**, p. 201, Springer-Verlag, N. Y.

Aubry, S., and Andre, G., (1980). Analyticity breaking and Anderson localization in incommensurate lattices, *Annals Israel Phys. Soc.* **3**, 133.

Aubry, S., (1983). Devil's staircase and order without periodicity in classical condensed matter, *J. Physique* **44**, 147.

Baesens, C., and Nicolis, G., (1983). Complex bifurcations in a periodically forced normal form, *Z. Phys.* **B52**, 345.

Bak, P., (1981). Chaotic behavior and incommensurate phases in the anisotropic Ising model with competing interactions, *Phys. Rev. Lett.* **46**, 791.

Bak, P., (1982). Commensurate phases, incommensurate phases and the devil's staircase, *Rep. Prog. Phys.* **45**, 587.

Bak, P., and Hogh Jensen, M., (1982). Bifurcations and chaos in the f**4 theory on a lattice, *J. Phys.* **A15**, 1983.

Bak, P., and Bruinsma, R., (1982). One-dimensional Ising model and the complete devil's staircase, *Phys. Rev. Lett.* **49**, 249.

Bak, P., (1983). "Universality and fractal dimension of mode-locking structure in systems with competing periodicities," in *Statics and Dynamics of Nonlinear Systems*, ed. G. Benedek, H. Bilz, R. Zeyher, p. 160, Springer-Verlag.

Baker, N. H., Moore, D. W., and Spiegel, A., (1971). Aperiodic behavior of a nonlinear oscillator, *Q. Journ. Mech. and Appl. Math.* **4**, 391.

Baive, D., and Franceschini, V., (1981). Symmetry breaking on a model of five-mode truncated Navier-Stokes equations, *J. Stat. Phys.* **26**, 471.

Bar-Eli, K., and Noyes, R. M., (1977). Model calculations describing bistability for the stirred flow oxidation of cerous ion by bromate, *J. Chem. Phys.* **61**, 1988.

Barrow, J. D., (1982a). General relativistic chaos and nonlinear mechanics, *Gen. Relativ. & Gravit.* **14**, 523.

Barrow, J. D., (1982b). Chaotic behavior in general relativity, *Phys. Reports* **85**, 1.

Beasley, M. R., and Huberman, B. A., (1982). Chaos in Josephson junctions, *Comments on Solid State Phys.* **10**, 155.

Beddington, J. R., Free, C. A., Lauton, J. H., (1975). Dynamic complexity in predator-prey models framed in difference equations, *Nature* **255**, 58.

Beiersdorfer, P., and Wersinger, J.-M., (1983). Topology of the invariant manifolds of period-doubling attractors for some forced nonlinear oscillators, *Phys. Lett.* **96A**, 269.

Bellissard, J., Besis, D., and Moussa, P., (1982). Chaotic states of almost periodic Schrödinger operator, *Phys. Rev. Lett.* **49**, 701.

Beloshapkin, V. V., and Zaslavsky, G. M., (1983). On the spectral properties of dynamical systems in the transition region from order to chaos, *Phys. Lett.* **97A**, 121.

Belyaev, Yu. N., Monakhov, A. A., Scherbakov, S. A., and Yavoskaya, I. M., (1979). Onset of turbulence in rotating fluid, *Sov. Phys. JETP Lett.* **29**, 295; Russian original, *Pis'ma JETP*, **29**, 329.

Benettin, G., Galgani, L., and Strelcyn, J.-M., (1976). Kolmogorov entropy and numerical experiments, *Phys. Rev.* **14A**, 2338.

Benettin, G., and Strelcyn, J.-M., (1978). Numerical experiments on the free motion of a point mass moving in a plane convex region: stochastic transition entropy, *Phys. Rev.* **A17**, 773.

Benettin, G., Froeschle, C., and Scheidecker, H. P., (1979). Kolmogorov entropy of a dynamical system with an increasing number of degrees of freedom, *Phys. Rev.* **A19**, 454.

Benettin, G., Cercignani, C., Galgani, L., and Giorgilli, A., (1978a). Tous les nombres caracteristiques de Lyapounov sont effectivement calculables, *C. R. Acad. Sc.* **286A**, 431.

Benettin, G., Casartelli, M., Galgani, L., Giorgilli, A., and Strelcyn, J.-M., (1978b). On the reliability of numerical studies of stochasticity: I. Existence of time averages, *Nuovo Cimento* **44B**, 183.

Benettin, G., Casartelli, M., Galgani, L., Giorgilli, A., and Strelcyn, J.-M., (1979). On the reliability of numerical studies of stochasticity: II. Identification of time averages, *Nuovo Cimento* **50B**, 211.

Benettin, G., Cercignani, C., Galgani, L., and Giorgilli, A., (1980). Universal properties in conservative dynamical systems, *Lett. Nuovo Cimento* **28**, 1.

Benettin, G., Galgani, L., and Giorgilli, A., (1980). Further results on universal properties in conservative dynamical systems, *Lett. Nuovo Cimento* **29**, 163.

Benettin, G., Galgani, L., Giorgilli, A., and Strelcyn, J.-M., (1980). Lyapunov characteristic exponents for smooth dynamical systems and for Hamiltonian systems; A method for computing all of them. Part 1: Theory, and Part II: Numerical application, *Meccanica* **15**, 9: **15**, 21.

Benettin, G., and Galgani, L., (1982). Transition to stochasticity in a one-dimensional model of a radiant cavity, *J. Stat. Phys.* **27**, 153.

Benettin, G., and Tenenbaum, A., (1983). Ordered and stochastic behavior in a two-dimensional Lenard-Jones system, *Phys. Rev.* **A28**, 3020.

Ben-Jacob, E., Braiman, Y., Shainsky, R., and Imry, Y., (1981). Microwave induced "devil's staircase" structure and "chaotic" behavior in current-fed Josephson junctions, *Appl. Phys. Lett.* **38**, 822.

Ben-Jacob, E., Goldhirsch, I., Imry, Y., and Fishman, S., (1982). Intermittent chaos in Josephson junctions, *Phys. Rev. Lett.* **49**, 1599.

Bennett, D., Bishop, A. R., and Trullinger, S. E., (1982). Coherence and chaos in the driven, damped sine-Gordon chain, *Z. Phys.* **B47**, 265.

Berge, P., (1979). "Experiments on hydrodynamic instabilities and the transition to turbulence," in *Dynamical Critical Phenomena and Related Topics*, ed. C. P. Enz, *Lect. Notes in Phys.* **104**, p. 288, Springer-Verlag, N. Y.

Berge, P., and Dubois, M., (1979). Time-dependent velocity in Rayleigh-Benard convection: a transition to turbulence, *Opt. Commun.* **19**, 129.

Berge, P., Dubois, M., Manneville, P., and Pomeau, Y., (1980). Intermittency in Rayleigh-Benard convection, *J. de Phys. Lett.* **41**, 341.

Berge, P., Pomeau, Y., (1980). La turbulence, *La Recherche* **11**, 422.

Berman, G. P., and Zaslavsky, G. M., (1977). Theory of quantum nonlinear resonance, *Phys. Lett.* **61A**, 295.

Berman, G. P., and Zaslavsky, G. M., (1982). Quantum mappings and the problems of stochasticity in quantum systems, *Physica* **111A**, 17.

Berman, G. P., and Kolovsky, A. R., (1983). Correlation function behavior in quantum systems which are classically chaotic, *Physica* **8D**, 117.

Berman, G. P., and Kagansky, A. M., (1983). On stochasticity for a system with a finite region of interaction of particles, *Physica* **9D**, 225.

Berry, M. V., (1978). "Regular and irregular motion," in *Topics in Nonlinear Dynamics*, ed. S. Jorna, *AIP Conf. Proc.* Vol. 46.

Bindal, V. N., Saksena, T. K., and Singh, G., (1980). Subharmonic emission produced by high power ultrasonic waves in air, *Acoust. Lett.* **4**, 89.

Bishop, A. R., Fesser, K., and Lomdahl, P. S., (1983). Coherent spatial structure versus time chaos in a perturbed Sine-Gordon system, *Phys. Rev. Lett.* **50**, 1095.

Bishop, A., (1983). "Chaos and solitons in dissipative nonlinear systems," in *Statics and Dynamics of Nonlinear Systems*, ed. G. Benedeck, H. Bilz, R. Zeyher, p. 197, Springer Verlag.

Block, L., Guckenheimer, J., Misiurewicz, M., and Young, L.-S., (1979). Periodic points of one-dimensional maps, *Lect. Notes in Math.*, Vol. **819**, 18. Springer-Verlag.

Bogoyavlenskii, O. I., (1981). "Geometrical methods of the qualitative theory of dynamical systems in problems of theoretical physics," in *Math. Phys. Rev.*, ed. S. P. Novikov, Vol. **2**, p. 117.

Boldrighini, C., and Franceschini, V., (1979). A five-dimensional truncation of the plane incompressible Navier-Stokes equations, *Commun. Math. Phys.* **64**, 159.

Boon, J.-P., (1981). "Hydrodynamic instability: Structure and chaos," in *Scattering Techniques Applied to Supramolecular and Non-equilibrium Systems*, ed. S. H. Chen, B. Chu, and R. Nossal, Plenum Press, N. Y.

Booty, M., Gibbon, J. D., and Fowler, A. C., (1982). A study of the effect of mode truncation on an exact periodic solution of an infinite set of Lorenz equations, *Phys. Lett.* **87A**, 261.

Born, M., (1955). Is classical mechanics in fact deterministic? *Phys. Bllater* 11: 9, 49; reprinted in *Physics in My Generation*, p. 164, Springer, 1969.

Bountis, T. C., (1981). Period doubling, bifurcations, and universality in conservative systems, *Physica* **3D**, 577.

Bountis, T., and Helleman, R. H., (1981). On the stability of periodic orbits of two-dimensional mappings, *J. Math. Phys.* **22**, 1867.

Bowen, R., (1970a). Markov partitions and minimal sets for Axiom A diffeomorphisms, *Amer. J. Math.* **92**, 907.

Bowen, R., (1970b). Markov partitions for Axiom A diffeomorphisms, *Amer. J. Math.* **92**, 725.

Bowen, R., (1973a). Topological entropy for noncompact sets, *Trans. Amer. Math. Soc.* **184**, 125.

Bowen, R., (1973b). Symbolic dynamics for hyperbolic flows, *Amer. J. Math.* **95**, 429.

Bowen, R., and Ruelle, D., (1975). The ergodic theory of Axiom A flows, *Invent. Math.* **29**, 181.

Bowen, R., (1975a). A horseshoe with positive measure, *Invent. Math.* **29**, 203.

Bowen, R., (1975b). "Equilibrium states and the ergodic theory of Anosov diffeomorphisms," in *Lect. Notes in Math.* Vol. **470**, Springer-Verlag.

Bowen, R., (1978). On Axiom A diffeomorphisms. *CBMS Regional Conference Series in Mathematics*. Vol. **35**, A. M. S. Publications, Providence.

Brandstater, A., Swift, J., Swinney, H. L., Wolf, A., Farmer, J. D., Jen, E., and Crutchfield, J. P., (1983). Low-dimensional chaos in a hydrodynamic system, *Phys. Rev. Lett.* **51**, 1442.

Bridges, R., Rowlands, G., (1977). On the analytic form of some strange attractors, *Phys. Lett.* **63A**, 189.

Brahic, A., (1971). Numerical study of a simple dynamical system, *Astron. and Astrophys.* **12**, 98.

Brindley, J., and Moroz, I. M., (1980). Lorenz attractor behavior in a continuously stratified baroclinic fluid, *Phys. Lett.* **77A**, 441.

Broomhead, D., McCreadie, G., and Rowlands, G., (1981). On the analytic derivation of Poincare maps — the forced Brusselator problem, *Phys. Lett.* **84A**, 229.

Broomhead, D. S., and Rowlands, G., (1982). A simple derivation of the Mel'nikov condition for the appearance of homoclinic points, *Phys. Lett.* **89A**, 63.

Broomhead, D. S., and Rowlands, G., (1983). On the analytic treatment of nonintegrable difference equations, *J. Phys.* **A16**, 9.

Brorson, S. D., Dewey, D., and Linsay, P. S., (1983). Self-replicating attractor of a driven semiconductor oscillator, *Phys. Rev.* **A28**, 1201.

Brunner, W., Paul, H., (1983). Regular and chaotic behavior of multimode gas lasers, *Opt. Quantum Electron.*, **15**, 87.

Budinsky, N., and Bountis, T., (1983). Stability of nonlinear modes and chaotic properties of 1D Fermi-Ulam-Pasta lattices, *Physica* **8D**, 445.

Bunow, B., and Weiss, G. H., (1979). How chaotic is chaos? Chaotic and other 'noisy' dynamics in the frequency domain, *Math. Biosciences* **47**, 221.

Bussac, M. N., and Meunier, C., (1982). Statistical properties of type I intermittency, *J. de Phys.* **43**, 585.

Busse, F. H., (1981). "Transition to turbulence in Rayleigh-Benard convection," in *Hydrodynamic instabilities and the transition to turbulence*, ed. H. L. Swinney and J. P. Gollub, Springer Verlag, N. Y.

Campanino, M., (1980). Two remarks on the computer study of differentiable dynamical systems, *Commun. Math. Phys.* **74**, 15.

Campanino, M., and Epstein, H. D., (1981). On the existence of Feigenbaum's fixed point, *Commun. Maths. Phys.* **79**, 261.

Carey, R. F., Rooney, J. A., and Smith, C. W., (1979). Subharmonic response in liquid helium, *J. Acoust. Soc. Am.* **66**, 1801.

Carmichael, H. J., (1983). "Chaos in nonlinear optical systems," in *Laser Physics, Proc. 3rd New Zealand Symposium*, p. 64, Springer-Verlag.

Carmichael, H. J., Savage, C. M., and Walls, P. F., (1983). From optical tristability to chaos, *Phys. Rev. Lett.* **50**, 163.

Caroli, B., Caroli, C., and Roulet, B., (1983). Effect of a small slow modulation on Pomeau-Manneville intermittencies, *Phys. Lett.* **94A**, 117.

Cartwright, M. L., and Littlewood, J. E., (1945). On non-linear differential equations of the second order, *J. London Math. Soc.* **20**, 180.

Cartwright, M. L., (1948). Forced oscillations in nearly sinusoidal systems, *J. Inst. Elec. Eng.* **95**, 88.

Casartelli, M., Diana, E., Galgani, L., and Scotti, A., (1976). Numerical computation on a stochastic parameter related to the Kolmogorov entropy, *Phys. Rev.* **A13**, 1921.

Casati, G., and Comparin, G., (1982). Decay of correlations in certain hyperbolic systems, *Phys. Rev.* **A26**, 1.

Casati, G., and Guarneri, I., (1983). Chaos and special features of quantum systems under external noise, *Phys. Rev. Lett.* **50**, 640.

Cascais, J., Dilao, R., and Costa, A. N., (1983). Chaos and reverse bifurcation in a RCL circuit, *Phys. Lett.* **93A**, 213.

Chang, S. J., and Wright, J., (1981). Transitions and distribution functions for chaotic systems, *Phys. Rev.* **A23**, 1419.

Chang, S. J., Wortis, M., and Wright, J., (1981a). Iterative properties of a one dimensional quartic map, *Phys. Rev.* **A24**, 2669.

Chang, S. J., Wortis, M., and Wright, J., (1981b). "Tricritical points and bifurcation in a quartic map," in *Nonlinear problems: present and future*, ed. A. Bishop, D. Campbell, B. Nicolaenko, North-Holland.

Chang, Y. F., Tabor, M., Weiss, J., and Corliss, G., (1981). On the analytic structure of the Henon-Heiles system, *Phys. Lett.* **85A**, 211.

Chernoff, D. F., and Barrow, J. D., (1983). Chaos in the mixmaster universe, *Phys. Rev. Lett.* **50**, 134.

Chirikov, B. V., (1969). Research concerning the theory of nonlinear resonance and stochasticity, *Nuclear Phys. Inst. of Siberian section of Acad. Sci., USSR*, No. **267**, *CERN Trans.* 71–40 by A. T. Sanders, Geneva, 1971.

Chirikov, B. V., (1979). A universal instability of many oscillator systems, *Phys. Reports* **52**, 265.

Chirikov, B. V., Ford, J., and Vivaldi, F., (1979). "Some numerical studies of Arnold diffusion in a simple model," in *AIP Conf. Proc.* No. **57**, p. 323.

Chirikov, B. V., Israilev, F. M., and Shepelyansky, D. L., (1981). "Dynamical stochasticity in classical and quantum mechanics," in *Math. Phys. Rev.* ed. S. P. Novikov, Vol. **2**, p. 209.

Chirikov, B. V., Shepelyansky, D. L., (1981). Stochastic oscillations of classical Yang-Mills fields, *Pis'ma JETP* **34**, 171.

Chirikov, B. V., and Shepelyansky, D. L., (1982). Diffusion during multiple passage through a nonlinear resonance, *Sov. Phys.-Tech. Phys.* **27**, 156; Russian original: *JTF* **52**, 238.

Chrostowski, J., Vallee, R., and Delisle, C., (1983). Self-pulsing and chaos in acousto-optic bistability, *Can. J. Phys.* **61**, 1143.

Chui, S. T., and Ma, K. B., (1982). Nature of some chaotic states for Duffing's equation, *Phys. Rev.* **A26**, 2262.

Churchill, R. C., (1982). "On proving the non-integrability of a Hamiltonian system," in *The Riemann Problem, Complete Integrability and Arithmetic Applications, Lect. Notes in Math.* Vol. **925**, 103, Springer Verlag.

Collet, P., and Eckmann, J.-P., (1980a). On the abundance of aperiodic behavior for maps on the interval, *Commun. Math. Phys.* **73**, 115.

Collet, P., and Eckmann, J.-P., (1980b). On the abundance of chaotic behavior in one dimension, *Ann. N. Y. Acad. Sci.* **357**, 337.

Collet, P., and Eckmann, J.-P., (1980c). "Properties of continuous maps of the interval to itself," in *Math. Problems in Theor. Phys.*, ed. K. Osterwalder, *Lect. Notes in Phys.* **116**, p. 331, Springer Verlag, N. Y.

Collet, P., Eckmann, J.-P., and Lanford, O. E., (1980). Universal properties of maps on an interval, *Commun. Math. Phys.* **76**, 211.

Collet, P., Eckmann, J.-P., and Koch, H., (1981a). Period doubling bifurcations for families of maps on Rn, *J. Stat. Phys.* **25**, 1.

Collet, P., Eckmann, J.-P., and Koch, H., (1981b). On universality for area-preserving maps of the plane, *Physica* **3D**, 457.

Collet, P., Eckmann, J.-P., and Thomas, L., (1981c). A note on the power spectrum of the iterates of Feigenbaum's function, *Commun. Math. Phys.* **81**, 261.

Cook, A. E., and Roberts, P. H., (1970). The Rikitake two-disc dynamo system, *Proc. Camb. Phil. Soc.* **68**, 547.

Coste, J., (1980). Iterations of transformation on the unit interval: approach to aperiodic attractor, *J. Stat. Phys.* **23**, 521.

Coullet, P., and Tresser, C., (1978). Iterations d'endomorphismes et groupe de renormalisation, *J. de Phys.* **39**, Coll. C5–25.

Coullet, P., Tresser, C., and Arneodo, A., (1979). Transition to stochasticity for a class of forced oscillators, *Phys. Lett.* **72A**, 268.

Coullet, P., and Tresser, C., (1980). Critical transition to stochasticity for some dynamical systems, *J. de Phys. Lett.* **41**, L255.

Coullet, P., and Tresser, C., (1981). Some universal aspects of the transition to stochasticity for non-conservative dynamical systems, in *Field Theory, Quantization and Statistical Physics*, ed. E. Tirapegui, p. 249, D. Riedel, Boston.

Coullet, P., and Vanneste, C., (1983). Scenarios for the onset of chaos, *Helv. Phys. Acta* **56**, 813.

Cramer, E., and Lauterborn, W., (1982). Acoustic cavitation noise spectra, *Appl. Sci. Res.* **38**, 209.

Croquette, V., and Poitou, C., (1981). Cascade of period doubling bifurcations and large stochasticity in the motions of a compass, *J. de Phys. Lett.* **42**, 537.

Crutchfield, J. P., and Huberman, B. A., (1980). Fluctuations and the onset of chaos, *Phys. Lett.* **77A**, 407.

Crutchfield, J. P., Farmer, J. D., Packard, N. H., and Shaw, R. S., (1980). Power spectral analysis of a dynamical system, *Phys. Lett.* **76A**, 1.

Crutchfield, J. P., Nauenberg, M., and Rudnick, J., (1981). Scaling for external noise at the onset of chaos, *Phys. Rev. Lett.* **46**, 933.

Crutchfield, J. P., Farmer, J. D., and Huberman, B. A., (1982). Fluctuations and simple chaotic dynamics, *Phys. Reports* **92**, 45.

Crutchfield, J. P., and Packard, N. H., (1982). Symbolic dynamics of 1D maps: entropies, finite precision, and noise, *Int. J. Theor. Phys.* **21**, 433.

Crutchfield, J. P., and Packard, N. H., (1983). Symbolic dynamics of noisy chaos, *Physica* **7D**, 201.

Curry, J. H., and Yorke, J. A., (1978). "A transition from Hopf bifurcation to chaos: computer experiments on maps on R2," in *Structure of Attractors in Dynamical Systems*, ed. N. G. Markley, J. C. Martin, and W. Perrizo, *Lect. Notes in Math.* **668**, p. 48, Springer Verlag, N. Y.

Curry, J. H., (1978). A generalized Lorenz system, *Commun. Math. Phys.* **60**, 193.

Curry, J. H., (1979a). On the Henon transformation, *Commun. Math. Phys.* **68**, 129.

Curry, J. H., (1979b). An algorithm for finding closed orbit, in *Global Theory of Dynamical Systems*, ed. Z. Nitecki, C. Robinson, p. 111, *Lect. Notes in Math.* Vol. **819**, Springer-Verlag.

Curry, J. H., (1979c). Chaotic response to periodic modulation of a model of a convecting fluid, *Phys. Rev. Lett.* **43**, 1013.

Curry, J. H., (1980). "On some systems motivated by the Lorenz equations: numerical results," in *Math. Problems in Theor. Phys.*, ed. K. Osterwalder, *Lect. Notes in Phys.* **116**, p. 316, Springer Verlag.

Curry, J. H., (1981). On computing the entropy of the Henon attractor, *J. Stat. Phys.* **26**, 683.

Curry, J. H., and Johnson, J. R., (1982). On the rate of approach to homoclinic tangency, *Phys. Lett.* **92**, 5.

Cvitanovic, P., and Myrheim, J., (1983). Universality for period n-tuplings in complex mappings, *Phys. Lett.* **94A**, 329.

Da Costa, L. N., Knobloch, E., and Weiss, N. O., (1981). Oscillations in double diffusive convection, *J. Fluid Mech.* **109**, 25.

Daido, H., and Tomita, K., (1979). Thermal fluctuation of a self-oscillating reaction system entrained by a periodic external force, *Prog. Theor. Phys.* **61**, 825.

Daido, H., (1980). Analytic conditions for the appearance of homoclinic and heteroclinic points of a 2-dimensional mapping: the case of the Henon mapping, *Prog. Theor. Phys.* **63**, 1190, 1831.

Daido, H., (1981a). Theory of the period-doubling phenomenon of one-dimensional mappings based on the parameter dependence, *Phys. Lett.* **83A**, 246.

Daido, H., (1981b). Universal relation of a band splitting sequence to a preceding period-doubling one, *Phys. Lett.* **86A**, 259.

Daido, H., (1982a). Period-doubling bifurcations and associated universal properties including parameter dependence, *Prog. Theor. Phys.* **67**, 1698.

Daido, H., (1982b). On the sealing behavior in a map of a circle onto itself, *Prog. Theor. Phys.* **68**, 1935.

Daido, H., (1983a). Nonuniversal accumulation of bifurcations leading to homoclinic tangency, *Prog. Theor. Phys.* **69**, 1304.

Daido, H., (1983b). Resonance and intermittent transition from torus to chaos in periodically forced systems near intermittency threshold, *Prog. Theor. Phys.* **70**, 879.

Decroly, P., and Goldbeter, A., (1982). Birthythmicity, chaos, and other patterns of temporal self-organization in a multiply regulated biochemical system, *Proc. Natl. Acad. Sci. USA.* **79**, 6917.

De Leon, N., and Berne, B. J., (1981). Intramolecular rate process: isomerization dynamics and the transition to chaos, *J. Chem. Phys.* **75**, 3495.

De Leon, N., and Berne, B. J., (1982a). Reaction dynamics in an ergodic system: the Siamese stadium billiard, *Chem. Phys. Lett.* **93**, 162.

De Leon, N., and Berne, B. J., (1982b). Reaction dynamics in a non-ergodic system: The Siamese stadium billiard, *Chem. Phys. Lett.* **93**, 169.

Derrida, B., Gervois, A., and Pomeau, Y., (1977). Iteration d'endomorphismes de la droite reele et representation des nombres, *C. R. Acad. Sc. Paris* **285A**, 43.

Derrida, B., (1978). Proprietes universelles de certains systems discrets dans le temps, *J. de Phys.* **39**, Colloq. C5–49.

Derrida, B., Gervois, A., and Pomeau, Y., (1978). Iteration of endomorphisms on the real axis and representation of numbers, *Ann. Inst. Henri Poincare* **29A**, 305.

Derrida, B., Gervois, A., and Pomeau, Y., (1979). Universal metric properties of bifurcations of endomorphisms, *J. Phys.* **A12**, 269.

Derrida, B., and Pomeau, Y., (1980). Feigenbaum's ratios of two-dimensional area preserving maps, *Phys. Lett.* **80A**, 217.

Derrida, B., (1980). "Critical properties of one-dimensional mappings," in *Bifurcation Phenomena in Mathematical Physics and Related Topics*, ed. C. Bardos and D. Bessis, p. 137, D. Riedel, Boston.

Dewel, C., Walgraef, D., and Borckmans, P., (1981). Layered structures in two dimensional nonequilibrium systems, *Journal de Phys. Lett.* **42**, 361.

D'Humieres, D., Beasley, M. R., Huberman, B. A., and Libchaber, A., (1982). Chaotic states and routes to chaos in the forced pendulum, *Phys. Rev.* **A26**, 3483.

Dias De Deus, J., Dilao, R., and Taborta Durate, J., (1982). Topological entropy and approaches to chaos in dynamics of the interval, *Phys. Lett.* **90A**, 1.

Dias De Deus, J., Dilao, R., and Taborta Durate, J., (1983). Topological entropy characteristic exponents and scaling behavior in dynamics of the interval, *Phys. Lett.* **93A**, 1.

Doolen, G. D., DuBois, D. F., Rose, H. A., and Hafizi, B., (1983). Coherence in chaos and caviton turbulence, *Phys. Rev. Lett.* **51**, 335.

Doveil, F., and Escande, D. F., (1981). Destabilization of an enumerable set of cycles in a Hamiltonian system, *Phys. Lett.* **84A**, 399.

Doveil, F., and Escande, D. F., (1982). Fractal diagrams for non-integrable Hamiltonians, *Phys. Lett.* **90A**, 226.

Dowell, E. H., (1982). Flutter of a buckled plate as an example of chaotic motion of a deterministic autonomous system, *J. Sound Vib.* **85**, 333.

Dubois, M., and Berge, P., (1978). Experimental study of the velocity field in Rayleigh-Benard convection, *J. Fluid Mech.* **85**, 641.

Dubois, M., and Berge, P., (1979). "Velocity field in Rayleigh-Benard instability: transitions to turbulence," in *Synergetics — far from equilibrium*, ed. A. Pacault and C. Vidal, Springer Verlag, N. Y.

Dubois, M., and Berge, P., (1980). Experimental evidence for the oscillators in a convective biperiodic regime, *Phys. Lett.* **76A**, 53.

Dubois, M., and Berge, P., (1981). Instabilites de couche limite dans un fluide en convection. Evolution vers la turbulence, *J. de Phys.* **42**, 167.

Dubois, M., Rubio, M. A., and Berge, P., (1983). Experimental evidence of intermittencies associated with a subharmonic bifurcation, *Phys. Rev. Lett.* **51**, 1446; Erratum, ibid, **51**, 2345.

Eckmann, J.-P., (1981). Roads to turbulence in dissipative dynamical systems, *Rev. Mod. Phys.* **53**, 643.

Eckmann, J.-P., Thomas, L., and Wittwer, P., (1981). Intermittency in the presence of noise, *J. Phys.* **A14**, 3153.

Eckmann, J.-P., and Thomas, L. E., (1982). Remarks on stochastic resonances, *J. Phys.* **A15**, L261.

Eckmann, J.-P., Koch, H., and Wittwer, P., (1982). Existence of a fixed point of the doubling transformation for area-preserving maps of the plane, *Phys. Rev.* **A26**, 720.

Elgin, J. N., and Forster, D., (1983). Mechanism for chaos in the Duffing's equation, *Phys. Lett.* **94A**, 195.

Englund, J. C., Snapp, R. R., and Schieve, W. C., (1983). Fluctuations, instabilities and chaos in the laser-driven nonlinear ring cavity, in *Progress in Optics XXI*, ed. E. Wolf, North-Holland.

Epstein, H., Lascoux, J., (1981). Analyticity properties of the Feigenbaum Equation, *Commun. Math. Phys.* **81**, 437.

Epstein, I. R., (1983). Oscillations and chaos in chemical systems, *Physica* **7D**, 47.

Erber, T., Johnson, P., and Everett, P., (1981). Cebysev mixing and harmonic oscillator models, *Phys. Lett.* **85A**, 61.

Escande, D. F., and Doveil, F., (1981a). Renormalization method for the onset of stochasticity in a Hamiltonian system, *Phys. Lett.* **83A**, 307.

Escande, D. F., and Doveil, F., (1981b). Renormalization method for computing the threshold of the large-scale stochastic instability in two degrees of freedom Hamiltonian systems, *J. Stat. Phys.* **26**, 257.

Escande, D. F., (1982a). Renormalization approach to nonintegrable Hamiltonians, *Phys. Lett.* **91A**, 327.

Escande, D. F., (1982b). Renormalization for stochastic layers, *Physica* **6D**, 119.

Escande, D. F., (1982c). Large-scale stochasticity in Hamiltonian systems, *Phys. Script.* **T2**, 126.

Escande, D. F., and Mehr, A., (1982). Link between KAM tori and nearby cycles, *Phys. Lett.* **91A**, 327.

Farmer, J. D., Crutchfield, J. P., Froehling, H., Packard, N. H., and Shaw, R. S., (1980). Power spectra and mixing properties of strange attractors, *Annals N. Y. Acad. Sci.* **357**, 453.

Farmer, J. D., (1981). Spectral broadening of period doubling bifurcation sequences, *Phys. Rev. Lett.* **47**, 179.

Farmer, J. D., (1982a). Chaotic attractors of an infinite dimensional systems, *Physica* **4D**, 366.

Farmer, J. D., (1982b). Information dimension and the probabilistic structure of chaos, *Z. Naturforsch.* **37a**, 1304.

Farmer, J. D., Ott, E., Yorke, J. A., (1983). The dimension of chaotic attractors, *Physica* **7D**, 153.

Farmer, J. D., (1982). "Dimension, fractal measures, and chaotic dynamics," in *Evolution of Order and Chaos in Physics, Chemistry, and Biology*, H. Haken, ed. Springer-Verlag.

Farmer, D., Hart, J., and Weidman, P., (1982). A phase space analysis of baroclinic flow, *Phys. Lett.* **91A**, 22.

Fatou, M. P., (1919–1920). Sur les equations fonctionelles, *Bull. Societe Math. de France* **47**, 161; **48**, 33; **48**, 208.

Fauve, S., Laroche, C., and Libchaber, A., (1981). Effect of a horizontal magnetic field on convective instabilities in mercury, *J. de Phys. Lett.* **42**, 455.

Fauve, S., and Libchaber, A., (1981). "Rayleigh-Benard experiment in a low Prandtl number fluid, mercury," in *Chaos and Order in Nature*, ed. H. Haken, p. 25, Springer Verlag.

Feigebaum, M. J., (1978). Quantitative universality for a class of nonlinear transformations, *J. Stat. Phys.* **19**, 25.

Feigenbaum, M. J., (1979). The universal metric properties of nonlinear transformations, *J. Stat. Phys.* **21**, 669.

Feigenbaum, M. J., (1980a). Universal behavior in nonlinear systems, *Los Alamos Sci. Summer*, 4.

Feigenbaum, M. J., (1980b). The onset spectrum of turbulence, *Phys. Lett.* **74A**, 375.

Feigenbaum, M. J., (1980c). The transition to aperiodic behavior in turbulent systems, *Commun. Math. Phys.* **77**, 65.

Feigenbaum, M. J., (1980d). The metric universal properties of period-doubling bifurcations and the spectrum for a route to turbulence, *Ann. N. Y., Acad. Sci.* **357**, 330.

Feigenbaum, M. J., (1981). "Tests of the period-doubling route to chaos," in *Nonlinear Phenomena in Chemical Dynamics*, ed. C. Vidal, A. Pacault, p. 95, Springer Verlag.

Feigenbaum, M. J., Kadanoff, L. P., and Shenker, S. J., (1982). Quasiperiodicity in dissipative systems: a renormalization group analysis, *Physica* **5D**, 370.

Feigenbaum, M. J., and Hasslacher, B., (1982). Irrational decimations and path intergrals for external noise, *Phys. Rev. Lett.* **49**, 605.

Feigenbaum, M. J., (1983). Universal behavior in nonlinear systems, *Physica* **7D**, 16.

Feingold, M., and Perez, A., (1983). Regular and chaotic motion of coupled rotators, *Physica* **9D**, 433.

Feit, S. D., (1978). Characteristic exponents and strange attractors, *Commun. Math. Phys.* **61**, 249.

Fenstermacher, P. R., Swinney, H. L., Benson, S. V., and Gollub, J. P., (1979a). Bifurcations to periodic quasiperiodic, and chaotic regimes in rotating and convecting fluids, *Ann. N. Y., Acad. Sci.* **316**, 652.

Fenstermacher, P. R., Swinney, H. L., and Gollub, J. P., (1979b). Dynamical instabilities and the transition to chaotic Taylor vortex flow, *J. Fluid Mech.* **94**, 103.

Firth, W. J., and Wrigth, E. M., (1982). Oscillation and chaos in a Fabry-Perot bistable cavity with Gaussian input beam, *Phys. Lett.* **92**, 211.

Fisher, M. E., and Huse, D. A., (1982). "Melting, order, flows, mappings and chaos," in *Melting, Localization and Chaos*, ed. by R. K. Kalia and P. Vashita, Elsevier Science Publishing Co.

Fishman, S., Grempel, D. R., and Prange, R. E., (1982). Chaos, quantum recurrences, and Anderson localization, *Phys. Rev. Lett.* **49**, 509.

Flaherty, J. E., and Hoppensteadt, F. C., (1978). Frequency entrainment of a forced van der Pol oscillator, *Stud. Appl. Math.* **58**, 5.

Ford, J., and Lunsford, G. H., (1970). Stochastic behavior of resonant nearly linear oscillator systems as the nonlinear coupling approaches zero, *Phys. Rev.* **A1**, 59.

Ford, J., and Lunsford, G. H., (1972). On the stability of periodic orbits for nonlinear oscillator systems in regions exhibiting stochastic behavior, *J. Math. Phys.* **13**, 700.

Ford, J., (1973). The transition from analytic dynamics to statistical mechanics, *Adv. Chem. Phys.* **24**, 155.

Ford, J., (1974). Stochastic behavior in nonlinear oscillator systems, in *Lect. Notes in Phys.* No. **28**, Springer-Verlag.

Ford, J., (1975). "The statistical mechanics of classical analytic dynamics," in *Fundamental problems in statistical mechanics III*, ed. E. G. D. Cohen, p. 215, North Holland, Amsterdam.

Ford, J., (1983). "How random is a coin toss?" *Phys. Today*, April, 40.

Fowler, A. C., Gibbon, J. D., and McGuinness, M. J., (1982). The complex Lorenz equations, *Physica* **4D**, 139.

Fowler, A. C., and McGuinness, M. J., (1982). Hysteresis in the Lorenz equations, *Phys. Lett.* **92A**, 103.

Fowler, A. C., (1983). Note on a paper by G. Rowlands (chaotic trajectories of ODE's), *J. Phys.* **A16**, 3139.

Fowler, A. C., Gibbon, J. D., and McGuinness, M. J., (1983). The real and complex Lorenz equations and their relevance to physical systems, *Physica* **7D**, 126.

Fradkin, E., and Hernandez, O., Huberman, B. A., Pandit, R., (1983). Periodic, incomcomensurate, and chaotic states in a continuum statistical mechanics model, *Nucl. Phys.* **B215 (FS7)**, 137.

Franceschini, V., and Tebaldi, C., (1979). Sequences of infinite bifurcations and turbulence in a five-mode truncation of the Navier-Stokes equations, *J. Stat. Phys.* **21**, 707.

Franceschini, V., (1980). A Feigenbaum sequence of bifurcations in the Lorenz model, *J. Stat. Phys.* **22**, 397.

Franceschini, V., and Tebaldi, C., (1981). A seven-mode truncation of the plane incompressible Navier-Stokes equations, *J. Stat. Phys.* **25**, 397.

Franceschini, V., and Russo, L., (1981). Stable and unstable manifolds of the Henon mapping, *J. Stat. Phys.* **25**, 757.

Franceschini, V., (1983). Bifurcations of tori and phase locking in a dissipative system of differential equations, *Physica* **6D**, 285.

Fraser, S., and Kapral, R., (1981). A resonance model for chaos near period three, *Phys. Rev.* **A23**, 3303.

Fraser, S., and Kapral, R., (1982a). Behavior of the Lyapunov number near period three, *Phys. Rev.* **A25**, 2827.

Fraser, S., and Kapral, R., (1982b). Analysis of flow hysteresis by a one-dimensional map, *Phys. Rev.* **A25**, 3223.

Fraser, S., Celarier, E., and Kapral, R., (1983). Stochastic dynamics of the cubic map: a study of noise-induced transition phenomena, *J. Stat. Phys.* **33**, 341.

Froehling, H., Crutchfield, J. P., Farmer, J. D., Packard, N. H., and Shaw, R. S., (1981). On determining the dimension of chaotic flows, *Physica* **3D**, 605.

Froyland, J., (1983). Lyapunov exponents for multidimensional orbits, *Phys. Lett.* **97A**, 8.

Fujisaka, H., and Yamada, T., (1977). Theoretical study of a chemical turbulence, *Prog. Theor. Phys.* **57**, 734.

Fujisaka, H., and Yamada, T., (1978a). Trajectory instability and strange attractors in a discrete model exhibiting chaotic behavior, *Phys. Lett.* **66A**, 450.

Fujisaka, H., and Yamada, T., (1978b). Theoretical study of time correlation functions in a discrete chaotic process, *Z. Naturforsch.* **33a**, 1455.

Fujisaka, H., and Yamada, T., (1980). Limit cycles and chaos in realistic models of the Belousov-Zhabotinskil reaction system, *Z. Physik* **B37**, 265.

Fujisaka, H., (1982). Multiperiodic flows, chaos and Lyapunov exponents, *Prog. Theor. Phys.* **68**, 1105.

Fujisaka, H., and Grossmann, S., (1982). Chaos-induced diffusion in nonlinear discrete dynamics, *Z. Phys.* **B48**, 261.

Fujisaka, H., (1983). Statistical dynamics generated by fluctuations of local Lyapunov exponents, *Prog. Theor. Phys.* **70**, 1264.

Fujisaka, H., and Yamada, T., (1983). Stability theory of synchronized motion in coupled oscillator systems, *Prog. Theor. Phys.* **69**, 32.

Fujisaka, H., Kamifukumoto, H., and Inoue, M., (1983). Intermittency associated with the breakdown of the chaos symmetry, *Prog. Theor. Phys.* **69**, 333.

Galavotti, G., (1982). A criterion of integrability for perturbed nonresonant harmonic oscillators. 'Wick ordering' of the perturbations in classical mechanics and invariance of the frequency spectrum, *Commun. Math. Phys.* **87**, 365.

Galgani, L., (1982). Planck's formula for classical oscillator with stochasticity thresholds, *Lett. Nuovo Cimento* **35**, 93.

Gambaudo, J. M., and Tresser, C., (1983). Simple models for bifurcations creating horseshoes, *J. Stat. Phys.* **32**, 455.

Gaspard, P., (1983). Generation of a countable set of homoclinic flows through bifurcation, *Phys. Lett.* **97A**, 1.

Gaspard, P., and Nicolis, G., (1983). What can we learn from homoclinic orbits in chaotic dynamics? *J. Stat. Phys.* **31**, 499.

Gao, J. Y., Yuan, J. M., and Narducci, L. M., (1983a). Instabilities and chaotic behavior in a hybrid bistable system with a short delay, *Opt. Commun.* **44**, 201.

Gao, J. Y., Narducci, L. M., Schulman, L. S., Squicciarini, M., and Yuan, J. M., (1983b). Route to chaos in a hybrid bistable system with delay, *Phys. Rev.* **A28**, 2910.

GBL, (1981). Period-doubling route to chaos shows universality, *Phys. Today*, March, 17.

Geisel, T., and Nierwetberg, J., (1981). A universal fine structure of the chaotic region in period-doubling systems, *Phys. Rev. Lett.* **47**, 975.

Geisel, T., Nierwetberg, J., and J. Keller, (1981). Critical behavior of the Lyapunov number at the period-doubling onset of chaos, *Phys. Lett.* **86A**, 75.

Geisel, T., and Nierwetberg, J., (1982). Onset of diffusion and universal scaling in chaotic systems, *Phys. Rev. Lett.* **48**, 7.

Ghikas, D. P., and Nicolis, G., (1982). Stochasticity from deterministic dynamics: an explicit example of generation of Markovian strings, *Z. Phys.* **B47**, 279.

Ghikas, D. P., (1983). A method of approximation of invariant measures for maps on the unit interval, *Lett. Math. Phys.* **7**, 91.

Gibbon, J. D., and McGuinness, M. J., (1980). A derivation of the Lorenz equations for some unstable dispersive physical systems, *Phys. Lett.* **77A**, 295.

Gibbon, J. D., and McGuinness, M. J., (1982). The real and complex Lorenz equations in rotating fluids and lasers, *Physica* **5D**, 8.

Gibbs, H. M., Hopf, F. A., Kaplan, D. L., and Shoemaker, R. L., (1981). Observation of chaos in optical bistability, *Phys. Rev. Lett.* **46**, 474.

Giglio, M., Musazzi, S., and Perini, U., (1981). Transition to chaotic behavior via a reproducible sequence of period-doubling bifurcations, *Phys. Rev. Lett.* **47**, 243.

Gilgio, M., Musazzi, S., and Perini, U., (1983). "Transition to deterministic chaos in a hydrodynamic system," in *Statics and Dynamics of Nonlinear Systems* ed. G. Benedeck, H. Bilz, R. Zeyher, p. 189, Springer-Verlag.

Glass, L., and Perez, R., (1982). Fine structure of phase locking, *Phys. Rev. Lett.* **48**, 1772.

Glass, L., Guevara, M. R., and Shrier, A., Perez, R., (1983). Bifurcation and chaos in a periodically stimulated cardiac oscillator, *Physica* **7D**, 89.

Golberg, A. I., Sinai, Ya. G., and Khanin, K. M., (1983). Universal properties of sequences of period-triplings, *Usp. Mat. Nauk.* **38**, 159.

Goldbeter, A., and Decroly, O., (1983). Temporal self-organization in biochemical systems: periodic behavior vs. chaos, *Amer. J. Physiology* **245**, R478.

Gollub, J. P., and Swinney, H. L., (1975). Onset of turbulence in a rotating fluid, *Phys. Rev. Lett.* **35**, 927.

Gollub, J. P., and Benson, S. V., (1978). Chaotic response to periodic perturbation of a convecting fluid, *Phys. Rev. Lett.* **41**, 948.

Gollub, J. P., Brunner, T. O., and Danly, B. G., (1978). Periodicity and chaos in coupled nonlinear oscillators, *Science* **200**, 48.

Gollub, J. P., and Benson, S. V., (1979). "Phase-locking in the oscillations leading to turbulence," in *Pattern Formation and Pattern Recognition*, ed. H. Haken, p. 74, Springer Verlag, N. Y.

Gollub, J. P., and Benson, S. V., (1980). Many routes to turbulent convection, *J. Fluid Mech.* **100**, 449.

Gollub, J. P., Benson, S. V., and Steinman, J. F., (1980). A subharmonic route to turbulent convection, *Ann. N. Y. Acad. Sci.* **357**, 22.

Gollub, J. P., Romer, E. J., and Socolar, J. E., (1980). Trajectory divergence for coupled relaxation oscillators: measurements and models, *J. Stat. Phys.* **23**, 321.

Gollub, J. P., (1980). The onset of turbulence: convection, surface waves, and oscillators, *Proc. Sitges Int. School on Stat. Mech.*, ed. L. Garrido, *Lect. Notes in Phys.* **132**, p. 162, Springer Verlag, N. Y.

Gollub, J. P., (1983). What causes noise in a convecting fluid, *Physica* **118A**, 28.

Gonzalez, D. L., and Piro, O., (1983). Chaos in a nonlinear driven oscillator with exact solution, *Phys. Rev. Lett.* **50**, 870.

Gorman, M., Swinney, H. L., and Rand, D. A., (1981). Doubly periodic circular Couette flow: experiments compared with prediction from dynamics and symmetry, *Phys. Rev. Lett.* **46**, 992.

Graham, R., and Scholz, H. J., (1980). Analytic approximation of the Lorenz attractor by invariant manifolds, *Phys. Rev.* **22A**, 1198.

Graham, R., (1983a). Exact solution of some discrete stochastical models with chaos, *Phys. Rev. Lett.* **A28**, 1679.

Graham, R., (1983b). Quantization of a two-dimensional map with a strange attractor, *Phys. Lett.* **99A**, 131.

Grasman, J., Veling, E. J. M., and Willems, G. M., (1976). Relaxation oscillations governed by a van der Pol equation with periodic forcing term, *SIAM J. Appl. Math.* **31**, 667.

Grassberger, P., (1981). On the Hausdorff dimension of fractal attractors, *J. Stat. Phys.* **26**, 173.

Grassberger, P., and Scheunert, M., (1981). Some more universal scaling laws for critical mappings, *J. Stat. Phys.* **26**, 697.

Grassberger, P., (1983a). New mechanism for deterministic diffusion, *Phys. Rev.* **A28**, 3666.

Grassberger, P., (1983b). On the fractal dimension of the Henon attractor, *Phys. Lett.* **97A**, 224.

Grassberger, P., (1983c). Generalized dimensions of strange attractors, *Phys. Lett.* **97A**, 227.

Grassberger, P., and Procaccia, I., (1983a). Measuring the strangeness of strange attractors, *Physica* **9D**, 189.

Grassberger, P., and Procaccia, I., (1983b). Characterization of strange attractors, *Phys. Rev. Lett.* **50**, 346.

Grassberger, P., and Procaccia, I., (1983c). Estimation of the Kolmogorov entropy from a chaotic signal, *Phys. Rev.* **A28**, 2591.

Graziani, R. R., Hudson, J. L., and Schmitz, R. A., (1976). The Belousov-Zhabotinskii reaction in a continuous flow reactor, *Chem. Eng. J.* **12**, 9.

Grebogi, C., and Kaufman, A. N., (1981). Decay of statistical dependence in chaotic orbits of deterministic mappings, *Phys. Rev.* **A24**, 2829.

Grebogi, C., Ott, E., and Yorke, J. A., (1982). Chaotic attractors in crisis, *Phys. Rev. Lett.* **48**, 1507.

Grebogi, C., Ott, E., and Yorke, J. A., (1983a). Crises, sudden changes in chaotic attractors, and transient chaos, *Physica* **7D**, 181.

Grebogi, C., Ott, E., and Yorke, J. A., (1983b). Are three-frequency quasiperiodic orbits to be expected in typical nonlinear dynamical systems? *Phys. Rev. Lett.* **51**, 339.

Grebogi, C., Ott, E., and Yorke, J. A., (1983c). Fractal basin boundaries, long-lived chaotic transients, and unstable-unstable pair bifurcation, *Phys. Rev. Lett.* **50**, 935.

Grebogi, C., McDonald, S. W., Ott, E., and Yorke, J. A., (1983). Final state sensitivity: an obstruction to predictibility, *Phys. Lett.* **99A**, 415.

Greene, J. M., (1968). Two-dimensional measure-preserving mappings, *J. Math. Phys.* **9**, 760.

Greene, J. M., (1979). A method for determining a stochastic transition, *J. Math. Phys.* **20**, 1183.

Greene, J. M., and Percival, I., (1981). Hamiltonian maps in the complex plane, *Physica* **3D**, 530.

Greene, J. M., (1980). The calculation of KAM surfaces, *Ann. N. Y. Acad. Sci.* **357**, 80.

Greene, J. M., Mackay, R. S., Vivaldi, F., and Feigenbaum, M. J., (1981). Universal behavior in families of area preserving maps, *Physica* **3D**, 468.

Greenside, H. S., Wolf, A., Swift, J., and Pignataro, T., (1982). Impracticality of a box-counting algorithm for calculating the dimensionality of strange attractors, *Phys. Rev.* **A25**, 3453.

Greenspan, B., and Holmes, P. J., (1982). "Homoclinic orbits, subharmonics and global bifurcations in forced oscillations," in *Nonlinear Dynamics and Turbulence*, p. 172, ed. G. Barenblatt, G. Iooss, D. D. Joseph, Pitman, London.

Gregorio, S. De., Scoppola, E., and Tirozzi, B., (1983). A rigorous study of periodic orbits by means of a computer, *J. Stat. Phys.* **32**, 25.

Grossmann, S., and Thomae, S., (1977). Invariant distributions and stationary correlation functions of the one-dimensional discrete processes, *Z. Naturforsch.* **32a**, 1353.

Grossmann, S., and Fujisaka, H., (1982). Diffusion in discrete nonlinear dynamical systems, *Phys. Rev.* **A26**, 1779.

Grossmann, S., (1983). Shape dependence of correlation times in chaos-induced diffusion, *Phys. Lett.* **97A**, 263.

Guckenheimer, J., (1976). "A strange, strange attractor," in *The Hopf Bifurcation and Its Application*, ed. J. E. Marsden and M. McCracken, *Appl. Math. Sci.* **19**, p. 368, Springer Verlag, N. Y.

Guckenheimer, J., Oster, G., and Ipaktchi, A., (1977). Dynamics of density dependent population models, *J. Math. Biol.* **4**, 101.

Guckenheimer, J., (1977). Bifurcations of maps of the interval, *Inventiones Math.* **39**, 165.

Guckenheimer, J., (1979a). Sensitive dependence on initial conditions for one-dimensional maps, *Commun. Math. Phys.* **70**, 133.

Guckenheimer, J., (1979b). The bifurcation of quadratic functions, *Ann. N. Y. Acad. Sci.* **316**, 78.

Guckenheimer, J., and Williams, R., (1979). Structural stability of the Lorenz attractor, *Publ. Math. IHES* **50**, 307.

Guckenheimer, J., (1980a). One-dimensional dynamics, *Ann. N. Y. Acad. Sci.* **357**, 343.

Guckenheimer, J., (1980b). Dynamics of the van der Pol equation, *IEEE Trans. on Circuits and Systems* **T-CAS 27**, 983.

Guckenheimer, J., (1980). Symbolic dynamics and relaxation oscillations, *Physica* **1D**, 227.

Guckenheimer, J., (1981a). "On a condimension two bifurcation," in *Lect. Notes Math.* **898**, ed. D. A. Rand and L. S. Young, p. 99, Springer, Berlin.

Guckenheimer, J., (1981b). "Instabilities and chaos in nonhydrodynamical systems," in *Hydrodynamic Instabilities and the Transition to Turbulence*, ed. H. L. Swinney, J. P. Gollub, p. 271, Springer Verlag.

Guckenheimer, J., (1982). Noise in chaotic systems, *Nature* **298**, 358.

Guckenheimer, J., (1983). Persistent properties of bifurcations, *Physica* **7D**, 105.

Guckenheimer, J., and Buzyma, G., (1983). Dimension measurements for geostrophic turbulence, *Phys. Rev. Lett.* **51**, 1438.

Guevara, M. R., Glass, L., and Shrier, A., (1981). Phase locking, period-doubling bifurcations, and irregular dynamics in periodically stimulated cardiac cells, *Science* **214**, 1350.

Guevara, M. R., and Glass, L., (1982). Phase locking, period-doubling bifurcations and chaos, *J. Math. Biol.* **14**, 1.

Gumowski, I., and Mira, C., (1980). Recurrences and discrete dynamic systems, *Lect. Notes in Math.* **809**, Springer Verlag, N. Y.

Gutkin, E., (1983). Propagation of chaos and the Burgers equation, *SIAM J. Appl. Math.* **43**, 971.

Haken, H., (1975). Analogy between higher instabilities in fluids and lasers, *Phys. Lett.* **53A**, 77.

Haken, H., and Mayer-Kress, G., (1981a). Chapman-Kolmogorov equation for discrete chaos, *Phys. Lett.* **84A**, 159.

Haken, H., and Mayer-Kress, G., (1981b). Chapman-Kolmogorov equation and path integrals for discrete chaos in presence of noise. *Z. Phys.* **43B**, 185.

Haken, H., and Wunderlin, A., (1977). New interpretation and size of strange attractor of the Lorenz model of turbulence, *Phys. Lett.* **62A**, 133.

Haken, H., and Wunderlin, A., (1982). Some exact results on discrete noisy maps *Z. Phys.* **B46**, 181.

Haken, H., (1983). At least one Lyapunov exponent vanishes if the trajectory of an attractor does not contain a fixed point, *Phys. Lett.* **94A**, 71.

Halas, N. J., Liu, S.-N., and Abraham, N. B., (1983). Route to mode locking in a three-mode He-Ne 3.39 μm laser including chaos in the secondary beat frequency, *Phys. Rev.* **A28**, 2915.

Hamilton, I., and Brumer, P., (1982). Relaxation rates for two-dimensional deterministic mappings, *Phys. Rev.* **A25**, 3457.

Hao, B.-L., (1981). Universal slowing-down exponent near period-doubling bifurcation points, *Phys. Lett.* **86A**, 267.

Hao, B.-L., (1982). Two kinds of entrainment-beating transitions in a driven limit cycle oscillator, *J. Theor. Biol.* **98**, 9.

Hao, B.-L., and Zhang, S.-Y., (1982a). Subharmonic stroboscopy as a method to study period-doubling bifurcations, *Phys. Lett.* **87A**, 267.

Hao, B.-L., and Zhang, S.-Y., (1982b). Hierarchy of chaotic bands and periodicities embedded in them in a forced nonlinear oscillator, *Commun. Theor. Phys.* **1**, 111.

Hao, B.-L., and Zhang, S.-Y., (1982c). Hierarchy of chaotic bands, *J. Stat. Phys.* **28**, 769.

Hao, B.-L., and Zhang, S.-Y., (1983). Subharmonic stroboscopic sampling method for study of period-doubling bifurcation and chaotic phenomena in forced nonlinear oscillators, *Acta Physica Sinica* **32**, 198.

Hao, B.-L., Wang, G.-R., Zhang, S.-Y., (1983). U-sequences in the periodically forced Brusselator, *Commun. Theor. Phys.* **2**, 1075.

Hao, B.-L., (1983). Bifurcation, chaos, strange attractor, turbulence and all that, *Progress in Physics* **3**, 329.

Harrison, R. G., Firth, W. J., Emshary, C. A., and Al-saidi, I. A., (1983). Observation of period-doubling in an all-optical resonator containing NH3 gas, *Phys. Rev. Lett.* **51**, 562.

Haucke, H., and Maeno, Y., (1983). Phase space analysis of convection in a He3- superfluid He4 solution, *Physica* **7D**, 69.

Heldstab, J., and Thomae, H., (1983). Linear and nonlinear response of discrete dynamical systems. I. Periodic attractor, *Z. Phys.* **B50**, 141.

Helleman, R. H. G., (1979). "Exact results for some linear and nonlinear beam-beam effects," in *Nonlinear Dynamics and the Beam-Beam Interaction*, ed. M. Month. and J. C. Herrera, *AIP Proc.* **57**, 236.

Helleman, R. H. G., and Bountis, T., (1979). "Periodic solutions of arbitrary period: variational methods," in *Stochastic Behavior in Classical and Quantum Hamiltonian Systems*, ed. G. Casati and J. Ford, *Lect. Notes in Phys.* **93**, Springer-Verlag, N. Y.

Helleman, R. H. G., (1980). "Self-generated chaotic behavior in nonlinear mechanics," in *Fundamental Problems in Stat. Mech. V*, ed. E. G. D. Cohen, p. 165, North-Holland, Amsterdam.

Helleman, R. H. G., (1982). "One mechanism for the onsets of large-scale chaos in conservative and dissipative systems," in *Long-Time Prediction in Dynamics*, ed. W. Horton, L. Reichl and V. Szebehely, p. 95, John Wiley.

Henon, M., and Heiles, C., (1964). The applicability of the third integral of the motion, some numerical experiments, *Astron. J.* **69**, 73.

Henon, M., (1976). A two-dimensional mapping with a strange attractor, *Commun. Math. Phys.* **50**, 69.

Henon, M., and Pomeau, Y., (1977). "Two strange attractors with a simple structure," in *Turbulence and the Navier-Stokes Equations*, ed. R. Teman, *Lect. Notes in Math.* **565**, Springer Verlag, N. Y. p. 29.

Henon, M., (1982). On the numerical computation of Poincare maps, *Physica* **5D**, 412.

Hentschel, H. G. E., and Procaccia, I., (1983). The infinite number of generalized dimensions of fractals and strange attractors, *Physica* **8D**, 435.

Herring, C., and Huberman, B. A., (1980). Dislocation motion and solid-state turbulence, *Appl. Phys. Lett.* **36**, 975.

Hirsch, J. E., Huberman, B. A., and Scalapino, D. J., (1982a). A theory of intermittence, *Phys. Rev.* **A25**, 519.

Hirsch, J. E., Nauenberg, M., and Scalapino, D. J., (1982b). Intermittency in the presence of noise: a renormalization group formulation, *Phys. Lett.* **87A**, 391.

Hirschman, S. P., and Whiteson, J. C., (1982). Spectral analysis of noisy linear maps, *Phys. Fluids* **25**, 967.

Hitzl, D. L., (1981). Numerical determination of the capture escape boundary for the Henon attractor, *Physica* **2D**, 370.

Hofstadter, D. R., (1981). Strange attractors: mathematical patterns delicately poised between order and chaos, *Science* **245**:5, 22.

Hogg, T., and Huberman, B. A., (1982). Recurrence phenomena in quantum dynamics *Phys. Rev. Lett.* **48**, 711.

Hogg, T., and Huberman, B. A. (1983). Quantum dynamics and nonintegrability, *Phys. Rev.* **A28**, 22.

Holmes, C., and Holmes, P. J., (1981). Second order averaging and bifurcations to subharmonics in Duffing's equation, *J. Sound and Vib.* **78**, 161.

Holmes, P., (1977). 'Strange' phenomena in dynamical systems and their physical implications, *Appl. Math. Modelling* **1**, 362.

Holmes, P., and Rand, D. A., (1978). Bifurcations of the forced van der Pol oscillator, *Quart. Appl. Math.* **35**, 495.

Holmes, P., (1979a). A nonlinear oscillator with a strange attractor, *Phil. Trans. Roy. Soc.* **A292**, 420.

Holmes, P. J., (1979b). Domains of stability in a wind induced oscillation problem, *Trans. ASME. J. Appl. Mech.* **46**, 672.

Holmes, P., and Marsden, J. E., (1979). Qualitative techniques for bifurcation analysis of complex systems, *Ann. N. Y. Acad. Sci.* **316**, 608.

Holmes, P., (1980a). Averaging and chaotic motions in forced oscillations, *SIAM J. Appl. Math.* **38**, 65; cf. ibid, **40**, 167.

Holmes, P., (1980b). Unfolding a degenerate nonlinear oscillator, *Ann. N. Y. Acad. Sci.* **357**, 473.

Holmes, P., (1980c). A strange family of three-dimensional vector fields near a degenerate singularity, *J. Diff. Eqns.* **37**, 382.

Holmes, P., (1981). "Space- and time-periodic perturbations of the sine-Gordon equations," in *Dynamical Systems and Turbulence*, ed. D. A. Rand and L.-S. Young, p. 164. *Lect. Notes in Math.* Vol. **898**. Springer-Verlag.

Holmes, P., and Marsden, J. E., (1981). A partial differential equation with infinitely many periodic orbits: chaotic oscillations of a forced beam, *Arch. Rat. Mech. Anal.* **76**, 135.

Holmes, P., (1982). Proof of non-integrability for the Henon-Heiles Hamiltonian near an exceptional integrable case, *Physica* **5D**, 335.

Holmes, P., and Marsden, J. E., (1982a). Melnikov's method and Arnold diffusion for perturbations of integrable Hamiltonian systems, *J. Math. Phys.* **23**, 669.

Holmes, P., and Marsden, J. E., (1982b). Horseshoes in perturbations of Hamiltonian systems with two degrees of freedom, *Commun. Math. Phys.* **82**, 523.

Holmes, P., and Marsden, J. E., (1983). Horseshoes and Arnold diffusion for Hamiltonian systems on Lie groups, *Indiana U. Math. J.* **32**, 273.

Holmes, P., and Whitley, D., (1983a). "On the attracting set for Duffing's equation I: Analytical methods for small force and damping," in *Proceedings of the Year of Concentration in PDE's and Dynamical Systems*, University of Houston.

Holmes, P., and Whitley, D., (1983b). On the attracting set for Duffing's equation II: A geometrical model for moderate force and damping, *Physica* **7D**, 111.

Hopf, E., (1948). A mathematical example displaying features of turbulence, *Commun. on Pure Appl. Math.* **1**, 303.

Hopf, F. A., Kaplan, D. L., Gibbs, H. M., and Shoemaker, R. L., (1982). Bifurcations to chaos in optical bistability, *Phys. Rev.* **A25**, 2172.

Hu, B., (1981). Dissipative bifurcation ratio in the area-preserving Henon map, *J. Phys.* **A14**, L423.

Hu, B., and Mao, J. M., (1982a). Third-order renormalization-group calculation of the Feigenbaum universal bifurcation ratio in the transition to chaotic behavior, *Phys. Rev.* **A25**, 1196.

Hu, B., and Mao, J. M., (1982b). Period doubling: universality and critical point order, *Phys. Rev.* **A25**, 3259.

Hu, B., (1982a). Introduction to real space renormalization group methods in critical and chaotic phenomena, *Phys. Reports* **91**, 233.

Hu, B., (1982b). A two dimensional scaling theory of intermittency, *Phys. Lett.* **91A**, 375.

Hu, B., and Rudnick, J., (1982a). Exact solutions to the Feigenbaum renormalization group equations for intermittency, *Phys. Rev. Lett.* **48**, 1645.

Hu, B., and Rudnick, L., (1982b). Exact solutions to the renormalization-group fixed-point equations for intermittency in two-dimensional maps, *Phys. Rev.* **A26**, 3035.

Hu, B., (1983). A simple derivation of the stochastic eigenvalue equation in the transition from quasiperiodicity to chaos, *Phys. Lett.* **98A**, 79.

Hu, B., Satija, I. I., (1983). A spectrum of universality classes in period doubling and period tripling, *Phys. Lett.* **98A**, 143.

Hu, B., and Mao, J. M., (1983). Universal metric properties of an approximate poincare map for Duffing's equation with negative stiffness, *Phys. Rev.* **A27**, 1700.

Hu, G., Hao, B.-L., (1983). A scaling relation for the Hausdorff dimension of the limiting sets in one-dimensional mappings, *Commun. Theor. Phys.* **2**, 1473.

Huberman, B. A., and Crutchfield, J. P., (1979). Chaotic states of anharmonic systems in periodic fields, *Phys. Rev. Lett.* **43**, 1743.

Huberman, B. A., Crutchfield, J. P., and Packard, N. H., (1980). Noise phenomena in Josephson junctions, *Appl. Phys. Lett.* **37**, 750.

Huberman, B. A., and Rudnick, J., (1980). Scaling behavior of chaotic flows, *Phys. Rev. Lett.* **45**, 154.

Huberman, B. A., and Zisook, A. B., (1981). Power spectra of strange attractors, *Phys. Rev. Lett.* **46**, 626.

Huberman, B. A., (1983). Mostly chaos, *Physica* **118A**, 323.

Hudson, J. L., Hart, M., and Marinko, D., (1979). An experimental study of multiple peak periodic and nonperiodic oscillators in the Belousov-Zhabotinskii reaction, *J. Chem. Phys.* **71**, 1601.

Hudson, J. L., and Mankin, J. C., (1981). Chaos in the Belousov-Zhabotinskii reaction, *J. Chem. Phys.* **74**, 6171.

Hunt, E. R., (1982). Comment on a driven nonlinear oscillator, *Phys. Rev. Lett.* **49**, 1054.

Huppert, H. E., and Moore, D. R., (1976). Nonlinear double-diffusive convection, *J. Fluid Mech.* **78**, 821.

Ibanez, J. L., Pomeau, Y., (1978). A simple case of non-periodic (strange) attractor, *J. Non-Equilib. Thermodyn.* **3**, 135.

Ikeda, K., (1979). Multiple-valued stationary state and its instability of the transmitted light by a ring cavity system, *Opt. Commun.* **30**, 257.

Ikeda, K., Daido, H., and Akimoto, O., (1980). Optical turbulence: chaotic behavior of transmitted light from a ring cavity, *Phys. Rev. Lett.* **45**, 709.

Ikeda, K., and Akimoto, O., (1982). Instability leading to periodic and chaotic self-pulsations in a bistable optical cavity, *Phys. Rev. Lett.* **48**, 617.

Ikeda, K., and Kondo, K., (1982). Successive higher-harmonic bifurcations in systems with delayed feedback, *Phys. Rev. Lett.* **49**, 1467.

Imada, M., (1983). Chaos caused by the soliton-soliton interaction, *J. Phys. Soc. Japan* **52**, 1946.

Imry, Y., (1983). "Chaos and solitons in Josephson junctions," in *Statics and Dynamics of Nonlinear Systems*, ed. G. Benedeck, H. Bilz, R. Zeyher, p. 170, Springer-Verlag.

Inoue, M., and Koga, H., (1982). Chaos and diffusion in a sinusoidal potential with a periodic external field, *Prog. Theor. Phys.* **68**, 2184.

Inoue, M., and Koga, H., (1983). Chaotic response of a self-interacting pseudospin model, *Prog. Theor. Phys.* **69**, 1403.

Iooss, G., and Langford, W. F., (1980). Conjectures on routes to turbulence via bifurcations, *Ann. N. Y. Acad. Sci.* **357**, 489.

Ito, A., (1979a). Perturbation theory of self-oscillating system with a periodic perturbation, *Prog. Theor. Phys.* **61**, 45.

Ito, A., (1979b). A perturbation theory of a quasiperiodic motion: an asymptotic expansion method, *Prog. Theor. Phys.* **62**, 620.

Ito, A., (1979c). Successive subharmonic bifurcations and chaos in a nonlinear Mathieu equation, *Prog. Theor. Phys.* **61**, 815.

Izrailev, F. M., Rabinovich, M. I., and Ugodnikov, A. D., (1981). Approximate description of three-dimensional dissipative systems with stochastic behavior, *Phys. Lett.* **86A**, 321.

Jaffe, C., and Reinhardt, W. P., (1982). Uniform semiclassical quantization of regular and chaotic classical dynamics on the Henon-Heiles surface, *J. Chem. Phys.* **77**, 5191.

Jakobson, M. K., (1981). Absolutely continuous invariant measure for one parameter families of one dimensional maps, *Commun. Math. Phys.* **81**, 39.

Jeffries, C., and Perez, J., (1982). Observation of a Pomeau-Manneville intermittent route to chaos in a nonlinear oscillator, *Phys. Rev.* **A26**, 2117.

Jeffries, C., and Perez, J., (1983). Direct observation of crises of the chaotic attractor in a nonlinear oscillator, *Phys. Rev.* **A27**, 601.

Jeffries, C., and Usher, A., (1983). Frequency division using diodes in resonant systems, *Phys. Lett.* **99A**, 427.

Jensen, M. H., Bak, P., and Bohr, T., (1983). Complete devil's staircase, fractal dimension, and universality of mode-locking structure in the circle map, *Phys. Rev. Lett.* **50**, 1637.

Jensen, R. V., and Oberman, C. R., (1982). Statistical properties of chaotic dynamical systems which exhibit strange attractors, *Physica* **4D**, 183.

Joseph, D. D., (1981). "Hydrodynamic stability and bifurcation," in *Hydrodynamic instabilities and the transition to turbulence*, ed. H. L. Swinney and J. P. Gollub, Springer Verlag, N. Y.

Julia, G., (1918). Memoires sur l'iteration des fonctions rationelles., *J. Math. Pures Appl.* **4**, 47.

Kadanoff, L. P., (1981a). Scaling for a critical Kolmogorov-Arnold-Moser trajectory, *Phys. Rev. Lett.* **47**, 1641.

Kadanoff, L. P., (1981b). "Critical behavior of a KAM surface, II. renormalization approach," in *Melting, Localisation and Chaos*, p. 209, North-Holland, 1982.

Kadanoff, L. P., (1983a). Supercritical behavior of an ordered trajectory, *J. Stat. Phys.* **31**, 1.

Kadanoff, L. P., (1983b). Roads to chaos, *Phys. Today* December, 46.

Kai, T., and Tomita, K., (1979). Stroboscopic phase portrait of a forced nonlinear oscillator, *Prog. Theor. Phys.* **61**, 54.

Kai, T., and Tomita, K., (1980). Statistical mechanics of deterministic chaos: the one-dimensional discrete process, *Prog. Theor. Phys.* **64**, 1532.

Kai, T., (1981). Universality of power spectra of a dynamical system with an infinite sequence of period-doubling bifurcations, *Phys. Lett.* **86A**, 263.

Kai, T., (1982). Lyapunov number for a noisy 2**n Cycle, *J. Stat. Phys.* **29**, 329.

Kaneko, K., (1982). On the period-adding phenomena at the frequency locking in a one-dimensional mapping, *Prog. Theor. Phys.* **68**, 669.

Kaneko, K., (1983a). Similarity structure and scaling property of the period-adding phenomena, *Prog. Theor. Phys.* **69**, 403.

Kaneko, K., (1983b). Transition from torus to chaos accompanied by frequency lockings with symmetry breaking, *Prog. Theor. Phys.* **69**, 1427.

Kaneko, K., (1983c). Doubling of torus, *Prog. Theor. Phys.* **69**, 1806.

Kaplan, H., (1983). New method for calculating stable and unstable periodic orbits, *Phys. Lett.* **97A**, 365.

Kaplan, J. L., and Marotto, F. R., (1977). "Chaotic behavior in dynamical systems," in *Proceedings Int. Conf. on Nonlinear Systems and Applications*, ed. V. Lakshmikantham, p. 199, Academic.

Kaplan, J. L., and Yorke, J. A., (1979a). Preturbulence: a regime observed in a fluid flow model of Lorenz, *Commun. Math. Phys.* **67**, 93.

Kaplan, J. L., and Yorke, J. A., (1979b). The onset of chaos in a fluid flow model of Lorenz, *Ann. N. Y. Acad. Sci.* **316**, 400.

Kaplan, J. L., and Yorke, J. A., (1979c). "Chaotic behavior of multidimensional difference equations," in *Functional Diff. Eq. and Approx. of Fixed Points*, ed. H.-O. Peitgen, H.-O. Walther, *Lect. Notes in Math.* No. **730**, Springer Verlag, p. 204.

Kapral, R., Schell, M., and Fraser, S., (1982). Chaos and fluctuations in nonlinear dissipative systems, *J. of Phys. Chem.* **86**, 2205.

Karney, C. F. F., Rechester, A. B., and White, R. B., (1982). Effect of noise on the standard mapping, *Physica* **4D**, 425.

Karney, C. F. F., (1983). Long-time correlations in the stochastic regime, *Physica* **8D**, 360.

Katok, A. B., (1980). Lyapunov exponents, entropy and periodic points for diffeomorphisms. *Publ. Math. IHES.* **51**, 137.

Kautz, R. L., (1981). Chaotic states of rf-biased Josephson junctions, *J. Appl. Phys.* **52**, 6241.

Keolian, R., Putterman, S. J., Turkevich, L. A., Rudnick, I., and Rudnick, J. A., (1981). Subharmonic sequences in the Faraday experiment: departures from period-doubling, *Phys. Rev. Lett.* **47**, 1133.

King, G., Swinney, H. L., (1983). Limits of stability and defects in wavy vortex flow, *Phys. Rev.* **A27**, 1240.

Kitano, M., Yabuzaki, T., and Ogawa, T., (1983). Chaos and period-doubling bifurcations in a simple diffusive acoustic system, *Phys. Rev. Lett.* **50**, 713.

Knobloch, E., (1979). On the statistical dynamics of the Lorenz model, *J. Stat. Phys.* **20**, 695.

Knobloch, E., (1981). Chaos in a segmented disk dynamo, *Phys. Lett.* **82A**, 439.

Knobloch, E., and Proctor, M. R. E., (1981). Nonlinear periodic convection in double diffusive systems, *J. Fluid Mech.* **108**, 291.

Knobloch, E., and Weiss, N. O., (1981). Bifurcations in a model of double-diffusive convection, *Phys. Lett.* **85A**, 127.

Kohyama, T., and Aizawa, Y., (1983). Orbital stability in a piece-wise linear map of the circle onto itself, *Prog. Theor. Phys.* **70**, 1002.

Kolmogorov, A. N., (1954). Preservation of conditionally periodic movements with small change in the Hamilton function, *Akad. Nayk SSSR Doklady* **98**, 527; English translation in *Stochastic Behavior in Classical and Quantum Hamiltonian Systems*, ed. G. Casati, J. Ford, p. 51, Springer Verlag, 1979.

Kozak, J. J., Musho, M. K., and Hatlee, M. D., (1982). Chaos, periodic chaos, and the random-walk problem, *Phys. Rev. Lett.* **49**, 1801.

Kuramoto, Y., and Yamada, T., (1976a). Turbulent state in chemical reactions, *Prog. Theor. Phys.* **56**, 679.

Kuramoto, Y., and Yamada, T., (1976b). Pattern formation in oscillatory chemical reactions, *Prog. Theor. Phys.* **56**, 724.

Kuramoto, Y., and Koga, S., (1982). Anomalous period-doubling bifurcations leading to chemical turbulence, *Phys. Lett.* **92A**, 1.

Kus, M., (1983). Integrals of motion for the Lorenz system, *J. Phys.* **A16**, L689.

Lafon, A., Rossi, A., and Vidal, C., (1983). The power of chaos measured through the spectral analysis of experimental data, *J. Physique* **44**, 505.

Landa, P. S., Stratonovich, R. L., (1982). Stationary probability distribution for one of the simplest strange attractors, *Sov. Phys.-Dokl.* **27**, 1032; Russian original; *Doklady* **267**, 832.

Landau, L. D., (1944). On the problem of turbulence, *C. R. Acad, Sci. URSS*, **44**, 311; in *Collected Papers of Landau*, ed. by D. ter Haar, p. 387, Pergamon, 1965.

Lanford, O. E., (1977). Computer pictures of the Lorenz attractor. *Appendix to Williams* (1977).

Lanford, O. E., (1980). "Remarks on the accumulation of period doubling bifurcations," in *Math. Problems in Theor. Phys.* ed. K. Osterwalder, *Lect. Notes in Phys.* **116**, p. 340, Springer Verlag, N. Y.

Lanford, O. E., (1981). "Strange attractors and turbulence," in *Hydrodynamics in Stabilities and the Transition to Turbulence*, ed. H. L. Swinney and J. P. Gollub, ed. Springer, Berlin, p. 7.

Lanford, O. E., (1982a). The strange attractor theory of turbulence, *Annual Review of Fluid Mechanics* **14**, 347.

Lanford, O. E., (1982b). A computer-assisted proof of the Feigenbaum conjectures, *Bull. Amer. Math. Soc.* **6**, 427.

Lanford, O., (1983). Period doubling in one and several dimensions, *Physica* **7D**, 124.

Lauterborn, W., and Cramer, E., (1981). Subharmonic route to chaos observed in acoustics, *Phys. Rev. Lett.* **47**, 1445.

Lauterborn, W., (1982). Cavitation bubble dynamics — new tools for an intricate problem, *Appl. Sci. Res.* **38**, 165.

Ledrappier, F., (1981). Some relations between dimension and Lyapunov exponents, *Commun. Math. Phys.* **81**, 229.

Lee, K.-C., (1983). The universality of period-doubling bifurcations in certain 2D reversible area-preserving mappings with quadratic nonlinearity, *J. Phys.* **A16**, L137.

Leven, R. W., and Koch, B. P., (1981). Chaotic behavior of a parametrically excited damped pendulum, *Phys. Lett.* **86A**, 71.

Levi, M., (1980). Periodically forced relaxation oscillations, in *Lect. Notes in Math.* No. **819**, Springer, p. 300.

Levi, M., (1981). Qualitative analysis of the periodically forced relaxation oscillations, *Memoirs Am. Math. Soc.* No. **244**.

Levinson, M. T., (1982). Even and odd harmonic frequencies and chaos in Josephson junctions: impact on parametric amplifiers? *J. Appl. Phys.* **53**, 4294.

Levy, Y. E., (1982). Some remarks about computer studies of dynamical systems, *Phys. Lett.* **88A**, 1.

Li, T. Y., and Yorke, J. A., (1975). Period three implies chaos, *Am. Math. Monthly* **82**, 985.

Li, T. Y., Misiurewicz, M., Pianigiani, G., and Yorke, J. A., (1982). Odd chaos, *Phys. Lett.* **87A**, 271.

Libchaber, A., and Maurer, J., (1978). Local probe in a Rayleigh-Benard experiment in liquid helium, *J. de Phys. Lett.* **39**, L369.

Libchaber, A., and Maurer, J., (1980). Une experience de Rayleigh-Benard de geometrie reduite: multiplication, accrochage, et demultiplication de frequences, *J. de Phys.* **41**, Coll. C3–51.

Libchaber, A., (1982). "Experimental study of hydrodynamic instabilities. Rayleigh-Benard experiment: helium in a small box," in *Nonlinear Phenomena at Phase Transitions and Instabilities*, ed. T. Riste, Plenum, N. Y., 259.

Libchaber, A., Laroche, C., and Fauve, S., (1982).Period doubling cascade in mercury, quantitative measurement, *J. de Phys. Lett.* **43**, L211.

Libchaber, A., Fauve, S., and Laroche, C., (1983). Two-parameter study of the routes to chaos, *Physica* **7D**, 73.

Lieberman, M. A., (1980). "Arnold diffusion in Hamiltonian systems with three degrees of freedom," in *Nonlinear Dynamics*, ed. R. H. G. Helleman, p. 119. New York Academy of Sciences, New York.

Linsay, P. S., (1981). Period doubling and chaotic behavior in a driven anharmonic oscillator, *Phys. Rev. Lett.* **47**, 1349.

Littlewood, J. E., (1957). On non-linear differential equations of second order, III, *Acta Mathematica* **97**, 268; **98**, 1.

Lorenz, E. N., (1963). Deterministic nonperiodic flow, *J. Atmos. Sci.* **20**, 130.

Lorenz, E. N., (1964). The problem of deducing the climate from the governing equations, *Tellus* **16**, 1.

Lorenz, E. N., (1979). "On the prevalence of aperiodicity in simple systems," in *Global Analysis*, ed. M. Grmela, J. E. Marsden, *Lect. Notes in Math.* **755**, p. 53, Springer Verlag, N. Y.

Lorenz, E. N., (1980a). Noisy periodicity and reverse bifurcation, *Ann. N. Y. Acad. Sci.* **357**, 282.

Lorenz, E. N., (1980b). Attractor sets and quasi-geostrophic equilibrium, *J. Atmos. Sci.* **36**, 1685.

Lorenzen, A., Pfister, G., and Mullin, T., (1983). End effects on the transition time-dependent motion in the Taylor experiment, *Phys. Fluids* **26**, 10.

Lozi, R., (1978). Un attracteur etrange (?) du type attracteur de Henon, *J. de Phys.* **39**, Colloq. C5–9.

Lucke, M., (1976). Statistical dynamics of the Lorenz model, *J. Stat. Phys.* **15**, 455.

Lugiato, L. A., Narducci, L. M., Bandy, D. K., and Pennise, C. A., (1983). Breathing, spiking and chaos in a laser with injected signal, *Opt. Commun.* **46**, 64.

Lunsford, G. H., and Ford, J., (1972). On the stability of periodic orbits for nonlinear oscillator systems in regions exhibiting stochastic behavior, *J. Math. Phys.* **13**, 700.

L'vov, V. S., and Predtechensky, A. A., (1981). On Landau and stochastic attractor pictures in the problem of transition to turbulence, *Physica* **2D**, 38.

Lyubimov, D. V., and Zaks, M. A., (1983). Two mechanisms of the transition to chaos in finite-dimensional models of convection, *Physica* **9D**, 52.

Mackay, M. C., and Glass, L., (1977). Oscillation and chaos in physiological control systems, *Science* **197**, 287.

Mackay, R. S., (1982). Islets of stability beyond period doubling, *Phys. Lett.* **87A**, 7.

Malomed, A. B., (1983). A simple dynamical system with stochastic behavior, *Physica* **8D**, 343.

Malraison, B., Atten, P., Berge, P., and Dubois, M., (1983). Dimension of strange attractors: an experimental determination for the chaotic regime of two convective systems, *J. Physique Lett.* **44**, 897.

Mandel, P., and Kapral, R., (1983). Subharmonic and chaotic bifurcation structure in optical bistability, *Opt. Commun.* **47**, 151.

Manneville, P., and Pomeau, Y., (1979). Intermittency and the Lorenz model, *Phys. Lett.* **75A**, 1.

Manneville, P., (1980a). Intermittency in dissipative dynamical systems, *Phys. Lett.* **79A**, 33.

Manneville, P., (1980b). Intermittency, self-similarity and 1/f spectrum in dissipative dynamical systems, *J. de Phys.* **41**, 1235.

Manneville, P., and Pomeau, Y., (1980). Different ways to turbulence in dissipative systems, *Physica* **1D**, 219.

Manneville, P., (1981). The transition to turbulence in nematic liquid crystals, *Mol. Cryst. & Liq. Cryst.* **70**, 1501.

Manneville, P., and Piquemal, J. M., (1982). Transverse phase diffusion in Rayleigh-Benard convection, *J. de Phys. Lett.* **43**, 253.

Manneville, P., (1982). On the statistics of turbulent transients in dissipative systems, *Phys. Lett.* **90A**, 327.

Manton, N. S., and Nauenberg, M., (1983). Universal scaling behavior for iterated maps in the complex plane, *Commun. Math. Phys.* **89**, 555.

Marotto, F. R., (1978). Snap-back repellers imply chaos in R**n, *J. Math. Anal. Appl.* **63**, 199.

Marotto, F. R., (1979). Chaotic behavior in the Henon mapping, *Commun. Math. Phys.* **68**, 187.

Martin, P. C., (1976). Instabilities, oscillations, and chaos, *J. de Phys.* **37**, Colloq. C1-57.

Marzec, C. J., and Spiegel, E. A., (1980). Ordinary differential equations with strange attractors, *SIAM J. Appl. Math.* **38**, 387.

Matinyan, S. G., Savvidy, G. K., and Ter-Arutynnyan-Savvidy, N. G., (1981). Stochastic classical mechanics of Yang-Mills and its elimination by the Higgs mechanism, *Pisma v JETP* **34**, 613; *JETP* **80**, 830.

Matkomsky, B. J., and Reiss, E. L., (1977). Singular perturbations of bifurcations, *SIAM J. Appl. Math.* **33**, 230.

Matsumoto, K., Tsuda, I., (1983). Noise-induced order, *J. Stat. Phys.* **31**, 87; Addendum, ibid, **33**, 757.

Maurer, J., and Libchaber, A., (1979). Rayleigh-Benard experiment in liquid He: frequency locking and the onset of turbulence, *J. de Phys. Lett.* **40**, 419.

Maurer, J., and Libchaber, A., (1980). Effect of Prandtl number on the onset of turbulence, *J. de Phys. Lett.* **41**, L515.

May, R. M., (1974). Biological populations with nonoverlapping generations: stable points, stable cycles, and chaos, *Science* **186**, 645.

May, R. M., and Oster, G. F., (1976). Bifurcations and dynamic complexity in simple ecological models, *Amer. Natur.* **110**, 573.

May, R. M., (1976). Simple mathematical models with very complicated dynamics, *Nature* **261**, 459.

May, R. M., and Oster, G. F., (1980). Period-doubling and the onset of turbulence, an analytic estimate of the Feigenbaum ratio, *Phys. Lett.* **78A**, 1.

Mayer, J. A., (1978). Sur la dynamigue des systems ecologigues nonlineares, *J. Physique* **39**, Colloq. C5–29.

Mayer-Kress, G., and Haken, H., (1981a). Intermittent behavior of the logistic system, *Phys. Lett.* **82A**, 151.

Mayer-Kress, G., and Haken, H., (1981b). The influence of noise on the logistic model, *J. Stat. Phys.* **26**, 149.

McCreadie, G. A., and Rowlands, G., (1982). An analytical approximation to the Lyapunov number for 1D maps, *Phys. Lett.* **91A**, 146.

McGuinness, M. J., (1983). The fractal dimension of the Lorenz attractor, *Phys. Lett.* **99A**, 5.

McGuire, J. B., and Thompson, C. J., (1981). On the universality and computation of Feigenbaum's delta, *Phys. Lett.* **84A**, 9.

McGuire, J. B., and Thompson, C. J., (1982). Asymptotic properties of iterates of nonlinear transformations, *J. Stat. Phys.* **27**, 183.

McKay, S. R., Berker, A. N., and Kirkpatrick, S., (1982). Spin-glass behavior in frustrated Ising models with chaotic renormalization-group trajectories, *Phys. Rev. Lett.* **48**, 767.

McLaughlin, D. W., Moloney, J. V., and Newell, A. C., (1983). Solitary waves as fixed points of infinite-dimensional maps in an optical bistable ring cavity, *Phys. Rev. Lett.* **51**, 75.

McLaughin, J. B., and Martin, P. C., (1975). Transition to turbulence in a statically stressed fluid system, *Phys. Rev.* **A12**, 186.

McLaughlin, J. B., (1976). Successive bifurcations leading to stochastic behavior, *J. Stat. Phys.* **15**, 307.

McLaughlin, J. B., (1979a). Stochastic behavior in slightly dissipative systems, *Phys. Rev.* **A20**, 2114.

McLaughlin, J. B., (1979b). The role of dissipation in a truncation of Henon's map, *Phys. Lett.* **72A**, 271.

McLaughlin, J. B., (1980). Connection between dissipative and resonant conservative nonlinear oscillators, *J. Stat. Phys.* **19**, 587.

McLaughin, J. B., (1981). Period-doubling bifurcations and chaotic motion of a parametrically forced pendulum, *J. Stat. Phys.* **24**, 375.

Meiss, J. D., and Cary, J. R., (1983). Correlations of periodic, area-preserving maps, *Physica* **6D**, 375.

Melnikov, V. K., (1963). On the stability of the center for time periodic perturbations, *Trans. Moscow Math. Soc.* **12**, 1.

Mendes, R. V., (1981). Critical-point dependence of universality in maps of the interval, *Phys. Lett.* **84A**, 1.
Metropolis, N., Stein, M. L., and Stein, P. R., (1967). Stable states of a nonlinear transformation, *Numer. Math.* **10**, 1.
Metropolis, N., Stein, M. L., and Stein, P. R., (1973). On finite limit sets for transformations on the unit interval, *J. Comb. Theor.* **A15**, 25.
Milonni, P. W., Ackerhalt, J. R., and Galbraith, H. W., (1983). Chaos and nonlinear optics: a chaotic Raman attractor, *Phys. Rev.* **A28**, 887.
Miracky, R. F., Clarke, J., and Koch, R. H., (1983). Chaotic noise observed in a resistively shunted self-resonant Josephson tunnel junction, *Phys. Rev. Lett.* **50**, 856.
Misiurewicz, M., and Szewc, B., (1980). Existence of a homoclinic point for the Henon map, *Commun. Math. Phys.* **75**, 285.
Misiurewicz, M., (1980). Strange attractors for the Lozi mapping, *Ann. N. Y. Acad. Sci.* **357**, 348.
Misiurewicz, M., (1981). Absolutely continuous measures for certain maps of an interval, *Math. Publ. IHES* **53**, 17.
Misiurewicz, M., (1981). The structure of mapping of an interval with zero entropy. *Publ. Math. IHES* **53**, 5.
Moiseyev, N., and Perez, A., (1983). Motion of wave packets in regular and chaotic systems, *J. Chem. Phys.* **79**, 5945.
Moloney, J. V., Hopf, F. A., and Gibbs, H. M., (1982). Effects of transverse beam variation of bifurcations in an intrinsic bistable ring cavity, *Phys. Rev.* **25A**, 3442.
Monin, A. S., (1978). On the nature of turbulence, *Sov. Phys. Usp.* **21**, 429; Russian original, *UFN* **125**, 94.
Moon, F. C., and Holmes, P. J., (1979). A magnetoelastic strange attractor. *J. Sound Vib.* **65**, 285; **69** (1980), 339.
Moon, F. C., (1980). Experiments on chaotic motion of forced nonlinear oscillator strange attractors, *Trans. ASME J. Appl. Mech.* **47**, 638.
Moon, F. C., and Shaw, S. W., (1983). Chaotic vibrations of a beam with nonlinear boundary conditions, *Int. J. Non-Linear Mech.* **18**, 465.
Moon, H. T., Huerre, P., and Redekopp, L. G., (1982). Three frequency motion and chaos in the Ginzburg-Landau equation, *Phys. Rev. Lett.* **49**, 458.
Moon, H. T., Huerre, P., and Redekopp, L. G., (1983). Transition to chaos in the Landau-Ginsburg equation, *Physica* **7D**, 135.
Moore, D. R., Toomre, J., Knobloch, E., Weiss, N. O., (1983). Period doubling and chaos in partial differential equations for thermosolutal convection, *Nature* **303**, 663.
Mori, H., and Fujisaka, H., (1980). Statistical dynamics of chaotic flows, *Prog. Theor. Phys.* **63**, 1931.
Mori, H., (1980). Fractal dimensions of chaotic flows of autonomous dissipative systems, *Prog. Theor. Phys.* **63**, 1044.
Mori, H., (1981). "Evolution of chaos and power spectra in one-dimensional map," in *Nonlinear Phenomena in Chemical Dynamics*, p. 88, ed. C. Vidal and P. Pacault, Springer-Verlag.
Mori, H., So, B.-C., and Ose, T., (1981). Time-correlation functions of one-dimensional transformations, *Prog. Theor. Phys.* **66**, 4.

Morioka, N., and Shimizu, T., (1978). Transition between turbulent and periodic states in the Lorenz model, *Phys. Lett.* **66A**, 447.

Moser, J., (1962). On invariant curves of area-preserving mappings of an annulus, *Nachr. Akad. Wiss. Gottingen Math. Phys. Kl.* **2,** 1.

Moser, J., (1968). Lectures on Hamiltonian systems, *Memoirs Am. Math. Soc.* **81**, 1.

Moser, J., (1978). Is the solar system stable? *Math. Interlligencer* **1**, 65.

Nagashima, T., and Shimada, I., (1977). On the C-system-like property of the Lorenz system, *Prog. Theor. Phys.* **58**, 1318.

Nagashima, T., and Haken, H., (1983). Chaotic modulation of correlation functions, *Phys. Lett.* **96A**, 385.

Nakamura, K., (1977). Nonlinear fluctuations associated with instabilities in dissipative systems, *Prog. Theor. Phys.* **57**, 6.

Nakamura, K., (1978a). Numerical experiments on trajectory instabilities, *Prog. Theor. Phys.* **59**, 64.

Nakamura, K., (1978b). Trajectory instabilities and stochastic behavior in dissipative systems with multiple steady states, *Suppl. Prog. Theor. Phys.* **64**, 378.

Nakamura, K., (1979). Stochastic instabilities and turbulence in nonlinear dissipative systems, *Proc. Inst. Nat. Sci.* Nihon Univ., **14**, 9.

Nakamura, K., Ohta, S., and Kawasaki, K., (1982). Chaotic states of ferromagnets in strong parallel pumping fields, *J. Phys.* **C15**, L143.

Nakatsuka, H., Asaka, S., Itoh, H., Ikeda, K., and Matzuoka, M., (1983). Observation of bifurcation to chaos in an all-optical bistable system, *Phys. Rev. Lett.* **50**, 109.

Nauenberg, M., and Rudnick, J., (1981). Universality and the power spectrum at the onset of chaos, *Phys. Rev.* **B24**, 493.

Nauenberg, M., (1982). On the fixed points for circle maps, *Phys. Lett.* **92A**, 7.

Newhouse, S. E., Ruelle, D., and Takens, F., (1978). Occurrence of strange axiom A attractors near quasi-periodic flows on Tm (m = 3 or more), *Commun. Math. Phys.* **64**, 35.

Newhouse, S., (1980). The abundance of wild hyperbolic sets and non-smooth stable sets of diffeomorphisms, *Publ. Math. IHES* **50**, 101.

Nicolis, J. S., Mayer-Kress, G., and Haubs, H., (1983). Nonuniform chaotic dynamics with implications to information processing, *Z. Naturforsch.* **38a**, 1157.

Normand, C. Y., Pomeau, Y., and Velarde, M. G., (1977). Convective instability: a physicist's approach, *Rev. Mod. Phys.* **49**, 581.

Nozaki, K., (1982). Stochastic instability of sine-Gordon solitons, *Phys. Rev. Lett.* **49**, 1883.

Nozaki, K., and Bekki, N., (1983a). Chaos in a perturbed nonlinear Schrodinger equation, *Phys. Rev. Lett.* **50**, 1226.

Nozaki, K., and Bekki, N., (1983b). Pattern selection and spatiotemporal transition to chaos in the Ginzburg-Landau equation, *Phys. Rev. Lett.* **51**, 2171.

Nozierres, P., (1978). Reversals of the earth's magnetic field: an attempt at a relaxation model, *Phys. of Earth and Planet. Interiors* **17**, 55.

Ogura, H., Ueda, Y., and Yoshida, Y., (1981). Periodic stationarity of a chaotic motion in the system governed by Duffing's equation, *Prog. Theor. Phys.* **66**, 2280.

Olsen, L. F., Degn, H., (1979). Chaos in an enzyme reaction, *Nature* **267**, 177.

Oono, Y., (1978a). A heuristic approach to the Kolmogorov entropy as a disorder parameter, *Prog. Theor. Phys.* **60**, 1944.

Oono, Y., (1978a). Period. NE. 2**n implies chaos, *Prog. Theor. Phys.* **59**, 1029.

Oono, Y., and Takahashi, Y., (1980). Chaos, external noise, and Fredholm theory, *Prog. Theor. Phys.* **63**, 1804.

Oono, Y., and Osikawa, M., (1980). Chaos in nonlinear differential equations, I — qualitative study of (formal) chaos, *Prog. Theor. Phys.* **64**, 54.

Oono, Y., Kohda, T., and Yamazaki, H., (1980). Disorder parameter for chaos, *J. Phys. Soc. Japan* **48**, 738.

Orszag, S. A., and McLaughlin, J. B., (1980). Evidence that random behavior is generic for nonlinear differential equations, *Physica* **1D**, 68.

Oseledec, V. I., (1968). A multiplicative ergodic theorem: Lyapunov characteristic numbers for dynamical systems, *Trans. Moscow Math. Soc.* **19**, 197.

Ostlund, S., Rand, D., Sethna, J., and Siggia, E., (1983). Universal properties of the transition from quasi-periodicity to chaos in dissipative systems, *Physica* **8D**, 303.

Otsuka, K., and Iwamura, H., (1983). Theory of optical multistability and chaos in a resonant-type semiconductor laser amplifier, *Phys. Rev.* **A28**, 3153.

Ott, E., (1981). Strange attractors and chaotic motions of dynamical systems, *Rev. Mod. Phys.* **53**, 655.

Ott, E., and Hanson, J. D., (1981). The effect of noise on the structure of strange attractors, *Phys. Lett.* **85A**, 20.

Packard, N. H., Crutchfield, J. P., Farmer, J. D., and Shaw, R. S., (1980). Geometry from a time series, *Phys. Rev. Lett.* **45**, 712.

Pakarinen, P., and Nieminen, R. M., (1983). Period-multiplying bifurcations and multifurcations in conservative mappings, *J. Phys.* **A16**, 2105.

Park, K., Crawford, G. L., and Donnelly, R. J., (1981). Determination of transition in Couette flow in finite geometries, *Phys. Rev. Lett.* **47**, 1448.

Park, K., and Donnelly, R. J., (1981). Study of the transition to Taylor vortex, *Phys. Rev.* **24A**, 2277.

Park, K., and Crawford, G. L., (1983). Deterministic transition in Taylor wavy-vortex flow, *Phys. Rev. Lett.* **50**, 343.

Parry, W., (1976). Symbolic dynamics and transformation of the unit interval, *Trans. Amer. Math. Soc.* **122**, 368.

Parry, W., (1977). "The Lorenz attractor and a related population model," in *Ergodic Theory Conf.*, ed. M. Denker, K. Jacobs, *Lect. Notes in Math.* No. **729** (1979), p. 169, Springer-Verlag.

Pedlovsky, J., and Frenzen, C., (1980). Chaotic and periodic behavior of finite amplitude baroclinic waves, *J. Atmos. Sci.* **37**, 1177.

Pederson, N. F., Soerenson, O. H., Dueholm, B., and Mygind, J., (1980). Half-harmonic parametric oscillations in Josephson junctions, *J. Low Temp. Phys.* **38**, 1.

Pederson, N. F., and Davidson, A., (1981). Chaos and noise rise in Josephson junctions, *Appl. Phys. Lett.* **39**, 830.

Percival, I. C., (1982). Chaotic boundary of a Hamiltonian map, *Physica* **6D**, 67.

Perez, J., and Jeffries, C., (1982). Effects of additive noise on a nonlinear oscillator exhibiting period-doubling and chaotic behavior, *Phys. Rev.* **B26**, 3460.

Perez, R., and Glass, L., (1982). Bistability, period doubling bifurcations and chaos in a periodically forced oscillator, *Phys. Lett.* **90A**, 441.

Perrin, B., (1982). Emergence of a periodic mode in the so-called turbulent region in a circular Couette flow, *J. de Phys. Lett.* **43**, 5.

Pesin, Ya. B., (1976). Lyapunov characteristic exponent and ergodic properties of smooth dynamical systems with an invariant measure, *Sov. Math. Dokl.* **17**, 196.

Pesin, Ya. B., (1977). Characteristic Lyapunov exponents and smooth ergodic theory, *Russ. Math. Surv.* **32:4**, 55.

Pesin, Ya. B., and Sinai, Ya. G., (1981). "Hyperbolicity and stochasticity of dynamical systems," in *Math. Phys. Rev.* ed. S. P. Novikov, Vol. **2**, p. 53.

Peters, H., (1982). Chaos in a time-delayed differential equation, *Z. Angew. Math. und Mech.* **62**, 297.

Piangiani, G., and Yorke, J. A., (1979). Expanding maps on sets which are almost invariant: decay and chaos. *Trans. Amer. Math. Soc.* **252**, 351.

Picard, G., and Johnston, T. W., (1982). Instability cascades, Lotka-Volterra population equations, and Hamiltonian chaos, *Phys. Rev. Lett.* **48**, 23.

Pieranski, P., (1983). Jumping particle model. Period doubling cascade in an experimental system, *J. Physique* **44**, 573.

Pikovsky, A. S., and Rabinovich, M. I., (1981a). Stochastic oscillations in dissipative systems, *Physica* **2D**, 8.

Pikovsky, A. S., and Rabinovich, M. I., (1981b). "Stochastic behavior in dissipative systems," in *Math. Phys. Rev.* ed. S. P. Novikov, Vol. **2**, p. 165.

Pikovsky, A. S., (1983). A new type of intermittent transition to chaos, *J. Phys.* **A16**, L109.

Pismen, L. M., (1982). Bifurcation sequences in a third-order system with a folded slow maniford, *Phys. Lett.* **89A**, 59.

Pomeau, Y., and Manneville, P., (1980). Intermittent transition to turbulence in dissipative dynamical systems, *Commun. Math. Phys.* **74**, 189.

Pomeau, Y., (1980). "Intermittency: a simple mechanism of continueons transition from order to chaos," in *Bifurcation Phenomena in Math. Phys. and Related Topics*, p. 155, Reidel.

Pomeau, Y., Roux, J. C., Rossi, A., Bachelart, and Vidal, C., (1981). Intermittent behavior in the Belousov-Zhabotionsky reaction, *Jour. de Phys. Lett.* **42**, 271.

Pounder, J. R., and Rogers, T. D., (1980). The geometry of chaos: dynamics of nonlinear second-order difference equation, *Bull. Math. Biol.* **42**, 551.

Prima, R. C., and Swinney, H. L., (1981). "Instabilities and transition in flow between concentric rotating cylinders," in *Hydrodynamic Instabilities and the Transition to Turbulence*, ed. H. L. Swinney and J. P. Gollub, p. 139, Springer, Berlin.

Rabinovich, M. I., (1978). Stochastic self-oscillations and turbulence, *Sov. Phys. Usp.* **21**, 443; Russian original, *UFN* **125**, 123.

Rabinovich, M. I., and Fabrikant, A. L., (1979). Stochastic self-modulation of waves in nonequilibrium media, *Sov. Phys. JETP*, **50**, 311; Russian original, *JETP*, **77**, 617.

Rabinovich, M. I., (1980). Strange attractors in modern physics, *Ann. N. Y. Acad Sci.* **375**, 435.

Rand, D., (1978). The topological classification of Lorenz attractors, *Math. Proc. Cambr. Phil. Soc.* **83**, 451.

Rand, D., (1982). Dynamics and symmetry. Predictions for modulated waves in rotating fluid, *Arch. Ration. Mech. & Anal.* **79**, 1.

Rand, D., Ostlund, S., Sethna, J., and Siggia, E. D., (1982). A universal transition from quasi-periodicity to chaos in dissipative systems, *Phys. Rev. Lett.* **49**, 132.

Rand, R. H., and Holmes, P. J., (1980). Bifurcation of periodic motions in two weakly coupled van der Pol oscillators, *Int. J. Nonlinear Mech.* **15**, 387.

Rayleigh, Lord, (1916). On convective currents in a horizontal layer of fluid when the higher temperature is on the underside, *Phil. Mag.* **32**, 529.

Rechester, A. B., and White, R. B., (1980). Calculation of turbulent diffusion for the Chirikov-Taylor model, *Phys. Rev. Lett.* **44**, 1586.

Rechester, A. B., and White, R. B., (1983). Invariant distribution on the attractors in the presence of noise, *Phys. Rev.* **A27**, 1203.

Reichl, L. E., de Fainchtein, R., Petrosky, T., and Zheng, W.-M., (1983). Field induced chaos in the Toda lattice, *Phys. Rev.* **A28**, 3051.

Riela, G., (1982a). Universal spectral property in higher dimensional dynamical systems, *Phys. Lett.* **92A**, 157.

Riela, G., (1982b). A new six-mode truncation of the Navier-Stokes equations on a two-dimensional torus: a numerical study, *Nuovo Cimento* **69B**, 245.

Riela, G., (1982c). Loss of stability and disappearance of two-dimensional invariant tori in a dissipative dynamical system, *Phys. Lett.* **91A**, 203.

Robbins, K. A., (1977). A new approach to subcritical instability and turbulent transitions in a simple dynamo, *Math. Proc. Camb. Phil. Soc.* **82**, 309.

Robbins, K. A., (1979). Periodic solutions and bifurcation structure at high R in the Lorenz model, *SIAM J. Appl. Math.* **36**, 457.

Röessler, O. E., (1976a). An equation for continuous chaos, *Phys. Lett.* **57A**, 397.

Röessler, O. E., (1976b). Chaotic behavior in simple reaction systems, *Z. Naturforsch.* **31a**, 259.

Röessler, O. E., (1976c). Chemical turbulence: chaos in a simple reaction-diffusion system, *Z Naturforsch.* **31a**, 1168.

Röessler, O. E., (1976d). Different types of chaos in two simple differential equations, *Z. Naturforsch.* **31a**, 1664.

Röessler, O. E., (1977a). Chaos in abstract kinetics: two prototypes, *Bull. Math. Biol.* **39**, 275.

Röessler, O. E., (1977b). Horseshoe-map chaos in the Lorenz equation, *Phys. Lett.* **60A**, 392.

Röessler, O. E., (1979a). "Chaotic oscillations — an example of hyperchaos," in *Nonlinear Oscillations in Biology*, ed. F. C. Hoppensteadt, *Lect. Notes in Appl. Math.* Vol. **17**, Amer. Math. Soc., Providence, RI.

Röessler, O. E., (1979b). An equation for hyperchaos, *Phys. Lett.* **71A**, 155.

Röessler, O. E., (1976c). Continuous chaos — four prototype equations, *Ann. N. Y. Acad. Sci.* **316**, 376.

Röessler, O. E., (1981). "Chaos and chemistry," in *Nonlinear Phenomena in Chemical Dynamics*, ed. by C. Vidal, A. Pacault, Springer Verlag.

Rollins, R. W., and Hunt, E. R., (1982). Exactly solvable model of a physical system exhibiting universal chaotic behavior, *Phys. Rev. Lett.* **49**, 1295.

Roux, J.-C., Rossi, A., Bachelart, S., and Vidal, C., (1980). Representation of a strange attractor from an experimental study of chemical turbulence, *Phys. Lett.* **77A**, 391.

Roux, J.-C., Rossi, A., Bachelart, S., and Vidal, C., (1981). Experimental observations of complex dynamical behavior during a chemical reaction, *Physica* **2D**, 395.

Roux, J.-C., and Swinney, H. L., (1981). "Topology of chaos in a chemical reaction, in *Nonlinear Phenomena in Chemical Dynamics*, ed. by A. Pacault & C. Vidal, Springer.

Roux, J.-C., Turner, J. S., McCormick, W. D., and Swinney, H. L., (1982). "Experimental observations of complex dynamics in a chemical reaction," in *Nonlinear Problems: Present and Future*, ed. A. R. Bishop et al, North-Holland, Amsterdam.

Roux, J.-C., (1983). Experimental studies of bifurcations leading to chaos in the Belousov-Zhabotinsky reaction, *Physica* **7D**, 57.

Roux, J.-C., Simoyi, R. H., and Swinney, H. L., (1983). Observation of a strange attractor, *Physica* **8D**, 257.

Rowlands, G., (1983). Chaotic trajectories of ordinary differential equations, *J. Phys.* **A16**, 585; cf. Fowler, (1983).

Rubenfeld, L. A., and Siegmann, W. L., (1977). Nonlinear dynamic theory for a double-diffusive convection model, *SIAM J. Appl. Math.* **32**, 871.

Ruelle, D., and Takens, F., (1971). On the nature of turbulence, *Commun. Math. Phys.* **20**, 167; Added note, *Commun. Math. Phys.* **23**, 343.

Ruelle, D., (1976). "The Lorenz attractor and the problem of turbulence," in *Quantum Dynamics: Models and Mathematics*, ed. L. Streit, *Acta Physica Austriaca Supp. XVI*, Springer Verlag, N. Y.

Ruelle, D., (1977a). Applications conservant une mesure absolument continue par rapport a dx sur [0,1], *Commun. Math. Phys.* **55**, 47.

Ruelle, D., (1978a). An inequality for the entropy of differentiable maps, *Bol. Soc. Bras. Math.* **9**, 83.

Ruelle, D., (1978b). "Dynamical systems with turbulent behavior," in *Math. Problems in Theor. Phys.*, ed. G. Dell'Antonio, S. Dopplicher, and G. J. Lasinio, *Lect. Notes in Phys.* **80**, p. 341, Springer Verlag, N. Y.

Ruelle, D., (1978c). What are the measures describing turbulence, *Prog. Theor. Phys. Supp.* **64**, 339.

Ruelle, D., (1979a). Ergodic theory of differential dynamical systems, *Publ. Math. IHES.* **50**, 275.

Ruelle, D., (1979b). Sensitive dependence on initial condition and turbulent behavior of dynamical systems, *Ann. N. Y. Acad. Sci.* **316**, 408.

Ruelle, D., (1980a). "Recent results on differentiable dynamical systems," in *Math. Problems in Theor. Phys.* ed. K. Osterwalder, *Lect. Notes in Phys.* **116**, p. 321, Springer Verlag, N. Y.

Ruelle, D., (1980b). Strange attractors, *Math. Intelligencer* **2**, 126; *La Recherche* **108**, 132.

Ruelle, D., (1981a). Differentiable dynamical systems and the problem of turbulence, *Bull. Am. Math. Soc.* **5**, 29.

Ruelle, D., (1981b). Measures describing a turbulent flow, *Ann. N. Y. Acad. Sci.* **357**, 1.

Ruelle, D., (1981c). Small random perturbations of dynamical systems and the definition of attractors, *Commun. Math. Phys.* **82**, 137.

Ruelle, D., (1982). Do turbulent crystals exist? *Physica* **113A**, 619.

Ruelle, D., (1983). Five turbulent problems, *Physica* **7D**, 40.
Russell, D. A., Hanson, J. D., and Ott, E., (1980). Dimension of strange attractors, *Phys. Rev. Lett.* **45**, 1175.
Russell, D. A., and Ott, E., (1981). Chaotic (strange) and periodic behavior in instability saturation by the oscillating two-stream instability, *Phys. Fluids* **24**, 1976.
Saltzman, B., (1962). Finite amplitude convection as an initial value problem, I., *J. Atmos. Sci.* **19**, 329.
Sanders, J. A., (1982). Melnikov's method and averaging, *Celestial Mechanics* **28**, 171.
Sano, N., and Sawada, Y., (1983). Transition from quasiperiodicity to chaos in a system of coupled nonlinear oscillators, *Phys. Lett.* **97A**, 73.
Sato, S., Sano, M., and Sawada, Y., (1983). Universal scaling property in bifurcation structure of Duffing's and of generalized Duffing's equations, *Phys. Rev.* **A28**, 1654.
Sarkovskii, A. N., (1964). Coexistence of cycles of a continuous map of a line into itself, *Ukranian Math. J.* **16**, 61.
Savage, C. M., Carmichael, H. J., and Walls, D. F., (1982). Optical multistability and self oscillations in three level systems, *Opt. Commun.* **42**, 211.
Scalapino, D. J., Hirsch, J. E., and Huberman, B. A., (1982). "Intermittency — another road to chaos," in *Melting, Localization, and Chaos*, ed. R. K. Kalia and P. Vashishta, p. 243, Elsevier Science Pub. Co.
Schell, M., Fraser, S., and Kapral, R., (1982). Diffusive dynamics in systems with translational symmetry: a one-dimensional-map model, *Phys. Rev.* **A26**, 504.
Schell, M., Fraser, S., and Kapral, R., (1983). Subharmonic bifurcation in the sine map: an infinite hierarchy of cusp bistabilities, *Phys. Rev.* **A28**, 373.
Schmidt, G., and Bialek, J., (1982). Fractal diagrams for Hamiltonian stochasticity, *Physica* **5D**, 397.
Schmitz, R. A., Graziani, K. R., and Hudson, J. L., (1977). Experimental evidence of chaotic states in the Belousov-Zhabotinskii reaction, *J. Chem. Phys.* **67**, 3040.
Schmitz, R. A., Renola, G. T., and Garrigan, P. C., (1979). Observation of complex dynamic behavior in the H_2-O_2 reaction on nickel, *Ann. N. Y. Acad. Sci.*, **316**, 638.
Scholz, H. J., Yamada, Y., Brand,H., and Graham, R., (1981). Intermittency and chaos in a laser system with modulated inversion, *Phys. Lett.* **82A**, 321.
Scholz, H. J., (1982). Markov chains and discrete chaos, *Physica* **4D**, 281.
Schreiber, I., and Marek, M., (1982a). Strange attractors in coupled reaction-diffusion cells, *Physica* **5D**, 258.
Schreiber, I., and Marek, M., (1982b). Transition to chaos via two-torus in coupled reaction-diffusion cells, *Phys. Lett.* **91**, 263.
Schulmann, N. J., (1983). Chaos in piecewise-linear systems, *Phys. Rev.* **A28**, 477.
Seifert, H., (1983). Intermittent chaos in Josephson junctions represented by stroboscopic maps, *Phys. Lett.* **98A**, 43.
Shapiro, M., and Child, M. S., (1982). Quantum stochasticity and unimolecular decay, *J. Chem. Phys.* **76**, 6176.
Shaw, R. S., (1981). Strange attractors, chaotic behavior, and information flow, *Z. Naturforsch.* **36a**, 80.
Shaw, R. S., Andereck, C. D., Reith, L. A., and Swinney, H. L., (1982). Superposition of traveling waves in the circular Couette system, *Phys. Rev. Lett.* **48**, 1172.

Shaw, S. W., and Holmes, P., (1983). Periodically forced linear oscillator with impacts: Chaos and long-period motions, *Phys. Rev. Lett.* **51**, 623.

Shenker, S. J., and Kadanoff, L. P., (1981). Band to band hopping in one-dimensional maps, *J. Phys.* **A14**, L23.

Shenker, S. J., and Kadanoff, L. P., (1982). Critical behavior of a KAM surface: I. Empirical results, *J. Stat. Phys.* **27**, 631.

Shenker, S. J., (1982). Scaling behavior in a map of a circle onto itself: Empirical results, *Physica* **5D**, 405.

Shepelyansky, D. L., (1981). Dynamic stochasticity in nonlinear quantum systems, *Theor. Math. Phys. (USA)* **49**, 925; Russian original, *TMF* **49**, 36.

Shepelyansky, D. L., (1983). Some statistical properties of simple classically stochastic quantum systems, *Physica* **8D**, 208.

Sherman, J., and McLaughlin, J. B., (1978). Power spectra of nonlinearly coupled waves, *Commun. Math. Phys.* **58**, 9.

Shigematsu, H., Mori, H., Yoshida, T., and Okamoto, H., (1983). Analytic study of the power spectra of the tent maps near band-splitting transitions, *J. Stat. Phys.* **30**, 649.

Shimada, I., and Nagashima, T., (1978). The iterative transition phenomenon between periodic and turbulent states in a dissipative dynamical system, *Prog. Theor. Phys.* **59**, 1033.

Shimada, I., and Nagashima, T., (1979). A numerical approach to ergodic problem of dissipative systems, *Prog. Theor. Phys.* **61**, 1605.

Shimada, I., (1979). Gibbsian distribution on the Lorenz attractor, *Prog. Theor. Phys.* **62**, 61.

Shimizu, T., and Morioka, N., (1978a). Chaos and limit cycles in the Lorenz model, *Phys. Lett.* **66A**, 182.

Shimizu, T., and Morioka, N., (1978b). Transient behavior in periodic regions of the Lorenz model, *Phys. Lett.* **66A**, 447.

Shimizu, T., and Morioka, N., (1978c). Transitions between turbulent and periodic states in the Lorenz model, *Phys. Lett.* **69A**, 148.

Shimizu, T., (1979). Analytic form of the simplest limit cycle in the Lorenz model, *Physica* **97A**, 383.

Shimizu, T., and Morioka, N., (1981). Period-doubling bifurcations in a simple model, *Phys. Lett.* **83A**, 243.

Shimizu, T., (1981). Asymptotic form of a strange attractor, *Phys. Lett.* **84A**, 85.

Shimizu, T., and Ichimura, A., (1982). Asymptotic solution of a chaotic motion, *Phys. Lett.* **91A**, 52.

Showalter, K., Noyes, R. M., and Bar-Eli, K., (1978). A modified Oregonator model exhibiting complicated limit cycle behavior in a flow system, *J. Chem. Phys.* **69**, 2514.

Shraiman, B., Wayne, C. E., and Martin, P. C., (1981). A scaling theory for noisy period-doubling transitions to chaos, *Phys. Rev. Lett.* **46**, 935.

Shtern, V. N., (1983). Attractor dimension for the generalized Baker's transformation, *Phys. Lett.* **99A**, 268.

Siegel, C. L., (1941). On the integrals of canonical systems, *Ann. Math.* **42**, 806.

Siegel, C. L., (1954). Uber die Existenz einer Normalform analytische Hamiltonischer

Differentialgleichungen in der Nahe einer Gleichgewichtslosung, *Math. Ann.* **128**, 144.

Siegmann, W. L., and Rubenfeld, L. A., (1975). A nonlinear model for double-diffusive convection, *SIAM J. Appl. Math.* **29**, 540.

Silnikov, L. P., (1965). A case of the existence of a denumerable set of periodic motions, *Sov. Math. Dokl.* **6**, 163; Russian original, *Doklady* **160**, 588.

Silnikov, L. P., (1969). On a new type of bifurcation of multidimensional dynamical systems, *Sov. Math. Dokl.* **10**, 1368; Russian original, *Doklady* **189**, 59.

Simo, C., (1979). On the Henon-Pomeau attractor, *J. Stat. Phys.* **21**, 465.

Simoyi, R. H., Wolf, A., and Swinney, H. L., (1982). One-dimensional dynamics in a multicomponent chemical reaction, *Phys. Rev. Lett.* **49**, 245.

Sinai, J. G., and Vul, E., (1981). Hyperbolicity conditions for the Lorenz model, *Physica* **2D**, 3.

Singer, D., (1978). Stable orbits and bifurcations of maps of the interval, *SIAM J. Appl. Math.* **35**, 260.

Singh, S., and Agarwal, G. S., (1983). Chaos in two-phonon coherent processes in a ring cavity, *Opt. Commun.* **47**, 73.

Smale, S., (1965). "Diffeomorphisms with many periodic points," in *Differential and Combinatorial Topology*, p. 63, Princeton.

Smale, S., (1967). Differentiable dynamical systems, *Bull. Am. Math. Soc.* **13**, 747.

Smith, C. W., Tejwani, M. J., and Farris, D. A., (1982). Bifurcation universality for first-sound subharmonic generation in superfluid helium-4, *Phys. Rev. Lett.* **48**, 492.

Smith, C. W., and Tejwani, M. J., (1983). Bifurcation and the universal sequence for first-sound subharmonic generation in superfluid helium-4, *Physica* **7D**, 85.

Snapp, R. R., Carmichael, H. J., and Schieve, W. C., (1981). The path to 'turbulence': Optical bistability and universality in the ring cavity, *Opt. Commun.* **40**, 1.

Steeb, W. H., Erig, W., and Kunic, A., (1983). Chaotic behavior and limit cycle behavior of anharmonic systems with periodic external perturbations, *Phys. Lett.* **93A**, 267.

Steen, P. H., and Davis, S. H., (1982). Quasiperiodic bifurcation in nonlinearly coupled oscillators near a point of strong resonance, *SIAM J. Appl. Math.* **42**, 1345.

Stefan, P., (1977). A theorem of Sarkovskii on the existence of periodic orbits of continuous endomorphisms of the real line, *Commun. Math. Phys.* **54**, 237.

Stefanski, K., (1982). Fluctuations and structure of attractors — simple tests on the Henon mapping, *Phys. Lett.* **92A**, 315.

Stratonovich, R. L., (1982). Correlators of processes in very simple systems with strange attractors, *Sov. Phys.-Dokl.* **27**, 942; Russian original, *Dokl.* **267**, 355.

Stuchl, I., and Marek, M., (1982a). Experiments on 'relative stabilities' in a chemical system, *J. Chem. Phys.* **77**, 1607.

Stuchl, I., and Marek, M., (1982b). Dissipative structures in coupled cells: experiments, *J. Chem. Phys.* **77**, 2956.

Sun, Y.-S., Froeschle, C., (1981). The dependence of the Kolmogorov entropy of mappings on the system of coordinates, *Acta Astron. Sinica* **22**, 168.

Sun, Y.-S., Froeschle, C., (1982). Kolmogorov entropy of 2D area-preserving mappings, *Scientia Sinica* No. **4**, 357.

Sun, Y.-S., (1983a). On the measure-preserving mappings with odd dimension, *Celestial Mechanics* **30**, 7.

Sun, Y.-S., (1983b). Stochasticity of the measure-preserving mappings with three dimensions, *Acta Astron. Sinica* **24**, 128.

Swinney, H. L., Fenstermacher, P. R., and Gollub, J. P., (1977a). "Transition to turbulence in a fluid flow," in *Synergetics: a workshop*, ed. H. Haken, p. 60, Springer Verlag, N. Y.

Swinney, H. L., Fenstermacher, P. R., and Gollub, J. P., (1977b). "Transition to turbulence in circular Couette flow," in *Symposium on turbulent shear flows*, Penn. State Univ., **2**, 17.1.

Swinney, H. L., (1978). Hydrodynamic instabilities and the transition to turbulence, *Prog. Theor. Phys. Supp.* **64**, 164.

Swinney, H. L., and Gollub, J. P., (1978). Transition to turbulence, *Phys. Today* **31**, 41.

Swinney, H. L., (1983). Observations of order and chaos in nonlinear systems, *Physica* **7D**, 3.

Szepfalusy, P., and Tel, T., (1982). Fluctuations in the limit cycle state and the problem of phase chaos, *Physica* **112A**, 146.

Tabor, M., (1981). The onset of chaotic motion in dynamical systems, *Adv. in Chem. Phys.* **46**, 73.

Tabor, M., and Weiss, J., (1981). Analytical structure of the Lorenz system, *Phys. Rev.* **A24**, 2157.

Tabor, M., (1982). Analytic structure and integrability of dynamical systems, *Int. J. Quantum Chem.* **16**, 167.

Takens, F., (1981). Detecting strange attractors in turbulence, in *Lect. Notes in Math.* **898**, ed. D. A. Rand and L. S. Young, p. 336, Springer, Berlin.

Takeyama, K., (1978). Dynamics of the Lorenz model of convective instabilities I, *Prog. Theor. Phys.* **60**, 613.

Takeyama, K., (1980). Dynamics of the Lorenz model of convective instabilities II, *Prog. Theor. Phys.* **63**, 91.

Takeyama, K., (1983). A parameter renormalization for orbit-splittings and band-mergings of iterated maps of an interval, *Z. Phys.* **B52**, 253.

Takeyama, K., and Satoh, K. K., (1983). A finite-order splitting-merging bifurcation of bands in iterated maps of an interval with a cusp maximum, *Phys. Lett.* **98A**, 313.

Tedeschini-Lalli, L., (1982). Truncated Navier-Stokes equations: continuous transition from a five-mode to a seven-mode model, *J. Stat. Phys.* **27**, 365.

Tel, T., (1982a). On the construction of invariant curves of period-two points in two-dimensional maps, *Phys. Lett.* **94A**, 334.

Tel, T., (1982b). On the construction of stable and unstable manifolds of two-dimensional invertible maps, *Z. Phys.* **B49**, 157.

Tel, T., (1983c). Fractal dimension of the strange attractor in a piecewise linear two-dimensional map, *Phys. Lett.* **97A**, 219.

Tennyson, J. L., Lieberman, M. A., and Lichtenberg, A. J., (1979). "Diffusion in near-integrable Hamiltonian systems with three degrees of freedom," in *Nonlinear Dynamics and the Beam-Beam Interaction*, ed. M. Moth and J. C. Herrera, *AIP Conf. Proc.* **57**, 272.

Termonia, Y., and Alexandrowicz, Z., (1983). Fractal dimension of strange attractors from radius versus size of arbitrary clusters, *Phys. Rev. Lett.* **51**, 1265.

Testa, J., Perez, J., and Jeffries, C., (1982). Evidence for universal chaotic behavior of a driven nonlinear oscillator, *Phys. Rev. Lett.* **48**, 714.

Testa, J., and Held, G. A., (1982). Study of a one-dimensional map with multiple basins, *Phys. Rev.* **A28**, 3085.

Thomae, S., and Grossmann, S., (1981a). A scaling property in critical spectra of discrete systems, *Phys. Lett.* **83A**, 181.

Thomae, S., and Grossmann, S., (1981b). Correlations and spectra of periodic chaos generated by the logistic parabola, *J. Stat. Phys.* **26**, 485.

Thomae, S., (1983). "Chaos-induced diffusion," in *Statics and Dynamics of Nonlinear Systems*, ed. G. Benedeck, H. Bilz, R. Zeyher, p. 204, Springer-Verlag.

Thompson, J. M. T., and Ghaffari, R., (1982). Chaos after period-doubling bifurcations in the resonance of an impact oscillator, *Phys. Lett.* **91A**, 5.

Thompson, J. M. T., Bokaian, A. R., and Ghaffari, R., (1983). Subharmonic resonances and chaotic motions of a billinear oscillator, *IMA J. Appl. Math.* **31**, 207.

Thompson, J. M. T., and Ghaffari, R., (1983). Chaotic dynamics of an impact oscillator, *Phys. Rev.* **A27**, 1741.

Tomita, K., Kai, T., and Hikami, F., (1977). Entrainment of a limit cycle by a periodic external excitation, *Prog. Theor. Phys.* **57**, 1159.

Tomita, K., and Kai, T., (1978a). Chaotic behavior of deterministic orbits: the problem of turbulent phase, *Prog. Theor. Phys. Supp.* **64**, 280.

Tomita, K., and Kai, T., (1978b). Stroboscopic phase portrait and strange attractors, *Phys. Lett.* **66A**, 91.

Tomita, K., and Kai, Y., (1979a). Stroboscopic phase portrait of a forced nonlinear oscillator, *Prog. Theor. Phys.* **61**, 54.

Tomita, K., and Kai, T., (1979b). Chaotic response of a nonlinear oscillator, *J. Stat. Phys.* **21**, 65.

Tomita, K., and Tsuda, I., (1979). Chaos in Belousov-Zhabotinskii reaction in a flow system, *Phys. Lett.* **71A**, 489.

Tomita, K., and Daido, H., (1980). Possibility of chaotic behavior and multi-basins in forced glycolytic oscillations, *Phys. Lett.* **79A**, 133.

Tomita, K., and Tsuda, I., (1980a). Towards the interpretation of Hudson's experiment on the Belousov-Zhabotinskii reaction: chaos due to delocalization, *Prog. Theor. Phys.* **64**, 1138.

Tomita, K., and Tsuda, I., (1980b). Towards the interpretation of the global bifurcation structure of the Lorenz system: a simple one-dimensional model, *Prog. Theor. Phys. Supp.* **69**, 185.

Tomita, K., (1982a). Chaotic response of nonlinear oscillators, *Phys. Reports* **86**, 113.

Tomita, K., (1982b). Structure and macroscopic chaos in biology, *J. Theor. Biol.* **99**, 111.

Tresser, C., and Coullet, P., (1978). Iterations d'endomorphismes et groupe de renormalisation, *C. R. Acad. Sc.* **287A**, 577.

Tresser, C., and Coullet, P., (1980). Preturbulent states and renormalization group for simple models, *Rep. on Math. Phys.* **17**, 189.

Tresser, C., Coullet, P., and Arneodo, A., (1980a). Transition to turbulence for doubly periodic flows, *Phys. Lett.* **77A**, 327.

Tresser, C., Coullet, P., and Arneodo, A., (1980b). On the existence of hysteresis in a transition to chaos after a single bifurcation, *J. de Phys. Lett.* **41**, L243.

Tresser, C., Coullet, P., and Arneodo, A., (1980c). Topological horseshoe and numerically observed chaotic behavior in the Henon mapping, *J. Phys.* **13A**, L123.

Tsuchiya, T., Szabo, A., and Saito, N., (1983). Exact solutions of simple difference equation systems that show chaotic behavior, *Z. Naturforsch.* **38a**, 1035.

Tsuda, I., (1981a). Multi-time expansion and universality in one-dimensional difference equation, *Prog. Theor. Phys.* **66**, 1086.

Tsuda, I., (1981b). Self-similarity in the Belousov-Zhabotinsky reaction, *Phys. Lett.* **85A**, 4.

Tsuda, I., (1981c). On the abnormality of period-doubling bifurcation — In connection with the bifurcation structure in the Belousov-Zhabotinsky reaction system, *Prog. Theor. Phys.* **66**, 1985.

Turner, J. S., Roux, J.-C., McCormick, W. D., and Swinney, H. L., (1981). Alternating periodic and chaotic regimes in a chemical reaction — experiment and theory, *Phys. Lett.* **85A**, 9.

Turner, J. S., (1980). "Complex periodic and nonperiodic behavior in the Belousov-Zhabotinskii reaction," in *Aspects of Chemical Evolution*, ed. G. Nicolis and M. Herzchkowitz-Kaufman.

Tyson, J. J., (1978). On the appearance of chaos in a model of the Belousov reaction, *J. Math. Biology* **5**, 351.

Ueda, Y., Hayashi, C., and Akamatsu, N., (1973). Computer simulation of nonlinear ordinary differential equations and nonperiodic oscillations, *Electronics and Commun. in Japan* **56A**, 27.

Ueda, Y., (1979). Randomly transitional phenomena in the system governed by Duffing's equation, *J. Stat. Phy.* **20**, 181.

Ueda, Y., (1980a). "Steady motions exhibited by Duffing's equation: a picture book of regular and chaotic motions," in *New Approaches to Nonlinear Problems in Dynamics*, ed. P. Holmes, p. 311, SIAM, Philadelphia.

Ueda, Y., (1980b). Explosion of strange attractors exhibited by Duffing's equation, *Ann. N. Y. Acad. Sci.* **357**, 422.

Ueda, Y., Akamatsu, N., (1981). Chaotically transitional phenomena in the forced negative-resistance oscillator, *IEEE Trans. Circuits and Systems* **CAS28**, 217.

Ueda, Y., and Noguch, A., (1983). A model that realizes the soliton and the chaos simultaneously, *J. Phys. Soc. Japan* **52**, 713.

Uezu, T., and Aizawa, Y., (1982a). Some routes to chaos from limit cycle in the forced Lorenz system, *Prog. Theor. Phys.* **68**, 1543.

Uezu, T., and Aizawa, Y., (1982b). Topological character of periodic solution in three-dimensional ODE system, *Prog. Theor. Phys.* **68**, 1907.

Ulam, S. M., and von Neumann, J., (1947). On combinations of stochastic and deterministic processes, *Bull. Amer. Math. Soc.* **53**, 1120.

Ushiki, S., (1982). Central difference scheme and chaos, *Physica* **4D**, 407.

Van Exter, M., and Lagendijk, A., (1983). Observation of fine structure in the phase locking of a nonlinear oscillator, *Phys. Lett.* **99A**, 1.

Velarde, M. G., and Normand, C., (1980). Convection, *Sci. Amer.* **243**, 79.

Velarde, M. G., (1981). "Steady states, limit cycles and the onset of turbulence: A few model calculations and exercises," in *Nonlinear Phenomena at Phase Transitions and Instabilities*, ed. T. Riste, Plenum, N. Y., 205.

Velarde, M. G., and Antoranz, J. C., (1981). Strange attractor (optical turbulence) in a model problem for the laser with saturable absorber and the two-component Benard convection, *Prog. Theor. Phys.* **66,** 717.

Velarde, M. G., (1982a). "Benard convection and laser with saturable absorber, oscillations and chaos," in *Evolution of order and chaos*, ed. H. Haken, p. 132, Springer-Verlag, N. Y.

Velarde, M. G., (1982b). "Dissipative structures and oscillations in reaction-diffusion model with or without time-delay," in *Stability of Thermodynamic System*, ed. J. Casas and G. Lebon, p. 248, Springer-Verlag.

Vidal, C., Roux, J.-C., (1980). La turbulence chimique existe-t-elle? *La Recherche* **11**, 66.

Vidal, C., Roux, J.-C., Bachelart, S., and Rossi, A., (1980). Experimental study of the transition to turbulence in the Belousov-Zhabotinskii reaction, *Ann. N. Y. Acad. Sci.* **357**, 377.

Vidal, C., Bachelart, S., and Rossi, A., (1982). Bifurcations en cascade conduisant a la turbulence dans la reaction de Belousov-Zhabotinskii, *J. de Phys.* **43,** 7.

Viet, O., Westfreid, J. E., and Guyon, E., (1983). Kinetic art and mechanical chaos, *Eur. J. Phys.* **4**, 72.

Vivaldi, F., Casati, G., and Guarneri, I., (1983). Origin of long-time tails in strongly chaotic systems, *Phys. Rev. Lett.* **51**, 727.

Vlasova, O. F., and Zaslavsky, G. M., (1983). Dissipation effect on chaos onset in a two-resonance overlap case, *Phys. Lett.* **99A**, 405.

Vul, E. B., and Khanin, K. M., (1982). On the unstable manifold of Feigenbaum fixed point, *Usp. Mat. Nauk.* **37**, 173.

Walden, R. W., and Donnelly, R. J., (1979). Reemergent order of chaotic circular Couette flow, *Phys. Rev. Lett.* **42**, 301.

Walker, G. H., Ford, J., (1969). Amplitude instability and ergodic behavior for conservative nonlinear oscillator systems, *Phys. Rev.* **188,** 416.

Wang, G.-R., (1983). The period-doubling bifurcation sequences of the trimolecular model with forced oscillation term, *Acta Physica Sinica* **32**, 960.

Wang, G.-R., Chen, S.-G., Hao, B.-L., (1983). Intermittent chaos in the forced Brusselator, *Acta Physica Sinica* **32**, 1139.

Wegmann, K., and Roessler, O. E., (1978). Different kinds of chaotic oscillations in the Belousov-Zhabotinskii reaction, *Z. Naturforsch* **33a**, 1179.

Weiss, C. O., and King, H., (1982). Oscillation period-doubling chaos in a laser, *Opt. Commun.* **44**, 59.

Weiss, C. O., Godone, A., and Olafsson, A., (1983). Routes to chaotic emission in a CW He-Ne laser, *Phys. Rev.* **A28**, 892.

Weissman, Y., and Jortner, J., (1982). Quantum manifestations of classical stochasticity, I. Energetics of some nonlinear systems. II. Dynamics of wave packets of bound states, *J. Chem. Phys.* **77**, 1469, 1486.

Wersinger, J.-M., Finn, J. M., and Ott, E., (1980a). Bifurcations and strange behavior in instability saturation by nonlinear mode coupling, *Phys. Rev. Lett.* **44**, 453.

Wersinger, J.-M., Finn, J. M., and Ott, E., (1980b). Bifurcation and "strange" behavior in instability saturation by nonlinear three-wave mode coupling, *Phys. Fluids* **23**, 1142.

Whiteman, K. J., (1977). Invariants and stability in classical mechanics, *Rep. Prog. Phys.* **40**, 1033.

Widom, M., and Kadanoff, L. P., (1982). Renormalization group analysis of area-preserving maps, *Physica* **5D**, 287.

Widom, M., Bensimon, D., Kadanoff, L. P., and Shenker, S. J., (1983). Strange objects in the complex plane, *J. Stat. Phys.* **32**, 443.

Wightman, A. S., (1981). "The mechanism of stochasticity in classical dynamical systems," in *Perspectives in Statistical Physics* ed. H. J. Raveche, p. 343, North-Holland.

Williams, R. F., (1977). "The structure of Lorenz attractors," in *Turbulence Seminar Berkeley 1976/77*, ed. P. Bernard and T. Ratiu, p. 94, Springer-Verlag.

Williams, R. F., (1979a). The structure of Lorenz attractors, *Publ. Math. IHES,* **50**, 321.

Williams, R. F., (1979b). The bifurcation space of the Lorenz attractor, *Ann. N. Y. Acad. Sci.* **316**, 393.

Wolf, A., and Swift, J., (1981). Universal power spectra for the reverse bifurcation sequence, *Phys. Lett.* **83A**, 184.

Yahata, H., (1977). Slowly-varying amplitude of the Taylor vortices near the instability point, I and II, *Prog. Theor. Phys.* **57**, 347, 1490.

Yahata, H., (1978). Temporal development of the Taylor vortex in a rotating fluid, I, *Prog. Theor. Phys. Suppl.* **64**, 176.

Yahata, H., (1979). Temporal development of the Taylor vortex in a rotating fluid, II and III, *Prog. Theor. Phys.* **61**, 791; **64** (1980), 782.

Yahata, H., (1983). Period-doubling cascade in the Rayleigh-Benard convection, *Prog. Theor. Phys.* **69**, 1802.

Yamada, T., and Kuramoto, Y., (1976a). Spiral waves in a nonlinear dissipative system, *Prog. Theor. Phys.* **55**, 2035.

Yamada, T., and Kuramoto, Y., (1976b). A reduced model showing chemical turbulence, *Prog. Theor. Phys.* **56**, 681.

Yamada, T., and Fujisaka, H., (1977). A discrete model exhibiting successive bifurcations leading to the onset of chaos, *Z. Physik* **B28**, 239.

Yamada, T., and Fujisaka, H., (1983). Stability theory of synchronized motion in coupled-oscillator systems, II. The mapping approach, *Prog. Theor. Phys.* **70**, 1240.

Yamaguchi, Y., (1983). Chaotic behavior of magnetization in superfluid He3 driven by external periodic field, *Prog. Theor. Phys.* **69**, 1377.

Yamaguchi, Y., Kometani, K., and Shimizu, H., (1978). Self-synchronization of nonlinear oscillations in the presence of fluctuations, *J. Stat. Phys.* **27**, 719.

Yamaguchi, Y., and Sakai, K., (1983). New type of 'crisis' showing hysteresis, *Phys. Rev.* **A27**, 2755.

Yamaguti, M., Ushiki, S., (1981). Chaos in numerical analysis of ODE's, *Physica* **3D**, 618.

Yamazaki, H., Oono, Y., and Hirakawa, K., (1978). Experimental study on chemical turbulence, I and II, *J. Phys. Soc. Japan* **44**, 335; **46** (1979), 721.

Yeh, W. J., and Kao, Y. H., (1982). Universal scaling and chaotic behavior of a Josephson-junction analog, *Phys. Rev. Lett.* **49**, 1888.

Yorke, J. A., and Yorke, E. D., (1979). Metastable chaos: the transition to sustained chaotic behavior in the Lorenz model, *J. Stat. Phys.* **21**, 263.

Yorke, J. A., and Yorke, E. D., (1981). "Chaotic behavior and fluid dynamics," in *'Hydrodynamic Instabilities and the Transition to Turbulence'*, ed. H. L. Swinney and J. P. Gollub, p. 77, Springer.

Yoshimura, K., and Watanabe, S., (1982). Chaotic behavior of nonlinear evolution equation with fifth order dispersion, *J. Phys. Soc. Japan* **51** 3028.

Yoshida, T., and Mori, H., and Shigematsu, H., (1983). Analytic study of chaos of the tent map: band structures, power spectra, and critical behavior, *J. Stat. Phys.* **31**, 279.

Young, L.-S., (1982). Dimension, entropy, and Lyapunov exponents, *Ergodic Theory and Dynamical Systems* **2**, 109.

Yuan, J. M., Tung, M., Feng, D. H., and Narducci, L. M., (1983). Instability and irregular behavior of coupled logistic equations, *Phys. Rev.* **A28**, 1662.

Zardecki, A., (1982). Noisy Ikeda attractor, *Phys. Lett.* **90A**, 274.

Zaslavsky, G. M., and Chirikov, B. V., (1972). Stochastic instabilities in nonlinear oscillations, *Sov. Phys. Usp.* **14**, 549; Russian original: *UFN* **105** (1971), 3.

Zaslavsky, G. M., (1978). The simplest case of a strange attractor, *Phys. Lett.* **69A**, 145.

Zaslavsky, G. M., and Rachko, Kh. R. Ya., (1978). Singularities of the transition to a turbulent motion, *Sov. Phys. JETP* **49**, 1039; Russian original, *JETP* **776**, 2052.

Zaslavsky, G. M., (1981). Stochasticity in quantum systems, *Phys. Reports* **80**, 157.

Zippelius, A., and Lucke, M., (1981). The effect of external noise in the Lorenz model of the Benard problem, *J. Stat. Phys.* **24**, 345.

Zisook, A. B., (1981). Universal effects of dissipation in two-dimensional mappings, *Phys. Rev.* **A24**, 1640.

Zisook, A. B., (1982). Intermittency in area-preserving mappings, *Phys. Rev.* **A25**, 2289.

Zisook, A. B., and Shenker, S. J., (1982). Renormalization group for intermittency in area-preserving mapping, *Phys. Rev.* **A25**, 2824.

Zuniga, J. E., and Luss, D., (1978). Kinetic oscillations during the isothermal oxidation of hydrogen on platinum wires, *J. Catal.* **53**, 312.